QUANTITATIVE ANALYSIS AND MODELING OF EARTH AND ENVIRONMENTAL DATA

QUANTITATIVE ANALYSIS AND MODELING OF EARTH AND ENVIRONMENTAL DATA

Space-Time and Spacetime Data Considerations

JIAPING WU
Zhejiang University, China

JUNYU HE
Zhejiang University, China

GEORGE CHRISTAKOS
Zhejiang University, China
San Diego State University, United States

ELSEVIER

Elsevier
Radarweg 29, PO Box 211, 1000 AE Amsterdam, Netherlands
The Boulevard, Langford Lane, Kidlington, Oxford OX5 1GB, United Kingdom
50 Hampshire Street, 5th Floor, Cambridge, MA 02139, United States

Library of Congress Cataloging-in-Publication Data
A catalog record for this book is available from the Library of Congress

British Library Cataloguing-in-Publication Data
A catalogue record for this book is available from the British Library

ISBN: 978-0-12-816341-2

For information on all Elsevier publications
visit our website at https://www.elsevier.com/books-and-journals

Publisher: Charlotte Cockle
Acquisitions Editor: Amy Shapiro
Editorial Project Manager: Lindsay Lawrence
Production Project Manager: Kumar Anbazhagan
Cover Designer: Victoria Pearson

Typeset by STRAIVE, India

They were good days, there have been good days.

Jiaping Wu, Junyu He, and George Christakos

Contents

9. Chronotopologic BME estimation

10. Studying physical laws

11. *CTDA* by dimensionality reduction

12. DIA models

13. Syntheses of *CTDA* techniques with *DIA* models

Preface

As its title dictates, the subject of this book is the chronotopologic analysis and modeling of natural phenomena, that is, phenomena that vary as functions of both their spatial (*topos*)[a] and the temporal (*chronos*)[b] coordinates, in realistic conditions of *in situ* uncertainty. The need to study such phenomena, in particular, has led to significant developments in chronotopologic data analysis (*CTDA*), which is an important part of the book.

The notions, techniques, and thinking modes discussed in the book aim at improving the understanding of the chronotopologic laws of change underlying the available numerical datasets, while taking into consideration all relevant core and site-specific knowledge bases, which are subject to multi-sourced uncertainties and associated measurement or observational errors (conceptual, technical, computational). The core knowledge bases include scientific theories, physical laws, and models, whereas site-specific knowledge bases may have various forms and sources, including hard measurements, soft observations, secondary information, and auxiliary variables (ground-level measurements, satellite observations, scientific instrument records, protocols and surveys, empirical graphs and charts). Understanding the spatial distribution and temporal dynamics of knowledge bases such as the above is a great challenge to the elucidation of crucial questions in many physical, health, and social disciplines, including geology, hydrology, geophysics, geography, environmental science, agronomy, ecology, public health, epidemiology, economics, public policy, and risk management.

The book views the difference between training and education in terms of the following metaphor: The way to get people build a ship is not to teach them carpentry, but to inspire them to long for the infinite immensity of the sea. Accordingly, the book's methodology has four stages: The first stage is *diagnosis* (knowing through), that is, identifying the basic features of the phenomenon of interest, the second stage is *cardiognosis* (knowing the heart of the phenomenon), that is, appreciating what is at stake and choosing the appropriate mode of thinking, the third stage is *prognosis* (foreknowledge), that is, accurate prediction of certain important aspects of the phenomenon, and the fourth stage is *epignosis* (improved knowledge), that is, drawing important conclusions, inferences, and interpretations.

This methodology has some interesting implications that become evident throughout the book, including the following: new routes open when one type of data crosses another; potentially difficult problems are considered from alternative standpoints, assessing how well different methods can handle them, and how an improved problem solution is obtained by a synthesis of methods acting in synergy; the presentation of worth noticing personal conclusions based on experience and insight is facilitated.

[a] The term "topos" (see topography, topology etc.) was the ancient Greek term for space.

[b] The term "chronos" (see chronology etc.) was the ancient Greek term for time.

To achieve its particular goals, the book's focus is the rigorous presentation of the supporting theory, followed by a comprehensive and balanced fusion of theory-driven and data-driven techniques. Accordingly, the assertions presented in the book have theory-based content (the expression of a proposition about the real phenomenon being in a certain way) and are made in a data-based context (including the available evidential support and real world conditions). *In melioribus annis,* this approach would have seemed natural. *In praesenti annis,* however, it is necessary to stress the importance of this approach, since the excessive use of technological "black-box" nowadays runs the risk to eventually create impoverished human "black-boxes." Several numerical examples and actual case studies are included so that the readers gain a hands-on experience and valuable insight concerning the implementation of the presented notions, models, and techniques in the real world. Practice exercises at the end of each chapter can help the readers learn more from the text and hone their critical and practical skills.

Individual topics addressed in the book may be probably found in separate publications, but, as far as we know, there is not any particular book that covers all these topics in a complete, systematic, and integrative manner as the present book does. Arguably, then, readers with scientific background or engineering training in the aforementioned disciplines would appreciate a systematic presentation in one volume of the most important quantitative concepts, models and techniques studying the chronotopologic behavior (combined spatial distribution and temporal dynamics) of natural phenomena under conditions of uncertainty and site-specific measurement errors (i.e., the vast majority of phenomena encountered in the various scientific and engineering disciplines).

We would like to thank our families for their patience, and Ms Lindsay C. Lawrence of Elsevier for her continuing encouragement during the book-writing project.

Jiaping Wu
Junyu He
George Christakos

CHAPTER

1

Chronotopologic data analysis

1 From topos to chronotopos

The first section of the book starts with a broad discussion of developments in the collection, analysis, and interpretation of data, from the case where the data values are assumed to vary as functions of the spatial (*topos*)[a] coordinates to the case in which they are assumed to vary as functions of both the spatial (topos) and the temporal (*chronos*)[b] coordinates. The latter case leads to the unified notion of *chronotopology*, where the data values may vary in a separate (space-time) or a composite (spacetime) manner.

1.1 Scientific paradigms

But, before focusing on chronotopology matters, and in order to put these matters in due perspective, the discussion should start with a reference, albeit a brief one, to a fundamental notion of human inquiry, as follows:

> A **scientific paradigm** is a framework of concepts, assumptions, theories, thinking modes, results, and practices that define a scientific discipline at any particular period of time.[c]

Historically, the evolution of the scientific paradigm has undergone certain major developmental phases:

① It started, a few thousand years ago, with the **purely empirical** paradigm that focused on the description of natural phenomena using purely empirical means.
② This was followed, a few hundred years ago, by the **theory-laden experimentation** paradigm when a theoretical component (involving theories, models, laws, generalizations) was added to the empirical component (in which systematic experimentation played a key role).
③ During the last few decades, the **computation-dominated** paradigm

[a] The term "topos" (see topography, topology, etc.) was the ancient Greek term for space.

[b] The term "chronos" (see chronology, etc.) was the ancient Greek term for time.

[c] The word παράδειγμα (*paradeigma*) has been used in famous texts, such as *Plato's Timaeus*, as the model or the pattern that God used to create the cosmos.

https://doi.org/10.1016/B978-0-12-816341-2.00005-8

emerged that is characterized by the addition of a significant computational component (since theories of complex phenomena became too complicated to solve analytically, numerical simulations needed to be generated).

Furthermore, it has been argued by many investigators (scientists and nonscientists alike) that we currently enter the era of yet another paradigm:

④ The **big data-driven** paradigm that seeks to apply computational models to breathtaking amounts of data obtained by instruments or generated by simulators, then processed by the software, resulting in information stored in computers.

According to this paradigm, insight is gained through a self-reinforcing loop between experimental data and statistical analysis (Succi and Coveney, 2019). As should be expected, serious objections to the purely data-driven paradigm exist. Among them, *Lewis H. Lapham* has suggested that:

✍ Data mining engineers have no use for the meaning and value of words. They come to bury civilization, not to praise it.

In response to these and similar concerns, a challenge to the purely data-driven perspective has emerged, as follows:

- The big data perspective outlined above makes extravagant claims, which, once properly discarded, a **synergistic synthesis** of large datasets with a sound scientific theory could plausibly lead to a sensical paradigm-shift.

This is a serious matter to anyone who takes a moment to examine it with an attentive mind. Therefore, we will revisit it in various parts of the book.

1.2 Interplay of science and mathematics

No readers of this book, most of whom are presumably *quants* (i.e., quantitative analysts), need to be reminded of the all-important relationship between two key components of human inquiry (see Fig. 1.1):

> **Mathematics**, which focuses on how to develop abstract representations of the real world, and **science**, which is concerned with the inverse process, i.e., with how to use these abstract representations to obtain useful knowledge about the real world.[d]

Admittedly, this book uses a considerable amount of mathematics in the form of *analysis*, *modeling*, and *estimation*[e] notions and techniques. Compared to the complexity of the real

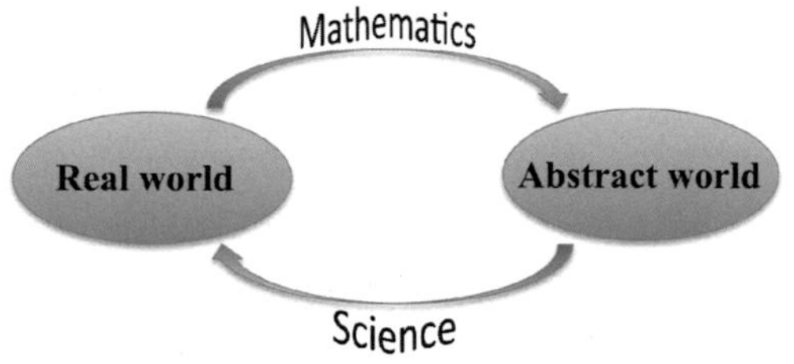

FIG. 1.1 The opposite directions of mathematics and science.

[d] Yet, the two have some interesting sociological differences, including the fact that in science each generation of investigators undoes the existing structure (replaces the theories of the previous generation with new ones etc.), whereas in mathematics each generation adds a new piece to the existing structure.

[e] The term "estimation" generally includes interpolation, extrapolation and prediction. The distinction between these three kinds of chronotopologic estimation is critical: in interpolation and extrapolation attribute estimates are sought at points, respectively, within and outside the chronotopologic sampling domain, and in prediction attribute estimates are sought at points within the spatial sampling area but outside the temporal sampling period. The issue is discussed in various parts of the book (e.g., Chapter 7).

world, these notions and techniques are not as complex as some investigators may seem to believe. Those who do not believe so, it is rather due to the fact that they may not realize how complicated real *life* actually is.

The links of mathematics with ordinary life are rather obvious. Many mathematical ideas are ways of mathematizing ordinary ideas, e.g., the idea of a derivative that is a straightforward mathematization of the ordinary idea of instantaneous change. Beyond this rather simple notion, many complex mathematical concepts eventually come down to direct perceptual experience.

Since science, *a natura eius*, is about finding the analogy and identity in the most remote parts, it should be noticed at the outset that this book's own interpretation of Fig. 1.1 is as follows:

> The **interworkings** of science and mathematics should be carefully considered in both directions: not only mathematics should be used in scientific investigations, but science should be used in mathematical investigations too.

Accordingly, investigators should closely fuse mathematical *symbology* with physical *meaning*, resulting in a powerful and productive structure. That is, scientific applications require us to integrate our understanding of the real world with symbolic relations of mathematics, thus adding meaning and structure to both. Science puts meaning to mathematics, adding additional levels of structure, interpretation, and even tools. This book, then, is more interested in the scientifically meaningful employment of mathematical tools than in the purely formal features of the tools. A direct consequence of the above considerations is the warning that looking at an equation and treating it purely as a matter of mathematical grammar, while neglecting to enrich its symbols with physical meaning, leaves one vulnerable to errors.

Example 1.1 Assume one is given the equation $E = \frac{F}{q}$, relating the electric field E with force F and a point charge q. If q is divided by 2, i.e., $\frac{q}{2}$, what is the corresponding E value? Focusing on the mathematical grammar of the equation, the obvious answer is $E = \frac{2F}{q}$. This answer is physically incorrect, though, since it neglects the essential knowledge that F is proportional to q, in which case the correct answer is that there is no change in the value of E. This error happens when one (who probably lacks the necessary background knowledge) looks at the equation and treats it purely as a matter of mathematical grammar, neglecting to enrich its symbols with physical meaning.

Since this book's perspective is that an equation's mathematical grammar should be integrated with physical meaning, our focus will not be only on technical deductions, but also on the rich variety of inductive inferences associated with our readers' ordinary experience. The same argument is obviously valid in terms of data, i.e., data should not be treated purely as numbers, neglecting to enrich it with the due physical meaning. In this respect, the following general conclusion is reached:

- Readers are encouraged to avoid falling into the trap of reasoning purely in terms of mathematical grammar and numerics, thus missing clues to a different **interpretation** given by physical meaning.

The payoff of this approach, integrating mathematical grammar and numerics with physical meaning, is comprehensible, since it can produce new knowledge, add to the existing body of knowledge in a particular scientific field, and, in many cases, scientific findings can be transferred into useful technology.

1.3 Natural attributes

Quantitative analysis, as practiced in sciences, relies on the tools of the discipline of applied mathematics, which studies the variational characteristics of empirical observations or measurements[f] associated with natural attributes. Before we proceed, we need a working definition of the latter:

A **natural attribute** is a measurable or observable entity characterizing selected aspects of a natural phenomenon (physical, biological, social) that vary over a region during a time period.

Simply put, everything that varies across space and/or time in nature and can be measured (using simple devices or sophisticated equipment of various kinds) or observed (via perceptual experience) can be considered a natural attribute. In this setting, known (measured or observed) attribute values are characterized as attribute *data*.

Example 1.2 There is a wide variety of such natural attributes, including pollutant concentration, topographic elevation, ocean surface temperature, rainfall intensity, human exposure, disease incidence, population mortality, land-use variables, commodity prices, and regional poverty indicators.

In cases such as the above, attribute characteristics of prime importance that need to be adequately defined and understood are the following:

- The **variational characteristics** describe attribute dynamics,[g] regional changes, and interrelationships due to the underlying mechanisms of the phenomenon under study (Section 2).

Indeed, data variation across space and time is a dominant characteristic that has tremendous impacts on the performance of certain quantitative analysis and modeling aspects. For example, the accuracy of attribute interpolation at unsampled points (discussed in Section 4) may decrease considerably with increasing data variation. Accordingly, understanding the attribute's variational characteristics based on the available data is a great challenge to the elucidation of central questions in many physical, social, and health disciplines, including environmentology, geology, agronomy, ecology, geography, public health, epidemiology, economics, social policy, and risk management.

1.4 Kinds of scientific data analysis

As regards the practice of data analysis in sciences, where there is a strong physical dependence of data values on location and time, three kinds of studies have historically emerged (in ascending order of modeling sophistication):

1. Studies concentrating solely on the variation of data across space, known as **spatial data analysis** (*SDA*), or solely on temporal data variation, known as **time data analysis** (*TDA*).
2. Studies focusing on some kind of **separable space-time data analysis** (hereafter denoted as *S-TDA*), in the sense that they consider space and time in isolation.
3. Studies favoring a **composite spacetime data analysis** (denoted as *STDA*), in the sense that space and time are considered as an integrated whole, meaning that the integration obeys the physics of the phenomenon.

In the above settings, space and time individually become three- and one-dimensional projections of the four-dimensional domain, respectively. Conversely, the classical three-dimensional geometry of *SDA* becomes a four-dimensional geometry in space-time analysis. Due to its wide popularity, let us comment a little further on *SDA*. Although it may be seen,

[f] The distinction between measurement and observation is discussed later in the book.

[g] The term dynamics here refers to the temporal change of the attribute.

methodologically, as a special case of *S-TDA*, there are considerable differences, as will become clear in the following. Moreover, *SDA* was historically developed many decades before *S-TDA*. Hence, one can find a plethora of *SDA* techniques—including inverse distance, splines, trend surface, statistical regression, and geostatistics Kriging (Chapters 7 and 8)—that have been successfully implemented in many disciplines.

1.4.1 Geostatistics and time-series statistics

SDA techniques did not develop *uno ictu temporis*. A case in point is the following well-known *SDA* field:

> **Geostatistics** consists of a collection of concepts and techniques that study spatially distributed attributes based on a space-dependent statistical theory that was built on pre-existing results formulated in different scientific disciplines with varying objectives in mind.

Indeed, the roots of geostatistics go back to the 1940s–60s as a group of spatial correlation, regression, and mapping techniques used in forestry and earth sciences with considerable success, see the pioneering work of Matérn (1947) and Matheron (1965). Remarkably, several decades later the field was rediscovered and renamed *spatial statistics*.[h] In any case, many important advances have been made since the 1960s, and the range of geostatistics or spatial statistics applications nowadays is vast (the interested readers are referred to the relevant literature).

TDA, on the other hand, generally focuses on the study of sequences of attribute data ordered in time (time is often considered the independent variable). In particular, the systematic statistical study of this kind of time data was the focus of a well-known *TDA* field:

> **Time series analysis** (***TSA***) is a collection of concepts and techniques for studying the statistical characteristics of time series data and making attribute forecasts for future times.

Historically, the earliest theoretical *TSA* developments can be traced back to the pioneering work of Yule (1927) and Walker (1931). The next major development was the work of Box and Jenkins (1976) presenting the complete time series modeling process (specification, estimation, diagnostics, and forecasting). One can find several *TSA* models in the relevant literature, including moving averages, exponential smoothing of various degrees, analysis of variance, autoregressive integrated moving average, among many other models. There have been numerous *TSA* applications in almost any scientific discipline, including econometrics, finance, geophysics, seismology, signal processing, pattern recognition, and meteorology. The interested readers are referred to the very rich literature on the subject.

Example 1.3 Both *SDA* and *TSA* capitalize on spatial and temporal correlations, respectively. To most scientists, the fact that closely spaced or temporally separated samples tend to be similar is not surprising, since such samples are likely to be influenced by similar physical processes.

An obvious disadvantage of *SDA* and *TSA* is that they both neglect any cross space-time data associations, and the reason of this neglect may be a physical justification, an apparently plausible approximation, or simply a matter of convenience. Yet, reality often does not strictly satisfy the *SDA* or *TSA* assumptions. Instead, strong correlations may occur simultaneously across space and time:

[h] A well-known spatial statistics book is that of Noel Cressie (*Statistics for Spatial Data*. J Wiley, NY, 1991).

- In reality, the links between attribute data values can be critically affected by space-time **cross-effects**,[i] and the corresponding data distributions can be highly skewed.

In a human sense, after all, to represent matters in a purely spatial manner is to freeze time and to deny the dynamism implicit in living. A similar criticism holds when representing matters in a purely temporal manner that denies spatial dependency. By way of a summary, then, the following conclusion is drawn based on the above considerations:

- Both *SDA* and *TSA* encounter profound **challenges** to their methodologies that have been based on notions and models that are, respectively, timeless-space (*SDA*) and space-independent (*TSA*).

It is reasonably anticipated that the adequate responses to these challenges could lie in richer approaches that account for both space and time in a systematic way. Such approaches are introduced in the remaining of this chapter and they are, in fact, the subject matter of this book.

1.4.2 S-TDA and STDA

For the above reasons, the advent of *S-TDA* has been seen as a significant development over *SDA* and *TSA* because it offers a more accurate quantitative representation of the real world, including a more realistic study of the space-time distribution of data from natural attributes. Indeed, many useful *S-TDA* models have been developed with emphasis on the quantitative expression of attribute properties and relationships that take into account the space-time localization of these properties and relationships in a direct way. The notions and methods discussed in the book address these and similar issues.

Example 1.4 Space-time adaptive processing extends adaptive antenna techniques to processors that simultaneously consider spatial domain information (in the form of signals received on multiple antenna array elements) and temporal domain information (in the form of multiple pulse repetition periods) of a coherent processing interval.

Although *S-TDA* is very useful in many practical applications, its key limitation is rather profound: *S-TDA* essentially suggests seeing the world in slices. I.e., the *S-TDA* models, by definition, consider space and time as separate arguments. In layman terms, then, the need for the composite spacetime perspective proposed by *STDA* is based on the following rather straightforward argument:

> In many cases, it is hard to pin down the truth when one is obliged to see the world in slices, because **snapshots** may conceal as much as they make plain.

Otherwise said, *STDA* also studies the distribution of natural attributes, but, while *S-TDA* views space and time as separate arguments, *STDA* considers a composite spacetime argument. Classical statistics assumes nonspatiality and that the samples are independent from one another, whereas *S-TDA* and *STDA* account for spatiality and temporality, although in different ways. *STDA* is obviously a more recent development than *S-TDA*.

Undoubtedly, *S-TDA* and *STDA* offer improved representations of reality, which is why one can find a considerable number of interesting real-world publications of *S-TDA* and *STDA* in a variety of scientific journals representing different and often unconnected

[i] This means that, due to space-time cross effects, data values that are further apart in space but closer in time may turn out to be more similar than those that are closer in space but further apart in time, and vice versa.

disciplines. Yet, there is a very limited number of books on *S-TDA*, and even fewer on *STDA* theory and applications.[j] This is somewhat surprising, since there is a great potential and need for comprehensive and systematic presentations of *S-TDA* and *STDA* in practice, due to the obvious fact that the vast majority of real data vary across both space and time. Moreover, improved technology is becoming available that facilitates the *in-situ* implementation of *S-TDA* and *STDA*, e.g., low-cost *Temporal Geographic Information Systems, TGIS*, with user-friendly interfaces (Christakos et al., 2002).

1.5 The unifying chronotopologic data analysis notion

It is now time to return to the notion of chronotopology briefly introduced at the beginning of this chapter. It should be brought to the reader's attention that, for methodological and presentation reasons, in this book a unifying notion will be favored, on occasion, depending on the context (e.g., when the properties of interest are shared by both *S-TDA* and *STDA*).

> It is methodologically appropriate and occasionally convenient to use the general term **chronotopologic**[k] **data analysis** (*CTDA*) that includes both cases of *S-TDA* and *STDA*.

Generally, *CTDA* aims at extracting implicit knowledge such as chronotopologic relations and patterns that may not be explicitly stored in the relevant databases. This extraction relies on the merging of scientific knowledge and logical principles. *CTDA* distinguishes itself from classical data analysis in that it associates with each attribute of interest physically meaningful chronotopologic features. In particular, while classical statistics assumes that the samples are independent from one another, *CTDA* considers sample dependence governed by physics, it is not tied to a distribution model that assumes that all samples of a population are normally distributed, and it is devoted to the interpretation of uncertainties caused by various *in-situ* conditions (Section 2 below).

Due to the increasing availability of datasets that are simultaneously location and time dependent, their systematic quantitative analysis and modeling provided by *CTDA* are of great importance to a variety of scientific and engineering disciplines, including earth, ocean and atmospheric sciences, environmental and ecological engineering, health and epidemiological studies, risk assessment, social policy and financial management.

1.5.1 CTDA conditions

As a matter of fact, it should be emphasized at the outset that *CTDA* can be used in the study of natural attributes that satisfy these conditions:

① They are measurable or observable.
② They vary within a well-defined chronotopologic domain.
③ They are characterized by chronotopologic dependence.
④ They occur in conditions of *in-situ* uncertainty.

In other words, the *CTDA* concepts and techniques should be used in the study of natural phenomena when meaningful values of the attribute may occur at every point in the chronotopologic domain of interest that can be potentially measured or observed.

[j] Spatiotemporal data analysis books include Chistakos G (*Spatiotemporal Random Fields*, Elsevier, the Netherlands, 2017; *Modern Spatiotemporal Geostatistics*. Oxford UP, NY, 2000); and Christakos G and Hristopulos DT (*Spatiotemporal Environmental Health Modeling*, Kluwer Acad, MA, 1998).

[k] The term "chronotopologic" includes the cases of space-time and spacetime data variability.

Example 1.5 While *CTDA* concepts and techniques can be implemented to study a large variety of natural attributes (Example 1.2 earlier), they may not be necessarily useful when points represent merely the presence of events (e.g., earthquake occurrence), people, or some physical object (e.g., volcanoes, buildings).

CTDA methods are based on a rigorous problem definition, which is then used to formalize a *CTDA* protocol in order to answer the questions of interest. In this respect, *CTDA* is a systematic inquiry to produce new knowledge, refine or validate the existing knowledge regarding the natural attributes considered, and generate attribute estimates.

- The **interdisciplinary** mantra of *CTDA* is to not work in silos and to value that which is cross-cutting.

In this setting, *CTDA* surely benefits from developments in *TGIS* technology. This technology allows the chronotopologic visualization of a variety of attributes such as individual populations, quality of life indices, pollutant distribution, disease spread, and company sales over time in the region of interest. To achieve that, it is enough to have a chronotopologic database, and the *TGIS* is capable of presenting an animated set of colored maps that provide informative visualizations of the chronotopologic pattern of the attributes of interest.

Example 1.6 *CTDA* applications currently span many disciplines, with their methods varying in relation to the specific questions being addressed, whether predicting air pollution, examining suspiciously high frequencies of disease events, or handling the vast data volumes being generated by the *Global Positioning System* (*GPS*) and *Satellite Remote Sensing* (*SRS*).

1.5.2 Generic CTDA notions

CTDA is a broad framework for studying natural attribute distributions, linking attribute coordinates, *in-situ* conditions and values. Considerable information exists individually on each of these items, which must be brought together to assure consistency in concepts and methods. *CTDA* provides some initial principles and notions, as follows.

- The **chronotopologic unit** is the basic building block for the *CTDA* of space- and time-specific natural attributes.

The choice of the chronotopologic unit should be based on a relevant set of criteria (physical, ecological, social, etc.) and it may affect the *CTDA* results.

Example 1.7 A chronotopologic unit may be associated with observable characteristics of terrestrial, marine and atmospheric domains. It may correspond to the smallest or elementary working unit that is representative of these characteristics.

Information on natural attributes may be available in several different scales (say, city, county and state), and an adequate *CTDA* may require conveying information from one scale to another (see, also, Section 4 of Chapter 3). This also includes the methods of transferring information from one chronotopologic point to another.

- **Chronotopologic scaling** is the process of conveying information from one chronotopologic scale to another.

Scaling may take different forms: Downscaling (conveying information from larger to smaller chronotopologic domains), upscaling (conveying information from smaller to larger domains), and transferring (conveying information from one point to another).

Example 1.8 A general existing ecosystem data analysis framework operates at the national level, whereas a site-specific study may require a much finer chronotopologic scale. Investigators need to understand when and how to convey information from the point (e.g., pollution monitoring) to the chronoregion (e.g., city population exposure) scale.

Adequate information gathering is important in understanding *CTDA* results as well as for communication purposes.

- **Chronotopologic aggregation** is the process of reducing certain measures characterizing the chronotopologic attribute distribution into simpler ones.

Aggregation may refer to chronotopologic units or the corresponding natural attribute values. Adequate aggregation relies on a satisfactory understanding of the underlying mechanisms, interrelationships, and their relative importance. These measures may be the same across different scales, or they may differ between scales, in which case the appropriate links between them need to be established. Also, different *CTDA* models and techniques may have their own chronotopologic units, scaling notions, and aggregation tools.

Example 1.9 Aggregation may consist in combining several data variability measures into a simpler set of measures. Aggregation may take the form of indicators, indices, summary statistics, and graphs. Disparate measures may be aggregated using a reference state or a common factor, such as unit of measurement.

1.5.3 *Fourfold CTDA objectives*

In light of the above deliberations, we chose to conclude this first section of the book by making some suggestions regarding the main objectives of an adequate *CTDA* study.

Whatever the specifics of the real-world case under investigation, the ultimate judgment of progress in *CTDA* should be **measurable results** obtained during a **reasonable time** based on **better science** and more **powerful technology**.

The adequacy of the above fourfold objective will become apparent throughout the book with varying emphasis on its individual objective, depending on the context. After all, it should be kept in mind that this is an instructional book that relies on sound theory, scientific reasoning, and substantive knowledge, rather than on mere data massaging techniques and popular quick fixes in challenging times.

2 Chronotopologic variability, dependency and uncertainty

As *CTDA* techniques become powerful tools in many fields (earth sciences, oceanography, geography, natural resource management, biological conservation, and societal processes), adequate quantitative characterizations of good quality chronotopologic datasets of natural attributes are increasingly needed.

2.1 Conceptual chronotopology

Accordingly, some further comments concerning the notions of chronotopology and its geometric dynamics are in order at this point. We start with the following two key notions:

Chronotopologic **variability** refers to the degree of the attribute's joint change (fluctuation) across space and time, whereas chronotopologic **dependency** refers to the degree of the attribute's joint smoothness (connectivity) across space and time.

That is, although both notions are concerned with differences between attribute values at different points, there is an inverse relationship between these two notions. In the real world, chronotopologic variability or dependency takes up particular forms that are functions of certain factors, like the nature of the phenomenon and the chosen representation of its chronotopologic domain. The attribute data used in a study may come from diverse, and sometimes unexpected, sources set apart by their distinct variability (dependency) features.

Example 2.1 When studying air temperature data all over the globe during 150 years in order to describe global climate change (e.g., Jones et al., 1986), one could calculate temperature averages at a given moment (say, day), but these numbers would not take into account the natural chronotopologic variability of these data (at any given moment, the range of temperatures around the world is huge due to different local climates, seasons, topography, whereas no temperature data exist over large portions of the oceans). In addition to the standard temperature datasets, recent studies of global warming relied on many other data sources, including data on the timing of the first appearance of tree buds in spring, greenhouse gas concentrations in the atmosphere, and measurements of isotopes of oxygen and hydrogen from ice cores (Egger and Carpi, 2011). The study of such diverse and multisourced datasets is essentially interdisciplinary, i.e., it requires the integrated efforts of scientists with a variety of backgrounds and expertise, see the report of the Intergovernmental Panel on Climate Change (IPCC, 2007).

The profound inverse relationship that exists between the notion of variability and that of dependency is described by the following relationship:

- **Complementarity relationship**: the presence of higher (lower) attribute variability in a chronotopologic domain would mean weaker (stronger) attribute dependency in the domain.

For reasons that will become evident later, complementarity assumes a key role in *CTDA* practice. Another essential property of chronotopology is as follows:

- **Dataset continuity**: in principle, data can be considered at any point of the chronotopologic domain.

It should be noticed that data continuity in the above sense refers to the attribute chronotopology rather than to the attribute value per se. In the latter case, attribute value continuity means that the attribute (e.g., chemical concentrations) can assume any value within a specified range to be distinguished from discrete attribute values that can be counted (e.g., number of days an individual has been exposed to a pollutant) and can take a finite number of possible values. Although chronotopologically continuous datasets play a pivotal role in investigations in the fields of earth monitoring, ocean research, environmental risk assessment and planning, public health management and decision-making, they are usually not readily available and are often difficult and/or expensive to acquire (e.g., in mountainous and deep marine regions).

Example 2.2 The following real-world situations illustrate the significance of continuous datasets and the need for *CTDA* techniques that can effectively compensate for their lack.

- Environmental managers require chronotopologically continuous data over a region to make effective and informed decisions concerning the prevention and control of severe pollution, and scientists need accurate data that are well-distributed in a chronotopologic domain to make justified interpretations.
- The marine environment in Australia is a typical case where seabed mapping, habitat classification, and prediction of marine biodiversity, essential for marine biodiversity conservation, need reliable continuous data of the marine environment. Yet, in most of the Australian marine regions such data are not available, and only sparsely and unevenly scattered point samples have been collected. Hence, chronotopologic interpolation techniques (Section 4) are essential for computing biophysical variables at the unsampled locations.
- Botanists collect information that is discrete at a particular set of sampling locations and time instances, but has to be converted in some way into useful dynamic map representations. To achieve this goal one

needs to manipulate attribute data in a way that accurately estimates attribute values in chronotopologic domains that have not been sampled. In essence, one needs from the information available to produce continuous maps at unknown domains.

2.2 Quantitative chronotopology

At the center of *CTDA* is the quantitative representation of a natural attribute that varies jointly across space and time.

A natural attribute is represented as a **mathematical function** $X(\mathbf{p})$ of the point $\mathbf{p}$ varying in a chronotopologic domain, where $\mathbf{p}$ consists of spatial coordinates $\mathbf{s} = (s_1,\ldots,s_n)$, the number n depending on the dimensionality of the region,[l] and a temporal coordinate (time instance) t.[m]

A digression may be appropriate at this point. To facilitate the writing, below we specify the notation to be used throughout the book, once and for all.

- **Regular English** letters are used to represent mainly natural attributes and chronotopologic arguments (space, time and spacetime); **bold English** letters denote vectors (e.g., assigned to points in a chronotopologic domain and locations in a topologic region); **Greek** letters represent attribute realizations; **bold Greek** letters denote vector realizations; and **subscripts** are used to identify knowledge bases, specific datasets, geographical coordinates and time instances.

As the following example illustrates, chronotopologically distributed data are often collected from point sources $\mathbf{p}$, which are defined in terms of their spatial locations and time instances throughout the domain of interest.

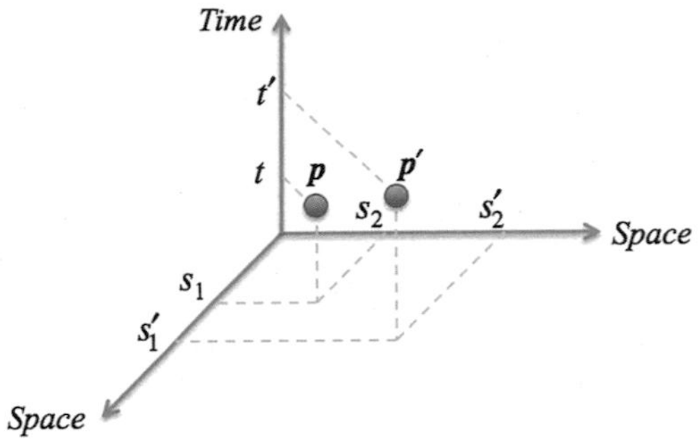

FIG. 2.1 An example of chronotopologic point data coordinates.

Example 2.3 Two points $\mathbf{p}$ and $\mathbf{p}'$ (dots) are shown in Fig. 2.1. The spatial coordinates of the point $\mathbf{p}$ are $\mathbf{s} = (s_1, s_2)$ and the time instance is t, which are always combined with chronotopologic coordinates; similar is the case of point $\mathbf{p}'$. Typical cases of chronotopologic point data can be found easily when natural attributes are studied such as, e.g., soil moisture, contaminant concentration, disease incidence, ocean salinity, sea surface chlorophyll concentration, temperature, and humidity. The distribution of natural attribute values varies over an infinite number of points, thus forming what is often called a chronotopologic pollutant distribution (enviromentalism), a disease spread pattern (epidemiology), or a moving surface (geography). At these points, the location on the earth's surface may change in elevation, orientation, or proximity to a feature, and the associated time instance may vary within a specified period (day, season, year, etc.) or frequency. The term "associated" is used here to emphasize the important issue that in the chronotopologic context space and time are linked and not independently varying.

As should be expected, a central objective of *CTDA* is the development of mathematical models that account for the chronotopologic

[l] Usually, $n = 2$ or 3.

[m] In the same *CTDA* setting, "location" is a topologic entity, whereas "point" is a chronotopologic entity.

variability (or dependency) of the natural attribute $X(\boldsymbol{p})$ or $X(\boldsymbol{s},t)$ in a rigorous and interpretable manner. As it turns out, the following mathematical operator plays a pivotal role in *CTDA*.

Quantitative models of attribute variability (dependency) represent dynamical (time-dependent) processes across space described by a mathematical **operator** $S[X(\boldsymbol{p}_i), X(\boldsymbol{p}_j)]$ for any points $\boldsymbol{p}_i$ and $\boldsymbol{p}_j$.

Based on the form of the *S*-operator,[n] these variability (dependency) models have different properties that guide their implementation in practice. Accordingly, some of them are data-depended whereas some others are data-free. We postpone a detailed consideration of the attribute variability (dependency) models until Chapters 4–6.[o]

2.2.1 *Revisiting Tobler's law*

Remarkably, *CTDA* analysis may generate some interesting *albeit* sometimes unexpected results. For illustration, one such result is related to the well-known 1st law of geographical science, which consists of an intuitively rather profound statement by *Waldo R. Tobler* Tobler (1970):

✍ Everything is related to everything else, but near things are more related than distant things.

As *CTDA* shows, this intuitively appealing empirical statement does not apply to all natural attributes, including those attributes exhibiting periodic or wave-like variability, where attribute values at longer distances and separations apart may be substantively more related than values at close distances and separations.

Example 2.4 Two typical cases of this situation are shown in Figs. 2.2 and 2.3.

- Fig. 2.2 (also, Example 3.21, Section 3 of Chapter 5) shows the variogram γ_X of sea surface heights that exhibits a wave pattern (the variogram is a function that measures attribute variability across space and time, Chapter 5). As is determined by the

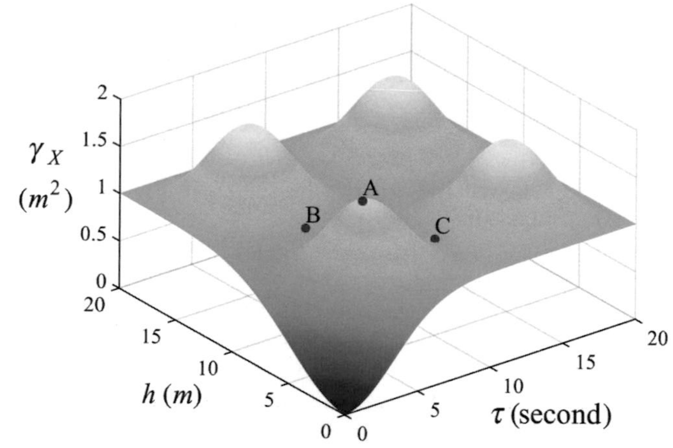

FIG. 2.2 The variogram of sea surface heights exhibiting a wave variation pattern.

[n] For technical details, see Eqs. (2.5a–2.5b), Section 2 of Chapter 5.

[o] The covariance and variogram are the mainstream chonotopologic variability models expressed by the *S*-operator. Additional models include the trivariance, contingogram and sysketogram.

[p] As will be discussed in Chapter 5, the lower the variogram value the higher the chronotopologic correlation (dependency) between the attribute values at the corresponding pair of points.

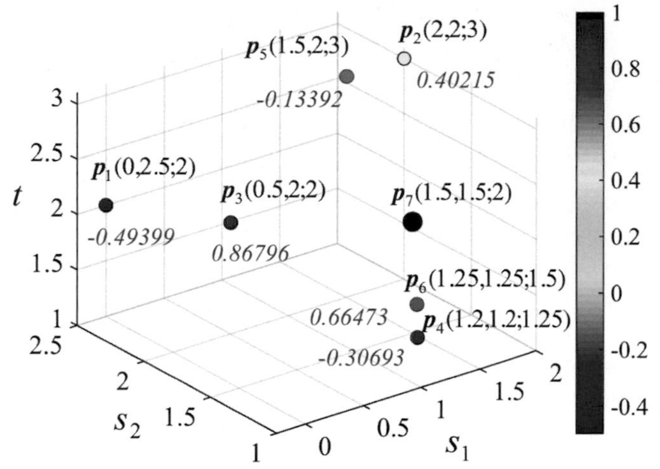

FIG. 2.3 Data points p_i ($i = 1, \ldots, 6$) and interpolation point p_7. The attribute value is indicated by the color level of the dot according to the scale legend on the right. The triplet within a parenthesis denotes coordinates $p_i = (s_{i1}, s_{i2}; t_i)$, and the italic number is the contribution of datum at p_i to interpolation at p_7. The contribution of p_6 is much larger than those of p_1, p_2, p_4 and p_5 that are farther away from p_7; but the contribution of p_3 is larger than those of p_4 and p_6, which are both closer to p_7 (this is called the "screen effect").

variogram γ_X values,[p] the sea surface heights at a space-time lag represented by C (distance 5 m and separation 10 s) with $\gamma_X = 0.9614\ \text{m}^2$ are more closely related than heights at a lag B (distance 10 m and separation 5 s) with $\gamma_X = 0.9945\ \text{m}^2$, and, in turn, heights at a space-time lag B are more closely related than heights at a lag A (distance 5 m and separation 5 s) with $\gamma_X = 1.6160\ \text{m}^2$.

- Chronotopologic interpolation would assign bigger weights to faraway points than to near ones.[q] This is, in fact, the case of Fig. 2.3 that represents the so-called *screen effect*, where p_i ($i = 1, \ldots, 6$) denote data points and p_7 the interpolated point (the triplet of numbers within the parentheses indicate space-time coordinates (s_{i1}, s_{i2}; t_i)). The italic numbers represent the contribution of each datum at p_i to the interpolated value p_7. A positive value represents a positive contribution, whereas a negative value represents a negative contribution, and the interpolated value was obtained using the space-time ordinary Kriging interpolation technique (to be discussed in Chapter 8). As indicated by these numbers, samples that are further away from the attribute interpolation point may have a greater influence on the interpolated value than points much closer to the interpolation point.

Rather not surprisingly, then, these examples seem to have illustrated numerically the following assertion:

- The first half of the empirical Tobler's law (i.e., "Everything is related to everything else") is valid more often in practice than the second half (i.e., "near things are more related than distant things").

[q] These violations of Tobler's law actually may be the starting point of valuable discoveries. After all, the progress of any scientific field is based on the derivation of unexpected and previously unknown results.

As it turns out, this assertion points out another significant issue that is worth revisiting in later chapters.

2.3 Concerning real-world uncertainty and its probabilistic description

In addition to chronotopologic attribute variability, another key characteristic of *CTDA* is the frequent lack of *in-situ* certainty. Every investigator and practitioner of almost every scientific field can testify to this fact. Otherwise said, humans do not seem capable of the kind of lasting immunity from doubt that the ideal of certainty requires.

Example 2.5 When studying changes in global mean surface air temperature from 1861 to 1984, Jones et al. (1986) used processing techniques to correct for uncertainties and inconsistencies in the historical data not related to climate. As they notice, "Sea surface temperatures were measured using water collected in uninsulated, canvas buckets, while more recent data come either from insulated bucket or cooling water intake measurements, with the latter considered to be 0.3-0.7 °C warmer than uninsulated bucket measurements." Accounting for these uncertainties is a complicated matter, since most sea surface temperature data do not include a description of what kind of bucket was used. Moreover, the air temperature measurements over the ocean were obtained aboard ships of unknown type and size. But ship type and size determined the height at which measurements were obtained, which introduced additional measurement uncertainties (temperatures can change rapidly with height above the ocean).

What is required, then, is a comprehensible and widely applicable notion of uncertainty. As it turns out, however, this is not necessarily the case in the real world.

2.3.1 Multifaceted notion

In fact, what makes matters of uncertainty so complicated is that the following argument can be made with actual certainty: There is not a unique definition of **uncertainty**, which is a multifaceted notion (ontic, epistemic, or technical), which may emerge in a number of distinct ways and may refer to different things.

Given, then, that there is considerable uncertainty about the definition of uncertainty itself, the associated *probability* notion may refer to more than one consideration:

❶ Probability as a **mathematical quantity**: a *number* belonging to the range between 0 and 1 (or 0 and 100%) expressing the frequency of occurrence of possible attribute values.

❷ Probability as a **state of mind**: a state of *doubt* or *lack of certainty* about the true attribute value, or a state of *incomplete understanding* of the phenomenon, or a state of *surprise* regarding the outcomes of experiments or future events.

❸ Probability as an **irreducible (inherent) property** of the natural phenomenon: representing, e.g., *Heisenberg*'s uncertainty principle of phenomena in microscales.

Example 2.6 Uncertainty in science is closely related to the fact that most data have a range of possible values as opposed to a precise single value). It might be enlightening to view matters from a layman's perspective. The layman's classification of the laws of uncertainty rather distinguishes between four kinds, as described in Table 2.1.

TABLE 2.1 Layman's laws of uncertainty.

Law	Description
Inevitability	Almost every investigation step contains a factor of uncertainty, it's an inevitable part of doing science in the real world
Invisibility	The absence of evidence is not evidence of absence
Obviousness	In many data collections, the figure most obviously correct could be the mistake
Interpretation	No matter what the result, there is always someone who could misinterpret it

Yet, despite the considerable and multifaceted difficulties caused by the above aspects of uncertainty, it is widely acknowledged that to understand anything about any phenomenon of nature, one must understand its uncertainty. Accordingly, the next suggestion is a reasonable one:

- Uncertainty-related assumptions about a phenomenon should be made on the basis of **rational reasons**, i.e., they should be justified assumptions.

Rational reasoning shows that data uncertainty does not necessarily imply that the data are wrong and useless. Quite the contrary, rigorous uncertainty quantification becomes very important in this respect, since the magnitude of uncertainty offers an assessment of how confident the investigators would be on the accuracy of their data (i.e., higher uncertainty implies lower confidence to the data).

Example 2.7 While the uncertainty level of historical air temperature data increases going further back in time, it has been reduced considerably, for all practical purposes, in more recent years starting around 1900 (Mann et al., 1998). In another case, although one cannot specify the exact amount of precipitation during next month, nevertheless, one can assume, as practically certain, that there will be no major weather pattern changes. Hence, the known weather conditions of previous years provide good grounds for reasonably assuming a certain probability rainfall distribution.

2.3.2 *The limited notion of practical certainty*

Surely, the level of rigor associated with uncertainty determination may also depend on the study conditions and goals. In this respect, the following notion could be useful in applications:

A notion of **practical certainty** would be appropriate (*a*) when the supporting argumentation is rational or justified enough to be relied upon in practical deliberation and in deciding what to do, or (*b*) when the stakes are low, whereas a more rigorous or stricter uncertainty notion may be appropriate when the stakes are higher.

In *CTDA*, one can distinguish between temporal *vs*. intemporal practical uncertainty (i.e., certainty at a specific time *vs*. lasting certainty), as well as, between context-dependent *vs*. context-independent practical certainty.

Example 2.8 While in certain applications it is often sufficient to focus on the conditions required for a phenomenon to be certain at a time or within a certain context, in other applications, it is recognized that a phenomenon might be certain at one time (when a clear and distinct perception of the phenomenon has been established) or within a particular context but not at another time (when one no longer has or is not attending to this clear and distinct perception) or within a different context.

2.3.3 *Sources of uncertainty*

Real-world case studies encounter considerable uncertainties from different *sources*, which is why *CTDA* studies put great emphasis on the reliability and quality of the data used. A list of key uncertainty sources is given in Fig. 2.4, and a brief review of them follows (see below items ❶ through ❹).

❶ One of the most common sources is ***in-situ* data** uncertainty, which is linked to experimental, observation and interpretation errors, heterogeneity and scale variation, as well as to limited or sparse data, see Example 2.1 of Section 2.

❷ One of the most important sources is **model** (or **conceptual**) uncertainty, i.e., whether or

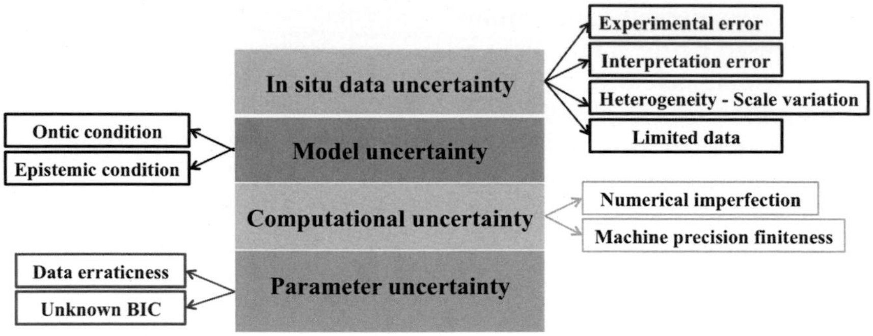

FIG. 2.4 Uncertainty sources considered in *CTDA*.

not a mathematical model used by *CTDA* is an adequate representation of the real phenomenon.

Model uncertainty may be linked to the ontic condition, which refers to an inherent feature of the natural phenomenon, and the epistemic[r] condition, which refers to incomplete understanding of the phenomenon.

Example 2.9 This kind of uncertainty may be due to imperfections and idealizations (due to ontic or epistemic conditions) made in the construction of a model representing a physical phenomenon, and to the choices of the relevant probability functions to describe statistically the model parameters.

Altera pars, perhaps not surprisingly, there are several *CTDA* studies in which uncertainty is linked to the opposite situation, as follows:

- Uncertainty may be linked to an **unmodeled** source that often can cause errors and biases that are chronotopologically correlated.

Example 2.10 Geomagnetic field studies consider both measurement errors, i.e., instrument accuracy (see data uncertainties from the Swarm mission; Tøffner-Clausen et al., 2016) and errors from unmodeled sources (even the most modern satellite field models do not fully capture the ionospheric processes, particularly in the auroral region; Finlay et al., 2017).

❸ **Computational** uncertainty could be the result of imperfection of a numerical technique and the finiteness of machine precision.

Example 2.11 Computational uncertainty may characterize the sensitivity of a numerical problem solution taking into account the errors and inaccuracies of inputs, algorithms, truncations, and roundings.

❹ **Parameter** uncertainty quantifies the lack of knowledge about the actual value of a parameter, and is linked to data erraticness, physical properties, process optimization, design variables, rate coefficients, and unknown boundary and initial conditions (*BIC*).

Example 2.12 Parameters that are inputs to computer models are uncertain when their exact values can neither be computed in physical experiments nor be inferred by statistical techniques. When more than one parameter is involved, their uncertainties are correlated and their proper merging is needed to fully assess uncertainty in an attribute derived from more than one parameter.

[r] The branch of philosophy that concerns itself with knowledge is called epistemology, or methodology, or philosophy of science. It is the focus of some important developments in *CTDA* (see, also, subsequent chapters).

2.3.4 Consequences of uncertainty

Uncertainty can have serious consequences in many areas of scientific inquiry. In particular, each of the above uncertainty sources affects basic issues of attribute modeling, estimation, and mapping,[s] as follows:

- The reliability and accuracy of the chronotopologic attribute **maps** that one derives depend to a considerable extent on the uncertainty conception that one uses.
- Another issue is the **propagation** of initial uncertainties by physical laws (chaos phenomena).
- Uncertainty has a profound influence on **risk analysis** (environmental impacts, health effects, social implications), **decision-making** (optimality criteria), and **planning** (multiple objectives, economic factors, public policy).

In view of the above concerns, in *CTDA* studies investigators take into consideration not only what the data suggest, but also the data quality and the consequences of uncertainty as regards various courses of action under the specific circumstances.

2.3.5 Quantification of uncertainty

Another central objective of *CTDA* is the development of quantitative models that describe the uncertainty sources of the natural attribute $X(\boldsymbol{s},t)$ or $X(\boldsymbol{p})$ in a rigorous and interpretive manner. This is, in fact, the role of a variety of probability models representing the corresponding multifaceted notion of probability (introduced in Section 2.3.1 above).

> Quantitative representations of attribute uncertainty include the **probability** models (probability density and distribution functions) describing the occurrence conditions of possible attribute values, and their **statistical** functions (mean, variance, covariance, variogram).

Example 2.13 When attribute uncertainty is fully characterized by a normal (Gaussian) probability model, it can be partially summarized in terms of the model's standard deviation (also called error bar) measuring model's width. Assessing the uncertainties of a posteriori probability model requires adequate observation error assessment and prior parameter information.

Remarkably, the underlying probability notion at the core of these models does not admit a unique interpretation. As a matter of fact, two essential interpretations may be considered, as follows:

> Probability may be a fundamentally **epistemic** concept, i.e., the probability of an attribute value is relative to some body of knowledge. If a body of knowledge is construed as a set of statements about the phenomenon, then probability becomes a **logical** concept.

Example 2.14 As a further classification of the twofold probability characterization above, Table 2.2 presents five commonly encountered cases of probability interpretation that depend on the context considered.

As it turns out, different probability models have distinct properties that guide their implementation in practice. Some of them are more informative but more difficult to develop, whereas some others are easier to compute but offer a limited representation of the actual attribute uncertainty.

Example 2.15 To mention a well-known case, there is a high demand for more informative probabilistic models of weather forecasting and extreme event warning (hurricanes, earthquakes, tsunamis, etc.); yet, admittedly, these models become increasingly complex too.

At the moment, it seems appropriate to postpone until a later chapter the presentation of a

[s] We will revisit each one of these three *CTDA* components, at various parts of the book.

TABLE 2.2 Probability interpretations.

Probability interpretation	Contextual description
Classical	The ratio of cases in which an attribute occurs over all possible cases
Frequentist	The value of classical ratio when number of cases is very large (tends to infinity)
Geometric	The ratio of the domain in which an attribute occurs over the entire domain of interest
Subjective	The result of personal experience, preferences and intuition
Scientific	The result of rational assessment based on logical reasoning and scientific knowledge

more detailed discussion of the quantitative uncertainty models employed in scientific applications.

3 Theory and evidence

A chief objective of *CTDA* is to adequately use the existing body of knowledge in a specified area of science in an efficient and systematic way in order to better understand a phenomenon and make predictions. Let us start with a brief yet concise description of what constitutes a "theory" and what constitutes "evidence".

Theory is the product of constructive imagination and logical thinking (mathematical models, scientific theories, physical laws, and hypotheses),[t] whereas **evidence** is the product of inquisitive observation and practical experience (evidential support, perceptual experience, empirical data, measurements, recordings, surveys) of various degrees of reliability and uncertainty.

The above definitions emphasize both the clear *distinction* between theory and evidence about the natural attribute of interest and their strong *complementarity*. Another interesting distinction should be made between theory, which is about predicting what has not been observed yet, and *inference*, which is based on evidence alone and cannot exceed the domain of the data it uses.

As regards *scientific* theories, they typically develop gradually over time into increasingly complex models representing concentrated human knowledge that often spans several centuries.

Example 3.1 In physics, the original Newton's theory was advanced considerably during the following three centuries, through the works of Bohr, Einstein, Heisenberg, Higgs, and many others to form today's Standard Model.

This gradual increase in complexity may be due to various reasons, i.e., as the understanding of the phenomenon improves, new variables need to be included in the theory, thus increasing its complexity, or, a phenomenon may manifest only at a higher level of complexity and not at lower levels, or, a theory that is valid in one scale maybe not so in a different scale. On the other hand, a deeper physical understanding can lead to theory simplification, i.e., by omitting variables that play a minor role in the phenomenon, or certain correlations are found to have negligible effects.

3.1 The value of theory

The value of theory has been emphasized throughout the centuries by many great thinkers. According to *Leonardo da Vinci*:

✍ He who loves practice without theory is like the sailor who boards a ship without a rudder and compass and never knows where he may cast.

[t] Classical examples of theories are Coulomb's law of electrical attraction and repulsion, Maxwell's electromagnetic equations, and the ideal gas equations describing relations between temperatures and pressures of enclosed gases.

In the words of *Auguste Comte*:

✍ If it is true that every theory must be based upon observed facts, it is equally true that facts cannot be observed without the guidance of some theory. Without such guidance, our facts would be desultory and fruitless; we could not retain them: for the most part, we could not even perceive them.

And, writing about *Charles Darwin*, Lionel Ruby (1966) notices:

✍ Darwin and a fellow scientist were searching for fossils in the north of England. They were not aware of the glacial theory at the time. Years later Darwin revisited the area, and he was now astonished to discover how clearly marked were the glacial ridges on the rocks. He had not noticed them on his earlier visit because he was not looking for them.... Darwin was able to appreciate the glacial markings only after he became aware of the glacial theory.

Sufficiently motivated by these words of wisdom, next we examine some other valuable uses of theory.

3.2 Against convenient slogans

Science-based *CTDA* ought to proceed with caution and avoid the oversimplifying slogans of purely *data-driven* analytics. Chief among them is the convenient slogan encouraging investigators to:

☞ Let the data speak.

Yet, surrendering oneself to the data is often a naive policy with many hidden dangers (see, also, discussion in Section 3 of Chapter 3). One of these dangers is that those surrendering themselves to this convenient slogan often fail to appreciate the fact that an attribute dataset is not just a set to be manipulated by statistical techniques, their numerical correlations computed, and purely number-driven (content-free) conclusions drawn. Yet, reality is more complicated and lies beyond the convenient slogan above.

A dataset has a natural "soul" that needs to be explored strictly within the specified **phenomenal context** so that **theory-driven** (content-dependent) conclusions can be drawn.

The knowledgeable investigators understand that the same numbers may have very different meanings, depending on the physical characteristics of the attribute they represent, and their possible ranges are restricted by these characteristics. Any number representing an attribute value has units that assign to it a physical meaning, and any data-processing technique should be conditioned by this meaning.

Example 3.2 A negative temperature value makes both numerical and physical sense. However, a negative permeability value, although makes numerical sense it surely does not have physical meaning. To declare that the ozone concentration at a certain location is, say, 5, is like to provide a meaningless number, unless physical units, *ppm*, are assigned to it. Furthermore, this number should be consistent with other numbers representing concentrations that are linked to it by a physical law.

Another hidden danger associated with the convenient slogan above is the consideration of apparent relationships among the data that do not actually exist.

- Currently, the only way to avoid the traps of such **spurious correlations** among the data is the protection provided by a sound theory.

If, instead, one focuses on correlation and ignores considerations of theory or mechanisms, one may be engaging in flawed reasoning. Therefore, if investigators really need a slogan, then a much more appropriate one should be to:

☞ Let the theory speak about the data.

Indeed, it is a theory that can take data analysis beyond mere numerics and assign the necessary meaning to the data and avoid common numerical traps like spurious data correlations. Also, theoretical developments may be more captivating than empirical evidence in cases where randomization and replication are difficult and several confounders exist.

3.3 Evidence unavailability

It is not surprising that the necessity of theory in many real-world investigations is also established on the basis of practical evidence-related reasons. Indeed, people need to understand the world, and when sufficient empirical evidence is not available, they naturally tend to rely on theory.

> Theoretical developments may be necessary because of **unavailable evidence**, i.e., in many real-world cases, evidence is not available at the current state.

Example 3.3 Evidence concerning the future occurrence of a phenomenon is usually unavailable, and empirically inaccessible attribute domains may exist.

As a matter of fact, logically the notion of "future evidence" may be a contradiction in terms, since the "evidence" expresses a current state (i.e., one definitely talks about "current evidence"), whereas the future refers to an unrealizable yet state (which may or may not be realized in the future, but this is not determined with certainty).

Example 3.4 When investigators study the weather conditions at a future time, they need to rely heavily on theoretical developments in the form of complicated weather models and hypotheses, surely supported by past evidence, whereas direct evidence of future events is profoundly absent. Similarly, when one attempts to forecast future developments in financial affairs or politics, future evidence is, indeed, a contradiction in terms, and one has no other option than to theorize or hypothesize about possible future scenarios.

3.4 Synergy

The crux of the matter is that to understand the world, one needs both theory and evidence. One should take advantage of the available evidence in all its forms (traditional and emerging) but without ignoring the treasure of core knowledge accumulated in scientific theories. And, the deeper meaning of the distinct yet complementary features of theory and evidence discussed above can be summarized as follows:

> Beyond the distinct yet crucial roles theory and evidence individually play in scientific investigation, perhaps even more important is the realization that they act in **synergy**, i.e., there is an intrinsic and essential relationship between the two.

Some key features of theory and evidence, together with some comments concerning their interplay, are presented in Table 3.1.

Example 3.5 Interesting cases of the synergy between theory and evidence in two different scientific disciplines are as follows:

- In modern health sciences, there is an urgent need to combine evidence bases with theoretical and computational models in order to adequately investigate the huge amounts of data from healthcare and medical research.
- In earth sciences, many applications merge the information contained into geophysical observations with that coming from dynamical models, a process often known as assimilation.

Although theory develops before any evidence becomes available in the form of data, yet it can be properly modified (updated or

TABLE 3.1 Theory, evidence, and their interplay.

Theory feature	Empirical evidence feature
Theory is a scientific and systematic description of knowledge that usually relies on a set of assumptions about the phenomenon of interest	Evidence is obtained by observing natural and experimentally generated events, objects, and effects
Theory is customarily represented as a collection of sentences, propositions, beliefs and formulas, and their logical consequences	Evidence sources are theory-laden observations and research findings. If evidence is separated from theory, it is disconnected from its origin and it may become devoid of scientific meaning
Theory uses evidence (real data) to calibrate its parameters	Evidence-based practice applies the most credible and reliable evidence available in order to understand and describe a phenomenon
Theory and theorizing are cornerstones of a scientific discipline. Research findings and conclusions are usually presented in the form of theories and hypotheses	Evidence is a conceptual term, and, thus, it is linked to knowledge fields that are conceptual. As such, the evidence does not look the same to observers with different conceptual resources[a]
A sound theory helps avoid both unanchored abstract thought and a narrow empiricism that concerns itself only with observation or measurement and unidirected data gathering	Evidence depends on the relevant context. By changing the context, the notion of evidence may change and even become problematic[b]
A theory can be challenged and defended in different ways and at different times as new empirical evidence and investigative techniques are introduced	Very few investigators would be convinced by a study drawing an apparently interesting conclusion without a plausible, at least, theory adequately fitting the relevant data

[a] *For example, to measure soil porosity one associates it with an appropriate concept; and proponents of a caloric account of heat do not understand heat experiment results in the same way as those who view heat in terms of mean kinetic energy or radiation.*
[b] *For example, if dark clouds are observed in the sky, there is a high likelihood of rain, but if dark clouds are observed when the temperature is far below freezing, there may be no rain.*

calibrated) when new and reliable evidence emerges in the process of scientific inquiry. Actually, this seems to be an appropriate procedure for combining theory with evidence. In practice, theory and evidence should rest comfortably with each other and they should also reinforce each other (detecting data correlations facilitates theory testing and refinement; and combined with the increasing computing power, the data may suggest different ways of developing a theory).

- When significant **differences** are found between theory and evidence, they can lead to improvements in both the way a theory is constructed and the way the evidence is obtained, including the type of evidence that counts in scientific reasoning.

If a theory suggests that the current evidence is inadequate, it implies that more research on that specific subject is necessary, whereas when hard data contradict a theory or the theory cannot explain it, the theory must be revised accordingly. Naturally, theory and practice have not been developed to the same degree in all disciplines.

Example 3.6 Weather forecasting is a discipline that is both theory- and practice-strong, whereas public health is considered a theory-weak and practice-strong discipline.

By way of a summary:

> Evidence is **necessary** for science but **not sufficient**. Evidence needs to be integrated with theory in order to have any explanatory and predictive value.

Ut ita dicam, a more appropriate goal should be seeking "content-rich data" or "content-deep data" rather than merely "big data" (Section 3 of Chapter 3). A detailed discussion of several important theory- *vs.* evidence-related topics will be presented in Chapters 3 and 5.

4 Chronotopologic estimation and mapping

While the study of attribute variation or dependence (introduced in Section 2) aims at a deep understanding of the chronotopologic structure and laws of change of the natural attribute of interest, another key *CTDA* objective is the following.

> **Chronotopologic estimation** seeks the derivation of estimates $\hat{X}(\boldsymbol{p})$ of an attribute $X(\boldsymbol{p})$ at unsampled points within the chronotopologic domain of interest, using all available information about the attribute.

Within the broad context of chronotopologic estimation, in a space-time or a spacetime setting, two very important special cases are considered:

- Chronotopologic **interpolation** that seeks to fill in a domain with attribute values at unsampled points.
- Chronotopologic **extrapolation** that seeks to obtain attribute values beyond this domain.

Interpolation is clearly a much easier problem than extrapolation.

4.1 Threefold conditions

It is worth stressing that in epistemic (knowledge-theoretic) terms, chronotopologic attribute estimation is not a categorical device, but only a *conditional* one. Accordingly, there are three main conditions of attribute estimation that are worth keeping in mind:

❶ Attribute estimation is **uncertain** and **approximate**.
❷ It depends on one's **knowledge** (models, theories, data) about the phenomenon of interest.
❸ It is warranted only if the **assumptions** made are reasonable and realistic.

Condition ❶ is an implication of the conditionality of estimation (i.e., the data are too weak to support categorical estimation of perfect reliability), in which case the estimated attribute values are uncertain and only approximations of the real *albeit* unknown ones. Condition ❷ is about how much data will support how much conditional estimation. Regarding condition ❸, one should distinguish between the assumptions one rationally and justifiably makes about the phenomenon and the modeling assumptions one uses in estimation. Otherwise said, if the characterization of the chronotopologic variability and the *in-situ* uncertainty of the natural attribute $X(\boldsymbol{p})$ is valid and meaningful, then the estimated values $\hat{X}(\boldsymbol{p})$ will be accurate and informative.

As it turns out, the restrictions imposed on estimation by the conditions ❶, ❷, and ❸ above may not necessarily keep estimation models and techniques from being informative and useful in practice. An instructive example is appropriate here.

Example 4.1 We offer our readers two illustrations of the above claim:

- That the law of gravity is not perfectly reliable, because of wind currents and other intervening factors, is not an excuse for one to ignore its predictions and decide to jump from a ravine. Any common-sensical human being understands that despite its imperfection, the law remains reasonably valid for all practical purposes.
- Similarly, to decline to make critical climate change estimates on the grounds that the data is imperfect (as many investigators have argued) would probably not be a good decision, especially given the gravity of the matter. As it turns out, investigators can indeed draw sufficiently informative inferences with the evidence we have.

4.2 Confidence and accuracy

On a relevant note, since chronotopologic estimation is subject to error, it is necessary to assess how much uncertainty can be tolerated and how much *confidence* can be placed in the generated attribute estimates. *Arthur Rudolph*, the scientist who developed the Saturn V moon rocket, expressed this critical issue very succinctly:

✍ You want a valve that doesn't leak, and you try everything possible to develop one. But the real world provides you with a leaky valve. Then, you have to determine how much leaking you can tolerate.

On practical grounds, then, the need for tolerance and confidence assessment leads to the development of several estimation *accuracy indicators*, which are absolutely necessary given the data heterogeneity and uncertainty (Section 2). After all, a general motto of scientific investigations is:

> Being aware of how much one **does not know** is often more important than being cognizant of how much one knows.

Remarkably, the kind of "quality control" introduced by the accuracy indicators is not part of deterministic interpolation (see, e.g., the inverse distance techniques to be discussed in Chapter 7). Moreover, some techniques do not take into account intrinsic properties of the interpolated phenomena (e.g., rapid changes in the data with distance *vs.* more gradual changes, or law-obeying variations), as they only take account of the location/instance of the data points. The *CTDA* techniques to be discussed in the book will try to address the above and similar issues.

4.3 Interpretation

We conclude with the following comment concerning the *interpretation* of the *CTDA* outcomes, like the chronotopologic attribute maps.

> In general, when investigators **interpret** the attribute maps, they attempt to explain the uncovered patterns and trends, bringing all of their background knowledge, experience and skills to bear on the question and relating their data to existing scientific ideas.

Example 4.2 Weather forecasting is an interpretation made by a meteorologist of data collected by satellites using climate models.

Given the personal nature of the knowledge the investigators rely upon, interpretation is bound to be subjective to some extent, in the sense that by bringing a different background, thinking mode and experience to bear on the same dataset, an investigator can come to very different conclusions. A widely publicized case in point is the global climate debate concerning the presence or absence of a temperature trend in time, which is the focus of the following example.

Example 4.3 The study of global warming turned out to be an arena of strong disagreements concerning data analysis and interpretation. Jones et al. (1986) interpreted the changes in global mean surface air temperature from 1861 to 1984 to imply a long-term warming trend (e.g., they emphasized that 1980, 1981, and 1983 were the three warmest years in the entire dataset). On the other hand, Lindzen (1990) opposed this interpretation using a number of arguments, like the inadequacy of the existing data collection network to correct for data uncertainty, substantial gaps in temperature coverage (especially over the oceans), errors in data analysis, and inappropriate interpretation of the global mean temperatures, thus, reaching the opposite conclusion that there is no trend in the data. This is a classical case where different investigators made very different interpretations and drew very different conclusions from the same datasets, as a result of their different backgrounds, experiences, and thinking modes.

Ut aequum, disagreement is not uncommon in scientific investigations. When it occurs, it routinely leads to more data collection and analysis, but perhaps more importantly it should also lead to the consideration and appreciation of the distinct *thinking modes* used by the different investigators.

5 A review of CTDA techniques

As already noted, there exist several *CTDA* techniques (both *S-TDA* and *STDA*) available in the relevant literature. Naturally, each technique has its own set of specific features and assumptions.

5.1 Classification

The *CTDA* techniques to be discussed in this book, in varying degrees of detail, belong to the following distinct groups:

1. **Pure chronogeographic** statistics techniques (point center, diffusion, inverse metrics, etc.; Chapters 4 and 7). They are concerned only with the chronotopological characteristics (coordinates, distances, instances) of the phenomenon.
2. **Attribute chronogeographic** statistics techniques (Moran's index, Geary's index, etc.; Chapters 4, 7, and 12). They are concerned both with the chronotopological characteristics of the phenomenon and its physical values (mass, weight, temperature, concentration, etc.). They include multiple linear regression (*MLR*), land-use regression (*LUR*), and linear mixed-effect models.
3. **Standard geostatistics** techniques (ordinary and simple Kriging, statistical regression, etc.; Chapters 5 and 8). They are popular techniques that focus on hard (uncertainty-free) data, and they usually rely on rather restrictive modeling assumptions (linear estimators, Gaussian data distribution, low-order statistics).
4. **Modern geostatistics** techniques (Bayesian maximum entropy, stochastic logic, factoras, etc.; Chapters 6 and 9). They are very versatile, incorporate core knowledge, assimilate both hard and soft (uncertain) data, and avoid the restrictive modeling assumptions of the standard techniques (nonlinear estimators and non-Gaussian data distribution are automatically considered).
5. **Machine learning** techniques (artificial neural networks, random forest, gradient boost machine, etc.; Chapter 12). They are often characterized as "black-box," because they do not involve confirmed mathematical expressions representing the substance of the natural attribute of interest. Yet, due to their high nonlinearity, self-adaptation, self-learning and other features, they can fit very well complex real world processes.

Each group of *CTDA* techniques has its *pros* and *cons*. Although many factors affect a technique (including sample size, scale, sampling design, data properties, and law of change), there are not always consistent findings about how exactly these factors affect the technique's performance (accuracy, informativeness, substantiveness). Therefore, it may not be always an easy task to select an appropriate chronotopologic modeling and interpolation technique for a given input dataset. It is noteworthy that different data interpolation techniques can generate different maps that are interpreted as representing very different phenomena.

In view of the classification introduced, it is often methodologically useful to make a key distinction:

> *CTDA* distinguishes between **defining** and **accompanying** properties of a class of techniques, where the former refer to properties that hold for any member of the class and the latter to properties that hold only for a certain subset of the class.

The test whether a property is a defining one or not consists of investigating whether its absence would exclude a technique from inclusion in the specific *CTDA* class. Chronotopology is a defining property of the *CTDA* class, whereas uncertainty is an accompanying property, since it holds for some members of the *CTDA* class (e.g., geostatistics, Chapter 8) but not for others (e.g., deterministic chronogeometric, Chapter 7). Depending on their defining and accompanying properties, one may distinguish between techniques that are global *vs.* local, exact *vs.* inexact, and formal *vs.* substantive.[u] *CTDA* techniques that have been developed for and applied to various disciplines (physical, biological, and social) are often data-specific or attribute-specific. In fact, there exist several chronotopologic estimation techniques available in the relevant literature (mainly interpolation rather than extrapolation techniques, Section 5 above).

Interestingly, of the several changes in the meaning of the term *CTDA* over the years, a certain proportion is due to a tendency of accompanying features to become gradually defining ones (in some cases in addition to the original defining properties, and some others displacing the original defining properties). It is also possible that the confusion between defining and accompanying properties could produce misconceptions that stand in the way of a realistic view of the phenomenon. Misconceptions about the physical world often originate in semantic shortsightedness, but are usually liable to correction through observation.

5.2 Computer technology

Naturally, one could not underestimate the important contribution of technology in computer-assisted *CTDA*. During the last few decades, *CTDA* underwent a profound transformation from being manual (and tedious) to predominantly computer-based (this was made possible mainly due to advances in computer hardware during the 1990s).

- **Computers** play a major role in the development and implementation of *CTDA* techniques.

Among the advantages of computer software is that it makes *CTDA* much more efficient and effective, allowing improved numerical accuracy, higher speed, time-savings, and handling complicated computations (see, also, Section 8). Technology advances of various

[u] The distinction between a formal and a substantive technique is that the former considers data as numbers assigning no physical meaning to them, whereas the latter views data as physical quantities assigning different interpretations to them, depending on the phenomenon.

forms have made it easier for investigators to collect and organize evidence in ways that allow systematic analysis and encourage worldwide collaboration among groups of *CTDA* experts and institutes.

We conclude our very brief review with a word of caution. Although the continuous development of more powerful computers in the past has facilitated improvements in real-world *CTDA* that would not have been possible otherwise, in the view of many scholars, this does not seem to necessarily be the case anymore. Instead, it has been argued that processors do not continue getting faster at the previous pace; instead, it is hoped that the continuation of the steady increase in power will come from massively parallel machines.

6 Chronotopologic visualization technology

After *CTDA* has been used to derive quantitative data dependency measures and generate numerical attribute estimates from isolated data, the next task is usually to portray these results on a continuous chronotopologic domain. In such a context, the following objective is of particular interest:

> Chronotopologic **mapping** is the act of representing the numerical *CTDA* results, involving data and attribute estimates, in a continuous and visually informative way.

Given its objective above, mapping is closely linked to the following form of information communication:

> **Visualization** in the form of maps, graphs, charts, diagrams, and animation is communication of the available *CTDA* information to the interested individuals or groups of individuals accurately and efficiently.

As such, visualization is enabling the act of mapping in a very effective and efficient manner. This brings to mind what *Albert Einstein* once said about scientists' ultimate goal:

> ✍ The natural scientists try to develop, in the fashion that suits him best, a simplified and intelligible picture of the world; he then tries to some extent to substitute this cosmos of his for the world of experience, and thus to overcome it. This is also what the painter, the poet, or the speculative philosopher try to do.

In the present case, the "intelligible picture of the world" is represented by the properly visualized attribute map. Surely, achieving intelligibility in mapping depends on choosing the kind of visualization tool that works best given the study objectives. In this respect, modern visualization is a very powerful technology that provides a large collection of tools that can generate chronotopologic maps of varying levels of sophistication (contour maps, domain coloring maps, animated maps, etc.). The idea is not so much to follow a formula, but rather to be aware of the many different aspects of visualizing a chronotopologic map or other *CTDA* results.

Generally, experience has shown that visualization is an incredibly effective medium for informing or conveying a viewpoint or a message across. Indeed, it is widely acknowledged that the informativeness and usefulness of *CTDA* results can be greatly improved by means of powerful visualization techniques. More specifically, based on the attribute data at the sampled points and the estimated attribute values at the unsampled points a detailed and, hopefully, highly enlightening map can be visualized. The emphasis is on the visualization of attribute properties and relationships that takes into account the chronotopologic localization/instantanization of these properties and relationships.

6.1 The emergence of visual thinking

Humans are inherently visual. The human brain has the ability to literally process and store visual information faster than text, symbols, numerics, and statistics. Visualization takes complex information that most people don't have the time, skills, or attention span to comprehend, and transforms it into easily understood, pleasant and visually focused presentations. In this setting, then, another important human cognition issue emerges that transcends mere technological considerations:

> The visualization of *CTDA* results requires the adaptation of a new way of thinking, viz. **visual thinking**, and, accordingly, to develop a deep understanding of what it means to think visually.

The modern scientists are expected to be able to think visually and use visualization technology to its full potential so that they can generate plots that allow them and others (the public, stakeholders, decision-makers) to grasp information in an efficient and physically meaningful manner. Given that *CTDA* involves a variety of knowledge bases, core and specificatory, shared by many different individuals with different backgrounds (scientific and cultural), this has made visualization a very important *common language*. One then comes to appreciate the claim that if data is information (of various kinds) about the world, then visualization is the creative communication of this information. Therefore:

- A good visualization technique does not only need to be **technically** adequate and **aesthetically** pleasant, but **contextually** rich and **functional** too.

The first two requirements focus on the eyes and follow certain presentation rules of visual grammar, whereas the last two requirements focus on the mind and respond to context (rather than set it). According to this perspective, visualization conveys facts and ideas clearly and also sets them in the appropriate context that can potentially influence people's views and decisions. Therefore, a good visualization technique should add to data's meaning and/or clarity.

More about this interesting topic will be presented in the following chapters. In the meantime, let us note that increasingly sophisticated and high-quality visualization technology products become available nowadays that are often both affordable and easy to use. The interested readers are referred to the relevant visualization literature (e.g., Tufte, 2001; Knaflic, 2015; Healy, 2018; Kirk, 2016; and references therein).

7 The range of CTDA applications

CTDA (in its *S-TDA* and *STDA* formulations) provides useful quantitative tools of chronotopologic dependency characterization and mapping. Besides the characterization and visual perception of attribute distributions, it is very important to translate the emerging chronotopologic patterns into useful objective and measurable entities, which is why *CTDA* can help the investigators answer key questions linked to real-world problems of numerous disciplines, such as those listed in Table 7.1.

The list in Table 7.1 is by no means exhaustive – the readers may propose many other *CTDA* issues that are clearly relevant to their fields of interest. On the other hand, it may come as no surprise to our readers that many fields of inquiry have contributed to *CTDA*'s rise in its modern forms, including cartography and surveying, botanical studies of global plant distributions and local plant locations, forestry, ethnological surveys of population movement, human exposure investigations, landscape ecological studies of vegetation blocks, environmental studies of population dynamics, biogeography research, epidemiology, and disease spread assessment combined with locational healthcare delivery data, and space-time econometrics.

TABLE 7.1 List of scientific disciplines and related questions.

Discipline	Questions
Environmental science: Investigators need to know the chronotopologic distribution of pollutant concentrations during peak days	• Does pollution exceed certain thresholds • In what parts of a region, and when and how frequently does threshold exceedance happen • Which are the most anticipated public health effects • Which are the most efficient prevention and control measures
Earth sciences: Investigators need to derive estimates of a wide range of physical attributes in large domains and during long time periods	• What is the fluctuation of water table elevation in a region during the summer months • How weather variables—such as temperature, air pressure, relative humidity, precipitation, cloud cover, wind speed, and direction—change from a region to the next, or from current to future
Mining: Data analysts need to estimate from a usually limited set of samples of different periods the extension of a mineral deposit in a site	• Can those samples be used to estimate the mineral distribution in that site or to forecast its deposit in the near future • What is the accuracy of the estimates to be used in decision-making and management
Agronomy: Scientists often must analyze a region for agricultural zoning purposes	• How to choose the independent variables—soil, vegetation, and geomorphology • What is the dependency structure of such variables within the domain • How to determine what the contribution of each variable is in order to define where the growth of each type of crop is more adequate
Epidemiology: Health professionals collect data about disease occurrence	• Does the distribution of disease cases develop a specific chronotopologic pattern and which are its major features • Is there any association with any pollution sources • Do geographical disease patterns vary with time • Should a state of emergency be declared, how long should it last, and which regional hospitals should participate
Social studies: People would like to know whether there is any chronotopologic concentration in regional theft distribution	• Are there any "hot spots" in this distribution • Are thefts that occur at certain regions or during specific seasons correlated to the socio-economic characteristics of these regions

Lastly, as noted earlier, advances in *TGIS* have contributed considerably to continuing *CTDA* developments and wide applicability. In return, *TGIS* popularity requires *CTDA* methods that allow the transformation of quantitative results into the kind of systematic procedures and qualitative representations that are tractable to nonexpert users.

8 Public domain software libraries

Admittedly, most *CTDA* techniques have been developed either for particular disciplines or even for specific attributes based on the data properties modeled. Several useful public domain and commercial software are listed in the Stochastic Analysis Software Library documentation (SANLIB, 1995).

More recent and in certain respects more advanced *CTDA* software libraries in the public domain include the following (the vast majority of the numerical examples and the real-world case studies in this book have been conducted using these libraries):

① **Spatiotemporal Epistemic Knowledge Synthesis Graphical User Interface** (*SEKS-GUI*): This is a library interface that was developed on a Matlab platform, and it provides an application for interactive *CTDA* analysis. The original space-time data, processed data, and final results are all available for users visualization. *SEKS-GUI* can be downloaded from:
http://homepage.ntu.edu.tw/~hlyu/software/SEKSGUI/SEKSHome.html.

Example 8.1 The interface of *SEKS-GUI* is shown in Fig. 8.1. In the "*BME* Spatiotemporal Analysis" part of the software library, the procedure is as follows: import hard data, specify soft data type and format, import soft data, specify predict locations or output grid, check for co-located data, data detrending with Gaussian kernel (optional) and exploratory analysis, data transformation (optional), spatial or spatiotemporal covariance computation and modeling, *BME* estimation and visualization.

② **Bayesian Maximum Entropy Graphical User Interface (*BMEGUI*)**: This is a Python-based *GUI* implementation for the *BMELIB* library, and has a variety of components for different types of analysis. *BMEGUI* can be downloaded from:
https://mserre.sph.unc.edu/BMEGUI_web/BMEGUI3.0.1_web/BMEGUI3.0.1_WEB_2014.htm.

③ **Space-Time Analysis Rendering** with **BME** (*STAR-BME*): This is, also, a Python-based plugin interacted with *Q* Geographic Information System software (*QGIS*, https://www.qgis.org/en/site/). It has a superior ability on rendering visualizations within the *GIS* environment, which makes it easier for mapping purposes. *STAR-BME* can be downloaded from:
https://stemlab.bse.ntu.edu.tw/wordpress/starbme/

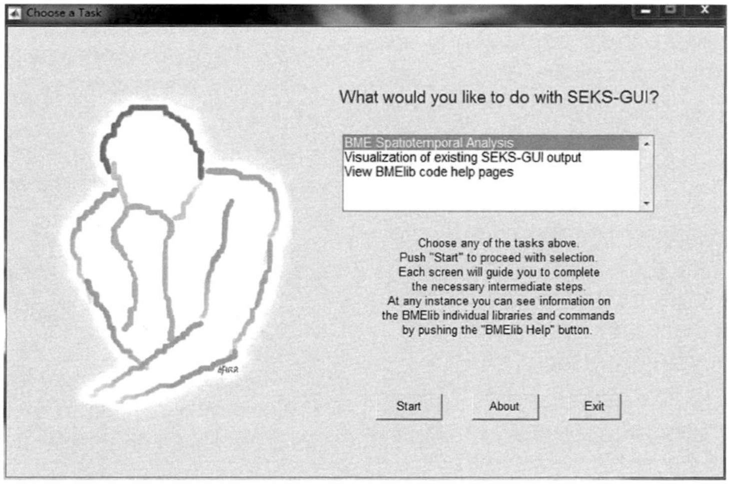

FIG. 8.1 Interface of the *SEKS-GUI* library.

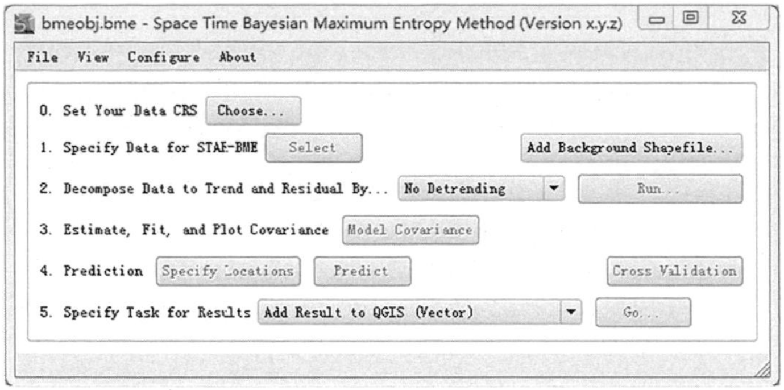

FIG. 8.2 Interface of the *STAR-BME* library.

Example 8.2 The interface of the *STAR-BME* software library is shown in Fig. 8.2. Briefly, the process introduced by *STAR-BME* is as follows: set the coordinate reference system, import hard and soft data, data detrending with Gaussian kernel or spatiotemporal mean (optional), spatiotemporal covariance modeling and estimation, prediction points, *BME* prediction and visualizations.

Ergo, a reasonable suggestion would be that the readers visit the above sites and decide which software library best fits their needs. Moreover, the users of the software tools should make sure that the underlying assumptions and limitations are well understood and not hesitate to question their applicability in the specific case study, if necessary. It is hoped that the readers will be sufficiently encouraged to apply one or more of these software tools into their *CTDA* data to make sense of theory and data.

Concluding, the discussion in this chapter is a skeletal description of the various *CTDA* elements and features, with the understanding that much of the ensuing discussion in Chapters 2–13 will serve to flesh out the skeleton.

9 Practice exercises

1. How a physical law is presented by an equation matters in determining what conceptualizations are activated. Thus, changes in symbolic structure or arrangement may cause changes in conceptualization. A difference in the arrangement of the symbols in an equation could imply a different organization of the body of knowledge, which could, in turn, lead to different interpretations. Newton's second law can be represented by two mathematically equivalent equations, $F = m\gamma$ and $\gamma = \frac{F}{m}$, where F is the total force on an object, m is its mass and γ its acceleration. Explain why although their mathematical formulations are the same, the conceptualization of the two equations could be quite different.
2. One can think of many natural quantities that exhibit chronotopologic variation, such as rainfall, vegetation, population density, human mortality, economic wealth, unemployment and crime rates. Suggest attributes in natural sciences that
   (a) do not vary in space,
   (b) do not vary in time,
   (c) vary neither in space nor in time.
3. Consider the chronotopologic coordinates associated with:
   (a) the occurrence of volcano eruption,
   (b) soil moisture samples,
   (c) household density in the state of California,
   (d) migration of swallows from Ireland to S. Africa, and
   (e) *SARS* disease incidences in East Asia.

Which ones of the above can be studied by the *CTDA* techniques?

4. In one sentence, what is *CTDA* all about and what kinds of problems can it solve, in your view?

5. Describe natural attributes in your field of interest characterized by chronotopologic dependence.

6. Consider the following phenomena:
(a) the occurrence of your father's birth,
(b) pollutant concentration samples,
(c) poverty levels in North Carolina counties,
(d) the battle of Marathon, and
(e) Plague mortality throughout India in the early 1900s.

Which ones of the above can be studied by *CTDA* techniques?

7. With which ones of the following descriptions do you agree, if any?

There are scientific fields that:
(a) capitalize on chronotopologic dependence to estimate attribute values at unsampled points,
(b) arc concerned with the study of phenomena that vary in a composite spacetime domain,
(c) focus on numerical techniques that deal with chronotopology in conditions of *in-situ* uncertainty,
(d) offer a way of describing the continuity of natural phenomena and include adaptations of regression techniques to take advantage of this continuity.

8. In what way do *S-TDA* and *STDA* distinguish themselves from classical data analysis, and in what way are they distinct from each other?

9. Fill the empty spaces (__) in the following statement:

> "Compared to the __ statistics that merely examine the statistical distribution of a set of sampled data, __ incorporates both the statistical distribution of the samples and the chronotopologic correlation among them. Because of this difference, many real-world problems are more effectively addressed using __ techniques."

10. Compare the chronotopologic variability and dependency of the temperature distribution in the countryside *vs.* the industrial city center.

11. Design an investigation of the chronotopologic variation of atmospheric humidity in a geographical region during a time period of your choice. List sources that may generate uncertainty.

12. Provide some special situations or data that cannot be analyzed by *CTDA* techniques.

13. List uncertainty sources associated with different types of sampling (random, regular, stratified, etc.).

14. Compare the variability of daily temperature in May at a city of your choice *vs.* the variability of daily temperature during the entire calendar year.

15. Compare the variability of daily humidity in May at a specific city of your country *vs.* the variability of daily humidity in May at several cities of your country.

16. Provide examples of attribute distributions that violate Tobler's law of geographical science.

17. Write a brief review of *CTDA* techniques.

18. Which is the branch of philosophy that concerns itself with knowledge, and why is it important for *CTDA* research purposes?

19. Do you consider implementing *CTDA* techniques, and in what specific fields?

20. Should *CTDA* be concerned with the representation and analysis of common-sense knowledge or with solving problems designed to test research-motivated axiomatic theories, or with both?

21. Like all types of human inquiry, *CTDA* too can be influenced by the underlying conceptual or methodological perspective. For illustration, Table 9.1 illustrates some

TABLE 9.1 Differences between the positivist and the interpretivist approach to human inquiry.

Issue	Positivist perspective	Interpretivist perspective
Access to real-world	Direct	Indirect
External reality	Single	Not single
Reality studied through	Objective knowledge	Perceived knowledge
Focus of human inquiry	On generalization and abstraction based on hypothesis and theory	On the specific and concrete seeking to understand the specific context
Investigation focus	On description and explanation	On understanding and interpretation
Investigator's role	• Detached, external observer • Make clear distinction between reason and feeling. • Aims to discover external reality rather than create the object of study • Use rational, consistent, and logical approach • Maintain a clear distinction between facts and value judgments • Distinguishes between science and personal experience	• Experiences what is studying • Allows feeling and reason to govern actions • Partially creates the object of study, the meaning of phenomena • Uses preunderstanding (intentional structure of feelings and thoughts) • Assumes a less clear distinction between facts and value judgments • Accepts influence from both science and personal experience
Techniques used	Predominantly quantitative (mathematical, statistical) methods	Primarily nonquantitative methods

key differences between the positivist and interpretivist approaches. What kind of approach do you use in your research? Justify.

22. Do you agree or not with the view that truth-seeking is the critical thinking skill that is defined as the practice of seeking knowledge, and following reason and evidence, *even if it disagrees* with a deep-stated belief system. Justify.

CHAPTER

2

Chronotopology theory

1 Introduction

Let us start by introducing an essential study perspective pertaining to the main focus of the present chapter regarding the role of *chronos* (*time*) and *topos* (*space*) and that of their relationship in *CTDA*:

> The goal of a theory of **chronotopology** is to link chronos and topos in a mathematically rigorous and physically meaningful manner.

Whether the natural attribute of interest concerns human populations or the physical environment, chronotopology is of fundamental importance. Historically, two distinct conceptions of chronotopology have been developed, around which all the other *CTDA* notions revolve:

- The **classical** conception, in which the chronotopologic domain serves as an arena for the natural phenomena to take place, and which is independent from the phenomena.
- The **modern** conception, according to which the chronotopologic domain does not claim existence on its own, but it is intimately linked to the natural phenomena and cannot exist independently of them.

The implementation of a particular chronotopologic framework may require theoretical and methodological commitments in different fields of human inquiry, including physics, mathematics, ontology, epistemology, cognition and sociology. Chronotopology, then, plays a key role in achieving certain key objectives.

> An **operational** chronotopology can achieve certain real-world key objectives: determining what kinds of natural attribute distributions exist, assessing how they vary *in-situ*, describing the relevant body of knowledge, and classifying the attributes into different scientific disciplines.

In this respect, some initial characterizations are in order regarding the main constituents of an operational chronotopology. These characterizations concern the correspondence between elements of the material (real or physical) world and these of the chronotopologic domain, as follows:

Quantitative Analysis and Modeling of Earth and Environmental Data
https://doi.org/10.1016/B978-0-12-816341-2.00012-5

1. A **material** point[a] (or particle) in the material (or real) world consisting of continuous matter and an **immaterial** (abstract or geometric) point[b] representing only position and time instance in the chronotopologic domain.
2. A material **body** (or object) in the material world and its abstract **configuration** in the chronotopologic domain within which the configuration resides and moves.
3. A **mobile** point in the chronotopologic domain representing the activity of a material particle in the material world and an **immobile** point in the chronotopologic domain that merely serves as an **attribute argument** (location and instance of the occurrence of an attribute value).

Example 1.1 Fig. 1.1A shows a point *P* and a body in the material world (consisting of a set of material points) that are mapped into a point ***p*** and a configuration, respectively, in the chronotopologic domain, i.e., the configuration of the body is obtained by fixing the material points of the body to abstract points on the chronotopologic domain. The study of this representation (i.e., a material point or body residing and moving in the chronotopologic domain) is the focus of

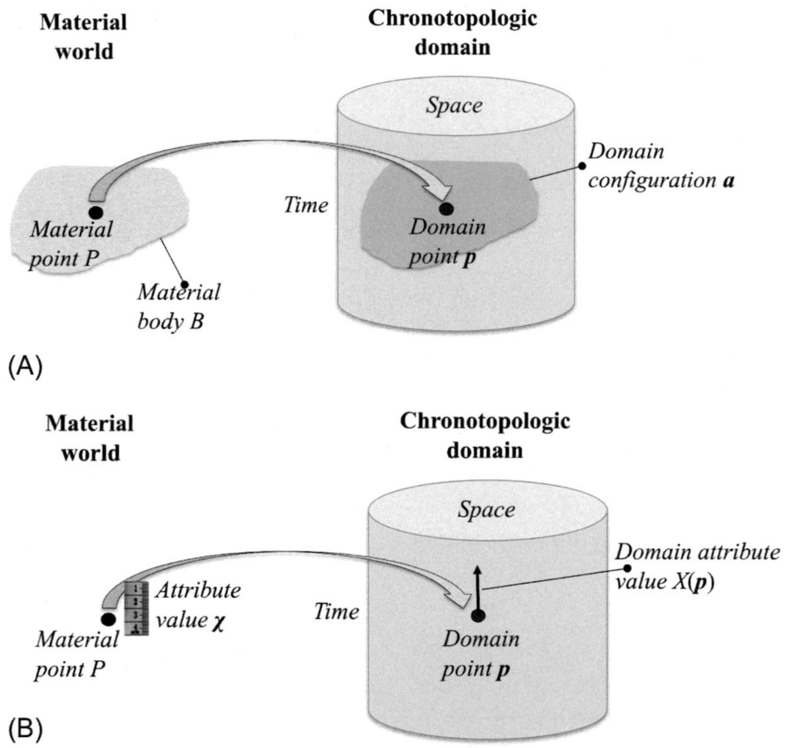

FIG. 1.1 (A) Elements of the material (or real) world and their mapping in the chronotopologic domain; and (B) attribute value and its mapping in the chronotopologic domain.

[a] I.e., an entity having size (width, no length and no depth).

[b] I.e., a sizeless entity in geometry having no size (no width, no length and no depth).

Section 2.2. In Fig. 1.1B, on the other hand, to the abstract point $\boldsymbol{p}$ of the chronotopologic domain one associates a natural attribute value χ in the material world. In this case, $\boldsymbol{p}$ is a fixed point in the chronotopologic domain that merely serves as the argument (location and instance) of an attribute value denoted as $X(\boldsymbol{p}) = \chi$; each point of the chronotopologic domain has its own attribute value that may change from one point to another. The study of this representation (i.e., a material world attribute mapped on the chronotopologic domain) is the concern of Section 2.5.

Ad pond om, the review of the foundations of a chronotopology theory in the remaining of this chapter is intended to illustrate a wide range of space and time concepts, and to draw out common themes between them. This will furnish a methodological platform on which to build a rigorous and systematic *CTDA*.

2 Basic chronotopologic notions

In chronotopology theory the relevant symbols serve as prompts for an array of conceptual operations and the recruitment of background knowledge. It follows from this view that assigning meaning to chronotopologic notions is a process rather than a discrete "thing" that can simply be "packaged" by language. Meaning construction draws upon physical knowledge and involves inferencing strategies that relate to different aspects of conceptual structure, organization and packaging. For instructional purposes, we continue our discussion with a closer look at what admittedly are the most fundamental notions of a chronotopology theory.

2.1 Space and time

Since long time, certain critical distinctions, both conceptual and practical, have been made between the notions of time and space, including the following:

① There are a few space coordinates, but there is precisely one time coordinate.

② There is a time arrow, but there is nothing like a space arrow.

③ One can travel in space in all directions, but one cannot do that in time.

④ One experiences time and space in completely different ways, e.g., one remembers the past and not the future, which refers to time, not to space.

⑤ There is a unidirectional causality principle that refers to time, not to space.

The interpretation of ①–⑤ has given rise to several conjectures concerning the nature of space and time. Furthermore, it has been recognized that despite the above distinctions between space and time, the two notions are closely *linked* in a certain way, as the following example reveals.

Example 2.1 Time is measured by certain arbitrary reference to space itself, such as the earth's journey around the sun, or the progression of a sundial's shadow. This fact reveals the existence of an important strand tying both space and time together.

Acceptance of the space and time *integrative* (or *synthetic*) perspective implies that the world can be regarded, represented, and understood as consisting of four-dimensional phenomena combined with their inter-relations. The readers would appreciate the viewpoint unfolded in this chapter according to which the way space and time are integrated—in theory and in practice—has a profound influence on the development of both *CTDA* epistemology and technology.

2.2 Chronotopologic domains

In order to express the above integrative perspective in mathematical terms, two kinds of chronotopologic domains may be considered for *CTDA* purposes (Christakos, 2017):

Space-time domain $R^{n,1}$ (used in *S-TDA*, see Section 1 of Chapter 1), which conceptually consists of a **series** of separate topological spaces R^n and a time axis R^1.

Spacetime domain R^{n+1} (used in *STDA*, see also Section 1 of Chapter 1), which consists of a **single** continuum, i.e., space and time mix together to form a new whole.

The above distinction is fundamental in *CTDA* and will be emphasized in our discussions throughout the book. The $R^{n,1}$ and R^{n+1} domains allow significantly different considerations of an object (a location, an area or a block). In the *empirical* three-dimensional ($n = 3$) world, simultaneity cannot be conventional. As the empirical world coincides with the present (everything that exists simultaneously at the time "now"), a conventional simultaneity readily implies that what exists is also conventional, which is unacceptable. On the other hand, in the four-dimensional ($n + 1 = 4$) world, simultaneity is unavoidably conventional, since it is really a matter of convention which three-dimensional cross-section of the four-dimensional spacetime one regards as one's empirical three-dimensional world (i.e., as a set of simultaneous events).

Furthermore, the topological space R^n in the space-time $R^{n,1}$ domain is usually, but not necessarily, Euclidean, whereas the time axis R^1 is uniquely singled out by natural laws, i.e., the laws of nature single out one dimension, the *lawful time*, over the others in some way. Lawful time should be distinguished from *causal time* (temporal relations are defined in terms of empirically accessible or primitive causal relations) and *metaphysical* time (unlike space, time is irreducibly passing, flowing, or becoming).

Separate space-time domain $R^{n,1}$ does not mean **no space-time interaction**, merely that this interaction is represented in a different way than the interaction in the composite spacetime domain R^{n+1}.

Often an investigator is interested about the history of a chronotopologic domain configuration representing a material body or object, in which case the following notion can be operationally useful:

The **worldtube *WT*** of a material body represented by its time-varying configuration is the part of the chronotopologic domain occupied by the configuration.

The *WT* stretches from the past to the future showing the spatial location of the object at every time instant. Once the complete *WT* of an object has been determined from the physics of the situation, its full history has become known. An immediate implementation of these ideas is that the *motion* of a material body can be properly represented in the chronotopologic domain as a *WT* consisting of a set of configurations parameterized by time t.

Next, the above lines of thought are further considered in terms of two examples in the convenient, for visual illustration purposes, $R^{2,1}$ and R^{2+1} domains. Each domain reveals certain coherent features of space and time.

Example 2.2 In Fig. 2.1, the space-time $R^{2,1}$ domain is seen as a series of horizontal space planes R^2 across time t, whereas the space-time R^{2+1} domain is seen as a single continuum within which spacetime planes at any angle are considered. I.e., while in $R^{2,1}$ we can consider only parallel planes at fixed times t as shown in Fig. 2.1A, in R^{2+1} we can consider any inclined plane as shown in Fig. 2.1B. If the $R^{2,1}$ description is chosen, we can only experience one horizontal plane at a time, whereas if the R^{2+1} description is chosen, we can experience any plane (horizontal or inclined) or even the entire spacetime history at once.

Example 2.3 Table 2.1 illustrates the different meanings that material bodies (objects) can have in the $R^{2,1}$ and the R^{2+1} domains. Fig. 2.2 offers visualizations of certain aspects of these meanings.

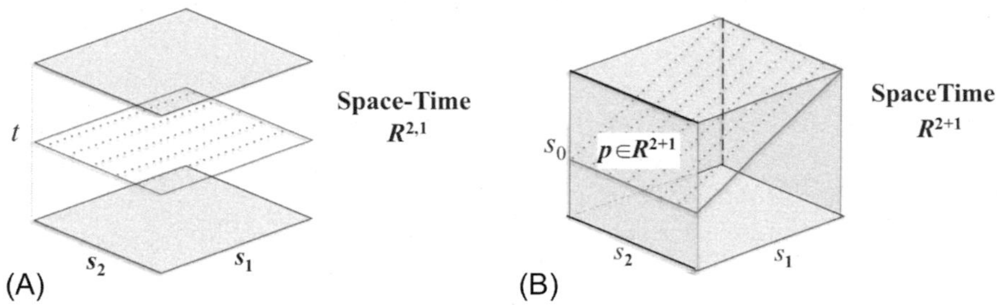

FIG. 2.1 (A) The space-time domain $R^{2,1}$, and (B) the spacetime domain R^{2+1}.

TABLE 2.1 The different meanings of a body according to the $R^{2,1}$ and R^{2+1} descriptions.

Meaning in $R^{2,1}$ (left of Fig. 2.2)	Meaning in R^{2+1} (right of Fig. 2.2)
A physical body retains its identity as a 2-dimensional (spatial) body, i.e., at all times of its history, it is the same body	A physical body retains its identity as a continuous (2 + 1)-*dimensional* body representing it at all moments of its history
The history of a body (i.e., its representation at all times) consists of distinct 2-dimensional cross-sections that are considered at different times	The history of a body is its *WT* that is not divisible into 2-dimensional cross-sections, i.e., at different times the body is a different body
A physical body preserves its identity in time, and it can undergo change in the sense that it is the same body that changes	A physical body does not preserve its identity in time, since at any time the body is a different cross-section of the spacetime *WT*
There is an ordinary flow of time in the sense that time instants are actualized one by one, i.e., there is the explicit division into past, present, and future	There is no ordinary flow of time, but all-time instances are given at once forming the 3rd dimension, i.e., there is no explicit division into past, present, and future
If a body exists only at the present time, it is a two-dimensional body	If a body exists at more than one time, it is a three-dimensional body
The time dimension does not exist in the same way the space dimensions do	The space and time dimensions are equally existent

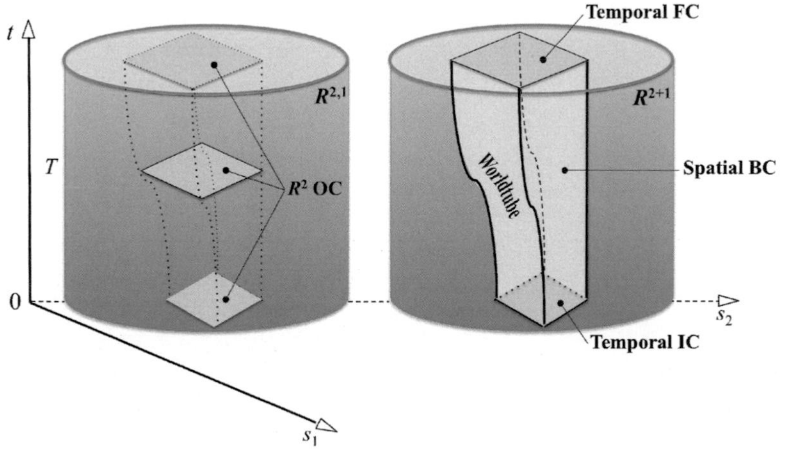

FIG. 2.2 A body (object) in the space-time $R^{2,1}$ domain (left) and in the spacetime R^{2+1} domain (right). The OC denotes object configuration in R^2, whereas the BC, IC and FC denote, respectively, boundary condition, initial condition and final condition in R^{2+1}.

Ad rem, the differences between the $R^{n,1}$ description *vs.* the R^{n+1} description are not necessarily ontologic, i.e., it does not mean that the $R^{n,1}$ ontology is true and the R^{n+1} ontology is false or vice versa. On the other hand, when it comes to investigatory and practical *CTDA* matters the following assessment can be made:

Operationally, the differences between the $R^{n,1}$ *vs.* the R^{n+1} descriptions have some considerable **modeling implications**: in some *CTDA* contexts the $R^{n,1}$ description may be a better choice and in some others the R^{n+1} description may be a more suitable one.

The next task is to introduce the necessary quantitative tools of chronotopologic characterization.

2.3 Equipping the domain with quantitative tools

According to the discussion in the introduction section above, the process underlying a chronotopologic characterization consists of three basic stages as presented in Fig. 2.3:

Stage ①: A **point** is considered in the domain of interest.
Stage ②: The point is equipped with **coordinates** $\boldsymbol{p}$ (with space and time units) that provide a quantitative description of the domain.
Stage ③: A **natural attribute** $X(\boldsymbol{p})$ (with physical units) is associated with $\boldsymbol{p}$ that assigns physical meaning to it. Otherwise said, in this stage physics is considered the space-time description of nature.

Building on this three-stage process, several important operational developments can be made.

Example 2.4 For illustration, Table 2.2 presents a classification of cases considered in the chronotopologic domains R^{2+1} and $R^{2,1}$. Specifically:

- The *static* **s**-*static t* representation of the phenomenon in Table 2.2 implies a fixed $\boldsymbol{p}$-grid in R^{2+1}, the attribute $X(\boldsymbol{p})$ values are simultaneously assigned to this grid, and a single spacetime attribute image is obtained.
- The *dynamic* **s**-*static t* representation implies a moving **s**-grid in R^{2+1} (i.e., the location **s** is dynamic and changes in time), the attribute $X(\boldsymbol{p})$ values are concurrently assigned to all **s**-grids, and a single spacetime attribute image is obtained.
- The *static* **s**-*evolving t* representation implies a fixed **s**-grid in $R^{2,1}$, the attribute $X(\boldsymbol{p})$ values are consecutively assigned to this grid as it passes through time *t*, and a sequence of spatial attribute images are obtained.
- The *dynamic* **s**-*evolving t* representation implies a moving **s**-grid in $R^{2,1}$, the attribute $X(\boldsymbol{p})$ values are consecutively assigned to this grid as it passes through time *t*, and a sequence of spatial attribute images are obtained.

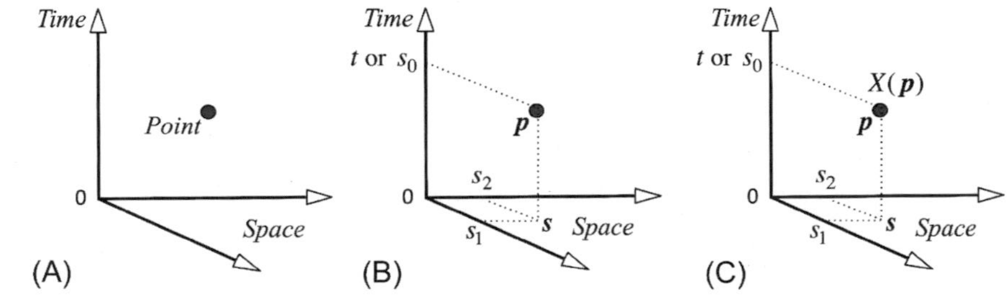

FIG. 2.3 The chronotopologic process: (A) the point; (B) its chronotopologic coordinates; and (C) the natural attribute associated with the point.

TABLE 2.2 Chronotopologic domain classification.

	Static time	Evolving time
Static locations	• Phenomenon represented as a single spacetime image in the R^{2+1} continuum • Recorded at the same spatial grid 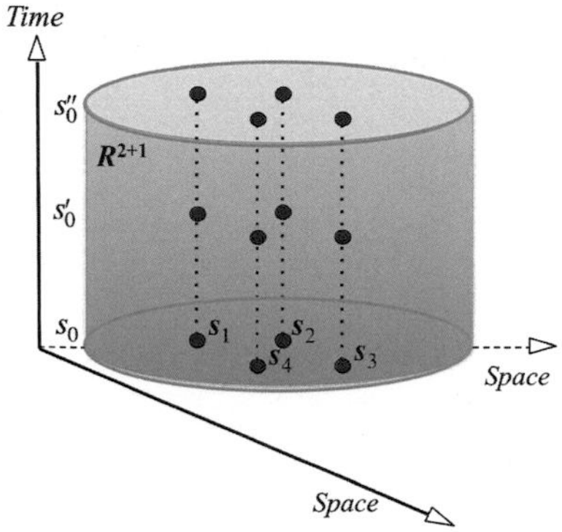	• Phenomenon represented as a temporal sequence in $T \subset R^1$ of its spatial images in the $D \subset R^2$ region • Recorded at the same spatial grid 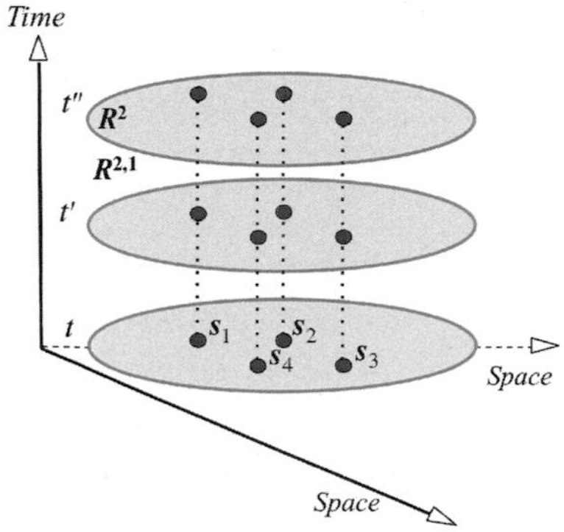
Dynamic locations	• Phenomenon represented as a single spacetime image in the R^{2+1} continuum. • Recorded at a changing spatial grid 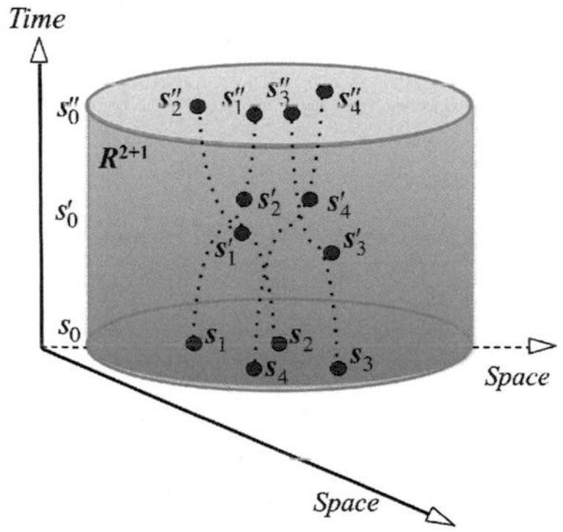	• Phenomenon represented as a temporal sequence in $T \subset R^1$ of its spatial images in the $D \subset R^2$ region • Recorded at a changing spatial grid 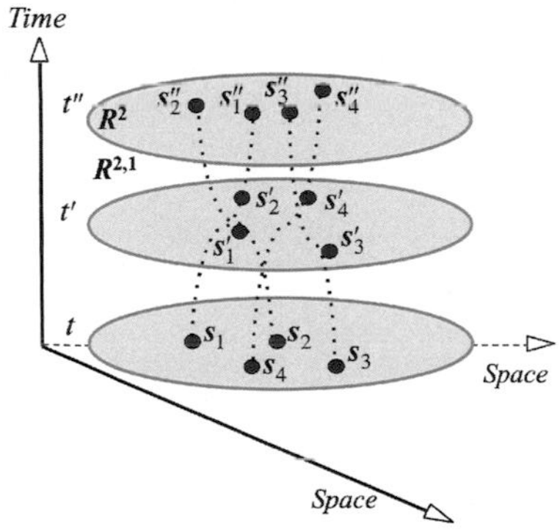

Special cases of the above representations may be considered. The static *s*-evolving *t* representation, i.e., includes the case where the phenomenon is represented only by its most recent spatial image, and the case where the phenomenon is represented by the complete temporal sequence (series) of images.

In a modeling setting, three natural attribute *arguments* characterize a chronotopologic domain: point coordinates, lags and metrics. Accordingly, the essential differences between the domains $R^{n,1}$ and R^{n+1} are described in terms of their corresponding threefold arguments, as follows:

1 The space-time domain $R^{n,1}$ is mathematically defined in a **biargumental** manner separating space from time so that the following conditions apply[c]:

Bi-point: this is a chronotopologic point defined in terms of a pair consisting of a vector and a scalar argument as

$$\boldsymbol{p} = (\boldsymbol{s}; t) \in R^{n,1} \tag{2.1a}$$

where the vector argument $\boldsymbol{s} = (s_1, \ldots, s_n) \in R^n$ denotes the spatial location and the scalar argument $t \in R^1$ denotes the time instance.

Bi-lag: this is a chronotopologic lag between two points $\boldsymbol{p}$ and $\boldsymbol{p}'$ determined by a pair of separately considered vector and scalar arguments as

$$\Delta\boldsymbol{p} = (\boldsymbol{h}; \tau) \in R^{n,1} \tag{2.2a}$$

where the vector argument[d] $\boldsymbol{h} = (h_1, \ldots, h_n) \in R^n$ denotes the spatial distance between the locations $\boldsymbol{s}$ and $\boldsymbol{s}'$ and the scalar argument $\tau \in R^1$ denotes the time lag between t and t'.[e]

Bi-metric: this is a chronotopologic distance between $\boldsymbol{p}$ and $\boldsymbol{p}'$ determined in terms of a pair of scalar arguments as

$$|\Delta\boldsymbol{p}| : |\boldsymbol{h}| \wedge \tau \tag{2.3a}$$

where the distance magnitude $|\boldsymbol{h}|$ is a function of space that depends on the phenomenon of interest, τ is the time lag, and the symbol $\wedge$ denotes that both $|\boldsymbol{h}|$ and τ are separately needed in the quantitative description of the bi-metric (the symbol $\wedge$ implies that both separable and composite functions of $|\boldsymbol{h}|$ and τ may be considered).

2 The spacetime domain R^{n+1} is mathematically defined in a **uniargumental** (or **monoargumental**) manner unifying space and time so that the following conditions apply:

Uni-point: this is a chronotopologic point determined in terms of a unique vector argument

$$\boldsymbol{p} = (s_1, \ldots, s_n, s_0) \in R^{n+1} \tag{2.1b}$$

where space is considered jointly with time, since in the R^{n+1} description the space coordinates s_i ($i = 1, \ldots, n$) and the time coordinate s_0 are assumed equally existent.[f]

Uni-lag: this is a chronotopologic difference between two points defined in terms of a unique vector argument

$$\Delta\boldsymbol{p} = (\Delta s_1, \ldots, \Delta s_n, \Delta s_0) \in R^{n+1} \tag{2.2b}$$

where the space lag is considered jointly with time separation.[g]

Uni-metric: this is a chronotopologic distance defined in terms of a unique scalar argument

$$|\Delta\boldsymbol{p}| = g(\Delta s_1, \ldots, \Delta s_n, \Delta s_0) \tag{2.3b}$$

where g is a function with shape depending on the features of the phenomenon of interest.

The reader may notice that the quantitative differences between the domains $R^{n,1}$ and R^{n+1} above are expressed notationally by their distinctive prefixes "uni-" *vs.* "bi-" corresponding to the uniargumental *vs.* the biargumental way they are defined. Moreover, unlike in the case of Eqs. (2.1a)–(2.3a), no semicolon is used in Eqs. (2.1b)–(2.3b) to separate space from time

[c] The semicolon will be occasionally used in the relevant formulas to emphasize the physical separation of space from time.

[d] As usual, $h_i = s_i' - s_i$, $i = 1, \ldots, n$.

[e] For convenience, it is usually assumed that $t' > t$, i.e., $\tau = t' - t > 0$.

[f] With the convention that the vector component with $i = 0$ denotes time separation.

[g] The order of the points in the chronotopologic lag expression (i.e., whether the $\Delta\boldsymbol{p}$ is defined as $\boldsymbol{p}' - \boldsymbol{p}$ or as $\boldsymbol{p} - \boldsymbol{p}'$) can affect mathematical manipulations and it can also be consequential in applications.

(in R^{n+1} the time coordinate is considered jointly with the space coordinates).

Example 2.5 With reference to the distinction between *S-TDA* and *STDA*:

- Each space-time point in *S-TDA* is denoted by $p_i = (s_{i,1}, s_{i,2}; t_i) \in R^{2,1}$, where $s_{i,1}$ and $s_{i,2}$ are the location coordinates and t_i is the time instant.
- Each spacetime point in *STDA* is denoted by $p_i = (s_{i,1}, s_{i,2}, s_{i,0}) \in R^{2+1}$, where $s_{i,0}$ is the time instant.

Let us have a more detailed look at the differences between the $R^{n,1}$ and the R^{n+1} descriptions introduced above. The following example illustrates the earlier consideration (Eqs. (2.1a) and (2.1b)) that according to the alternative domain descriptions $R^{n,1}$ *vs.* R^{n+1}, different ways may be used to denote a point for modeling purposes.

Example 2.6 One may distinguish between two notations to present the chronotopologic points, see Fig. 2.4.

- In the first notation (Fig. 2.4A) the spatial and temporal coordinates are separately displayed as in Eq. (2.1a), i.e., in the $R^{2,1}$ domain one can write that,

$$\left.\begin{array}{l}(s_1;t_1)=(2,0;0),(s_1;t_2)=(2,0;1),(s_1;t_3)=(2,0;2),(s_1;t_4)=(2,0;3)\\ (s_2;t_1)=(3,1;0),(s_2;t_2)=(3,1;1),(s_2;t_3)=(3,1;2),(s_2;t_4)=(3,1;3)\\ (s_3;t_1)=(1,3;0),(s_3;t_2)=(1,3;1),(s_3;t_3)=(1,3;2),(s_3;t_4)=(1,3;3)\end{array}\right\} \tag{2.4a}$$

- In the second notation (Fig. 2.4B), the spatial and temporal coordinates are jointly displayed as in Eq. (2.1b), i.e., in the R^{2+1} domain one can write that,

$$\left.\begin{array}{l}p_1=(2,0,0),p_2=(2,0,1),p_3=(2,0,2),p_4=(2,0,3)\\ p_5=(3,1,0),p_6=(3,1,1),p_7=(3,1,2),p_8=(3,1,3)\\ p_9=(1,3,0),p_{10}=(1,3,1),p_{11}=(1,3,2),p_{12}=(1,3,3)\end{array}\right\} \tag{2.4b}$$

∴ As noted earlier, although the formal differences between Eqs. (2.4a) and (2.4b) are merely notational (space and time coordinates are distinguished by a **semicolon** in Eq. (2.4a) *vs.* a **comma** in Eq. (2.4b)), the differences in physical meaning are significant (the semicolon signifies that space and time are separate entities, whereas the comma means that they form a unique whole).

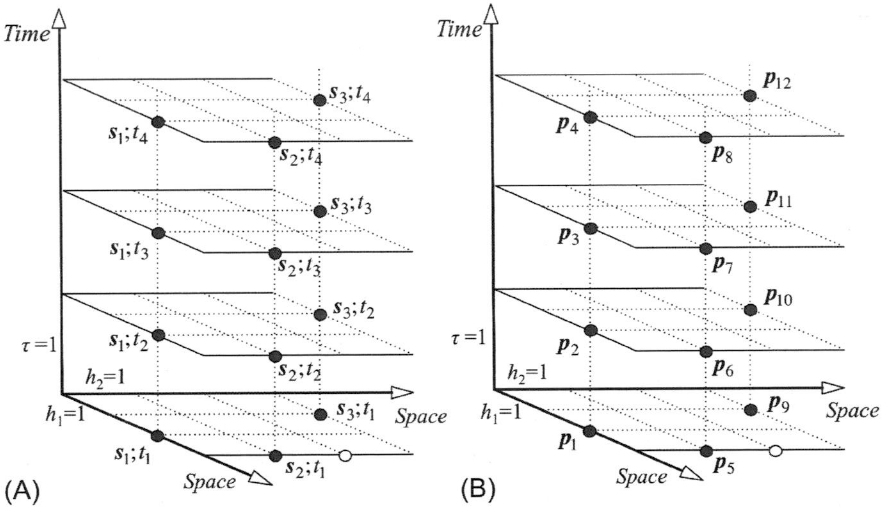

FIG. 2.4 Alternative ways of chronotopologic point determination: (A) the biargument setting and (B) the uniargument setting.

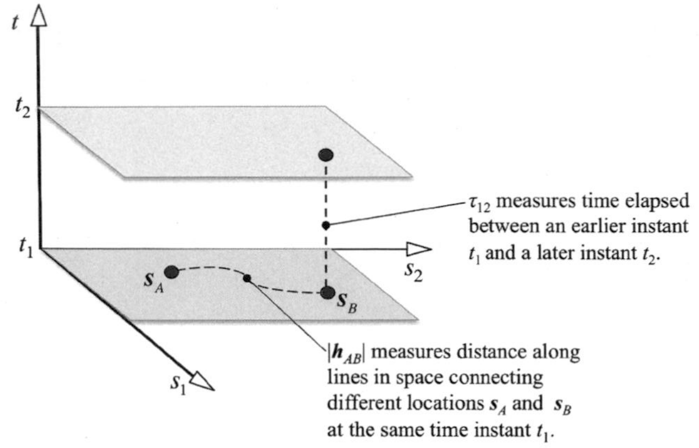

FIG. 2.5 The role of space and time in the space-time $R^{2,1}$ domain.

The notion of the chronotopologic lag between two points, Eqs. (2.2a) and (2.2b), emerges naturally, since the study of a phenomenon in its domain of occurrence requires the simultaneous consideration of pairs of points. The $R^{n,1}$ description explicitly distinguishes time lag (separation) τ from space lag $\boldsymbol{h}$, whereas the R^{n+1} description does not explicitly distinguish time lag from space lag.[h]

The metrical structure of Eqs. (2.3a) and (2.3b) is arguably the most fundamental chronotopologic feature of a phenomenon. So, $|\Delta \boldsymbol{p}|$, or, equivalently, the distance $|\boldsymbol{h}|$ and the time separation τ, are among the most salient aspects of the chronotopologic description of a phenomenon affecting the *CTDA* results.

Example 2.7 Fig. 2.5 gives a visual illustration of the spatial distance $|\boldsymbol{h}|$ (not necessarily Euclidean, see Section 3 later in this chapter) and the time separation τ. Points $\boldsymbol{p}$ with varying locations $\boldsymbol{s}$ (say, $\boldsymbol{s} = \boldsymbol{s}_A, \boldsymbol{s}_B, \ldots$) at the same time instance (say, fixed t_1) form a plane that is a snapshot of space at one instance (t_1). Points $\boldsymbol{p}$ with varying time t at the same location (say, fixed $\boldsymbol{s}_B$) form a vertical line that is the time line of one location ($\boldsymbol{s}_B$).

2.4 Worldtube, worldsequence and worldlines

The *worldtube* (*WT*) of an object was introduced qualitatively earlier in Section 2.2. Here the *WT* is expressed quantitatively in terms of an equation of the corresponding time-varying configuration and the associated point coordinates. We start with the R^{n+1} domain.

The **worldtube** $WT_{\boldsymbol{a}}$ of a material object or body B (Fig. 1.1A) is the part of spacetime R^{n+1} occupied by the configuration $\boldsymbol{a} \in R^{n+1}$, i.e.,

$$WT_{\boldsymbol{a}} = \cup_{\boldsymbol{p} \in R^{n+1}} \{\boldsymbol{p} = (\boldsymbol{s}, s_0) \in \boldsymbol{a}\} \tag{2.5a}$$

where $WT_{\boldsymbol{a}} \subset R^{n+1}$.

In the $R^{n,1}$ domain the equivalent of the continuous *WT* notion is a discrete notion, as follows:

The **worldsequence** $WS_{\boldsymbol{a}}$ of a material object or body B is the discrete sequence of configurations $\boldsymbol{a}_t \in R^n$ for all $t \in T$ in the space-time $R^{n,1}$ domain, i.e.,

$$WS_{\boldsymbol{a}} = \cup_{\boldsymbol{p} \in R^{n,1}} \{\boldsymbol{s} \in \boldsymbol{a}_t \wedge t \in T\} \tag{2.5b}$$

[h] $R^{n,1}$ considers a Newtonian space-time with the spatial lag $\boldsymbol{h}$ having the signature (1, …,1;0) and the time lag τ having the signature (0, …,0;1), whereas in the R^{n+1} description the corresponding $\Delta \boldsymbol{p}$ signature is (1, …,1,1).

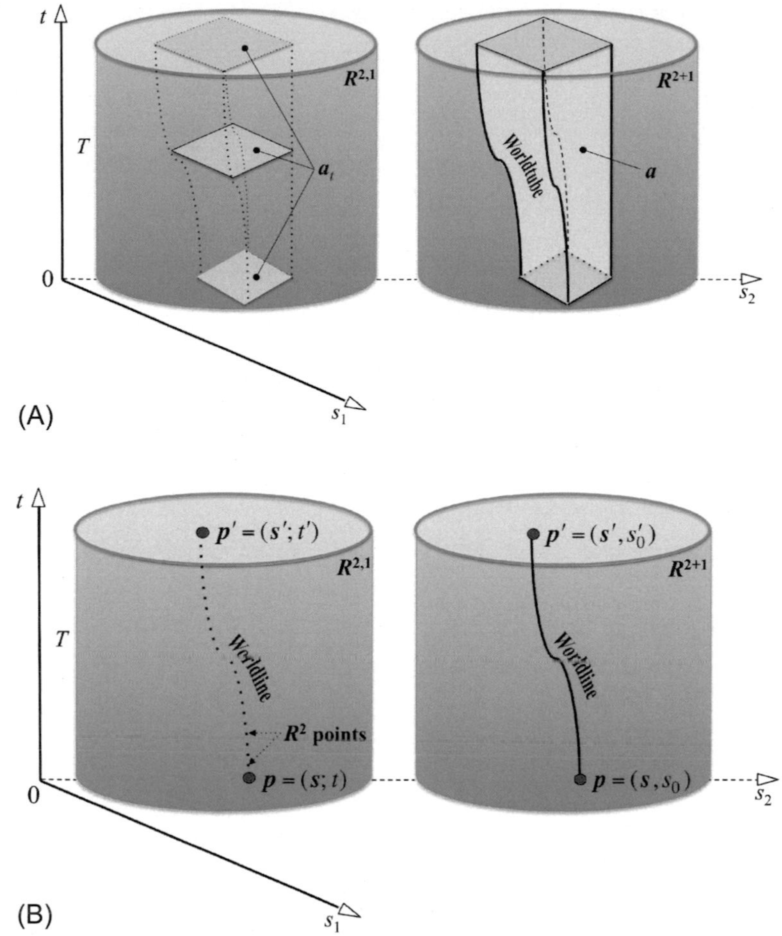

FIG. 2.2(CONT.) (A) Worldtube $WT_{\boldsymbol{a}}$ of configuration $\boldsymbol{a}$ in the spacetime R^{2+1} domain (right), and world sequence $WS_{\boldsymbol{a}}$ of discrete configurations $\boldsymbol{a}_t$, $t \in T$, in the space-time $R^{2,1}$ domain (left). (B) Continuous worldline $WL_{\boldsymbol{p}}$ of point $\boldsymbol{p}$ in the spacetime R^{2+1} domain (right), and discrete worldline $WL_{\boldsymbol{p}}$ in the space-time $R^{2,1}$ domain (left).

Example 2.2(cont.) Fig. 2.2A(cont.) visualizes the continuous *WT* of configuration $\boldsymbol{a}$ in the R^{2+1} domain, and the discrete *WS* of an ensemble of configurations $\boldsymbol{a}_t \in R^2$, $t \in T$, in the $R^{2,1}$ domain.

In view of the above notions, some relevant characterizations of a point in the chronotopologic domain representing a material point in the real world emerge naturally, such as the next one.

The **worldline** $WL_{\boldsymbol{p}}$ of a material point $\boldsymbol{p}$ (Fig. 1.1A) is represented in the spacetime R^{n+1} domain as the *continuous* line occupied by the abstract point $\boldsymbol{p} = (\boldsymbol{s}, s_0) \in R^{n+1}$, and in the space-time $R^{n,1}$ domain as the *dotted line* occupied by the point $\boldsymbol{p} = (\boldsymbol{s}; t) \in R^{n,1}$.

Example 2.2(cont.) Fig. 2.2B(cont.) visualizes the continuous worldline of point $\boldsymbol{p}$ in the R^{2+1} domain (right), and the discrete worldline of point $\boldsymbol{p}$ in the $R^{2,1}$ domain (left). A thing to notice is that the worldline in Fig. 2.2B(cont.) may be seen essentially as the worldtube of Fig. 2.2A (cont.) for which the object consists of just a single point. Two simple special *WL* cases of Fig. 2.2B(cont.) are as follows: a line with constant spatial coordinates (i.e., a line perpendicular to these coordinates) represents a point-object at rest; and a line inclined from the time axis represents a point-object with a constant speed (constant change in space coordinates with increasing time coordinate), where the more the line is inclined, the larger the speed is.

Generally, the history of a point-object (e.g., a particle) in the real world is represented by the corresponding *WL* in the chronotopologic domain (i.e., each worldline point may be seen as an event labeled with the time instant and the spatial position of the object). In the case of the discrete *WL*, in particular, the unobservable part of the *WL* in-between the discrete-time instances is obtained by *interpolation*. Interpolation in $R^{n,1}$ may differ from that in R^{n+1} because of the different characterizations of the two domains discussed above.[i] Since the *WL* can be viewed as a special case of the *WT*, the latter can be expressed in terms of the former, see Exercise 13 at the end of the chapter.

Another empirical issue is that although a worldline traces out the path of a point in the chronotopologic domain,[j] the concept of a "worldline" should be distinguished from concepts such as the "trajectory" or the "orbit" of an object. An illustration of the matter follows, in terms of the Earth's worldline *vs.* the Earth's orbit.

Example 2.8 The Earth's orbit is a three-dimensional closed curve in a space domain (often approximated by a circle) that returns yearly to the same position in space relative to the Sun but it does so at a different (later) time. On the contrary, the Earth's worldline is a helical curve in a four-dimensional chronotopologic domain that does not return to the same position.

Investigators would carefully study the worldlines (paths or trajectories) and extract useful information about the **point distribution** patterns, such as important parts of the worldlines or ensembles of worldlines with common features.[k]

The next natural step in the *CTDA* context is to examine ways to assign physical meaning to the above notions concerning domain points and their configurations (including *WL* and *WT*).

2.5 Assigning physical meaning to points, worldlines, and worldtubes

It was argued earlier (Fig. 2.3) that a point in the chronotopologic domain represents a sizeless position in time, chronotopology assigns coordinates (with space and time units) to the point, and science assigns physical meaning to the point by associating a natural attribute (with physical units) to it. The last stage is discussed next.

As the readers recall, the notion of a natural attribute was introduced qualitatively in Chapter 1, and several real world examples were discussed. Following the presentation of the chronotopology terminology in this chapter, the natural attribute can be expressed in quantitative terms as a function of its chronotopologic arguments, i.e., the chronos (time) and topos (space) of its occurrence. These arguments jointly constitute what was earlier called a point in the attribute's domain of interest.

A **natural attribute** X may be indexed by space $\boldsymbol{s}$ and time t, or more concisely, by spacetime $\boldsymbol{p}$. As such, it can be viewed either as a function $X(\boldsymbol{s};t)$ varying in the space-time domain $R^{n,1}$, or as a function $X(\boldsymbol{p})$ varying in the spacetime domain R^{n+1}.

[i] I.e., in $R^{n,1}$ the interpolation uses the biargumental domain characterization, whereas in R^{n+1} it uses the uniargumental domain characterization.

[j] Which is why the worldline is also known as *point-path* in time-geography.

[k] See, also, point clusters in Section 4 of Chapter 4.

Following the formal introduction of $X(s;t)$ or $X(\boldsymbol{p})$ above, the concept of the worldtube of an attribute (e.g., pollutant concentration) can be considered so that instead of representing an object or a configuration $\boldsymbol{a}(\boldsymbol{p})$ of points moving in time, the worldtube represents a set of *attribute points*, i.e., points to which the natural attribute assigns a certain physical meaning.

The **attribute worldtube** WT_X is the part of the R^{n+1} domain in which the attribute values satisfy certain conditions C, i.e.,

$$WT_X = \cup_p \{\boldsymbol{p} \in R^{n+1} : X(\boldsymbol{p}) \text{ satisfies } C\} \quad (2.6a)$$

The **attribute worldsequence** WS_X in $R^{n,1}$ represents the ensemble of attribute values such that

$$WS_X = \cup_t \{\boldsymbol{s} \in R^n : X(\boldsymbol{s}, t) \text{ satisfies } C\} \quad (2.6b)$$

i.e., the WS_X is the discrete union in the $R^{n,1}$ domain of all distinct spatial locations in R^n where the conditions C are satisfied during a time period T.

The WT_X is viewed as a whole that occupies a subdomain of R^{n+1}, whereas the WS_X is rather considered as a stack of spatial locations considered during the time period T within the $R^{n,1}$ domain.

Example 2.9 Illustrations of the above conditions C imposed on the values of the attribute $X(\boldsymbol{p})$ include the following:

- The $X(\boldsymbol{p})$ values exceed a certain (environmental or public health) threshold ζ, i.e., $C : X(\boldsymbol{p}) \geq \zeta$. More than one attribute worldtube may describe the threshold-exceeding phenomenon, as shown in Fig. 2.6.
- The $X(\boldsymbol{p})$ values vary within a specified range I, i.e., $C : X(\boldsymbol{p}) \in I$.
- The $X(\boldsymbol{p})$ values obey a probability law P, i.e., $C : X(\boldsymbol{p}) \sim P$.

Thus, the WT_X and WS_X notions are very useful in the study of threshold exceeding phenomena (e.g., hazardous states like over-contamination, epidemics, earthquakes, and floods). An attribute-related (i.e., physical) version of a worldline is introduced next.

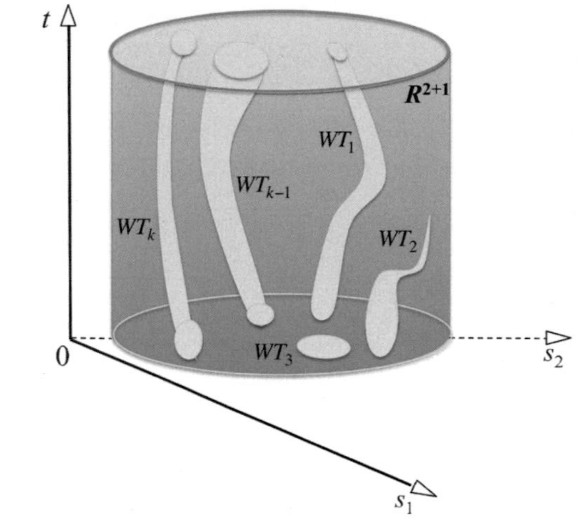

FIG. 2.6 Visualization of the ζ-exceeding attribute worldtubes indicated as WT_i, $i = 1, \ldots, k$. The worldtube WT_2 expires before the end time t', whereas that of WT_3 consists of a single time instant t.

The **attribute worldline** WL_X is a continuous line in the R^{n+1} domain or a dotted line connecting spatial points at different times in the $R^{n,1}$ domain along which the conditions C are satisfied.

In this setting, the attribute worldline may be viewed as a special case of the attribute worldtube or worldsequence. In terms of configurations, the WT_X is defined so that at all points $\boldsymbol{p}$ belonging to the configuration $\boldsymbol{a} \in R^{n+1}$ the $X(\boldsymbol{p})$ satisfies C, whereas the WS_X is defined so that at all $\boldsymbol{s}$ belonging to the configurations $\boldsymbol{a}_t \in R^n$ with $t \in T$ the corresponding $X(\boldsymbol{s}, t)$ satisfies C.

- Each point of WT_X or WS_X is seen as the point at which the C-conditioned attribute value occurs, in which case one could write that

$$\left.\begin{array}{l} WT_X = WT_p \\ WS_X = WS_p \end{array}\right\} + \text{physical meaning}$$

and refer to the worldtube and worldsequence of a natural attribute as **chronotopologic histories** of *C*-conditioned natural attribute values.

As we will see in subsequent chapters, attribute values assigned at the data points of the chronotopologic domain may need to be interpolated to generate continuous attribute distributions, attribute worldlines, worldtubes, etc. In light of the preceding discussion, one would anticipate that the different chronotopologic domain descriptions have different implications, as follows:

Attribute modeling implication: The $R^{n,1}$ *vs.* the R^{n+1} representations of a natural attribute use different descriptions of the fundamental chronotopologic arguments (points, lags, and metrics), which may affect the mathematical formulations of the natural attribute and the *CTDA* applications based on them.

Discere faciendo, we conclude this section with an illustrative example of the different chronotopologic characteristics of the $R^{n,1}$ *vs.* the R^{n+1} domains, and their potential implications in rigorous attribute representation, explanation and estimation.

Example 2.10 Suppose that the attribute X of interest is sea level height (units in m). One can model this attribute chronotopologically in two ways, as follows (Fig. 2.1(cont.)):

- According to the $R^{2,1}$ convention, any location $\boldsymbol{s}$ (in km units) on a horizontal plane at a fixed time instance t (in days units) in Fig. 2.1A (cont.) belongs to R^2. At any space-time point $(\boldsymbol{s},t)$ in the same figure, the assigned sea level height is denoted as $X(\boldsymbol{s},t)$, and the space-time sea level distribution is represented as a set of two-dimensional maps in Fig. 2.1C(cont.).
- According to the R^{2+1} convention, on the other hand, any point $\boldsymbol{p}$ on the inclined plane of Fig. 2.1B(cont.) belongs to R^{2+1}. The sea level height at any point $\boldsymbol{p}$ in the same

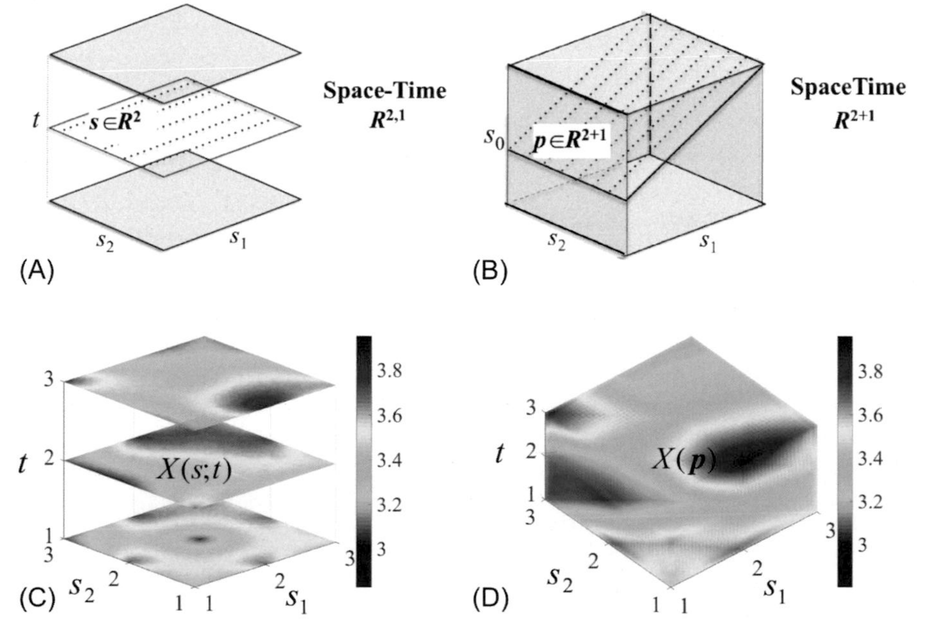

FIG. 2.1(CONT.) (A) Locations $\boldsymbol{s} = (s_1, s_2)$ on parallel planes (fixed t) of the space-time domain $R^{2,1}$, (B) points $\boldsymbol{p} = (s_1, s_2, s_0)$ on inclined planes of the spacetime domain R^{2+1}, (C) a set of two-dimensional attribute $X(\boldsymbol{s};t)$ realizations in $R^{2,1}$, and (D) a continuous three-dimensional $X(\boldsymbol{p})$ attribute realization in R^{2+1}.

figure is denoted as $X(\boldsymbol{p})$, and the spacetime sea level height distribution is represented as a continuous three-dimensional map in Fig. 2.1D(cont.).

☞ As will be shown in later chapters, this difference in mathematical formulation may have considerable consequences in attribute modeling and estimation (i.e., different chronotopologic dependence models and estimation techniques may be developed, depending on the specific mathematical formulation considered; Chapters 4–9).

In view of the above considerations, the following conclusions about operational chronotopology are drawn:

❶ Representing a natural attribute in the R^{n+1} domain means that a **static** view is assumed, i.e., all the spatial and temporal attribute values are considered *simultaneously* (e.g., Fig. 2.1(cont.)B and D). That is, a single spacetime snapshot is considered of all the attribute values assigned to the locations and the corresponding time instances where and when, respectively, the values were recorded.

❷ Representing a natural attribute in the $R^{n,1}$ domain means that a **dynamic** view is assumed that unfolds in the form of *sequences* of attribute spatial maps or images, i.e., the time-changing attribute values on a spatial grid are followed consecutively (e.g., Fig. 2.1(cont.)A and C).[l] That is, the $R^{n,1}$ viewpoint focuses on the evolution of the entire attribute history.

Our discussion of the chronotopologic domain arguments above still leaves a few, mainly technical, questions still unanswered. Addressing these questions is the goal of the remaining sections of this chapter.

3 Chronotopologic metric modeling

Metric modeling depends on whether a bi-metric $(|\boldsymbol{h}|;\tau)$ or a uni-metric $|\Delta\boldsymbol{p}|$ argument is considered. In both cases, the starting point is to understand the physics of the phenomenon, since this will allow one to account for important aspects of the phenomenon and achieve a science-based metric modeling.

A sound **chronotopologic metric** should be consistent with the body of scientific knowledge about the phenomenon (physical law and/or empirical evidence). In practice, this implies either of two things:

① The metric is derived directly from the physical law (**lawful** metric).[m]

② Different metric models are considered, and the one that provides the best fit to the available evidence is selected (**empirical** metric).

Otherwise said, the contribution of the time dimension in a chronotopologic metric is mostly application-dependent. In the following, it is methodologically appropriate that any investigation consistently distinguishes between bi-metric and uni-metric modeling.

3.1 Bi-metric modeling: *S-TDA*

In real world situations, many *S-TDA* bi-metrics $(|\boldsymbol{h}|;\tau)$ in the $R^{n,1}$ domain are used that involve the spatial distance types listed in Table 3.1. By considering space and time separately, bi-metric modeling is often reduced to the selection of an adequate $|\boldsymbol{h}|$ model.

The **distance** $|\boldsymbol{h}|$ component of a bi-metric $(|\boldsymbol{h}|;\tau)$ model can be determined **empirically** by considering various $|\boldsymbol{h}|$ models (like those in Table 3.1) and then choosing the one that offers the best fit to the actual distance of the real phenomenon.

[l] These attribute values (e.g., ground temperatures) may be remotely sensed from satellites.

[m] Examples of lawful metrics can be found in Christakos (2017).

TABLE 3.1 Examples of distance forms, $|\boldsymbol{h}|$, in *S-TDA* metrics $(|\boldsymbol{h}|;\tau)$ $(R^{n,1})$.

Distance type	Distance form $\lvert\boldsymbol{h}\rvert$
Euclidean, $\lvert\boldsymbol{h}\rvert_E$	$\sqrt{\sum_{i=1}^{n} h_i^2}$
Manhattan, $\lvert\boldsymbol{h}\rvert_M$	$\sum_{i=1}^{n} \lvert h_i \rvert$
Minkowski, $\lvert\boldsymbol{h}\rvert_{Mi}$	$(\sum_{i=1}^{n} h_i^{\lambda})^{1/\lambda}$ $(\lambda \in [1,\ 2])$
Spherical (radius ρ), $\lvert\boldsymbol{h}\rvert_{Sp}$	$2\rho \sin^{-1}\left(\frac{\lvert\boldsymbol{h}\rvert_E}{2\rho}\right)$

In practice, these four metrics provide sufficient flexibility and generality in a variety of realistic situations, and, in addition, they may be related to each other.

Example 3.1 The Minkowski distance (Table 3.1), $|\boldsymbol{h}|_{Mi}$, in particular, defines a very flexible metric, given that various λ values can be considered and the one that offers the best fit to the empirical evidence should be chosen. Note that as $\lambda \to 1$, $|\boldsymbol{h}|_{Mi} \equiv |\boldsymbol{h}|_M$, and as $\lambda \to 2$, $|\boldsymbol{h}|_{Mi} \equiv |\boldsymbol{h}|_E$.

In layman terms, a distance between two points is what the investigator defines it to be depending on the *in-situ* conditions and the objectives of the study. Such a distance model has to be well-defined and unambiguous.

Example 3.2 On the Cartesian plane, the Euclidean distance between two points is defined as the straight line joining them, but the Manhattan distance between them is the shortest path if, say, a particle moves from one location to the other only along the lines parallel to the coordinate axes.

Due to the availability of powerful *TGIS* software, the distance $|\boldsymbol{h}|$ and time separation τ between two locations (say, along a road network) can be computed in a rather straightforward manner (using, e.g., a shortest path algorithm). Yet, an analytical expression of the distance $|\boldsymbol{h}|$ (like the models in Table 3.1) is needed for quantitative modeling purposes.[n] The following example considers some descriptive aspects of bi-metric modeling.

Example 3.3 Epidemic transmission is seriously controlled by the available transportation and travel means, which, in turn, affects the way travel distances and times are conceived and calculated. Perhaps, the most common distance type is the Euclidean distance, i.e., $|\boldsymbol{h}|_E$ in Table 3.1. A typical case of such a space-time metric in $R^{2,1}$ is

$$(|\boldsymbol{h}|_E;\tau) = \left(\sqrt{h_1^2 + h_2^2};\tau\right) \tag{3.1}$$

Although $|\boldsymbol{h}|_E$ is intuitive and straightforward to calculate, in many real-world situations it does not provide an accurate estimate of the actual distance, say, in the case of cities with their usual rectangular block patterns and grid-like road networks, the accurate estimation of the actual distance between patient homes and hospitals cannot be made in terms of $|\boldsymbol{h}|_E$. In such cases, a more realistic distance may be the Manhattan distance $|\boldsymbol{h}|_M$ in Table 3.1, so that

$$(|\boldsymbol{h}|_M;\tau) = (|h_1| + |h_2|;\tau) \tag{3.2}$$

which is a metric that can measure, say, the actual home-hospital distance along a rectangular path with right angle turns.

The following synthetic (simulation) example presents the *step-by-step* **computational** process of selecting a bi-metric.

Example 3.4 Consider flow in a porous medium within which particles are moving from different locations $\boldsymbol{s}_i = (s_{i,1},\ s_{i,2})$ $(i = 1,\ldots,100)$ towards the target location $\boldsymbol{s}_0 = (s_{0,1},\ s_{0,2}) = (3,4)$ in space units.

[n] As a matter of fact, *TGIS* is a central element of an ongoing debate that is re-examining and re-inventing space and time and their representations.

Step 1: As noted above, a sound metric modeling needs to account for the important physical features of the phenomenon. Basically, each particle follows a different path $\boldsymbol{s}_i \rightarrow \boldsymbol{s}_0$ within the medium, which makes it a particular travel time τ to cover. The corresponding particle velocity $\boldsymbol{v}$ depends on the path followed.

Step 2: Based on the physics of the situation, a total of 100 such real particle paths were simulated (Fig. 3.1) with coordinates $s_{i,1}$, $s_{i,2} \in [1,2]$ space units, and the corresponding distances $|\boldsymbol{h}_{i0}|$ between $\boldsymbol{s}_i$ and $\boldsymbol{s}_0$ were calculated. These $|\boldsymbol{h}_{i0}|$ were considered the empirical (actual) distances of the particle paths. Given the geometry of the medium, the simulated $|\boldsymbol{h}_{i0}|$ distances ranged between the Euclidean distance $|\boldsymbol{h}_{i0}|_E$ and the Manhattan distance $|\boldsymbol{h}_{i0}|_M$.

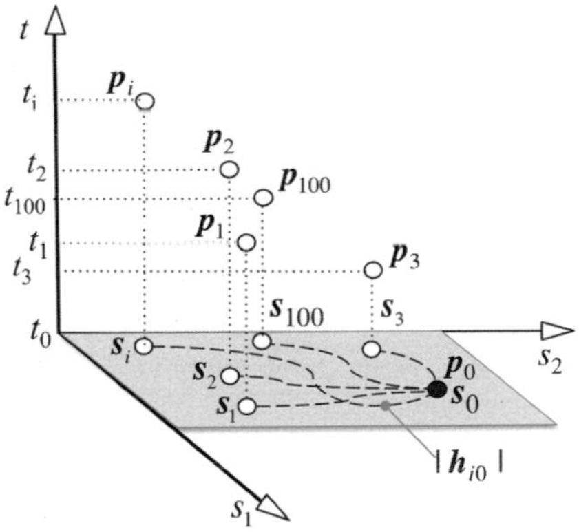

FIG. 3.1 Simulated distances $|\boldsymbol{h}|_{i0}$ along the indicated paths between points $\boldsymbol{p}_i = (\boldsymbol{s}_i; t_i)$ $(i = 1,\ldots,100)$ and $\boldsymbol{p}_0 = (\boldsymbol{s}_0; t_0)$ (blue dashed lines).

Step 3: The travel times τ_i $(i = 1,\ldots,100)$ are the simulated times during which each distance $|\boldsymbol{h}_{i0}|$ was covered by the corresponding particle at initial location $\boldsymbol{s}_i$.

Step 4: The Minkowski model of Table 3.1 was fitted to the empirical distances $|\boldsymbol{h}_{i0}|$ of the particle paths for various λ values in the range [1,2] using a $\delta\lambda = 0.05$ step, thus leading to the Minkowski distance estimates $\left|\hat{\boldsymbol{h}}_{i0}\right|_{Mi} = \left(\hat{h}_1^{\lambda} + \hat{h}_2^{\lambda}\right)^{1/\lambda}$.

Step 5: Linear regression modeling was performed between the empirical $|\boldsymbol{h}_{i0}|$ values and the Minkowski estimates $\left|\hat{\boldsymbol{h}}_{i0}\right|_{Mi}$. The regression R^2 coefficient was calculated for each $\lambda \in [1,2]$, and it was found that the value $\lambda = 1.36$ gives the $\left|\hat{\boldsymbol{h}}_{i0}\right|_{Mi}$ with the highest $R^2 = 0.634$, i.e., the following Minkowski bi-metric model

$$\left(\left|\hat{\boldsymbol{h}}_{i0}\right|_{Mi}; \tau\right) = \left(\left(\hat{h}_1^{1.36} + \hat{h}_2^{1.36}\right)^{0.74}; \tau\right) \tag{3.3}$$

was selected to represent the particle path distances $|\boldsymbol{h}_{i0}|$ $(i = 1,\ldots,100)$. Table 3.2 displays the statistics of the empirical and model distances. As was expected, the values of the Minkowski ($\lambda = 1.36$) model (mean, min, max, range) vary between those of the Euclidean and the Manhattan models.

∴ The **Minkowski** model ($\lambda = 1.36$) offers the best estimate of the empirical mean distance, followed by the Euclidean and the Manhattan model.

TABLE 3.2 Statistics of the empirical and model distances.

	$\|\boldsymbol{h}_{i0}\|$	$\|\boldsymbol{h}_{i0}\|_E$ ($\lambda = 1$)	$\|\hat{\boldsymbol{h}}_{i0}\|_{Mi}$ ($\lambda = 1.36$)	$\|\boldsymbol{h}_{i0}\|_M$ ($\lambda = 2$)
Mean distance, $\bar{h}$	3.424	3.975	3.354	3.510
Min distance, $h_{\min}$	2.407	3.144	2.671	2.339
Max distance, $h_{\max}$	4.593	4.844	4.069	3.510
Range, $h_{\max}$-$h_{\min}$	2.186	1.700	1.398	1.171

3.2 Uni-metric modeling: *STDA*

Some essential cases of *STDA* uni-metrics in R^{n+1} are listed in Table 3.3. In general, a metric is uniquely singled out by the laws of nature. Christakos (2017) and Christakos et al. (2017) discuss an approach to derive uni-metric as follows:

> A **spacetime metric** for the phenomenon of interest can be selected among formally valid uni-metric $|\Delta p|$ models so that it is optimally consistent with the **physics** of the phenomenon.

The uni-metric models listed in Table 3.3 can be considered for this purpose. Unlike the bi-metric models, the uni-metric models include coefficients (ε_{ii}, ε_{0i}, ε_{00}) with physical meaning assigned to them, since they can be directly linked to the parameters of the physical law of the phenomenon under investigation. Furthermore, these coefficients may be related. In the case of the Riemann metric, e.g., the coefficient $2\sqrt{\varepsilon_0\varepsilon_i}$ of the spacetime product term is twice the square root of the product of the coefficient ε_i of the space term and the coefficient ε_0 of the time term.[o]

TABLE 3.3 Examples of *STDA* uni-metrics, $|\Delta p|$ (R^{n+1}).

Uni-metric type	Uni-metric form $\|\Delta p\|$
Manhattan, $\|\Delta p\|_M$	$\sum_{i=1}^{n,0} \varepsilon_i \|h_i\|$
Pythagorean, $\|\Delta p\|_P$	$\sqrt{\sum_{i=1}^{n,0} \varepsilon_i h_i^2}$
Riemann, $\|\Delta p\|_R$	$\sqrt{\sum_{i=1}^{n} \left(\varepsilon_i h_i^2 + 2\sqrt{\varepsilon_0\varepsilon_i}\|h_i\| h_0\right) + \varepsilon_0 h_0^2}$
Minkowski, $\|\Delta p\|_{Mi}$	$\sqrt[\lambda]{\sum_{i=1}^{n,0} \varepsilon_i \|h_i\|^\lambda}$ ($\lambda \in [1,\ 2]$)
Traveling, $\|\Delta p\|_T$	$\sqrt{\sum_{i=1}^{n} (\|h_i\| + \varepsilon_i \|h_0\|)^2}$
Plane wave, $\|\Delta p\|_W$	$\sqrt{\left(\sum_{i=1}^{n} \frac{\varepsilon_i}{\sqrt{\varepsilon_0}} \|h_i\| + \sqrt{\varepsilon_0}\|h_0\|\right)^2}$

☝ The following examples present the *step-by-step* **computational** process of selecting a uni-metric.

Example 3.5 In the study of the $PM_{2.5}$ concentration distribution over the Shandong province (China) during the period Jan 1–31, 2014 (Fig. 3.2A, Christakos et al., 2018), the uni-metric determination followed the procedure used by the *SEKSGUI* library (Yu et al., 2007) to search for points neighboring the interpolation point of interest, see Fig. 3.3, as follows:

Step 1: To choose a sound metric one needs to account for the available physical information, which in the Shandong case included the $PM_{2.5}$ concentration variogram plots of Fig. 3.2B.

Step 2: Based on this information, the selected uni-metric in R^{2+1} was the Pythagorean one,

$$|\Delta p|_P = \sqrt{h_1^2 + h_2^2 + \varepsilon\tau^2} \tag{3.4}$$

where the parameter ε was of the form

$$\varepsilon = \left(\frac{\text{spatial correlation range}}{\text{temporal correlation range}}\right)^2 = \left(\frac{200{,}000\ \text{m}}{20\ \text{days}}\right)^2 = 10^8\ (\text{m/day})^2$$

so that the terms of Eq. (3.4) have the same units.

∴ In view of Eq. (3.4), the $PM_{2.5}$ chronotopologic domain may be considered analogous to a three-dimensional Euclidean space if the time coordinate is scaled by the physical parameter ε.

Example 3.6 A different case of uni-metric selection was discussed in Christakos (2017) and Christakos et al. (2017), as follows:

[o] For further technical, interpretational and notational details, see Christakos (2017); Table 3.3 presents some simplified cases.

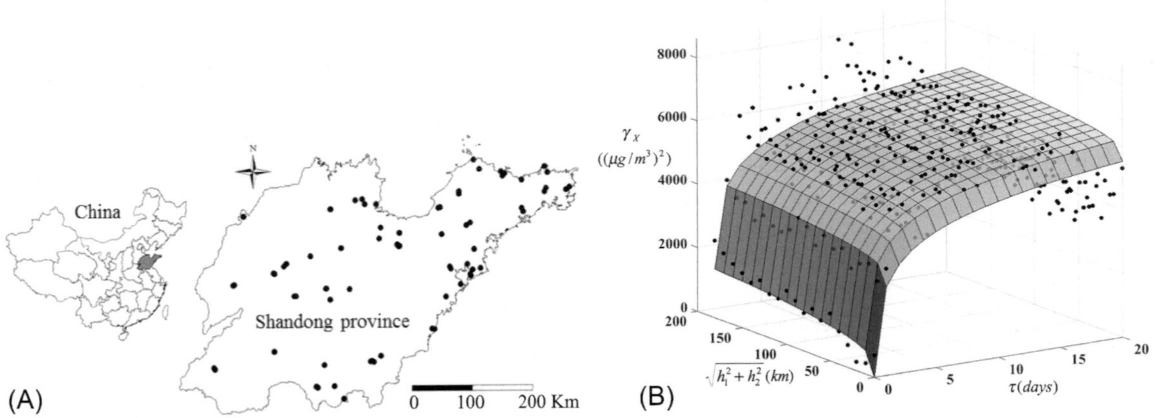

FIG. 3.2 (A) Locations of $PM_{2.5}$ samples ($\mu g/m^3$) in the Shandong province during the period January 1–31, 2014; and (B) the empirical variogram (denoted with black dots, in $(\mu g/m^3)^2$) and the theoretical variogram (continuous line) of the $PM_{2.5}$ distribution.

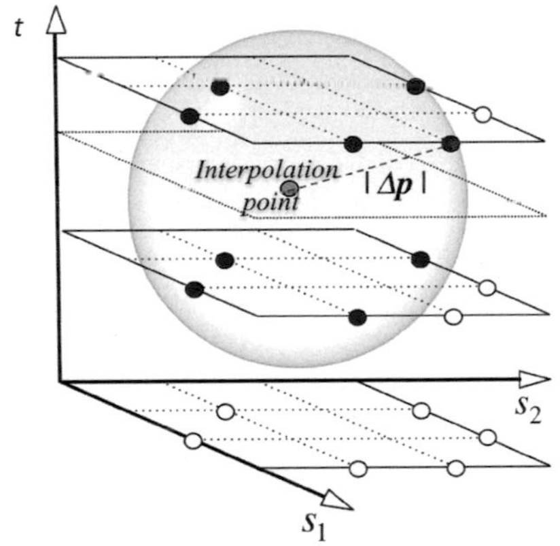

FIG. 3.3 A uni-metric used in the nearest neighbor search.

Step 1: The available physical information included the subsurface flow law.

Step 2: Based on this information, the selected uni-metric in R^{2+1} was that of the type,

$$|\Delta \boldsymbol{p}|_M = \sqrt{\sum\nolimits_{i=1}^{2} h_i + \xi^{-1}\left(\sum\nolimits_{i=1}^{2} k_i - 2\kappa\right)\tau} \quad (3.5)$$

where the uni-metric coefficients are expressed in terms of the physical parameters of this law ξ, κ and k_i, $i = 1, 2$.

The following example demonstrates that in some cases the bi-metric and the uni-metric models associated with the same phenomenon may lead to different conclusions.

Example 3.7 Consider the same phenomenon as in Example 3.4 above.

Steps 1–2: The same as in Example 3.4.

Step 3: A total of 100 locations $\boldsymbol{s}_i = (s_{i1}, s_{i2})$, $i = 1, \ldots, 100$, were simulated at time t with their coordinates, s_{i1} and s_{i2}, ranging from 1 to 2 units (Fig. 3.1). The actual spatial distance h_S between the simulated points $\boldsymbol{s}_i$ at t and the target point $\boldsymbol{s}_0 = (3,4)$ at time t' was assumed to range between the corresponding $|\boldsymbol{h}|_E$ and $|\boldsymbol{h}|_M$ values; and the actual temporal distances h_T between the $(\boldsymbol{s}_i, t)$ points and the $(\boldsymbol{s}_0, t')$ point ranged from 1 to 1.2 units.

Step 4: The velocity υ of the 100 simulated cases ranged from 1 to 1.4 units, and the travel times for each one of the 100 simulations were calculated as $\tau = h_T/\upsilon$. The actual chronotopologic distance can be calculated as $|\Delta \boldsymbol{p}|^\lambda = h_S^\lambda + h_T^\lambda$.

Step 5: Then, various λ values were selected ranging between 1 and 2 with a 0.01 interval, and the formula $|\Delta \boldsymbol{p}|_{Mi} = \left(h_1^\lambda + h_2^\lambda + \bar{\upsilon}^\lambda \tau^\lambda\right)^{1/\lambda}$ was used to get a Minkowski distance.

Step 6: A linear regression model was used to fit the calculated Minkowski $|\Delta \boldsymbol{p}|_{Mi}$ to the real $|\Delta \boldsymbol{p}|$, and the R^2 coefficient was calculated for each λ. The solution with the highest $R^2 = 0.6347$ was obtained for $\lambda = 1.5$, i.e., the selected metric was

$$|\Delta \boldsymbol{p}|_{Mi} = \left(h_1^{1.5} + h_2^{1.5} + \overline{v}^{1.5}\tau^{1.5}\right)^{0.667} \tag{3.6}$$

∴ The above uni-metric model is different than the bi-metric model of Eq. (3.3) derived in Example 3.4 for the same phenomenon.

Concluding, in empirical data analysis the different descriptions of the chronotopologic domains (space-time $R^{n,1}$ *vs.* spacetime R^{n+1}) have some considerable attribute modeling implications.

> ***S-TDA*** uses the $R^{n,1}$ description, which is based on the space-time point of Eq. (2.1a), the bi-lag of Eq. (2.2a) and the bi-metric of Eq. (2.3a) in Section 2, whereas ***STDA*** uses the R^{n+1} description based on the spacetime point of Eq. (2.1b) in Section 2, the uni-lag of Eq. (2.2b) and the uni-metric of Eq. (2.3b) in Section 2.

In the next section we discuss the possibility that a chronotopologic metric can be used in different contexts than the one considered above and for other useful purposes.

3.3 Accessibility indicators

Beyond determining the fundamental characteristics of the chronotopologic domain of interest, the metric can have some other useful applications too, like the one described next.

> The chronotopologic metric can be a useful **indicator of accessibility** between key points in the chronotopologic domain of a phenomenon by adequately assessing its geographical pattern and time evolution.

This important role of the metric emerges in various applications, as illustrated in the following public health example.

Example 3.8 Chronotopologic bi-metrics (distance magnitude $|\boldsymbol{h}|$ and time separation τ) play a vital role in public health studies assessing spatial disease patterns and time evolution. Specifically, both $|\boldsymbol{h}|$ and τ are important measures of accessibility to hospital services during the critical period of an epidemic. In today's complex health care environments, the slightest micro-geographic and micro-temporal differences in service availability can decisively affect access to health care (e.g., Shahid et al., 2009). It is in this setting that the distance $|\boldsymbol{h}|$ and/or travel time τ between patient homes and the nearest hospitals can serve as valuable indicators of geographic accessibility to hospital services.

In situations like that discussed in Example 3.8, the calculation of the bi-metric ($|\boldsymbol{h}|;\tau$), where $|\boldsymbol{h}|$ and τ are the actual distance and time on the existing road network, can be key parameters of models and techniques assessing the effective accessibility to public health services, in general. Yet, perhaps not surprisingly, empirical data also show that in some cases in practice the distance $|\boldsymbol{h}|$ and the time separation τ are not equally reliable indicators of accessibility. Both $|\boldsymbol{h}|$ and τ can be computed using *TGIS* software. Although the τ between any two points is easy to calculate with *TGIS* in large-scale domains (e.g., airplane routes, missile and satellite paths), *experientia docet*, it is more difficult to compute it in small-scale domains (e.g., road networks characterized by local conditions, including delays due to traffic, accidents and road construction), in which cases the $|\boldsymbol{h}|$ could be a more reliable accessibility indicator.

4 Metric effects on chronotopologic attribute interpolation

The metric choice can have significant effects on chronotopologic attribute interpolation and the associated interpolation error. The following remark is based on real-world experience.

The **metric effect** is demonstrated by the fact that the same data interpolation techniques[p] using the same dataset and assuming the same chronotopologic dependence structure[q] but different metrics can generate different attribute maps, which may be physically interpreted as representing different phenomena.

To address these potential metric effects, an example involving two illustrative simulation experiments is discussed next.

Example 4.1 The spatial and temporal units in this example are km and days. In light of the simulated sea level dataset (in m) presented in Fig. 4.1A, two different distance types of the space-time metric ($|\boldsymbol{h}|;\tau$) were initially assumed: one with the Manhattan distance

$$|\boldsymbol{h}| = |\boldsymbol{h}|_M = \sum\nolimits_{i=1}^{2} |h_i|$$

and another one with the Euclidean distance

$$|\boldsymbol{h}| = |\boldsymbol{h}|_E = \sqrt{\sum\nolimits_{i=1}^{2} h_i^2}$$

Then, two numerical synthetic (simulation) experiments were performed, as follows.

Experiment 1

The above two different $|\boldsymbol{h}|$ types, together with the same dataset of Fig. 4.1A and a space-time Exponential covariance model with fixed parameters (sill $c_0 = 2\,\text{m}^2$, spatial correlation range $\varepsilon_s = 4\,\text{km}$, and temporal correlation range $\varepsilon_t = 1\,\text{day}$, Chapter 5) were assumed. Then, using the space-time ordinary Kriging data interpolation technique (*STOK*, Chapter 8), two different space-time maps were generated in Fig. 4.1B (Manhattan distance) and C (Euclidean distance). For better visualization the corresponding spatial map sets are shown in Fig. 4.1D (Manhattan distance) and E (Euclidean distance). It is easily seen that the two maps in Fig. 4.1D and E exhibit very different spatial patterns at the fixed time instance.

∴ The informed investigator will resolve the issue concerning which is the better metric by choosing the one that generates the map that is the most meaningful in **physical** terms.

Experiment 2

The same distance type (Euclidean, $|\boldsymbol{h}|_E$), simulated dataset (Fig. 4.1A), and a space-time Gaussian covariance model but with two different sets of parameters (sill $c_0 = 5\,\text{m}^2$ *vs.* $2\,\text{m}^2$, spatial correlation range $\varepsilon_s = 5\,\text{km}$ *vs.* $1\,\text{km}$, and a common temporal correlation range $\varepsilon_t = 1\,\text{day}$) were assumed. Then, using the STOK interpolation technique two different space-time map sets were generated in Fig. 4.1F and G, depending on the model parameter values selected. It is seen that the two maps exhibit very different variational features and that they also have different attribute value ranges (i.e., the scale of Fig. 4.1F ranges from 2.8 to 4.0 m, whereas that of Fig. 4.1G ranges from 1.0 to 4.0 m).

∴ As before, the informed investigator will resolve the issue concerning which is the better space-time attribute dependency model by choosing the one that generates the map that is the most meaningful in **physical** terms.

[p] Such techniques are discussed in Chapters 8 and 9.

[q] Chronotopologic dependence is discussed in Chapter 5.

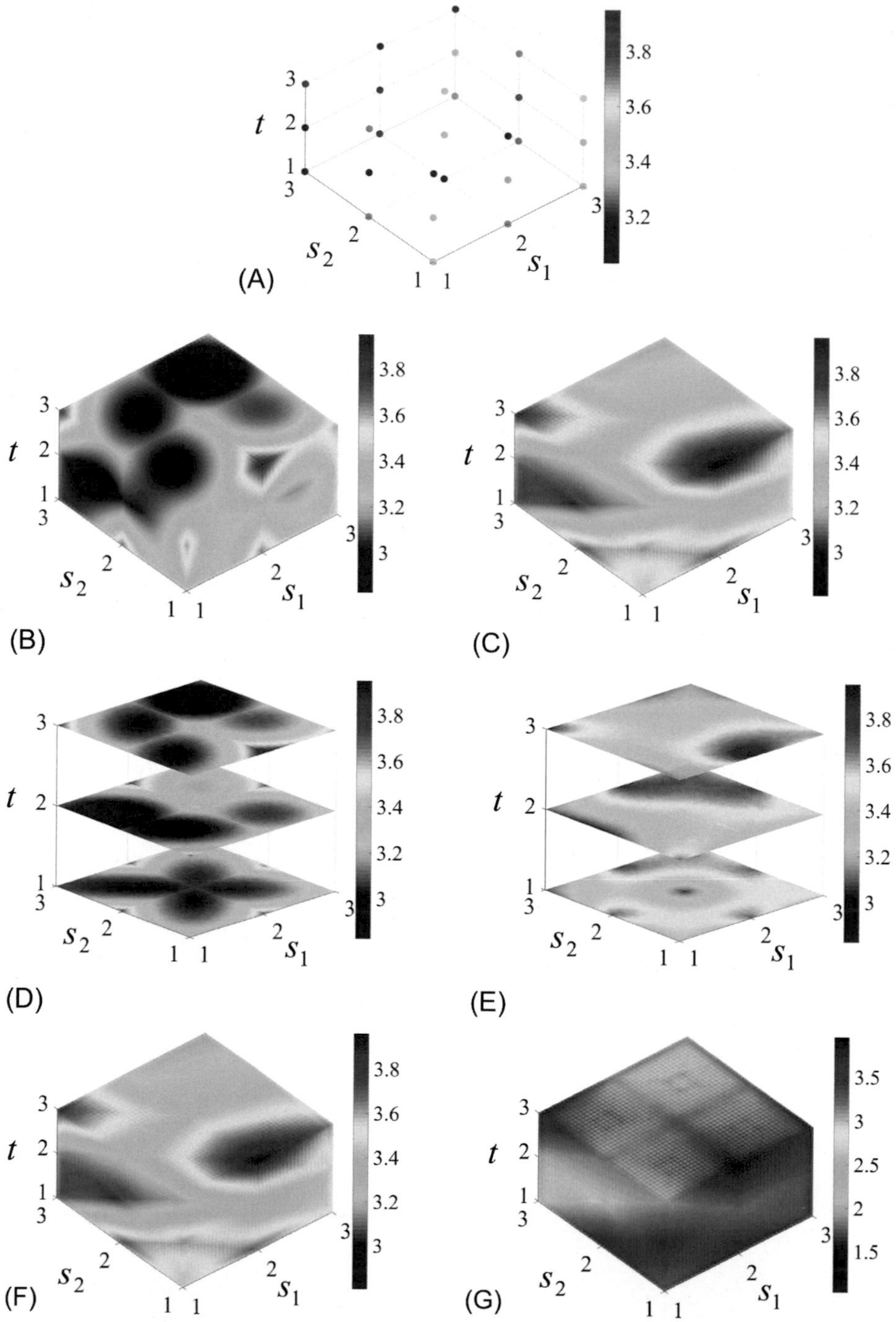

FIG. 4.1 (A) Simulated attribute data (color legend denotes attribute values in suitable units). (B)–(E) *STOK* interpolation maps produced using the data of (A) and the same jointly Exponential covariance model (sill 2 m^2, spatial correlation range 4 km, temporal correlation range 1 day) but different metrics ($|\boldsymbol{h}|_M$ *vs.* $|\boldsymbol{h}|_E$). Distinct interpretations are assigned to the maps in (B), (D) *vs.* (C), (E). Different interpretations are assigned to the *STOK* maps in (F) and (G) generated using the data in (A) with $|\boldsymbol{h}|_E$ and the same jointly Gaussian covariance model, but with varying parameters (sill 5 m^2 *vs.* 2 m^2, spatial range 5 km *vs.* 1 km, common temporal range 1 day).

We leave this chapter with a concluding remark:

The goal of the wide-ranging review of chronotopologic notions and techniques discussed in this chapter was to illustrate the rich and diverse nature of one's thinking about an **integrated** perspective of space and time that is physically realistic.

In the following chapters, these notions and techniques will be considered in various *CTDA* settings.

5 Practice exercises

1. Discuss the main differences between the conception of a space-time domain and that of a spacetime domain.
2. Describe natural phenomena that should be modeled in a space-time domain and phenomena that should be modeled in a spacetime domain, instead.
3. Among the following, which ones can be regarded as space-time domain phenomena *vs.* spacetime domain phenomena?
   (a) Digital camera photos detecting fish location in a pool every 5 min.
   (b) Digital camera photos detecting fish location in a pool from 4 a.m. to 11 a.m.
   (c) Recording the noise level on a city's main roads from 6 a.m. to 12 p.m. at 2-h intervals.
   (d) Recording the noise level on a city's main roads continuously from 11:30 a.m. to 12:30 a.m.
   (e) Sea level variation over the Aegean Sea.
   (f) Daily temperature variations in various rooms of a house.
   (g) Daily temperature variations in a room of a house.
   (h) Monthly incidence of a disease in each county of a city.
   (i) Daily mean humidity in each city of a country.
4. Formally transfer the separately displayed spatial and temporal coordinates below ($R^{2,1}$) into jointly displayed spatiotemporal coordinates (R^{2+1}):
   (a) $(s_{1,1}, s_{1,2}; t_1) = (3,4;15) \in R^{2,1}$.
   (b) $(s_{2,1}, s_{2,2}; t_2) = (6,1;7) \in R^{2,1}$.
   What physical reason would justify such a transformation?
5. Formally transfer the jointly displayed spatiotemporal coordinates below (R^{2+1}) into separately displayed spatial and temporal coordinates ($R^{2,1}$).
   (a) $\boldsymbol{p}_1 = (3,1,6) \in R^{2+1}$.
   (b) $\boldsymbol{p}_2 = (6,1,8) \in R^{2+1}$.
   What physical reason would justify such a transformation?
6. What are the key differences between a chronotopologic uni-metric and a chronotopologic bi-metric?
7. Calculate the Euclidean, Manhattan, Minkowski ($\lambda = 1.4$) and Spherical ($\rho = 6$) metrics between points *A* and *B* coordinates displayed in Table 5.1 using bi-metric modeling.
8. Repeat Experiment 1 of Example 4.1 of Section 4 using a Minkowski ($|\Delta\boldsymbol{p}|_{Mi}$, $\lambda = 1.3$) metric.
9. Repeat Experiment 2 of Example 4.1 of Section 4 using a Minkowski ($|\Delta\boldsymbol{p}|_{Mi}$, $\lambda = 1.3$) metric.
10. Describe circumstances in the real world where it is necessary to think of a unified spacetime *vs.* separable space-time.

TABLE 5.1 Spatial and temporal coordinates of point A and B.

No.	*A*	*B*
1	(1,4; 1)	(5,7; 3)
2	(2,4; 2)	(10,11; 3)
3	(7,4; 4)	(2,4; 5)
4	(3,6; 9)	(4,6; 15)

11. Argue how descriptions of reality emerging from common sense can be combined with descriptions derived from different scientific disciplines.

12. Write a brief essay expanding on the following metaphor regarding metric structures.

(a) The separate space-time metric structure would be suitable to represent one's common sense view of space as a container (within which all events take place) and time as an absolute entity (that registers the successive or simultaneous occurrences of these events). I.e., space and time exist independently of natural processes and laws, as a kind of a theatre for the natural processes and laws to enact their drama.

(b) On the other hand, the basic idea underlying the composite spacetime metrical structure is that the theatre (spacetime continuum) is intimately linked to its actors (natural processes and laws) and cannot exist independently of them.

(c) A famous legal principle of Roman law was *"da mihi facta, dabo tibi ius,"* i.e., the parties to a suit should present the facts and the judge will rule on the law that governs them. Why can't this principle be applied in scientific investigations?

13. Derive analytical expressions of the worldtube $WT_{\boldsymbol{a}}$ of a configuration $\boldsymbol{a}$ in the spacetime R^{n+1} domain and the space-time $R^{n,1}$ domain in terms of the worldline $WT_{\boldsymbol{p}}$.

CHAPTER

3

CTDA methodology

1 Methodologic chain

So far, the discussion in the previous chapters has gradually come to the following broad conclusion:

> The principal concern of ***CTDA*** is the study of **natural attributes** that vary in a specified **chronotopologic domain** based on the availability of a relevant body of **knowledge** and using suitable quantitative **models** and **visualization** techniques.

An illustration of these different links of the *CTDA* methodologic chain is portrayed in Fig. 1.1. The investigation of these links may well begin by considering five relevant questions:

① **Why** should the selected natural attribute be studied?
② **What** chronotopologic domain should be considered?
③ **Which** knowledge bases should be collected and utilized?
④ **How** should the attribute be represented quantitatively?
⑤ **By** means of what kind of tools should the attribute be visualized?

When the five questions above are considered in the context of a realistic *CTDA*, the concern of an insightful investigation is as follows:

- In addition to asking "What is the **answer** to this question?" one should also ask "Is this question a **worthy** one to investigate?"

Indeed, what constitutes a useful question and how investigators can identify such questions are crucial issues of *CTDA* and scientific inquiry, in general. For example, the *pros* and *cons* of asking the question may be weighted and a decision is made based on the anticipation of its answer leading to some worthy outcome. So, a question may be worth asking if it deepens understanding, it is relevant to the issues of concern, and it is purposeful.

According to this line of thought, addressing the above five basic questions should simultaneously address the corresponding five key *CTDA* links of Fig. 1.1. Next, we throw more light on these links, some of which have already been introduced in previous chapters, and all of which will be revisited in subsequent chapters of the book.

Quantitative Analysis and Modeling of Earth and Environmental Data
https://doi.org/10.1016/B978-0-12-816341-2.00010-1

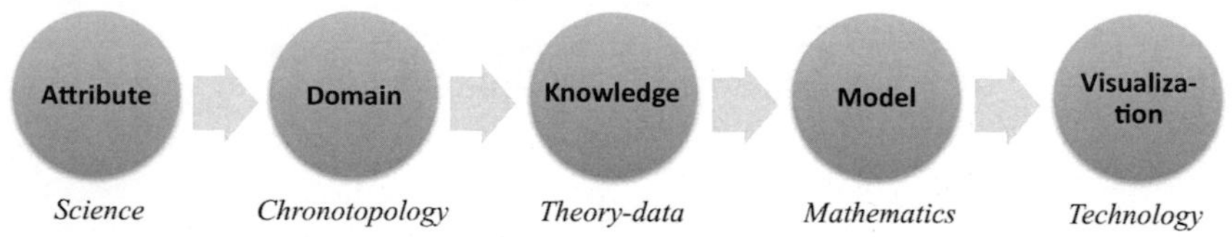

FIG. 1.1 The methodologic *CTDA* chain.

1.1 Study attributes

Concerning question ①, the readers may recall the following characterization of the subject matter from our previous discussions (see, e.g., Section 1 of Chapter 1):

> A **natural attribute** characterizes selected measurable or observable aspects of a natural phenomenon (physical, biological, or social) that vary over a region during a particular time period.

A basic fact about any natural attribute is that its chronotopologic variation obeys a law of nature,[a] in the presence of which uncertainty is constrained causally.

- **Lawful** variation and ***in-situ* uncertainty** are, thus, the main traits of a natural attribute.

It is with these two traits (lawfulness and uncertainty), disparate as they may be in some respects, that *CTDA* is chiefly concerned.

Example 1.1 The natural attributes occur in the environment (atmosphere, hydrosphere, biosphere, lithosphere, and anthroposphere), and include pollutant concentration, atmospheric pressure, topographic elevation, sea temperature, rainfall intensity, human exposure level, population mortality, land use, regional poverty index, commodity prices, and income indicators, among many others.

Deciding, then, on which natural attributes to focus is a critical starting point of a *CTDA* study. This decision clearly depends on the appreciation of a combination of factors on behalf of the investigator:

❶ Sufficient understanding of the underlying phenomenon (its physical, biological, or social mechanisms and principles).
❷ Careful outline of the study's objectives (deepening understanding, information acquisition, hypothesis testing, explanation, prediction, and discovery).
❸ Adequate resources (datasets, observation instruments, computational equipment, time availability, and financial means).

Since a phenomenon may be characterized by several attributes, the selection of the limited number of natural attributes to study should account for the governing principles and offer informative manifestations of the phenomenon. Moreover, these attributes should enable the fulfillment of the study objectives, and their study should be possible based on the current resources.

- In sum, *CTDA* investigators typically make deliberation and select the attributes to focused on, based on the factors ❶, ❷, and ❸ (i.e., they do not follow a "black box" approach where the inputs and outputs can be observed, but not the internal workings).

1.2 Chronotopologic domain

Question ② concerns the appropriate choice of the domain of the natural attribute, which is defined next (see, also, Section 1 of Chapter 2).

[a] Unlike the *auctoritas non veritas facit legem* of politics and social policy, i.e., where, according to Thomas Hobbes, it is authority, not truth, that makes laws (*Leviathan*, 1651), the physical sciences attempt to discover the laws imposed by Nature.

TABLE 1.1 Distinct ways of describing chronotopologic arguments, depending on the data analysis type.

Type	Point p	Lag Δp	Metric $\|\Delta p\|$
S-TDA	Separate space-time point determination in the domain $R^{n,1}$: Spatial vector coordinates s_i $(i=1,\ldots,n)$ independent of time scalar coordinate t	Separate space-time lag (bi-lag) in the domain $R^{n,1}$: Spatial vector distances h_i $(i=1,\ldots,n)$ independent of time scalar separation τ (also known as the vector-scalar lag)	Pair of scalar arguments (bi-metric): Spatial vector magnitude $\|\boldsymbol{h}\|$ independent of time separation τ
STDA	Joint spacetime point determination in the domain R^{n+1}: Vector coordinates s_i $(i=1,\ldots,n,0)$ form a unique spacetime vector	Composite spacetime vector lag (uni-lag) in the domain R^{n+1}: Vector elements Δs_i $(i=1,\ldots,n,0)$ form a unique spacetime vector	Single scalar argument (uni-metric): $g(\Delta p)$

An attribute **domain** is the chronotopologic physical field to which the considered natural attribute belongs or occurs.

Chapter 2 introduced two distinct ways of describing chronotopologic arguments (points, lags, and metrics) depending on the kind of *CTDA* considered, see also the relevant summary in Table 1.1. Beyond the physical domains, *CTDA* may also consider a variety of techniques that focus on technically convenient domains, such as the "time-frequency" and the "space-time projected" domains.

Example 1.2 The following are illustrations of attribute domains (physical and technical) considered in particular scientific studies.

- *S-TDA* was used to study $PM_{2.5}$ concentrations that varied within the space-time physical domain, "China-period 2015–16" (He et al., 2019a; Xiao et al., 2020).
- When the wavelet transformation technique was used to analyze surface chlorophyll-a concentration time series at a reservoir monitoring station during the period Nov. 2011 to Aug. 2012, the wavelet spectrum of chlorophyll-a concentration lies in the "time-frequency" technical domain, whereas the chlorophyll-a concentration data itself belongs to the "time" physical domain (Xiao et al., 2017).

1.3 Relevant body of knowledge

As regards question ③, the following description of what constitutes knowledge[b] is useful in the *CTDA* context.

Knowledge base (*KB*) denotes the collection of information sources on the natural attribute of interest. This includes theoretical developments (physical laws, scientific theories, etc.) together with practical experience and empirical support (data, evidence, etc.) of various degrees of reliability and uncertainty.

The totality of available *KB* about a phenomenon possesses both theoretical and empirical content. The choice of the *KB* to use depends on:

❶ The specific study characteristics, needs, and requirements.
❷ The quality of the *KB*.

[b] Going all the way back to Plato's *Meno*, knowledge is considered more valuable than mere belief. *Inter alia*, knowledge is better suited to guide action than belief. For example, if one knows rather than merely believes that this is the way to San Diego, then one might be less likely to be perturbed by the fact that the road initially seems to be going in the wrong direction.

❸ The close relevancy of *KB* to the study.

Thinking based on *KB* that are stripped of any meaningless information is considered "distilled." In this respect, one refers to the body of knowledge that satisfies conditions ❶, ❷, and ❸ as the *rational corpora*, that is, the body of knowledge that is considered relevant and of interest to the problem at hand.

Example 1.3 The goal of the Beijing (China) study was the dynamic (time-dependent) assessment of the PM_{10} concentration levels over the city. The empirical support of this kind of study is provided by the *KB* consisting of PM_{10} concentration data at a set of monitoring stations. Table 1.2 lists the PM_{10} measurements in the specified space-time domain: "four monitoring stations in Beijing—at 1:00 a.m. of January 3, 2018." The displayed "Latitude" and "Longitude" of each monitoring station are regarded as the location's spatial coordinates; "Date" depicts the exact moment that a PM_{10} concentration was measured, which is regarded as the temporal coordinate.

Section 2 continues the discussion of several *KB* aspects of wide *CTDA* interest.

1.4 Attribute modeling

Addressing the question ④ above requires introducing the notion of a "model," initially in a rather broad sense.

A **model** is a useful representation of a phenomenon that may involve symbols, plots, and pictures, without necessarily representing the true nature of reality.

As a representation in the broad sense, a model can be expressed in different ways of varying degrees of formalism and ambiguity. Generally, a model could take the form of a set of mathematical equations, diagrams, or pictures, a physical construct, a computer program that describes a phenomenon and predicts its future behavior. Models allow investigators to capture essential assumptions and to undertake sound reasoning. By accounting for certain attribute features of interest, whereas (intentionally or unintentionally) ignoring some others, a model is not a perfect image of the actual phenomenon, yet it is a comprehensive one that is supposed to work.

1.4.1 From Altamira paintings to quantum physics

It is intriguing to start with a historical incident which, in contemporary terms, could be seen as the oldest known model of aspects of reality that were vital to the lives of the people who created them.

Example 1.4 In the view of many scholars, one of the oldest examples of a "model" is the famous *Altamira paintings*. During the Paleolithic period (15,000–10,000 BC), the inhabitants

TABLE 1.2 PM_{10} concentration measurements at four monitoring stations in Beijing, China.

Latitude	Longitude	Name of station	Date	Time	PM_{10} concentration ($\mu g/m^3$)
116.417	39.929	Dongsi	01/03/2018	1:00 a.m.	17
116.407	39.886	Tiantan	01/03/2018	1:00 a.m.	19
116.339	39.929	Guanyuan	01/03/2018	1:00 a.m.	24
116.352	39.878	Wangshouxigong	01/03/2018	1:00 a.m.	21

of Altamira created the now-famous cave paintings of the hunting of animals on which they fed. The representations evoke a complex relationship between the creators and their creations with regard to the represented reality. It has been hypothesized that people created these representations of reality because of their belief in the power inherent in them to influence aspects of that reality.

After the above intriguing historical example of a model in the broad sense, let us investigate some modern model conceptions starting with some general cases.

Example 1.5 Modeling constitutes an indispensable component of modern science, as the following cases amply demonstrate.

- Continuously improved models of the atmosphere (including general-circulation models of global climate) are required in order to predict the weather with increasing accuracy and to assess climate change.
- Air quality models are necessary in order to assess human exposure to hazardous pollutants and prevent severe damage to the population.
- Realistic models of disease spread play a crucial role in any effort to control an epidemic and avoid considerable population mortality.
- Evolutionary models and the double-helix model of DNA play major roles in biologic investigations.
- Agent-based models in the social sciences are of similar importance.

Due to their great importance in scientific investigations, different model classifications that vary in degrees of formalism have been considered in the scientific literature, including visual, mathematical, and computer models. Indeed, regarding the numerous models considered in sciences nowadays, the following classifications have been used:

- The first classification depends on modeling language, syntax, and semantics. As described above, modeling generally refers to a certain form of symbolic representation of an investigator's assumptions about reality, in **informal** modeling the symbols are mental, verbal, or pictorial, whereas in **formal** modeling the symbols are mathematical.[c]
- Another classification is between **mechanistic** models that focus on *how* a phenomenon behaves (i.e., they are physically detailed and causal based on mathematical descriptions of the underlying natural mechanisms), and **phenomenologic** models that merely seek to describe the data as accurately as possible (seeking a close data fit by means of statistical regression, etc.).
- Some models have deeply **explanatory** power (i.e., they explain *why* a natural attribute adopts a certain value at a specified chronotopologic point and can be used in an *ex ante*[d] analysis), and some others are only **descriptive** (they do not offer any explanation, and can only be used in an *ex post* analysis and not in a predictive analysis).
- Some models are purely **conceptual** models (i.e., they offer a rather abstract description of a situation that may not have a concrete physical existence), and some others are models that are expressed **quantitatively** (i.e., they involve mathematical descriptions of a real-world phenomenon).

[c] While formal models rely on precise statements (involving mathematical and physical relationships), informal models may lack precision and be rather ubiquitous, but, nevertheless, they can guide the investigator's thinking during the initial study stages (e.g., experimentalists often rely on informal models to design experiments).

[d] *Ex ante* means "before the event" (e.g., prediction is always *ex ante*). The opposite of *ex ante* is *ex post*, which means "after the event."

Example 1.6 Below we present some illustrative cases of models belonging to the different classifications described above.

- Conceptual models of aquatic ecosystems include hydro-ecological models describing the components and processes of key aquatic ecosystem asset types, and pressure-stressor models identifying the impacts specific to key activities (pressures) and the mechanisms through which these activities may cause stress to the environment (stressor).
- In biogeography, one encounters quantitative representations involving a species growth equation of the form

$$\frac{d}{dt}X(\boldsymbol{s},t) = aX(\boldsymbol{s},t), \tag{1.1}$$

  where the attribute $X(\boldsymbol{s},t)$ is the species population at the time t and geographical location $\boldsymbol{s} = (s_1,s_2)$, with s_1 and s_2 denoting spatial coordinates and a being an experimental parameter. The model of Eq. (1.1) is a mathematical representation of the actual but incompletely known phenomenon "species growth." Many other models of natural systems can be found in the relevant scientific literature.
- According to the Newtonian gravity model, the gravity force F between two bodies is proportional to the product of the masses m_1 and m_2 of the two bodies divided by the distance h squared, i.e., the model output F is determined by the three input values m_1, m_2, and h. While the model explains the *how* (i.e., how m_1, m_2, and h interact to produce the F), it does not explain the *why* (i.e., why these should be the factors that influence gravity and why they interact in this way). Thus, the model is mechanistic (it merely provides a parsimonious explanation that allows us to predict gravity for any situation), but it is not a deeply explanatory model.

Generally, for a scientific model to be adequate and useful, it should fully incorporate the current state of knowledge regarding the natural phenomenon of interest, and it should account for the needs of the real world investigation. Typically, a model should capture the assumptions one makes about a phenomenon (e.g., what are the inputs and outputs, how the inputs might interact to generate the outputs), with further details about its parameters to be obtained from empirical investigations.

Modeling, then, is both about one's *subjective* representation of the world and about the *objective* reality of the world. In this respect, one may distinguish between two kinds of modeling assumptions: The *epistemic* assumptions made by the investigator that reflect one's level of understanding of the world, and the *ontic* assumptions about what is characterized as the natural world. The former assumptions are much looser and much more subject to modification in the light of new evidence than the latter.

Naturally, various combinations of the above model classifications may be considered in the pragmatic world of scientific modeling (e.g., the quantitative models are formal models). Of special interest is the following advanced modeling conception.

A **theoretical** model is based on a scientific theory that explains a phenomenon starting from the first principles,[e] and it is also able to predict it. It can lead to new findings and, in a few cases, even to the development of new sciences.

As a matter of fact, theoretical modeling investigations have been responsible for some of the greatest scientific and technological advances in human history. This is the case of the following example.

[e] I.e., it is deeply explanatory.

Example 1.7 By examining experimental data on radiation energy density ρ *vs.* cavity temperature T and electromagnetic radiation frequency υ, Max Planck fitted to the data the empirical formula:

$$\rho(\upsilon,T) = a\upsilon^3\left(e^{\frac{b\upsilon}{T}} - 1\right)^{-1}, \quad (1.2)$$

where a and b are coefficients determined from data fitting. Planck was not satisfied with the excellent fit that Eq. (1.2) provided to the data. By using scientific insight, theoretical concepts, and hypotheses he established the now famous theoretical radiation model:

$$\rho(\upsilon,T) = \frac{8\pi h\upsilon^3}{c^2}\left(e^{\frac{h\upsilon}{k_B T}} - 1\right)^{-1}, \quad (1.3)$$

where c is the speed of light and h, k_B are constants. By comparing the empirical curve of Eq. (1.2) with the theoretical model of Eq. (1.3), Planck derived expressions for the fitting coefficients a and b, and obtained a powerful interpretation of Eq. (1.3), thus, gaining a much better physical understanding of the phenomenon.

∴ The **thought process** involved in the derivation of Eq. (1.3) led to the establishment of new physics that is responsible for some of the greatest scientific and technological advances of the 20th century.

Ultimately, all the kinds of models discussed above should be grounded in their interaction with the physical world. Modeling is a central element of *CTDA* and will be considered in detail in various parts of the book.

1.4.2 *Interpretation*

Models are of central importance in many scientific contexts. Yet, rigorous model building is not the end of the story. A physically meaningful interpretation of the constructed model is equally important.

The **interpretation** of a model is the act of understanding the relationship between the physical quantities of the model and how the phenomenon it represents is expected to behave under certain conditions.

How an investigator construes *abstract meaning* might be constrained by and often derived from the investigator's very *concrete experiences* in the physical world. A related interpretation issue is that *ancillary* knowledge can be critical to determining the meaning of a scientific model.

- To understand a model and properly interpret, it is to understand its **relation** to the entire knowledge body considered.

Example 1.8 Let us examine the possible interpretations of the model $\hat{X} = \hat{\theta}_0 + \hat{\theta}_1 Y$ used in chronotopologic linear regression interpolation (*CTLR*, Section 3.2 of Chapter 7) by considering different perspectives:

- According to the *minimalistic* perspective of abstract mathematics, the model simply expresses the quantity $\hat{X}$ in terms of the sum of the quantity $\hat{\theta}_0$ and the product of the quantities $\hat{\theta}_1$ and Y. The minimalistic perspective is usually not sufficient in *CTDA* applications as in reality the above model surely carries more meaning than the formal relation between its four symbols.
- According to the *CTDA* perspective, the model takes on the meaning of a regressive relation between the independent variable Y and the dependent variable $\hat{X}$, where $\hat{\theta}_0$ and $\hat{\theta}_1$ are constants, i.e., the model represents a straight line. The constants gain additional, geometrical, meaning: $\hat{\theta}_1$ is the slope of the line and $\hat{\theta}_0$ is its intercept on the $\hat{X}$-axis. The meaning of the model understood in terms of regression notions and plots is much richer than the minimalistic perspective merely expressing the symbol $\hat{X}$ in terms of other symbols.

- According to a **physical** perspective discussed in Exercise 16 at the end of the chapter, an even richer interpretation of the above model is possible.

1.4.3 *Content-free vs. content-driven modeling*

There are clear limitations to the specificity of inferences that may be drawn about the physical mechanisms from the behavior of abstract quantitative models. One of them is that it is a theoretical and practical fact that often not one but *several* models may honor the same set of data points. This is because when the "data speak," what they actually do is to give the investigators several choices, among which they pick the right one based on a deep understanding of the phenomenon, scientific introspection, and sound logical reasoning. The following example illustrates this crucial point.

Example 1.9 Fig 1.2A presents the plot of an oscillating natural attribute together with the available data points in spacetime (20 dots), where h and τ denote spatial distance and temporal separation, respectively. The amplitude of oscillation increases with h and τ. In a real-world study, one often tries to describe the attribute distribution usually based on a limited number of samples. *Least squares* and *thin-plate splines* models were both fitted to the data points in Fig 1.2A. For further illustration, Fig 1.2B shows a slice cut at an angle of 45° across the actual attribute variation of Fig 1.2A, together with the two fitted models, where the spacetime metric is defined as $|\Delta p| = \sqrt{h^2 + \tau^2}$. The splines model was favored over the least-squares model, in this case, because it better captured the fluctuation structure of the attribute, i.e., the selection of the splines model relied on the basis of the natural properties of the attribute.

The take-home message, then, is that several quantitative models may offer good fits to a particular dataset, in which case the appropriate way to decide among them is not in terms of **content-free** (technical fit) rules but rather by means of **content-driven** (substantive) arguments.

1.4.4 *A modeling process*

Building on ideas such as the ones discussed above, the *modeling process* in real-world *CTDA* applications undergoes two major phases, as described in Table 1.3. The premodeling phase

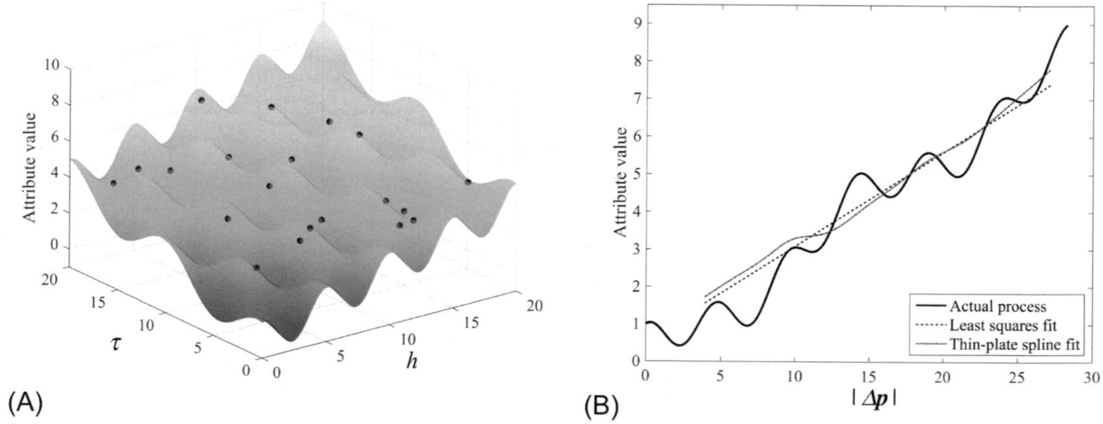

FIG. 1.2 Comparison of model-fitting results: (A) actual natural attribute (shaded color surface) with data locations (red dots), and (B) a slice of the actual plot (black line) in (A) that is cut at an angle of 45°, together with the two models (blue dot line represents least-squares model, and the purple line represents thin-plate spline model) fitted to the data.

TABLE 1.3 The model building process.

❶ *Premodeling phase*	1*a*. Identification of the proper *desirables*, i.e., the attributes, variables, and parameters that can be quantified by means of some process of measurement or observation (at least in principle, if not always in practice)
	1*b*. Intuitive understanding of the *physical relations* between the measurables or observables
❷ *Modeling phase*	2*a*. The real-world measurables or observables are expressed in mathematical symbols so that the physical relations between them are "translated" into mathematical *equations* (i.e., mathematical relations between symbols)
	2*b*. Mathematical tools are used to *solve* the equations and find answers to problems that could not be answered by direct intuitive thinking about the physical system
	2*c*. The mathematical solution is *interpreted*, i.e., its meaning is understood in the context of the physical system of interest. The physical interpretation of the equation parameters helps with developing intuition
	2*d*. The solution and its interpretation are *evaluated* in terms of both their agreement with observation and the logical consequences they lead to. If they fail to pass either of these two tests, a modification of the selected model may be necessary

is concerned with the careful selection of *desirables*, i.e., the particular substantive aspects of a physical system that need to be described, measured or observed, explained, and predicted. The main modeling phase is concerned with *equations* and their *solutions*, *interpretations*, and *evaluations*. That is, the physical links between the phenomenal aspects that can be measured or observed are expressed in terms of equations that must be interpreted, and their validity and physical meaningfulness are duly assessed.

Example 1.10 A natural forest assessment study needs to focus on those physical attributes required to generate the target information (desirables in Table 1.3). Although the decision as to which attributes to measure or observe (measurables or observables in Table 1.3) may be relatively simple, sometimes it is not easy to implement in practice.

Even when a model's shortcomings are profound, it could still reveal something useful about the underlying mechanism of the natural attribute of interest: it may not provide specific attribute predictions but it may indicate the presence of significant trends, or it may not accurately measure data correlations but it may reveal their existence.

1.5 Visualization means

Ceteris paribus, the intention of the last question ⑤ is to bring to the readers' attention certain important technological aspects of visualization, which play an increasingly important role in modern *CTDA* applications.

> Chronotopologic **visualization** aims to display the data and/or the model outputs by an easily understandable figure or movie rather than directly listing the numerical data.[f] In this sense, visualization can generate valuable new information.

Certain visualization aspects were already discussed in Section 6 of Chapter 1, and the topic

[f] As, say, in Table 1.2.

will be revisited in Section 6 of the present section. At this point, it is appropriate to conclude our review with a visualization of a real-world climate-disease association.

Example 1.11 Fig. 1.3 presents a chronotopologic visualization of the monthly strength of the association between global climate dynamics (multivariate El Niño-Southern Oscillation index, *MEI*) and the hemorrhagic fever with renal syndrome (*HFRS*) across east China during the year 2012. These maps offer an informative visualization of the geographical spread and monthly dynamics of the corresponding coherency values (for more details, the readers are referred to He et al., 2018).

2 About knowledge

Real-world *CTDA* involves a variety of *KB*, some of which are directly applicable in practice whereas some others need further verification before being used in real-world applications.

2.1 General and specificatory

Methodologically, it is considered appropriate for *CTDA* purposes to distinguish between two major *KB*[g]:

> **General** or **core** *KB*, *G*, which includes fundamental natural laws, primitive equations, scientific models, and theories. The *G-KB* is distilled knowledge about a phenomenon, meaning that it has been known for a long time and is well-established.

> **Specificatory**, or **case-** or **site-specific** *KB*, *S*, which includes hard data (measurements and observations at different chronotopologic scales and resolutions) and soft information (uncertain information, noisy measurements, data intervals and histograms, auxiliary variables, and empirical charts).

The union of these two *KB* leads to what can be considered as the wholeness of what is known:

- The **total** or **full** *KB*, *K*, currently available about the phenomenon of interest is symbolically denoted as $K = G \cup S$.

The fundamental laws of classical and modern physics, as well as the scientific theories and the primitive equations, belong to the *G-KB* because they are distilled knowledge for the corresponding disciplines. One cannot assign meaning to the data and adequately interpret it without theoretical support, both of which are the focus of the *G-KB*.

The *S-KB*, on the other hand, is empirical evidence that involves a collection of measurements and observations, as well as other relevant information sources on a natural attribute. One cannot know without experiencing, and data brainstorming can deliver valuable insights, both of which are the focus of the *S-KB*.

2.1.1 *KB interplay*

The distinction between the *G-KB* and the *S-KB* outlined above is often based on the *interplay* between two major stages of the knowledge process:

[g] Justifications of the distinction between general and specificatory *KB* can be found in disciplines like *evolutionary biology* (endosomatic knowledge is incarnated in organisms though evolution, and exosomatic knowledge is encoded in new experience with environment), *neuroscience* (compressed knowledge in the left brain hemisphere enables an individual to deal with familiar problems relatively comfortably, whereas the right brain hemisphere enables one to explore new data about the specific situation often with certain discomfort), and *sociobiology* (considers the distinction between stored knowledge and sensory data).

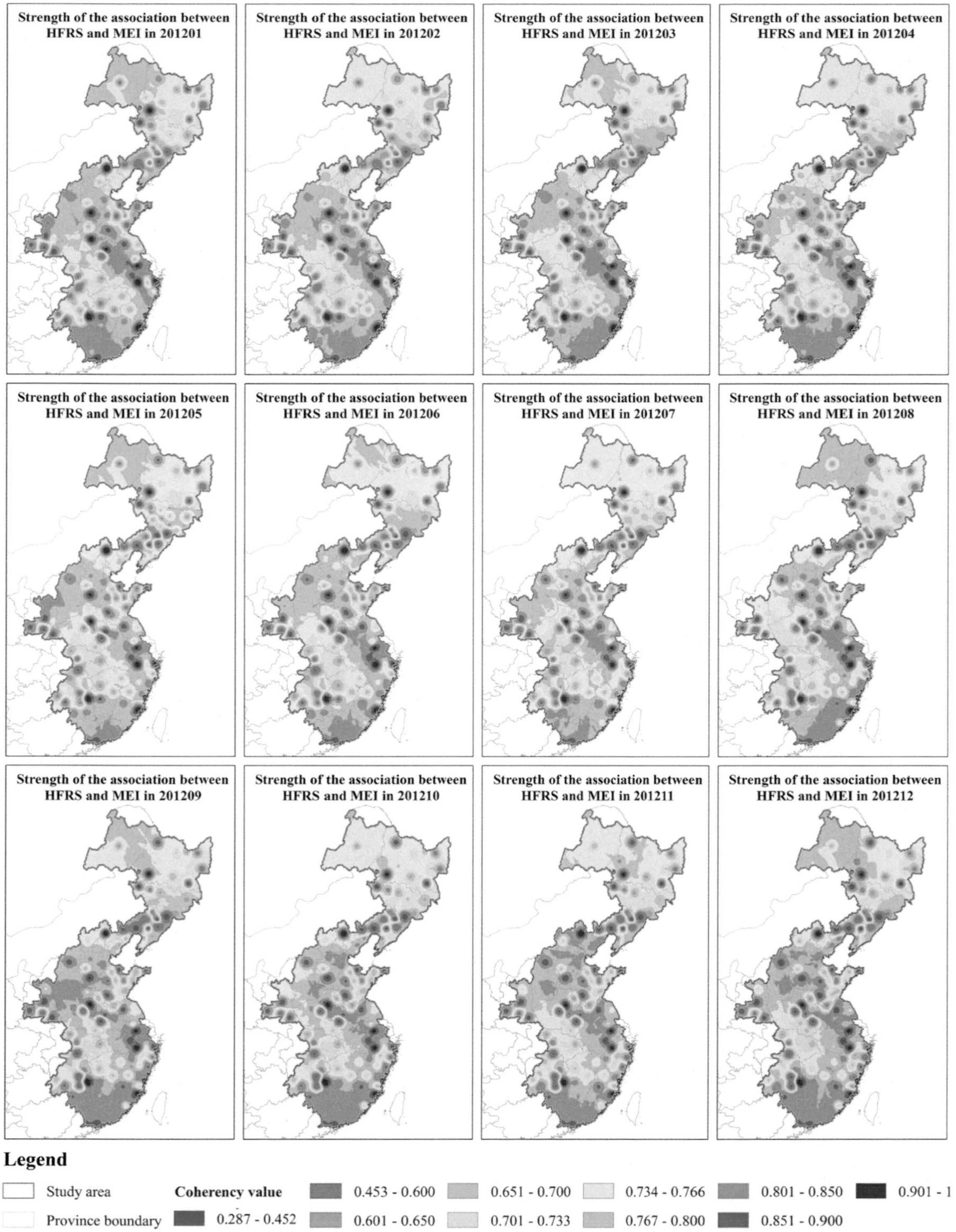

FIG. 1.3 Strength of the climate-*HFRS* association (coherency) in the Eastern China study region during the year 2012.

❶ The stage leading from perceptual knowledge (i.e., empirical evidence, sensory experience and experimentation considered in S) to conceptual knowledge (i.e., concepts and theories considered in G), and

❷ The stage leading from conceptualization (G) back to perception (S) in order to ascertain whether the concepts and theories meet with the anticipated success in practice.

The adequate understanding of a natural phenomenon is usually arrived at after several repetitions of the ❶-❷-❶ process. Then, a valuable conclusion is reached:

- ***KB* intersection** is an essential consequence of the above process. New routes open when one type of *KB* crosses another.

In some cases, the interplay between the *G-KB* and the *S-KB* may take a particularly interesting form: information that was initially thought to be part of the *S-KB* of a phenomenon, may later turn out to be part of its *G-KB*. An illustrative case is discussed next.

Example 2.1 Based on site-specific data in three Chinese regions (Zhejiang, Guangdong, and Xinjiang), He et al. (2020b) found that at its early stage the coronavirus 2019 (COVID-19) spread obeyed the log-linear differential equation.

$$\log\frac{dT(t)}{dt} = 0.759\log T(t) + 0.055, \tag{2.1}$$

where $T(t)$ represented the infected and removed case numbers at time t (including cured and death cases). Eq. (2.1) links the log-number of daily new infected cases and the log-accumulated number of infected cases, thus providing a measure of the COVID-19 transmission speed. Since Eq. (2.1) was derived based on data available in the three Chinese regions, it was originally considered as part of the *S-KB* of these regions. However, it was subsequently found that Eq. (2.1) was also obeyed by the COVID-19 spread in other countries, including South Korea, Italy, and Iran, at which point Eq. (2.1) could be potentially seen as an empirical COVID-19 growth law that belonged to the *G-KB* of the COVID-19 spread (i.e., it may be applicable to COVID-19 outbreaks in other world regions).

The particular notions and methods presented in the book will cover a wide range of *G-* and *S-KB* including data that describe a variety of attributes characterizing physical phenomena and systems including earth, ocean, and atmospheric variables, environmental and ecological parameters, population health states, disease indicators, social and economic characteristics.

2.1.2 KB symbolization

Datasets and other kinds of information belonging to the *S-KB* are represented in practice using mathematical symbols to convey meaning. An illustrative example of how this is typically done follows.

Example 2.2 Let T represent the attribute of the *daily highest temperature* (*DHT*, in °C) in the Zhejiang Province (China). The numerical T data during March 1–7, 2019 at 11 cities of the province are listed in Table 2.1. The name of each city is also listed. Overall, this dataset consists of $n = 77$ measurements at $k = 11$ cities. Accordingly, the following symbolic representation can be used in this case:

- The spatial vector $\boldsymbol{s} = (s_1, s_2)$ symbolizes city location, where the s_1- and s_2-coordinates are longitude and latitude, respectively, and the t-coordinate is the particular date during the time period of interest.
- The symbol $T(\boldsymbol{s}, t)$ denotes the *DHT* at city $\boldsymbol{s} = (s_1, s_2)$ and time t. For illustration, $T(\mathbf{1}, 1) = 10$°C is the *DHT* at the 1st city (Hangzhou City with geographical coordinates 120.15 and 30.28) and at the 1st time instance (March 1, 2019); similarly, the *DHT* at the 3rd city (Wenzhou City with geographical coordinates 120.70 and 28) and at the 7th time instance (March 7, 2019) is denoted as $T(\mathbf{3}, 7) = 15$°C.

TABLE 2.1 Daily highest temperatures (*DHT*) in Zhejiang Province (March 2019).

s_1	s_2	City	t	*DHT*	t	*DHT*
120.15	30.28	Hangzhou	March 1	10	March 2	9
			March 3–5	10	March 6	14
			March 7	12		
121.55	29.88	Ningbo	March 1	8	March 2–4	11
			March 5	10	March 6	12
			March 7	11		
120.70	28.00	Wenzhou	March 1	11	March 2–3	14
			March 4	15	March 5	13
			March 6	17	March 7	15
120.75	30.75	Jiaxing	March 1	12	March 2	9
			March 3–4	10	March 5	11
			March 6	14	March 7	12
120.08	30.90	Huzhou	March 1	12	March 2	8
			March 3	10	March 4	9
			March 5	10	March 6	14
			March 7	11		
120.57	30.00	Shaoxing	March 1	10	March 2	9
			March 3	11	March 4	10
			March 5	11	March 6	13
			March 7	12		
119.65	29.08	Jinhua	March 1	10	March 2–5	11
			March 6–7	13		
118.87	28.93	Quzhou	March 1–2	9	March 3	14
			March 4	12	March 5	10
			March 6	14	March 7	13
122.20	30.00	Zhoushan	March 1–2	10	March 3	8
			March 4	10	March 5	13
			March 6	11	March 7	9
121.43	28.68	Taizhou	March 1	11	March 2–3	12
			March 4	14	March 5	11
			March 6	16	March 7	12

Continued

TABLE 2.1 Daily highest temperatures (*DHT*) in Zhejiang Province (March 2019)—cont'd

s_1	s_2	City	*t*	*DHT*	*t*	*DHT*
119.92	28.45	Lishui	March 1	12	March 2	17
			March 3	14	March 4	13
			March 5	12	March 6	14
			March 7	13		

CTDA investigators need really efficient ways to collect, represent, and evaluate the data and other information sources belonging to the *S-KB*. This is a procedure that involves appropriate *measurement protocols* and *field manuals* assuring that the natural attributes to be measured or observed are defined in an unambiguous manner, the data collection procedure can be described in detail, etc.

2.1.3 Measurement vs. observation

The chronotopologic context includes several types of data that exhibit different properties. Each data type offers distinct opportunities for extracting useful knowledge. Measurement and observation are basic components of all data collection protocols and field manuals. They may be gathered either in the field (*in-situ*) or the laboratory. Before proceeding further, a fundamental distinction between the two should be made.

A **measurement** is distinguished from an **observation** in the sense that the former denotes an active process (it involves an experimental equipment, instrument, or apparatus), whereas the latter is a rather passive process (it is based on direct monitoring, surveillance, or inspection of the phenomenon).

To experiment is to isolate, set up, and manipulate conditions (in the lab or *in-situ*) to produce epistemically useful evidence, whereas observing is about noticing and attending to interesting features of phenomena perceived under natural conditions. There are, of course, cases where the two coincide. Nevertheless, in practice the terms "measurement" and "observation" are often used interchangeably, and the exact meaning is inferred from the context.

Example 2.3 Measuring the maximum heat that a metal can sustain before it starts melting is the result of a certain action on the agent's behalf involving the experimental setup (instrument, device, etc.) for the measurement to be made possible. On the other hand, an observational statement of the kind "the sun rises every morning" neither requires nor is depending on any instrument or device in order to occur. Lastly, measuring the number of migrating birds in a region is essentially the same as observing this number.

2.1.4 S-KB structural features

When dealing with the various *S-KB* encountered in applications, initially investigators must be aware of certain of their structural characteristics:

Inclusivity: *S-KB* may include various combinations of *measurements* and *observations*—the two may need to be distinguished from each other (see above).

Support: *S-KB* may be collected from sources with different supports. Although mainly *point* sources are considered, other kinds of sources may also be used, such as *chronoarea* and *chronovolume* data supports (Section 3.1 of Chapter 8).

Performance: The choice and organization of tools (techniques, instruments, equipment) influence data *representability*, i.e., how well *S-KB* represents the attribute it is intended to, and *reliability*, i.e., how well *S-KB* maintains its consistency or repeatability in measurements (Section 2.3 below).

Multidisciplinarity and **multiscality**: Data of different kinds of *attributes* from various scientific disciplines may be analyzed and processed, which are obtained at different *resolutions*, from coarse to fine (Section 4.2 later).

Example 2.4 For illustration, some typical scientific applications involving different data characteristics are as follows.

- In *public health*, investigators compare group scores on some dependent variable (e.g., a disease), and based on the analysis, they draw a conclusion about whether the independent variable (say, disease treatment) had a causal effect on the dependent variable.
- In *earth* and *atmospheric sciences*, large-scale monitoring systems usually operate at coarse space-time resolutions that are expensive but still incapable of capturing the complex atmosphere-land surface interactions with sufficient precision that would make possible the development of realistic models of the natural system.
- *Remote sensing* can generate measurements on a dense space-time grid, but it is incomplete in that not all the attributes are measured that are required in state determination.

2.1.5 *Emerging skepticism*

It is commonly claimed that one starts with datasets, from which a model is constructed and inferences are drawn. Then, a major issue naturally emerges:

Why are these particular datasets considered?

Accordingly, data gathering should not be a mere recording of facts, but it should penetrate the mystery of their origin and the assumptions (sometimes informal, implicit, or even hidden) that precede data acquisition. The use of an *S-KB* in *CTDA* is not a straightforward affair, but requires a healthy dose of skepticism concerning the issues raised by the following questions:

① **Origin**: Where did the data and other information come from?
② **Collection**: What was the data collection procedure?
③ **Quality**: What is the level of data quality?
④ **Peer-review**: Has the data and other information been peer-reviewed?
⑤ **Correlation**: Are the data correlations robust?
⑥ **Context** and **physical meaning**: Are numbers reported in context and what is their physical meaning?
⑦ **Confounding factors**: What factors possibly causing spurious associations might be involved?
⑧ **Hypothesis testing** and **logical reasoning**: Are the hypotheses underlying data collection testable and the resulting conclusions logical?
⑨ **Replication**: Have the results been replicated by others?
⑩ **Commissioning**: Who ordered the study?

Data and other information sources should be thought of not as "givens" but as a thoroughly examined and tested sensory experience.

2.2 Data collection issues

Important data-related activities include collecting, structuring, and cleaning the dataset. Actually, the question ② above concerning data collection refers to basic information about the data.

> It is rather a matter of common sense that to study and interpret the data adequately, and subsequently derive useful physical conclusions from it, one needs to know how the data was collected, that is, the **data-specific collection technique**.

2.2.1 *Techniques*

Below we consider four common techniques of data collection that are implemented, to varying degrees, in many scientific disciplines.

Census: This technique relies on the possibility to obtain data from every member of a *population*.

In most real world studies, a census may not be a practical approach, because of the cost and/or time required.

Example 2.5 One wants to know the IQ and EQ of all children between 8 and 12 years old in the USA. Thus, one must implement a census for all children, which is the population in this case.

Sample survey: This is a study that obtains data from a *subset* of a population, in order to estimate population attributes.

Because of its practical feasibility, this technique is favored in a large number of applications.

Example 2.6 One only chooses five representative elementary schools from each state to investigate the IQ and EQ of the children between 8 and 12 years old in the USA.

Experiment and **measurement**: An experiment is a controlled study in the *lab* or *in-situ* that produces measurements at different scales.

The goal of experimentation is to help investigators understand the underlying mechanisms and cause-and-effect relationships. It is a "controlled" study, in the sense that the investigator controls (i) the kind and conditions of the experiment, (ii) which aspects of the phenomenon the experiment accounts for, (iii) what the theoretical support of the experimental setup is, (iv) how entities are assigned to groups, and (v) which treatments each group receives.

Example 2.7 Storm data considered in Cobos et al. (2019) covered the Spanish and North African territorial waters with a 65×10^4 km^2 area and perimeter larger than 4×10^3 km. Hard datasets included spectral wave heights from several sources: REDCOS and REDEXT datasets recorded at shallow and deep waters providing hourly values in 24 and 26 min intervals, and the SIMAR database from the Puertos del Estado forecasting system updated with wind and pressure data evolution and consistent with observations (Table 2.2). The spatial data distribution is shown in Fig 2.1A. Soft datasets (Fig 2.1B) included the empirical relation of spectral wave height at REDEXT-COS (H_R, in m) and wave height at SIMAR locations (H_S, in m).

We already stressed the difference between a measurement and an observation. Here we further elaborate on this difference.

Inspection: It attempts to understand mechanisms and cause-and-effect relationships by means of passive observation rather than active measurement.[h]

Unlike experiments, in an observational study the investigator may not be able to control (i) what the observation conditions are, (ii) how subjects are assigned to groups, and/or (iii) which treatments each group receives.

TABLE 2.2 Informative parameters of several sources of hard data.

Hard data source	Interval (min)	No.	Duration (yrs)	Comments
REDCOS	24	14	Depends on device	Measurements in shallow waters
SIMAR	–	121	1958–2017[a]	Forecasting consistent with observation
REDEXT	26	7	Depends on device	Measurements in deep waters

[a] *The analysis of data consistent with observation started on January 1, 2000.*

[h] See discussion earlier, for a distinction between the two.

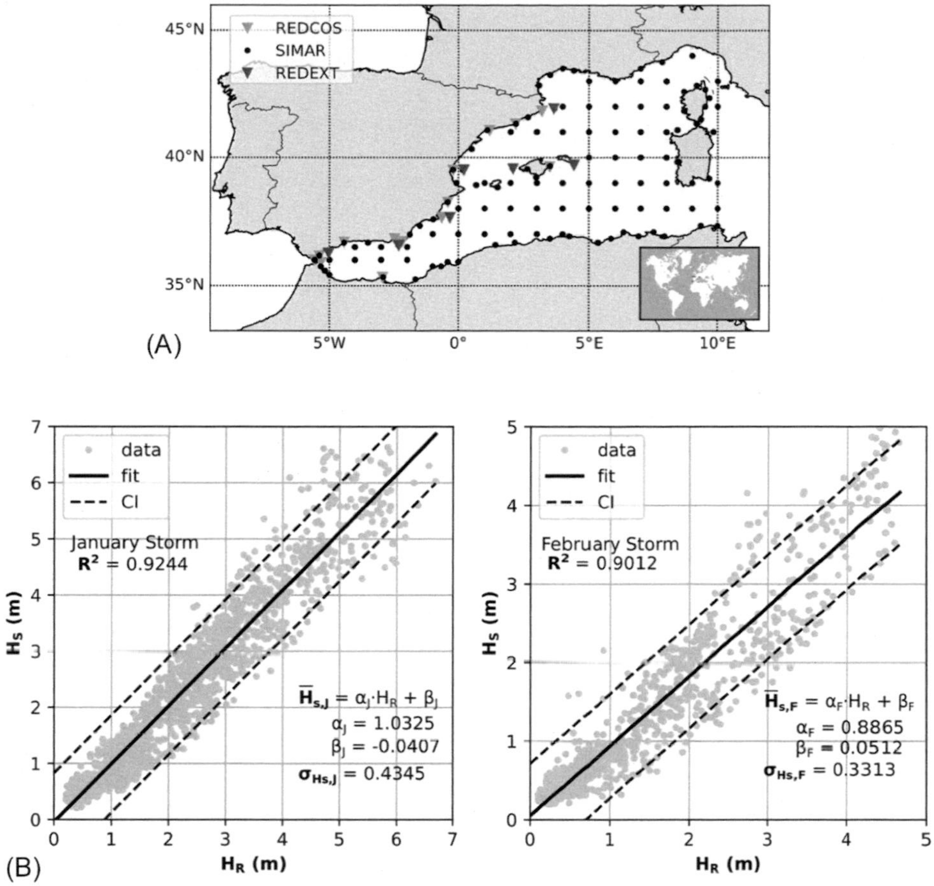

FIG. 2.1 (A) Hard spectral data locations on the Spanish Mediterranean Sea. (B) Soft data based on empirical laws between H_R and H_S.

Example 2.8 Some typical scientific applications involving observational studies are as follows:

- One observes the path of a hurricane under conditions decided by nature and is beyond the observer's control.
- Observed hourly sea-heights along the coastal region of Los Angeles depend on weather conditions beyond human control.

In recent years, data collection activities have made considerable gains from technologies like the *Global Positioning System* (*GPS*) and *Information and Communication Technology* (*ICT*) devices. We will revisit this important matter in the later chapters of the book.

2.2.2 Selection criteria

Concluding our discussion of the data collection issues, it is worth noting that, as is the case with the application of every technique in practice, each data collection technique has its *pros* and *cons*. Let us review three relevant criteria for selecting a data collection technique:

Resources: It refers to the requirements of a technique concerning cost, speed, materials, energy, staff, or other assets.

Example 2.9 When the population is large, a sample survey has a big resource advantage over a census. A well-designed sample survey can provide very precise estimates of population parameters—quicker, cheaper, and with less manpower than a census.

Generalizability: It refers to the appropriateness of applying the findings generated by the technique to a larger population,[i] and it may require random selection.

Example 2.10 If participants in a study are randomly selected from a larger population, it is appropriate to generalize study results to the larger population; if not, it is not appropriate to generalize. Observational studies do not feature random selection.[j] Hence, it may not be appropriate to generalize from the results of an observational study to a larger population.

Causal inference: Cause-and-effect relationships can be established when a natural law connects the relevant attributes, or when subjects are randomly assigned to groups.

Example 2.11 Experiments that involve the conditional application of natural laws or allow the controlled assignment of subjects to treatment groups may be the best ways for investigating causal relationships.

The closer to satisfying the above three **criteria** a technique is, the more appropriate it is for the study.

2.3 Data quality issues

We will now focus on question ③ above regarding data quality. We will start with a relevant definition.

Data quality is the degree to which data features fulfill certain requirements associated with a particular purpose and specific use. Such data features include data availability, completeness, validity, accuracy, and consistency. Study requirements are defined as needs or expectations that are stated for the application of interest.[k]

As noted earlier, essential data-related activities include collecting, structuring, and cleaning a dataset. Here, we must add that the specification of these activities is a twofold affair involving:

- Data **quality assurance** principles, including (i) detecting, correcting or removing corrupt, noisy, or inaccurate data, (ii) identifying incomplete, incorrect, or irrelevant data parts, (iii) replacing, modifying or deleting dirty, noisy, or coarse data, and, lastly, (iv) making sure that the data correctly represents the real-world phenomenon to which it refers.
- Data **analysis objectives**, i.e., data is fit for its intended uses, internal data consistency is achieved, and data validity is assured.

Example 2.12 Data quality is a major focus of public health investigations, especially as demand for accountability increases. Data quality assurance programs should establish strong monitoring and evaluation procedures that produce high-quality data, determine quality indicators, and assess the underlying data management and

[i] A reasoning process known as *induction*.

[j] Random selection, e.g., of floods or earthquakes to observe is rather nonsensical.

[k] However, as will be discussed below, there may also exist other standards related to data quality.

reporting processes for different diseases, like AIDS, tuberculosis (*TB*), and malaria.

2.3.1 *Error as negative information*

Quality assurance requires efficient data collection and management techniques in conditions of *in-situ* uncertainty. In the *S-KB* context, one should distinguish between appearance and reality by introducing the possibility of being wrong to a certain degree.

- With the introduction of measurement or observation **error**, the empirical content now connects appearance and reality, since an error is, in a sense, useful information.

Example 2.13 When mapping the seafloor depth, echo soundings (sonar measurements) are taken at regular intervals from a vessel traversing a system of paths over the ocean surface. This yields a set of depth readings, $D(s_i) = D_i$ $i = 1, 2, \ldots, n$, Fig 2.2A. As the ocean is not a homogeneous medium, such echo soundings are influenced by the local zooplankton concentration in the region of each sounding, Fig 2.2B. Zooplankton clouds create interference called ocean volume reverberation. This interference pattern tends to vary from location to location, and even from day to day. So, actual readings are random variables of the form:

$$D(s_i) = X(s_i) + e(s_i),$$

where $X(s_i)$ is the actual (but unknown) depth and $e(s_i)$ is the measurement error (due to interference). Errors $e(s_i)$ are statistically dependent as plankton concentrations at nearby locations and will tend to be more similar than at locations widely separated in space.

The error being viewed as negative information, its detection points to what could potentially have hazardous effects on *CTDA*. This information offers a valuable warning to an investigator. Following information gain is action, i.e., the error needs to be eliminated or reduced. This is an important goal of the following activity.

Data cleansing is a data quality assurance activity seeking error detection and elimination, which can be combined with other data processing stages like integration and maintenance.

In nuce, the ultimate goal of data cleansing is producing a dataset that is free of errors, although such an objective may not be easy to achieve.

Example 2.14 The cleansing of data may have to be performed every time the data is accessed, which increases the response time considerably and lowers efficiency. As a result, in many cases, it may be preferable that data cleansing becomes an iterative process, involving significant exploration and interaction, which avoids the recleansing of data in its entirety after some values in data collection change and be repeated only on values that have changed.

Data cleansing can be an expensive and time-consuming process during which the data is often audited with the use of statistical and database methods to detect and remove potential

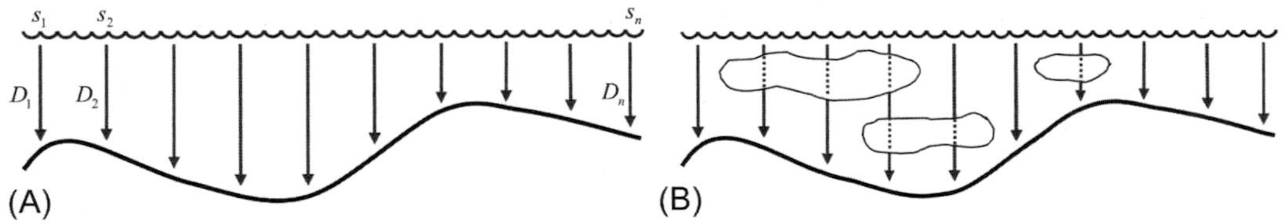

FIG. 2.2 Mapping the sea floor depth. (A) Sonar measurements of depth readings $D(s_i) = D_i$ $(i = 1, \ldots n)$ taken at regular intervals from vessel traversing paths over the ocean surface, and (B) echo soundings are influenced by the local concentration of zooplankton.

anomalies, including measurement or observation errors, their locations and/or instances, and possible data contradictions. The causes of these anomalies in the data must be assessed in an efficient manner. In the case of large datasets, this inevitably involves a trade-off because the execution of the data quality assurance process can be computationally expensive.

2.3.2 *Seven key requirements*

Adequate data quality assurance is not an arbitrary process, but it must satisfy the seven essential requirements that are listed in Table 2.3. As regards requirement ❶, *real* accuracy (in a deterministic sense) may be hard to achieve, as, in principle, it requires knowing the true attribute value, which is often unavailable. *Statistical* accuracy, on the other hand, is usually achievable by means of a paradigm shift from deterministic to stochastic (probabilistic) thinking.

The latter involves statistical functions (e.g., minimum variance estimation error) and probabilistic functions (e.g., maximum likelihood).

Example 2.15 In public health studies, accuracy may be achieved in some cases, such as confirming a patient's contact data by means of databases that match up zip codes to geographical locations, verify the existence of street addresses, etc.

In the setting of requirement ❷, incompleteness results when the investigator cannot infer datasets that were not recorded.

Example 2.16 In the case of interview data, it may be possible to fix the incompleteness of the interview by going back to the original source of data and reinterview the subject. Yet, even this does not guarantee success because of recall problems (e.g., in an interview seeking to gather data on human exposure, most probably one will not remember exactly how often and for how long one was exposed to a pollutant during the last 6 months). When certain data columns should not be left empty, one may designate a default value as "unknown" or "missing."

In the setting of the requirement ❸, inconsistency occurs when two data items in the study dataset contradict each other.

Example 2.17 Inconsistency can emerge when an attribute is recorded by two different devices as having two different values, only one of which can be correct. Fixing inconsistency requires a variety of strategies, e.g., deciding which data source is likely to be most reliable, or by testing both data items in another way.

TABLE 2.3 Data quality assurance requirements.

Requirement	Description
❶ Accuracy	The degree of conformity of an attribute datum to a standard or a true value
❷ Completeness	The degree to which all datasets necessary for an adequate *CTDA* are available
❸ Consistency	The degree to which data items are equivalent within the various components of a *CTDA* study
❹ Uniformity	The degree to which the data values are specified using the same units of measure, if the data have different sources, or were obtained at different scales
❺ Duplicate detection	The determination of duplicate representations of the same entity in a dataset
❻ Data transformation	The mapping of a dataset from its initial format into one that is appropriate for the *CTDA* application of interest
❼ Statistics	By calculating data statistics (means, variances, range, clustering indicators), it is possible for an expert to detect values that are considered unexpected or erroneous

Requirement ❹ refers to the possibility of different data sources and scales.

Example 2.18 Typical cases include the following:

- ❖ The weight may be recorded either in pounds or kilos and must be converted to a single measure using a suitable arithmetic transformation.
- ❖ Disease incidence data may be obtained in both the county and the state scale, pooled from different locales or recorded during different times.

Requirement ❺ is concerned with duplicate data detection, which is often a very challenging problem of data cleansing.

Example 2.19 There are several techniques seeking to detect duplicate entries. In many cases, faster identification of duplicate entries is achieved by sorting the data in a way that could bring duplicate entries closer together.

Yet, there are cases in practice that the transformation of the requirement ❻ cannot be determined. Transformations may involve value conversions or translation functions, as well as the normalization of numerical values to conform to minimum and maximum limits.

Example 2.20 The Gaussian anamorphosis can transform a numerical dataset that is not normally distributed into normally distributed values. In certain cases, the necessary transformations or corrections cannot be determined because the available information on data anomalies is insufficient, leaving the deletion of such anomalies as the only option. However, the deletion of data can lead to particularly costly information loss, especially if there is a large amount of deleted data.

In the case of the requirement ❼, although the real correction of such data is difficult, as the true value is unknown, the issue can be resolved, at least in part, by setting the values to an average or other statistical value.

Example 2.21 Statistical techniques are used to handle missing values that can be replaced by one or more plausible values, which are usually obtained by means of extensive data augmentation.

The availability of good quality data is an essential prerequisite of any rigorous and efficient *CTDA* implementation.

> In a very real sense, then, the ultimate goal of quality assurance is **purposefulness**, i.e., the totality of data features should bear on its ability to satisfy a given purpose.

In this respect, quality assurance should not be just a matter of implementing strong validation checks. It should rather seek a state of completeness, validity, consistency, and accuracy that makes the dataset appropriate for a specific use.

2.4 Theoretical evaluation of empirical evidence

There exist key epistemic issues concerning the structure of empirical evidence represented by the *S-KB*, i.e., what it means when an argument or suggested evidence about a phenomenon lacks a definitive proof. As it will become apparent in the following, these issues not only have theoretical consequences, but they can have significant practical *CTDA* implications, as well. This is because empirical evidence is sometimes directly linked to potential logical fallacies that directly affect practical thinking.[1] One of them concerns the meaning of the "absence of confirmatory evidence" about a phenomenon.

[1] As was noted earlier, practical thinking usually refers to thinking adapted to one's environment so that one is able to pursue a particular everyday goal (this is sometimes referred to as common sense).

2.4.1 *Warnings and fallacies*

In *CTDA* studies, the absence of confirmatory evidence was initially linked to the following statement in rather strong terms (Stephens, 2011):

> **Argument from Ignorance (*AIg*)**: The absence of evidence is evidence of absence.

Let us consider the two parts of *AIg* in some detail: "Absence of evidence" means that one searched for confirmatory evidence about something during a specified time period but failed to find it. "Evidence of absence," on the other hand, means that there is evidence that something does not exist. According to the *AIg* perspective, then, because one does not possess evidence that something exists, this implies that one has evidence that it does not exist. Here is a typical example.

Example 2.22 That no traces of contaminants have been found in the water samples of an aquifer (absent evidence) is considered evidence by the environmental authorities that the aquifer is not contaminated.

Because it is expressed in the most definite terms, the *AIg* may be viewed as a statement of deterministic mathematics. This is precisely the problem *AIg* encounters when applied in a real-world context, where its validity has been under serious criticism. A typical example where *AIg* seems not to be valid in reality is as follows.

Example 2.23 That an ecologist does not have, at the moment, evidence that a species exists in a region (absence of evidence), it does not mean that the species does not exist (evidence of absence), but, simply that the species may not yet have been detected in the region.

Cases like the one discussed in the above example have led to the view that a key issue regarding the role of evidence or its lack, at least as it is viewed by *AIg*, is the following statement:

> **Presence of Evidence (*PEv*)**: If something is true, then one should have evidence that it is true.

According to this line of thought, the *PEv* can determine whether an argument from ignorance concerning evidence is valid or not. There are cases in the real world, however, in which no evidential support can be obtained regarding a scientific proposition (very difficult *in-situ* conditions, extremely costly and noise-interrupted experiments, etc.), nevertheless, it cannot be argued with certainty that the proposition is not true. Hence, in real-world terms *PEv* can be a fallacy.[m] Moreover, relevant to the case described by *AIg* (i.e., that the absence of confirming evidence demonstrates that something is untrue) is the argument that the lack of falsifying evidence against something means that it must be true. This can be a fallacy too.

The above considerations have led many investigators to adopt the rather negative way of thinking proposed by the following equally strong (deterministic) statement:

> **Non-argument from Ignorance (*NAIg*)**: The absence of evidence is not evidence of absence.

For one thing, it has been argued, it is rather common in practice that the claim about "absence of evidence" in *AIg* could be based only on the limited studies that have been done and one cannot comment on those that have not been

[m] To give the readers a more complete picture, the supporters of *PEv* have also used the following line of thought: There are cases in the real world in which no evidential support can be obtained about a scientific proposition, in which case even if the proposition was true, it would be anyway impossible to have evidence that it is true and, therefore, the proposition should be considered not to be true.

done due to practical reasons (e.g., lack of funding). Also, it is too often assumed that if proof about something is lacking (even if hypothetically obtainable), it must be false. It is this common misconception that makes it important that *NAIg* emphasizes, in the strongest terms, that absence of evidence is not evidence of absence.

Although *NAIg* offers a sound warning, it is a rather strong kind of a statement, and not a very constructive one in the real world, where more positive thinking is rather needed. Indeed, while investigators should be aware of empirical evidence-based fallacies, like the above ones, they may also draw some constructive conclusions from these fallacies. A way to achieve such a goal is provided by a paradigm shift from the strict deterministic thinking adopted by *AIg*, *PEv*, and *NAIg* to a more flexible and realistic sort of thinking.

2.4.2 *Probabilistic thinking*

On a closer look, the validity of the above deterministic arguments has been questioned not so much when it is considered in a purely mathematical (deductive) context, but rather when they are used in scientific studies characterized by real-world conditions of uncertainty.[n] This situation can be resolved, to a satisfactory degree, by properly extending the investigation of empirical evidence in the probabilistic domain. In this domain of thinking, the absence of evidence that something is true is not definite proof that something is absent but it merely increases the probability of being absent. In this case, the *AIg* and *NAIg* can be replaced by a more realistic and useful claim, as follows.

> **Probabilistic Argument from Ignorance (*PAIg*)**: The absence of evidence increases the probability of absence.

Example 2.23 (cont.): The existing evidence decreases the initial probability that a species exists in a region, which counts as partial "evidence of absence" (i.e., a certain degree of evidence that the species does not exist).

Continuing along the lines of the probabilistic thinking mode that applies in real world cases in the presence of *in-situ* uncertainty, let us assume that H is a scientific hypothesis or proposition about a phenomenon and e is empirical evidence that is part of the *S-KB* about the phenomenon. Also, let $P_G[H] = q < 1$ be the *prior* probability of H when only the *G-KB* is considered (P_G denotes probability based on G). The $q < 1$ implies that H is an uncertain hypothesis, whereas if $q = 1$ there will be no reason to search for or evaluate evidence. The *posterior* probability $P_G[H|e]$ is the conditional (updated) probability given to evidence e. Table 2.4 describes certain significant evidence-related cases with their associated probabilistic conditions and implications. The conditions in Table 2.4 establish probabilistic correlations between e and H (and/or their negations $\neg H$ and $\neg e$) rather than deductive relationships between them. In view of these probabilities, Cases 1 and 2 both suggest that the presence of evidence (e) supports H to a certain degree. Cases 3–5 are linked to *NAIg* (role of $\neg e$). Case 3 essentially claims that the absence of evidence ($\neg e$) is evidence of absence ($\neg H$), which also implies that if the probability inequality of Case 3 is violated, then *NAIg* may be valid. In Case 4 the conditional probability of H given the absence of evidence ($\neg e$), $P_G[H|\neg e]$, should not decrease, which agrees with *NAIg*; whereas, if "the absence of evidence is evidence of absence," then the $P_G[H|\neg e]$ should decrease (Case 5). Hence, it is possible that, under certain conditions, the absence of evidence could indicate evidence of possible absence, which properly invalidates *NAIg*.

[n] The readers may recall that in mathematical reasoning, because of its deductive structure, if the premises are true, the conclusions are always true. In probabilistic reasoning, because of its inductive structure, the premises may be true and the conclusions wrong.

TABLE 2.4 Focus, condition, and implication of evidence assessment according to *PAIg*.

Case	Focus	Condition	Implication
1	Presence of evidence (e)	$P_G[H\|e] > P_G[H]$	e supports H
2	The relative impact of the presence of evidence (e) *vs.* absence of evidence ($\neg e$) on H *vs.* $\neg H$	$P_G[e\|H] > P_G[e\|\neg H]$	e offers more support for H
3		$P_G[\neg e\|\neg H] > P_G[\neg e\|H]$	$\neg e$ offers more support for $\neg H$
4	Absence of evidence ($\neg e$)	$P_G[H\|\neg e] > P_G[H]$	$\neg e$ supports H
5		$P_G[H\|\neg e] < P_G[H]$	$\neg e$ does not support H

Some more examples follow.

Example 2.24 Some illustrations of the cases presented in Table 2.4 are examined next:

- As regards Case 1 of Table 2.4, if the conditional probability that flood occurs (H) given the available evidence e is greater than the prior flood probability, then the e supports H. It does not mean that flood will surely occur, merely its probability of occurrence is $P_G[H|e] = q$, and the larger the difference $P_G[H|e] - P_G[H]$ is the higher the support of e on H.
- Regarding Case 2, if it is more likely that evidence e is found that a species exists in a region when it indeed exists (H) than that e is found when the species does not exist ($\neg H$), then e provides more support to the hypothesis that species exists than to its negation, and *NAIg* is not valid.
- According to Case 3, the absence of evidence of soil over-contamination supports the hypothesis of no over-contamination, i.e., Case 3 claims that the absence of evidence ($\neg e$) is evidence of absence ($\neg H$).
- As regards Case 4, if it is more likely that a species exists in a region given that no supporting evidence is found ($\neg e$) than when evidence e is found, then the absence of evidence is not evidence that no species exists (evidence of absence), and *NAIg* holds.

Some stronger conditions, in the probabilistic logic sense, may be introduced based on the total or full *KB* available ($K = G \cup S$).

Example 2.25 In Cases 2 and 3 one could use, instead, the conditions $P_K[e|H] > P_K[e|\neg H]$ and $P_K[\neg e|\neg H] > P_K[\neg e|H]$, respectively, where P_K denotes probability based on the total available *K-KB*.

2.4.3 *Misleading evidence*

So far, the discussion of Table 2.4 does not consider the cases that the available evidence e is potentially misleading (or even false). However, in real life, there is always the possibility of previously unknown evidence, say e', which may contradict or invalidate e.

Example 2.22 (cont): After collecting empirical evidence e that no traces of contaminants have been found in the water samples of an aquifer, new evidence e' was revealed that showed that e was collected under inappropriate conditions (malfunctioning apparatus, unprofessional *in-situ* operators, etc.). Then, it can be argued that e' invalidated e.

There are other cases when new evidence e' does not necessarily cancel out the general validity of e but rather explains why the conditions in Table 2.4 may fail in the presence of e'.

Example 2.26 Assume, *arguendo*, that the presence of an indicator generally confirms the existence of a cancerous tumor (e). The indicator was not detected during a medical test ($\neg e$), so it was considered that the patient was cancer-free

(¬*H*). Yet, unknown to the examiner, at the time of the exam the patient was under the influence of a drug that is known to temporarily hide the presence of the indicator (this is new evidence *e′*). In this case, the condition of Case 3 (Table 2.4) may fail.

3 Big data: Why learn, if you can look it up?

With the great improvements in computational power in recent years, perhaps not surprisingly, a new conception of data has been introduced, as follows:

Big Data (*BD*) refers to extremely large datasets analyzed computationally to reveal patterns-trends associations, supposedly, without the need of any theory or model of Nature. Accordingly, the focus of the so-called **Data Intensive Analysis** (*DIA*)[o] is on *BD* and massive computing power.

BD has attracted considerable attention worldwide, including from governments, industries, and academia, and it is applied to an increasing number of fields (Boyd and Crawford, 2012). Noticeably, *BD*-driven analysis or *DIA* is not merely about huge *volumes* of data (terabytes or petabytes), it is also about high *velocity* (in or near real-time data generation), *diversity* (structured and unstructured), *exhaustivity* (seeking to capture entire populations or systems), fine-grained *resolution*, *relationality* (enables the conjoining of different datasets), and *flexibility* (new fields can be easily added, and its size can be expanded rapidly) (Kitchin, 2014). Thus, while *conventional* data generation consists of limited sets of measurements or observations seeking to be representative of a population, *BD* is exhaustive, seeking to cover nearly the entire population, continuously generated, fine-grained, and flexible.[p]

3.1 The new analytics

As existing technologies could not handle efficiently the volume and variety of most *BD* cases, to achieve the *BD* goals, *DIA* relies on high-powered computation and on new analytics techniques designed to cope with data abundance.

- New analytics techniques were originated mostly in **Artificial Intelligence** (*AI*) seeking to produce **Machine Learning** (*ML*) that can computationally and automatically mine and detect patterns in huge datasets, develop predictive tools, and optimize the analysis results.

In the *ML* setting (Mitchell, 1997; Hastie et al., 2001; Bishop, 2006), although *how* a system works may be completely transparent, *why* it works the way it does can be potentially unfathomable (i.e., it is a difficult and rather unrealistic expectation to make sense of the purpose of the various components in an *ML* model). This is because *ML* models rely on nonlinearities and complex interactions between inputs (e.g., the effect of changing one input may depend critically on the values of other inputs). Although it is very hard to figure out why the *ML* setup works, the relevant details are fully transparent.

Based on *ML* models, *DIA* enables a radically different approach compared to the traditional ones where the investigator seeks to select an appropriate technique for the case based on their knowledge of available techniques and existing

[o] Also termed *BD-driven analysis*.

[p] It may be worth noting that the conventional analysis of what is considered small data currently accounts for about 95% of all data analysis, often using Microsoft Excel (Brodie, 2015).

datasets. Instead, *DIA* considers an ensemble sort of approach that applies many different algorithms to a dataset to determine the best or a composite model.

3.2 The *DIA* epistemology

Being the operational component of the *BD*-driven paradigm of scientific inquiry (Section 1 of Chapter 1), *DIA* is currently used in a variety of scientific and nonscientific fields, including health care, geosciences, manufacturing, education, finances, policing, and marketing. This introduces an alternative to the traditional way of learning and knowing.

> While the traditional scientific method involves testing human-generated hypotheses or theories by analyzing relevant data, *DIA* creates a new **epistemology** that involves data-generated hypotheses and seeks to gain insights born from the data (e.g., significant correlations identified algorithmically in the data).

Example 3.1 Some fields of inquiry are listed below where the implementation of *DIA* notions and techniques is rapidly increasing:

- In earth and atmospheric sciences, as a result of new technologies (e.g., earth observation, deep exploration, computer simulations), the volume of the acquired geospatial[q] data has increased tremendously. Well over two thousand satellites currently orbit the earth, which generate large amounts of remote sensing images at different geographical, temporal, and spectral resolutions.
- Electronic health records produce *BD* health care environments in which large amounts of data are captured and transferred digitally. It is hoped that *BD*-driven analysis will accelerate treatment discoveries that can manage and prevent chronic diseases more precisely and cost-effectively.
- *DIA* can also be useful in many cases, including collaborative filtering, group recommendations, and personalization, in which the correlations detected by data-crunching are actually sufficient to draw interesting conclusions.
- In marketing, *DIA* may help to discover new laws of macrobehavior that were overlooked in the past due to lack of data. Also, establishing correlations is the main goal when evaluating geo-location data for mobile phone providers. Data correlations may provoke situational interpretations that can improve one's making sense of the data, whereas comparing different interpretations may generate new insights and help discover nonintuitive correlations in consumer buying patterns.
- Real-time economic indicators (like credit card purchases, package trucking, and cell phone usage) are useful bodies of information that become available on a real-time basis. "Now-casting" is an innovative use of *BD* in business decision- and revenue-making strategies that use real-time data to describe contemporaneous activities before official data sources become available.

3.2.1 The symbiosis of science and analytics

Several decades ago, *John von Neumann* famously warned investigators about the perils of purely data-based analysis and related data fitting (regression) techniques:

✍ With four free (independent) parameters, one can fit an elephant to the data, and with five parameters one can make him wiggle his trunk.

[q] "Geospatial" data refers to organization, visualization and multisource integration of big earth data related to geographical locations.

By this Neumann meant that one should not be impressed when a computational technique fits a dataset well, as with enough parameters you can fit any dataset. Although successful in many nonscientific (commercial, etc.) applications, *DIA* has had rather limited applicability in scientific applications involving complex natural phenomena. This is due, at least in part, to the fact that pure *DIA* seems to neglect the essential existence of a symbiotic environment in which scientists do science and analysts do analysis.[r]

> In *CTDA*, the direct implication of the science-analysis **symbiosis** is that theory and data act in synergy, i.e., instead of purely data-driven *DIA* that employs automatic hypotheses, **theory-driven *DIA*** applies human-generated scientific hypotheses to large amounts of data.

Along these lines of thought, *Enrico Fermi* expressed a rather strong view concerning this symbiosis:

✍ When one does a calculation, there are two ways of doing it. Either one should have a clear physical model in mind or one should have a rigorous mathematical basis.

Computationally intensive theoretical and numerical models of natural systems (atmosphere, land surfaces, oceans, biotechnology, genome-wide analysis, health care) are progressively refined by continual comparison with large datasets flowing from the real world, which makes it possible to improve the prediction of the future behavior of these immensely complex natural systems.

Example 3.2 Below we mention two representative cases of the above synergy.

- In *oceanography*, data are generated using co-located arrays of many sensor types operated over long periods of time (decades to centuries). These data are archived, visualized, and compared to theoretical and numerical model simulations that are configured to address complexity at scales comparable in time and space to the actual measurements (Delaney and Barga, 2009).
- In *public health*, on the other hand, while electronic health records increasingly produce big health care data, the theoretical and computational models required to study this *BD* are currently scarce (Buchan et al., 2009).

3.3 Concerns with *DIA*

A number of concerns have been associated with the application of *DIA* in various fields (scientific, commercial, etc.). Among them, two prime concerns are as follows:

① The majority of *DIA* deployments in **marketing** and **finance** focus on mainstream and even trivial issues rather than seeking to discover previously unanticipated aspects in these fields, or to make fundamental advances and new inventions.

② At the same time, new **social perils** arise, like the risk of privacy violations: tracking the websites consumers are visiting and the purchases they are making, collecting data about citizens' health state without their consent, or monetizing people's private data.

In health care, a well-publicized case of *BD* failure is the so-called *Google Flu Trends* (*GFT*) described in the following example.

Example 3.3 As discussed in Lazer et al. (2014), in February 2013, *GFT* made headlines but not for reasons that Google executives would have hoped. It was reported that *GFT* was predicting more than double the proportion of doctor visits for influenza-like illness than the *Centers for Disease Control and Prevention* (*CDC*) estimates based on surveillance reports from laboratories across

[r] See discussion of theory and evidence in Section 3 of Chapter 1.

the U.S. This happened despite the fact that *GFT* was built specifically to predict the *CDC* reports. Given that *GFT* is often held up as an exemplary use of *BD*, some cautionary lessons should be drawn from this error.

By way of a summary, major concerns associated with *DIA* or *BD*-driven analysis (in scientific and nonscientific applications alike) include but are not limited to the issues listed in Table 3.1. Closely related to the assertions of Table 3.1 are the specific illustrations presented in the following example.

Example 3.4 In connection to Table 3.1, some of the potentially serious real-world problems of *DIA* or *BD*-driven analysis are as follows.

- Companies analyze *mobile device* data to make inferences about people's lives and the economy. Yet, some issues of concern emerge naturally. Massive piles of data imply increasing data noise and uncertainty, usually of unknown origins and beyond control, in which case the validity of the motto "garbage in garbage out" finds its prime application here.
- While *BD* is often complex, imprecise, and noisy, the associated uncertainty and accuracy are not adequately quantified, say, in terms of error bars, likelihood, or probability of the results.

TABLE 3.1 Issues with *DIA* (*BD*-driven analysis).

Data noise	Most data include *errors* of various kinds (measurement, technical, and conceptual). Increasing the amount of data also increases noise, as data errors and noisy datasets add up. There is a mismatch between the messy randomness and noise of *BD* and the human intuition of events. Data noise can contribute significantly to the toxicity of data, which may increase much faster than its benefits[a]
Data size	Appropriate handling of large amounts of data becomes increasingly more difficult as the dataset size increases. In several cases, too much data can alter a phenomenon. *BD* often means the detection of *inexistent* or *false patterns* (this case is growing rapidly as a side effect of the techno-information age). As a result, a good data-user may be one who occasionally manages to be blind to most of what others see in the *BD*
Uncertainty principle	*BD* involves a *trade-off*, i.e., it improves the knowledge of one physical attribute at the cost of another (e.g., increasing the number of particle position measurements decreases position uncertainty but increases velocity uncertainty)
Discovery	The bigger the database is the more difficult the discovery of unique hidden items. *BD* typically lives in databases that are defined too narrowly to allow for the unexpected. Yet, new ideas typically come from *juxtaposition*, i.e., by combining things that previously seemed unrelated or by contrasting things in order to highlight the differences
Theory-dependence	All meaningful data have *theoretical support* (i.e., they are not just numbers). A theory or a model is needed to distinguish between good and bad data and to guide discovery. Data are like a ventriloquist's dummy, which may be made to utter words of wisdom or may talk nonsense, depending on the person who uses it
Extrapolation	*BD* reveals statistical correlation, but not necessarily causation. It may even result in a cause-and-effect illusion, which is called *epiphenomenalism*
Data security	Mobile phones and tablets facilitate the collection, reporting, and analysis of *BD* in near real time. However, in the case of health datasets, e.g., this accessibility leaves them vulnerable to *security risks* and data breaches that could jeopardize data confidentiality
Existential risks	These are risks facing *humanity* from *DIA* involving *AI*. Relevant concerns have been voiced by world-leading authorities and institutions aiming at safeguarding life[b]

[a] *As Brodie (2015) reports, 80% of the errors in DIA may arise during the data management processes.*
[b] *See, e.g., the Organization for Economic Co-operation and Development and the legal communities (Bohannon, 2015; Gershman et al., 2015; Horvitz and Mulligan, 2015; Brodie, 2015).*

- Although collecting large amounts of data on the consumption of a certain product is not going to affect the relevant phenomenon, installing too many boreholes in the case of groundwater contamination can severely alter the phenomenon, often in unpredictable ways.
- In *risk assessment* and *natural hazard science* one needs to be alert to unexpected forms of threat to people and the environment, and their unplanned consequences, not to the ones whose correlations are established using computers and are often trivial and repetitive.
- Another potentially serious problem with *BD* is *outlier identification*, i.e., determining what is an outlier and what is valuable information, e.g., a subversive outlier that could have a big upside.
- The deployment of densely arranged sensors in *harsh* and *remote environments* is challenging. The sensors must remain operational for a long time-period using restrictive power sources (small batteries, solar panels, or other types of environmental energy) so that the combined process of sensing, computation, and communication is energy efficient. Sensors and their communication links must be sufficiently robust to ensure reliable data acquisition in harsh outdoor environments. Moreover, invalid sensor data due to system failures or environmental impacts must be identified and treated accordingly.
- It has been argued that in many cases, *DIA* is not adequately understood to allow verification and assessment, like quantification of the probability or likelihood that the analysis results will occur within estimated error bounds (Brodie, 2015).
- Although *electronic health records* greatly increase the amount of available health care data, the clinical coding behavior introduces new potential biases. Incentives for primary care professionals to tackle specific conditions lead to fluctuations in the coding amount of new cases. The falling costs of devices for remote monitoring and near-patient testing, on the other hand, are leading to further capture of objective measures in electronic health records, which can provide less biased signals but may create the illusion of an increase in disease prevalence simply due to more data becoming available (data noise issue). Biases can also arise from artifacts in biotechnical data processing (Buchan et al., 2009).
- Also, *security* experts are aware that organized crime worldwide seeks to come up with patterns that will not be detected by correlations of *BD*. Naturally, the same data technology is available to the organized crime that enables it to become more actively random and undetected by correlations.

3.3.1 BD threshold

Based on the past experience, the following suggestion seems reasonable in many scientific investigations of the environment:

- A **data-threshold** may exist beyond which further data-crunching can obscure one's understanding of the underlying phenomenon and it can even change it.

Properly determining the *BD* threshold is a critical *albeit* rather underappreciated issue. It could depend, *inter alia*, on the particular phenomenon of interest and on a deeper understanding of the underlying mechanisms. Neglecting this threshold may turn a data-rich analysis into a data-buried analysis.

In the end, the real challenge is asking the data the right questions and understanding what kind of empirical evidence is needed to form a sound theory and make good decisions.

This important topic will be revisited later in this chapter.

3.4 The unbearable naivety of *BD*-crunchers

Some proponents of *BD* go too far by advocating that absolutely no theory is necessary. Instead, for them data-crunching is always a sufficient approach to identify emerging trends and novel patterns of information and inference. Nevertheless, the available evidence offers only conditional support to such a perspective, which may be occasionally sufficient in nonscientific fields like marketing and related fields, where *DIA* could improve business decision making and help develop new revenue-making strategies.[s] Yet, even in these nonscientific fields, there are serious concerns about the current state of *BD*-driven analysis.

- One of the most important issues is the impossibility of **theory-free** prediction, extrapolation, and causality.

Example 3.5 Some direct expressions of the above concern even in marketing and related fields are listed below:

- The derivation of *BD* correlations can be functional and useful in stimulating sales, but they can also be highly imperfect, given that correlations are inherently limited as predictors.
- *BD* is about exactly right now, with no significant historical context possessing a predictive value. The pace of cultural change is so significant that one cannot extrapolate from present data-driven expectations of what currently makes sense, to what may make sense to a younger generation in the future.
- The search for correlations could be a trap as in marketing one is often looking for unpredictable things that one can use opportunistically. In this setting, when anticipating future events, it may be advisable that one does not rely on old models that come from the old data.
- A prime goal of data-crunching is to establish correlation, yet correlation does not establish causality. The latter requires theories and models, which are often based on restrictive assumptions and have distinct prediction limits. Even when all one cares about is correlations, one has to make certain that core theory decisions are considered at the very beginning of data analysis and processing.

On the other hand, the theory-free perspective is profoundly inadequate as far as scientific inquiry is concerned. There are various reasons for this inadequacy, one of which is as follows:

One cannot **measure** anything scientifically unless one has a clear idea or conceptual construction of what one is going to measure.

Example 3.6 Typical illustrations of the last claim are as follows:

- In hydrogeology, measuring hydraulic conductivity requires a sound theory of

[s] For example, mobile phone providers collecting data on the consumption of certain everyday life products, such as soaps, detergents, tacos or enchiladas, predicting people's movements and social interactions.

groundwater flow. The better the theory, the more realistic the measurement of hydraulic conductivity.

- In public health, data alone will not suffice; instead, a considerable overhaul of theory and methodology is needed to handle the realistic complexity of health, ultimately leading to greatly improved public health care standards.

The above concerns do not seem to bother *BD* radicals who see *DIA* as a panacea introducing new forms of empiricism that declare the creation of *data-driven* analysis against a *knowledge-driven* science, and to strongly support extreme views like that of Anderson (2008):

☞ The end of theory: the data deluge makes the scientific method obsolete.

Some other *BD* radicals have expressed a set of views that Brodie (2015) collectively has called the "point-and-click self-service claim," i.e.:

☞ Point us at the data and we will find the patterns of interest for you.

Such rather *ad absurdum* claims ignore both the unique features of theoretical modeling and the pitfalls of purely data-driven analysis, as stated by many experts (Bollier, 2010), such as *Hal Varian*:

☞ Theory is what allows you to extrapolate outside the observed domain.

Shirley Coleman:

☞ Unless you create a model of what you think is going to happen, you cannot ask questions about the data.

John Timmer:

☞ Correlations are a way of catching a scientist's attention, but the models and mechanisms that explain them are how we make the predictions that not only advance science but generate practical applications.

And, Brodie (2015):

☞ Too much human-in-the-loop leads to errors; too little leads to nonsense.

Lastly, Succi and Coveney (2019) have presented four critical objections to radical *BD* claims:

① **Complex** systems are strongly correlated, hence, they do not obey Gaussian statistics, in general.
② No data are big enough for systems with a strong sensitivity to **data inaccuracies**.
③ **Correlation** does not imply **causation**, the link between the two becoming exponentially fainter at increasing data size.
④ In a **finite-capacity** world, too much data is just as bad as no data.

Assertion ④ is linked to the *BD* threshold issue discussed earlier. Beyond this link, this assertion expresses a broader societal concern according to which treating the world's resources as if they were unlimited can lead to disasters (environmental, social, and economic). In sum, although there exists plenty of useful information to be obtained from *BD* correlations, there is a strong consensus among experts that extreme views like those of Anderson go way too far and are largely nonsensical.

3.4.1 Not BD alone

The upshot of the above discussion is that human inquiry will not improve through *BD* alone; even worse, it cannot go far with the "point-and-click self-service" perspective of *BD* radicals. A more drastic and thoughtful approach is needed:

The investigators should be able to explore **rich-content data** and seek **deep-content data** that enable the detection of those tiny clues which, combined with sound theory, can uncover essential features and real trends of the phenomenon and facilitate deducing substance and function from form.

In physical sciences, this kind of data concerning the observatory infrastructure of natural systems is obtained by taking advantage of current technological advancements (e.g., sensor technologies, robotic systems, real-time data, high-speed communication, eco-genomics, and nanotechnology). As a result, the approaches that scientists, educators, technologists, and policymakers take in interacting with these systems are substantially transformed and improved. Arguably, as the theoretical and numerical models of these systems become more sophisticated and data become routinely available via the *Internet*, the latter is emerging as a very powerful research tool.

Example 3.7 Meaningful uses of rich-content or deep-content data exploration take place in geospatial modeling, where efforts are made to use *BD* processing and simulation models to identify data problems and discover scientific laws governing space-time distribution patterns (Hey et al., 2009). Yet, the application of geospatial models requires highly efficient computational infrastructure, and as the complexity and simulation accuracy of input data increase, geospatial models require increasing amounts of accurate input data, which is a time consuming and costly effort. Current challenges include the automatic match of shared data for use in geospatial modeling and the promotion of the integrated sharing of "data-models" to benefit the interpretation of the model calculation results.

Concluding, rich-content and deep-content data offer further possibilities. They could give investigators teasing insights about the phenomenon of interest, which can be explored using computational induction tools to consider multiple alternatives and assess the plausibility of competing explanatory models. Furthermore, using inferential tools, investigators could identify the most valuable information to collect in conditions of *in-situ* uncertainty, to select the best test to perform and to inspect potentially implications visually, depending on the current objectives. Such inferential tools can guide exploration and confirmation of physical connections among chronotopologic attribute values.

4 Attribute data scales

Data (observation and measurement) scales are systematically used in *CTDA* to categorize and quantify natural attributes. There exist different kinds of scale classifications considered in *CTDA*, some of which are discussed next.

4.1 Property-oriented classification

According to one kind of scale classification, each data scale may satisfy one or more of the following four key properties:

① **Identity**: Each value on the data scale has a unique meaning.

② **Magnitude**: Values on the scale have an ordered relationship to one another (that is, some values are larger and some are smaller).

③ **Equal intervals**: Units along the scale are equal to one another (this means, e.g., that the difference between 1 and 2 physical units is equal to that between 19 and 20 units).

④ **Absolute zero**: The scale has a true zero point, below which no values exist.

In view of the four above properties, the present section describes four data scales that are commonly used in *CTDA*, as follows:

- The **nominal** scale only satisfies the identity property of measurement. Values assigned to attributes represent a descriptive category but have no inherent numerical value with respect to magnitude.

Example 4.1 Gender is an example of an attribute that is measured on a nominal scale. Individuals may be classified as "male" or "female," but neither value represents more or less "gender" than the other. Religion and political affiliation are other examples of attributes that are normally measured on a nominal scale.

- The **ordinal** scale has the properties of identity and magnitude. Each value on this scale has a unique meaning, and it has an ordered relationship to every other value on the scale.

Example 4.2 An ordinal scale in action would be the results of a horse race, reported as "win", "place", and "show". One knows the rank order in which horses finished the race. The horse that won finished ahead of the horse that placed, and the horse that placed finished ahead of the horse that showed. However, one cannot tell from this ordinal scale whether it was a close race or whether the winning horse won by a mile.

- The **interval** scale has the properties of identity, magnitude, and equal intervals. With an interval scale, one knows not only whether different values are bigger or smaller, but also how much bigger or smaller they are.

Example 4.3 A perfect case of an interval scale is the Fahrenheit scale measuring temperature. The scale is made up of equal temperature units so that the difference between 40°F and 50°F (degrees Fahrenheit) is equal to the difference between 50°F and 60°F. Now, suppose it is 60°F on Monday and 70°F on Tuesday; then, one knows not only that it was hotter on Tuesday, but also that it was 10°F hotter.

- The **ratio** scale satisfies all four properties: identity, magnitude, equal intervals, and an absolute zero.

Example 4.4 The weight of an object is a typical illustration of a ratio scale. Each value on the weight scale has a unique meaning, weights can be rank-ordered, units along the weight scale are equal to one another, and there is an absolute zero (objects at rest can be weightless, but they cannot have negative weight).

4.2 Attribute-oriented classification

Beyond the property-oriented classification above, the attribute-oriented classification is another classification that includes the following eight scales:

- The **physical** scale characterizes the mechanisms underlying natural attributes. It is determined by the intrinsic properties of the natural attribute and the structure of the medium within which it takes place, and it may involve varying resolutions.

Example 4.5 Dispersion of an injected tracer in a heterogeneous porous medium with a resident fluid. The mixing length is a physical scale determining the extent of the zone within which the two fluids are mixed. Medium correlations are important in determining the mixing length. If correlations (structures and patterns) exist at different scales more than one length scales may be relevant.

- The **support** scale refers to the experimental or empirical support of the sample space in the chronotopologic domain.

Example 4.6 In *geostatistics* studies, one distinguishes between point-, areal-, and block-support scales. In *land use* studies, land cover change may be summarized at the national level, but a land cover change matrix requires smaller units to assess what land cover types have changed and what they have changed into.

- The **sampling** scale refers to the average spacing and/or time separation between measurements.

Sampling density is, on the average, proportional to the inverse of the sampling scale. A sampling scale may involve a *regular*, a *random*, or a *stratified* sampling grid, depending on the case (topography, financial resources, etc.). Economic considerations and time constraints severely limit sampling density in practice.

Example 4.7 For chronotopologic estimation purposes, a large sampling density is desirable, for it improves predictability.

- The **measurement** or **instrument** scale refers to the averaging of the measured quantity over the instrument window.

This averaging leads to a loss of variability, and hence limits the resolution of the instrument. Measurements can be viewed either as point attribute values or the coarse-graining of the instrument must be taken into account. The effects of the instrument or measurement process on the physical law must also be considered. The measurement scale is determined by the available technological resources and the resourcefulness of the experimentalists.

Example 4.8 If the data are sufficiently homogenized by the measurement process, nonlocal relations may lead to local empirical expressions.

- The **modeling** scale is the scale at which a specific physical model is valid.

Example 4.9 Consider the modeling of fluid flow within a porous medium. At the atomic scale, the internal motion of the fluid particles is determined by the laws of quantum mechanics. At the molecular scale, the collection of fluid molecules is organized according to the laws of statistical mechanics. At a higher scale, the molecular fluctuations in density are averaged out and the fluid behaves as a continuum. At the pore scale, fluid flow is governed by Navier-Stokes equations. For relatively large porous media blocks it gets too expensive to solve these equations, which are hence replaced by coarse-grained partial differential equations (e.g., Darcy's law or some extension). At the field scale, local permeability fluctuations at the Darcian scale are homogenized, leading—under certain conditions—to an effective flow equation.

- The **observation** scale defines the domain within which a specific variability or correlation measure is valid and a particular estimation technique applies.

Within this domain, attribute estimates are obtained using a chronotopologic dependency or variability measure (e.g., covariance or variogram, respectively) and an interpolation technique (e.g., Kriging or regression). The observation scale is usually not known a priori but it is determined from the natural variability of the data using an optimization criterion.

Example 4.10 The observation scale may refer to a subset of the total available dataset that includes the data used in determining the covariance or the variogram and obtaining Kriging interpolations (Fig. 4.1).

- The **mapping** scale (or **estimation** scale) defines the spacing of the lattice on which estimates are generated. As such, it also defines the size of the domain within which estimates are sought.

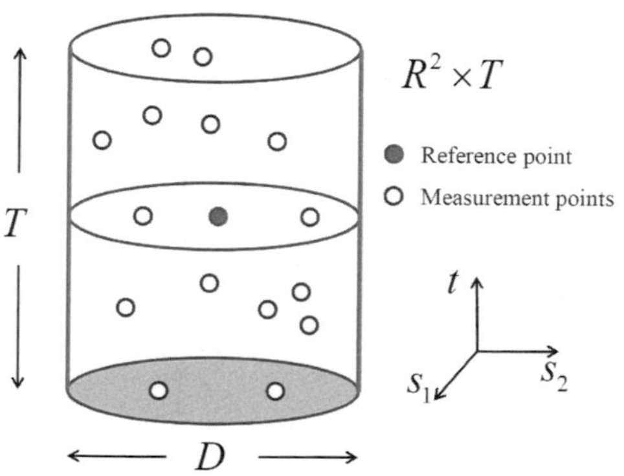

FIG. 4.1 An illustration of the observation scale.

In cases in which the physical laws are not known, attribute estimation procedures are essentially educated interpolation techniques. The mapping scale is constrained by the observation scale, in which case the interpolation accuracy at points that lie outside the observation scale decreases fast.

Example 4.11 In Fig. 4.2 the mapping scale in $R^{2,1}$ is expressed by the resolution of the numerical estimation grid.

- The **discretization** scale defines the resolution of the numerical grid, that is, the spacing of the lattice on which a numerical solution of a physical law or a scientific model is obtained and estimates are generated.

How a continuous physical law is presented in the discrete domain is directly linked to the discretization scale chosen. In many cases, the discretization scale may coincide with the mapping scale.

Example 4.12 Two common expressions of the discretization scale are as follows:

- When maps of a natural attribute are generated, the discretization scale determines the size of the smallest features that can be resolved on the map.
- The effect of discretization scale on flow modeling has been demonstrated by comparing flows simulated at regional and subregional scales by models having equivalent hydraulic parameters.

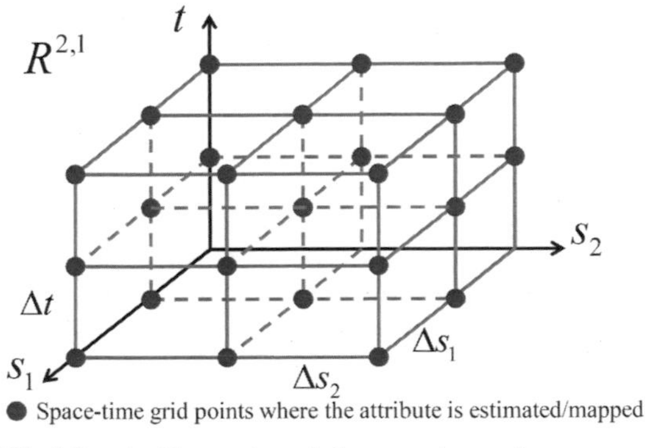

FIG. 4.2 An illustration of the mapping scale.

The point here is that in the real world, the phenomenon of interest may be characterized by several kinds of scales involving a variety of resolutions (fine to coarse), which may complicate the *CTDA*.

Operating at the **appropriate scale** and using an **adequate resolution** are crucial prerequisites for the improved understanding and accurate mapping of the attribute of interest, especially when the physical scale or the resolution differs from the model or the mapping scales.

Example 4.13 Below we examine two typical scale-related real-world phenomena involving more than one scale:

- The highly complex features of mountainous surfaces (bare rocks, debris, patchy snow cover, sparse trees, shallow and deep soils with varying vegetation) can occur within a single kilometer, which is a resolution that is typically not reached by current weather forecasting and climate change models, which still operate at a grid resolution that is far too coarse (multiple kilometers) to correctly map mountainous surface heterogeneity (Lehning et al., 2009).
- The merging of different data scales (say, from molecular-level to population-level) is a great challenge for data-intensive health science.

Arguably, a complete theoretical understanding of the mechanisms that relate natural attributes at different chronotopologic scales has not been achieved yet. In practical applications, one is usually seeking a scale level that reveals important chronotopologic variability characteristics of the specific attribute considered. It is possible that different models with various levels of discretization may need to be constructed, and the appropriate model for the given task will have to be selected using suitable criteria.

5 Emergence of chronotopology-dependent statistics

Due to the vast range of statistical applications in the real world (news, surveys, weather forecasting, disease incidence, crime rate, finances, politics, sports, and governments), scientists cannot avoid becoming familiar with the subject and understand the basic ideas behind statistics.

What we are talking about here is *Classical Statistics* (*CS*), which is generally devoted to the analysis and interpretation of uncertainties caused by the limited sampling of the attribute of interest. Accordingly, *CS* is tied to a population distribution model that assumes that the samples of a population are independent from one another. Generally speaking, a statistic is a number that summarizes some characteristics of a dataset, such as the mean, median, mode, midrange, standard deviation, coefficient of variation, skewness, and kurtosis.

Example 5.1 The mean is a measure of the data central tendency, the standard deviation measures outcome spread (uncertainty), and skewness measures asymmetry, where positive skewness describes distributions with occasional returns well-above average (say, risky investments with high potential winnings), and negative skewness describes distributions with occasional poor outcomes (e.g., rare catastrophic occurrences during hurricane data collection).

Statistical tables and graphics are very often presented in the news. *CS* techniques are advantageous in many ways. A few of their uses are:

① To organize information and interpret observations.
② To condense large amounts of information into figures and/or statements.
③ To describe and quantify sources of uncertainty and/or error in experimental data, and to assess the usefulness of the data.
④ To deduce and infer properties of a population, based on the information we can derive from a sample of the population.
⑤ To analytically reason experiments during their planning, data collection, and analysis stages, in order to design and develop more effective research designs.

Accordingly, it is critical that scientists know (and practice) statistics. The readers who want to refresh their knowledge of *CS* can do so by consulting the voluminous literature on the subject.

5.1 The CS inadequacy

A fundamental fact one should keep in mind is that *CS* does not consider the location or the time of an attribute value. As a result, there are several situations in applied sciences demonstrating the inadequacy of *CS*.

We start with a simple yet illustrative example.

Example 5.2 Assume that there are two datasets characterizing a natural attribute in a chronotopologic domain, see Tables 5.1 and 5.2. As it turns out, these datasets have the same histogram and summary statistics (i.e., mean 10.05 and variance 3.46 in proper units), but exhibit very different physical behaviors: the natural attribute underlying Dataset #1 shows a random character (Fig 5.1A) but that of Dataset #2 shows an increasing temporal pattern at each location (Fig. 5.1B). One naturally concludes that there exists more to a physical dataset than what is suggested by *CS*.

5.2 Blending chronotopology with uncertainty assessment

As we saw above, in *CS* one is interested in the statistical distribution of the attribute *values*, i.e., how much dispersion or variation is there. Yet, attribute *locations* and *instances* should be also regarded as physical features of equal interest, i.e., the dynamic distribution of

TABLE 5.1 Dataset #1.

s_1-coord	s_2-coord	t-coord	Attribute value	s_1-coord	s_2-coord	t-coord	Attribute value
1.8497	1.8086	1	9.4201	1.8497	1.8086	4	12.3482
1.1378	0.8248	1	7.0372	1.1378	0.8248	4	9.4148
1.7571	1.5653	1	7.8068	1.7571	1.5653	4	10.0917
1.8486	0.8133	1	10.2186	1.8486	0.8133	4	13.6090
1.2433	1.6992	1	9.4795	1.2433	1.6992	4	5.6280
0.1208	1.5531	1	10.8037	0.1208	1.5531	4	11.6246
1.8497	1.8086	2	12.5231	1.8497	1.8086	5	10.2539
1.1378	0.8248	2	10.3110	1.1378	0.8248	5	8.9184
1.7571	1.5653	2	9.0140	1.7571	1.5653	5	9.8724
1.8486	0.8133	2	13.6280	1.8486	0.8133	5	8.5538
1.2433	1.6992	2	11.2003	1.2433	1.6992	5	7.3459
0.1208	1.5531	2	12.9404	0.1208	1.5531	5	11.0911
1.8497	1.8086	3	10.9508	1.8497	1.8086	6	8.6864
1.1378	0.8248	3	11.6371	1.1378	0.8248	6	9.3827
1.7571	1.5653	3	9.6385	1.7571	1.5653	6	11.2227
1.8486	0.8133	3	10.6240	1.8486	0.8133	6	11.0531
1.2433	1.6992	3	11.1879	1.2433	1.6992	6	7.1180
0.1208	1.5531	3	9.3464	0.1208	1.5531	6	7.8967

TABLE 5.2 Dataset #2.

s_1-coord	s_2-coord	t-coord	Attribute value	s_1-coord	s_2-coord	t-coord	Attribute value
1.8497	1.8086	1	7.1180	1.8497	1.8086	4	11.6246
1.1378	0.8248	1	7.8967	1.1378	0.8248	4	11.0531
1.7571	1.5653	1	5.6280	1.7571	1.5653	4	10.8037
1.8486	0.8133	1	8.5538	1.8486	0.8133	4	10.2539
1.2433	1.6992	1	8.6864	1.2433	1.6992	4	9.8724
0.1208	1.5531	1	7.0372	0.1208	1.5531	4	10.2186
1.8497	1.8086	2	8.9184	1.8497	1.8086	5	12.9404
1.1378	0.8248	2	9.0140	1.1378	0.8248	5	11.1879

Continued

TABLE 5.2 Dataset #2—cont'd

s_1-coord	s_2-coord	t-coord	Attribute value	s_1-coord	s_2-coord	t-coord	Attribute value
1.7571	1.5653	2	7.8068	1.7571	1.5653	5	10.9508
1.8486	0.8133	2	9.3464	1.8486	0.8133	5	11.2227
1.2433	1.6992	2	9.4201	1.2433	1.6992	5	10.6240
0.1208	1.5531	2	7.3459	0.1208	1.5531	5	11.0911
1.8497	1.8086	3	10.3110	1.8497	1.8086	6	13.6280
1.1378	0.8248	3	9.6385	1.1378	0.8248	6	11.6371
1.7571	1.5653	3	9.4148	1.7571	1.5653	6	11.2003
1.8486	0.8133	3	9.3827	1.8486	0.8133	6	12.3482
1.2433	1.6992	3	9.4795	1.2433	1.6992	6	13.6090
0.1208	1.5531	3	10.0917	0.1208	1.5531	6	12.5231

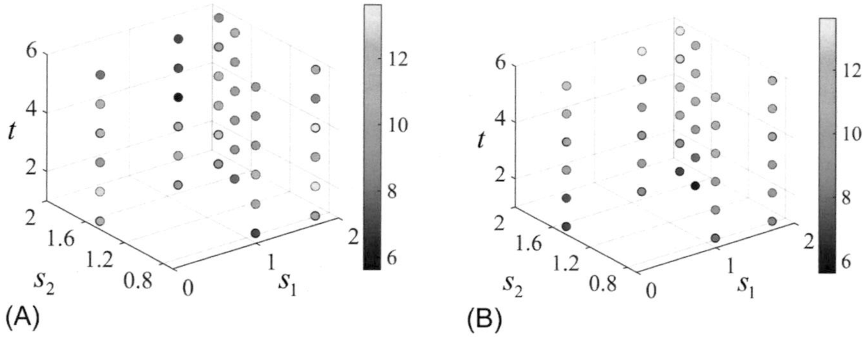

FIG. 5.1 Three-dimensional plots of (A) Dataset #1 (Table 5.1), and (B) Dataset #2 (Table 5.2).

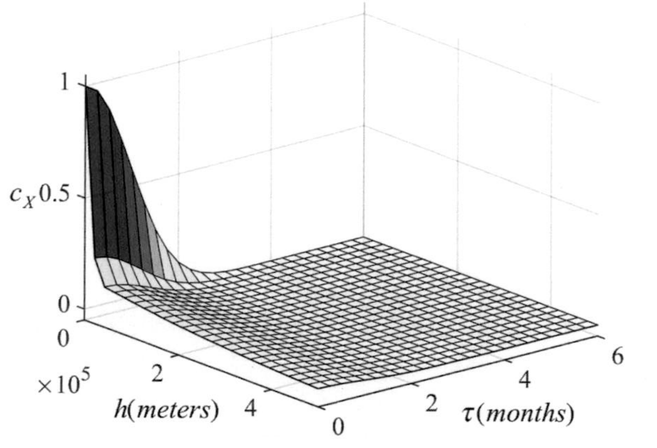

FIG. 5.2 Chronotopologic covariance c_X of the normalized *HFRS* incidence of the 4th class as a function of the spatial distance h and the time separation τ between pairs of points in the domain of interest (Heilongjiang Province during 2005–13).

locations is also of importance when exploring the physical characteristics of a natural attribute in conditions of *uncertainty*. Hence, a major research and development field of wide applicability has emerged within the general framework of *CTDA*, as follows:

Chronotopologic statistics introduces notions and techniques that describe statistical relationships between attribute values that simultaneously depend on space and time in conditions of *in-situ* uncertainty.

One encounters a large number and variety of chronotopologic statistics applications, including environmental, ecological, social, and public health studies. Accordingly, important chronotopologic statistics developments have been associated with the fields of chronogeography (Chapter 4), interpolation (Chapter 7), krigology (Chapter 8), etc.

Example 5.3 For illustration purposes, two typical cases of real-world chronotopologic statistics applications are as follows.

- Botanists collect information that is discrete at particular sampling points but has to be converted in some way into useful map representations. In order to achieve this, the botanists need to use chronotopologic statistics to represent the data in a way that allows adequate assessment of attribute variability and accurately predicts the occurrence of similar attribute features in domains where one has not sampled.
- Environmental managers often require accurate maps over the region of interest obtained by chronotopologic statistics in order to make effective and informed decisions, and environmental scientists need accurate datasets that are well-distributed across a region and in time so that justified interpretations can be made.

In real-world situations, chronotopologic statistics considers the distribution of an attribute as the synthesis of its dependence *structure* (e.g., the laws governing the attribute change in nature) and *in-situ uncertainty* (which may be the result of insufficient information and incomplete understanding of the natural mechanisms underlying the attribute distribution, or predictions attribute values are sought at unknown points from a usually limited set of data).

Example 5.4 The observation points in an area on the earth's surface vary at different elevations. Yu and Wang (2013) employed background stations built on the mountains to set up a number of imaginary stations with the same physical statistics at nonmonitored places in the mountains. The CO and PM_{10} distributions of the entire Taiwan were studied by integrating the chronotopologic CO and PM_{10} structures with *in-situ* uncertainties sources. The uncertainties characterizing these data were due to imperfections of the measurement instruments and the fact that it is often difficult and expensive to acquire these measurements, especially in mountainous and deep marine regions.

Chronotopologic statistics research and development include advances in the fields of *descriptive space-time statistics* (Chapters 4 and 7), *classical geostatistics* (Chapters 5 and 8), and *modern geostatistics* (Chapters 6, 9, and 10). Capitalizing on chronotopology-dependence, techniques have been developed in these fields for analyzing attribute values, locations, times, and correlations.

Example 5.5 Using chronotopologic geostatistics, He et al. (2018) studied the highly heterogeneous distribution of *HFRS* incidences in Heilongjiang province (China) during 2005–13. *HFRS* incidences spread with geographical location and time, i.e., their distribution exhibits chronotopologic dependence. For this reason, the *HFRS* incidence at the 130 counties of Heilongjiang Province was divided into four classes in terms of percentiles. For illustration, Fig. 5.2 plots the chronotopologic covariance function c_X[t] of the normalized *HFRS* incidence $X(\boldsymbol{p})$ of the 4th class

[t] Covariance function is a measure of chronotopologic dependence to be introduced in Chapter 5.

(i.e., the one with the highest *HFRS* incidence among the four classes) as a function of the spatial distance h and the time separation τ between pairs of points $\boldsymbol{p}$ in the domain of interest (Heilongjiang Province during 2005–13).

Given the considerable advances in information science, *TGIS*-based chronotopologic statistics techniques are becoming powerful tools in natural resource management, biological conservation, and social processes, informative chronotopologic maps of natural attributes are increasingly required. Besides the visual perception of attribute distributions, it is very important to translate the existing patterns into useful objective and measurable entities, which is why *TGIS* can help an investigator answer key questions, like in the next example (see, also, Section 7 of Chapter 1).

Example 5.6 Fig 5.3A displays a simulated chronotopologic log-transformed disease incidence dataset. This dataset is documented at 361 geographical locations and 10 time instances (3610 data in total). Using *TGIS*-based interpolation techniques, a continuous disease incidence distribution is obtained in Fig 5.3B.

It comes then as no surprise that many fields of inquiry have contributed to chronotopologic statistics' rise in its modern form: cartography and surveying, botanical studies of global plant distributions and local plant locations, forestry, ethnological surveys of population movement, econometrics, human exposure investigations, landscape ecological studies of vegetation blocks, environmental studies of population dynamics, biogeography and medical geography research, epidemiology and disease spread assessment combined with locational health care delivery data. In this broad sense, chronotopologic statistics is naturally a branch of *CTDA*.

6 More on chronotopologic visualization

As mentioned earlier, visualization aims at displaying the data or model outputs by means of an easily understandable figure or movie rather than directly showing the data values. In the same setting, visualization may offer a more intuitive manner to understand the chronotopologic data distribution than statistics.

6.1 The epistemic value of visualization

This brief introduction emphasizes the following fine point as regards human cognitive faculties:

> The epistemic value of visualization is based on the idea that humans are able to sense chronotopologic variation primarily in a **visual** manner.

In more words, the apparent epistemic value of visualization is a combination of the following five elements:

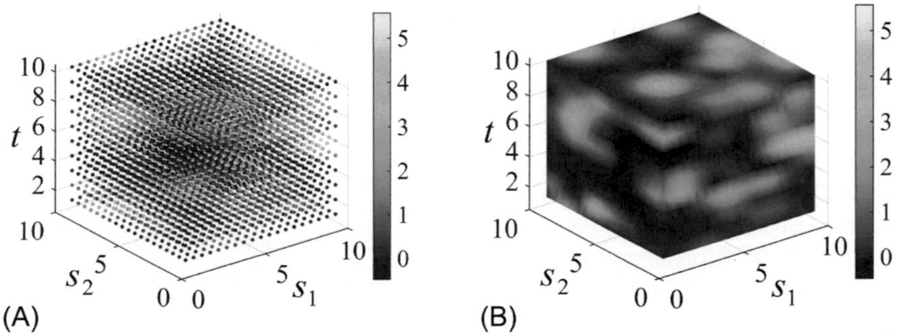

FIG. 5.3 (A) Chronotopologic distribution of log-transformed disease incidence, (B) generated continuous disease incidence distribution in a 3-D map.

① **Intuitiveness** (including the communication of key ideas and notions).
② **Interpretation** (correcting factual misperception and improving context understanding).
③ **Aesthetics** (superseding numerical comprehension).
④ **Presentation** (making qualitative judgments about the phenomenon on the basis of its presentation).
⑤ **Effectiveness** (including memorability, bolstering the presentation of salient aspects of the phenomenon, removing distracting elements and detecting hidden correlation changes).

Example 6.1 Computer-aided data visualization is a very effective tool for identifying meaningful correlations among data and then exploring them in order to develop new models or theories. The way to improve understanding is linked to improving the way visualization and *CTDA* tools can be used together in all scientific fields (in fact, visualization and *CTDA* challenges are more common across scientific fields than they are different).

6.2 Producer and viewer

Visualization involves two agents: the *producer* and the *viewer*. The adequate visualization depends on what aspects of the phenomenon the map producer seeks to emphasize, say, global pattern *vs.* local trends, temporal change *vs.* spatial variation, big picture *vs.* individual performance, or general outline *vs.* individual features. The map viewer's objectives are finding meaning, making connections, prioritizing information, and forming narratives. At its best, visualization achieves focus, clarity, understanding, and simplicity, making it possible to experience even quite complex data arrangements as a coherent whole. This enables the detection of broad data trends and the elucidation of data inconsistencies.

A good way to start *thinking visually* is to consider two key questions about the nature of visualization:

❶ Is the information **conceptual** (visualizing concepts and qualitative information) or **data-driven** (plotting data and information)?
❷ Is the purpose of visualization **declarative, confirmatory,** or **exploratory**. This involves further targeted questions like:
- Is one seeking to declare or explore something?
- Is one seeking to show pollutant data distribution over a region, or to confirm whether or not a hypothesis holds?
- Is one seeking to explore what unknown disease patterns, trends, and anomalies may emerge?
- Is visualization presenting, researching, or seeking ideas?

The answers to these questions can help the investigator plan what visualization resources and tools to use. Note that as one moves from declarative toward exploratory, certainty about what one knows tends to increase.

6.3 Epistemic conditions on visualization

At the same time, visualization is, to a considerable extent, epistemically restricted, which means the following:

> Visualization is conditioned by one's **ability** to visualize, and the **way** one is taught or accustomed to viewing reality, including conventions, expectations, and metaphors concerning data meaning.

Example 6.2 It has been expertly suggested that humans cannot distinguish more than eight different colors in a map. Also, on the basis of one's previous experience one mentally stores different kinds of conventions about data meaning and interpretation, and disrupting what one expects to see may affect one's ability to find meaning in a map.

Viewers make qualitative judgments about the information on the basis of its presentation. If the proposed description of a phenomenon is difficult to perceive visually, it makes it hard to search for its physical meaning and can construct false narratives. As a consequence, this description is often judged unfavorably or even rejected. Even worse, the available data on the basis of which this description was based may also be assessed as inadequate, not credible, or even useless (which could be unfair, in some cases).

Also, being aware of the way humans see and conceive visually offers valuable guidance on *what* should be shown in a map and *how* it should be shown. One's visual examination of a map may start with a quick overview followed by stimulated concentration on certain parts of the map that stand out (e.g., the map viewer's attention is often attracted by sharp changes and distinctions, such as peaks, steep slopes, outliers, and intense colors). People's expertise and background can also play a major role in visualization. Surely, people with expertise in the subject matter of the map will be able to read through it and interpret it more accurately and efficiently than others.

7 Practice exercises

1. The government of California, USA wants to understand the chronotopologic situation of ambient ozone situations during the summer season. Provide an adequate methodological chain of studying the problem to help the local government accomplish the purpose.
2. The manager of Zhejiang University (China) wants to develop a strategy on road light control in the evening by taking into consideration the chronotopologic distribution of students in the campus. Provide an appropriate methodological chain of studying the problem to help the manager make a decision.
3. Consider point-source pollution at one specific location in a lake. Assume you want to study the pollutant distribution in the lake across time. What kind of model might you use?
4. Assume you want to study the chronotopologic distribution of ambient NOx in the center of a city, where many vehicles (one source of NOx emissions, etc.) drive during working hours. What kind of general or core knowledge bases (including fundamental natural laws, primitive equations, scientific models and theories) and specific knowledge base (including measurements, auxiliary variables, etc.) will you consider?
5. To investigate an environmental problem several kinds of data are required. Suggest an appropriate way to collect the following:
   - **(a)** Indoor temperature in the morning.
   - **(b)** Local people's attitude concerning climate change at various times.
   - **(c)** Alkalinity of the local soil.
   - **(d)** Meteorologic data in the region.
   - **(e)** Impact of varying levels of rainfall on local fruits.
6. Which soft data cases represent better quality among the pairs compared below:
   - **(a)** Interval data [1,2] or interval data [1,1.5]?
   - **(b)** Gaussian distribution data $N(1,1)$ or Gaussian distribution data $N(1,0.5)$?
   - **(c)** Uniform distribution data $U(1,2)$ or uniform distribution data $U(1,4)$?
   - **(d)** Lowest limit of $PM_{2.5}$ measurement $= 5\ \mu g/m^3$ or lowest limit of $PM_{2.5}$ measurement $= 4\ \mu g/m^3$?
7. Assume you have obtained a daily temperature dataset (in °C) in one month; please find a way to do data cleaning.
   22, 23.4, 24, 24, 24, 24.5, 24.2, 25, 25, 26, 50, 27, 26, 24, 23, 23, 23.5, 24.2224, 26.148, 24, 20, 21.8, 1, 32, 28.5, 2.48, 25, 25.5, 24, 23, 22
8. An investigator wants to record the room temperature during an entire day. The lower and higher limits of the thermometer are 10 and 40°C, respectively;

and the precision of the thermometer is 0.1° C. The recorded temperature is as follows:

34.5, 36.2, 36.12, 36.5, 37.2, 37.5, 37.6, 37.7, 40.1, 40.2, 39.25, 39.4, 35, 37.5, 37.2, 36.5, 36.2, 36.2, 36.1, 36.1, 36.1, 36.0, and 35.8.

Perform a data cleaning procedure.

9. Explain which properties are used in the data classification schemes below:
   (a) 1, 3, 5, 7, 9, 11, 13, 15.
   (b) 0: female, 1: male.
   (c) Tree, house, horse, lake, flower, boy, girl.
   (d) 1, 10, 100, 1000, 100,001.
   (e) $\{x|x > 1.1\}$.

10. Explain what kind of support scales apply in the following cases:
   (a) 30 soil samples are obtained randomly at the Zhejiang University campus.
   (b) Flu disease cases reported at each county of Zhejiang Province during winter.
   (c) 150 water samples collected during a cruise.

11. Describe situations in your field of interest that may be characterized by a change of support or scale effects.

12. Describe natural systems and/or attributes characterized by chronotopologic dependence in your field of interest.

13. Do you agree with the dictum that "data analysis is based on the assumption that one cannot know before experiencing," and why?

14. An investigator claims that the particular database is dreadfully weak and cannot support the extrapolation involved. What does the investigator mean in *CTDA* terms by this claim?

15. Do you agree that the ultimate judgment of progress in *CTDA* is measurable results in a reasonable time? Explain your rationale.

16. Revisit Example 1.7 of Section 1, and explain why a richer interpretation of the model considered in that example is possible if one's knowledge can assign specific physical meaning to the model's symbols by letting $Y = t$ denote time, $\hat{\theta}_1 = \gamma$ a particle's acceleration, $\hat{\theta}_0 = v_0$ its initial velocity, and $\hat{X} = v$ its current velocity.

17. The *CTDA* methodology discussed in this chapter was concerned with the constraints that the physical world imposes on potential solutions to the problem we face. Discuss any concern about the social problems of implementing this methodology, and explore the additional constraints imposed by social realities.

18. Discuss different forms of explanation depending on the scale used.

19. Discuss which of the following scales are associated with deterministic *vs.* stochastic *vs.* contingent:
   (a) Small spatial and short time scales.
   (b) Large spatial scales and long time scales.

20. Do you support or not the view that many of the sophisticated ideas and formulations in mathematics are intricately entwined with the physicality of our being? Explain.

21. Consider the following statements concerning scientific inquiry you agree and with which you do not:
   (a) It draws inferences that are consistent with the data and scientific reasoning.
   (b) It uses evidence, logic, and imagination to develop explanations about the natural world.
   (c) It progresses through a continuous process of questioning, data collection, analysis, and interpretation.
   (d) It is a thoughtful and coordinated attempt to search out, describe, explain and predict natural phenomena.
   (e) It requires the sharing of findings and ideas for critical review by colleagues and other scientists.

   With which of the above do you agree and with which do you not agree? Explain.

22. Consider the arguments:
   (a) One does not really understand a physical concept until one can describe it quantitatively in mathematical terms.

(b) One does not really understand a physical concept until one can describe how it behaves without having to resort to mathematics.

Do you agree with argument (a) or with argument (b)? Or, do you suggest a third argument? Explain.

23. Would you use a model that, despite its shortcomings, (a) reveals something about the mechanism of the phenomenon (say, indicates trends and reveals the interaction of forces), and (b) tends to lead to errors on the side of conservatism? Explain.

CHAPTER

4

Chrono-geographic statistics

1 Introduction

This chapter introduces certain statistical notions and techniques that were developed originally in geographical sciences, with the time component being added later in order to enable the study of attributes and events that occur in a space-time framework. That is, the time addition led to the extended *chrono-geographic* versions of the original purely geographic techniques.

Ut aequum, it serves our purpose to give a broad description of the subject matter of this chapter.

Chrono-geographic statistics (*CGS*) is concerned with the statistical description of attribute point distributions and values that are functions of dynamic (time-dependent) geographical coordinates.

A characteristic of the *CGS* techniques is that they consider space and time separately. In this respect, they belong to the *S-TDA* field introduced in Section 1 of Chapter 1. In practice, the *CGS* techniques have been used mainly in the $R^{2,1}$ domain, where they introduce new factors, both of knowledge and uncertainty, and offer a transdisciplinary quantitative perspective on spatial and temporal objects and processes, such as attribute locations, instances and values, species distributions, ecological interactions, societal and environmental activities, and public health events.

Historically, the so-called *time-geography* is a field that was developed in order to study the coordination of human activities in society and nature (Hägerstrand, 1975; Parkes and Thrift, 1980). The quantitative *CGS* techniques can be very useful in this field, especially since a rigorous analytical framework of the field's basic entities and relations is still lacking or is rather limited (see, e.g., discussion in Miller, 2005). Although there are a number of issues in the time-geography field that can be illuminated by *CGS*, the latter suggests a broader perspective that includes many applications outside this field. Nowadays, the *CGS* techniques are applied in multiple fields related to geosciences, ecology, environmentology, regional planning, ecology, human geography, and public health.

https://doi.org/10.1016/B978-0-12-816341-2.00006-X

Example 1.1 *CGS* can study the coordinates that define the worldline or worldtube[a] of disease movement in the $R^{2,1}$ domain (Chapter 2). In this way, valuable insight could be gained into the epidemic spread and the daily activities of infected individuals (duration of contacts, the direction of movements, etc.), and it can also reveal logical problems in reasoning (e.g., a group of infected individuals cannot simultaneously be at two locations or leave out a time period).

In actu, some of the *CGS* techniques focus solely on space-time data coordinate information (these techniques are discussed in Section 2 below), whereas some others also include physical data values in their quantitative analysis of the phenomenon (these techniques are the focus of Section 3).

2 CGS of data point information

For obvious reasons, many of the developments in this chapter are directly linked to the chronotopology theory discussed in Chapter 2.

2.1 Fundamental domain distinction

A point distribution or pattern consists of a set of data points $\boldsymbol{p}$ in the $R^{n,1}$ or the R^{n+1} domain. In common with previous discussions, here the following terminology is used (Section 2.2 of Chapter 2):

In $R^{n,1}$, the point distribution is considered as a **series** of spatial entities at specified times, and the **separate** space-time domain of interest is denoted as $D \times T \subset R^{n,1}$, with $D \subset R^n$ and $T \subset R^1_{+,\{0\}}$. Points $\boldsymbol{p}$ within $D \times T$ have distinct spatial and temporal coordinates, $\boldsymbol{s} = (s_1, \ldots, s_n) \in R^n$ and $t \in T$, respectively.

In R^{n+1}, the point distribution is considered as a **single** spacetime entity, and the **composite** domain of interest is denoted as $\Omega \subset R^{n+1}$. Points $\boldsymbol{p}$ within Ω have joint spacetime coordinates $(s_1, \ldots, s_n, s_0)$.

Each perspective has its own merits. Analysis in the $R^{n,1}$ domain allows the *sequential* study of a point distribution during a particular time period, i.e., it follows the spatial point arrangement as it moves with time. The $R^{n,1}$ analysis may be appropriate when the investigator seeks to understand the dynamic (i.e., time-dependent) evolution of the geographic point arrangement. Analysis in the R^{n+1} domain allows the study of the point distribution during the time frame of interest as a *whole*. The R^{n+1} analysis may be appropriate when the investigator is interested in the point distribution as a continuum representing it simultaneously at all moments of its history (the readers may recall the discussion in Table 2.1 of Chapter 2).

Example 2.1 The $R^{2,1}$ study of groups of infected individuals within a geographical region D that move from one location to another during the time period of interest can offer valuable hints to public health managers in order to better deal with an epidemic. In this case, a *neighborhood* of points in the separate domain $D \times T \subset R^{2,1}$ may be represented by a cylinder with a circular geographic basis with radius $|\boldsymbol{h}| \in D$ (Table 3.1 of Chapter 2) and height corresponding to the time lag $\tau \in T$. On the other hand, the R^{2+1} study of an epidemic as a single entity can give a picture of its overall spacetime spread and help understand the behavior of the epidemic spread as a single entity. In this case, a neighborhood of points in the composite domain $\Omega \subset R^{2+1}$ may be represented by a sphere of radius $|\Delta\boldsymbol{p}|$ (Table 3.3 of Chapter 2).

[a] In time-geography the term time-geographic *path* (or *trajectory*) has been used, instead (Hägerstrand, 1975).

As noted earlier, separate space-time does not mean that no space-time interaction exists, but merely that this interaction is represented in a different way than the composite spacetime interaction. An investigator would study the point paths, trajectories, or worldlines and extract useful information about the point distribution patterns, such as important parts of the worldlines or clustering of worldlines with common features.

Example 2.2 No space-time interaction may imply that the intensity function (average number of points $\boldsymbol{p}$ per unit chronotopologic domain) is a product of one function depending only on the spatial location s and another function depending only on t.

In this section, the attention of the chrono-geographic statistics measures is restricted to information about data point configurations, ignoring the physical values of the corresponding measurements or observations at these points.

The focus of the present *CGS* is the computation of the statistics of **chrono-geographic coordinates** of data point distributions.

Technically, the extension of Classical Statistics (*CS*, see Section 5 of Chapter 3) proposed by *CGS* consists of introducing a set of space-time equivalents of the standard *CS* measures. An illustrative list of coordinate-independent *CS* measures and the corresponding coordinate-dependent measures is given in Table 2.1.

The *CGS* provides rigorous analytical expressions of space-time point distributions. Describing the statistics of point locations and time instances is very useful in the study of attributes in nature and society (e.g., how these attributes are arranged and coordinated with geographical location and time).

2.2 Descriptive measures

The *CGS* point distribution measures to be discussed below are collectively denoted as descriptive measures $DM(\boldsymbol{p})$.[b] Another significant issue to stress is that, in accordance with the domain distinction above, and depending on the objectives of the study, the investigators may consider the development of such measures in three possible ways:

1. As a **separate space-time** or **dynamic** measure in the $R^{n,1}$ domain,

$$DM(\boldsymbol{s}_t) = DM[g_1(s_1), \ \ldots, g_n(s_n); t] \qquad (2.1a)$$

for each $t \in T$, g is a suitable function, i.e., as a time-varying set of geographically-averaged values.

2. As a **time-averaged** $DM(\boldsymbol{s}_t)$ measure in the $R^{n,1}$ domain,

$$DM_T = \frac{1}{T}\int_T dt DM(\boldsymbol{s}_t) \qquad (2.1b)$$

over all $t \in T$, i.e., it is obtained by averaging the $DM(\boldsymbol{s}_t)$ values over time.

3. As a **composite spacetime** measure in R^{n+1},

$$DM(\boldsymbol{p}) = DM[g_1(s_1), \ \ldots, g_n(s_n), g_0(s_0)], \qquad (2.1c)$$

i.e., a single $DM(\boldsymbol{p})$ value is considered over all points of the Ω domain.

TABLE 2.1 *CS* measures and their *CGS* counterparts.

Measure	Central tendency	Absolute dispersion	Relative dispersion
CS	Mean	Standard deviation	Coefficient of variation
CGS	Mean center	Standard distance	Relative distance

[b] Rather obviously, "*DM*" here means "descriptive measure."

Note that the case ❸ above the *DM* is a function g_0 of time s_0, unlike the case ❶, where an individual time t is considered and the *DM* values change with time. All of the above cases will be discussed in the following sections while, on occasion, we will be particularly focusing on the $R^{2,1}$ and R^{2+1} domains.

2.3 Point distribution center

We begin with the definition of the center of a dynamic (time-dependent) geographic data point distribution. As above, one should distinguish between separate space-time point distribution centers (corresponding to cases ❶ and ❷) and composite spacetime centers (corresponding to case ❸).

2.3.1 *Separate space-time center measures*

Assume that m points $\boldsymbol{p}_i$ ($i = 1, \ldots, m$) are considered at time t, which is a subset of the total number of points in the $R^{n,1}$ domain (i.e., including all times $t \in T$). This definition uses the minimum information available at each point $\boldsymbol{p}_i$, i.e., its spatial coordinates $(s_{i,1}, \ldots, s_{i,n})$ and time instance t.[c]

Chrono-geographic mean center (*CGMC*) of the separate space-time data point distribution consisting of the locations $\boldsymbol{s}_i$ ($i = 1, \ldots, m$) at time t is an artificial summary point representing its "center" and calculated in the $R^{n,1}$ domain as

$$\overline{\boldsymbol{s}}_t = (\overline{\boldsymbol{s}}; t) = (\overline{s}_1, \ldots, \overline{s}_n; t) = \left(\frac{1}{m}\mathbf{1}^{\mathrm{T}}\boldsymbol{s}_1, \ \ldots, \frac{1}{m}\mathbf{1}^{\mathrm{T}}\boldsymbol{s}_n; \ t\right), \tag{2.2}$$

see Fig. 2.1, $\mathbf{1}^{\mathrm{T}} = [1 \ldots 1]$, $\boldsymbol{s}_k^T = [s_{1,k} \ldots s_{m,k}]$ ($k = 1, \ldots, n$). The *CGMC* is the time-dependent vector $\overline{\boldsymbol{s}}_t$ expressed as the pair "space vector $\overline{\boldsymbol{s}}$, time scalar t."

Sensu lato, CGMC belongs to the group of measures described by Eq. (2.1a) with $g(\boldsymbol{s}_k) = \frac{1}{m}\mathbf{1}^{\mathrm{T}}\boldsymbol{s}_k = \frac{1}{m}\sum_{i=1}^{m} s_{i,k} = \overline{s}_k, \quad k = 1, \ldots, n.$ A visualization of the *CGMC* $\overline{\boldsymbol{s}}_t$ is presented in Fig. 2.1 in a cylindrical space-time $D \times T \subset R^{2,1}$ setting.

An interesting property of the *CGMC* Eq. (2.2) is that it satisfies the minimization condition,

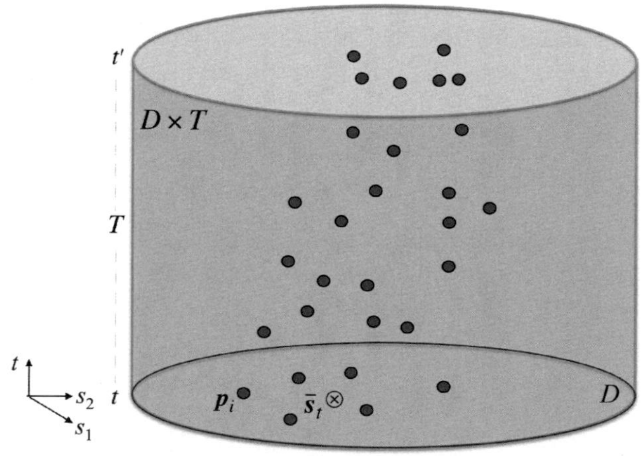

FIG. 2.1 A visualization in $R^{2,1}$ of the *CGMC* $\overline{\boldsymbol{s}}_t$ (⊗) at time t, where $\boldsymbol{p}_i$ ($i = 1, \ldots, m$) are those points within the circle at time t that is considered when determining $\overline{\boldsymbol{s}}_t$.

[c] The space-time point notion was introduced in earlier chapters (see, e.g., Section 1 of Chapter 2).

$$\bar{s}_t = (\bar{s}_1, \ldots, \bar{s}_n; t)$$
$$: \min\left\{d^2 = \sum_{i=1}^{m}\sum_{k=1}^{n}(s_{i,k} - \bar{s}_k)^2\right\}, \quad (2.3)$$

i.e., it minimizes the sum of the squared coordinate deviations from *CGMC*, d^2, at time t. This is an alternative, sometimes useful, way to define the point distribution center.

The interpretation of *CGMC* occurs through the fruitful interaction of mathematical formulation and conceptual content, where the former is highly symbolic and the latter is rich in meaning. In $R^{2,1}$, given that $\bar{s}_t$ is the geometric point at which the mean of the $s_{i,1}$ coordinates meets with the mean of the $s_{i,2}$ coordinates at the time instance t, its physical meaning can be seen as the "gravity center," "the balancing point," or the "center of action" of a *space-time* point pattern. The *CGMC* is used to monitor or track changes in a data point distribution or to compare the point distributions of different datasets.

Example 2.3 Monitoring the space-time evolution of an epidemic's *CGMC* may offer the starting point for further investigations concerning its origin. The *CGMC* is used to compare the distributions of different epidemic types in the same chrono-geographic domain or the same epidemic at different domains.

Potential problems with *CGMC* are that it is sensitive to outliers (which can be a problem in the case of skewed data point distributions), it is sensitive to movements in the point distribution, and it could be outside the domain of interest. Accordingly, occasionally it makes sense to consider other ways to define the central tendency of point distribution. Some of these ways are presented next.

The chrono-geographic **Euclidean median** center (***CGEMC***) at time t is defined as

$$\bar{s}_{Em,t} = (\bar{s}_{Em}; t), \quad (2.4a)$$

where the time-dependent vector $\bar{s}_{Em}$ in $R^{n,1}$ satisfies the minimization condition:

$$\bar{s}_{Em} = (\bar{s}_{Em,1}, \ldots, \bar{s}_{Em,k})$$
$$: \min \sum_{i=1}^{m}\sqrt{\sum_{k=1}^{n}(s_{i,k} - \bar{s}_{Em,k})^2}. \quad (2.4b)$$

CGEMC belongs to the group of measures of Eq. (2.1a). Eq. (2.4a) may be seen as a possible space-time extension of the median measure of *CS* that minimizes the aggregate distance to the distribution center at time t.

The chrono-geographic **Manhattan median** center (***CGMMC***) at time t is defined as the time-dependent vector in $R^{n,1}$,

$$\bar{s}_{Mm,t} = (\bar{s}_{Mm}; t) = (\bar{s}_{Mm,1}, \ldots, \bar{s}_{Mm,n}; t). \quad (2.5)$$

The *CGMMC* also belongs to the group of Eq. (2.1a). Eq. (2.5) determines the location at time t that is the intersection of two mutually perpendicular lines each dividing the population into equal groups. A less frequently used measure of central tendency of a point distribution is as follows.

The chrono-geographic **modal** center (***CGMoC***) at time t is the point with the maximum surface point density at a given time.

CGMoC is useful when one seeks to illustrate high-density effects with reference to a specific point.

Example 2.4 The *CGEMC* is the center of convergence where the entire point distribution of exposed individuals could be assembled with the minimum aggregate travel distance, whereas the *CGMMC* is particularly useful when seeking to compare different point distributions of infected *vs.* noninfected individuals within the same space-time domain. Among the *CGMoC* features are that it is sensitive to the size of the areal unit considered, and it is the least sensitive measure to population movement. Yet, it is useless in the analysis of changing areal patterns (*CGMoC* is a spatially microscopic measure).

The measures of a chrono-geographic point distribution's central tendency have their own characteristics: *CGMC* is affected by all the points of the distribution, and it is easy to compute. *CGEMC* defines the shortest path, it is less influenced by outliers, and it is harder to compute (it has no exact solution). *CGMMC* minimizes the absolute distance deviations, it is equally influenced by each point in the distribution, it is the shortest distance when traveling only NS and/or EW, it is rather easy to compute, and it has no exact solution for an even number of points. Although the interpretation of *CGEMC* is easier than that of *CGMC*, its calculation is more difficult (this is because *CGMC* minimizes the sum of Eq. (2.3), whereas *CGEMC* minimizes the sum of Eq. (2.4b)).

In practice, computational techniques are used to calculate the particular kind of measure that has been selected to describe the central tendency of chrono-geographic point distribution. For example, the *CGEMC* is usually found by using an iterative procedure so that the Euclidean distance in Eq. (2.4b) is minimized (e.g., Bart and Barber, 1996). Next, we focus on the *CGMC*, which is the most commonly used center measure in practice.

The next example presents a *step-by-step* *CGMC* **computation** procedure in practice.

Example 2.5 The goal of this study (see, Fei et al., 2016b) was to investigate the associations between two related diseases, female thyroid (*FT*) and breast cancer (*BC*), in Hangzhou province, China during the period 2008–2012. For illustration, the recorded occurrences of the two diseases in one county are presented in Table 2.2 and plotted in Fig. 2.2A in the chrono-geographic domain of interest. In particular, there are 69 *FT* and 72 *BC* cases in the

TABLE 2.2 Geographic coordinates of recorded female thyroid (*FT*) and breast cancer (*BC*) cases.

	Female thyroid						Breast cancer					
Time	No.	$s_{i,1}$	$s_{i,2}$	No.	$s_{i,1}$	$s_{i,2}$	No.	$s_{i,1}$	$s_{i,2}$	No.	$s_{i,1}$	$s_{i,2}$
0	1	1.9232	0.9962	7	2.6398	0.3235	1	2.6013	1.1595	8	2.8166	1.2639
	2	3.5042	2.3128	8	3.2940	1.3815	2	2.9797	1.9383	9	2.6746	1.1357
	3	2.4672	0.8567	9	3.4436	0.7753	3	2.8118	1.6843	10	2.6633	1.2023
	4	3.0601	2.0766	10	4.0271	3.2448	4	3.4601	1.3860	11	3.5432	1.2639
	5	2.6733	1.5744	11	2.8835	1.9951	5	3.0177	1.5646	12	3.1956	1.4590
	6	3.1821	2.3620	12	1.7170	2.3274	6	3.5928	1.4420	13	3.1598	1.0546
							7	2.7182	1.0836			
1	1	3.3226	1.4622	8	3.2056	1.9029	1	3.3918	1.4826	8	2.5554	1.7109
	2	4.1003	1.9288	9	5.1470	0.7755	2	2.9707	1.4197	9	2.6166	1.7803
	3	1.6447	1.3770	10	4.6617	1.9303	3	3.2866	1.8421	10	3.4881	1.8302
	4	3.5173	1.4255	11	2.1901	2.4781	4	2.7054	1.8747	11	3.3338	1.3036
	5	3.1913	2.3938	12	4.8210	1.7933	5	3.3473	1.2056	12	2.8805	1.7477
	6	2.2154	2.3454	13	3.4352	2.1208	6	2.5382	1.5387	13	3.6403	1.7206
	7	2.7398	2.3503				7	2.8323	1.4901	14	2.5413	1.1789

TABLE 2.2 Geographic coordinates of recorded female thyroid (*FT*) and breast cancer (*BC*) cases—cont'd

	Female thyroid						Breast cancer					
Time	**No.**	$s_{i,1}$	$s_{i,2}$	**No.**	$s_{i,1}$	$s_{i,2}$	**No.**	$s_{i,1}$	$s_{i,2}$	**No.**	$s_{i,1}$	$s_{i,2}$
2	1	2.3319	0.8631	9	3.0195	0.6466	1	3.6006	1.8737	10	3.4350	1.4956
	2	2.9959	2.9103	10	3.3315	1.7929	2	2.8430	1.3423	11	3.6208	1.0922
	3	3.9196	1.1306	11	3.6604	1.3936	3	3.4086	1.5814	12	2.6559	1.2519
	4	2.5382	1.9488	12	3.9265	1.3824	4	3.4045	1.1822	13	3.1826	2.0047
	5	3.2228	1.3845	13	3.0516	2.3516	5	2.9565	1.6622	14	3.0633	1.1676
	6	2.8646	2.0332	14	2.1050	1.6750	6	3.1814	1.2893	15	2.5143	1.9084
	7	3.6704	1.0411	15	2.5546	1.6187	7	2.5910	1.7195	16	2.9045	1.5922
	8	2.3466	0.6586				8	2.5647	1.7581	17	2.6946	2.0957
							9	3.1370	1.8230			
3	1	1.8235	1.6692	10	3.6591	1.3286	1	3.0890	1.0658	9	2.7900	1.8049
	2	2.8814	1.5201	11	2.8333	1.0012	2	3.0871	1.2583	10	2.9847	1.7125
	3	2.2753	0.6998	12	3.4209	0.9125	3	2.9053	1.3885	11	2.6157	1.4960
	4	4.7448	2.1765	13	1.7689	0.8062	4	3.5801	1.9033	12	2.6584	1.6017
	5	3.4951	1.7101	14	2.7877	1.1799	5	2.9431	1.0169	13	3.6305	1.3260
	6	3.8274	1.3206	15	2.5058	0.2984	6	2.6334	1.0473	14	3.6474	1.8192
	7	2.3651	1.5137	16	2.0538	2.0785	7	3.4363	1.1859	15	3.1903	1.2079
	8	2.7188	1.3428	17	3.3048	1.8120	8	2.9677	1.7140			
	9	2.8365	0.4499									
4	1	3.2730	1.5138	7	3.7564	1.8751	1	3.2470	2.0157	8	2.7337	1.2839
	2	2.4908	1.5308	8	3.3961	1.6099	2	3.2045	1.4732	9	2.7711	1.4496
	3	2.7991	1.9956	9	2.9593	0.8821	3	2.7493	1.2033	10	2.7048	1.6544
	4	3.3317	2.4162	10	2.8829	2.0695	4	2.8615	1.9954	11	2.7732	1.2884
	5	3.6235	1.7801	11	2.8694	1.6842	5	3.0651	2.0777	12	3.0228	1.6631
	6	2.3294	1.3742	12	2.8181	1.5811	6	2.7766	1.4828	13	2.8733	1.7823
							7	3.5132	1.1222			

county during a 5-month period (i.e., time instances range from 0 to 4 in Table 2.2).

Step 1: For the *FT* disease, the *CGMC* of female thyroid at the time instance $t = 0$ is calculated as

$$\bar{\mathbf{s}}_{t=0,FT} = (\bar{s}_1, \bar{s}_2; t=0)_{FT} = (2.9013,\ 1.6855;\ 0),$$

where $\bar{s}_1 = \frac{1}{12}\sum_{i=1}^{12} s_{i,1} = 2.9013$, and $\bar{s}_2 = \frac{1}{12}\sum_{i=1}^{12} s_{i,2} = 1.6855$. Similarly, the *CGMC* of *BC* spread at $t = 0$ is

$$\bar{\mathbf{s}}_{t=0,BC} = (\bar{s}_1, \bar{s}_2; t=0)_{BC} = (3.0181, 1.3567; 0).$$

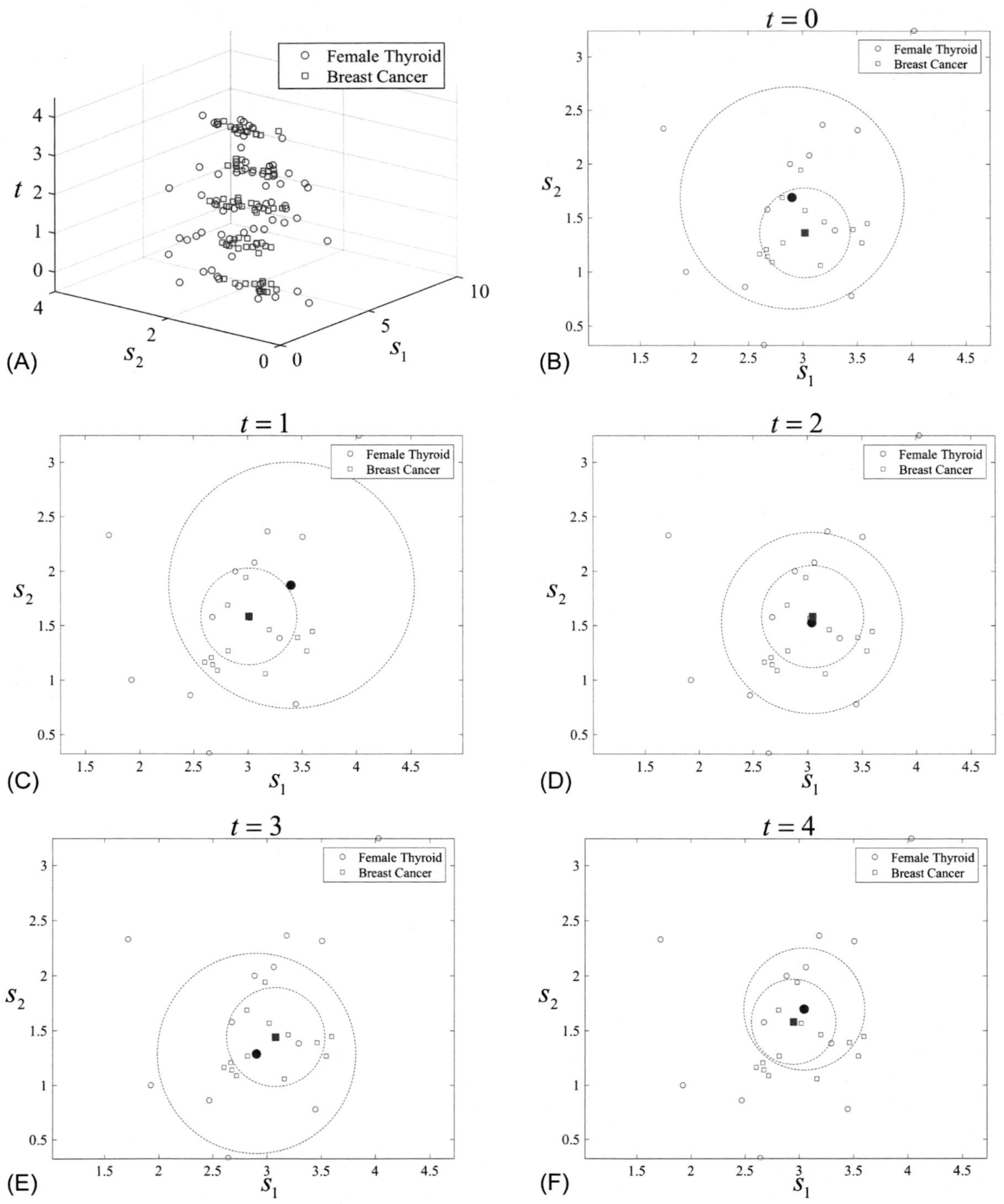

FIG. 2.2 (A) The chrono-geographic distribution of the *FT* (blue small circle) and *BC* (red rectangle) occurrences at various times, (B–F) the corresponding *CGMC* (solid circle or rectangle) and *CGSD* (big circles).

Step 2: By comparing the two *CGMC* in Fig. 2.2B (the solid blue circle and the red rectangle represent the *CGMC* of *FT* and *BC* distributions, respectively) one would conclude that the mean center of the *FT* distribution is located to the NW of the mean center of the *BC* distribution at $t = 0$.

Step 3: The *CGMC* of the two diseases at times $t = 1, 2, 3$ were also computed at various times, as shown in Table 2.3 and Fig 2.2C–F. Interestingly, the relative locations of the *FT* and *BC* centers vary in time.

Sine dubio, there is considerably more into *CGMC*-based analysis, which extends beyond the centrographic statistics discussed above. Indeed, starting from the point distribution information (chrono-geographic coordinates), other useful deliverables are developed, as follows:

The *CGMC* ***path***, also called the *CGMC* ***worldline*** $WL_{\bar{s}_t}$,[d] represents the *CGMC* movement at various times $t \in T$ as

$$Path_{\bar{s}_t} = WL_{\bar{s}_t} = \{\bar{s}_t : t \in T\}. \quad (2.6)$$

TABLE 2.3 Chrono-geographic statistics of the female thyroid (*FT*) and breast cancer (*BC*) dataset.

	FT			*BC*		
Time instance	***CGMC*** $\bar{s}_t$	***CGSD*** $\delta\bar{p}$	***CGRD*** $\delta\bar{p}_R$	***CGMC*** $\bar{s}_t$	***CGSD*** $\delta\bar{p}$	***CGRD*** $\delta\bar{p}_R$
0	(2.9013,1.6855;0)	(1.0306;0)	(0.6089;0)	(3.0181,1.3567;0)	(0.4158;0)	(0.2457;0)
1	(3.3994,1.8680;1)	(1.1308;1)	(0.6681;1)	(3.0092,1.5804;1)	(0.4439;1)	(0.2623;1)
2	(3.0359,1.5221;2)	(0.8322;2)	(0.4917;2)	(3.0446,1.5788;2)	(0.4691;2)	(0.2771;2)
3	(2.9001,1.2835;3)	(0.9167;3)	(0.5416;3)	(3.0773,1.4365;3)	(0.4526;3)	(0.2674;3)
4	(3.0441,1.6927;4)	(0.5574;4)	(0.3293;4)	(2.9459,1.5763;4)	(0.3891;4)	(0.2299;4)

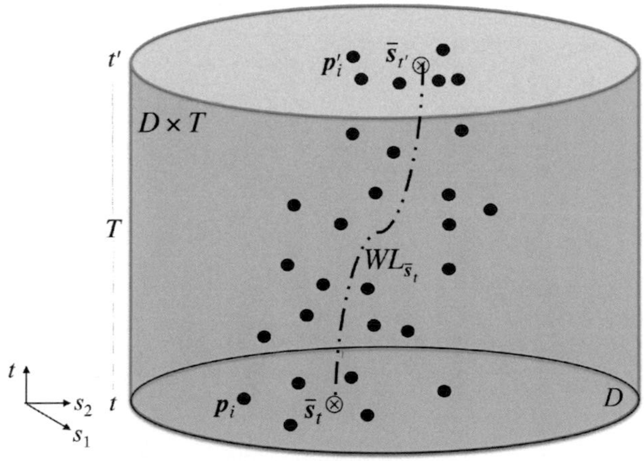

FIG. 2.3 A visualization in $R^{2,1}$ of the $WL_{\bar{s}_t}$ of the *CGMC* $\bar{s}_t$ from times t to t' (indicated as a dashed-dotted line — ··).

[d] See the worldline definition in Section 2 of Chapter 2.

In other words, the *CGMC path* (or *CGMC worldline, WL*) is the path (or *WL*) of the *CGMC* point distribution over time, i.e., it is a plot of $\overline{s} = (\overline{s}_1, \overline{s}_2)$ *vs. t*. Fig. 2.3 presents a visualization of the $WL_{\overline{s}_t}$ or $Path_{\overline{s}_t}$ in $R^{2,1}$ of $\overline{s}_t$ from time *t* to time *t'* in terms of the plot of $\overline{s} = (\overline{s}_1, \overline{s}_2)$ *vs. t*.

Example 2.5 (cont.) Given that chrono-geographic disease datasets are purposefully divided into a specified number of geographical data subsets in terms of specified time instances, the mean centers of the data point subsets at various times can be computed, their evolution interpreted, and future developments inferred. Proceeding further with the *FT* and *BC* study of Example 2.5 above, one more step is added, as follows:

Step 4: By integrating the mean centers of the *FT* and *BC* distributions, the movements of the geographical patterns of the two diseases are plotted in Fig. 2.4. These are the time paths or the *WL* of the *CGMC* of the *FT* and *BC* diseases. The mean center paths or *WL* of the two diseases have different origins and different endings. The *WL* of the *FT* disease is considerably more diffused geographically than the *WL* of the *BC* disease, traveling at longer distances and covering a much wider region in the Hangzhou province.

The *CGMC* path can also be matched to background information (e.g., transportation and communication networks). In practice, the *CGMC* path at unobserved points is constructed using an interpolation technique. It is useful to examine some additional quantities that are useful in point distribution studies in the $R^{2,1}$ domain.

The vector $\overline{h}_{tt'}$ represents the **geographic distance** of the *CGMC* move between times *t* and *t'*, i.e.,

$$\overline{h}_{tt'} = \overline{s}'_t - \overline{s}_t = (\overline{s}_{1,t'} - \overline{s}_{1,t}, \overline{s}_{2,t'} - \overline{s}_{2,t}) = (\overline{h}_{1,tt'}, \overline{h}_{2,tt'}). \quad (2.7)$$

Combined with $\overline{s}_t$, the $\overline{h}_{tt'}$ gives the investigator valuable information about the dynamic

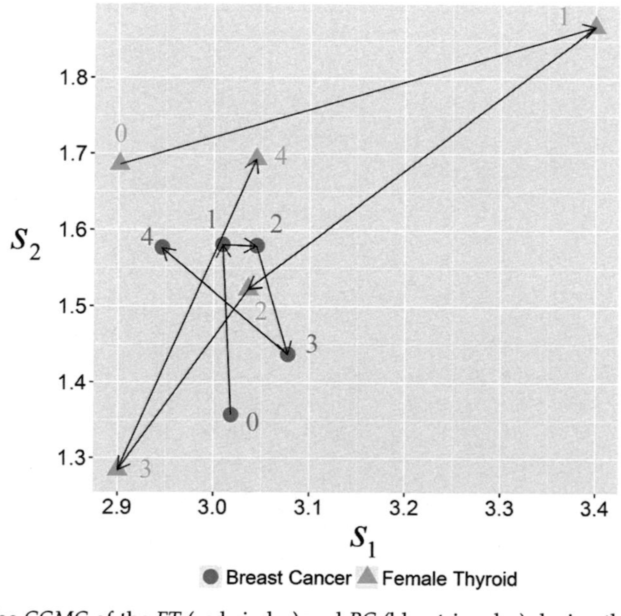

FIG. 2.4 Paths or *WL* of the *CGMC* of the *FT* (red circles) and *BC* (blue triangles) during the 5-month period of interest.

(time-dependent) evolution of the data point distribution. The further away $\overline{s}_{t'}$ is from $\overline{s}_t$, the further away the point distribution center of the action is at time t' from the center at time t. And the larger the $\overline{h}_{tt'}$ is, the faster the diffusion of the distribution with time. Also very useful is information about the direction of the dynamic evolution of the point distribution.

The *CGMC* **direction** θ at time t is defined as the angle of the vector $\overline{\boldsymbol{h}}$ with the positive h_1-axis, which is computed as

$$\cos\theta = \arctan\frac{\overline{h}_{2,tt'}}{\overline{h}_{1,tt'}}. \tag{2.8}$$

The following is a temporal characteristic of the traveled distance.

The $\tau = t' - t$ is the **travel time** between $\overline{s}_t$ and $\overline{s}_{t'}$, i.e., the duration of *CGMC* movement.

Obviously, the longer it takes for *CGMC* to move between two specified geographical locations (as determined by τ), the slower it takes for the point distribution to travel along a certain direction θ (e.g., the slower a species moves from a region to another, or the slower the contamination moves from the soil to the aquifer).

The *CGMC* **velocity** $\boldsymbol{v}$ at time t is defined as the ratio of the distance $\overline{\boldsymbol{h}}_{tt'}$ the *CGMC* covered during the period τ, over the period τ, i.e.,

$$\boldsymbol{v} = \frac{\overline{\boldsymbol{h}}_{tt'}}{\tau}. \tag{2.9}$$

In view of Eq. (2.7), the velocity in Eq. (2.9) can also be written as $\boldsymbol{v} = \left(\frac{\overline{h}_{1,tt'}}{\tau}, \frac{\overline{h}_{2,tt'}}{\tau}\right)$, i.e., in terms of the dynamic (time-dependent) geographic distance coordinates.

The *CGMC* **speed** v at time t is the magnitude of the velocity $\boldsymbol{v}$ at this time, i.e.,

$$v = |\boldsymbol{v}| = \sqrt{\frac{\overline{h}^2_{1,tt'} + \overline{h}^2_{2,tt'}}{\tau^2}}. \tag{2.10}$$

The description of the velocity $\boldsymbol{v}$ vector in Eq. (2.9) includes both the speed or velocity magnitude v and the direction θ. A real-world study implementing the *CGMC* velocity is discussed in Example 2.6 below. Basically, the use of mathematics in Eqs. (2.6)–(2.10) above is illustrative, aiming at getting the intuition of the deliverables. A summary of the above concepts is provided in Table 2.4.

TABLE 2.4 CGMC-based chrono-geographic deliverables.

Deliverable	Mathematical definition
CGMC worldline $WL_{\overline{s}_t}$, or CGMC path	Eq. (2.6) of this section
Geographical distance (vector) $\overline{\boldsymbol{h}}_{tt'}$ of CGMC between times t and t'	Eq. (2.7) of this section
CGMC direction θ (scalar)	Eq. (2.8) of this section
Space-time bi-lag $\Delta\overline{\boldsymbol{p}} = (\overline{\boldsymbol{h}}; \tau)$	Eq. (2.2a) of Section 2, Chapter 2
Space-time bi-metric $\lvert\Delta\overline{\boldsymbol{p}}\rvert = (\lvert\overline{\boldsymbol{h}}\rvert; \tau)$	Eq. (2.3a) of Section 2, Chapter 2
CGMC velocity (vector) $\boldsymbol{v}$	Eq. (2.9) of this section
CGMC speed (scalar) v	Eq. (2.10) of this section

The *CGMC* is also amenable to modifications, as required by the real-world study. In certain applications, it is appropriate to assign weights to chrono-geographic points, in which case a more general formulation of Eq. (2.2) has been proposed, as follows.

The **weighted *CGMC*** is defined as

$$\bar{\mathbf{s}}_{t,w} = (\bar{s}_{1,w}, \ldots, \bar{s}_{n,w}; t)_w = \left(\frac{1}{W}\boldsymbol{w}^{\mathrm{T}}\boldsymbol{s}_1, \ldots, \frac{1}{W}\boldsymbol{w}^{\mathrm{T}}\boldsymbol{s}_n;\ t\right), \tag{2.11}$$

where $\boldsymbol{w}^{\mathrm{T}} = [w_1 \ldots w_m]$, w_i is the weight of point $\boldsymbol{p}_i = (s_{i,\,1}, \ \ldots, s_{i,\,n}, t)$, and $W = \sum_{i=1}^{m} w_i$.[e]

An interesting issue is that the weights w_i may have different forms with distinct interpretations assigned to them, such as:

(a) The w_i may be analogous to frequencies in the analysis of grouped data (e.g., weighted mean); the mean center, then, serves as a chronotopologic analog to the statistical mean, in that it is the point that minimizes the sum of squared deviations of a set of points.

(b) The weights w_i may be assigned (physical, ecological, or social) meaning; in the case of an epidemic, e.g., w_i may represent the disease feature (incidence, mortality, etc.) at location $\boldsymbol{s}_i$ and time t.

Example 2.6 The 14th century Black Death epidemic in Europe was arguably the deadliest epidemic in human history (Christakos et al., 2005, 2007).

- The Black Death *CGMC* (i.e., centroids defined as the centers of the local areas struck by Black Death at given times) were computed using Eq. (2.11), where w_i was taken to be the mortality at a point $\boldsymbol{p}_i$. By connecting the series of *CGMC* across time, an epidemic *CGMC* path was plotted. Because of the different transportation methods and epidemic origins, multiple Black Death paths were initiated at different locations and propagated along with different directions throughout the continent.
- In addition, the spread velocity of Black Death, see Eq. (2.9), was obtained by using the spatial distances and temporal lags between selected *CGMC* at various time instances. Then, a mean epidemic speed, Fig. 2.5, was calculated as the arithmetic average of the speeds of the various epidemic *CGMC* along with different directions throughout Europe (Christakos et al., 2007). A conspicuous feature of the plot in Fig. 2.5 is the intense fluctuations of the *CGMC* speed with time. It was finally concluded that the spread velocity of Black Death fluctuated mainly around 1.5–6 km/day. Lastly, Fig. 2.6A and B present the coordinates (longitude and latitude) for the various *CGMC* as a function of time. Most *CGMC* show a similar trend that moves northward with time. In the early years of Black Death, the mean *CGMC* latitude was about 42°; in the final stage, the mean latitude was close to 55°. The longitude plot shows that the epidemic started at about 15°, then it moved westward, where it stayed for a while, and finally, it moved towards the East. In the final years of the epidemic the mean longitude was also at about 15°.

2.3.2 *Time-averaged CGMC*

In some applications, one may find it useful to consider the *CGMC* $\bar{\mathbf{s}}_t$ of a point distribution at different times, and then compute its temporal average, as follows:

[e] Obviously, when $w_i = 1$ $(i = 1, \ldots, m)$, it implies that $W = m$ and Eq. (2.11) reduces to Eq. (2.2).

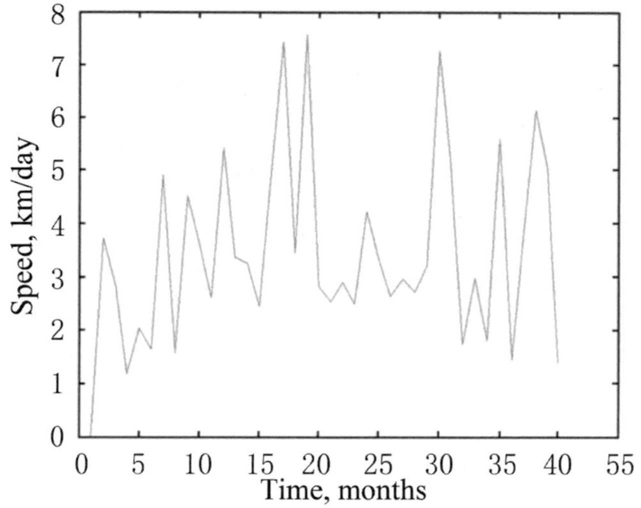

FIG. 2.5 *CGMC* speed of Black Death epidemic in 14th century Europe.

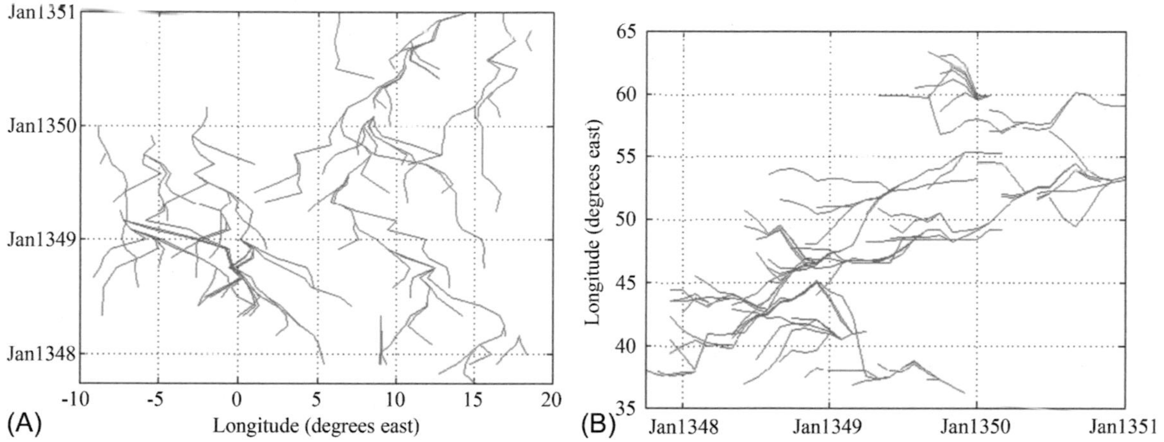

FIG. 2.6 Plots of the (A) longitude and (B) latitude coordinates for the Black Death *CGMC* (centroid) in 14th century Europe.

The ***T-CGMC*** is defined in the $R^{n,1}$ domain as the time-averaged $\bar{\mathbf{s}}_t$ within a period T, i.e.,

$$\bar{\bar{\mathbf{s}}}_t = \left(\bar{\bar{s}}_1, \ldots, \bar{\bar{s}}_n\right) = \frac{1}{T}\int_T dt\, \bar{\mathbf{s}}_t \qquad (2.12)$$

over all $t \in T$, where the double bar implies double averaging (one over space and then over the considered time period). The *T-CGMC* belongs to the group of Eq. (2.1b).

The $\bar{\bar{\mathbf{s}}}_{Em,t}$ and $\bar{\bar{\mathbf{s}}}_{Mm,t}$ may also be averaged over the time period T in a manner similar to that of Eq. (2.12) leading to the corresponding ***T-CGEMC*** and ***T-CGMMC*** measures.

While the $\bar{\mathbf{s}}_t$ is useful for summarizing spatial shifts in the point distribution over time, the corresponding $\bar{\bar{\mathbf{s}}}_t$, $\bar{\bar{\mathbf{s}}}_{Em,t}$ and $\bar{\bar{\mathbf{s}}}_{Mm,t}$ define the balancing points of these shifts.

Example 2.7 The *T-CGEMC* may be useful as an initial choice for the location of health facilities to serve a population in the case of an epidemic (it could also be updated by accounting for other factors like transportation and communication networks).

2.3.3 Composite spacetime center measure

As mentioned earlier, depending on the objectives of the study, a composite spacetime definition of the point distribution center may be appropriate.

In the R^{n+1} domain, a **composite *CGMC*** measure can be computed as

$$\overline{p} = (\overline{s}_1, \ldots, \overline{s}_n, \overline{s}_0) = \left(\frac{1}{m}\mathbf{1}^T s_1, \ \ldots, \frac{1}{m}\mathbf{1}^T s_n, \frac{1}{m}\mathbf{1}^T s_0\right). \quad (2.13)$$

A visualization of Eq. (2.13) in the R^{2+1} domain is shown in Fig. 2.7, where the composite *CGMC* provides a unique spacetime point $\overline{p}$ that is representative of the entire domain (i.e., it does not vary with t). The *CGMC* measure of Eq. (2.13) belongs to the group of Eq. (2.1c), with $g(s_k) = \frac{1}{m}\sum_{i=1}^{m} s_{i,k}$, $k = 1, \ldots, n, 0$.

Yet, the *CGMC* expression of Eq. (2.2) is usually preferred to that of Eq. (2.13) because it allows the consideration of other useful notions, like the *CGMC* worldline or path and the *CGMC* velocity.

2.4 Point dispersion

The above measures of central distribution tendency suggest *where* the chrono-geographic center of the point distribution would be. An investigator, however, wants to know something about *how* the data points are distributed around the distribution's center. In this case, one needs a measure similar to the sample standard deviation considered in *CS*, but in the chrono-geographic context. These measures are defined as follows.

2.4.1 Separate space-time dispersion measures

The space-time dispersion measures may come in different forms, depending on the physical characteristics of the real-world phenomenon. The following dispersion measures distinguish between anisotropic and isotropic variation.

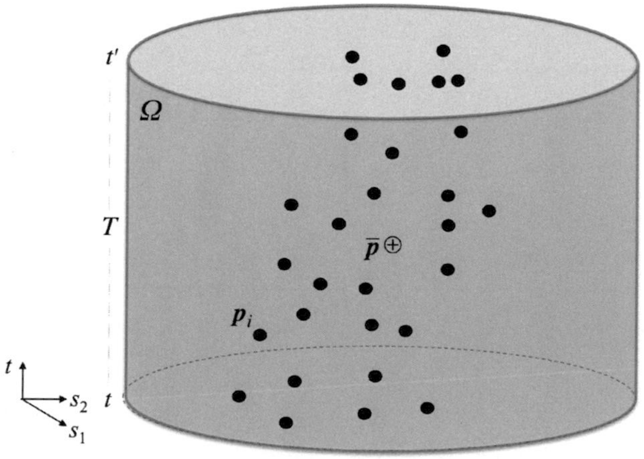

FIG. 2.7 The composite *CGMC* $\overline{p}$ (⊕) of the point distribution in R^{2+1}.

The chrono-geographic **standard distance** (***CGSD***)[f] of a point distribution in the $R^{2,1}$ domain is a measure of the space-time data point dispersion around the *CGMC* at time *t*, and is defined as

$$\delta\bar{\boldsymbol{h}}_t = (\delta\bar{\boldsymbol{h}}; t) = (\delta\bar{h}_1, \delta\bar{h}_2; t) = \left(\sqrt{\frac{1}{m}d_1^2}, \sqrt{\frac{1}{m}d_2^2}; t\right), \quad (2.14)$$

where $d_k^2 = \sum_{i=1}^{m}(s_{i,k} - \bar{s}_k)^2$, $k = 1, 2$, is the sum of the squared coordinate deviations from *CGMC* at time *t*, and the **dispersion area** of the point distribution is denoted as $A_t = A(\bar{\boldsymbol{s}}_t, \delta\bar{\boldsymbol{h}}_t)$, i.e., it is an ellipse with center $\bar{\boldsymbol{s}}_t$ and axes determined by $\delta\bar{\boldsymbol{h}}_t$.

The *CGSD* $\delta\bar{\boldsymbol{h}}_t$ of Eq. (2.14) is expressed as "the distance vector $\delta\bar{\boldsymbol{h}} = (\delta\bar{h}_1, \delta\bar{h}_2)$, and time scalar *t*," and is used in the *anisotropic* case (i.e., when directional trends are present). It is a point distribution diffusion measure that belongs to the group described by Eq. (2.1a) with $g(s_k) = \sqrt{\frac{1}{m}\sum_{i=1}^{m}(s_{i,k} - \bar{s}_k)^2}$, $k = 1, 2$. The dispersion area A_t represents the extent of spatial data diffusion or scattering at time *t* of the point distribution.

A special case of Eq. (2.14) emerges when $\delta\bar{h}_1 = \delta\bar{h}_2 = \delta\bar{h}$, i.e., the *isotropic* case when the ellipse reduces to a *circle*, in which case Eq. (2.14) reduces to the following.

The ***CGSD*** of a point distribution at time *t* in the $R^{2,1}$ domain when no directional trends in the point distribution exist is given by

$$\delta\bar{h}_t = (\delta\bar{h}; t) = \left(\sqrt{\frac{1}{m}d^2}; t\right) \quad (2.15)$$

where d^2 is the sum of the squared coordinate deviations from *CGMC* at time *t*,[g] and $\delta\bar{h}$ is the radius of the circle that outlines the time-dependent **dispersion area** $A_t = A(\bar{\boldsymbol{s}}_t, \delta\bar{h})$ of the point distribution centered at $\bar{\boldsymbol{s}}_t$.

Visual representations of the $\delta\bar{h}$ of Eq. (2.15) and the associated A_t at various times during the period *T* (from *t* to *t'*) are shown in Fig. 2.8 in the context of a cylindrical $D \times T$ domain representation. The worldline $WL_{\bar{\boldsymbol{s}}_t}$ of the *CGMC* $\bar{\boldsymbol{s}}_t$ is also shown.

In the following, unless stated otherwise the isotropic *CGSD* will be considered. Interpretatively, the isotropic *CGSD* is the average distance the points spread from the *CGMC* and measures the *compactness* of point distribution, i.e., the degree to which the points are concentrated or dispersed around their *CGMC*. In other words, the $\delta\bar{\boldsymbol{h}}_t$'s value is that it measures the amount of absolute dispersion in a point pattern at a given time *t*. A large isotropic *CGSD* value means that the points are relatively scattered, whereas a small value means that they are relatively clustered.

Example 2.8 Among other things, the *CGSD* can be used to assess which species occupies the wider territory in time, to compare changes over time such as the weekly dispersion of auto thefts, or what is the proportion of the domain points where pollution exceeds certain thresholds.

The scalar quantity $\delta\bar{h}$ in Eq. (2.15) considered at time *t* is different than the magnitude $|\bar{\boldsymbol{h}}_{tt'}|$ of the vector $\bar{\boldsymbol{h}}_{tt'}$ of Eq. (2.7) considered between two distinct times *t* and *t'*.[h] Generally, if

[f] Also known as chrono-geographic *standard distance deviation*.

[g] See Eq. (2.3), $d = \sqrt{\sum_{i=1}^{m}(s_{i,1} - \bar{s}_1)^2 + \sum_{i=1}^{m}(s_{i,2} - \bar{s}_2)^2} = \sqrt{m[g^2(s_1) + g^2(s_2)]}$.

[h] I.e., $|\bar{\boldsymbol{h}}| = \sqrt{\left(\frac{1}{m}\sum_{i=1}^{m}s'_{i,1} - \bar{s}_1\right)^2 + \left(\frac{1}{m}\sum_{i=1}^{m}s'_{i,2} - \bar{s}_2\right)^2}$.

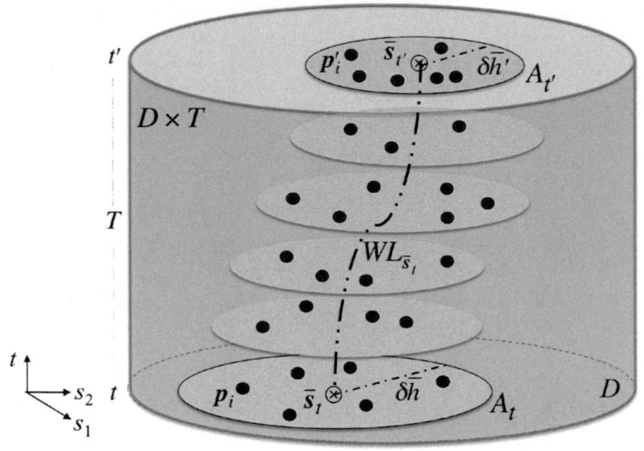

FIG. 2.8 An $R^{2,1}$ visualization of the *CGMC* $\bar{\mathbf{s}}_t$, its $WL_{\bar{\mathbf{s}}_t}$, the *CGSD* $\delta\bar{h}$, and the time-dependent dispersion areas A_t of point distribution during the time period from t to t'; $\boldsymbol{p}_i$ are the points considered at each time.

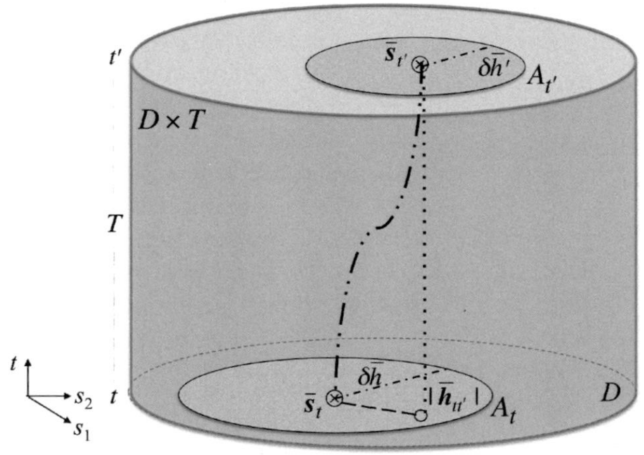

FIG. 2.9 A visualization of the difference between the geographic distance magnitude $|\bar{\boldsymbol{h}}_{tt'}|$ between the *CGMC* $\bar{\mathbf{s}}_t$ and $\bar{\mathbf{s}}_{t'}$, and the composite isotropic *CGSD* $\delta\bar{h}$ of the data point distribution in $R^{2,1}$.

$|\bar{\boldsymbol{h}}_{tt'}| < \delta\bar{h}$, the $\boldsymbol{s}$ at time t' falls within the A_t, where if $|\bar{\boldsymbol{h}}_{tt'}| > \delta\bar{h}$, it does not. The greater the difference $|\bar{\boldsymbol{h}}_{tt'}| - \delta\bar{h}$ the further away the dispersion $A_{t'}$ of the point pattern at time t' is from the dispersion A_t of the point patterns at time t.

Example 2.9 For illustration, Fig. 2.9 gives a geometrical visualization of the difference between $|\bar{\boldsymbol{h}}_{tt'}|$ and $\delta\bar{h}$. In geometric terms, the vector $\bar{\boldsymbol{h}}_{tt'} = \bar{\mathbf{s}}_t' - \bar{\mathbf{s}}_t = (\bar{s}_{1,t'} - \bar{s}_{1,t}, \bar{s}_{2,t'} - \bar{s}_{2,t})$ is the projection on the point distribution plane at time t of the chrono-geographic vector from the *CGMC* $\bar{\mathbf{s}}_t$ to the *CGMC* $\bar{\mathbf{s}}_{t'}$. The scalar $\delta\bar{h}$, on the other hand, is the radius of the dispersion circle surrounding the data.

Just as the *CGMC* $\bar{\mathbf{s}}_t$ of Eq. (2.2) serves as a chrono-geographic analog to the *CS* mean (that measures data central tendency), the *CGSD* $\delta\bar{\boldsymbol{h}}_t$ is the chrono-geographic equivalent of the *CS* standard deviation (which measures data spread) that provides a single summary measure of a point spread around their center (similar to the way the *CS* standard deviation measures the distribution of data values around the statistical mean).

Example 2.10 As we saw above, together, the $\bar{s}_t$ and the $\delta\bar{h}_t$ define the **dispersion area** A_t at time t with center $\bar{s}_t$ and radius $\delta\bar{h}_t$. If the point distribution obeys a Gaussian law (i.e., the point distribution of the input features is concentrated in the *CGMC* with fewer points towards the periphery), then 68.27%, 95.45%, and 99.73% of the points fall within a distance of $\delta\bar{h}$, $2\delta\bar{h}$ and $3\delta\bar{h}$, respectively, from the *CGMC*.

Like the standard deviation of *CS*, *pro rata*, the *CGSD* is strongly influenced by extreme or peripheral points. Because distances about the *CGMC* are squared, "uncentered" or atypical points may have a dominating impact on the magnitude of the *CGSD*.

The following example illustrates a step-by-step *CGSD* **computational** procedure in practice.

Example 2.5(cont.) Step 1 is the same as in Example 2.5 above.

Step 2: For the *FT*, the *CGSD* at temporal instant 0 is calculated as

$$\delta\bar{\boldsymbol{h}}_{t,FT} = (\delta\bar{h}_{FT}, t) = \left(\sqrt{\delta\bar{h}_{1,FT} + \delta\bar{h}_{2,FT}}, 0\right)$$
$$= (1.0306, 0),$$

where

$$\delta\bar{h}_{1,FT} = \frac{1}{12}\sum\nolimits_{i=1}^{12}(s_{i,1} - \bar{s}_{A,1})^2$$
$$= \frac{1}{12}\left[(1.9232 - 2.9013)^2 + \ldots + (1.7170 - 2.9013)^2\right]$$
$$= 0.4043,$$

$$\delta\bar{h}_{2FT} = \frac{1}{12}\sum\nolimits_{i=1}^{12}(s_{i,2} - \bar{s}_{A,2})^2$$
$$= \frac{1}{12}\left[(0.9962 - 1.6855)^2 + \ldots + (2.3274 - 1.6855)^2\right]$$
$$= 0.6578.$$

Similarly, for the *BC* we find that

$$\delta\bar{\boldsymbol{h}}_{t,BC} = (0.4158, 0).$$

Step 3: The center and radius of the blue dashed circle in Fig. 2.2B represent, respectively, the *CGMC* and the *CGSD* of the recorded *FT* disease at time 0. Similarly, the center and radius of the red dashed circle in Fig. 2.2B represent, respectively, the mean center and the standard distance of the recorded *BC* disease.

Step 4: Furthermore, the *CGSD* of the two diseases at the remaining times were calculated as shown in Table 2.3; the corresponding plots are shown in Fig. 2.2C–F.

As noted earlier, the isotropic *CGSD* $\delta\bar{\boldsymbol{h}}_t$ is a dispersion measure usually visualized as a circle with a center at the *CGMC* radius equal to the *CGSD*. We remind the readers that the $\delta\bar{\boldsymbol{h}}_t$ in the form of Eq. (2.15) works optimally in the case of distributional isotropy (i.e., when no strong directional trends are present in the point pattern). In the case when directional trends exist a better approach may be to use the anisotropic *CGSD* of Eq. (2.14). Other forms of *CGSD* may be considered. Here are some possibilities. Instead of d^2 in Eq. (2.14), other geographic metrics forms $\boldsymbol{h}^2$ may be considered, see a list of metrics in Table 3.1 of Section 3 of Chapter 2. Moreover, in applications that require a weighted measure, the *CGSD* of Eq. (2.15) can be modified as follows.

The **weighted CGSD** in $R^{2,1}$ is given by the time-dependent function

$$\delta\bar{\boldsymbol{h}}_{t,w} = (\delta\bar{h}_w; t)$$
$$= \left(\sqrt{\frac{1}{W}\sum\nolimits_{k=1}^{n}\sum\nolimits_{i=1}^{m} w_i(s_{i,k} - \bar{s}_k)^2}; t\right), \tag{2.16}$$

where w_i denotes the weight of point $\boldsymbol{p}_i = (s_{i,1}, \ldots, s_{i,n}, t)$ that may assume different forms and interpretations and are such that $W = \sum_{i=1}^{m} w_i$.

The dispersion area $A(\bar{s}_t, \delta\bar{h})$ at each time t together with the time period T of interest outline the chrono-geographic domain of a distribution of points $\boldsymbol{p}$, i.e., this is the dispersion area at time t with center $\bar{s}_t$ and radius $\delta\bar{h}$.

The **worldtube** WT_p of a data point distribution is defined as the union of the corresponding time-dependent dispersion areas, i.e.,

$$WT_p = \cup_{t \in T} A(\bar{s}_t, \delta\bar{h}). \tag{2.17}$$

The WT_p is visually represented as the chrono-geographic domain within which the points $\boldsymbol{p}$ are scattered.

Example 2.11 Fig 2.10A presents a worldtube in $R^{2,1}$ consisting of circular dispersion areas of varying radius $\delta\bar{h}$ over the time period T, i.e., the spatial location of the *CGMC* varies with time. Fig 2.10B presents a trivial case consisting of a worldtube that is a cylinder with a constant radius $\delta\bar{h}$ and height T. The spatial location of the *CGMC* does not change with time (i.e., it is a vertical line) and the chrono-geographic extend of the worldtube is equal to $2\pi\delta\bar{h}T$.

2.4.2 Time-averaged CGSD

One may also consider the *CGSD* of a point distribution at different times, and then focus on its time average, as follows:

The ***T-CGSD*** is defined as the temporally averaged $\delta\bar{h}_t$ over a period T, i.e.,

$$\delta\bar{\bar{h}}_t = \frac{1}{T}\int_T dt\, \delta\bar{h}_t, \tag{2.18}$$

where $t \in T$.

Apparently, the *T-CGSD* belongs to the group of measures of Eq. (2.1b). The physical meaning of $\delta\bar{h}_t = (\delta\bar{h}; t)$ is that it is a time-dependent function that summarizes the diffusion changes of the time-dependent spatial shifts in the point distribution at each given time, whereas the $\delta\bar{\bar{h}}_t$ is a unique time-averaged value of these changes.

2.4.3 Composite spacetime dispersion measure

As with other point distribution measures, a composite spacetime definition of the point distribution standard distance may be appropriate, depending on the objectives of the study.

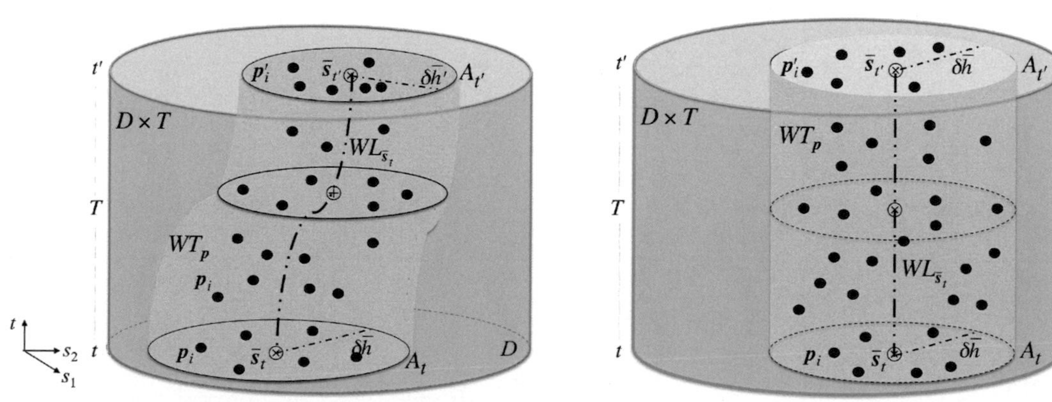

FIG. 2.10 (A) A general WT_p with a curved worldline $WL_{\bar{s}_t}$ and time-dependent dispersion area $A(\bar{s}_t, \delta\bar{h})$. (B) A special case of a cylindrical WT_p with a vertical worldline and time-independent dispersion area.

The **composite *CGSD*** in R^{n+1} is given by

$$\delta\overline{p} = |\Delta\overline{p}_i| = \sqrt{\frac{1}{m}\sum\nolimits_{i=1}^{m}\Delta p_i^2}, \tag{2.19}$$

where $\Delta p_i^2 = \sum\nolimits_{k=1}^{n,0}\varepsilon_{ik}(s_{i,k} - \bar{s}_k)^2$, and the coefficients $\varepsilon_{i,k}$ ($k = 1, \ldots, n, 0$) are determined empirically.[i]

In Eq. (2.19), the $|\Delta\overline{p}_i|$ is of the uni-metric form of Eq. (2.3b), Section 2 of Chapter 2. Related to the composite *CGMC* of Eq. (2.13) and the *CGSD* of Eq. (2.19) is the following spacetime notion.

The **dispersion sphere** is described by

$$S_{\overline{p}} = S(\overline{p}, \delta\overline{p}) \tag{2.20}$$

i.e., its center is the composite *CGMC* $\overline{p}$ and its radius the composite *CGSD* $|\Delta\overline{p}_i|$.

Visualization of $\delta\overline{p}$ and $S_{\overline{p}}$ is shown in Fig. 2.11. Many other chronotopologic metric forms Δp_i^2 may be considered, see the list of metrics in Table 3.3 of Section 3 of Chapter 2.

2.5 Normalized point distribution diffusion

The coefficient of variation (*COV*, i.e., standard deviation divided by the mean) is the *CS* measure of relative dispersion. Certain extensions of this notion in the chrono-geographic domain are considered next.

2.5.1 Separate space-time relative dispersion measures

The corresponding relative dispersion measures in the separate space-time domain $R^{n,1}$ are defined as follows:

The **chrono-geographic relative distance** (***CGRD***) of a point pattern at time *t* is defined as the ratio of the corresponding anisotropic *CGSD* over a reference measure of regional magnitude at time *t*, i.e.,

$$\delta\overline{h}_R = \left(\frac{\delta\overline{h}_1}{\rho_1}, \frac{\delta\overline{h}_2}{\rho_2}; t\right), \tag{2.21}$$

where ρ_k ($k = 1, 2$) are the axes of an ellipse with the same area as the region of interest at time *t*. In the case of an isotropic *CGSD*, the *CGRD* is defined as

$$\delta\overline{h}_R = \left(\frac{\delta\overline{h}}{R}; t\right). \tag{2.22}$$

where *R* is the radius of a circle with the same area as the region of interest at time *t*.

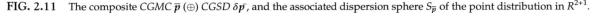

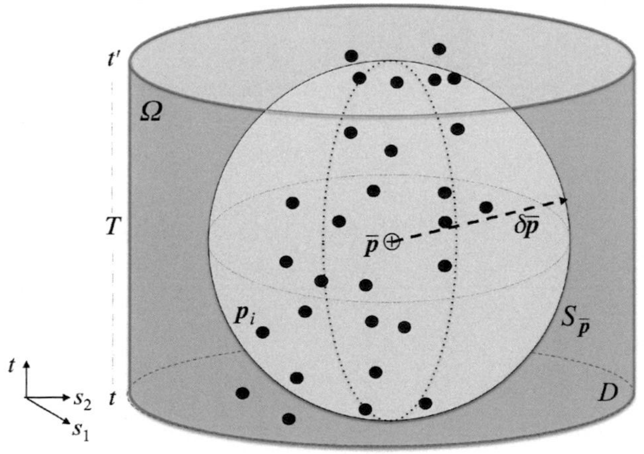

FIG. 2.11 The composite *CGMC* $\overline{p}$ ($\oplus$) *CGSD* $\delta\overline{p}$, and the associated dispersion sphere $S_{\overline{p}}$ of the point distribution in R^{2+1}.

[i] In several applications, $\varepsilon_{i,1} = \varepsilon_1$, $\varepsilon_{i,2} = \varepsilon_2$ and $\varepsilon_{i,0} = \varepsilon_0$.

The *CGRD* is a descriptive measure of relative dispersion that allows direct comparison of the dispersions of distinct point patterns at different regions and at the same time t, or at different times t and t'. Otherwise said, the *CGRD* of Eq. (2.22) is a chrono-geographic measure belonging to the group defined by Eq. (2.1a), with $g(s_k) = \frac{1}{\rho_k}\sqrt{\frac{1}{m}\sum_{i=1}^{m}(s_{i,k} - \bar{s}_k)^2}$ $k = 1, 2$. This *CGRD* allows direct comparison of the dispersion of different point patterns from different areas (even if the areas are of varying sizes) at a specified time t or at different times t and t'.

Example 2.12 For comparison purposes, Fig. 2.12 displays a certain data point distribution and the associated study areas at different times. Regions A, B, and C have the same *CGSD* (same amount of absolute dispersion), but the area increases from region A to C, i.e., the radius R of the region increases from region A to C. As a consequence, in the small region A the point distribution has high *CGRD* (high degree of relative dispersion), but the larger regions B and C have low *CGRD*. On the other hand, regions D and E have the same *CGRD*, but the point distribution in D has a larger *CGSD* than the point distribution in E, because of its larger size.

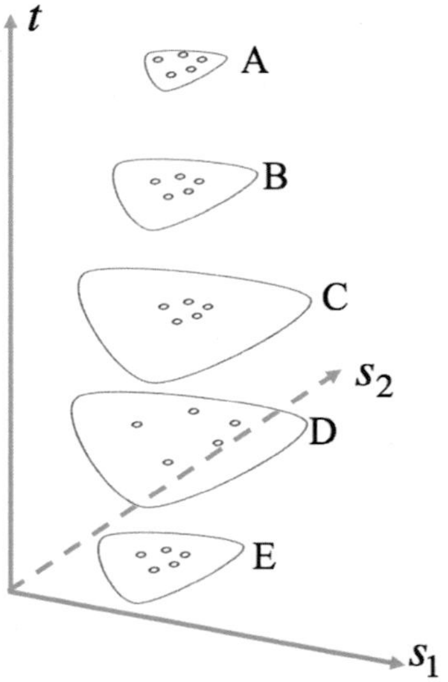

FIG. 2.12 Five arrangements of the data point distributions (red circle) and the study area (black boundary) at a specified time t.

The following example presents the *step-by-step* **computational** process of *CGRD* in practice.

Example 2.5(cont.) Let us revisit Example 2.5, as follows:

Steps 1–2: The same as in Example 2.5(cont) above.

Step 3: Assume that the area of the study region at time $t = 0$ is 9 units^2; then, the radius at time $t = 0$, i.e., R_0, can be calculated as $\pi R_0^2 = 9$, which gives $R_0 = 1.6926$ units.

Step 4: The *CGRD* of *FT* and *BC* at $t = 0$ are calculated as

$$\delta\bar{h}_{R,FT} = \left(\frac{1.0306}{1.6926}, 0\right) = (0.6089, 0), \text{and}$$

$$\delta\bar{h}_{R,BC} = \left(\frac{0.4158}{1.6926}, 0\right) = (0.2457, 0),$$

respectively. Similarly, the *CGRD* of the two diseases at various temporal instants can be calculated as in Table 2.3. The conclusion is drawn that the *FT* dispersion is considerably larger than that of the *BC* disease.

2.5.2 Time-averaged CGRD

As usual, one may obtain a single number measuring the relative dispersion of a point distribution for the entire chrono-geography of interest, as follows:

The ***T-CGRD*** is defined as the time-averaged $\delta\bar{h}_R$ within a period T, i.e.,

$$\delta\bar{h}_{R,T} = \frac{1}{T}\int_T dt\,\delta\bar{h}_R \qquad (2.23)$$

for all $t \in T$, i.e., the *T-CGRD* belongs to the group of measures of Eq. (2.1b).

The *T-CGRD* measures the averaged relative dispersion of a point distribution throughout the entire chrono-geographic domain considered. As such, it makes possible the direct comparison of the overall dispersions of different point patterns over the same period T, or during different periods T and T'.

2.5.3 Composite spacetime relative dispersion measure

As with other point distribution measures, a composite spacetime definition of the point distribution standard distance may be appropriate, depending on the objectives of the study.

The **composite** ***CGRD*** in R^{n+1} is given by

$$\delta\overline{p}_R = \frac{\delta\overline{p}}{\Delta\rho}, \tag{2.24}$$

where $\Delta\rho$ is the radius of a spacetime sphere with the same size as the spacetime domain of interest.

As in previous cases, the investigators may also assume $\Delta\rho$ that have an appropriate spacetime metric form (see, e.g., the spacetime bi-metrics in Table 3.3, Section 3 of Chapter 2).

Concluding, the *CGS* techniques discussed in this section seek to develop macro-level (overall point distribution arrangements) descriptions, interpretations, and explanations, starting from micro-level (individual points) information processing.

3 CGS of chrono-geographic attribute values

In the previous section, we focused on the chrono-geographic statistics of *point sets* distributed within the domain of interest. Here, we study chrono-geographic statistics of *attribute values* (such as temperature, air pollution, disease incidence or mortality, household income, and poverty level) distributed within the same domain. Because these chrono-geographic statistics are concerned with data values they may be characterized as physical statistics.

3.1 Assessing attribute correlation

Being able to transform data into useful correlation measures for *CGS* purposes takes particular knowledge and insight. The first measure of this kind to be considered assesses data value similarity within a chrono-geographic domain $\Omega \subseteq R^{n,1}$. In the following definitions, Ω is divided into m units U_i so that $\cup_{i=1}^{m} U_i \subset \Omega$. Let $\chi_i = \chi(\boldsymbol{p}_i)$ be attribute values at points $\boldsymbol{p}_i = (\boldsymbol{s}_i, t_i) \in U_i$, which are the centroids of the units U_i ($i = 1, \ldots, m$), and let

$$\overline{X} = \frac{1}{m}\sum\nolimits_{i=1}^{m} \chi_i,$$

$$\sigma_X^2 = \frac{1}{m}\sum\nolimits_{i=1}^{m} \left[\chi_i - \overline{X}\right]^2$$

be the corresponding data mean and variance. Then, $\widetilde{\chi}(\boldsymbol{p}_i) = \chi(\boldsymbol{p}_i) - \overline{X}$ conveniently denotes attribute fluctuations ($i = 1, \ldots, m$). Furthermore, the *connectivity*, *contiguity*, or *weight* matrix is given by

$$\mathbf{W} = \{w_{ij};\ i,j = 1, \ldots, m\} = \begin{bmatrix} w_{11} \cdots w_{1m} \\ \cdots \\ w_{m1} \cdots w_{mm} \end{bmatrix},$$

with the connectivity weights w_{ij} assigned to pairs of points $\boldsymbol{p}_i$ and $\boldsymbol{p}_j$ depending on whether the corresponding units U_i and U_j are chronotopologic neighbors or not, so that[j]

$$W = sum\mathbf{W} = \mathbf{1}^T\mathbf{W}\mathbf{1} = \sum\nolimits_{i=1}^{m}\sum\nolimits_{j=1}^{m} w_{ij}.$$

[j] More technical details on the meaning of the term "chronotopologic neighbors" will be given a little later.

Apparently, the w_{ij} are seen as a measure of contiguity between units. A typical case is to assume that $w_{ij}=1$ if the units U_i and U_j are neighbors ($i \neq j$), and $w_{ij}=0$ otherwise (this includes $w_{ii}=0$), but this is, obviously, not the only possibility.

The attribute correlation measures may be divided into two groups: the *global* measures that consider all chronotopologic lags between pairs of points in the domain of interest and provide an average correlation value of the entire domain, and the *local* measures that are functions of the chronotopologic lags (i.e., they distinguish between lags) and provide lag varying correlation values.

3.1.1 Spacetime global correlation measure

A global measure of correlation across space and time that is commonly used by *CGS* is the following.

The **chrono-geographic Moran's Index** (*CGMI*) is a composite spacetime global measure of attribute correlation within Ω expressed in terms of attribute deviations from the mean at the point pairs $\boldsymbol{p}_i, \boldsymbol{p}_j \in \Omega$, i.e.,

$$c_{\Omega}=\frac{1}{W\sigma_X^2}\widetilde{\boldsymbol{\chi}}_p^T \mathbf{W} \widetilde{\boldsymbol{\chi}}_p, \tag{3.1}$$

where $\widetilde{\boldsymbol{\chi}}_p^T=[\widetilde{\chi}(\boldsymbol{p}_1)\ldots\widetilde{\chi}(\boldsymbol{p}_m)]$.

Eq. (3.1) measures the weighted correlation between χ_i and χ_j, assuming attribute homostationarity.[k] An interesting case of a *CGMI* is examined next.

Example 3.1 The following notation emphasizing time dynamics can be considered in the discrete space-time domain Ω, with data values denoted as $\chi_{i,t}$ at $\boldsymbol{p}_i=(\boldsymbol{s}_i, t_i=t)$ and $\chi_{j,t-\tau}$ at $\boldsymbol{p}_j=(\boldsymbol{s}_j, t_j=t_i-\tau)$. A constant attribute mean $\overline{X}$ is assumed, and $\widetilde{\chi}_{i,t}=\chi_{i,t}-\overline{X}$ and $\widetilde{\chi}_{j,t-\tau}=\chi_{j,t-\tau}-\overline{X}$ are attribute fluctuations. Then, the *CGMI* can be written as (Griffith, 1981)

$$c_{\Omega}=\frac{m(T-\tau)}{W}\frac{\sum_{t=\tau+1}^{T}\sum_{i=1}^{m}\sum_{j=1}^{m} w_{ij,t-\tau}\widetilde{\chi}_{i,t}\widetilde{\chi}_{j,t-\tau}}{\sum_{t=1}^{T}\sum_{i=1}^{m}\widetilde{\chi}_{i,t}^2}, \tag{3.2}$$

where m is the number of locations, T is the time period considered, $\overline{X}=\frac{1}{mT}\sum_{t=1}^{T}\sum_{i=1}^{m}\chi_{i,t}$ is the mean, $w_{ij}=1$ if the $\boldsymbol{s}_i$ and $\boldsymbol{s}_j$ are neighboring locations during the specified time, and $w_{ij}=0$ otherwise, and $W=\sum_{t=1+\tau}^{T}\sum_{i=1}^{m}\sum_{j=1}^{m} w_{ij,t-\tau}$ is the total weight summation in space-time.

3.1.2 Space-time global correlation measures

Next, we focus on a separate space-time *CGMI* formulation motivated by the classical *S-TDA* conception that time is distinguished from space, which implies that the attribute distribution is considered at separate time instances. As Moran's mainstream notion of neighborhood was originally conceived in a purely geographic setting, of special practical interest is a dynamic definition of *CGMI* at specified times t that constrain simultaneous attribute values (i.e., the attribute representation consists of a series of temporal states). For this purpose, at time $t \in T$ (T is the time axis) the spatial region $S \subseteq R^n$ is considered so that $\Omega = S \times T$. The $\chi_{i,t}=\chi(\boldsymbol{s}_i,t)$ and $\chi_{j,t}=(\boldsymbol{s}_j,t)$ are data values at locations $\boldsymbol{s}_i \in S$ and $\boldsymbol{s}_j \in S$ and at the same time $t_i=t_j=t$, m_t is the data number at time t, and $\overline{X}_t=\frac{1}{m_t}\sum_{i=1}^{m_t}\chi_{i,t}$ and $\sigma_{X_t}^2=\frac{1}{m_t}\sum_{i=1}^{m_t}\widetilde{\chi}_{i,t}^2$ are the data mean[l] and variance, respectively. Under these conditions, the following correlation measure is used.

[k] I.e., an attribute that is spatially homogeneous and temporally stationary (for a formal definition, see Section 3 of Chapter 5).

[l] Assuming ergodicity, i.e., the ensemble mean is equal to the sample mean $\overline{X}_t$.

The **time-specific** ***CGMI*** is a separate space-time measure of attribute correlation within a global region S at time t expressed in terms of deviations from the attribute mean at all locations s_i, $s_j \in S$ at time t, which is computed by

$$c_{S,t} = \frac{1}{W_t \sigma_{X_t}^2} \tilde{\chi}_{s,t}^{\mathrm{T}} \mathbf{W}_t \tilde{\chi}_{s,t}, \tag{3.3}$$

where $\tilde{\chi}_{s,t}^{\mathrm{T}} = [\tilde{\chi}(s_1,t) \dots \tilde{\chi}(s_m,t)]$, m is the "number of data locations at time t" (sometimes concisely denoted as m_t), and the elements $w_{ij,t}$ of matrix $\mathbf{W}_t$ indicate whether locations s_i and s_j are spatial neighbors or not at time t such that $W_t = sum\ \mathbf{W}_t = \mathbf{1}^T\mathbf{W}_t\mathbf{1}$.[m]

Eq. (3.3) introduces a straightforward way of space-time correlation analysis by developing time *snapshots* of the spatial data distribution, computing the *CGMI* value for each time frame, and then presenting the results by means of an animation technique or spatial map multiples (presenting a series of maps, one for each time instance).

Example 3.2 Eq. (3.3) is essentially a special case of Eq. (3.2). Indeed, Eq. (3.3) is obtained from Eq. (3.2) by letting $\tau = 0$, $m_t = m$, and the summation $\Sigma_{t=\tau+1}^{T}$ being reduced to a single time instance t (conversely, Eq. (3.2) is a generalization of Eq. (3.3) by considering several time instances).

As regards the key *CGMI* values, just as in the case of the *CS* correlation measure, the normalized *CGMI* values too range between −1 and +1.

- A positive *CGMI* sign represents positive data correlation, a negative sign denotes negative correlation, and the zero *CGMI* value means no correlation (data that tend to be randomly scattered).

There have been proposed certain alternatives to *CGMI* over the years. Perhaps, the most noticeable is the one discussed next.

3.1.3 Spacetime local correlation measure

The *CGMI* of Eq. (3.1), and the special case of Eq. (3.2), provide a general expression between data that accounts for all pairs of chronotopologic points in the domain of interest, which link attribute values in an integrated manner that is simultaneously based on both spatial and temporal proximity. Eq. (3.1) can be extended to yield a composite spacetime correlation measure that is a function of the chronotopologic lags. Some of these possibilities are briefly discussed next.

The **local** *CGMI* is a rather straightforward extension of the composite *CGMI* of Eq. (3.1) defined as

$$c_{\Delta p} = \frac{1}{W_{\Delta p} \sigma_X^2} \tilde{\chi}_p^{\mathrm{T}} \mathbf{W}_{\Delta p} \tilde{\chi}_p, \tag{3.4}$$

where the connectivity matrix $\mathbf{W}_{\Delta p}$ includes only a limited number of weights associated with attribute points within the chronotopologic lag $\boldsymbol{\Delta p}$.

3.1.4 Space-time local correlation measure

Separate space-time local versions of Eq. (3.3) can be found in the literature, such as the following (e.g., a space-time extension of results by Anselin, 1995).

The **local** *CGMI* for an attribute value at location s_i and time t is given by

$$c_{i,t} = \tilde{\chi}_{i,t} w_{i,t}^{\mathrm{T}} \tilde{\chi}_{s,t}, \tag{3.5}$$

where now $w_{i,t}^{\mathrm{T}} = [w_{i1,t} \dots w_{im,t}]$.

The local *CGMI* above reflects the level of clustering of similar values around s_i and time t within a specified sub-domain of Ω.

[m] The notation in Eq. (3.3) is further clarified in the numerical Example 3.5 below.

- In light of the above expressions, it is valid that

$$c_{S,t} = \frac{1}{W_t \sigma_{X_t}^2} \sum\nolimits_{i=1}^{m_t} c_{i,t}, \tag{3.6}$$

i.e., the sum of the local *CGMI* overall points considered is proportional to the global *CGMI*, with a scale factor $\lambda_{c,t} = W_t \sigma_{X_t}{}^2$.

Otherwise said, Eq. (3.6) decomposes the global *CGMI* into a summation of contributions from each datum.

In practice, several useful modifications may apply. One of the best known modifications is as follows: for each point $\boldsymbol{p}_i$ the weights w_{ij} are replaced by $\omega_{ij} = \frac{w_{ij}}{\sum_{j=1}^{q} w_{ij}}$, i.e., the weights w_{ij} are normalized by the local summation of the weights of all points $\boldsymbol{p}_j$ ($j = 1, \ldots, q$) that are neighbors of $\boldsymbol{p}_i$. Note that, generally, $\omega_{ij} \neq \omega_{ji}$.

Example 3.3 Assume that at time t the arrangement of points in Fig. 3.1 occurs. In this case, for point $\boldsymbol{s}_1$ the local ω_{1j}-weights ($j = 2, 4, 5$) are $\omega_{12} = \frac{w_{12}}{\sum_{j=2}^{4} w_{1j}} = \frac{1}{1+1} = \frac{1}{2}$, $\omega_{14} = \frac{w_{14}}{\sum_{j=2}^{4} w_{1j}} = \frac{1}{1+1} = \frac{1}{2}$, but $\omega_{15} = 0$. Note that $\omega_{21} = \frac{w_{21}}{\sum_{j=1}^{3,5} w_{2j}} = \frac{1}{1+1+1} = \frac{1}{3} \neq \omega_{12}$, $\omega_{41} = \frac{w_{41}}{\sum_{j=1}^{5,7} w_{4j}} = \frac{1}{1+1+1} = \frac{1}{3}$, but $\omega_{51} = 0 = \omega_{15}$.

The statistic represented by the local *CGMI* is similar to the notion of the *spatiotemporal covariance* (or *correlation*) of geostatistics (see, e.g., Section 2.3 of Chapter 5).[n] We do not elaborate further on the geostatistical covariance here, but we refer the interested readers to the relevant theory and applications presented in Chapter 5.

3.2 Assessing attribute decorrelation

Unlike *CGMI*, the next physical statistic globally measures data dissimilarity. Yet, like correlation measures, the decorrelation measures may be divided into two groups: the *global* decorrelation measures and the *local* decorrelation measures.

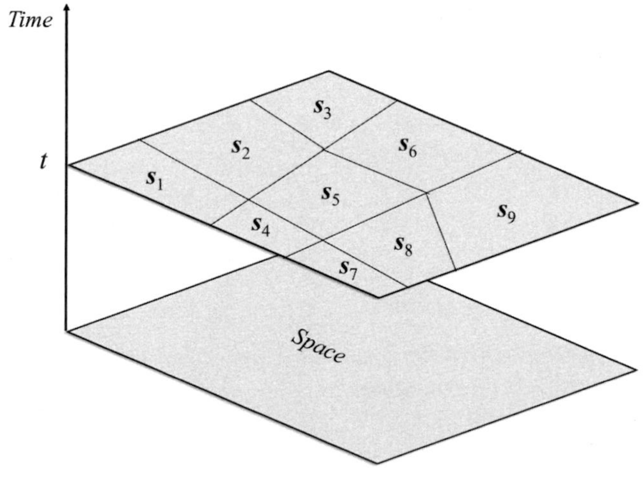

FIG. 3.1 Calculation of ω weights at time t.

[n] Although some key differences exist, e.g., the *CGMI* considers all chronotopologic lags between pairs of points, whereas the geostatistical covariance is a function of the lags (i.e., it distinguishes between lags).

3.2.1 Spacetime global decorrelation measure

In particular, the following decorrelation alternative to the global *CGMI* measure of Eq. (3.2) has been suggested that is expressed in terms of the squared differences of attribute values at all points $p_i = (s_i,t_i)$, $p_j = (s_j,t_j) \in \Omega$.

The **chrono-geographic Geary's Index (*CGGI*)** is a composite measure of attribute decorrelation within a global domain $\Omega \subseteq R^{n,1}$ defined as

$$\gamma_\Omega = \frac{m-1}{2mW\sigma_X^2} trW\left(\Delta\widetilde{\chi}_p^2\right)^{\mathrm{T}}, \tag{3.7}$$

where the elements of the matrix $\Delta\widetilde{\chi}_p^2$ are $\Delta\widetilde{\chi}_{p_ip_j}^2 = \left[\widetilde{\chi}(p_i) - \widetilde{\chi}(p_j)\right]^2$

Otherwise said, the *CGGI* measures decorrelation in terms of the squared attribute fluctuation differences between all pairs of points p_i, $p_j \in \Omega$ considered. The γ_Ω values range from 0 to 2 so that the following assertion holds:

- A *CGGI* value approaching 0 indicates that similar attribute values tend to cluster, a value approaching 2 indicates that dissimilar values tend to cluster, and a value approaching 1 indicates a random pattern.

Somewhat confusingly, then, positive autocorrelation corresponds to *CGGI* values less than 1, zero autocorrelation corresponds to a *CGGI* value equal to 1, and negative autocorrelation corresponds to a *CGGI* value greater than 1. One also notices that unlike *CGMI*, in *CGGI* the interaction is not the product of the attribute value deviations from the mean, but the squared deviations between attribute values. *CGMI* is believed to give a more global correlation index than *CGGI*, whereas *CGGI* is more sensitive to differences in small neighborhoods.

3.2.2 Space-time global decorrelation measure

The separate space-time alternative to Eq. (3.7) is as follows.

The **time-specific** *CGGI* is a measure of separate space-time attribute decorrelation within a global region $S \subseteq R^n$ at time t defined as

$$\gamma_{S,t} = \frac{m_t-1}{2m_tW_t\sigma_{X_t}^2} trW_t\left(\Delta\widetilde{\chi}_{s,t}^2\right)^{\mathrm{T}}, \tag{3.8}$$

where now the elements of the matrix $\Delta\widetilde{\chi}_{s,t}^2$ are $\Delta\widetilde{\chi}_{s_i,s_j,t}^2 = \left[\widetilde{\chi}(s_i,t) - \widetilde{\chi}(s_j,t)\right]^2$.

Apparently, the *CGGI* is expressed in terms of squared differences of attribute fluctuations at all pairs of locations s_i, $s_j \in S$ considered at time t.

3.2.3 Spacetime local decorrelation measure

The *CGGI* can be extended to yield a correlation measure that is a function of the chronotopologic lags. Some of these extensions are briefly discussed next.

The **local** *CGGI* is a rather straightforward extension of the *CGMI* of Eq. (3.7) defined as

$$\gamma_{\Delta p} = \frac{m-1}{2mW_{\Delta p}\sigma_X^2} trW_{\Delta p}\left(\Delta\widetilde{\chi}_p^2\right)^{\mathrm{T}}, \tag{3.9}$$

where the connectivity matrix $W_{\Delta p}$ includes only a limited number of weights associated with attribute points within the chronotopologic lag Δp.

3.2.4 Space-time local decorrelation measure

As with *CGMI*, other local versions of *CGGI* can be found in the literature, such as the following, which is a chronotopologic extension of the purely spatial local *GI* (for a discussion of the local *GI*, see, e.g., Anselin, 1995).

The **local** *CGGI* is a measure of separate space-time attribute decorrelation at location s_i and time t that is computed as

$$\gamma_{i,t} = w_{i,t}^{\mathrm{T}} \left(\Delta \widetilde{\chi}_{s_i,t}^{2} \right)^{\mathrm{T}}, \tag{3.10}$$

where the elements of the vector $\Delta \widetilde{\chi}_{s_i,t}^{2}$ are $\Delta \widetilde{\chi}^{2}_{s_i,s_j,t} = [\widetilde{\chi}_{s_i,t} - \widetilde{\chi}_{s_j,t}]^2$.

Again, the local *CGGI* above reflects the level of clustering of similar values around s_i and time t within a specified subdomain of Ω.

- In light of the above expressions, it is valid that

$$\gamma_{S,t} = \frac{m_t - 1}{2 m_t W_t \sigma^2_{X_t}} \sum\nolimits_{i=1}^{m_t} \gamma_{i,t}, \tag{3.11}$$

i.e., the sums of the local *CGGI* overall points considered are proportional to the global *CGGI*, with a scale factor $\lambda_{\gamma,t} \propto \frac{m_t W_t \sigma^2_{X_t}}{m_t - 1}$.

To put it in more words, Eq. (3.11) decomposes the global *CGGI* into a summation of contributions from each datum.

3.3 Practical issues

As one might expect, a number of practical issues emerge as regards the implementation of the global *CGMI* and *CGGI* statistics in practice.

3.3.1 *Connectivity weights*

The calculation of the connectivity weights depends on whether the formulations of Eqs. (3.1) for *CGMI* and (3.7) for *CGGI* or that of Eqs. (3.3) for *CGMI* and (3.8) for *CGMI* are chosen.[o]

① For the calculation of w_{ij} in Eqs. (3.1) and (3.7) one needs to decide in what *chronotopologic* sense the $\boldsymbol{p}_i = (\boldsymbol{s}_i, t_i)$ and $\boldsymbol{p}_j = (\boldsymbol{s}_j, t_j)$ are neighbors or not, when $\boldsymbol{s}_i \neq \boldsymbol{s}_j$ and $t_i \neq t_j$, in general. One approach is that the weight $w_{ij} = 1$ should be assigned when points $\boldsymbol{p}_i$ and $\boldsymbol{p}_j$ fall within the same chronotopologic lag, say, $\boldsymbol{h} = (h_1 = |s_{i1} - s_{j1}|, h_2 = |s_{i2} - s_{j2}|; \tau = |t_i - t_j|)$,[p] otherwise $w_{ij} = 0$.

② For the calculation of $w_{ij,t}$ in Eqs. (3.3) and (3.8) matters are rather more obvious, as one needs to determine whether locations $\boldsymbol{s}_i$ and $\boldsymbol{s}_j$ are *spatial* neighbors at the same time t or not. This means that the notion of "spatially neighboring locations" must be technically introduced. For this purpose, a structure (say, grid or unit) is imposed on the data locations that contains the number of neighbors considered. Typically, then, $w_{ij,t} = 1$ if $\boldsymbol{s}_i$ and $\boldsymbol{s}_j$ are neighbors at time t, and $=0$, otherwise.

The weight sum W of Eq. (3.1) (case ① above) can be expressed in terms of the weight sum W_t of Eq. (3.3) (case ②) as $W = \sum_{t \in T} W_t$, where T is the entire time period considered in the particular case study. Also, as regards the corresponding number of data, it holds that $m = \sum_{t \in T} m_t$.

Example 3.4 Illustrations of cases ① and ② above are presented in Fig 3.2A and B, respectively. Case ① breaks up the data into a series of time snapshots, whereas in case ② all the data is considered (data values $\chi(\boldsymbol{s}_i, t)$ that are spatially and temporally close to each other are analyzed together, the relationships between $\chi(\boldsymbol{s}_i, t)$ are assessed relative to the location $\boldsymbol{s}_i$ and time instance t).

Other possibilities concerning the choice of connectivity weights could be considered, as well.

- Generally, an operational notion of "chronotopologically neighboring locations" should be introduced for w_{ij}; similarly, an

[o] The readers are suggested to review the notions of chronotopologic (space-time and spacetime) lags and neighborhoods presented in Chapter 2.

[p] E.g., within $h_1 \times h_2 = 5 \times 5$ km^2 area and a $\tau = 3$ days window.

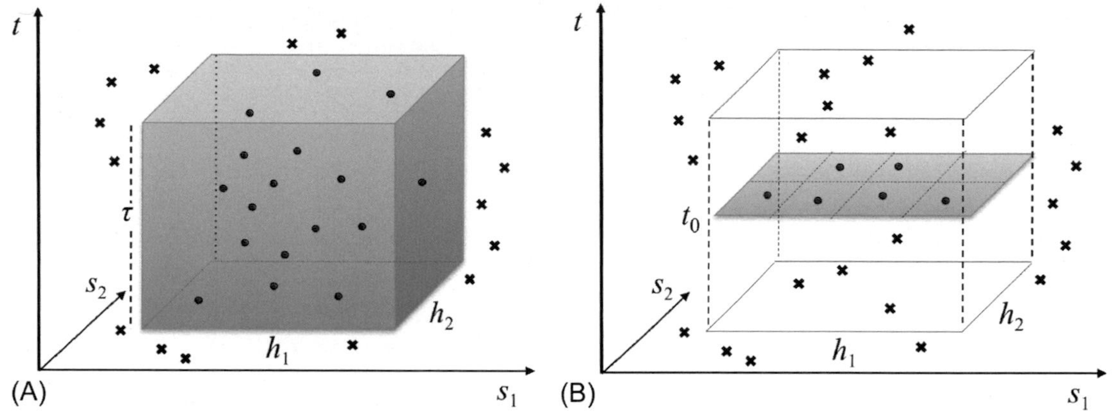

FIG. 3.2 Illustrations of (A) case ① with the solid circles indicating points that fall within the chronotopologic domain (h_1, h_2, τ), and crosses indicating points that fall outside, and (B) case ② with the solid circles indicate locations that belong to the area grid at time t_0, and crosses indicate locations that fall outside.

operational notion of "topologically neighboring locations" over time should be considered for $w_{ij,t}$

Furthermore, significant practical issues include the plotting of the *CGMI* c_Ω as a function of τ using Eq. (3.2) and the scatterplots of $w_{ij}\widetilde{\chi}_{i,t}$ on $\widetilde{\chi}_{j,t-\tau}$ for a set of τ values. These plots are useful in the visualization of space-time correlation patterns in the data distribution (e.g., geographical trends that persist in time *vs.* solely temporary geographical trends may be detected).

The following example outlines the *step-by-step* process of *CGMI* and *CGGI* **computation**.

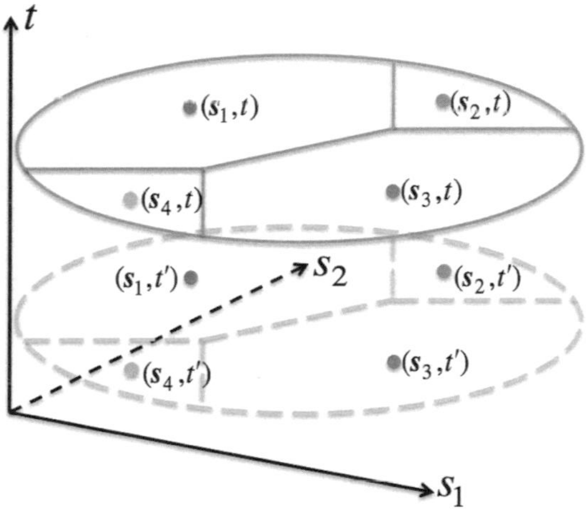

FIG. 3.3 A simple data pattern of four points at two different temporal instances.

Example 3.5 Consider the spatial data pattern at a specified time shown in Fig. 3.3, where the attribute data at time t are $\chi_{1,t} = 3$, $\chi_{2,t} = \chi_{3,t} = 2$ and $\chi_{4,t} = 1$ *units*. The computational steps of the time-specific *CGMI* of Eq. (3.3) are as follows:

Step 1: The binary $w_{ij,t}$ terms represent the chrono-geographic proximity of $(\boldsymbol{s}_i, t)$ and $(\boldsymbol{s}_j, t)$, and they are calculated based on adjacencies,[q] i.e.,

$$\mathbf{W}_t = \begin{bmatrix} w_{11,t}\, w_{12,t}\, w_{13,t}\, w_{14,t} \\ w_{21,t}\, w_{22,t}\, w_{23,t}\, w_{24,t} \\ w_{31,t}\, w_{32,t}\, w_{33,t}\, w_{34,t} \\ w_{41,t}\, w_{42,t}\, w_{43,t}\, w_{44,t} \end{bmatrix} = \begin{bmatrix} 0 & 1 & 1 & 1 \\ 1 & 0 & 1 & 0 \\ 1 & 1 & 0 & 1 \\ 1 & 0 & 1 & 0 \end{bmatrix}$$

so that $W_t = \mathbf{1}^{\mathrm{T}}\mathbf{W}_t\mathbf{1} = \sum_{i=1}^{m_t}\sum_{j=1}^{m_t} w_{ij,t} = \sum_{i=1}^{4}\sum_{j=1}^{4} w_{ij,t} = 0 \times 6 + 1 \times 10 = 10.$

[q] Recall that $w_{ij,t} = 1$ if $\boldsymbol{s}_i$ and $\boldsymbol{s}_j$ are neighbors and $=0$, otherwise.

Step 2: The sample mean and variance at time t are, respectively, $\overline{X}_t = \frac{1}{4}\sum_{i=1}^{4}\chi_{i,t} = 2$ and $\sum_{i=1}^{4}\left[\chi_{i,t} - \overline{X}_t\right]^2 = 2$, and $\sum_{i=1}^{4}\sum_{j=1}^{4} w_{ij}[\chi_{i,t} - 2][\chi_{j,t} - 2] = -2$.

Step 3: Finally, from Eq. (3.3) the numerical *CGMI* is found to be

$$c_{S,t} = \frac{4}{10} \times \frac{-2}{2} = -0.4,$$

which represents negative spatial correlation at time t. One can also calculate the *CGGI* as follows:

Step 1: As in Example 3.5 above.

Step 2: As above $\overline{X}_t = 2$ and $\sum_{i=1}^{4}\left[\chi_{i,t} - \overline{X}_t\right]^2 = 2$, but $\sum_{i=1}^{4}\sum_{j=1}^{4} w_{ij,t}[\chi_{i,t} - \chi_{j,t}]^2 = 14$. Obviously, in the case of a constant mean, the attribute fluctuation difference $\widetilde{\chi}_{i,t} - \widetilde{\chi}_{j,t}$ can be simply written as the attribute value difference $\chi_{i,t} - \chi_{j,t}$.

Step 3: Then, the *CGGI* of Eq. (3.8) is

$$\gamma_{S,t} = \frac{4-1}{2 \times 10} \times \frac{14}{2} = 1.05,$$

which, as the *CGMI* value above, also represents negative spatial correlation at time t.

Example 3.6 *CGMI* can be used to analyze a chrono-geographic dataset by dividing it into several spatial data sub-sets in terms of temporal instances. Fei et al. (2016a,b) found that the *CGMI* of Breast Cancer (*BC*) incidence in Hangzhou City (China) during the years 2008–2012 were equal to 0.353, 0.345, 0.352, 0.422 and 0.380, respectively. These values indicated positive *BC* correlation and geographically clustered characteristics in the corresponding years.

3.3.2 *CGMI vs. CGGI*

As noted earlier, the *CGGI* provides an alternative to *CGMI* for the same data arrangement, yet, in most applications both indexes are equally satisfactory. Perhaps, an advantage of *CGMI* is that it is arranged so that its extremes match our earlier intuitive notions of zero, positive, and negative correlation, whereas the *CGGI* uses a more confusing scale. These relationships are summarized in Table 3.1 (i.e., correspondence between *CGMI*, *CGGI*, and conceptual scales of correlation).

TABLE 3.1 Comparison between *CGMI* and *CGGI* of Eqs. (3.1) and (3.7).

Conceptual scale of correlation	*CGMI*	*CGGI*
Similar, regionalized, smooth	>0	∈(0,1)
Independent, uncorrelated, random	≈0	=1
Dissimilar, contrasting, checkerboard	<0	>1

3.3.3 *Global vs. local indexes*

The *CGMI* and *CGGI* expressions of Eqs. (3.1)–(3.3), (3.7)–(3.8) assess the overall degree of chronotopologic correlation (CGMI) and decorrelation (CGGI) based on the assumption of attribute homostationarity. If this assumption does not hold, the *CGMI* and *CGGI* expressions can be modified and implemented to each local unit to measure local attribute correlation at time t, as in Eqs. (3.4)–(3.6), (3.9)–(3.11). Specifically, in the local case the *CGMI* and *CGGI* for each datum $\chi_{i,t}$ provide an assessment of the extent of clustering of similar values around $\chi_{i,t}$ $(i = 1, \ldots, m_t)$. The global *CGMI* and *CGGI* compute only one value for the entire domain Ω at time t, whereas the local *CGMI* and *CGGI* are computed for each location $\boldsymbol{s}_i$ at time t within each sub-domain of Ω. Interpretationally, in a chronotopologic context the local *CGMI* and *CGGI* may be viewed as indexes of local units of heterogeneity, or hotspots (see Section 4 later) or used to assess the influence of individual points on the magnitude of the global *CGMI* and *CGGI* and to identify "outliers."

3.4 Assessing diversity

The next chrono-geographic indices originate in the idea of *biodiversity* on Earth. High biodiversity, in particular, is considered as an indication of ecosystem health. If an ecosystem comes under stress from overexploitation or pollution, it will show low diversity. In order to assess biodiversity, scientists take two factors into consideration: species richness (number of different kinds of species in a region), and species evenness (similarity of the population size of each species).

Example 3.7 A habitat where there are many different species is likely to be more diverse than one with only a few different species. A habitat where there are even numbers of individuals in each species is likely to be more diverse than one in which individuals of one species outnumber all the others.

The idea of biodiversity can be extended to other kinds of natural diversity, say, in ecology, environment, and epidemiology, in which case one may refer to eco-diversity, enviro-diversity, and epi-diversity, respectively. Accordingly, the following general description may be adopted:

The **natural diversity** of a space-time data distribution is characterized by the degree of variability of the data values considered. In this context, then, the **natural concentration** may be viewed as a complementary notion.

Example 3.8 An epidemic with a mortality rate distribution that includes the categories of "high," "medium-high," "medium," "medium-low" and "low" rates is more (epi-)diverse than one that includes only the categories of "high" and "low."

To describe diversity in a formal manner, consider a natural attribute $X(\boldsymbol{s},t)$, and assume that the available dataset is divided into l_t classes. At each time t, the region of interest A_t is divided into k_t areas $a_{j,t}$ ($j = 1, \ldots, k_t$) so that $A_t = \cup_{j=1}^{k_t} a_{j,t}$. Let $m_{ij,t}$ denote the number of data of class $i = 1, \ldots, l_t$ found in area $a_{j,t}$. Then, $m = \sum_{i=1}^{l_t}\sum_{j=1}^{k_t} m_{ij,t}$ is the total number of attribute data available in A_t, and $m_j = \sum_{i=1}^{l_t} m_{ij}$ is the number of data found in $a_{j,t}$. A quantitative diversity index should take into account the *richness* of the dataset (number of different classes in each sub-domain) and its *evenness* (similarity in size of each class). Then, indices addressing attribute value diversity over each $a_{j,t}$ at a specified time t are developed as follows.

The **probabilistic diversity indices** consider the random selection at time t of a pair of attribute values $X(\boldsymbol{s}_q, t) = \chi_{q,t}$ ($q = 1,\ 2$) at locations $\boldsymbol{s}_q \in a_{j,t}$, and define the diversity index by means of the complementary probability

$$\iota_{j,t}^{\mathfrak{I}} = 1 - P_j\left[\vee_{i=1}^{l_t}\left(\chi_{1,t}\in C_i\, \mathfrak{I}\, \chi_{2,t}\in C_i\right)\right], \tag{3.12a}$$

or by means of the inverse probability

$$\iota_{j,t}^{\mathfrak{I},*} = P_j^{-1}\left[\vee_{i=1}^{l_t}\left(\chi_{1,t}\in C_i\ \mathfrak{I}\, \chi_{2,t}\in C_i\right)\right], \tag{3.12b}$$

where C_i ($i = 1, \ldots, l_t$) are specified attribute value classes, and $\mathfrak{I}$ is a logical connective such as $\mathfrak{I} \equiv \wedge$ (conjunction), $\mathfrak{I} \equiv \vee$ (disjunction), $\mathfrak{I} \equiv \rightarrow$ (implication) and $\mathfrak{I} \equiv \leftrightarrow$ (equivalence).[r]

Using the standard probability addition rule,[s] the probability function in Eqs. (3.12a) and (3.12b) can be written as

$$\begin{aligned}
P_j\left[\vee_{i=1}^{l_t}\left(\chi_{1,t}\in C_i\, \mathfrak{I}\chi_{2,t}\in C_i\right)\right] &= \sum\nolimits_{i=1}^{l_t} P_j\left[\chi_{1,t}\in C_i\ \mathfrak{I}\chi_{2,t}\in C_i\right] \\
&\quad + (-1)^{l_t-1}\sum\nolimits_{i=i_1,\ldots,i_{l_t}} P_j\left[\wedge_{i=i_1}^{i_{l_t}}\left(\chi_{1,t}\in C_i\ \mathfrak{I}\chi_{2,t}\in C_i\right)\right] \\
&= \sum\nolimits_{i=1}^{l_t} P_j\left[\chi_{1,t}\in C_i\ \mathfrak{I}\chi_{2,t}\in C_i\right] \\
&\quad - \sum\nolimits_{i_1<i_2} P_j\left[\left(\chi_{1,t}\in C_{i_1}\mathfrak{I}\chi_{2,t}\in C_{i_1}\right)\wedge\left(\chi_{1,t}\in C_{i_2}\mathfrak{I}\chi_{2,t'}\in C_{i_2}\right)\right] \\
&\quad + \sum\nolimits_{i_1<i_2<i_3} P_j[\left(\chi_{1,t}\in C_{i_1}\mathfrak{I}\chi_{2,t}\in C_{i_1}\right)\wedge\left(\chi_{1,t}\in C_{i_2}\mathfrak{I}\chi_{2,t}\in C_{i_2}\right) \\
&\quad \wedge\left(\chi_{1,t}\in C_{i_3}\mathfrak{I}\chi_{2,t}\in C_{i_3}\right)] - \ldots \\
&\quad + (-1)^{l_t-1} P_j\left[\wedge_{i=1}^{i_{l_t}}\left(\chi_{1,t}\in C_i\mathfrak{I}\chi_{2,t}\in C_i\right)\right].
\end{aligned} \tag{3.13}$$

[r] See, also, Section 3.3 of Chapter 6 on stochastic logic.

[s] According to the probability addition rule, $P_j[\vee_{i=1}^{l_t} A_i] = \sum_{i=1}^{n} P_j[A_i] - \sum_{i_1<i_2} P_j[A_{i_1} \wedge A_{i_2}] + \sum_{i_1<i_2<i_3} P_j[A_{i_1} \wedge A_{i_2} \wedge A_{i_3}] - \ldots + (-1)^{n-1} P_j[\wedge_{i=1}^{l_t} A_i]$, with $A_i \equiv \chi_{1,t} \in C_i\ \ \mathfrak{I}\chi_{2,t} \in C_i$, in this case.

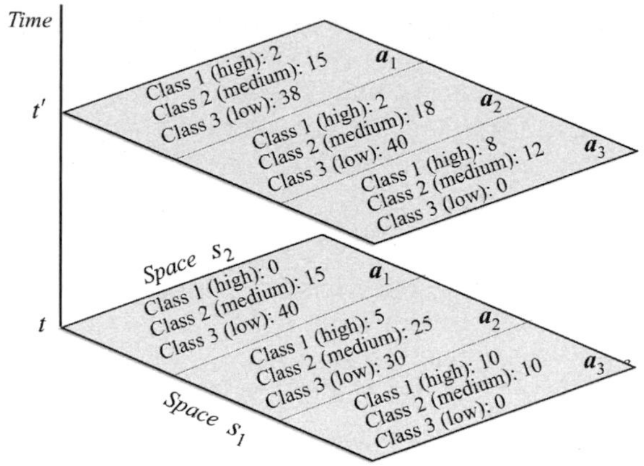

FIG. 3.4 Chrono-geographic patterns of *Cd* content in a domain.

Example 3.9 Fig. 3.4 presents chrono-geographic cadmium (*Cd*) content distribution in a region $A_t = \cup_{j=1}^{3} a_{j,t}$ consisting of the areas a_j at any time t. The *Cd* contents are classified as belonging to the *high*, *medium* and *low* class C_i corresponding to $i = 1, 2, 3$, respectively. The number of areas and data classes is constant, i.e., $l_t = k_t = 3$, at any time t. Then, at any time t, Eqs. (3.12a) and (3.13) reduce to

$$\begin{aligned} P_j\left[\vee_{i=1}^{l_t}\left(\chi_{1,t}\in C_i\ \Im\chi_{2,t}\in C_i\right)\right] &= \sum_{i=1}^{3} P_j\left[\chi_{1,t}\in C_i\ \Im\chi_{2,t}\in C_i\right] \\ &- P_j\left[\left(\chi_{1,t}\in C_1\ \Im\chi_{2,t}\in C_1\right)\wedge\left(\chi_{1,t}\in C_2\ \Im\chi_{2,t}\in C_2\right)\right] \\ &- P_j\left[\left(\chi_{1,t}\in C_1\ \Im\chi_{2,t}\in C_1\right)\wedge\left(\chi_{1,t}\in C_3\ \Im\chi_{2,t}\in C_3\right)\right] \\ &- P_j\left[\left(\chi_{1,t}\in C_2\ \Im\chi_{2,t}\in C_2\right)\wedge\left(\chi_{1,t}\in C_3\ \Im\chi_{2,t}\in C_3\right)\right] \\ &+ P_j\left[\left(\chi_{1,t}\in C_1\ \Im\chi_{2,t}\in C_1\right)\wedge\left(\chi_{1,t}\in C_2\ \Im\chi_{2,t}\in C_2\right)\right. \\ &\left.\wedge\left(\chi_{1,t}\in C_3\ \Im\chi_{2,t}\in C_3\right)\right]. \end{aligned} \quad (3.14)$$

The physical meaning of the diversity index $\iota_{j,t}^{\Im}$ naturally depends on the choice of logical connective $\Im$. The truth table of the standard connectives $\Im \equiv \wedge$ (conjunction), $\Im \equiv \vee$ (disjunction), $\Im \equiv \rightarrow$ (implication), and $\Im \equiv \leftrightarrow$ (equivalence) is shown in Table 3.2. Then, the corresponding index $\iota_{j,t}^{\Im}$ is computed in terms of the quadruples of possible combinations of T- and F-values for the elementary events $\chi_{q,t} \in C_i\,(q = 1,2)$ that cause a T-value for the event $(\chi_{1,t} \in C_i\ \Im\chi_{2,t} \in C_i)$ with $\Im = \wedge$, $\leftrightarrow$ and $\rightarrow$.

Eqs. (3.12a), (3.12b) and (3.13) are subject to the conditions of the particular random selection considered. In this sense, the conditions of mutual exclusiveness or independency may hold, such that, respectively, $P_j[\wedge_{i=i_1}^{i_{lt}}(\chi_{1,t} \in C_i \wedge \chi_{2,t} \in C_i)] = 0$ or $P_j[\wedge_{i=i_1}^{i_{lt}}(\chi_{1,t} \in C_i \leftrightarrow \chi_{2,t} \in C_i)] = \prod_{i=i_1}^{i_{lt}} P_j[\chi_{1,t} \in C_i \leftrightarrow \chi_{2,t} \in C_i]$ for all combinations of $i_1, \ldots, i_{l_t}$.

Example 3.9(cont.) For $\Im \equiv \wedge$, the selections $\chi_{1,t} \in C_i \wedge \chi_{2,t} \in C_i$ are mutually exclusive for $i \neq i'$,[t] and

$$\sum_{i=1}^{3} P_j\left[\chi_{1,t}\in C_i\ \wedge\chi_{2,t}\in C_i\right] = \sum_{i=1}^{3} \frac{m_{i1,t}(m_{i1,t}-1)}{m_{1,t}(m_{1,t}-1)}.$$

In the case of $\Im \equiv \leftrightarrow$,

$$\sum_{i=1}^{3} P_j\left[\chi_{1,t}\in C_i \leftrightarrow \chi_{2,t}\in C_i\right] = \sum_{i=1}^{3} P_j\left[\left(\chi_{1,t}\in C_i \wedge \chi_{2,t}\in C_i\right) \vee \left(\chi_{1,t}\notin C_i \wedge \chi_{2,t}\notin C_i\right)\right],$$

and with reference to Table 3.2, the last equation gives

TABLE 3.2 Truth table of the logical connectives of the diversity indices.

$\chi_{1,t} \in C_i$	$\chi_{2,t} \in C_i$	$\chi_{1,t} \in C_i \wedge \chi_{2,t} \in C_i$	$\chi_{1,t} \in C_i \leftrightarrow \chi_{2,t} \in C_i$	$\chi_{1,t} \in C_i \rightarrow \chi_{2,t} \in C_i$
T	*T*	*T*	*T*	*T*
T	*F*	*F*	*F*	*F*
F	*T*	*F*	*F*	*T*
F	*F*	*F*	*T*	*T*

[t] The fact that $\chi_{q,t} \in C_i$ $(q = 1, 2)$ excludes that $\chi_{q,t} \in C_{i'}$ too for $i \neq i'$.

$$\sum_{i=1}^{3} P_j\left[\chi_{1,t}{\in}C_i \leftrightarrow \chi_{2,t}{\in}C_i\right]$$
$$= \sum_{i=1}^{3} P_j\left[\chi_{1,t}{\in}C_i \wedge \chi_{2,t}{\in}C_i\right]$$
$$+ \sum_{i=1}^{3} P_j\left[\chi_{1,t}{\notin}C_i \wedge \chi_{2,t}{\notin}C_i\right]$$
$$= \sum_{i=1}^{3} \frac{m_{i1,t}(m_{i1,t}-1)}{m_{1,t}(m_{1,t}-1)}$$
$$+ \sum_{i=1}^{3}\left(1-\frac{m_{i1,t}}{m_{1,t}}\right)\left(1-\frac{m_{i1,t}-1}{m_{1,t}-1}\right).$$

In the case of area a_1:

$$\sum_{i=1}^{3} P_j\left[\chi_{1,t}{\in}C_i \wedge \chi_{2,t}{\in}C_i\right] = \frac{15\times 14}{55\times 54} + \frac{40\times 39}{55\times 54} = \frac{1770}{2970} = 0.596, \text{and}$$

$$\sum_{i=1}^{3} P_1\left[\chi_{1,t}{\in}C_i \leftrightarrow \chi_{2,t}{\in}C_i\right] = \tfrac{1770}{2970} + 1 + \left(1-\frac{15}{55}\right)\left(1-\frac{14}{54}\right) + \left(1-\frac{40}{55}\right)\left(1-\frac{39}{54}\right) = 1 + \frac{3595}{2970} = 2.21.$$

Based on the truth Table 3.2, the corresponding Table 3.3 of the logical connective probabilities associated with the diversity indices is developed.

Qualitatively, $\iota_{j,t}^{\mathfrak{I}}$ and $\iota_{j,t}^{\mathfrak{I}*}$ express the same diversity assessment: higher $\iota_{j,t}^{\mathfrak{I}}$ or $\iota_{j,t}^{\mathfrak{I}*}$ imply greater physical diversity. Numerically, $\iota_{j,t}^{\mathfrak{I}}$ and $\iota_{j,t}^{\mathfrak{I}*}$ may take different values. Since $P_j \in [0,1]$, Eqs. (3.12a) and (3.12b) give $\iota_{j,t}^{\mathfrak{I}} \in [0,1]$ and $\iota_{j,t}^{\mathfrak{I}*} > 0$, respectively.

A $\iota_{j,t}^{\mathfrak{I}}$ value close to 1 indicates a highly diverse (heterogeneous) system, and a value close to 0 indicates a less diverse (rather homogeneous) system. On the other hand, the lowest $\iota_{j,t}^{\mathfrak{I}*}$ value is 1 (representing, say, a community consisting of only one species).

Depending on the particular choice of the logical connective $\mathfrak{I}$ ($\wedge$, $\leftrightarrow$ and $\rightarrow$), the diversity indices above may include different random selections of attribute-value pairs $\chi_{q,t}$ ($q = 1, 2$) at locations $s_q \in a_{j,t}$. Indeed, below we consider the three logical connectives $\mathfrak{I}$, and the corresponding distinct diversity indices.

3.4.1 *Conjunction or Simpson index*

The next index is obtained by letting $\mathfrak{I} \equiv \wedge$ in Eq. (3.12a), i.e., the index considers diversity in terms of the occurrence probability of a pair of attribute values that belong to a specified class.

The **chrono-geographic conjunction index** (***CGCI***), also known as **chrono-geographic Simpson's index** (***CGSI***), of attribute $X(\boldsymbol{s},t)$ in sub-area $a_{j,t}$ at time t is expressed as

$$\iota_{j,t}^{\wedge} = 1 - P_j\left[\vee_{i=1}^{l_t}\left(\chi_{1,t}{\in}C_i \wedge \chi_{2,t}{\in}C_i\right)\right] = 1 - \sum_{i=1}^{l_t} P_j\left[\chi_{1,t}{\in}C_i \wedge \chi_{2,t}{\in}C_i\right], \quad (3.15a)$$

where

$$\sum_{i=1}^{l_t} P_j\left[\chi_{1,t}{\in}C_i \wedge \chi_{2,t}{\in}C_i\right] = \frac{\sum_{i=1}^{l_t} m_{ij,t}(m_{ij,t}-1)}{m_{j,t}(m_{j,t}-1)} \quad (3.15b)$$

TABLE 3.3 Logical connective probabilities of the diversity indices.[a]

	$P_{j,\,(i,\chi_1)}$			$P_{j,\,(i,\chi_2)}$			$P_{j,\,(i,\chi_1\wedge\chi_2)}$			$P_{j,\,(i,\chi_1\leftrightarrow\chi_2)}$			$P_{j,\,(i,\chi_1\rightarrow\chi_2)}$		
	$i=1$	**2**	**3**	**1**	**2**	**3**	**1**	**2**	**3**	**1**	**2**	**3**	**1**	**2**	**3**
$j=1$	0	$\frac{15}{55}$	$\frac{40}{55}$	0	$\frac{14}{54}$	$\frac{39}{54}$	0	$\frac{7}{99}$	$\frac{52}{99}$	1	$\frac{181}{297}$	$\frac{357}{594}$	1	$\frac{237}{297}$	$\frac{237}{297}$
2	$\frac{1}{12}$	$\frac{25}{60}$	$\frac{1}{2}$	$\frac{4}{59}$	$\frac{24}{59}$	$\frac{29}{59}$	$\frac{1}{177}$	$\frac{30}{177}$	$\frac{29}{118}$	$\frac{609}{708}$	$\frac{365}{708}$	$\frac{1}{2}$	$\frac{653}{708}$	$\frac{533}{708}$	$\frac{132}{177}$
3	$\frac{1}{2}$	$\frac{1}{2}$	0	$\frac{9}{19}$	$\frac{9}{19}$	0	$\frac{9}{38}$	$\frac{9}{38}$	0	$\frac{1}{2}$	$\frac{1}{2}$	1	$\frac{14}{19}$	$\frac{14}{19}$	1

[a] *Where* $P_{j,(i,\chi_q)} = P_j[\chi_{q,t} \in C_i]$, $q = 1, 2$, $P_{j,(i,\chi_1\wedge\chi_2)} = P_j[\chi_{1,t} \in C_i \wedge \chi_{2,t} \in C_i]$, $P_{j,(i,\chi_1\leftrightarrow\chi_2)} = P_j[\chi_{1,t} \in C_i \leftrightarrow \chi_{2,t} \in C_i]$ *and* $P_{j,(i,\chi_1\rightarrow\chi_2)} = P_j[\chi_{1,t} \in C_i \rightarrow \chi_{2,t} \in C_i]$.

is the probability at time t to randomly select two attribute data $\chi_{1,t}$ and $\chi_{2,t}$ of $X(\boldsymbol{s},t)$ in sub-area $a_{j,t}$ that belong to the same attribute class C_i ($i = 1, \ldots, l_t$).

Example 3.9(cont.) In this case, the *CGSI* of Eq. (3.15a) reduces to

$$\begin{aligned}\iota_{j,t}^{\wedge} &= 1 - \sum_{i=1}^{3} P_j[\chi_{1,t}{\in}C_i \wedge \chi_{2,t}{\in}C_i] \\ &= 1 - \frac{\sum_{i=1}^{3} m_{ij,t}(m_{ij,t}-1)}{\sum_{i=1}^{3} m_{ij,t}\left(\sum_{i=1}^{3} m_{ij,t}-1\right)}. \qquad (3.16)\end{aligned}$$

Index $\iota_{j,t}^{\wedge}$ includes the diversity selections $\chi_{1,t} \in C_i \wedge \chi_{2,t} \notin C_i$ and $\chi_{1,t} \notin C_i \wedge \chi_{2,t} \in C_i$, as well as the vague combination $\chi_{1,t} \notin C_i \wedge \chi_{2,t} \notin C_i$. The last combination is considered "vague", in the sense that includes the possibilities that both $\chi_{1,t}$ and $\chi_{2,t}$ may belong to any combination of classes that are different from C_i and/or from each other.[u] By definition, the *CGEI* is a probability, i.e., $\iota_{j,t}^{\wedge} \in [0,1]$ (the higher the $\iota_{j,t}^{\wedge}$ value, the greater the diversity, so that $\iota_{j,t}^{\wedge} \approx 1$ indicates a most diverse system, and $\iota_{j,t}^{\wedge} \approx 0$ indicates a least diverse system).

Example 3.10 As the *CGSI* was originally developed to study ecosystems, the $a_{j,t}$ may represent a specific sub-area j at time t, and $m_{ij,t}$ denotes the number of organisms of species i found in sub-area j. Then, the $\iota_{j,t}$ expresses the probability at time t to randomly select two organisms in sub-area j that do not belong to the same species i. For illustration, the probability of, say, two trees picked at random from a tropical rainforest being of the same species is relatively small (high Simpson's index), whereas in a boreal forest it is relatively large (low Simpson's index). Table 3.4 presents some examples of *CGSI* interpretations in an ecosystem context.

CGSI assigns more weight to the more abundant species recorded in a sample, whereas the addition of rare species to a sample would cause minor changes in the index's value. If $m_{j,t}$ is very large, the expression

$$\iota_{j,t} = 1 - \frac{\sum_{j=1}^{l_t} m_{ij,t}^2}{m_{j,t}^2}$$

may be used, instead of Eq. (3.15b). As *CGSI* assumes that the proportion of data values in a domain indicates their importance to diversity, it is concerned not only with diversity but also with dominance (it weights towards the

TABLE 3.4 Possible interpretations of CGSI.

Low species diversity (low *CGSI*) suggests	High species diversity (high *CGSI*) suggests
Relatively few successful species in the habitat	A greater number of successful species and a more stable ecosystem
The environment is quite stressful with relatively few ecological niches and only a few organisms are well adapted to that environment	More ecological niches are available and the environment is less likely to be hostile
Food webs are relatively simple	Complex food webs
Changes in the environment would probably have quite serious effects	Environmental change is less likely to be damaging to the ecosystem as a whole

[u] This includes the concentration combinations $\chi_{1,t}, \chi_{2,t} \in C_{i'} \neq C_i$ for all $i' \neq i$, and the diversity combinations $\chi_{1,t} \in C_{i'} \neq C_i \wedge \chi_{2,t} \in C_{i''} \neq C_i, C_{i'}$ for all $i' \neq i''$ and $i', i'' \neq i$.

abundance of the most common data values). *CGSI* is used in a large number of different fields, including environmental, health, social, ecological, vegetation, and animal studies.

Example 3.11 He et al. (2019b) studied the monthly Hemorrhagic fever with renal syndrome (HFRS) population-standardized incidence data in Heilongjiang province during the period 2005–2013. They introduced a class-dependent data analysis that divided the original dataset into discrete incidence classes that overcome data heterogeneity and skewness effects and can produce improved space-time HFRS incidence estimates together with their estimation accuracy.

3.4.2 *Equivalence index*

Useful chronotopologic attribute inferences can be drawn using stochastic logic, which may be preferable under certain conditions, like when what happens at one point affects what happens at the next point (i.e., $\chi_{1,t}$ and $\chi_{2,t}$ are related physically). In such cases, two measures are worth examining here. The first one is obtained below by letting $\mathfrak{I} \equiv \leftrightarrow$ in Eq. (3.12a).

The **chrono-geographic equivalence index** (***CGEI***) is the probability at time t to randomly select two attribute values in sub-area $a_{j,t}$ ($j = 1, \ldots, k_t$) such that $\chi_{1,t}$ belongs to class C_i ($i = 1, \ldots, l_t$) if and only if $\chi_{2,t}$ does so, i.e.,

$$\iota_{j,t}^{\leftrightarrow} = 1 - P_j^{\leftrightarrow}\left[\vee_{i=1}^{l_t}(\chi_{1,t}\in C_i \leftrightarrow \chi_{2,t}\in C_i)\right], \tag{3.17}$$

which is Eq (3.12a) with $\mathfrak{I} \equiv \leftrightarrow$.

Interpretively, since the stochastic logical equivalence ($\leftrightarrow$) in Eq. (3.17) can be expressed as

$$P_j[\chi_{1,t}\in C_i \leftrightarrow \chi_{2,t}\in C_i] = P_j[(\chi_{1,t}\in C_i \wedge \chi_{2,t}\in C_i) \vee (\chi_{1,t}\notin C_i \wedge \chi_{2,t}\notin C_i)], \tag{3.18a}$$

the *CGEI* involves the complementary probability of the union of randomly selecting two attribute values $\chi_{1,t}$ and $\chi_{2,t}$ in $a_{j,t}$ at time t so that either both values belong to class C_i (i.e., $\chi_{1,t} \in C_i \ \wedge \chi_{2,t} \in C_i$) or both values do not belong to C_i ($\chi_{1,t} \notin C_i \ \wedge \chi_{2,t} \notin C_i$). Accordingly, with reference to Table 3.2, the index $\iota_{j,t}^{\leftrightarrow}$ includes both diverse selections $\chi_{1,t} \in C_i \ \wedge \chi_{2,t} \notin C_i$ and $\chi_{1,t} \notin C_i \ \wedge \chi_{2,t} \in C_i$.

The probability in Eq. (3.18a) can be expressed in terms of the conjunction-based probability function as

$$\begin{aligned} P_j[\chi_{1,t}\in C_i \leftrightarrow \chi_{2,t}\in C_i)] &= 2P_j[\chi_{1,t}\in C_i \wedge \chi_{2,t}\in C_i] \\ &\quad + P_j[\chi_{2,t}\notin C_i\,] - P_j[\chi_{1,t}\in C_i\,] \\ &= 1 - 2P_j[\chi_{1,t}\in C_i \wedge \chi_{2,t}\in C_i] \\ &\quad + P_j[\chi_{2,t}\in C_i\,] - P_j[\chi_{1,t}\in C_i\,], \end{aligned} \tag{3.18b}$$

and, hence, it can be linked to the *CGSI* $\iota_{j,t}^{\wedge}$ as

$$\begin{aligned} &\sum\nolimits_{i=1}^{l_t} P_j[\chi_{1,t}\in C_i \leftrightarrow \chi_{2,t}\in C_i] \\ &= 2\sum\nolimits_{i=1}^{l_t} P_j[\chi_{1,t}\in C_i \wedge \chi_{2,t}\in C_i] \\ &\quad + l_t - \sum\nolimits_{i=1}^{l_t} P_j[\chi_{2,t}\in C_i\,] \\ &\quad - \sum\nolimits_{i=1}^{l_t} P_j[\chi_{1,t}\in C_i\,] = 2\iota_{j,t}^{\wedge} + l_t \\ &\quad - \sum\nolimits_{i=1}^{l_t} P_j[\chi_{2,t}\in C_i\,] - 1 \\ &= 2\frac{\sum\nolimits_{i=1}^{l_t} m_{ij,t}(m_{ij,t}-1)}{\sum\nolimits_{i=1}^{l_t} m_{ij,t}\left(\sum\nolimits_{i=1}^{l_t} m_{ij,t}-1\right)} \\ &\quad + l_t - \frac{\sum\nolimits_{i=1}^{l_t}(m_{ij,t}-1)}{\sum\nolimits_{i=1}^{l_t} m_{ij,t}-1} - 1 \\ &= \frac{\sum\nolimits_{i=1}^{l_t}(m_{ij,t}-1)}{\sum\nolimits_{i=1}^{l_t} m_{ij,t}-1}\left(2\frac{m_{ij,t}}{\sum\nolimits_{i=1}^{l_t} m_{ij,t}} - 1\right) \\ &\quad + l_t - 1 \geq 0, \end{aligned} \tag{3.19}$$

since $\sum_{i=1}^{l_t} P_j[\chi_{1,t}\in C_i\,] = \frac{\sum_{i=1}^{l_t} m_{ij,t}}{\sum_{i=1}^{l_t} m_{ij,t}} = 1$, and $\sum_{i=1}^{l_t} P_j[\chi_{2,t}\in C_i] = \frac{\sum_{i=1}^{l_t}(m_{ij,t}-1)}{\sum_{i=1}^{l_t} m_{ij,t}-1}$.

Example 3.9(cont.) In this case, assuming that the selections $\chi_{1,t} \in C_i \leftrightarrow \chi_{2,t} \in C_i$ and $\chi_{1,t} \in C_{i'} \leftrightarrow \chi_{2,t} \in C_{i'}$ are independent for $i \neq i'$, the *CGEI* of Eq. (3.17) reduces to

$$\begin{aligned} \iota_{j,t}^{\leftrightarrow} = 1 - \sum_{i=1}^{3} P_j[\chi_{1,t}{\in}C_i \leftrightarrow \chi_{2,t}{\in}C_i] \\ + P_j[\chi_{1,t}{\in}C_1 \leftrightarrow \chi_{2,t}{\in}C_1]P_j[\chi_{1,t}{\in}C_2 \leftrightarrow \chi_{2,t}{\in}C_2] \\ + P_j[\chi_{1,t}{\in}C_1 \leftrightarrow \chi_{2,t}{\in}C_1]P_j[\chi_{1,t}{\in}C_3 \leftrightarrow \chi_{2,t}{\in}C_3] \\ + P_j[\chi_{1,t}{\in}C_2 \leftrightarrow \chi_{2,t}{\in}C_2]P_j[\chi_{1,t}{\in}C_3 \leftrightarrow \chi_{2,t}{\in}C_3] \\ - \prod_{i=1}^{3} P_j[\chi_{1,t}{\in}C_i \leftrightarrow \chi_{2,t}{\in}C_i]. \end{aligned} \quad (3.20)$$

3.4.3 *Implication index*

Another interesting stochastic logic diversity measure is proposed below, by letting $\Im \equiv \rightarrow$ in Eq. (3.12a).

The **chrono-geographic Implication Index** (*CGII*) is the probability at time t to randomly select two attribute data in sub-area $a_{j,t}$ ($j=1, \ldots, k_t$) so that if $\chi_{1,t}$ belongs to class C_i ($i=1, \ldots, l_t$) it implies that $\chi_{2,t}$ does so too, i.e.,

$$\iota_{j,t}^{\rightarrow} = 1 - P_j\left[\vee_{i=1}^{l_t}(\chi_{1,t}{\in}C_i \rightarrow \chi_{2,t}{\in}C_i)\right], \quad (3.21)$$

which is Eq (3.12a) with $\Im \equiv \rightarrow$.

Example 3.9(cont.) Assuming that the selections $\chi_{1,t} \in C_i \rightarrow \chi_{2,t} \in C_i$ and $\chi_{1,t} \in C_{i'} \rightarrow \chi_{2,t} \in C_{i'}$ are independent for $i \neq i'$, the *CGII* of Eq. (3.21) reduces to

$$\begin{aligned} \iota_{j,t}^{\rightarrow} = 1 - \sum_{i=1}^{3} P_j[\chi_{1,t}{\in}C_i \rightarrow \chi_{2,t}{\in}C_i] \\ + P_j[\chi_{1,t}{\in}C_1 \rightarrow \chi_{2,t}{\in}C_1]P_j[\chi_{1,t}{\in}C_2 \rightarrow \chi_{2,t}{\in}C_2] \\ t \quad + P_j[\chi_{1,t}{\in}C_1 \rightarrow \chi_{2,t}{\in}C_1]P_j[\chi_{1,t}{\in}C_3 \rightarrow \chi_{2,t}{\in}C_3] \\ + P_j[\chi_{1,t}{\in}C_2 \rightarrow \chi_{2,t}{\in}C_2]P_j[\chi_{1,t}{\in}C_3 \rightarrow \chi_{2,t}{\in}C_3] \\ - \prod_{i=1}^{3} P_j[\chi_{1,t}{\in}C_i \rightarrow \chi_{2,t}{\in}C_i]. \end{aligned} \quad (3.22)$$

Interpretively, it is worth noticing that the logical implication terms of Eq. (3.21) can also be expressed as[v]

$$P_j[\chi_{1,t}{\in}C_i \rightarrow \chi_{2,t}{\in}C_i] = P_j[\neg[(\chi_{1,t}{\in}C_i) \wedge \neg(\chi_{2,t}{\in}C_i)]]. \quad (3.23)$$

The logical argument in the brackets, $\neg[(\chi_{1,t} \in C_i) \wedge \neg(\chi_{2,t} \in C_i)]$, is necessarily *truth-preserving*, as follows:

- A diversity argument is **valid** if it is impossible for the premise $\chi_{1,t} \in C_i$ to be true and the conclusion $\chi_{2,t} \in C_i$ false simultaneously (alternatively, it must be the case that when the premise is true, the conclusion is true).

With reference to Table 3.2, the index $\iota_{j,t}^{\rightarrow}$ includes one diversity combination $\chi_{1,t} \in C_i \wedge \chi_{2,t} \notin C_i$ but excludes the diversity combination $\chi_{1,t} \notin C_i \wedge \chi_{2,t} \in C_i$. In this sense, it refers to stricter diversity situations in the real world, keeping in mind that as important as it is, a valid logical argument does not necessarily imply a *sound* one.[w]

3.4.4 *Comparative analysis of the diversity indices*

As regards the ranking of the three diversity indices above, it holds that $\iota_{i,t}^{\rightarrow} \leq \iota_{i,t}^{\leftrightarrow} \leq \iota_{i,t}$ ($i=1, \ldots, l_t$).

By comparing Eqs. (3.15a), (3.17), (3.21) we see that $\iota_{j,t}^{\leftrightarrow}$ includes both diverse selections ($\chi_{1,t} \in C_i \wedge \chi_{2,t} \notin C_i$ and $\chi_{1,t} \notin C_i \wedge \chi_{2,t} \in C_i$), $\iota_{j,t}^{\wedge}$ includes these two diverse selections and also the vague selection $\chi_{1,t} \notin C_i \wedge \chi_{2,t} \notin C_i$, and $\iota_{j,t}^{\rightarrow}$ includes only one diverse selection $\chi_{1,t} \in C_i \wedge \chi_{2,t} \notin C_i$. In this sense, $\iota_{j,t}^{\leftrightarrow}$ seems to be the most explicit diversity index, whereas $\iota_{j,t}^{\wedge}$ is a broader index that also includes a vague selection, and $\iota_{j,t}^{\rightarrow}$ is a rather limited choice, in the random selection setting considered above. This

[v] This can also be written as $P_j[\chi_{1,t} \in C_i \rightarrow \chi_{2,t} \in C_i] = P_j[\neg(\chi_{1,t} \in C_i) \vee (\chi_{2,t} \in C_i)]$.

[w] The readers may recall that an argument is *valid* if and only if it is impossible for the premise to be true and the conclusion false, whereas an argument is *sound* if and only if it is both valid and its premise is actually true.

comparative interpretation will be assessed numerically in terms of an example below.

3.4.5 Computational diversity issues

In actu, to calculate the diversity indices in a specific case study, the area of interest must first be sampled, and the number of individuals in the sample belonging to each species present in the samples must be recorded. Naturally, several samples should be collected to assure an adequate estimate of the overall diversity.

The following example presents the *step-by-step* **computational** procedure of the *CGSI, CGEI,* and *CGII* diversity indices in practice.

Example 3.9(cont.) The step-by-step computation of the diversity indices of the areas a_j ($j = 1, 2, 3$) at time t is described in Table 3.5. In this numerical example, it was found that $\iota_{i,t}^{\rightarrow} \le \iota_{i,t}^{\leftrightarrow} \le \iota_{i,t}$ ($i = 1, 2, 3$), which confirms numerically what was expected in theory. Since higher *CGSI* implies a more *Cd*-diverse sub-area, it is concluded that it is most likely to select two *Cd* content values of the same class in the area a_1, followed by a_3 and, lastly, by a_2. Accordingly, a_2 is the most *Cd*-diverse area, followed by a_3 and then by a_1, according to *CGSI*. The *CGEI* and *CGII* lead to similar qualitative conclusions as *CGSI*. Yet, there is a difference in magnitude. *CGEI* indicates a much less diverse *Cd* distribution for the area a_2 than *CGSI* suggested; and *CGEI* also assessed the least possible diversity for areas a_1 and a_3. Lastly, the *CGII* suggested the least diversity for a_2 among the three indices; and, like *CGEI*, *CGII* computed the least possible diversity for a_1 and a_3.

3.5 Distributional segregation

There are a number of *segregation* indices that can be used for explaining the chrono-geographic data distribution at hand. One of them is as follows:

TABLE 3.5 Numerical computation of the diversity indices.

Diversity index	Computation
CGSI $\iota_{j,t}^{\wedge}$, Eq. (3.16)	$1-\frac{0\times(0-1)+15\times 14+40\times 39}{55\times 54}=0.404$ for a_1
	$1-\frac{5\times 4+25\times 24+30\times 29}{60\times 59}=0.579$ for a_2
	$1-\frac{10\times 9+10\times 9+0\times(-1)}{20\times 19}=0.526$ for a_3
CGEI $\iota_{j,t}^{\leftrightarrow}$, Eq. (3.20)	$1-0-1-\frac{1810}{2970}-\frac{1785}{2970}+1\times\frac{1810}{2970}+1\times\frac{1785}{2970}+\frac{1810\times 1785}{2970\times 2970}-1\times\frac{1810\times 1785}{2970\times 2970}=0$ for a_1
	$1-\frac{3045}{3540}-\frac{1825}{3540}-\frac{1770}{3540}+\frac{3045\times 1825}{3540\times 3540}+\frac{3045\times 1770}{3540\times 3540}+\frac{1825\times 1770}{3540\times 3540}-\frac{3540\times 1825\times 1770}{3540\times 3540\times 3540}=0.034$ for a_2
	$1-\frac{1}{2}-\frac{1}{2}-1+\frac{1}{2}\times\frac{1}{2}+\frac{1}{2}\times 1+\frac{1}{2}\times 1-\frac{1}{2}\times\frac{1}{2}\times 1=0$ for a_3
CGII $\iota_{j,t}^{\rightarrow}$, Eq. (3.22)	$1-1-\frac{2370}{2970}-\frac{2370}{2970}+1\times\frac{2370}{2970}+1\times\frac{2370}{2970}+\frac{2370\times 2370}{2970\times 2970}-1\times\frac{2370\times 2370}{2970\times 2970}=0$ for a_1
	$1-\frac{3265}{3540}-\frac{2665}{3540}-\frac{2640}{3540}+\frac{3265\times 2665}{3540\times 3540}+\frac{3265\times 2640}{3540\times 3540}+\frac{2665\times 2640}{3540\times 3540}-\frac{3265\times 2665\times 640}{3540\times 3540\times 3540}=0.005$ for a_2
	$1-\frac{280}{380}-\frac{280}{380}-1+\frac{280}{380}\times\frac{280}{380}+\frac{280}{380}\times 1+\frac{280}{380}\times 1-\frac{280\times 280}{380\times 380}\times 1=0$ for a_3

The **chrono-geographic dissimilarity index** (***CGDI***), which assesses how dissimilar the distributions of two datasets α and α' at time t are, is defined as

$$\eta_{\alpha\alpha' t} = {}^1/_2 \sum_{i=1}^{l_t} \left| \frac{m_{i\alpha,t}}{A} - \frac{m_{i\alpha',t}}{A'} \right|, \tag{3.24}$$

where $m_{i\alpha,t}$ and $m_{i\alpha',t}$ represent the number of data in α and α', respectively, belonging to class C_i ($i = 1, \ldots, l_t$) at time t, A and A' represent the total number of data belonging to α and α', respectively, and l_t is the number of data subclasses at time t.

The closer the *CGDI* is to 1, the more dissimilar two distributions are, whereas a *CGDI* equal to 0 reflects two identical distributions. *CGDI* can be also interpreted as the proportion of one of the distributions that would have to be geographically reallocated in order to achieve perfect similarity.

Example 3.9(cont.) At time t, consider the datasets $\alpha = 1$ and $\alpha' = 3$; in this case, these are the datasets belonging to the class of low *Cd* content (i.e., $\alpha = i = 1$) and the class of high *Cd* content ($\alpha' = i = 3$), respectively. For these datasets, the dissimilarity index is calculated as

$$\eta_{13,t} = \frac{1}{2} \sum_{i=1}^{3} \left| \frac{m_{i1,t}}{\sum_{i=1}^{3} m_{i1,t}} - \frac{m_{i3,t}}{\sum_{i=1}^{3} m_{i3,t}} \right|,$$

where $m_{i1,\,t}$ and $m_{i3,\,t}$ are the number of data values belonging to the class $\alpha = 1$ and $\alpha' = 3$; respectively, that are found in sub-areas $j = 1$, 2 and 3 (a_1, a_2 and a_3) at time t. The numerical value $\eta_{13,\,t}$ is

$$\begin{aligned}
\eta_{13,t} &= \frac{1}{2} \left| \frac{m_{11,t}}{m_{11,t} + m_{21,t} + m_{31,t}} - \frac{m_{13,t}}{m_{13,t} + m_{23,t} + m_{33,t}} \right| \\
&+ \frac{1}{2} \left| \frac{m_{21,t}}{m_{11,t} + m_{21,t} + m_{31,t}} - \frac{m_{23,t}}{m_{13,t} + m_{23,t} + m_{33,t}} \right| \\
&+ \frac{1}{2} \left| \frac{m_{31,t}}{m_{11,t} + m_{21,t} + m_{31,t}} - \frac{m_{33,t}}{m_{13,t} + m_{23,t} + m_{33,t}} \right| \\
&= \frac{1}{2} \left| \frac{0}{15} - \frac{40}{70} \right| + \frac{1}{2} \left| \frac{5}{15} - \frac{30}{70} \right| + \frac{1}{2} \left| \frac{10}{15} - \frac{0}{70} \right| = 0.667,
\end{aligned}$$

i.e., the distributions of low ($\alpha = 1$) and high ($\alpha' = 3$) *Cd* contents are more dissimilar than not.

4 Chrono-geographic clustering and hotspot (coldspot) analysis

The clustering phenomenon has already been addressed in various parts of this chapter. The main conclusion is that it is often necessary for *CTDA* applications to adequately identify the clustering of points in a physical domain associated with natural attributes, events, or objects. In this setting, hotspot (or coldspot) is a point clustering domain defined so that the attribute values are high (or low) at the specified point and the surrounding points. This important situation gave rise to the twofold subject matter of this section.

Chrono-geographic clustering aims at grouping points together based on their chronotopologic similarity features, and **hotspot** (or **coldspot**) **analysis** is a set of *CGS* notions and tools seeking to detect high- (or low-) valued clusters.

Example 4.1 Beyond groups of points that are sufficiently close to each other both in space and in time, clustering may include groups of worldlines that are correlated, or groups of attribute values that satisfy certain conditions (e.g., exceed a threshold) in the chronotopologic domain.

Due to the dominant role of space and time interactions, some clarification is appropriate concerning the matter of chrono-geographic clustering as considered above.

- Chrono-geographic clustering refers to **geographical clustering** with varying **dynamic character** (i.e., occurring in different ways as a function of time) and not to the separate consideration of spatial and temporal clusters.

Although the above description is rather clear-cut, two observations can be made about it. As regards "temporal clusters," it would not make much sense to consider a statistical measure of a purely temporal nearest neighbor at a geographical location, since a point's nearest neighbor in time is uniquely defined, i.e., it is the preceding or the following time instant (and it is often a constant unit time interval). As a consequence, the term "varying dynamic character" denotes that the distinct kinds of interactions between space and time may lead to different types of clustering, such as follows:

① **Continuous dynamic** clustering: Geographical clustering that happens all the time (including the case that clustering happens at certain regions continuously in time).
② **Moving dynamic** clustering: Clustering that occurs only within specific time periods (i.e., clustering may move with time from one area to another).
③ **Sporadic dynamic** clustering: Clustering that leads to hotspots (or coldspots) that appear at specific times and last certain periods.

Generally, identifying hotspots (or coldspots) is a key step in many decision-making and planning studies.

Example 4.2 It is illustrative to review a few applications of hotspot analysis in environmental and health sciences.

- Knowing where and when threshold exceeding pollution occurs in a city during the period of interest is very important in decision-making and risk assessment.
- At which regions of a country and during what times did the reassortment between human influenza A subtype H3N2 and avian influenza subtype H5N1 generate a novel influenza virus?
- When characterizing bacteria outbreaks samples close in time and space would be most likely to be associated with the same outbreak.
- How can crime hotspot maps most effectively guide police action (allocating scarce resources geographically and temporally, where and when demands for police are highest and where and when they are lowest)?
- A map showing locations of regional wildfire hotspots during a time period could reveal that certain hotspot areas are seasonal. Understanding such wildfire features can have important implications for fire resources allocation.

There are different techniques for analyzing chronotopologic patterns and detecting hotspots (coldspots). Two issues are worth emphasizing regarding hotspots (coldspots) techniques:

❶ A point clustering subdomain may be technically considered a **hotspot** (or **coldspot**) if it has a larger concentration of high (or low) attribute values compared to the expected number given a **random** distribution of values.
❷ Some hotspots (coldspots) techniques belong to the **point information** *CGS* group (Section 2 above), whereas some others belong to the **attribute information** *CGS* group (Section 3 above).

For instructional purposes, two techniques representing these two different groups are selected and discussed next, namely, the nearest neighbor point technique (point information group) and the autocorrelation technique (attribute information group).

4.1 Nearest neighbor (*NN*) point techniques

As with other *GCS* notions, the arguments about the nature of the nearest neighbor have traditionally divided along $R^{n,1}$-domain *vs.* R^{n+1}-domain lines. Let m points $\boldsymbol{p}_i$ ($i = 1, \ldots, m$) in the chronotopologic domain of interest be associated with natural attribute values $X(\boldsymbol{p}_i)$.[x] It would be interesting to obtain a measure of how far, on average, a point in a distribution is from its closest neighbor, since this can give the investigator a perception of the density of the point distribution at a particular time. We start with a global nearest neighbor approach described as follows.

In **chrono-geographic nearest neighbor** (*CGNN*) analysis, for each point $\boldsymbol{p}_i$ ($i = 1, \ldots, m$), the **observed mean** *CGNN* is defined as

$$\delta p_O = \frac{1}{m}\sum\nolimits_{i=1}^{m} |\Delta \boldsymbol{p}_i|, \tag{4.1}$$

where the metric $|\Delta \boldsymbol{p}_i|$ determines the nearest to $\boldsymbol{p}_i$ neighbor among the remaining m-1 points.

For a given point distribution consisting of m points, the smaller the δp_O is, the more likely it is that clustering occurs. Accordingly, Eq. (4.1) provides a measure of clustering in terms of the chronotopologic metric $|\Delta \boldsymbol{p}|$.

- As has been emphasized on various occasions, it is mostly application-dependent on how the weight of the time dimension should be considered in combination with that of the space dimension in a chronotopologic metric $|\Delta \boldsymbol{p}|$.

In common with our discussion in Chapter 2, the $|\Delta \boldsymbol{p}_i|$ may have the bi-metric form of Eq. (2.3a) or the uni-metric form of Eq. (2.3b), both in Section 2 of Chapter 2. Accordingly, the concept of chrono-geographic nearest-neighborship would depend on the form of $|\Delta \boldsymbol{p}_i|$ used:

① In $R^{n,1}$ it may include cases that are nearest-neighbors in a **separate** sense, i.e., both the $|\boldsymbol{h}|$ and τ metric components in Eq. (2.3a) in Section 2 of Chapter 2 are minimized.

② In R^{n+1} it may include cases that are nearest-neighbors in a **composite** sense, i.e., the joint metric $|\Delta \boldsymbol{p}|$ in Eq. (2.3a) in Section 2 of Chapter 2 is minimized, and not necessarily both the corresponding individual spatial and temporal lags.

The *CGNN* δp_O is rather an absolute measure and, therefore, a direct comparison of results from different domains is difficult. As a consequence, alternative global *CGNN* clustering indicators have been proposed that basically are computed by comparing the observed data value distribution represented by δp_O against an expected random distribution of these values.

The **standardized** ***CGNN*** of a point distribution in the study domain is defined as

$$\eta = \frac{\delta p_O}{\delta p_R}, \tag{4.2}$$

where δp_R refers to the theoretical mean *CGNN* of a totally random point distribution in the study domain.

The η considers a unique spacetime point cluster that is linked to the entire composite domain Ω and, thus, it simultaneously relates to past and future instances (technically, the cluster consists of uni-points, recall Eq. (2.1b), Section 2 of Chapter 2). The minimum η value is zero (perfectly clustered point pattern), but the maximum value (perfectly separated point pattern) depends on the point density. Generally, if $\eta < 1$, the empirical δp_O is shorter than the theoretical δp_R, and there is a tendency towards clustering; and if $\eta > 1$, there is a tendency towards regularly arranged points. The η value depends on the study domain involved in the computation of δp_R (i.e., as with δp_O, the theoretical expression of δp_R varies depending on the form and the kind of the study domain considered, e.g., $R^{n,1}$ *vs.* R^{n+1}).

[x] Say, pollutant concentration, disease incidence and crime rate.

A commonly used std *CGNN* expression in the $R^{n,1}$ domain is a dynamic (time-depending) formulation that allows the spatial nearest neighbor focus to move with time. This expression, which is useful when the investigator wants to detect if a space-time cluster exists at the specified time t, is defined as follows.

The **standardized (std)** ***CGNN*** of a point distribution in the study region $D \subset R^n$ at time t can be defined as

$$\boldsymbol{\eta}_t = \left(\frac{h_O}{h_R}; t\right), \tag{4.3}$$

where h_O is the arithmetic mean of the observed distances between each distribution location and its nearest neighbor at time t, and h_R is the theoretical mean nearest neighbor distance of a completely random point distribution over region D.[y]

There is a difference between η of Eq. (4.2) in the R^{n+1} domain and $\boldsymbol{\eta}_t$ of Eq. (4.3) in the $R^{n,1}$ domain. In the η case, the nearest neighborship is determined by the composite spacetime metric $|\Delta \boldsymbol{p}|$, whereas $\boldsymbol{\eta}_t$ focuses on geographical clustering as a function of time, given that a point's nearest neighbor in time is the preceding or the following time instant. Accordingly, unlike the η that focuses on a unique jointly chrono-topologic cluster in R^{n+1}, the $\boldsymbol{\eta}_t$ considers several purely geographical clusters in $D \subset R^n$ different time instances of T.[z] Some useful visualizations may be found in earlier Fig 2.10A and B.

4.1.1 Theoretical mean nearest neighbor distance computation

A useful closed-form formula of computing the theoretical mean nearest neighbor distance h_R of a completely random point distribution over the region is discussed in the following example.

Example 4.3 In $R^{n,1}$, consider a region D of unit size, $|D| = 1$, with m randomly located points per unit volume. In this case, one finds that the theoretical mean nearest neighbor distance of a completely random point distribution over region D is

$$h_R = \frac{\Gamma\left(\frac{n}{2}+1\right)^{\frac{1}{n}} \Gamma\left(\frac{n+1}{n}\right) \Gamma(m)}{\sqrt{\pi}\,\Gamma\left(m+\frac{1}{n}\right)}, \tag{4.4}$$

where Γ is the gamma function, and if m is very large, then the following approximation is obtained,

$$h_R \approx \frac{\Gamma\left(\frac{n}{2}+1\right)^{\frac{1}{n}} \Gamma\left(1+\frac{1}{n}\right)}{\sqrt{\pi}} \left(\frac{1}{m}\right)^{\frac{1}{n}}, \tag{4.5}$$

These equations can be easily extended to any region size $|D|$. So, for $n = 2$ the above approximation gives the familiar expression

$$h_R = \frac{1}{2}\sqrt{\frac{|D|}{m}}, \tag{4.6}$$

which is the expected value of the mean nearest neighbor h_{nn} of random points that are uniformly distributed in the study region D at time t, i.e., $h_R = \overline{h}_{nn}$, where $h_{nn} = \frac{1}{m}\sum_{i=1}^{m} h_i$ (h_i is the distance between location $\boldsymbol{s}_i$ and its nearest neighbor); and the associated h_{nn} variance is $Var(h_{nn}) = 0.068\frac{|D|}{m^2}$.

4.1.2 Numerical CGNN scale

In comparative terms, the $\boldsymbol{\eta}_t$ generally attempts to characterize a point distribution according to whether it is clustered, random or regular at each time instance t. In this setting, Eq. (4.3) produces a result between 0 and

[y] The η_t belongs to the class of measures defined by Eq. (2.1a) of Section 2.

[z] These clusters relate to bi-points, recall Eq. (2.1a), Section 2 of Chapter 2.

2.149. While interpreting the η_t measure, an investigator should keep in mind the following notable numerical values:

① The more closely the points are **clustered** together, the closer to 0 the η_t will be at the specified time t (since the average nearest point distance in the point distribution decreases).
② The closer η_t gets to 1, the more **randomly** spaced the points are (this follows from the definition of η_t).
③ The value of *std CGNN* approaches 2.149 for perfectly uniformly spaced points; hence, the closer *std CGNN* is to 2.149, the more **uniformly spaced** the data are.

In terea, the significance of the difference between the observed and the random expectation in Eq. (4.3) can be tested using a Z score. The primary issue η_t is its strong dependence on the study area D at time t. If you take a given point distribution and insert it into a larger D, the η_t will decrease until, for a sufficiently large area, it goes to 0. Conversely, if one decreases D, the η_t value will increase. Hence, the careful determination of D at the time t of interest is very important. The η_t works well for homogeneous point patterns (the points are all relatively dispersed), but interpretation becomes more difficult for heterogeneous patterns in which distributions of *clustered* points exist. The η_t offers a simple way for testing data clustering with the potential disadvantage that it does not account for chronotopologic attribute autocorrelation.

The following example illustrates the *step-by-step* **computational** process of *std CGND* in practice.

Example 4.4 This is a continuation of Example 2.5 of Section 2 above. The goal is to calculate the *std CGNN* of breast cancer (*BC*) in Hangzhou (China) at the initial time $t = 0$ (i.e., the year 2008). We notice that in practice it is often useful to implement the dynamic geographic notation in Eq. (4.3), which then can be replaced by $\eta_t = \frac{h_{O,t}}{h_{R,t}}$, where $h_{O,t}$ and $h_{R,t}$ are, respectively, the h_O and h_R of Eq. (4.3) at the given time t.

Steps 1–3: Same as in Example 2.5 of Section 2.

Step 4: The total study area at a time $t = 0$ is $D_0 = 9$, and the distances between the $m_0 = 13$ locations are calculated in Table 4.1. The shortest nonzero distances are shown in bold letters. (See Table 4.1.)

Step 5: Introducing them into Eq. (4.2) at $t = 0$, we can obtain the *CGNN* for *BC*,

$$\eta_0^{BC} = 2\sqrt{\frac{1}{m_0 D_0}} \sum_{i=1}^{m} h_{i,0} = 2\sqrt{\frac{1}{13 \times 9}} \times 2.2379$$
$$= 0.4138.$$

Since the *CGNN* is considerably less than 1, the point distribution is considered clustered. The *std CGNN* at other times, 2009–2012, are calculated in a similar fashion. It is worth noticing that when using the *std CGNN* in practice, one does not have to compare a location $\boldsymbol{s}_i$ at time t to all other locations of the distribution at the same time—only the ones that look like potential nearest neighbors. Just because $\boldsymbol{s}_j$ is the nearest neighbor to $\boldsymbol{s}_i$ does not mean that $\boldsymbol{s}_i$ will always be the nearest neighbor to $\boldsymbol{s}_j$. A point $\boldsymbol{s}_i$ may be the nearest neighbor to several $\boldsymbol{s}_j$, or to no points at all.

4.1.3 Other CGNN measures

As with other measures discussed earlier, one may obtain a single number measuring the relative dispersion of a point distribution for the entire chrono-geography of interest, as follows:

The **T-std** *CGNN* is defined as the time-averaged $\overline{\eta}_t$ within a period T, i.e.,

$$\overline{\overline{\eta}} = \frac{1}{T} \int_T dt \overline{\eta}_t \tag{4.7}$$

for all $t \in T$.

The *T-std CGNN* offers an average quantitative assessment of how far, on average, a point

TABLE 4.1 The distances between various locations of breast cancer at the time $t = 0$.

	1	2	3	4	5	6	7	8	9	10	11	12	13
1	0.0000	0.8659	0.5655	0.8882	0.5809	1.0309	0.1394	0.2392	0.0771	**0.0753**	0.9476	0.6655	0.5683
2	0.8659	0.0000	**0.3045**	0.7319	0.3757	0.7888	0.8939	0.6939	0.8587	0.8012	0.8788	0.5257	0.9019
3	0.5655	0.3045	0.0000	0.7135	**0.2381**	0.8176	0.6080	0.4204	0.5655	0.5043	0.8435	0.4450	0.7194
4	0.8882	0.7319	0.7135	0.0000	0.4771	**0.1440**	0.8012	0.6550	0.8244	0.8177	0.1477	0.2743	0.4472
5	0.5809	0.3757	0.2381	0.4771	0.0000	0.5880	0.5666	0.3617	0.5492	0.5068	0.6054	**0.2069**	0.5294
6	1.0309	0.7888	0.8176	**0.1440**	0.5880	0.0000	0.9452	0.7964	0.9679	0.9599	0.1848	0.3975	0.5809
7	0.1394	0.8939	0.6080	0.8012	0.5666	0.9452	0.0000	0.2054	**0.0679**	0.1308	0.8444	0.6074	0.4426
8	0.2392	0.6939	0.4204	0.6550	0.3617	0.7964	0.2054	0.0000	0.1913	**0.1652**	0.7266	0.4263	0.4020
9	0.0771	0.8587	0.5655	0.8244	0.5492	0.9679	0.0679	0.1913	0.0000	**0.0676**	0.8779	0.6132	0.4919
10	0.0753	0.8012	0.5043	0.8177	0.5068	0.9599	0.1308	0.1652	**0.0676**	0.0000	0.8820	0.5910	0.5180
11	0.9476	0.8788	0.8435	**0.1477**	0.6054	0.1848	0.8444	0.7266	0.8779	0.8820	0.0000	0.3985	0.4368
12	0.6655	0.5257	0.4450	0.2743	**0.2069**	0.3975	0.6074	0.4263	0.6132	0.5910	0.3985	0.0000	0.4060
13	0.5683	0.9019	0.7194	0.4472	0.5294	0.5809	0.4426	**0.4020**	0.4919	0.5180	0.4368	0.4060	0.0000

in a specified distribution is from its nearest neighbor in space during the entire time period of interest T. This is the average overall possible times of the corresponding spatial nearest neighbors, each one of them associated with different times. In this sense, $\bar{\bar{\eta}}$ is a temporal average of the spatial nearest neighbors $\bar{\eta}_t$, each one of them associated with a different time.

Another *std CGNN* measure is also considered that focuses on the lag varying correlation between points separated by Δp.

The ***CGNN* correlation** function of a point distribution in the study domain is defined as

$$\rho(\Delta p) = \frac{N_O(\Delta p)}{N_R(\Delta p)} - 1, \tag{4.8}$$

where $N_O(\Delta p)$ and $N_R(\Delta p)$ are the number of observed and random point pairs, respectively, separated by $|\Delta p|$.

Interpretatively, $\rho(\Delta p)$ provides a measure of the excess of probability to observe a pair of points separated by Δp compared to a random arrangement. One can calculate $\rho(\Delta p)$ for different Δp and estimate the corresponding probabilities.

Lastly, the point information statistics discussed in Section 2 (*CGMC*, *CGSD*, etc.) can also serve as valuable auxiliary tools of hotspot analysis, as they offer measures of data point centrality and dispersion that are useful for describing data patterns or for comparing two data different distributions.

4.2 The issue of statistically significant attribute hotspots (coldspots)

Unlike the previous section in which point information *CGS* were considered, the present section focuses on attribute information *CGS* (see, also, the relevant distinction in item ❷ of Section 4.3 above). In particular, chronotopologic attribute autocorrelation techniques examine how similar those attribute values that are closer to each other are. Such techniques based on the *CTMI* and *CTGI* were introduced in Section 3. As was observed in this section, the *CTMI* and *CTGI* typically measure the tendency of attribute values to cluster or the extent to which points close together have similar values on average than those farther apart.

Example 4.5 Typically, *CGMI* values near +1 indicate that similar data tend to cluster, whereas values near −1 indicate that dissimilar values tend to cluster.

Both global and local versions of the *CTMI* and *CTGI* attribute statistics were considered in Section 3. Even if there is no global correlation (or no global clustering), nevertheless, clusters may still be found at a smaller scale using local *CTMI* and *CTGI* statistics (the local indexes differ from the global indexes in that the former are applied to each local unit). Otherwise said, in this case, clustering is detected by downscaling from global to local statistics.

However, while the *CTMI* and *CTGI* can be used to find out whether the considered attribute values are clustered in general (autocorrelated), a quantitative tool would be useful that can specify whether the attribute values are clusters of high (or low) values in particular. This is the objective of the following tool, which is a straightforward chronotopologic extension in the $R^{n,1}$ domain of the original purely spatial *G* statistic (Getis and Ord, 1992).[aa]

The **chrono-geographic G statistic (*CGGS*)** of hotspot analysis is expressed as the ratio

$$G^*(s_i;t) = \frac{\sum_{j=1}^{m} w_{ij;t}\left[\chi(s_j,t) - \overline{X}(t)\right]}{S(t)\sqrt{\frac{n\sum_{j=1}^{m} w_{ij;t}^2 - \left(\sum_{j=1}^{m} w_{ij;t}\right)^2}{n-1}}}, \quad (4.9)$$

where $\chi(s_j,t)$ is the attribute value at location s_j and time t, $w_{ij;\,t}$ is the weight between locations s_i and s_j at t, m is the total number of locations, and

$$\overline{X}(t) = \frac{\sum_{j=1}^{m} \chi(s_j,t)}{m},$$
$$S(t) = \sqrt{\frac{\sum_{j=1}^{m} \chi(s_j,t)^2}{m} - \overline{X}(t)^2} \quad (4.10\text{a,b})$$

are the spatial attribute mean and variance, respectively, as functions of time.

The *CGGS* identifies statistically significant hotspots (or coldspots). As in previous cases, underlying $G^*(s_i;t)$ is the idea that the density of points within a certain subdomain may be compared against a completely random model (i.e., in which attribute values occur randomly at these points). Particularly, the null hypothesis that "there is no clustering of high or low attribute values within the specified distance of a location s_i at time t (the *CGGS* being close to zero)" *vs.* the hypothesis that "there is a clustering of high or low values within the specified distance of location s_i at time t (a significant positive *CGGS* value implies a clustering of high values, and a significant negative *CGGS* value indicates a clustering of low values)."

Hotspot analysis in terms of Eq. (4.9) can be used to delimit and identify attribute subdomains that are strongly or weakly associated. In this respect, the *CGGS* at each time t of interest measures the degree of attribute association that results from the concentration of weighted points or region centroids and all other weighted points within a certain distance from the point of interest. Hotspots (or coldspots) at 99%, 95%, and 90% significance level can be obtained accordingly. Therefore, beyond assessing the density of points in a given subdomain (point clustering), hotspot analysis in terms of the *CGGS* statistic may also assess the extent of

[aa] Hotspot analysis in terms of the *CGGS* measure provides information about local heterogeneity at a fixed time, whereas stratified heterogeneity could be tested by the q-statistic (Wang et al., 2016).

attribute value interaction at these points, thus improving one's understanding of the relevant chronotopologic patterns. On the other hand, clusters composed of only a few data may inflate the *CGGS* value.

Example 4.6 Let us revisit the application considered in Example 1.11 of Section 1 of Chapter 3. This application is concerned with the study of the monthly strength of the association between global climate dynamics (multivariate El Niño-Southern Oscillation index, *MEI*) and the *HFRS* across east China during 2012. Using the hotspot analysis described above, a suitable *HFRS* dataset was created, data autocorrelation testing was performed, and the hotspot maps of the corresponding *HFRS-MEI* associations were computed and plotted as in Fig. 4.1 (for more details, the readers are referred to He et al., 2018).

In Exercise 23, an extension of the *CGGS* formulation of Eq. (4.9) in a composite spacetime context is considered.

4.3 Concluding remarks

Chrono-geographic clustering can determine whether points that are close in space are also close in time and vice versa. This kind of determination is important in many fields, such as, e.g., the decision whether a disease is infectious or not, the control of excess contamination incidents, and the study of clusters of stars.

As a matter of fact, during the last decade or so, the application of hotspot analysis has increased considerably mainly due to the advent of *TGIS*-based software. In this setting, different mapping techniques are used to depict hotspots, including the following (Silverman, 1986; Chainey et al., 2002; NIJ, 2005; MSPH, 2019).

① Point mapping uses dots to display the distribution of objects, events, or attribute values in a chronotopologic domain.
② Geometrical surfaces around point clusters are also used to display hotspots on a map.
③ Thematic maps use geographic boundaries or quadrats to aggregate data and attribute details according to thematic areas.
④ The kernel density estimation technique interpolates (smooths) discrete data points so that a continuous map is produced and visualized.

Concerns associated with the implementation of the above mapping techniques include the following: Cluster identification in technique ① based on the observation of points within a domain could be troublesome as it is directly affected by the observer's visual perception. The geometrical surfaces of technique ② may not be very informative if the hotspot type under consideration does not follow the surface form. Detecting hotspots by means of technique ③ are restricted to the shape of the thematic units, which can lead to misinterpretations, when important domain features (e.g., population density) are not taken into consideration. Although aesthetically pleasing and producing easily understandable outputs, technique ④ is more subjective than other techniques, it involves a certain level of arbitrariness and many trial-and-error steps (several user-defined parameters must be established), and it is affected by the investigator's experience. Lastly, a point worth stressing is that when selecting a hotspot detection technique it should be kept in mind that several of the existing tools only measure gross space-time interactions rather than assessing chronotopologic clustering.

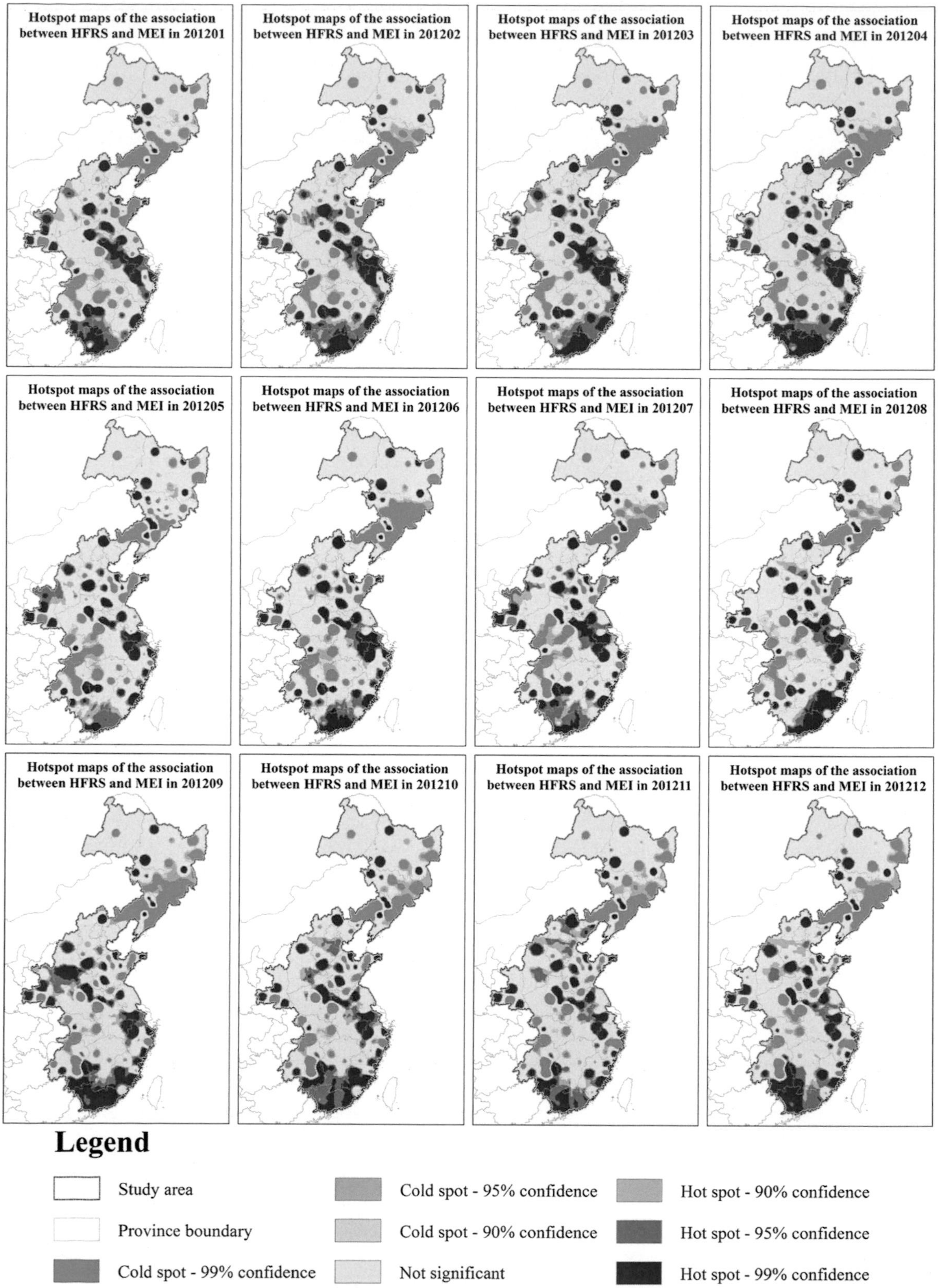

FIG. 4.1 Hot spot map of *HFRS-MEI* association in East China.

5 Practice exercises

1. Use the dataset shown in Table 5.1 below to calculate the *CGMC, CGSD, CGRD* (assume an area of 2 *units*^{2}), *CGNN;* describe the data distribution across space and time.
2. Calculate the *CGMI* and *CGGI* at various time instants of the dataset displayed in Fig. 5.1 below. The numbers in the figure are the data values at various space-time points.
3. Calculate the *CGSI* at various time instants of the dataset displayed in Table 5.2. Then, calculate the *CGDI* by comparing Woodrush and Holly at the two time instants.
4. Follow the steps of Example 2.5 in Section 2 above, and use the data provided in Table 5.3 to calculate *CGMC, CGSD, CGRD,* and *CGNN.*
5. Consider the point distribution in Fig. 2.2 of Section 2. Calculate the $|\overline{h}|$ and $\delta\overline{h}$ in terms of the corresponding coordinates $(s_{i,1}, s_{i,2})$ and $(s_{i,1}', s_{i,2}')$.

TABLE 5.1 Dataset.

s_1	s_2	t	s_1	s_2	t
1.3162	4.8003	1	1.0746	4.1966	2
1.0488	4.1419	1	1.1288	4.2511	2
1.1392	4.4218	1	1.4204	4.6160	2
1.2734	4.9157	1	1.1271	4.4733	2
1.4788	4.7922	1	1.1758	4.5678	3
1.4824	4.9595	1	1.4154	4.0759	3
1.0788	4.6557	1	1.2926	4.0540	3
1.4853	4.0357	1	1.2749	4.5308	3
1.4786	4.8491	1	1.4586	4.7792	3
1.2427	4.9340	1	1.1429	4.9340	3
1.4455	4.8143	2	1.3786	4.1299	3
1.4796	4.2435	2	1.3769	4.5688	3
1.2736	4.9293	2	1.1902	4.4694	3
1.0693	4.3500	2			

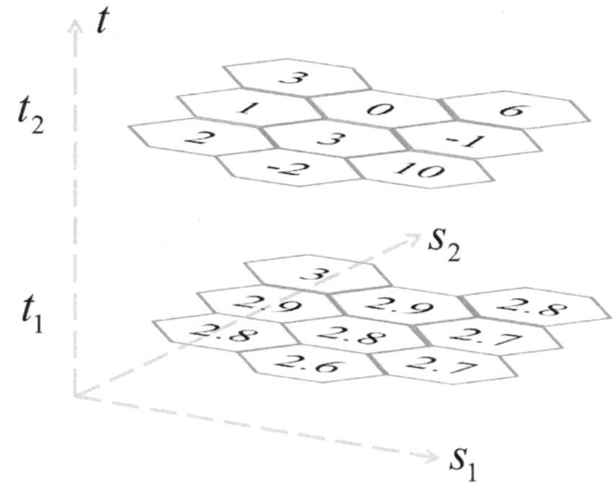

FIG. 5.1 Dataset.

TABLE 5.2 Dataset.

	Number (n)	
Species	t = **1**	t = **2**
Woodrush	2	3
Holly (seedlings)	8	7
Bramble	1	2
Yorkshire Fog	1	2
Sedge	3	1

TABLE 5.3 Dataset.

Time	**No**	$s_{i,1}$	$s_{i,2}$	**No.**	$s_{i,1}$	$s_{i,2}$
0	1	1.9331	1.2197	7	3.0526	0.9995
	2	4.1779	3.0562	8	3.7537	2.1539
	3	3.1681	1.6722	9	3.9612	1.1388
	4	3.7497	2.9709	10	4.6725	3.9028
	5	3.6399	2.5348	11	3.4821	2.3201
	6	3.4402	3.3171	12	2.1495	2.4955

Continued

TABLE 5.3 Dataset—cont'd

Time	No	$s_{i,1}$	$s_{i,2}$	No.	$s_{i,1}$	$s_{i,2}$
1	1	3.3494	1.6960	8	3.9532	2.4442
	2	4.1468	2.6865	9	5.8861	0.8704
	3	2.1738	1.7817	10	5.5745	2.0902
	4	4.4266	2.1670	11	2.6468	2.7894
	5	3.2100	3.0779	12	4.9586	2.5968
	6	3.0298	3.1026	13	4.2818	2.5117
	7	3.6240	2.4869			
2	1	2.8606	1.2649	9	3.8701	0.6701
	2	3.7616	3.5243	10	4.0591	2.4446
	3	4.6813	1.9828	11	4.0411	1.9596
	4	3.2890	2.0276	12	4.6314	2.3758
	5	3.9724	2.2739	13	4.0021	3.1129
	6	3.0357	2.3923	14	3.0720	2.6439
	7	4.4229	1.4235	15	3.5499	2.4978
	8	3.0316	0.9865			
3	1	2.6948	2.6402	10	4.3633	1.4330
	2	3.8578	1.9051	11	3.7366	1.2769
	3	2.6497	0.7633	12	4.3081	1.2207
	4	5.1965	2.9706	13	2.0050	1.3180
	5	4.2378	2.3628	14	3.5932	1.2936
	6	4.8069	2.2917	15	2.7860	0.8073
	7	2.5509	2.1361	16	2.1744	2.2597
	8	2.7395	2.0431	17	3.9849	2.0386
	9	3.6438	0.9124			
4	1	3.5881	1.9776	7	4.3310	2.5778
	2	2.8891	2.0481	8	3.4876	1.8490
	3	2.8219	2.4772	9	3.9422	1.3696
	4	3.8033	2.9498	10	2.9828	2.7582
	5	3.9568	1.8771	11	3.7375	1.8555
	6	3.5881	1.9776	12	4.3310	2.5778

6. Revisit Example 3.9 of Section 3 and calculate the corresponding *CGSI* at time t'.
7. Revisit Example 3.9 of Section 3 and calculate the corresponding *CGEI* at time t'.
8. Describe natural attributes characterized by chronotopologic dependence in your field of interest.
9. Which of the following statements are true?
   - **(a)** A sample survey is an example of an experimental study.
   - **(b)** An observational study requires fewer resources than an experiment.
   - **(c)** The best method for investigating causal relationships is an observational study.

   Justify your answers.
10. Consider the mapped pattern (cell points and values) (Fig. 5.2): the corresponding attribute values $X(\mathbf{s}_j, t)$ are:

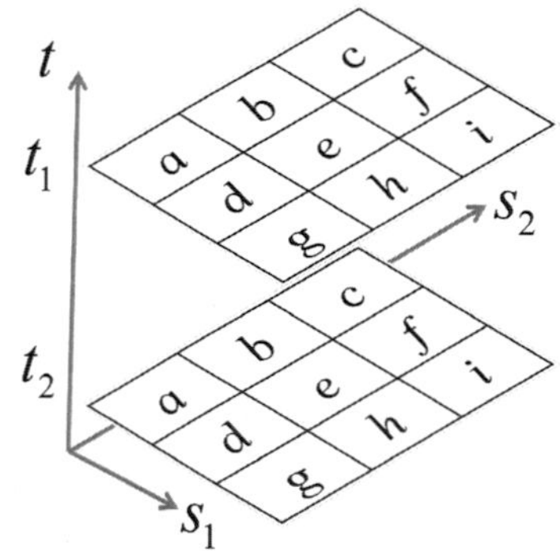

FIG. 5.2 Dataset.

$j =$	a	b	c	d	e	f	g	h	i
$X(s_j,t_1)$	9	6	3	8	5	2	7	4	1
$X(s_j,t_2)$	10	6	9	5	1	2	3	4	2

Calculate the *CTMI* and *CTGI* for the above data.

11. Calculate the chronotopologic mean center, standard distance, *CTMI*, and *CTGI* of the PM_{10} observations in Beijing City, China on Jan 05, 2018, shown in Table 5.4.

12. Express in double summation form:

(a) Eq. (3.1) of Section 3, i.e., the equation $c_{\Omega} = \frac{1}{W\sigma_X^2}\widetilde{\chi}^{T}\mathbf{W}\widetilde{\chi}$.

(b) Eq. (3.5) of Section 3, i.e., the equation $c_{i,t} = \widetilde{\chi}_{i,t}\boldsymbol{w}_{i,t}^{T}\widetilde{\chi}_{t}$.

(c) Eq. (3.7) of Section 3, i.e., the equation $\gamma_{\Omega} = \frac{m-1}{2mW\sigma_X^2} tr\mathbf{W}\left(\Delta\widetilde{\chi}^2\right)^{T}$.

(d) Eq. (3.10) of Section 3, i.e., the equation $\gamma_{s_i,\,t} = \boldsymbol{w}_{i,\,t}^{T}(\Delta\chi_{i,\,t}^{T})^{T}$.

TABLE 5.4 Spatial coordinates of the PM_{10} monitoring stations and the observations at various time instants.

Longitude	Latitude	07:00 a.m.	08:00 a.m.	09:00 a.m.
116.417	39.929	70	63	49
116.407	39.886	52	57	71
116.339	39.929	71	69	73
116.352	39.878	60	66	62
116.397	39.982	102	51	41
116.461	39.937	76	65	43
116.287	39.987	50	50	60
116.174	40.09	64	55	88
116.279	39.863	56	74	105
116.146	39.824	42	43	42
116.184	39.914	42	50	78
116.136	39.742	39	47	72
116.404	39.718	138	112	108
116.506	39.795	66	65	78
116.663	39.886	50	40	45
116.655	40.127	30	23	24
116.23	40.217	47	28	31
116.106	39.937	25	35	58

Continued

TABLE 5.4 Spatial coordinates of the PM_{10} monitoring stations and the observations at various time instants—cont'd

Longitude	Latitude	07:00 a.m.	08:00 a.m.	09:00 a.m.
117.1	40.143	24	31	34
116.628	40.328	28	28	35
116.832	40.37	38	49	48
115.972	40.453	112	110	126
116.22	40.292	28	28	30
115.988	40.365	46	42	55
116.911	40.499	21	25	20
116.783	39.712	115	118	118
116.3	39.52	89	95	112
116	39.58	96	81	103
116.395	39.899	56	59	52
116.394	39.876	67	72	60
116.349	39.954	94	73	60
116.368	39.856	85	87	112
116.483	39.939	106	113	90

13. Prove:

(a) Eq. (3.6) of Section 3, i.e., the equation $c_{S,t} = \frac{1}{W_t\sigma_{X_t}^2}\sum_{i=1}^{m_t} c_{i,t}$.

(b) Eq. (3.11) of Section 3, i.e., the equation $\gamma_{S,t} = \frac{m_t-1}{2m_tW_t\sigma_{X_t}^2}\sum_{i=1}^{m_t}\gamma_{i,t}$.

14. Think of a few real-world cases that render the assumptions of the following techniques so unrealistic as to be perniciously misleading: (*a*) *CTMI*, (*b*) *CTGI*, (*c*) *CTNN*, and (*d*) *CTGS*.

15. Do you agree that the *CGSD* is the average distance all features vary from the *CGMC* and measures the compactness of a distribution? Explain.

16. Concerning Eq. (2.16) of Section 2, show that $\delta\overline{h}_t = \left(\sqrt{g^2(s_1)+g^2(s_2)}; t\right)$.

17. Consider that an investigator is looking for clusters along a unidimensional region of length L. Assuming that m random points are located in this region:
    (a) What is the probability p that the minimum distance between any pair of points is larger than some value d?[ab]
    (b) What is the value of the expected minimum distance h_R as a function of m and L?
18. In R^3, what is the mean (straight-line) distance between two randomly chosen points:
    (a) On the unit sphere of radius r?
    (b) In the unit sphere?
19. Consider an n-dimensional Euclidean sphere D of unit volume $|D| = 1$ with m randomly located points per unit volume. Assuming a certain random point as the reference, there are $m - 1$ other random points within D with the reference point at its center. A point is the kth neighbor of another (the reference point) if there are exactly $k - 1$ other points that are closer to the latter than the former. What is the mean nearest distance h_R between a given reference point and its kth neighbor $(k < m)$?[ac]
20. Consider the stochastic equivalence logic expression.

$$P_j[\chi_{1,t}\in C_i \leftrightarrow \chi_{2,t}\in C_i] = P_1[(\chi_{1,t}\in C_i \wedge \chi_{2,t}\in C_i)\vee (\chi_{1,t}\notin C_i \wedge \chi_{2,t}\notin C_i)].$$

What is the corresponding expression for the $CGEI$ $\iota_{j,t}^{\leftrightarrow} = P_j[\cup_{i=1}^{l_t}(\chi_{1,t} \in C_i \leftrightarrow \chi_{2,t} \in C_i)] = 1$ when one class is empty, i.e., $C_i = \varnothing$, and how would you interpret it?
21. In the case of Example 3.9 of Section 3, show that

$$\sum_{i=1}^{3} P_j[\chi_{1,t}\in C_i \leftrightarrow \chi_{2,t}\in C_i] = 2\sum_{i=1}^{3} P_j[\chi_{1,t}\in C_i \wedge \chi_{2,t}\in C_i] + 3 - \sum_{i=1}^{3} P_j[\chi_{2,t}\in C_i] - 1,$$

$(j = 1, 2, 3)$, where $P_1[\vee_{i=1}^{3}(\chi_{1,t} \in C_i \wedge \chi_{2,t} \in C_i)] = \sum_{i=1}^{3} P_1[\chi_{1,t} \in C_i \wedge \chi_{2,t} \in C_i]$.
22. The selections $\chi_{1,t} \in C_i \leftrightarrow \chi_{2,t} \in C_i$ and $\chi_{1,t} \in C_{i'} \leftrightarrow \chi_{2,t} \in C_{i'}$ being independent for $i \neq i'$, show that Eq. (3.17) of Section 3 can be written as

$$\iota_{j,t}^{\leftrightarrow} = 1 - \sum_{i=1}^{l_t} P_j[\chi_{1,t}\in C_i \leftrightarrow \chi_{2,t}\in C_i] - (-1)^{l_t-1}\sum_{i=i_1,\ldots,i_l} \prod_{i=i_1}^{i_l} P_j[\chi_{1,t}\in C_i \leftrightarrow \chi_{2,t}\in C_i];$$

similarly, the selections $\chi_{1,t} \in C_i \rightarrow \chi_{2,t} \in C_i$ and $\chi_{1,t} \in C_{i'} \rightarrow \chi_{2,t} \in C_{i'}$ being independent for $i \neq i'$, show that Eq. (3.21) of Section 3 can be written as

$$\begin{aligned}\iota_{j,t}^{\rightarrow} = 1 &- \sum_{i=1}^{l_t} P_j[\chi_{1,t}\in C_i \wedge \chi_{2,t}\in C_i] + \sum_{i=1}^{l_t} P_j[\chi_{2,t}\in C_i] - l_t \\ &+ \sum_{i<i'} P_j[\chi_{1,t}\in C_i \rightarrow \chi_{2,t}\in C_i] \times P_j[\chi_{1,t}\in C_{i'} \rightarrow \chi_{2,t}\in C_{i'}] \\ &- \sum_{i<i'<i''} P_j[\chi_{1,t}\in C_i \rightarrow \chi_{2,t}\in C_i] \times P_j[\chi_{1,t}\in C_{i'} \rightarrow \chi_{2,t}\in C_{i'}] P_j[\chi_{1,t}\in C_{i''} \rightarrow \chi_{2,t}\in C_{i''}] + \ldots - (-1)^{l_t-1}\prod_{i=1}^{l_t} P_j[\chi_{1,t}\in C_i \rightarrow \chi_{2,t}\in C_i],\end{aligned}$$

which are expressions that are useful in computational applications?
23. Extend the $CGGS$ formulation of Eq. (4.9) in Section 4.2 in a composite spacetime context, assuming an attribute $X(\boldsymbol{p})$, $\boldsymbol{p} = (s,t)$, and defining the corresponding $G^*(\boldsymbol{p}_i)$.

[ab] The p should be a function of m, L, and d.

[ac] It should be a function of m, n, and k.

CHAPTER

5

Classical geostatistics

1 Historical introduction

Geostatistics is a well-established branch of applied mathematics *in the broad sense*,[a] which ultimately offers a collection of *CTDA* tools aimed at understanding and modeling natural attribute variability. A brief introduction into some fundamental geostatistical concepts and ideas is presented in this introductory section, which serve to organize our thinking around the more technical description of the geostatistical models and methods to follow in later sections. These models and methods are important in a wide variety of *CTDA* applications.

1.1 Spatial geostatistics: Basic postulates

As was discerned in Chapter 1, historically, nowadays geostatistics is considered a rather mature discipline.

In its purely spatial form, geostatistics can be probably traced back to the forestry studies of the 1940s (in Sweden and elsewhere), and the meteorological and mineral studies of the 1950–60s (in Russia, France, South Africa and elsewhere).

The five basic *postulates* of spatial geostatistics are listed in Table 1.1. The importance of the 1st postulate stems from the fact that it has close links with both geostatistical *inquiry* (drawing inferences about an unknown quantity in terms of known ones) and *inference* (identifying elements needed to draw reasonable conclusions, to form conjectures and hypotheses, to process relevant information, and to deduce the consequences). Both the known quantities and the identifiable elements involve data that are adequately configured across space.

Example 1.1 In environmental geostatistics, typically the known quantities are the available pollutant data at a number of monitoring

[a] *Sensu lato*, this refers to applied mathematics that rely on physical insight and interpretive reasoning to understand the analytical expressions representing the phenomenon. This viewpoint has certain significant consequences, e.g., the analytical expressions can be potentially reduced to simpler ones, free of any unnecessary complications, and then one can proceed with the solution of the simpler yet equally meaningful expression.

Quantitative Analysis and Modeling of Earth and Environmental Data
https://doi.org/10.1016/B978-0-12-816341-2.00001-0

TABLE 1.1 Postulates of spatial geostatistics.

Postulate	Description
1st: Data-based delineation	For spatial data analysis purposes, a natural attribute is characterized by the regional distribution of a sufficient number of measurements or observations
2nd: Spatial variation	A particular structure of attribute distribution representing an elemental relationship between attribute values that depends on space
3rd: Spatial homogeneity	Spatial attribute dependence structure characterized by a correlation that remains the same between any pair of points separated by the same vector distance, regardless of the locations of the two points
4th: Spatial isotropy	Special case of the 3rd postulate where correlation remains the same between any pair of points separated by the same distance magnitude, regardless of direction
5th: Mathematical modeling	A natural attribute occurring in conditions of uncertainty is mathematically represented as a **spatial random field** (*SRF*) that considers several possible attribute realizations with varying probabilities assigned to them

locations, and the unknown quantity is the pollutant value at an unmonitored location. Another immediate logical consequence of the previously mentioned postulate is that a limited environmental dataset cannot adequately support spatial estimation.

As regards the 2nd postulate, real-world attribute variability is present along every direction in space, and it gets higher as data dispersion increases.[b] It may vary horizontally and/or with depth, continuously or abruptly (*exempli causa*, certain physical features of spatial variability for different attributes are listed in Table 1.2). As it will be shown in Section 2.3, a useful quantitative expression of spatial variability is the so-called *variogram* function. *In-situ*, the 2nd postulate has a definite physical significance worthy of further attention in subsequent sections.

Example 1.2 A few common cases of attribute variability are as follows:

TABLE 1.2 Attribute variation features of different natural systems: short-range spatial variation (*SRSV*), vertical variation (*VV*), temporal variation (*TV*), std sampling density (*SSD*), remote-sensing detectability (*RSD*); high (*H*), medium (*M*), low or nonexistent (*L*) variability.

Natural system	Attribute examples	*SRSV*	*VV*	*TV*	*SSD*	*RSD*
Air pollution	PM, Ozone, and SO_2 concentrations	*H*	*H*	*H*	*H*	*L*
Soil contamination	Heavy metal and radioactive contamination	*H*	*L*	*M*	*H*	*L*
Water quality	Arsenic, phosphorus, and ammonium concentrations	*M*	*M*	*M*	*M*	*L*
Weather forecasting	Wind velocity, temperature, precipitation, cloud density	*M*	*M*	*H*	*M*	*H*
Public health	Susceptibility, infection, and mortality rates	*L*	*L*	*M*	*H*	*L*
Socioeconomic status	Population, poverty, and life expectancy rates	*M*	*L*	*M*	*H*	*H*

[b] Naturally, the opposite is the case with attribute value dependency, which gets weaker as data dispersion increases.

- If pollution is detected at a sample spot across a lake, it is very probable that places close to this spot are also polluted.
- When the presence of an adult tree inhibits the development of others, such inhibition decreases with distance, and beyond a certain radius, other big trees will be found.

In the 3rd postulate, homogeneity is associated with a variogram shape (Section 2.3) that is only a function of the vector distance between the location pairs, and not of the locations themselves. In practice, homogeneity is assumed in the largest number of geostatistical applications. In the 4th postulate isotropy is associated with a variogram shape that is only a function of the distance magnitude between the location pairs.

Example 1.3 Since soil data are the result of geological processes of medium and long duration, the homogeneity hypothesis is justified on the basis of the relative stability of these processes.

According to the 4th postulate, physically the *SRF* model integrates *spatial structure* (based on natural laws, physical mechanisms, and empirical models) with *uncertainty* (epistemic or ontic, conceptual or technical). Mathematically, the *SRF* is denoted as

$$X(\boldsymbol{s}) \sim f_X(\chi),$$

where $\boldsymbol{s} = (s_1, \ldots, s_n)$ are spatial coordinates, χ is a spatial attribute realization with probability density function (*PDF*) $f_X(\chi)$, and $\sim$ means that the *SRF* at $\boldsymbol{s}$ obeys f_X.

- In most real-world situations, it is appropriate to consider the spatial dataset not as a set of independent samples, but as one **space-dependent** realization of an *SRF*.

In other words, contrary to the usual independent samples assumption of classical statistics (*CS*), where each observation carries an independent information, *sine loco et anno*, in the *SRF* case all the observations are combined to describe the spatial pattern of the studied attribute.

Example 1.4 According to the *SRF* model, the physical prediction process is not deterministic but conditional, i.e., given the boundary and initial conditions of the phenomenon of interest, and assuming that certain events do not intervene in the process, then, a particular outcome (realization) may be expected—probably and approximately and within the limits of error. Such conditional predictions can be useful, even when one knows in advance that the conditions may not be satisfied.

While the *timeless*-space notions and techniques of spatial geostatistics will continue to be useful, the field faces a number of profound challenges whose solutions lie in the adoption of a richer approach based on attribute representations that account for natural variations across both space and time. This is based on the following sound standpoint:

- Knowledge of the real world is concerned with an **integrated** representation of space-time rather than with the spatialization of the temporal.

Over the last two decades, this standpoint gave rise to a considerable generalization of the geostatistical concepts and techniques.

1.2 Spatiotemporal geostatistics: Methodological notions

The first systematic introduction to spatiotemporal geostatistics was presented about 20 years ago in a book by Christakos (2000). The eight basic methodological notions underlying these early developments are listed in Table 1.3.

According to the 1st notion of Table 1.3, and as we saw in Chapter 2, a continuum of points may be seen either

(a) as a spatial arrangement of events combined with their temporal order where time dimension does not exist in the same way space dimensions do ($R^{n,1}$ domain), or

TABLE 1.3 Methodological notions of spatiotemporal geostatistics.

Notion	Description
1st: Chronotopologic continuum	Natural attribute occurs in a chronotopologic continuum of points enabling a structure imposed on it by means of *physical relationships* among its values uniquely linked to *geometrical relationships* among the points
2nd: Field	Attribute values are assigned to chronotopologic continuum of points so that the latter produces a *field* representing the attribute value distribution
3rd: Complementarity	Multiple chronotopologic *realizations* (potentialities) that are in agreement with factual knowledge are considered in order to achieve a complete understanding of the attribute distribution
4th: Total knowledge	For a geostatistical approach to be sound and useful, it should fully incorporate the current state of *knowledge* about the natural attribute
5th: Stochastic knowledge support	Quantification of the support that the current body of knowledge offers to hypothesis H in conditions of real-world uncertainty is made *stochastically* in terms of a *probability* function
6th: Correspondence	A new development accounts for the validity of its *predecessor* under the conditions the predecessor applies and has been confirmed empirically
7th: Information and probability	Geostatistics balances two requirements: High prior *information* about the initial attribute distribution given core knowledge and high posterior *probability* for its final distribution given specificatory knowledge
8th: Relativity	A geostatistics theory possesses no absolute truth or validity unequivocally established once and for all. What counts is the relative superiority of a theory to its rivals, this superiority being an objective feature

(b) as a single spacetime continuum where the space and time dimensions are equally existent (R^{n+1} domain).

Ut ita dicam, the continuum idea is paramount in representing the evolution of a natural attribute that assumes values at any point, thus requiring continuously varying chronotopologic coordinates.

- **Chronotopologic continuity** implies a substantive integration of space with time, and is a fundamental property of the mathematical formalism of phenomena in nature.

Example 1.5 Instead of relying on a static view coupled with arbitrary temporal parameters, the intuition behind the notion of the chronotopologic notion is that physical phenomena leave distinct yet complementary signatures across space and in time. An object can only exist if its body is spatially consistent—all its molecules are tightly packed in the proper form; simultaneously, an object must be temporally consistent—it can only be at one place at a time. We call this insight that objects have distinct and complimentary signatures in space and time as chronotopologic *consistency*.

According to the 2nd notion of Table 1.3, when a natural attribute is considered in a chronotopologic continuum, it needs to be described in terms of quantities related to its occurrence, behavior, and interactions with surroundings.

- A **field** may associate mathematical entities such as a scalar, a vector, or a tensor with chronotopologic points.

Field-based representations of an attribute may involve a hierarchy of fields. There is no unique field representing every aspect of reality, instead, one field describes one feature of reality and another field some other feature.

Example 1.6 Among the many kinds of fields, some represent materialistic variables (say, space-time distribution of some material in the soil or a contaminant in the water), and others express nonmaterialistic regions of influence (say, the earth's gravitational field or the electromagnetic field).

Given the real-world experience of most investigators and practitioners, there seems no reason to doubt that the chronotopologic distributions of most natural attributes are not sharply defined but, instead, they have an uncertain or indeterminate structure. A theoretical as well as a practical need to account for this uncertainty gave rise to complementarity, i.e., the 3rd notion of Table 1.3, which considers the multiple parallel processing of field realizations that are diverse yet necessary for a complete understanding of the phenomenon of interest.

- In light of complementarity, **uncertainty** manifests itself as an ensemble of possible field realizations that are all in agreement with what is known about the phenomenon, so that these different realizations are not contradictions but rather represent complementary aspects of a seamless unity.

Example 1.7 There seems no reason to doubt that an investigator may face several obstacles in doing real-world research. This happens in cases like: (a) when confronted with complex nonlinearities of nature, lack of symmetry, and a large number of degrees of freedom; and (b) when faced with experiments that are either too difficult or even impossible to make, with physical conditions that are not easily controlled in the laboratory, or with exponentially increasing costs of realistic observations. In such cases, complementarity turns out to be a very powerful idea that materializes in *computer simulation* producing multiple realizations, thus allowing the evaluation of the importance of the numerous parameters of a complex physical system, before expensive technology is used to perform an experiment. Moreover, multiple parallel processing of realizations can use modern animation to visualize abstract pictures, to study the complexities of the mathematical analysis and to provide valuable guidance regarding future theoretical investigations.

Complementarity may offer an interesting interpretation of the *cause-effect* concept: The use of *logical conditionals*, i.e., propositions of the form "if E_1 then E_2" or "$E_1 \rightarrow E_2$" where E_1 is the antecedent and E_2 is the consequent, [c] requires us to admit various possible realizations besides the actual (but unknown) one. A conditional then may be true not in terms of how things are, but of how they would be in an appropriate field realization. That is, a conditional $E_1 \rightarrow E_2$ is true in a realization if E_2 is true in the same realization that E_1 is also true.

- An event E_1 may be considered as **causing** an event E_2 if both E_1 and E_2 occur in the observed realization, but in the vast majority of the other realizations in which E_1 does not occur, E_2 does not occur either.

The implication of the 4th notion of Table 1.3, is that, in their application of mathematics to physical systems, investigators use total knowledge, i.e., both *core* knowledge (well-established knowledge) and *ancillary* knowledge (which may be implicit, tacit, or unstated): this is a dynamic process whereby mathematical symbols serve as prompts for conceptual and logical operations and the recruitment of background knowledge, which includes knowledge that is directly applicable for practice or most recent knowledge that may need further verification before application.

[c] E.g., E_1 may represent the attribute value at one point and E_2 its implied value at a different point (see, also, Chapters 4 and 6).

Example 1.8 We remind the readers that several knowledge bases that may be part of this principle can be found in Section 2 of Chapter 3.

The 5th notion views probability as an introspective field of inquiry that plays a critical role in the theory of knowledge, also called *epistemology* or *methodology of science*.

Example 1.9 Scientific reasoning is generally concerned with: (a) knowledge bases, K, obtained from experiments, observations, experience, and reasoning; and (b) hypotheses, H, regarding a phenomenon. Then, we have two expressions that describe the reasoning process from K to H, i.e.,

Deterministic argument: K is available, therefore H is valid.

Stochastic argument: K is available, therefore probably H is valid.

The term "probably" above weights the support that K gives to H (level of certainty about H given K). Assessment of this support leads to some kind of scale, like $p \in [a,b]$, where the a and b correspond to evidence completely opposing H and completely supporting H, respectively. Then, the stochastic argument mentioned earlier may be rewritten as

$$K \text{ is available, therefore } H \text{ is valid with probability } P[H|\,K] = p \in [0, 1]. \tag{1.1}$$

In the vast majority of practical situations, the available knowledge is not perfect, which is why the situation is represented by means of a $f_X(\chi)$ expressing the probability of occurrence of the possible attribute values $\chi_1, \chi_2, \ldots$ Given the *PDF*, several uncertainty assessments can be made. Among them, from the range of χ values in $f_X(\chi)$, the *uncertainty width* $L = \chi_{\max} - \chi_{\min}$ would express investigator's extent of incomplete knowledge. Probabilistic considerations play an important role in the quantitative theories of science. A physical theory that describes the chronotopologic behavior of an attribute, together with a theory that describes how observation errors are distributed statistically, can yield a probability law of the attribute observations. Beyond providing support to a hypothesis or a proposition, knowledge can guide decision-making and action. Decisions that produce optimal outcomes are those that are based on the most timely, accurate and relevant knowledge support.

An interesting feature of the 6th notion is that it may be expressed in various ways, one of which is that correspondence can be seen both as a constraint on *CTDA* developments and as an account of how more general developments (like, geostatistics) are related to their predecessors (like, deterministic data analysis). Then, an important implementation of the correspondence notion is that deterministic analysis is overtaken and incorporated into modern *stochastics*.[d] In this respect, in the *zero uncertainty* limit—i.e., when all uncertainties are practically negligible—the stochastic laws reduce to those of deterministic analysis.

Example 1.10 When $p = 1$ in Eq. (1.1), the stochastic argument of Example 1.9 reduces to a deterministic argument, i.e., Eq (1.1) becomes

$$K \text{ is available, therefore } H \text{ is valid with probability } P[H|\,K] = 1. \tag{1.2}$$

Hence, the deterministic argument is a special case of the stochastic argument.

The 7th notion underlies the scientific perspective that probability and information ought to be treated in the same manner. Accordingly, a mistake in judgment should not be determined *a*

[d] *Stochastics* is a discipline concerned with the mathematical modeling of natural phenomena in a way that accounts for (*a*) the mathematical structure of the underlying physical laws as well as (*b*) their probabilistic and statistical features. Component *a* is a generalization of deterministic mathematics fields (like calculus, differential equations, geometry, and logic) under conditions of uncertainty, and component *b* is a generalization of the fields of classical probability and statistics (Christakos and Hristopulos, 1998, Christakos, 2010, 2017; and references therein).

posteriori (after the event), but in the light of the *prior* information (before the event).

Example 1.11 The prior-posterior stages may be connected by information processing rules that are based on Bayes law. The prior-posterior principle states that from all prior probabilities $Prob_1$, $Prob_2$, etc. only the one that maximizes prior information while satisfying the given knowledge and data is the appropriate prior probability.

According to the 8th notion, questions like "which model is the true one?" have no meaning. What is of real value in scientific investigations rather is answers to questions like:

☞ Which model performs best among the various ones currently available?

Example 1.12 As we saw in Chapter 3, mechanistic models are more powerful than statistical models since they tell you about the underlying processes and driving patterns, and they are more likely to work correctly when extrapolating beyond the observed conditions.

Brevi manu, when appropriately used, the mentioned earlier eight methodological notions offer deeper insights and facilitate the solution of difficult problems. Although intuition and common sense will be of paramount importance for such a task, sophisticated mathematics justify their use by yielding useful and sometimes quite unobvious results. After all, what counts is a correct *CTDA*, and investigators should never view a correct analysis as unnecessarily theoretical and complicated.

It is generally appreciated that while the two spatial dimensions of a natural attribute are relatively manageable, their combination with time results in a number of challenges. Among other developments, these challenges have motivated the development of the random field theory.

2 Random field theory

In this section, the *spatiotemporal random field* (*S/TRF*) theory will be briefly reviewed,[e] including its underlying theses, postulates, and classifications in a broad *CTDA* context, as is the subject of this book. In common with previous discussions about *S/TRF* theory, here *S/TRF* modeling refers to the quantitative assessment and prediction of the chronotopologic attribute behavior in conditions of uncertainty by means of statistical and mathematical tools.

> In the *S/TRF* setting, the term "random" refers to *in-situ* attribute **uncertainties**, and the term "spatiotemporal" to the attribute's **physical pattern**, i.e., *S/TRF*'s concern is not merely the way in which attributes are located in space and occur in time, but also the manner in which they are essential components of the physical world.

Example 2.1 Sea surface salinity is a natural attribute to which geographic and time coordinates are assigned, but it is also an essential part of the sea surface, inheriting features from this surface (hydrodynamic, climatic, etc.). The temporal *S/TRF* modeling component determines the extent to which the phenomenon is captured by the available information, while the spatial component determines whether the phenomenon is associated with a set of fixed sensors or collected by moving sensors.

As happens with all notions considered in applied sciences, the *S/TRF* notion should be considered in both an intuitive and a mathematical setting.

2.1 Intuitive setting: The four theses

In an intuitive setting, the *S/TRF* notion admits that in many real-world phenomena the generator of reality is often not observable,

[e] For a detailed theoretical background, see Christakos et al. (2017).

and, as a consequence, it proposes the following *methodological* theses:

① When thinking about a phenomenon, the investigator should take into account both the observed realization and the **unobserved** possible realizations (potentialities, possibilities) about the phenomenon. In fact, one cannot adequately visualize a realized outcome without reference to the **non-realized** ones.

A literary digression may be refreshing, at this point. According to *Salman Rushdie*:

☞ All stories are haunted by the ghosts of the stories they might have been.

Mutatis mutandis, the observed attribute realizations of thesis ① are seen as the worldly stories that occurred, which are implicitly but decisively affected by the nonobserved attribute realizations that are seen as the ghosts of the stories that might have occurred.

② Possible realizations are usually **nonequiprobable**, whereas the observed realization (also called the actualization, because its probability of occurrence is one) is a **partial realization**, in the sense that it reveals the attribute values only at the sampled points.

It maybe worth noticing that sometimes in practice, when a very large number of realizations can be simulated, the equi-probability assumption may turn out to be a reasonable approximation.

③ What the observed and the possible realizations have in common is that they all satisfy the same **science-** and **logic-based conditions**.

The implication of thesis ③ is that these are not just arbtrarily generated realizations but substantive realizations that must be in agreement with the scientific knowledge of the phenomenon and do not violate any rules of logical thinking.

④ What needs to be determined is the limited number of realizations (ultimately although rarely it may be a single realization) across which the most important **properties** of the actual phenomenon hold.

We do not merely introduce these methodological theses –we have rational reasons to introduce them. These are justified assertions; there are, *exempli causa*, a number of issues in the random field theory that are illuminated by the viewpoint introduced by the previously mentioned theses. Taken to their logical conclusion, the methodological theses ①-④ lead to the following assertion:

> **Intuitive conception** of *S/TRF*: A collection of science- and logic-based realizations regarding the phenomenon of interest, with various probabilities assigned to them, where the observed realization is seen as the only certain albeit partial (incomplete) realization.

Remarkably, the insight gained by considering those realizations that did not happen to materialize in the physical world under the current specific conditions may lead to an improved understanding of the phenomenon under more general conditions.

Example 2.2 In many ecological studies, an investigator may need to question the possible realizations of the phenomenon of interest that did not take place, and not only the ones that actually happened.

2.1.1 Links between physics and geometry

The readers may find it interesting that some fundamental links between physics (attribute values) and chrono-geometry (attribute domain) underlie the intuitive *S/TRF* notion previously mentioned:

① The *S/TRF* domain consists of a set of points associated with a **continuous** arrangement of attribute values.

② Since attribute values are assigned to domain points, the associations between attribute values are essentially **relationships** between points in the chronotopologic domain.

③ Since **physical** relationships between attribute values are associated with geometrical relationships between points, a chronotopologic structure is imposed by means of these physical relationships.

④ The choice of an appropriate **geometry** to describe an *S/TRF* depends on whether one adopts an intrinsic or an extrinsic visualization of the chronotopologic domain.

> Understanding the **link** between physics and chrono-geometry is a prerequisite for an investigation to advance from purely descriptive to explanatory.

Example 2.3 A model that elucidates why a natural attribute at a specific location and time (chrono-geometry) adopts a certain value (physics) is explanatory. That is, the evolution of an attribute at one location and time is caused from another attribute value at a previous location and time.

2.1.2 *Separate space-time and composite spacetime*

As strongly indicated in the discussion of the previously mentioned link, a phenomenon's domain arguments (space and time coordinates) are key elements of a mathematically rigorous and scientifically meaningful *S/TRF* modeling, which is chiefly concerned with the following key distinction:

❶ When space and time are viewed **separately** ($R^{n,1}$ perspective), they are both continua sharing all the properties that the abstract notion of a continuum possesses. But there are salient differences too:[f] time has *extra-continua* physical properties that it does not share with any other continuum and by virtue of which time is specifically time and not just a continuum (e.g., recursivity is not a property that continua have in general, it is an extra-continua property of time but not of space). The same is true with space features (e.g., space evolves along many directions and it can be anisotropic).

Example 2.4 A typical case of a phenomenon represented by the *S/TRF* model is the outbreak of flu that first happens in one core region and over time spreads from this region to the neighbor regions. Such a flu outbreak lasts for sometime, i.e., the outbreak level of flu usually decreases with the distance to the core region and with the time separation to the highest outbreak period (this case belongs to the $R^{n,1}$ perspective mentioned earlier).

❷ When bringing space and time **together** (R^{n+1} perspective), the extra-continua physical properties of space integrate with those of time producing a holistic chronotopology in which the whole is greater than the sum of its parts. In such a holistic framework, chronotopologic connections and cross-effects control natural variations. It can also happen that only the empirical investigation of spacetime as a whole can disclose the nature of the geometry that best describes it. Physical equations may arise that put restrictions on the geometrical features of spacetime.

Example 2.5 The subsurface flow case discussed in Example 3.6, Section 3 of Chapter 2 was characterized by a composite spacetime metric that was expressed in terms of the physical

[f] These differences have been already discussed in some detail in Chapter 2.

parameters of the flow law (this case belongs to the R^{n+1} perspective above).

When investigators give to the issue its due weight, they realize something crucial: the perspective choice ($R^{n,1}$ *vs.* R^{n+1}) is linked to the relative role that the phenomenon domain arguments play in metric development, i.e., to the difference between the degree of importance of space and time in the phenomenon depends on their extra-continua properties. In sum:

> The contribution of the time dimension relatively to that of the space dimension in a chronotopologic metric as expressed by the metric is mostly **application**-dependent.

In the previously mentioned setting, the attribute covariance can be instructive in determining the appropriate metric of a chronotopologic continuum (Christakos, 2017). This is not surprising, since the covariance is a function of the chronotopologic metric that describes the attribute distribution.

2.1.3 *Basic operational rules*

As is commonly the case when proceeding from conceptualization to actualization, *a posse ad esse*, there are three basic operational rules to be followed as regards the implementation of the *S/TRF* notion in *CTDA* applications:

① The focus is shifted from a single attribute realization (traditional deterministic view) to an **ensemble** of possible realizations representing all potentialities restricted only by the investigator's knowledge of the attribute's properties. This means that instead of studying directly single attribute features, the focus turns on the stochastic properties of the ensemble realizations (probability laws, expected values).

② **Predictions** of any kind about the behavior of the single actual realization are derived on the basis of the ensemble of possible realizations.

③ The ultimate **judgment** of *S/TRF* modeling progress is measurable results in reasonable time.

Example 2.6 When the choice is made to represent the chronotopologic pollutant distribution by the *S/TRF* model, the operational implications are that (*a*) the modeler considers all possible pollutant realizations conditioned by the data and the physics of the phenomenon, (*b*) the focus is on the probability characterization and the statistics of these pollutant realizations, and (*c*) pollutant predictions are generated on the basis of *a* and *b*.

Next, based on the previously mentioned methodological and operational considerations, we will introduce the mathematical setting of the *S/TRF* model.

2.2 Mathematical setting

The notion of alternative realizations can be extended considerably and subjected to all manner of technical refinement. In particular, we start with the definition of a *random variable* (*RV*), x, which is a nondeterministic multivalued function that assigns a unique real number (*RV* realization) χ to each event u of a set of possible events U regarding the phenomenon, i.e.,

$$u \in U : x(u) = \chi. \tag{2.1}$$

Since the study of physical data is often interested in attribute values that change across space and time, the kind of generalization of the *RV* theory that is of interest is the one that associates a chronotopologic point $\boldsymbol{p}$ to each *RV* x, i.e., in the new theoretical setting Eq. (2.1) is replaced by

$$u \in U, \boldsymbol{p} \in R^{n,1} \text{ or } R^{n+1} : x(u, \boldsymbol{p}) = \chi(\boldsymbol{p}). \tag{2.2}$$

Eq. (2.2) represents the chronotopologic variation of an attribute in conditions of *uncertainty*, as reflected in the multiple $\boldsymbol{p}$-dependent *RV* realizations introduced by the *S/TRF* theory.

One of the main objectives of this theory is to suggest a formal model that can be used to describe the chronotopologic variation of the natural attribute represented by the $\boldsymbol{p}$-*RV* in conditions of uncertainty. This is a *PDF* model, $f_X(\chi, \boldsymbol{p})$, which measures the probability that the *RV* takes on a specific value (realization) χ associated with the event u, i.e.,

$$x(u, \boldsymbol{p}) \sim f_X(\chi, \boldsymbol{p}), \tag{2.3}$$

where the $\boldsymbol{p}$-*RV* is a function of the random event u, whereas the corresponding *PDF* is not. Then, a mathematical definition of an *S/TRF* is possible, as follows:

An *S/TRF* $X(\boldsymbol{p})$ is a collection of *RV* $x(u, \boldsymbol{p}_i)$ at points p_i, $i = 1, 2, \ldots$, where a different, in general, *PDF* $f_X(\chi, \boldsymbol{p}_i)$ is assigned at each point.

Example 2.7 According to the previous discussion, in Fig 2.3 of Chapter 1 the correlated *RV*s $x(\boldsymbol{p}_1), \ldots, x(\boldsymbol{p}_7)$ can be associated with points $\boldsymbol{p}_1, \ldots, \boldsymbol{p}_7$. This chronotopologic association defines the *S/TRF* $X(\boldsymbol{p})$.

A number of *CTDA* issues can be illuminated in the *S/TRF* setting. Interpretively, the *S/TRF* notion refers to the variation of events across space combined with their dynamic unfolding with the course of time. Naturally, the next step should be the characterization of the *S/TRF*.

2.2.1 *S/TRF characterization*

As it turns out, there are two levels of characterization: a full (or complete), yet more difficult to achieve in the real world, characterization; and a partial (or incomplete), albeit often adequate for all practical purposes characterization that is easier to execute in the real world.

The **fullest** possible description of $X(\boldsymbol{p})$ is provided by the infinite sequence of *PDF* covering all points of the domain of interest, i.e.,

$$f_X(\boldsymbol{\chi}, \boldsymbol{p})d\boldsymbol{\chi} = \text{Prob}[\boldsymbol{\chi} \le \boldsymbol{x} \le \boldsymbol{\chi} + d\boldsymbol{\chi}], \tag{2.4}$$

where $\boldsymbol{p} = (\boldsymbol{p}_1, \ldots, \boldsymbol{p}_m)$, $\boldsymbol{\chi} = (\chi_1, \ldots, \chi_m)$, and $d\boldsymbol{\chi} = d\chi_1 \ldots d\chi_m$, so that $\boldsymbol{\chi} \le \boldsymbol{x} \le \boldsymbol{\chi} + d\boldsymbol{\chi}$ denotes $\chi_1 \le x_1 = X(\boldsymbol{p}_1) \le \chi_1 + d\chi_1, \quad \ldots, \quad \chi_m \le x_m = X(\boldsymbol{p}_m) \le \chi_m + d\chi_m$.

The univariate *PDF* ($m = 1$) describes the behavior of the amplitude of the *S/TRF* $X(\boldsymbol{p})$, but it amounts to a very sparse description of $X(\boldsymbol{p})$'s general behavior. For a complete description of $X(\boldsymbol{p})$'s general behavior, the sequence of the multivariate *PDF* in Eq. (2.4) is needed. Moreover, by connecting selected values generated by $f_X(\chi, \boldsymbol{p})$ of the random field $X(\boldsymbol{p})$ any one of several possible *S/TRF* realizations obeying Eq. (2.4) can be determined.

Example 2.8 The plots in Figs. 2.1C (cont.) and 2.1D (cont.) of Chapter 2 may be viewed, respectively, as a set of two-dimensional *S/TRF* realizations in the $R^{2,1}$ domain and as a continuous three-dimensional *S/TRF* realization in the R^{2+1} domain.

As it turns out, the following *S/TRF* characterization is predominantly used in most practical applications.

In practice, a **partial** yet usually adequate *S/TRF* characterization is implemented in terms of low-order chronotopologic statistics.

The term "low-order" usually refers to 1st-order statistics (attribute mean value function) and 2nd-order statistics (attribute covariance and variogram functions). In particular, the covariance and the variogram are the two main geostatistics tools that define the chronotopologic structure of the data, see Eqs. (2.7a, b) and (2.8a, b) below.

Example 2.9 Tables 3.4 and 3.5 in Section 3 list various variogram models that provide partial *S/TRF* characterizations commonly used in many practical applications.

2.2.2 *S/TRF classifications*

Ut aequum, the classification discussion will start by considering two essential types of *S/TRF*, namely, discrete-valued and continuous-valued fields:

① **Discrete** *S/TRF* can take only a finite number of values that can be counted using integers, whereas **continuous** *S/TRF* can take any value on the real number line.

Generally speaking, discrete *S/TRF* representations are used in *computational* geostatistics (e.g., calculation of empirical mean, covariance and variogram values), whereas continuous *S/TRF* representations are used in *theoretical* geostatistics (e.g., the derivation of theoretical mean, covariance and variogram models, and the solution of stochastic differential equations).

Example 2.10 Some cases of discrete *vs*. continuous *S/TRF* representations are presented next.

- Discrete random fields may include the number of people in a household at a specific location during a particular year, the number of tornadoes in a county during the course of a year, and the number of flu occurrences in San Diego City on September 4, 2018. Discrete *S/TRF*, such as the previously mentioned, are generally used to measure the *counts* of some event or phenomenon.
- Other types of discrete random fields are used to represent qualitative characteristics of observations. Consider, say, the gender of the head of a household, where the discrete field can be considered as equal to 1 if the gender is male, and equal to 0 otherwise.
- Continuous random fields, on the other hand, include the temperature at Saint Petersburg City on September 4, 2018, the average monthly meat consumption at Sparta City during the last decade, the flu rate in Shanghai City during 2013, and the 2018 house prices as a function of their distance from the San Francisco Bay.

Beyond the physical value-based classification, ①, other kinds of *S/TRF classifications* may be established on the basis of the distinct random field features assumed, like: [g]

② Its **space** and **time arguments** (e.g., scalar *vs*. vector arguments).
③ Its **probability law** (e.g., Gaussian *vs*. non-Gaussian).
④ Its **variability pattern** (e.g., homostationary *vs*. heterogeneous variability).[h]
⑤ Its **dimensionality** (scalar *vs*. vector fields).

In general, investigators are interested in those particular *S/TRF* features that best represent the considered real case study. These features allow them to match the most appropriate models to the particular natural attributes they seek to study.

Example 2.11 For illustration purposes, let us consider three distinct cases.

- Due to global warming, the concerns in the adverse effects of high *temperature* on human *mortality* have increased, so that the extreme heat events are considered as a leading cause of weather-related fatality worldwide. To investigate the temperature impacts on mortality in a specific region during the time of interest, one should have a good idea of the chronotopologic distributions of both temperature-mortality. If, after collecting temperature

[g] More technical and interpretive details can be found in Christakos et al. (2017).

[h] See, also, the modeling assumptions discussed in Section 2.3 below.

measurements (from meteorological monitoring stations) and the relevant mortality records (from health departments), it is found that their joint variation exhibit distinct trends across space and time, the temperature-mortality distributions can be jointly modeled as a heterogeneous vector *S/TRF*. Then, the geostatistics tools can be utilized to accurately map the distributions of temperature and mortality, and gain valuable insight concerning the temperature-mortality *association*.

- With the rapid growth of industrialization and urbanization, more and more lands are contaminated by *hazardous materials*, especially heavy metals, many of which are carcinogenic. Environmental agencies need to remediate the contaminated land, especially the cropland. If the available heavy metal data follow a bell-shaped distribution and no observable trends in their variation exist, the heavy metal distribution can be modeled as a Gaussian and homogeneous *S/TRF*. This modeling decision allows heavy metal concentrations to be estimated across space and time by geostatistical tools. Then, by studying the chronotopologic distribution map, environmental managers select their remediation targets and areas.
- In recent years, *infectious diseases* have emerged as serious threats to public health, e.g., COVID-19, Ebola virus, and avian influenza H5N1 virus. To optimize medical recourse allocation, public health departments and hospitals need to comprehend the disease outbreak level and the disease incident scale. If the observed disease incidences at various locations and times follow an asymmetric law, the disease may be modeled as a non-Gaussian and heterogeneous *S/TRF*, and the chronotopologic incidence is mapped using geostatistics tools (in some cases, short-range forecasting may be possible).

Before proceeding further with the technical issues of random field modeling, our readers should be reminded of the following assertion:

> When implementing the *S/TRF* theory in *CTDA* applications, the focus should not be merely on the **mathematical grammar** of the relevant equations but also on their **physical meaning** and practical consequences.

In this setting, the real-world phenomena are studied as they unfold naturally, that is, in a nonmanipulatively and noncontrolling manner (no predetermined constraints exist), so that the investigators are open to whatever phenomenal features may emerge.

2.3 Chronotopologic statistics

The notion of "chronotopologic statistics" was first introduced in Section 5.2 of Chapter 3 to describe statistical relationships between attribute values that depend on both space and time in conditions of uncertainty. The notion can be rigorously developed in the *S/TRF* setting. As it turns out, in many applications, it is practically adequate to use the chronotopologic characteristics of an *S/TRF* expressed by its trend function (mean) and the dependence or variability functions (covariance or variogram). A brief presentation of the relevant mathematical formulas is provided here, in both the discrete domain (*DD*) and the continuous domain (*CD*).[i]

2.3.1 General operator

It is always convenient to consider a general mathematical formulation on the basis of which many useful chronotopologic statistics results can be derived as special cases. This is the role of the following operator.

[i] Basically, in the *CD* expressions the summations of the *DD* are replaced with integrals.

The general operator expresses the chronotopologic attribute variation in terms of the **stochastic expectation**

$$\overline{S[X(p_1),\ldots,X(p_m)]} = \begin{cases} \sum_{1_i}\cdots\sum_{m_i} S[\chi_{1_i},\ldots,\chi_{m_i}] f_X(\chi_{1_i},\ldots,\chi_{m_i}), & DD \\ \int d\chi_1\cdots\int d\chi_m S[\chi_1,\ldots,\chi_m] f_X(\chi_1,\ldots,\chi_m), & CD \end{cases} \quad (2.5\text{a, b})$$

where S is an operator that describes quantitatively the attribute characteristics of interest based on the available knowledge bases.

The choice of the S operator depends on both the physical characteristics of the phenomenon considered and the study objectives. The formal distinction in Eq. (2.5a) *vs.* Eq (2.5b) can be used to evaluate quantitatively the meaning of concepts like discreteness and continuity with respect to space and time. Table 2.1 displays some noticeable cases of the operator S above that are commonly encountered in *CTDA* applications.[j,k]

2.3.2 *One-, two-, and three-point chronotopologic statistics*

CTDA is chiefly concerned with the study of chronotopologic trends, correlations and variabilities between data. Among the various tools used for this purpose, the ones discussed next play a prominent role in *CTDA*.

- The **mean** attribute function at each point p_i is expressed as

$$\overline{X(\boldsymbol{p}_i)} = \begin{cases} \sum_i \chi_i f_X(\chi_i), & DD \\ \int_{-\infty}^{\infty} d\chi_i \chi_i f_X(\chi_i), & CD \end{cases} \quad (2.6\text{a, b})$$

where, as usual, the bar denotes statistical average, $\chi_i = \chi(\boldsymbol{p}_i)$, and $f_X(\chi_i)$ is the univariate *PDF* of the $X(\boldsymbol{p})$.[l]

The stochastic expectation operator also yields *two-point* variability assessment tools that are widely used in *CTDA* applications.

- The **covariance** and **variogram** attribute functions between the points $\boldsymbol{p}_i$ and $\boldsymbol{p}_j = \boldsymbol{p}_i + \Delta\boldsymbol{p}$ are given by, respectively,[m]

TABLE 2.1 Common cases of the stochastic expectation operator of Eqs. (2.5a, b).

S	$\overline{S}$	Name	Interpretation
$X(\boldsymbol{p})$	$\overline{X(\boldsymbol{p})}$	Mean	Describes chronotopologic trends
$\left[X(\boldsymbol{p}_i)-\overline{X(\boldsymbol{p}_i)}\right]\left[X(\boldsymbol{p}_j)-\overline{X(\boldsymbol{p}_j)}\right]$	$c_X(\boldsymbol{p}_i,\boldsymbol{p}_j)$	Covariance	Measures chronotopologic dependence between $X(\boldsymbol{p}_i)$ and $X(\boldsymbol{p}_j)$
$\frac{1}{2}\left[X(\boldsymbol{p}_i)-X(\boldsymbol{p}_j)\right]^2$	$\gamma_X(\boldsymbol{p}_i,\boldsymbol{p}_j)$	Variogram	Measures chronotopologic variation between $X(\boldsymbol{p}_i)$ and $X(\boldsymbol{p}_j)$
$\frac{f_X(\boldsymbol{p}_i,\boldsymbol{p}_j)}{f_X(\boldsymbol{p}_i)f_X(\boldsymbol{p}_j)}$	$\psi_X(\boldsymbol{p}_i,\boldsymbol{p}_j)$	Contingogram	Measures information on $X(\boldsymbol{p}_i)$ contained in $X(\boldsymbol{p}_j)$
$\log\frac{f_X(\boldsymbol{p}_i,\boldsymbol{p}_j)}{f_X(\boldsymbol{p}_i)f_X(\boldsymbol{p}_j)}$	$\beta_X(\boldsymbol{p}_i,\boldsymbol{p}_j)$	Sysketogram	Measures information on $X(\boldsymbol{p}_i)$ contained in $X(\boldsymbol{p}_j)$

[j] See Section 4 of Chapter 6.

[k] See Section 4 of Chapter 6.

[l] The $\overline{X(\boldsymbol{p}_i)}$ is directly obtained from Eq (2.5a-b) by letting $m = 1$ and $S[\chi_1] = \chi_1$.

[m] I.e., the covariance and variogram are obtained from Eq (2.5a-b) by letting $m = 2$ and $S[\chi_i,\chi_j] = \left[\chi_i - \overline{X(\boldsymbol{p}_i)}\right]\left[\chi_{i+\Delta p} - \overline{X(\boldsymbol{p}_i+\Delta\boldsymbol{p})}\right]$ and $S[\chi_i,\chi_j] = ½[\chi_i - \chi_{i+\Delta p}]^2$, respectively.

$$c_X\left(p_i,p_j\right)=c_X(p_i,p_i+\Delta p)=\overline{\left[X(p_i)-\overline{X(p_i)}\right]\left[X(p_i+\Delta p)-\overline{X(p_i+\Delta p)}\right]}$$
$$=\begin{cases}\sum_i\left[\chi_i-\overline{X(p_i)}\right]\left[\chi_{i+\Delta p}-\overline{X(p_i+\Delta p)}\right]\quad f_X\left(\chi_i,\ \chi_{i+\Delta p}\right), & DD\\ \int_{-\infty}^{\infty}\int_{-\infty}^{\infty}d\chi_i d\chi_{i+\Delta p}\left[\chi_i-\overline{X(p_i)}\right]\left[\chi_{i+\Delta p}-\overline{X(p_i+\Delta p)}\right]f_X\left(\chi_i,\ \chi_{i+\Delta p}\right), & CD\end{cases}\tag{2.7a, b}$$

$$\gamma_X\left(p_i,p_j\right)=\gamma_X(p_i,p_i+\Delta p)=\frac{1}{2}\overline{[X(p_i)-X(p_i+\Delta p)]^2}$$
$$=\frac{1}{2}\begin{cases}\sum_i\left[\chi_i-\chi_{i+\Delta p}\right]^2 f_X\left(\chi_i,\ \chi_{i+\Delta p}\right), & DD,\\ \int_{-\infty}^{\infty}\int_{-\infty}^{\infty}d\chi_i d\chi_{i+\Delta p}\left[\chi_i-\chi_{i+\Delta p}\right]^2 f_X\left(\chi_i,\ \chi_{i+\Delta p}\right), & CD,\end{cases}\tag{2.8a, b}$$

where $\chi_i=\chi(p_i)$, $\chi_{i+\Delta p}=\chi(p_i+\Delta p)$, $f_X(\chi_i,\ \chi_{i+\Delta p})$ is the bivariate *PDF* at $p=p_i$ and $p=p_j=p_i+\Delta p$.

The chronotopologic lag may assume a separate space-time or a composite spacetime form as defined in Section 2 of Chapter 2, i.e., $\Delta p=(h_1,\dots,h_n;\tau)\in R^{n,1}$ and $\Delta p=(\Delta s_1,\dots,\ \Delta s_n,\Delta s_0)\in R^{n+1}$, respectively.

- A specific case of the covariance is that with $i=j$ yielding the attribute **variance**

$$\sigma_X^2(p)=c_X(p,p).$$

Example 2.12 The separate space-time lag in $R^{2,1}$ is $\Delta p=(h_1,h_2;\tau)$, in which case in Eqs. (7a-b) and (8a-b) it holds that $p_i=(s_{i1},s_{i2};t_i)$ and $p_j=(s_{j1},s_{j2};t_j)=(s_{i1}+h_1,s_{i2}+h_2;\ t_i+\tau)$.

In the case that the physics of the situation requires the consideration of *three-point* variability, the following tool can be used.

- The **trivariance** attribute function between the points p_i, $p_j=p_i+\Delta p$ and $p_k=\ p_i+\Delta p'$ is given by

$$\zeta_X\left(p_i,p_j,p_k\right)=\overline{\left[X(p_i)-\overline{X(p_i)}\right]\left[X(p_i+\Delta p)-\overline{X(p_i+\Delta p)}\right]\left[X(p_i+\Delta p')-\overline{X(p_i+\Delta p')}\right]}$$
$$=\begin{cases}\sum_i\left[\chi_i-\overline{X(p_i)}\right]\left[\chi_{i+\Delta p}-\overline{X(p_i+\Delta p)}\right]\left[\chi_{i+\Delta p'}-\overline{X(p_i+\Delta p')}\right]\\ f_X\left(\chi_i,\ \chi_{i+\Delta p},\ \chi_{i+\Delta p'}\right),\quad DD\\ \int_{-\infty}^{\infty}\int_{-\infty}^{\infty}\int_{-\infty}^{\infty}d\chi_i d\chi_{i+\Delta p}d\chi_{i+\Delta p'}\left[\chi_i-\overline{X(p_i)}\right]\left[\chi_{i+\Delta p}-\overline{X(p_i+\Delta p)}\right]\left[\chi_{i+\Delta p'}-\overline{X(p_i+\Delta p)}\right]\\ f_X\left(\chi_i,\ \chi_{i+\Delta p},\ \chi_{i+\Delta p'}\right),\quad CD\end{cases}\tag{2.9a, b}$$

where $\chi_i=\chi(p_i)$, $\chi_{i+\Delta p}=\chi(p_i+\Delta p)$, $\chi_{i+\Delta p'}=\chi(p_i+\Delta p')$, $f_X(\chi_i,\ \chi_{i+\Delta p},\ \chi_{i+\Delta p'})$ is the trivariate *PDF* of $X(p)$ at $p=p_i$, $p=p_j=p_i+\Delta p$ and $p=p_k=p_i+\Delta p'$.[n]

Trivariance could be useful in practical applications where two-point statistics are inadequate to model curvilinear or long-range continuous physical formations (e.g., unrealistic

[n] Some interesting trivariance properties are discussed in Exercise 24 at the end of this chapter.

variogram–based depositional elements simulations that are poor representations of long-range continuous geologic bodies).

2.3.3 *Modeling assumptions*

There are certain assumptions concerning the *S/TRF* models used in geostatistical applications that are both mathematically convenient and also physically reasonable:

According to the **homostationarity** assumption, the mean is constant across space and time,

$$\overline{X} = \overline{X(\boldsymbol{p}_i)} \tag{2.10a}$$

for all $\boldsymbol{p}_i$, and the covariance, variogram and trivariance are functions of the lag,

$$\begin{aligned} c_X\left(\boldsymbol{p}_i, \boldsymbol{p}_j\right) &= c_X(\Delta\boldsymbol{p}), \\ \gamma_X\left(\boldsymbol{p}_i, \boldsymbol{p}_j\right) &= \gamma_X(\Delta\boldsymbol{p}), \\ \zeta_X\left(\boldsymbol{p}_i, \boldsymbol{p}_j, \boldsymbol{p}_k\right) &= \zeta_X(\Delta\boldsymbol{p}, \Delta\boldsymbol{p}'). \end{aligned} \tag{2.10b-d}$$

The following is a special, albeit very useful in practice, case of the homostationarity assumption previously mentioned.

According to the **isostationarity** assumption, the constant mean Eq. (2.10a) still holds but the covariance, variogram and trivariance are now only functions of the metric,

$$\begin{aligned} c_X\left(\boldsymbol{p}_i, \boldsymbol{p}_j\right) &= c_X(|\Delta\boldsymbol{p}|), \\ \gamma_X\left(\boldsymbol{p}_i, \boldsymbol{p}_j\right) &= \gamma_X(|\Delta\boldsymbol{p}|), \\ \zeta_X\left(\boldsymbol{p}_i, \boldsymbol{p}_j, \boldsymbol{p}_k\right) &= \zeta_X(|\Delta\boldsymbol{p}|, |\Delta\boldsymbol{p}'|). \end{aligned} \tag{2.11a-c}$$

As usual, the metric may assume a separate space-time form ($R^{n,1}$ domain) or a composite spacetime form (R^{n+1} domain) as was defined in Chapter 2; i.e.,

$$|\Delta\boldsymbol{p}| = \begin{cases} (|\boldsymbol{h}|, \tau) = \left(\sqrt{\sum_{i=1}^{n} h_i^2}, \tau\right), \text{ or} \\ g(\boldsymbol{h}, \tau) \end{cases}$$

where g is a function defined by the scientific knowledge of the situation. In many problems the separate metrical structure will suffice. In other cases, however, the more involved composite structure may be necessary (one may ask, e.g., which mathematical chronotopologic metrical structure, i.e., function g, describes physical reality, as there is no need to single out a particular metrical structure). The physical meaning of isostationarity is discussed in a following section.

- Note that the isostationarity implies homostationarity, but not necessearily *vice versa*.

The following assumptions enable the chronotopologic moments to be deduced from a single realization.

According to the **ergodicity** assumption, the 1st- and 2nd-order ensemble moments coincide with the 1st- and 2nd-order sample moments of the *S/TRF*.°

Otherwise said, the ergodicity assumption is assumed valid if the statistical properties of a sufficiently long chronotopologic realization are similar to the properties of an average of shorter realizations. In practice, ergodicity usually remains a purely modeling decision, i.e., an *S/TRF* is frequently assumed to be ergodic simply when it seems reasonable to do so. In view of ergodicity, under certain conditions it is valid that

$$\gamma_X(\Delta\boldsymbol{p}) = c_X(\boldsymbol{0}) - c_X(\Delta\boldsymbol{p}), \tag{2.12}$$

° This kind of ergodicity is called *wide sense* ergodicity. If all the ensemble and sample multivariate probability laws are equal to each other, the *S/TRF* is called ergodic in the *strict sense*, which, though, is a condition that has limited applicability.

i.e., the covariance and variogram are formally equivalent (although they have different interpretations).

There may be several reasons that one or more of the previously mentioned modeling assumptions could be violated in the real world. A prime one is the effect of the physical environment, i.e., the modeled attributes are embedded within this environment. Then, the resulting covariance and variogram functions may not be homostationary, but depend on the local physical conditions (ecology, hydrogeology, etc.). A more appropriate modeling assumption in such cases could be as follows:

> **Heterogeneity** assumes a combination of spatial nonhomogeneity and temporal nonstationarity, where the dependency or variability functions (covariance, variogram or trivariance) explicitly depend on the individual coordinates $\boldsymbol{p}_i$.

Chronotopologic heterogeneity can assume several forms, the detailed study of which is beyond the scope of this book (the readers are referred to detailed technical studies of heterogeneity in Christakos, 1992, 2010, 2017 and references therein).

2.3.4 Model permissibility

Not all functions can serve as dependency or variability models. On the contrary, there are well-defined technical conditions that apply. Specifically:

> Depending on the situation at hand (homostationary or heterogeneous variation, scalar or vector attributes, etc.), the functions that are candidate covariance or variogram models must satisfy certain **criteria of permissibility (*COP*)**.

For efficiency purposes, *COP* classifications have been proposed in the literature that distinguish between the following classes of *COP*:

- *COP* that are necessary and sufficient.
- *COP* that are only necessary or only sufficient.
- *COP* that are defined in the physical domain.
- *COP* defined in the frequency domain.

Naturally, the choice of the most suitable and convenient *COP* to use should depend on the particular study under consideration. A detailed exposition of the different mathematical *COP* classes can be found in Christakos et al. (2017). We can now put these mathematical issues aside and concentrate on implementation and inferential matters.

3 Covariography and variography

The difficulty of determining the complete *PDF* of Eq. (2.4) of Section 2 in most practical situations confronts investigators with a choice: either choose the *PDF-based* approach, with the associated significant theoretical and practical difficulties, or search for an alternative.

The dominant response is rather typical: choose the alternative, i.e., limit analysis to an approximation that offers a satisfactory, for all practical purposes, characterization of the attribute variability in terms of up to 2nd-order chronotopologic statistics. This turns out to be a satisfactory characterization, as long as the analysis is done in a rigorous and systematic manner, as follows:

> **Covariography** and **variography** are parallel procedures that are concerned with the systematic computation, adequate modeling, and meaningful interpretation of the covariance and variogram functions in practice.

To put it in more words, covariography and variography are hands-on procedures that have three main components:

① A **computational** component, where the empirical covariance and/or variogram values are calculated.

② An **analytical** component, where an adequate theoretical model is fitted to the empirical values.

③ An **intuitive** component, where all the features of the fitted covariogram and/or variogram models are interpreted and linked to known physical processes.

Covariography and variography are indispensable tools in the quantification of the chronotopologic attribute variability. In addition, they are the necessary inputs to most geostatistical estimation techniques. As it turns out, covariography is more common in cases involving sophisticated mathematical modeling, whereas variography is typically used in cases based mainly on datasets. We continue with a salient observation:

> The covariance and variogram functions offer quantitative and visual assessments of, respectively, physical **correlation** (similarity) and **decorrelation** (dissimilarity) as functions of spatial distance and time separation.

In this section, we will discuss several important covariography and variography aspects. The five main steps of these procedures are as follows:

① Gathering an adequate **dataset** (quality, number, locations and instances, Chapter 3) and other relevant **knowledge bases** (core or specificatory, also Chapter 3).
② Examining which **modeling assumptions** (Section 2.3) apply in the case study of interest. To ensure the validity of these assumptions a **data transformation** may be necessary.
③ Computing the **empirical**[p] covariance or variogram values at selected space and time lags, and identifying the relevant attribute variability features.
④ Fitting a **model** to the empirical covariance or variogram values.
⑤ Derive meaningful **interpretations** of the covariance or variogram plots.

When properly implemented, the covariography and the variography can account for evidential support and scientific laws, and enable an adequate physical interpretation. Then, the crux of the matter is that:

> An investigator cannot know before **experiencing**, and, at the same time, multi-sourced data brainstorming can deliver valuable **insights**.

In the following lines, the principles of covariography and variography are developed and illustrated with the help of a number of practical examples and real-world case studies.

3.1 Data selection and preprocessing

Not surprisingly, covariography and variography start with the selection of an adequate chronotopologic dataset. This means gathering, structuring, assessing, and cleaning the dataset before using it for covariography or variography purposes. The specification of these activities is a threefold affair:[q]

① Data **quality assurance**. Including, (*a*) detecting, correcting, or removing corrupt, noisy or inaccurate data, (*b*) identifying incomplete, incorrect or irrelevant data parts, (*c*) replacing, modifying, or deleting the dirty, noisy or coarse data, and, lastly, (*d*) making sure that the data correctly represents the real phenomenon to which it refers.
② Data analysis **objectives**: Including internal data consistency is achieved, data validity is assured, and data is fit for its intended uses.

[p] Sometimes also termed raw, sample, or experimental.

[q] A detailed discussion of the subject was given in Section 2 of Chapter 3.

③ Computational **needs**: Data should be available in sufficient numbers, as well as at physically appropriate locations and time instances, so that accurate calculations of the covariance or variogram values are possible.

3.1.1 *Covariography and variography prerequisites*

Before calculating a covariance or a variogram, a careful data analysis should ensure that certain prerequisites are *de facto* fulfilled:

Gaussian data transformation of the original data distribution into a normal pattern is usually performed. This includes normal and uniform score transform techniques.

Data trend removal may need to be performed prior to transformation to make sure, as much as technically possible, that the data distribution is **homostationary**, and it comes from a single statistical population.

Untransformed data is rarely considered in *CTDA*. The reason is that commonly used Gaussian methods require the transformation of the original data and the subsequent derivation of the variability models of the transformed data. Also, Gaussian transformation removes any data outliers. Covariography and variography are performed on the residuals (original data values minus the trend), whereas the chronotopologic trend is added back to the interpolated attribute values at the mapping stage of a study.

Brevi manu, a crucial covariography and variography prerequisite is the availability of a dataset that possesses a threefold character: it is in good order, of adequate quality, and offers a satisfactory coverage of the chronotopologic domain. In this context, being able to transform data into useful information for decision-making takes particular knowledge and skills.

3.2 Empirical covariance and variogram computations

To reiterate, the empirical covariance and variogram values are calculated using the basic expressions of Eqs. (2.7b)–(2.8b) in Section 2 or one of their convenient computational modifications involving specific choices of the *PDF*. In the chronotopologic case, one is actually dealing with covariance and variogram *surfaces* that are functions of the lags $\boldsymbol{\Delta p}$ (consisting of a spatial distance $\boldsymbol{h} = (h_1, h_2, \ldots, h_n)$ and a time separation τ) rather than *lines* that are functions purely of space $\boldsymbol{h}$. Since the variogram lags $\boldsymbol{\Delta p} = (\boldsymbol{h}, \tau)$ are defined both in space and in time, each lag considers all pairs of points that are separated by $\boldsymbol{h}$ in space (e.g., $|\boldsymbol{h}| \in (5, 10]$ km) and by τ in time (e.g., $\tau \in (1,3]$ hours). It is then hardly an understatement to suggest that:

Successful geostatistical modeling depends, to a large extent, on the adequacy of the computed **empirical** covariance or variogram values.

3.2.1 *Computational techniques*

The following standard computational formulas are often used to compute the empirical covariance, variogram and mean values:[r]

$$\hat{c}_X(\boldsymbol{\Delta p}) = \frac{1}{N(\boldsymbol{\Delta p})} \sum\nolimits_{i=1}^{N(\Delta p)} \left[\chi_i - \overline{X}\right] \left[\chi_{i+\Delta p} - \overline{X}\right], \tag{3.1}$$

$$\hat{\gamma}_X(\boldsymbol{\Delta p}) = \frac{1}{2N(\boldsymbol{\Delta p})} \sum\nolimits_{i=1}^{N(\Delta p)} \left[\chi_i - \chi_{i+\Delta p}\right]^2, \tag{3.2}$$

[r] As usual, the "ˆ" indicates that these formulas give the computed (empirical, raw, sample or experimental) values of the covariance, variogram and mean functions.

$$\hat{\bar{X}} = \frac{1}{N}\sum_{i=1}^{N} \chi(\boldsymbol{p}_i), \quad (3.3)$$

where $\Delta\boldsymbol{p}$ is the space-time lag, $\chi_i = \chi(\boldsymbol{p}_i)$, $\chi_{i+\Delta p} = \chi(\boldsymbol{p}_i + \Delta\boldsymbol{p})$, N is the number of data points considered, and $N(\Delta\boldsymbol{p})$ is the number of point pairs with lags $\Delta\boldsymbol{p}$. Obviously, the empirical covariance and variogram values are computed only for those distances, directions and time separations at which data pairs exist.

Eqs. (3.1) and (3.2) recognize that there is insight to be gained from both spatial and temporal data variations. A covariance or a variogram value at a given space-time lag $\Delta\boldsymbol{p}$ is the average product of attribute value fluctuations around the mean or the squared difference between the attribute values at the paired points, respectively. If the points $\boldsymbol{p}_i$ and $\boldsymbol{p}_i + \Delta\boldsymbol{p}$ are close to each other chronotopologically (this means that the metric size $|\Delta\boldsymbol{p}|$ is small), the corresponding attribute values are similar, so their difference $\chi_i - \chi_{i+\Delta p}$ will be small. As $\boldsymbol{p}_i$ and $\boldsymbol{p}_i + \Delta\boldsymbol{p}$ get farther apart, the corresponding attribute values become less similar, so the difference $\chi_i - \chi_{i+\Delta p}$ gets larger. Accordingly, the variance of $\chi_i - \chi_{i+\Delta p}$ increases with $\Delta\boldsymbol{p}$, which is why the variogram is assessing dissimilarity in the attribute distribution.

As usual in such cases, there are several practical issues associated with the calculation of the empirical covariance and variogram values, beyond the obvious assumption that the available dataset and other knowledge bases are sufficient to reliably compute these empirical values.

Dictum factum, the following two examples present the *step-by-step* process of computing the empirical covariance and variogram values using two different techniques, the **traditional technique** and the **cloud technique**.

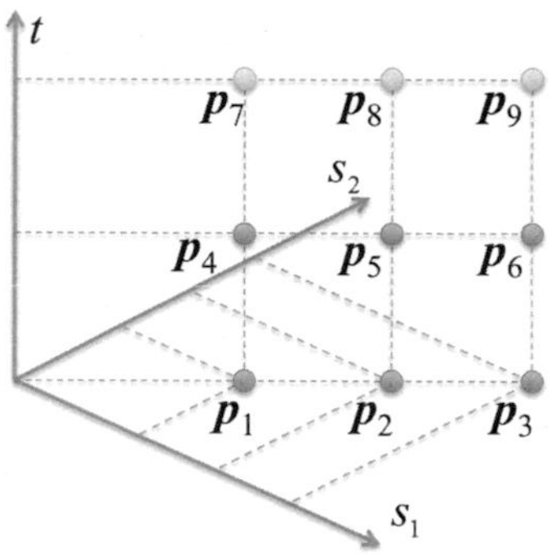

FIG. 3.1 Chronotopologic distribution of the nine points.

Example 3.1 Temperature is measured at three monitoring stations at three different time instances, see the nine chronotopologic points in Fig. 3.1. Table 3.1 describes the corresponding nine temperature data coordinates and measured temperature values.

Step 1: We start by considering the point pairs to be used in empirical covariance and variogram calculation. It is seen in Table 3.1 and Fig. 3.1 that for the paired points occurring at the same time instance,

$$(\boldsymbol{p}_1,\boldsymbol{p}_2),(\boldsymbol{p}_2,\boldsymbol{p}_3),(\boldsymbol{p}_4,\boldsymbol{p}_5),(\boldsymbol{p}_5,\boldsymbol{p}_6),(\boldsymbol{p}_7,\boldsymbol{p}_8),(\boldsymbol{p}_8,\boldsymbol{p}_9),$$

the spatial distances between them are the same, whereas for the paired points at the same location,

$$(\boldsymbol{p}_1,\boldsymbol{p}_4),(\boldsymbol{p}_4,\boldsymbol{p}_7),(\boldsymbol{p}_2,\boldsymbol{p}_5),(\boldsymbol{p}_5,\boldsymbol{p}_8),(\boldsymbol{p}_3,\boldsymbol{p}_6),(\boldsymbol{p}_6,\boldsymbol{p}_9),$$

the time separations between them are the same. Moreover, we consider spatial distances between paired points for various time instances,

$$(\boldsymbol{p}_1,\boldsymbol{p}_5),(\boldsymbol{p}_2,\boldsymbol{p}_6),(\boldsymbol{p}_2,\boldsymbol{p}_4),(\boldsymbol{p}_3,\boldsymbol{p}_5),(\boldsymbol{p}_4,\boldsymbol{p}_8),(\boldsymbol{p}_5,\boldsymbol{p}_7),$$
$$(\boldsymbol{p}_5,\boldsymbol{p}_9),(\boldsymbol{p}_6,\boldsymbol{p}_8),(\boldsymbol{p}_1,\boldsymbol{p}_8),(\boldsymbol{p}_2,\boldsymbol{p}_7),(\boldsymbol{p}_2,\boldsymbol{p}_9),(\boldsymbol{p}_3,\boldsymbol{p}_8),$$

As usual, the space-time lag is denoted[s] as $\Delta\boldsymbol{p} = (h_1, h_2, \tau)$, where h_1 and h_2 are spatial

[s] The readers may recall that usually a comma "," is used in between the spatial and temporal lags, whereas the semicolon ";" is favored when a distinction of the spatial from the temporal lags is explicitly made in the mathematical expression of the covariance or variogram functions.

TABLE 3.1 Temperature measurements.

Points (p_i)	($s_{i,1}, s_{i,2}; t_i$)	Temperature °C (χ_i)	Points (p_i)	($s_{i,1}, s_{i,2}; t_i$)	Temperature °C (χ_i)
p_1	(3,4;1)	27	p_6	(5,6;2)	28
p_2	(4,5;1)	29	p_7	(3,4;3)	24
p_3	(5,6;1)	30	p_8	(4,5;3)	27
p_4	(3,4;2)	26	p_9	(5,6;3)	29
p_5	(4,5;2)	27			

distances along the directions s_1 and s_2, respectively, and τ is the temporal separation. For illustration, we examine the numerical case $\boldsymbol{\Delta p} = (h_1, h_2, \tau) = (1,1,0)$, where the paired points involved are

$$(\boldsymbol{p}_1, \boldsymbol{p}_2), (\boldsymbol{p}_2, \boldsymbol{p}_3), (\boldsymbol{p}_4, \boldsymbol{p}_5), (\boldsymbol{p}_5, \boldsymbol{p}_6), (\boldsymbol{p}_7, \boldsymbol{p}_8), (\boldsymbol{p}_8, \boldsymbol{p}_9).$$

Step 2: The corresponding empirical covariance value is calculated using the point pairs previously mentioned as Eq. (3.1)

$$\begin{aligned}\hat{c}_X(\boldsymbol{p}_i, \boldsymbol{p}_i + \boldsymbol{\Delta p}) &= \hat{c}_X(1,0)\\ &= \frac{1}{6}[(\chi_1 - \overline{X})(\chi_2 - \overline{X}) + (\chi_2 - \overline{X})(\chi_3 - \overline{X}) + (\chi_4 - \overline{X})(\chi_5 - \overline{X}) + (\chi_5 - \overline{X})(\chi_6 - \overline{X})\\ &+ (\chi_7 - \overline{X})(\chi_8 - \overline{X}) + (\chi_8 - \overline{X})(\chi_9 - \overline{X})] = \frac{1}{6}[(27 - 27.44)(29 - 27.44)\\ &+ (29 - 27.44)(30 - 27.44) + \ldots + (27 - 27.44)(29 - 27.44)] = 0.75,\end{aligned}$$

where the temperature mean is (Eq. 3.3)

$$\begin{aligned}\hat{\overline{X}} &= \frac{1}{9}\sum\nolimits_{i=1}^{9}\chi_i\\ &= \frac{1}{9}(27 + 29 + 30 + 26 + 27 + 28 + 24 + 27 + 29)\\ &= 27.44.\end{aligned}$$

And, the empirical variogram value is (Eq. 3.2)

$$\begin{aligned}\hat{\gamma}_X(\boldsymbol{p}_i, \boldsymbol{p}_i + \boldsymbol{\Delta p}) &= \hat{\gamma}_X(1,0)\\ &= \frac{1}{2\times 6}\left[(\chi_1 - \chi_2)^2 + (\chi_2 - \chi_3)^2 + (\chi_4 - \chi_5)^2 + (\chi_5 - \chi_6)^2 + (\chi_7 - \chi_8)^2 + (\chi_8 - \chi_9)^2\right]\\ &= \frac{1}{12}\left[(27 - 29)^2 + (29 - 30)^2 + (26 - 27)^2 + (27 - 28)^2 + (24 - 27)^2 + (27 - 29)^2\right] = 1.67.\end{aligned}$$

Step 3: Working along the same lines, several other empirical covariance and variogram values at different space-time lags $\boldsymbol{\Delta p} = (h_1, h_2, \tau)$ are computed in Table 3.2. Fig. 3.2A and B shows the calculated empirical variogram $\hat{\gamma}_X$ and covariance $\hat{c}_X$ values, respectively, at all considered spatial distances h and temporal separations τ above.

TABLE 3.2 Empirical covariance and variogram values at different space-time lags.

Space-time lag $\Delta p = (h_1, h_2; \tau)$	Paired space-time points	Empirical $\hat{c}_X(h_1, h_2; \tau)$ and $\hat{\gamma}_X(h_1, h_2; \tau)$ values
(1,1;0)	(p_1,p_2), (p_2,p_3), (p_4,p_5), (p_5,p_6), (p_7,p_8), (p_8,p_9)	0.75, 0.67
(1,1;1)	(p_1,p_5), (p_2,p_4), (p_2,p_6), (p_3,p_5), (p_4,p_8), (p_5,p_7), (p_5,p_9), (p_6,p_8)	−0.14, 2.83
(0,0;1)	(p_1,p_4), (p_4,p_7), (p_2,p_5), (p_5,p_8), (p_3,p_6), (p_6,p_9)	1.235, 1.17
(0,0;2)	(p_1,p_7), (p_2,p_8), (p_3,p_9)	1.61, 2.33
(1,1;2)	(p_1,p_8), (p_2,p_7), (p_2,p_9), (p_3,p_8)	−0.97, 4.25
(2,2;0)	(p_1,p_3), (p_4,p_6), (p_3,p_9)	−2.43, 2.33
(2,2;1)	(p_1,p_6), (p_4,p_3), (p_4,p_9), (p_7,p_6)	−2.03, 5.25
(2,2;2)	(p_1,p_9), (p_7,p_3)	−4.75, 10

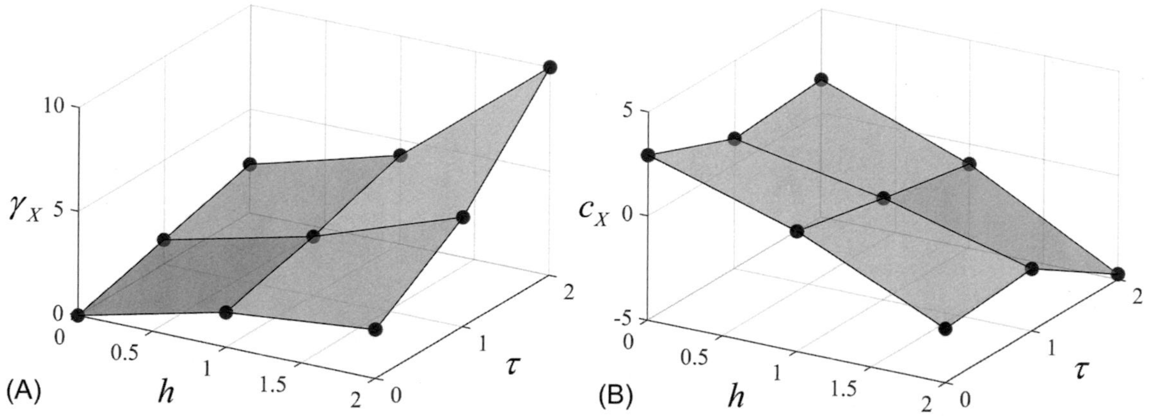

FIG. 3.2 Empirical space-time (A) variogram and (B) covariance values.

Step 4: Figs. 3.2A and B show that the variogram values increase with increasing spatial distances and time separations, whereas the covariance values decrease with increasing spatial distances and temporal separations (variogram and covariance exhibit inverse shapes).

Step 5: Lastly, theoretical variogram and covariance models (e.g., combinations of "Gaussian," "Spherical," or "Exponential" models) were fitted to these empirical variogram and covariance values (see discussion in Section 3.3 later).

It has been argued with considerable justification that the calculation of empirical covariogram and variogram values is as much a thinking mode as it is a computational method. In the following example, an alternative technique based on a different than the traditional way of thinking is presented for the computation of empirical covariogram and/or variogram values.

Example 3.2 Fig. 3.3A displays a simulated chronotopologic sea level dataset (in m) along a coastal area (2166 data, in total). The spatial coordinates of the sea level observation locations range from 1 to 10 km with a 0.5 km interval and the temporal instances range from 5 to 10 s with

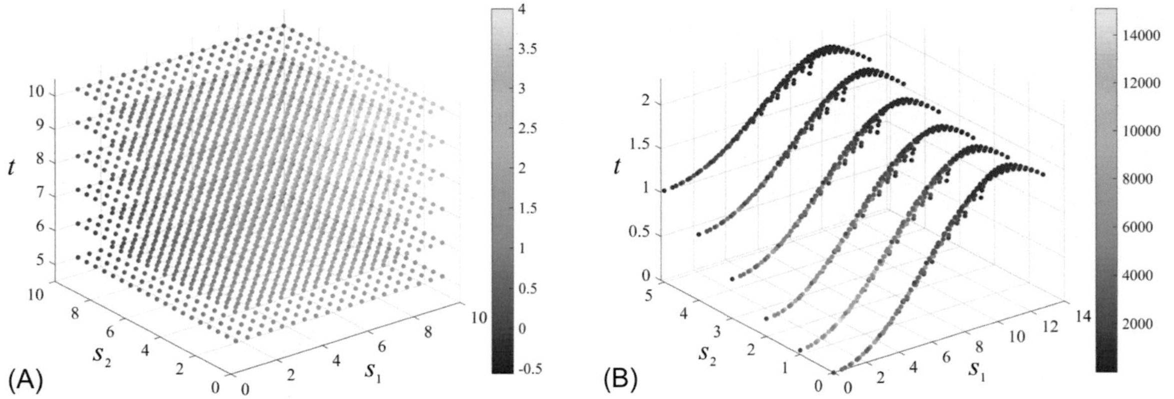

FIG. 3.3 (A) Space-time dataset with the legend representing sea level (in m), and (B) cloud of empirical variogram values; the color legend represents the number of pairs between two points at the specified spatial distance and time separation.

a 1 s interval. The legend in Fig. 3.3A represents the observed sea levels at the corresponding space-time points. We seek to calculate the empirical variogram values and to subsequently fit a theoretical variogram model to them.

Step 1: The possible temporal separations and spatial distances between any pair of points in Fig. 3.3A include, respectively, the group {0, 1, 2, 3, 4, 5 s) and {0, 0.5, 0.7071, 1, 1.118, 1.4142, 1.50 km}.[t]

Step 2: Fig. 3.3B shows the cloud of the calculated empirical variogram values with the corresponding legend representing the number of pairs between two points at the specified spatial distance and time separation. The variogram values increase with increasing spatial distance and temporal separation. Visually, the points in Fig. 3.3B form six smooth curves at various time separations. This is because the simulated sea level dataset is characterized by strong chronotopologic dependency. In real-world situations, however, the variogram values usually exhibit much higher fluctuations (than the smooth values shown in Fig. 3.3B) with a rather irregular shape.

Step 3: In order to make variogram model fitting easier, a set of lag classes along the spatial and temporal axes were selected in Fig. 3.3B, and then the vertical surfaces were drawn dividing the entire chronotopologic domain into several blocks: 165 spatial distances and 6 temporal separations were considered (the distance and separation intervals were equal to 2 km and 1 s, respectively). Variogram values at zero distance or zero separation were not included. The domain of interest was divided into various chronotopologic blocks using the spatial classes 0, (0,2], (2,4], (4,6], (6,8], (8,10], (10,12] and (12,14] km, and the temporal classes 0, (0,1], (1,2], (2,3], (3,4], (4,5] *s*, see Fig. 3.4A.

Step 4: The average value of the cloud of empirical variogram values was calculated for each space-time block. E.g., when spatial distance and temporal separation were (0,2] and (0,1], respectively, the average variogram value was equal to 0.1101. This value was calculated by averaging all variogram values located in this specific block; the coordinates of this average value were set equal to 1 (space) and 1 (time). After averaging all values within the various blocks, a series of empirical variogram values

[t] Because this is a simulated dataset, as one sees in Fig 3.3A, the simulated points are (1,1;5) (2,1;5) (2,2;5) (1,2;5), etc., the spatial resolution in the s_1 and s_2 axis is 0.5 km, and, then, the distances between these points are 0, 0.5, 0.71 etc.

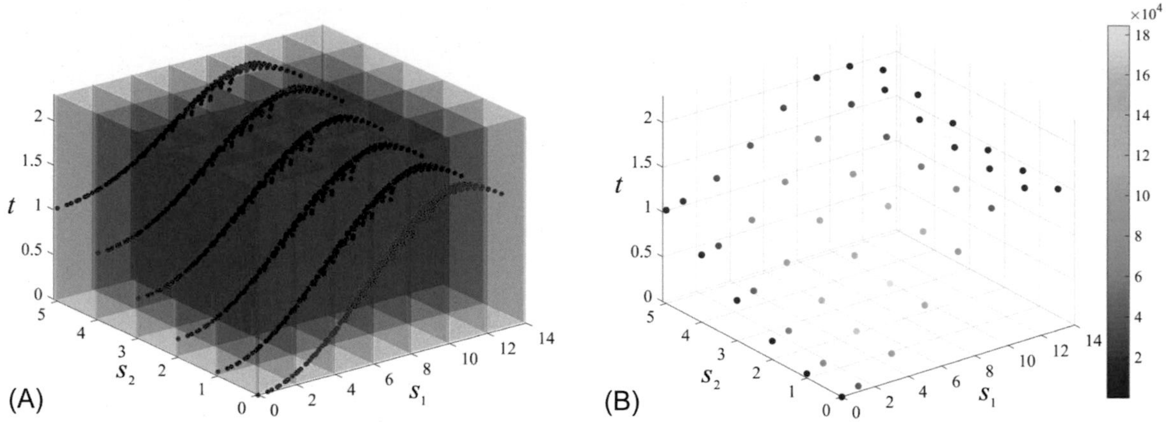

FIG. 3.4 Empirical variogram calculation: (A) Selected space-time blocks, and (B) average variogram values within these blocks.

were obtained at various distances and separations, as shown in Fig. 3.4B (the shades represent the number of pairs between points at each specified distance and separation).

3.3 Covariance and variogram modeling

Once the empirical covariance or variogram values have been computed and all key covariance or variogram parameters have been linked to known physical processes, one may proceed to covariance or variogram model selection.

> An adequate **theoretical model** satisfies the requirement of geostatistical analysis that the covariance or variogram values are needed for subsequent calculations at all distances, directions, and time separations within the neighborhood.[u]

This is because, as noted earlier, the empirical covariance and variogram values are calculated only for specific distances, directions, and time separations at which data pairs exist. Accordingly, the empirical covariance and variogram values must be interpolated at those distances, directions, and time separations where too few or no attribute data pairs are available.

Example 3.3 Three common real-world cases that make modeling necessary are as follows.

- In Examples 3.1 and 3.2, temperatures and sea levels, respectively, were calculated at point pairs determined by data availability. Yet, chronotopologic sea level interpolation and mapping require temperatures and sea level values at points in between those determined by the data.
- The empirical covariance and variogram values of petrophysical attributes are usually calculated along the principal (horizontal and vertical) directions, but chronotopologic estimation requires them in other directions containing contributions from both principal directions.
- Covariance and variogram models may reveal environmental gradients that illuminate animal–habitat relationships (either masked or impractical to investigate using a purely patch-based perspective) by quantifying spatial dependence among point samples (Sadoti et al., 2014).

[u] The covariance and the variogram have a significant effect, e.g., on chronotopologic estimation or mapping.

To satisfy the interpolation and mapping requirements illustrated in Example 3.3, and given the empirical covariance and variogram values between pairs of data points, adequate theoretical models are subsequently fitted to them. These covariance and variogram models have the following mathematical properties:

- They are functions of the **lag** $\boldsymbol{\Delta p}$ (location vector $\boldsymbol{h} = (h_1, h_2)$ and time separation τ) in the homostationary case, or of the **metric** $|\boldsymbol{\Delta p}|$ in the isostationary case. The independent variables, $\boldsymbol{h}$ and τ, of a covariance or a variogram plot have the same units as the location $\boldsymbol{s}$ and the instance t, respectively. The dependent variable, covariance or variogram axis, has units that are equal to the units of $X(\boldsymbol{p})$ squared.
- They must be **permissible** (i.e., positive definite or conditionally positive definite, see *COP* in Section 2.3). *Inter alia*, permissibility is necessary so that the Kriging interpolation error variance (Chapter 8) is positive. Hence, it has been a common practice to use covariance and variogram models that are known to be permissible (like those in Tables 3.4 and 3.5).

3.3.1 Variogram visualization

In the rest of this section, for instructional purposes it is convenient to focus on variogram models—the corresponding covariance models can be derived directly from the variogram models, when necessary. Fig. 3.5 presents a typical plot of a variogram model, $\gamma_X(\boldsymbol{\Delta p})$, where $h = |\boldsymbol{h}|$. Some general variogram features are listed in Table 3.3 that can help organize one's thinking concerning the information conveyed by a variogram model.

The plot of Fig. 3.5 may represent visually the variogram γ_X at a certain spatial direction, or an averaged variogram in the sense that its values are the averages of variogram values along different directions of the vector h but at the same magnitude $|\boldsymbol{h}| = h$. Some of the descriptive variogram features are as follows: At each set of lags $\boldsymbol{h}$ and τ, the chronotopologic variogram is $\gamma_X(\boldsymbol{h}, \tau)$ with marginal variograms $\gamma_X(\boldsymbol{h}, 0)$ and $\gamma_X(\boldsymbol{0}, \tau)$; the chronotopologic range is $\boldsymbol{\varepsilon}_{st} = (\varepsilon_s, \varepsilon_t)$ with marginal ranges ε_s and ε_t; and the chronotopologic sill is $c_{0,st} = \gamma_X(\varepsilon_s, \varepsilon_t)$ with marginal sills $c_{0,s} = \gamma_X(\varepsilon_s, 0)$ and $c_{0,t} = \gamma_X(0, \varepsilon_t)$. When variograms at different spatial directions are considered, they may be assumed to have the same

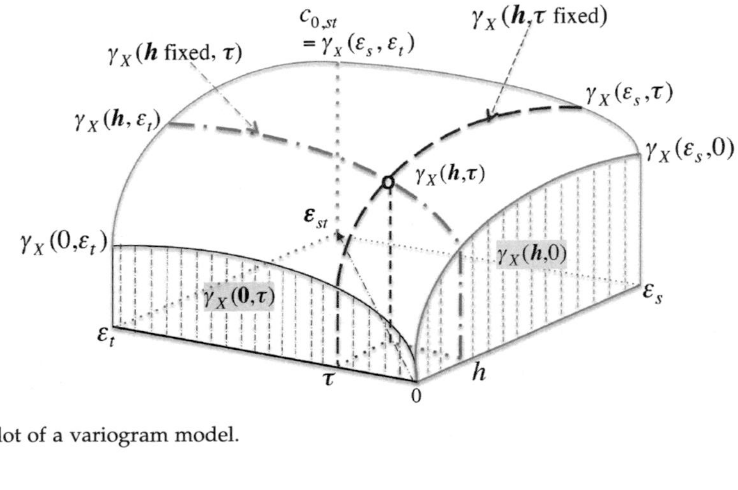

FIG. 3.5 Typical plot of a variogram model.

TABLE 3.3 Key variogram features.

Feature	Description
1	At zero space and time limits $\boldsymbol{h} = \boldsymbol{0}$ and $\tau = 0$, the **marginal temporal** and the **marginal spatial** variograms $\gamma_X(\boldsymbol{0}, \tau)$ and $\gamma_X(\boldsymbol{h}, 0)$ are obtained, respectively.
2	Variogram values generally increase with distance $\boldsymbol{h}$ and separation τ until a lag $\boldsymbol{\Delta p} = (\boldsymbol{h}, \tau)$ is reached at which the variogram levels out approximately at the **range vector** $\boldsymbol{\varepsilon}_{st} = (\varepsilon_s, \varepsilon_t)$.
3	Sample points separated by lags $\boldsymbol{\Delta p} < \varepsilon_{st}$ are spatially or temporally **correlated**, whereas sample points separated by lags $\boldsymbol{\Delta p} \geq \varepsilon_{st}$ are considered **independent**.
4	Purely **spatial sill**, $c_{0,s} = c_s(\boldsymbol{0})$, is the variogram value reached at a spatial range ε_s, while purely **temporal sill** $c_{0,t} = c_t(0)$ is the value reached at the temporal range ε_t.[a]
5	A **cross-sectional spatial** variogram, $\gamma_X(\boldsymbol{h}, \tau$ fixed), at a fixed nonzero separation ($\tau \neq 0$) exhibits a so-called **nugget**[b] that increases with higher τ values. Similar is the behavior of the **cross-sectional temporal** variogram, γ_X ($\boldsymbol{h}$ fixed; τ), at fixed nonzero distances ($h \neq 0$), which has a nugget that increases with higher h values.[c]

[a] *As we will see later, there may exist different kinds of sills depending on the phenomenon.*
[b] *Its origins will be discussed later.*
[c] *The $\gamma_X(h, 0) = \delta_{0,t}(h)$ and $\gamma_X(0, \tau) = \delta_{0,S}(\tau)$ can be seen as the space- and time-varying nugget effects of the corresponding cross-sectional variogram, $\gamma_X(\tau \mid h \neq 0)$, and cross-sectional spatial sea level variogram, $\gamma_X(h \mid \tau \neq 0)$, respectively.*

marginal temporal variogram $\gamma_X(\boldsymbol{0}, \tau)$. When variograms at different spatial directions are considered, they may be assumed to have the same marginal temporal variogram $\gamma_X(\boldsymbol{0}, \tau)$.

Among the major variography challenges are (*i*) the acquisition of the right data to be used in empirical variogram computations at the available $\boldsymbol{h}$ and τ, and (*ii*) the selection of the adequate models to be used for the final variogram representation at any $\boldsymbol{h}$ and τ. The nature of the attribute and its distributional features determine which variogram model type will be used to characterize the attribute's chronotopologic variability in the *CTDA* context. For practical reasons, a distinction is made between two major classes of variability characterization models, as follows.[v]

3.3.2 1st class: Separable variability models

The 1st class of variogram models is often the simplest way of looking at chronotopologic variation, and can be used with relative ease in practical applications. This class is defined as follows.

> Space-time **separable** models can be constructed by properly combining purely spatial with purely temporal models, which are thus considered as the main components of the synthetic chronotopologic variogram model obtained.

The term "properly" above means that not every possible combination of models is allowed. Examples of combinations leading to permissible space-time separable variogram models are:

$$\begin{aligned}
\gamma_X(\boldsymbol{\Delta p}) &= \gamma_S(\boldsymbol{h}) + \gamma_T(\tau),\\
\gamma_X(\boldsymbol{\Delta p}) &= c_S(0)\gamma_T(\tau) + \gamma_S(\boldsymbol{h})c_T(0) - \gamma_S(\boldsymbol{h})\gamma_T(\tau),\\
\gamma_X(\boldsymbol{\Delta p}) &= \gamma_S(\boldsymbol{h}) + \gamma_T(\tau) - a\gamma_S(\boldsymbol{h})\gamma_T(\tau),
\end{aligned} \tag{3.4a-c}$$

where $a_i \geq 0\,(i = 1,2,3)$ and $a \in \left(0, \frac{1}{\max(c_{0,s},\ c_{0,t})}\right]$.[w]

In practice, to ensure mathematical rigor and consistency, theoretical variograms are modeled in terms of suitable analytical functions:

[v] Other classifications exist, some of which will be mentioned in following sections and chapters.

[w] For future reference, the corresponding covariance relationships are: $c_X(\boldsymbol{\Delta p}) = c_S(\boldsymbol{h}) + c_T(\tau)$, $c_X(\boldsymbol{h}, \tau) = c_S(\boldsymbol{h})c_T(\tau)$, and $c_X(\boldsymbol{\Delta p}) = a_1c_S(\boldsymbol{h}) + a_2c_T(\tau) + a_3c_S(\boldsymbol{h})\gamma_T(\tau)$.

TABLE 3.4 Separable variogram models in space and time ($\zeta = |\boldsymbol{h}|$ or $= \tau$, $c_0 \geq 0$).

Model	Mathematical form
Exponential	$c_0\left(1-e^{-\frac{\zeta}{a_\zeta}}\right)(a_\zeta>0)$
Gaussian	$c_0\left(1-e^{-\frac{\zeta^2}{a_\zeta^2}}\right)(a_\zeta>0)$
Expo-Cosine	$c_0\left[1-e^{-\frac{\zeta}{a_\zeta}}\cos(b_\zeta\zeta)\right]\ (a_\zeta,b_\zeta>0)$
Spherical	$\begin{cases}\frac{c_0\zeta}{2a_\zeta}\left[3-\frac{\zeta^2}{a_\zeta^2}\right], & \zeta\in[0,a_\zeta] \\ 1, & \zeta>a_\zeta\end{cases}\ (a_\zeta>0)$
Wave	$c_0\left(1-\frac{\sin a_\zeta\zeta}{\zeta}\right)\ (a_\zeta>0)$
De Wijsian	$\frac{\ln\zeta^{a_\zeta}}{\ln(1+\zeta^{a_\zeta})}\ (a_\zeta\in(0,2])$
Cubic	$\begin{cases}c_0\left[\frac{7}{a_\zeta^2}\zeta^2-\frac{35}{4a_\zeta^3}\zeta^3+\frac{7}{2a_\zeta^5}\zeta^5-\frac{3}{4a_\zeta^7}\zeta^7\right], & \zeta\leq a_\zeta \\ c_0, & \zeta>a_\zeta\end{cases}$
Pentaspherical	$\begin{cases}c_0\left[\frac{15}{8a_\zeta}\zeta-\frac{5}{4a_\zeta^3}\zeta^3+\frac{3}{8a_\zeta^5}\zeta^5\right], & \zeta\leq a_\zeta \\ c_0, & \zeta>a_\zeta\end{cases}$
Pure nugget[y]	$c_0[1-H(0)]$

- Commonly used theoretical **spatial** variogram models are listed in Table 3.4 with $\zeta = |\boldsymbol{h}|$ and parameters directly related to the sill and range.[x]
- Variogram models of Table 3.4 can also be assumed in the **temporal** domain by simply letting $\zeta = \tau$.
- New **synthetic** chronotopologic variogram models can be constructed by choosing their purely spatial and temporal components from the models of Table 3.4.

Following the previously mentioned process, variogram models of a mixed space-time structure are conveniently produced in two ways:

① A **direct** space-time separable variogram model involves a single spatial and a single temporal model.

Sometimes, the real-world attribute variability is rather complex, and the simple combination of a single spatial model and a single temporal model cannot adequately fit the empirical values. This potential inadequacy leads to the following model building process.

② A **nested** space-time separable variogram model involves more than one spatial and/or more than one temporal model.

The previously mentioned distinction, ① *vs.* ②, allows considerable flexibility in the construction of variogram models in practice. That is, standard model shapes can be combined in many ways to build more detailed variogram model shapes.

The following examples present the *step-by-step* process of constructing **synthetic separable** variogram models in practice using Eqs. (3.4a–c), while emphasizing the procedural distinction between direct and nested models.

Example 3.4 For instructional purposes, we first discuss the direct attribute variogram model construction process.

Step 1: From Table 3.4 the Exponential component model is chosen to represent the spatial part and the Gaussian component model to represent the temporal part of the attribute variability.

[x] A special case is the pure nugget model that can be seen as a limiting case of some of the other models when the range is infinitesimally small.

[y] Heaviside function, $H(0) = 1$ at lag 0 and $=0$ otherwise.

Step 2: In light of Eq. (3.4b) the composite space-time theoretical variogram model can be written as

$$\gamma_X(\Delta \boldsymbol{p}) = c_0^2\left[\left(1-e^{-\frac{h}{\alpha}}\right)+\left(1-e^{-\frac{3\tau^2}{\beta^2}}\right) - \left(1-e^{-\frac{h}{\alpha}}\right)\left(1-e^{-\frac{3\tau^2}{\beta^2}}\right)\right], \quad (3.5)$$

which is an additive and multiplicative combination of the features of its Exponential and Gaussian components.

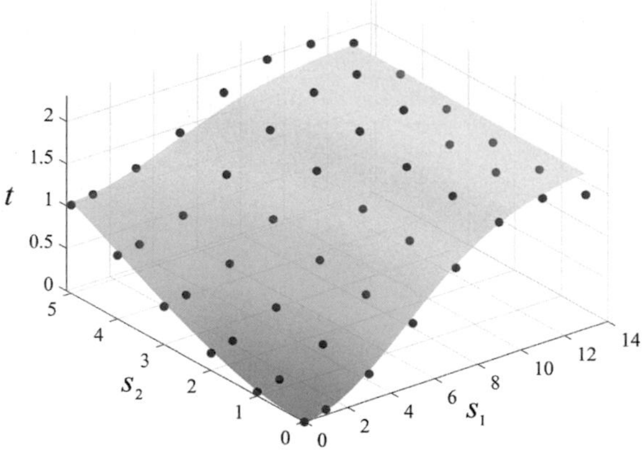

FIG. 3.6 Empirical variogram values (dots) of sea surface, and the fitted theoretical variogram model.

Example 3.2 (cont.): Going back to the chronotopologic sea level dataset (in m) of Example 3.2, the direct model fitting proceeds can be implemented as follows.

Steps 1-4: As in Example 3.2.

Step 5: A theoretical space-time separable Gaussian variogram model (i.e., a purely spatial Gaussian and a purely temporal Gaussian model selected from Table 3.4) is fitted to the experimental values (dots in Fig. 3.4B and Fig. 3.6 later), as follows

$$\gamma(\Delta \boldsymbol{p}) = 1.87\left(1-e^{-\frac{3h^2}{11.54^2}-\frac{3\tau^2}{9.36^2}}\right), \quad (3.6)$$

see Fig. 3.6, with estimated model parameters $c_0 = 1.87$ (common sill), $a_h = 6.66$ and $a_\tau = 5.40$. As one observes, at large distances h, the cross-sectional time variogram loses its Gaussian behavior and becomes rather linear; this is not valid at large τ, where the cross-sectional spatial variogram retains its Gaussian shape. The marginal spatial and temporal sea level variograms are, respectively,

$$\gamma_X(h;0) = 1.87\left(1-e^{-\frac{3h^2}{11.54^2}}\right), \text{ and } \gamma_X(0;\tau) = 1.87\left(1-e^{-\frac{3\tau^2}{9.36^2}}\right).$$

These functions are the space- and time-varying nugget effects of the cross-sectional temporal sea level variogram, $\gamma_X(\tau|h \neq 0)$, and cross-sectional spatial sea level variogram, $\gamma_X(h|\tau \neq 0)$.

Example 3.5 The implementation of nested variogram modeling can be illustrated in two distinct steps, as follows.

Step 1: The Exponential component model is selected from Table 3.4 to represent the spatial part and the Gaussian and Spherical component models to represent the temporal part of the space-time attribute variability.

Step 2: Using Eqs. (3.4a) and (3.4b), we get

$$\gamma_X(\Delta \boldsymbol{p}) = c_0^2\left[\left(2-e^{-\frac{h}{\alpha_1}}-e^{-\frac{3\tau^2}{\beta_1^2}}\right)+\left(2-e^{-\frac{h}{\alpha_2}}-e^{-\frac{3\tau^2}{\beta_2^2}}\right) - \left(1-e^{-\frac{h}{\alpha_2}}\right)\left(1-e^{-\frac{3\tau^2}{\beta_2^2}}\right)\right], \quad (3.7a)$$

$$\gamma_X(\Delta \boldsymbol{p}) = c_0\left[\left(1-e^{-\frac{h}{\alpha_1}}\right)+\left(1-e^{-\frac{h}{\alpha_2}}\right)+\left(1-e^{-\frac{3\tau^2}{\beta_1^2}}\right) + \frac{\tau}{2\alpha}\left(3-\frac{\tau^2}{\alpha^2}\right)\right]. \quad (3.7b)$$

The examples above suggest a straightforward yet general approach for constructing new space-time variogram models:

Suitable combinations of spatial and temporal models can be developed,[z] and then fitted to the empirical variogram values. The best-fitted combination, which is decided by merging insight with technical means (e.g., the least square error technique), leads to a new **synthetic** variogram model that depicts the real phenomenon and can serve further *CTDA* objectives.

Although this approach facilitates structural analysis considerably, it has some *in actu* drawbacks that are caused by the strict separation of spatial and temporal structures assumed, which may turn out to be physically inappropriate in many applications. Such a separation may imply, e.g., that the spatial behavior is the same at all times and that the temporal behavior is the same at all locations. This is not, however, what investigators often encounter in practice, where different spatial patterns emerge at different times, and time series at different locations show distinct behavior (Vyas and Christakos, 1997; Snepvangers et al., 2003).

Therefore, because in the real world there is not a single modeling approach that can act as a panacea to all conceivable applications, in the literature one encounters a variety of theoretical variogram models that are not conveniently separable in space and time, but they can offer a better fit to the empirical evidence available in real-world applications. Such variogram models are examined next.

3.3.3 2nd Class: Nonseparable or composite variability models

The 2nd class of variogram models may turn out to be a more substantive way of looking at the chronotopologic attribute variation than the 1st class of models, in the sense that it offers a more realistic perspective justified by the physics of the phenomenon of interest.

Space-time **nonseparable** or **composite** theoretical variogram models are those that cannot be decomposed into purely spatial and purely temporal models.

Analytical expressions of commonly used composite variogram models are listed in Table 3.5.[aa]

TABLE 3.5 Nonseparable variogram models ($n = 1, 2, 3$; $c_0 \geq 0$).

Model	Mathematical expression
1	$c_0 - (4\alpha\pi\tau)^{-0.5n} e^{-\frac{h^2}{4\alpha\tau}}$ $(c_0,\ \alpha > 0)$
2	$c_0 - (\beta\tau^{2\gamma} + 1)^{-0.5n} e^{-h^2(\beta\tau^{2\gamma}+1)^{-1}}$ $(c_0 > 0,\ \beta \in [0,1],\ \gamma \in (0,1])$
3	$c_0 - e^{-\frac{\lvert \boldsymbol{h} \pm \boldsymbol{v}\tau \rvert}{\alpha}}$ $(c_0, \alpha > 0)$
4	$c_0 - e^{-\frac{(\boldsymbol{h} \pm \boldsymbol{v}\tau)^2}{\alpha^2}}$ $(c_0, \alpha > 0)$
5	$c_0 - \left[1 + \frac{(\boldsymbol{h} \pm \boldsymbol{v}\tau)^2}{\beta^2}\right]^{0.5\nu} e^{-\frac{\lvert \boldsymbol{h} \pm \boldsymbol{v}\tau \rvert}{\alpha}}$ $(c_0, \alpha, \beta, \nu = \lvert \nu \rvert > 0)$
6	$c_0 \left[1 - (1+\tau^2)^{-\frac{1}{2}} e^{-\frac{h^2}{1+\tau^2}}\right]$ $(c_0 > 0)$
7	$c_0 - \left(1 + \frac{\tau^2}{\nu^2}\right)^{-0.5\nu} e^{-\frac{h}{a}}$ $(c_0, a, \nu > 0)$
8	$c_0 \left[1 - \left(1 + \frac{r^{\lambda_1}}{a_1^2} + \frac{\tau^{\lambda_2}}{a_2^2}\right)^{-\lambda_3}\right]$ $(\lambda_1,\ \lambda_2 \in (0,2], c_0, a_1, a_2, \lambda_3 > 0)$
9	$c_0 - \frac{\sin(\beta\tau)}{\pi^4\tau} \prod_{i=1}^{3} \frac{\sin(a_i h_i)}{h_i}$ $(c_0, a_i, \beta > 0)$
10	$(c_0 + c) - c\left[1 + \frac{1}{\beta^2}(h + \alpha\tau)^2\right]^{-\frac{\nu}{2}} e^{-\frac{1}{\lambda}\lvert \boldsymbol{h} + \alpha\tau \rvert}$ $(c_0, c, a, \lambda > 0)$

[z] E.g., the permissible combinations in Eq. (3.4a–c).

[aa] For a more detailed list, see Christakos et al. (2017).

- In all these models, space and time are **intimately linked** rather than considered separately.

In a certain respect, the space-time separable modeling class considered earlier can be seen as a special case of nonseparable (composite) modeling under limited conditions concerning the structure of chronotopologic variability.

Example 3.6 Consider the study of $PM_{2.5}$ concentration data (μg/m^3) at 19345 points covering the Jiangsu region (China) during 2014 (Yang et al., 2018).

Step 1: For the original $PM_{2.5}$ concentration data in the southern Jiangsu, the calculated Kolmogorov-Smirnov test value was 0.0001 ($P<0.005$), indicating that the null hypothesis of a normal data distribution was rejected. Accordingly, the original $PM_{2.5}$ data were logarithmically transformed to obey a normal distribution. Of course, when back-transformed the resulting $PM_{2.5}$ maps included the existing space-time trends of the raw data (Yang et al., 2018).

Step 2: The empirical variogram values of the log$PM_{2.5}$ concentration data (μg/m^3) are denoted by black dots in Fig. 3.7.

Step 3: Based on the configuration of the empirical values previously mentioned, a theoretical variogram model of the 2nd class (in particular, the nonseparable model No. 10 of Table 3.5) was selected. The weighted least square technique with weights equal to $(\boldsymbol{h}^2+\tau^2)^{-1}$ was used to calculate the model parameters as $c_0 = 0.022$ (μg/m^3)^2, $c = 0.506$ (μg/m^3)^2, $\beta = -1.87 \times 10^5$m, $\alpha = 4.941 \times 10^5$m/day, $v = -2.229$, and $\lambda = 1.629 \times 10^5$m. The final log$PM_{2.5}$ variogram model was

$$\gamma_X(\boldsymbol{\Delta p}) = 0.528 - 0.506\left[1 + 0.29\boldsymbol{\Delta p}^2\right]^{1.115} e^{-0.614|\boldsymbol{\Delta p}|} \tag{3.8}$$

in units of (μg/m^3)^2, where $|\boldsymbol{\Delta p}| = 10^{-5}|\boldsymbol{h}| + 4.94\tau$. The theoretical variogram model

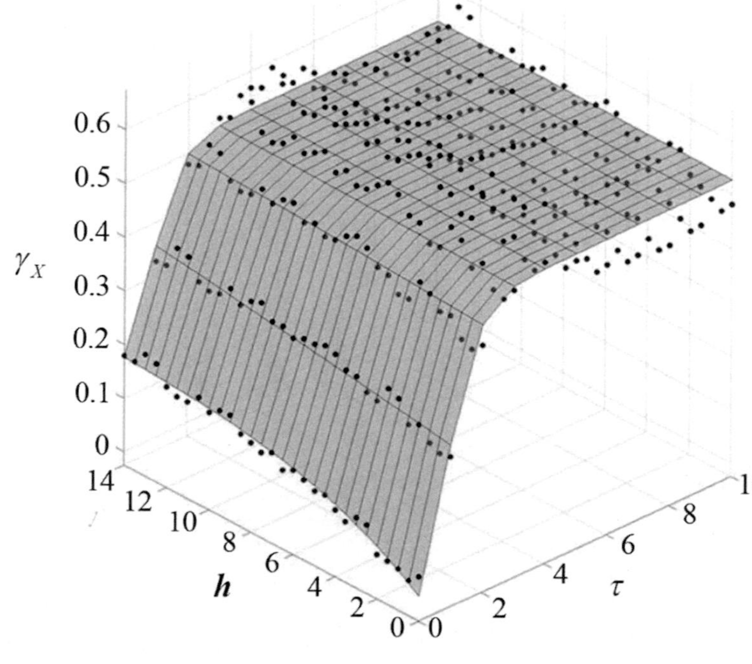

FIG. 3.7 Empirical log$PM_{2.5}$ variogram values and the fitted theoretical model ($h = |\boldsymbol{h}|$ in m × 10^4, τ in days).

plotted in Fig. 3.7 closely follows the pattern of the empirical values (black dots in Fig. 3.7), indicating that the fitted model of Eq. (3.8) adequately represents the actual chronotopologic $PM_{2.5}$ decorrelations. The marginal variograms,

$$\gamma_X(h,0) = 0.53 - 0.51\left[1 + 0.29\left(\frac{1}{10^5}h\right)^2\right]^{1.12} e^{-\frac{0.61}{10^5}h},$$

and

$$\gamma_X(0,\tau) = 0.53 - 0.51\left[1 + 0.29(4.94\tau)^2\right]^{1.12} e^{4.94\tau},$$

imply that the nugget effect $\delta_{0,t}(h) = \gamma_X(h,0)$ of the cross-sectional temporal variogram is generally smaller than the nugget effect $\delta_{0,s}(\tau) = \gamma_X(0,\tau)$ of the cross-sectional spatial variogram.

Example 3.7 The variogram model No. 6 in Table 3.5 has the interesting feature that it satisfies the following partial differential equation

$$\frac{\partial}{\partial h}\gamma_X(\boldsymbol{\Delta p}) = \frac{2h}{\tau}\frac{\partial}{\partial \tau}\gamma_X(\boldsymbol{\Delta p}), \tag{3.9a}$$

where the parameter $\frac{2h}{\tau}$ determines the mixing between the temporal variability change and the spatial variability change. Also, one may notice that the models Nos 3–5 in Table 3.5 are associated with a special class of random fields exhibiting a wave pattern so that

$$X(\boldsymbol{p}) = \widehat{X}(\boldsymbol{s} - \boldsymbol{v}t), \tag{3.9b}$$

where $\boldsymbol{v}$ is a vector parameter to be determined from the data (Christakos, 2017).

3.3.4 Metric-dependent variogram models

Beyond fitting an adequate model to the empirical data, nonseparable modeling offers an interesting way of building new models. As was shown in Chapter 2, the metric used in a variogram may take various forms, depending on the phenomenon. This suggests that new chronotopologic variogram models can be constructed as follows:

> Select a variogram γ_X **shape** (e.g., from Table 3.5) and a **metric** $|\boldsymbol{\Delta p}|$ form (Tables 3.1 and 3.3, Section 3 of Chapter 2), then insert $|\boldsymbol{\Delta p}|$ into γ_X to obtain an expression for the $\gamma_X(|\boldsymbol{\Delta p}|)$ model.

The following example illustrates how a nonseparable variogram model can be constructed in practice based on metric selection.

Example 3.8 According to the previous discussion, a valid chronotopologic variogram model of subsurface flow can be constructed as follows.

Step 1: Consider a nonseparable variogram model γ_X. Any known variogram shape like those listed in Table 3.5 may be used for this purpose.

Step 2: Suppose that the available data imply that the temporal metric component should be rescaled to match the spatial one in terms of the chronotopologic anisotropy correction $\sqrt{\varepsilon}$. Then, a suitable metric choice would be $|\boldsymbol{\Delta p}| = \sqrt{h^2 + (\sqrt{\varepsilon}\tau)^2}$, which leads to the variogram model

$$\gamma_X(|\boldsymbol{\Delta p}|) = \gamma_X\left(\sqrt{h^2 + (\sqrt{\varepsilon}\tau)^2}\right). \tag{3.10}$$

Step 3: A suitable subsurface flow metric is

$$|\boldsymbol{\Delta p}| = \sqrt{\sum\nolimits_{i=1}^{2} h_i + \xi^{-1}\left(\sum\nolimits_{i=1}^{2} k_i - 2\kappa\right)\tau},$$

which yields the physically interesting variogram model

$$\gamma_X(|\boldsymbol{\Delta p}|) = \gamma_X\left(\sqrt{\sum\nolimits_{i=1}^{2} h_i + \xi^{-1}\left(\sum\nolimits_{i=1}^{2} k_i - 2\kappa\right)\tau}\right), \tag{3.11}$$

where ξ, κ, and k_i ($i = 1,2$) are flow parameters.

3.3.5 Case-dependent variogram models

Despite the usefulness of the variogram models listed in Tables 3.4 and 3.5, they should not be used in a black-box manner in practice. As a matter of fact, there are real-world situations where it makes more sense to rely on a different approach:

Case-dependent variogram models are derived directly from the available *KB*s about the phenomenon of interest[ab] rather than relying solely on the black-box implementation of the models listed in Table 3.4 or 3.5.

Hence, models need to be chosen so that they represent two major kinds of knowledge of chronotopologic attribute variability:

- **Quantitative** knowledge (say, in the form of mathematical equations representing physical laws, and empirical models).
- **Qualitative** knowledge (say, in the form of site-specific morphology, stratigraphy and topography, thematic data using remote sensing techniques and stored in *TGIS*).

It may be worth noticing that statuses other than knowledge, such as justification or understanding, are also distinctively valuable for variogram modeling purposes. To illustrate some of the previously mentioned considerations, the following example presents the process leading to the construction of a case-dependent variogram model in practice.[ac]

Example 3.9 It is instructive to examine the implementation of this process, first in terms of an empirical algebraic equation representing a social case.

Step 1: The following empirical law has been found to link homicide rate X and unemployment rate Y in a specific geographical region during a time period,

$$X(\boldsymbol{p}) = 1.52Y(\boldsymbol{p}) - 4.55, \tag{3.12}$$

where the societal meaning of the coefficient 1.52 is that at location $\boldsymbol{s}$ within the region the temporal homicide change is about 50% higher than that of unemployment. The unemployment-statistics in the region during the time period of interest are well documented by the unemployment variogram $\gamma_Y(\Delta\boldsymbol{p})$, which is the product of two of the models in Table 3.4: the Exponential model in space and the Wave model in time with $c_0 = 1$, $\alpha = 0.5$, $\beta = 0.25$.

Step 2: The homicide variogram model is then directly derived from the empirical law of Eq. (3.12), which yields

$$\begin{aligned}\gamma_X(\Delta\boldsymbol{p}) &= 2.3\gamma_Y(\Delta\boldsymbol{p}) \\ &= 2.3\left(1 - e^{-2h}\right)\left[1 - \frac{\sin(0.25\tau)}{\tau}\right],\end{aligned} \tag{3.13}$$

see the plot in Fig. 3.8.

Step 3: Interpretatively, as one observes the cross-sectional temporal homicide variograms show a dampening periodic behavior, i.e., at fixed h the variogram varies as $\approx 1.75 + a\frac{\sin(0.25\tau)}{\tau}$. When, e.g., $h = 50, \tau = 20$, then $a = -18.1452$; and when $h = 50$, $\tau = 10$, $a = 10.0319$. A similar periodic behavior is exhibited by the sill levels of the cross-sectional spatial homicide variograms for fixed τ.

Occasione data, at this point we briefly reiterate the main results of this section, for future reference:

❶ The **separable** approach of theoretical variogram model building assumes individual spatial and temporal variation structures (1st class models), and obtains

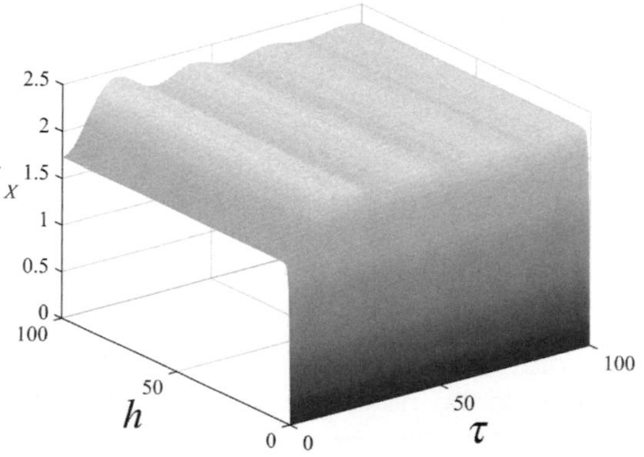

FIG. 3.8 The homicide variogram model in Example 3.9.

[ab] Situation-dependency directly implies that, beyond numerical data, general or core *KB*s, including physical and biological laws, scientific theories, or health and social relationships (Chapter 2), can be used in the development of variogram models.

[ac] This observation is confirmed by considering the temporal derivative of the model, $\frac{\partial}{\partial t}X(\boldsymbol{p}) = 1.52\frac{\partial}{\partial t}Y(\boldsymbol{p})$.

the final variogram as their permissible combinations.

❷ The **composite** (**nonseparable**) approach uses 2nd-class models, and can overcome some of the drawbacks of the separable approach.

❸ The **metric** approach selects a metric form (Tables 3.1 and 3.3 of Chapter 2) and a variogram shape (Tables 3.4 and 3.5), and then inserts the former into the latter.

❹ The **case-specific** approach builds models internally from the core knowledge (mainly physical laws and scientific models) about the phenomenon of interest.

❺ After the final variogram model γ_X has been selected via ❶–❹, it is fitted to the empirical values $\hat{\gamma}_X$, and the γ_X parameters are computed accordingly.

3.4 Variogram features and their interpretation

To remind the readers: discrete variogram values are calculated empirically (Section 3.2), a continuous mathematical model is fitted to the empirical values (Section 3.3), the model is physically understood and interpreted (Section 3.4), and then the variogram model is used in geostatistical estimation and mapping (Chapters 7 and 8).

Generally, the interpretation of the variogram model is the understanding of the relation between the model parameters and how the phenomenon it represents is expected to behave chronotopologically under certain conditions. When interpreting variograms, there are certain rules that should be followed. We recall the following facts concerning the interpretation and measurement of the variogram function.[ad]

> Variogram interpretation is phenomenological. It offers a measure of the attribute's chronotopologic **decorrelation** or **dissimilarity** that is computed in terms of the mean squared attribute value differences.

Example 3.10 The variogram model may offer a quantitative probabilistic description of the following phenomena:

- A dynamic, i.e., time varying, land surface roughness, i.e., in this case, dissimilarity is interpreted as roughness.
- Gold accumulation that shows an erratic pattern with considerable value dissimilarity at shorter distances, as compared to reef width variability that is far more continuous exhibiting higher value similarity.
- In oil and gas field studies, the main role of variograms is to reflect one's geologic understanding of the reservoir geometry and continuity of reservoir properties along different directions, which can have an important effect on volumetric estimation, predicted flow behavior, and reservoir-management decisions. All variogram-based techniques must give outcomes that honor data at wells. However, the use of different variogram techniques or vertical and lateral correlation lengths will result in potentially radial differences in model properties, as one moves away from the wells (Hand et al., 1994; Krishnan and Journel, 2003).

Having interpreted the variogram function in attribute dissimilarity terms, the next important point to be made is as follows:

> It is not only the **technical** (statistical) inferences that are of concern to *CTDA*, but also the **substantive** inferences regarding one's ordinary experience and interpretations about the physical meaning of the statistical inferences.

Indeed, this perspective raises certain interesting questions about the variogram and its role in a realistic *CTDA* study:

[ad] Correspondingly, the covariance is interpreted as the attribute's chronotopologic similarity, and measured in terms of the mean attribute fluctuation product.

☞ If a variogram represents a general chronotopologic variability model of a dataset, what does this mean physically?

Investigators attempt to answer this and similar questions by relating the variogram information to real-world conditions and actual data seen in the field in a way that this information has some kind of a *logical structure*. Awareness of the issue at hand has motivated the efforts to develop a set of correspondence links between model and reality that can be used efficiently in practice.

3.4.1 Variogram shape—Real-world attribute correspondence

The existence of certain key links between the observed variogram shape (form, structure, slope magnitude, etc.) and the *in-situ* chronotopologic behavior of the real-world attribute should be adequately stressed.

The links between the variogram shape and the natural attribute behavior are summarized by the **fundamental duality relation** (***FDR***) that can be symbolized as

Mathematical variogram shape
⇔ *In-situ* chronotopologic attribute distribution.

Interpretation of the variogram shape provides useful insight into certain aspects of attribute distribution.

Naturally, reliable variogram interpretation requires the establishment of appropriate *FDR* links between attribute variations and variogram behavior. Since a variogram model is fully characterized by the set of its key parameters (which are estimated by fitting a theoretical model to the empirical values), the *FDR* links are expressed in terms of these parameters and their interpretation. Otherwise said, the variogram parameters assess quantitatively different aspects of the chronotopologic attribute variability.

A list of key *FDR* links is presented in Table 3.6. But, before proceeding with a detailed discussion of each variogram feature listed in Table 3.6, a few typical cases from different scientific fields that address different FDR aspects are worth investigating.

Example 3.11 Let us consider a few illustrative cases of variogram interpretion from different scientific disciplines.

- Revisit Example 3.2. As noted earlier, the interpretation of these sea level variogram plots is basically phenomenological. I.e., based on these variogram plots, the meaning of any claim about the real sea level phenomenon is to be analyzed into the confirming or disconfirming empirical evidence out of which it was logically constructed.
- The variogram of zip-code Tuberculosis (*TB*) incidences is different than the variograms of the county-scale and the state-scale TB incidence data. By comparing variograms at different scales essential information about the spread of the disease is obtained by health scientists and managers.
- A variogram model for a stockwork-vein gold deposit should be different from a typical low-grade epithermal gold deposit (the ranges of the latter are expected to be longer along any lateral-direction than the former case).
- In the case of layered deposits (e.g., coal), the variogram of the data with periodic variation with depth (e.g., carbon content in coal) may show a cyclic-pattern of the hole-effect model along a vertical direction, which is not common in the case of a massive sulfide type mineral deposit, such as a porphyry-copper deposit.
- The short scale structure of certain aquifer properties (like permeability and porosity) often plays the most significant role: the nugget effect due to measurement error needs to be accounted for, as well as the size of the geological modeling units.

TABLE 3.6 A list of variogram *FDR* aspects (∴ denotes "implies").

Behavior at origin	Correlation ranges	Behavior at large lag	Shape characteristics
Parabolic behavior ∴ Continuous and regular attribute pattern	Long range ∴ High attribute dependency and low attribute variation	Asymptotic with sill (max variogram value) ∴ Homostationary attribute distribution with sill representing the full attribute variation at correlation range	Periodic shape that oscillates between positive and negative correlation values at the length scale of physical cycle ∴ Cyclic attribute variation
Linear shape ∴ Continuous attribute distribution but with greater attribute variability	Short range. ∴ Low attribute dependence and high variation	Linear increase beyond range exhibiting above the sill (negative) correlation values ∴ Heterogeneous attribute distribution with physical trend	Nested shape ∴ Superposition of different attribute variation scales
Nugget (discontinuity at origin) ∴ Small portion of noncontinuous and irregular attribute variability explained as random behavior		Cyclic variation among positive and negative values dampen out over large distances (*hole effect*) ∴ High attribute value domains surrounded by low value domains with physical cycle length being not perfectly regular	Directional variograms with different ranges ∴ *Geometric* attribute anisotropy Directional variograms with different sills ∴ *Zonal* attribute anisotropy
Pure nugget ∴ Fully random attribute variability		Exhibiting below the sill (positive) values, not encountering the full attribute variability ∴ Physical trend with zonal anisotropy	Flat shape ∴ Completely random attribute variation

Since real-world situations need to be studied as they unfold naturally, the *FDR* implementation often needs to be adapted as understanding deepens and/or situations change.

Establishing on rigorous grounds the "variogram shape-*in-situ* attribute" association expressed by *FDR* may provide a sort of a **verifiability** criterion that can be used to constrain theoretical speculation about the variogram form.

As these cases demonstrate, the variogram model has the potential to closely represent the chronotopologic structure of the actual dataset. The matter is, hence, worthy of further investigation in the following lines.

3.4.2 Nugget

Nugget is a discontinuity feature present in the variogram plot of many natural attributes, which generally implies increased randomness in the chronotopologic distribution of the attribute values. Understanding the origins of the nugget and modeling it properly are essential for *CTDA* purposes.

The **nugget** δ is a discontinuity at the variogram origin visualized as a function that, while it starts with a zero value at zero lags, it increases abruptly at very small lags.

The δ indicates a positive variogram value observed at a space-time lag with magnitude in the vicinity of zero and remaining constant in all directions. The term "in the vicinity" is selected to underline that δ is actually the intercept of the variogram at lags very close to zero, but not exactly zero, because at exactly zero lags the variogram has a zero value, indeed. Since a very small number of data pairs are usually available at very small lags (in the vicinity of zero), the δ is often either estimated by extrapolating the variogram surface to intercept the variogram axis or chosen based on the investigator's experience and insight. Also, in practice the closest spaced data is typically used to model the nugget value (in geophysics, e.g., such data are available along the downhole direction).

Concerning the possible origins of δ, it may be associated with a multiplicity of reasons, each one of which has its own significance:

① A small portion of attribute variability explained as **random** or **aleatory** behavior.
② **Sampling errors** at lags smaller than the minimum sample spacing/separation.
③ **Microstructure** with ranges less than the lags between the two closest data.[ae]
④ **Measurement noise** inherent in the measurement apparatus or due to human factor (location/time measurement error, attribute value measurement error, etc.).
⑤ **Data scale** issues (e.g., short-scale noise).

A limited number of available data (sparse data) may also be the cause of a higher nugget value. The δ is practically chosen to be equal in all directions, often selected as the smallest nugget among the study's different empirical variograms along these directions.[af] An omnidirectional δ can be calculated by using the isotropic variogram. But this is not necessarily a general rule. In certain cases, instead of the original data, composited (e.g., averaged) data may be considered, since the nugget effect reduces significantly when the data are composited. This case is illustrated in the following example.

Example 3.12 Some characteristic nugget cases are noticed as follows that offer useful insight into the nugget phenomenon.

- If equal-length composited (e.g., averaged) data are available along vertical drill-holes, the nugget is often calculated by modeling the down-the-hole vertical variogram. This normally provides the closest pairs for analysis around the zero spatial distances and temporal separations.
- In geological sciences, the nugget describes the expected difference between samples when the lags are almost zero and consists of two parts: the inherent small scale (physical) variability together with sampling and database errors.
- When grade continuity models in the three orthogonal directions are used, they often have the same nugget, the same sill values and the same model types, although the ranges can be different for each direction.

The selected nugget δ is a variogram model parameter that has serious consequences in *CTDA*, including the chronotopologic mapping of natural attributes. The δ usually has the biggest impact on the sample weights (and hence the chronotopologic estimates obtained). A nugget implies higher short scale variability (i.e., attribute data located close to each other are more dissimilar), resulting in a greater distribution of the estimation (Kriging) weights assigned to further away data, a considerable averaging of the attribute estimates and a smoother attribute map. Accordingly, if δ is over-estimated, the Kriging values will become unrealistically over-smoothed, whereas if δ is

[ae] Variation at microscales smaller than the sampling spacing may be considered as part of the nugget effect.

[af] The nugget effect has been represented by adding to the variogram function a term with sufficiently small ranges along all spatial directions and in time.

under-estimated, the Kriging values will appear excessively continuous, both cases having potentially serious *CTDA* consequences (also, Chapter 8).

3.4.3 Correlation (or dependence) range vector

One of the most critical parameters of a variogram is linked to lags with the very significant property of attribute variance maximization across space and time.

> The **space-time correlation** or **dependence lag** $\boldsymbol{\varepsilon}_{st}$ is the space-time lag where the variogram first flattens out.

The $\boldsymbol{\varepsilon}_{st}$ may be considered in the R^{n+1} or the $R^{n,1}$ domains. In theory, in the R^{n+1} domain, attribute values at sample points separated by lags $\boldsymbol{\Delta p}$ larger than $\boldsymbol{\varepsilon}_{st}$ are independent (after $\boldsymbol{\varepsilon}_{st}$, the variogram is invariant of the $\boldsymbol{\Delta p}$ value). Contingent on the application, spatial and temporal ranges may be represented quantitatively in more than one way in the $R^{n,1}$ domain, as described in Table 3.7 (for a visualization see Fig. 3.5).[ag]

As regards Eq. (3.14), a representative ε_s value may be an averaged spatial range value along several directions. Of special interest is the case where the ranges ε_s and ε_t correspond to the distance and time separation at which the marginal spatial and temporal variograms reach their maximum values, respectively.[ah] In Eq. (3.16), the range along each different direction in space is considered individually. Samples separated by distance $\boldsymbol{h}$ larger than the spatial correlation ranges are spatially independent, while samples separated by a time separation larger than the temporal range are temporally independent. In some applications, the inverse ranges, ε_s^{-1}, ε_t^{-1}, etc. are used to determine the *decay* rate of the spatial and temporal correlations, respectively.

Example 3.13 Real-world cases illustrating certain range features are as follows.

- For many geological attributes, the horizontal ranges, say, $\boldsymbol{\varepsilon}_h = (\varepsilon_1, \varepsilon_2)$ are usually larger than the vertical range ε_3.

TABLE 3.7 Determination of space-time correlation ranges and sills.

Representation as	Chronotopologic range	Chronotopologic sill
① Scalar-scalar	$\boldsymbol{\varepsilon}_{st} = (\varepsilon_s, \varepsilon_t)$ (3.14) ε_s, ε_t: representative attribute spatial and temporal ranges	$c_{0,st} = (c_{0,s}, c_{0,t})$ (3.17) $c_{0,s}$, $c_{0,t}$: representative attribute spatial and temporal ranges
② Function-scalar	$\boldsymbol{\varepsilon}_{st} = (\varepsilon_s(\phi), \varepsilon_t)$ (3.15) ε_s: spatial range (function of directional angle ϕ across space)	$c_{0,st} = (c_{0,s}(\phi), c_{0,t})$ (3.18) $c_{0,s}$: spatial sill (a function of directional angle ϕ across space)
③ Vector-scalar	$\boldsymbol{\varepsilon}_{st} = (\boldsymbol{\varepsilon}_s, \varepsilon_t)$ (3.16) $\boldsymbol{\varepsilon}_s = (\varepsilon_{s,1}, \ldots, \varepsilon_{s,n})$: spatial correlation range vector linked to lag vector $\boldsymbol{h} = (h_1, \ldots, h_n)$	$c_{0,st} = (\boldsymbol{c}_{0,s}, c_{0,t})$ (3.19) $\boldsymbol{c}_{0,s} = (c_{0,s,1}, \ldots, c_{0,s,n})$: spatial sill vector linked to lag vector $\boldsymbol{h} = (h_1, \ldots, h_n)$

[ag] Since the range and sill are closely related, both their determinations are presented in the same Table 3.7.

[ah] The ε_s and ε_t are such that, respectively, $\gamma_X(\varepsilon_s, 0) = \max_h \gamma_X(h, 0)$ and $\gamma_X(0, \varepsilon_t) = \max_\tau \gamma_X(0, \tau)$, see Fig. 3.5.

- In the case of a reservoir with anisotropic features, the range strongly depends on the direction considered.

The ranges $\boldsymbol{\varepsilon}_s$ and ε_t provide information about the size of the search window used in chronotopologic estimation techniques (Chapters 8 and 9). That is, the spatial and temporal ranges constitute a structural aspect of the variogram model: short dependence ranges may imply greater variability in the chronotopologic distribution of an attribute, whereas long ranges may indicate a distinct chronotopologic structure with clusters of similar attribute values. To gain additional insight into the *in-situ* meaning of the range, let us look into another example.

Example 3.14 Beijing is the capital city of China with more than 20 million inhabitants. Mohe is a county located in northeastern China (next to Russia) with 80 thousand inhabitants. Beijing and Mohe cover similar area sizes, i.e., 17 thousand km^2; however, Mohe is not as developed as Beijing in many sectors, such as economy, urbanization, and transportation. Therefore, the transmission of an infectious disease will be rather different in these two places. The high population and the high probability of contact with infected people can lead to serious disease outbreak in large areas of Beijing during long time periods, i.e., the disease pattern in Beijing is characterized by low spatial and temporal variability. The chronotopologic dependence is high, as a result of which the spatial and temporal ranges (ε_s and ε_t) are long. Mohe, on the other hand, is a county that is not as urbanized as Beijing and the infection rate among people is low. Hence, the infected area and period in Mohe should be smaller than those in Beijing, and the disease probably will not be a big concern for Mohe county. The chronotopologic variability is high in Mohe, whereas the spatiotemporal dependency is low and the correlation ranges ε_s and ε_t are short.

We will revisit the nugget effect notion and its features a little later in this chapter, particularly in our discussion of the directional anisostationarity.

3.4.4 Sill

Another important variogram parameter closely related to the notion of the correlation range is as follows.

> The **sill** $c_{0,st}$ refers to variogram values that correspond to zero attribute correlation. Variogram values smaller than the sill imply positive correlation between attribute values along a spatial direction and/or in time.[ai]

Accordingly, sill may be seen as the maximum variogram value representing the full physical variability associated with the attribute's correlation ranges. For computational purposes, the sill is often taken to be equal to the data variance. Just as in the case of the correlation ranges, the description above indicates that more than one sill notions and values may be potentially encountered in practice. The vector $\boldsymbol{c}_{0,st}$ may be considered in the R^{n+1} or the $R^{n,1}$ domains as the maximum variogram value beyond which the variogram behaves asymptotically. In the R^{n+1} domain, the sill is the variogram value reached at sample points separated by lags equal or greater than $\boldsymbol{\varepsilon}_{st}$. Depending on the application, spatial and temporal sills may be defined quantitatively in more than one way in the $R^{n,1}$ domain, as presented in Table 3.7.

Occasione data, it should be noted that a particular case of Eq. (3.17) is when $c_{0,s}$ and $c_{0,t}$ are, respectively, the spatial and temporal marginal sills corresponding to the maximum values of the marginal spatial and temporal variograms

[ai] Interestingly, certain practical studies have associated the existence of variogram values higher than the sill with negative correlation.

in $R^{n,1}$.[aj] The sill may have two additive components: a nugget effect δ_s (and/or δ_t) and a partial sill $c^*_{0,s}$ (and/or $c^*_{0,t}$), so that

$$c_{0,s} = \delta_s + c^*_{0,s}, \quad c_{0,t} = \delta_t + c^*_{0,t}. \qquad (3.20a, b)$$

In the case of nested variogram models, there may exist a partial sill for its corresponding variogram curve.

If the ratios of the spatial and temporal sills over the nugget, $\frac{c_{0,s}}{\delta_s}$ and $\frac{c_{0,t}}{\delta_t}$, are close to 1, then most of the variability is nonspatial and nontemporal, respectively. Generally, the nugget and sill parameters indicate random aspects of the data distribution, while the ε_s, ε_t, $c_{0,s} - \delta_s$, and $c_{0,t} - \delta_t$ indicate its structural aspects.

Example 3.15 Some characteristic cases of "variogram sill-real-world variability" associations are worth noticing.

- The vertical soil permeability variogram may have a higher sill than the horizontal one, possibly due to additional stratification and layering variances. On the other hand, the vertical variogram may exhibit a lower sill than the horizontal one due to significant differences in average soil property values within each well (vertical variogram includes additional variances between wells).
- Horizontal variograms are usually computed at six angles: 0°, 30°, 60°, 90°, 120° and 150°. In sedimentary formations, spatial dependency between soil properties is generally much greater along the horizontal plane than vertically (*inter alia*, this is due to the fact that data is sparse horizontally). As a result, horizontal wells, seismic and analog data, as well as conceptual geological models are properly incorporated.

Example 3.16 Rainfall intensity $X(\boldsymbol{p})$ is measured in cm/h; $\boldsymbol{p} = (s_1, s_2; t)$, where s_1 and s_2 represent the distance (measured in km) east and north of a benchmark, respectively, and t denotes time (in h). Assume that the theoretical variogram model of $X(\boldsymbol{p})$ is anisotropic and given by

$$\gamma_X(\Delta \boldsymbol{p}) = 0.5 + 2e^{-\frac{3h^2}{5^2} - \frac{3\tau^2}{3^2}}, \qquad (3.21)$$

which is plotted in Fig. 3.9. The nugget parameter of this model is $\delta = 0.5$ (cm/h)^2, the sill is $c_0 = 2.5$ (cm/h)^2, the spatial correlation range is $\varepsilon_s = \sqrt{3}\frac{5}{\sqrt{3}} = 5$km, and the temporal correlation range is $\varepsilon_t = \sqrt{3}\frac{3}{\sqrt{3}} = 3$h. The marginal spatial variogram model and the marginal temporal variogram model are both plotted in Fig. 3.10.

In some cases of heterogeneous attribute distribution, the variogram does not level off at large space-time lags, or, even it levels off, its value is not equal to the data variance. In such cases, using the data variance could be a more practical approach. We will revisit the sill notion and its practical implications in our discussion of directional anisostationarity.

As noted earlier, since the range and sill are closely related, their computation and interpretation can be considered jointly in practice. In

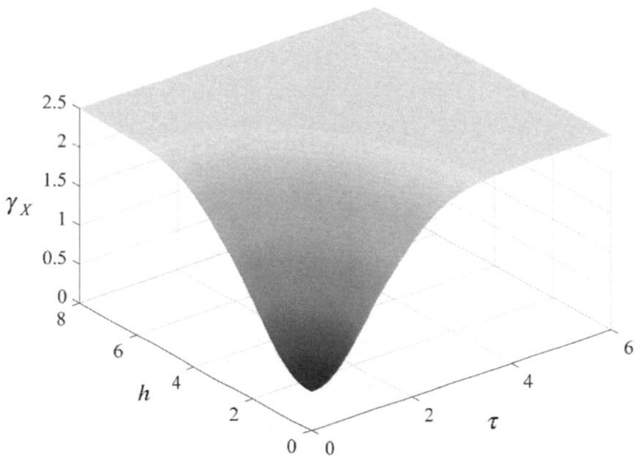

FIG. 3.9 Anisotropic variogram model. The h-axis of the variogram $\gamma_X(h,\tau)$ plot has units of km, the τ-axis has units of h, and the vertical axis has units of (cm/h)^2.

[aj] I.e., $c_{0,S} = \max_h \gamma_X(h,0) = \gamma_X(\varepsilon_S, 0)$ and $c_{0,t} = \max_\tau \gamma_X(0,\tau) = \gamma_X(0,\varepsilon_t)$.

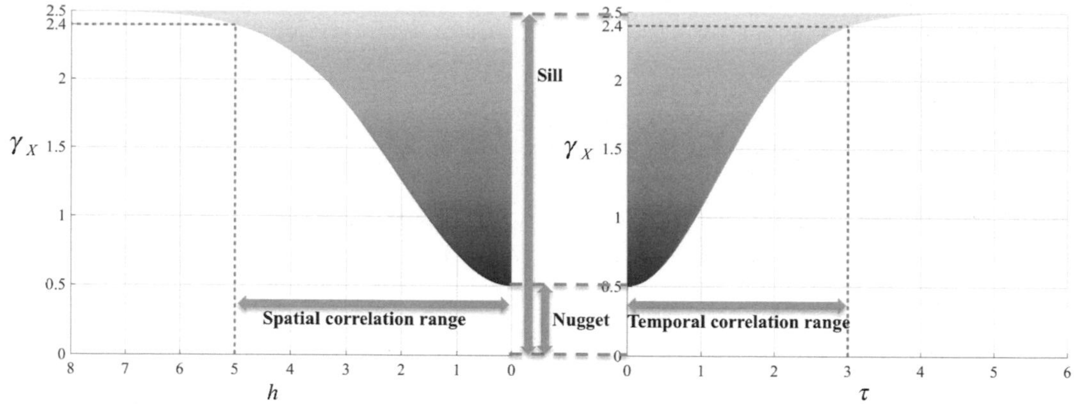

FIG. 3.10 Marginal purely spatial (left) and purely temporal (right) variogram models associated with Fig. 3.9. Note that for $h=0,\ \tau=3$ and $h=5,\ \tau=0$ one gets the marginal variogram value 2.4~2.5 in practice.

the case of Eqs. (3.15) and (3.18), it is usually an efficient approach to determine the corresponding spatial and temporal ranges and sills for each ϕ considered. *Ut supra*, the variogram plot in Fig. 3.5 may correspond to a particular direction (ϕ-value) associated with its range and sill. In this figure, for each fixed $h>0$ the corresponding temporal variogram has a nugget equal to the marginal spatial variogram $\gamma_X(h,0)$, and for each fixed $\tau>0$ the corresponding spatial variogram has a nugget equal to the marginal temporal variogram $\gamma_X(0,\tau)$. The sill can be initially identified numerically using the omnidirectional empirical variogram. If samples are properly distributed in an equidimensional grid, the omnidirectional empirical variogram can help estimate the nugget with reasonable accuracy.

3.4.5 *Behavior at the origin*

As noted earlier, another key variogram feature that also represents a structural aspect of the attribute distribution is its behavior at the space-time origin.

> The behavior of the variogram at the origin may be quantitatively characterized by its **slope**.

In general terms (also, Table 3.6):

- The **sharper** the slope, the higher the anticipated variability of the *in-situ* attribute distribution.
- A sharp slope combined with the presence of a **nugget** may indicate a considerably high attribute variability.

Other useful inferences are possible based on the variogram's shape at the origin. As a practical rule, under certain rather common conditions (nonnested and nonfluctuating directional variogram shape with a linear slope at the origin), the tangent at the variogram's origin may meet the corresponding sill at approximately 95% of the variogram range. This property is particularly helpful in estimating the range directly from the first few experimental variogram points, and it can be also used in chronotopologic estimation to properly ignore irrelevant samples.

3.4.6 *Directional anisostationarity*

A major issue of directional variogram modeling is the examination of its continuity features along different directions in space.

> **Directional anisostationarity** refers to different variogram shapes and parameters along different directions in space.

The main concern, of course, is spatial anisotropy, since time cannot be anisotropic. *Isostationary* variograms are used to characterize data with the same degree of continuity in all directions. On the other hand, when the data does not display the same degree of continuity in all directions, *anisostationary* variograms should be used. The latter show different behavior (e.g., changing variogram parameters, like correlation ranges and sills) along different spatial directions, which implies that the attribute distribution is spatially anisotropic.

Specifically, anisotropy implies that the variogram sills and dependence (correlation) ranges change with direction, see Eqs. (3.15) and (3.18). There are several techniques to address this situation. Two of them are analyzed here.

❶ If the variogram sills along different directions are comparable but their ranges change with direction, a model with **geometric** anisotropy components should be considered. In this case the nugget should remain constant in all directions.

Example 3.17 For illustration purposes, assume that the temporal variogram component is fixed so that we can focus on the spatial component. Fig. 3.11A presents the spatial variograms along two prime directions with the same sill. One has a spatial correlation range ε_1 and a directional angle ϕ_1, and the other has a spatial range ε_2 and an angle ϕ_2. After calculating the spatial correlation ranges along various directions, a function $\varepsilon(\phi)$ can be fitted to these ranges, see the ellipse in Fig. 3.11A. Also, the spatial variograms can be represented by the same model type (e.g., Gaussian model) but with different parameters.

❷ If the variogram ranges along different directions are comparable but their sills change with direction, a variogram model with **zonal** anisotropy components should be fitted to the empirical values.

Zonal anisotropy may be seen as an extreme case of geometric anisotropic where the correlation range along a specified direction exceeds the domain size, which leads to a directional variogram that does not reach the sill. Such directional variograms reach a plateau at a value that is lower than the sill, i.e., the attribute variability may not be observable along these directions.

Example 3.18 Assume that the spatial variograms along two directions not only have different correlation ranges but also different sills (Fig. 3.11B). One cannot fit the same model to these variograms. Hence, it is possible that to the values of the first empirical variogram a Gaussian (say) model can be fitted, but to the values of the second empirical variogram an Exponential (say) model can be fitted, instead. Yet, experience shows that natural phenomena usually do not exhibit this kind of anisotropy.

3.4.7 *Periodicity or cyclicity*

Generally, the variogram establishes a link between physical variability and the chronotopologic lag. Although, in practice, it seems that

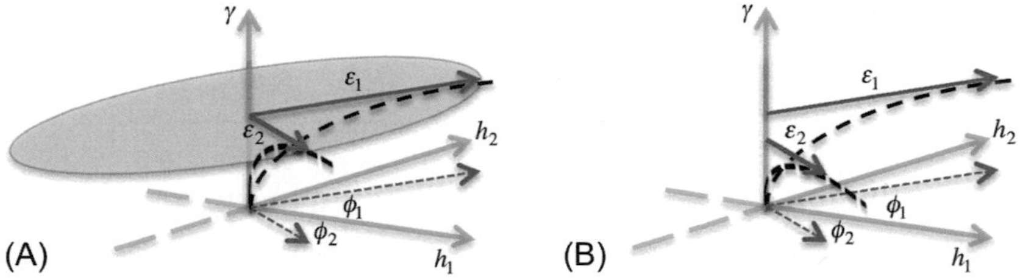

FIG. 3.11 (A) The variograms along the two directions have the same sill but different ranges; (B) The variograms along the two directions have different sills and different ranges.

the majority of variogram models have a smooth shape with monotonic increase, physically meaningful nonmonotonic configurations have been also observed.

Periodic or cyclic attribute variations are often detected by similar variogram behaviors: the variogram oscillates between positive and negative correlations at the length scales of the physical cycles. Such a variogram is sometimes termed a **hole effect**, and it may or may not have a sill, and be dampened (most often) or undampened (in rare cases)

Nonmonotonic variograms represent *in-situ* situations when high attribute value periods in time are suddenly interrupted by low value ones, when clustered or random lenses exist along specified spatial directions in the subsurface, and, similarly, when a stratified structure of varying bending thicknesses occurs along certain directions. Cyclic variations usually, but not always, dampen out over large distances or periods since the lengths of the physical cycles are not perfectly regular.

Example 3.19 The following are cases encountered in geosciences that illustrate the fruitful use of variogram modeling.

- Variogram cyclicity is linked to geological periodicity, i.e., the continuous process during which rocks are created, changed from one type to another, destroyed, and then formed again. Specifically, repetitive variations are observed in the reservoir properties as a result of geo-phenomena often occurring repetitively over geological time. This kind of phenomena is reflected in a cyclic variogram shape, that is, alternating positive and negative correlations shown in the variogram at the geologic cycle scale. Given that the length scale of the geological cycles is not perfectly regular, these cyclic variations dampen out at long distances.
- Using variogram models that account for geometric and zonal anisotropy, trends, and cyclic or periodic geologic variations is a key prerequisite for the adequate representation of facies and property distribution in reservoir modeling. It can improve understanding of reservoir heterogeneities, rock and fluid property distribution, and reservoir performance, and it can enable reliable production forecasting.

3.4.8 *Behavior at large space-time lags*

As regards *CTDA*, another important aspect of the variogram function concerns its behavior at large space-time lags.

The behavior of the variogram at large space-time lags provides information about the attribute variability in the real world, such as its **homostationary** or **heterogeneous** distribution characteristics.

As it turns out, based on extensive experience in the field of geostatistics, at larger $\boldsymbol{h}$ and τ lags the variogram may exhibit three possible configurations:

① An **asymptotic** shape (reaching a constant level), which seems to indicate that the attribute distribution is homostationary.
② A **fluctuating** shape, which indicates a periodic attribute distribution.
③ A **linear** shape, which means that the attribute variation is heterogeneous (i.e., a spatial nonhomogeneous and temporally nonstationary attribute distribution).

Example 3.20 In most cases, if the trend has not been removed from the data, it can be identified by a variogram shape that keeps increasing above the sill.

Interpretationally, physical attribute trends are associated with directional variograms increasing beyond their range, i.e., exhibiting above the sill (negative) correlation at large distances, or with directional variograms not encountering the full attribute variability, i.e., implying below the sill (positive) correlation at large distances (a behavior also associated with zonal anisotropy, see above).

Example 3.21 Sea surface heights $X(\boldsymbol{s},t)$ in Fig. 3.12A exhibit a periodic pattern, in which case a nonstandard form of height correlation applies. Height peaks have similar shapes, and the same is true for the height troughs. One notices the dampening in the variogram $\gamma_X(h,\tau)$ of Fig. 3.12B, that is, height peaks are isolated in space-time, and the same is true for the *in-situ* height troughs. Another interesting observation is that because of the wave pattern of the phenomenon, the wave height at the origin 0 is considerably more correlated to the height at point B than to that at point A, despite the fact that A is closer to 0 (Fig. 3.12B). When a variogram plot exhibits the shape of Fig. 3.12B, the inverse chronotopologic metric (*ICTM*) technique of attribute estimation (Chapter 7) ignores the periodic pattern of the *in-situ* phenomenon, and may generate inaccurate height estimates at unsampled locations and time instants. The spatiotemporal Kriging estimation method (Chapter 8), on the other hand, will properly take this periodic structure into account.

Brevi manu, real-world variograms usually involve a combination, to varying extents, of the distinct features described earlier, including *nugget, correlation ranges, sills, directional anisotropies, physical trends*, and *periodicities*. A brief summary of the previously mentioned *FDR* cases was given earlier in Table 3.6, and it is also visualized in Fig. 3.13. Both Table 3.6 and Fig. 3.13 describe the correspondence between

(a) the behavior of the variogram (at small and large distances, directionality, shape, etc.) with

(b) the properties of the natural attribute whose chronotopologic dependence is modeled by the variogram function.

Not surprisingly, real-world variograms usually exhibit a combination of the variogram features presented in Table 3.6 and Fig. 3.13.

Example 3.22 In light of Fig. 3.13, the key geometrical features of the variogram models of Figs. 3.6–3.10 are interpreted as follows.

- Fig. 3.6 is a rather typical example of sea surface variogram measured in Western Pacific Ocean, which possesses two properties: (a) its shape is linear at long time separations but asymptotic at large distances, in which case the physical insight

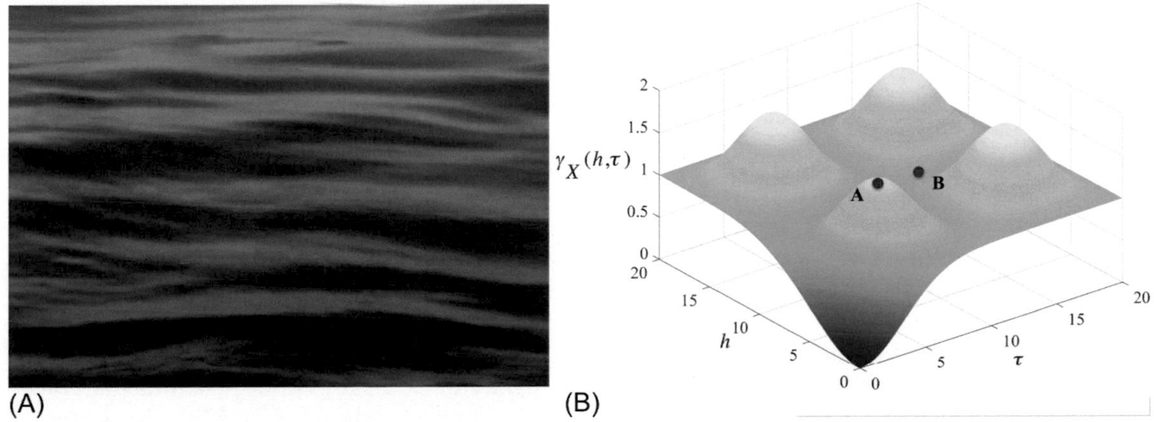

FIG. 3.12 (A) Wave pattern $X(\boldsymbol{s},t)$ and (B) its theoretical variogram model $\gamma_X(h,\tau)$.

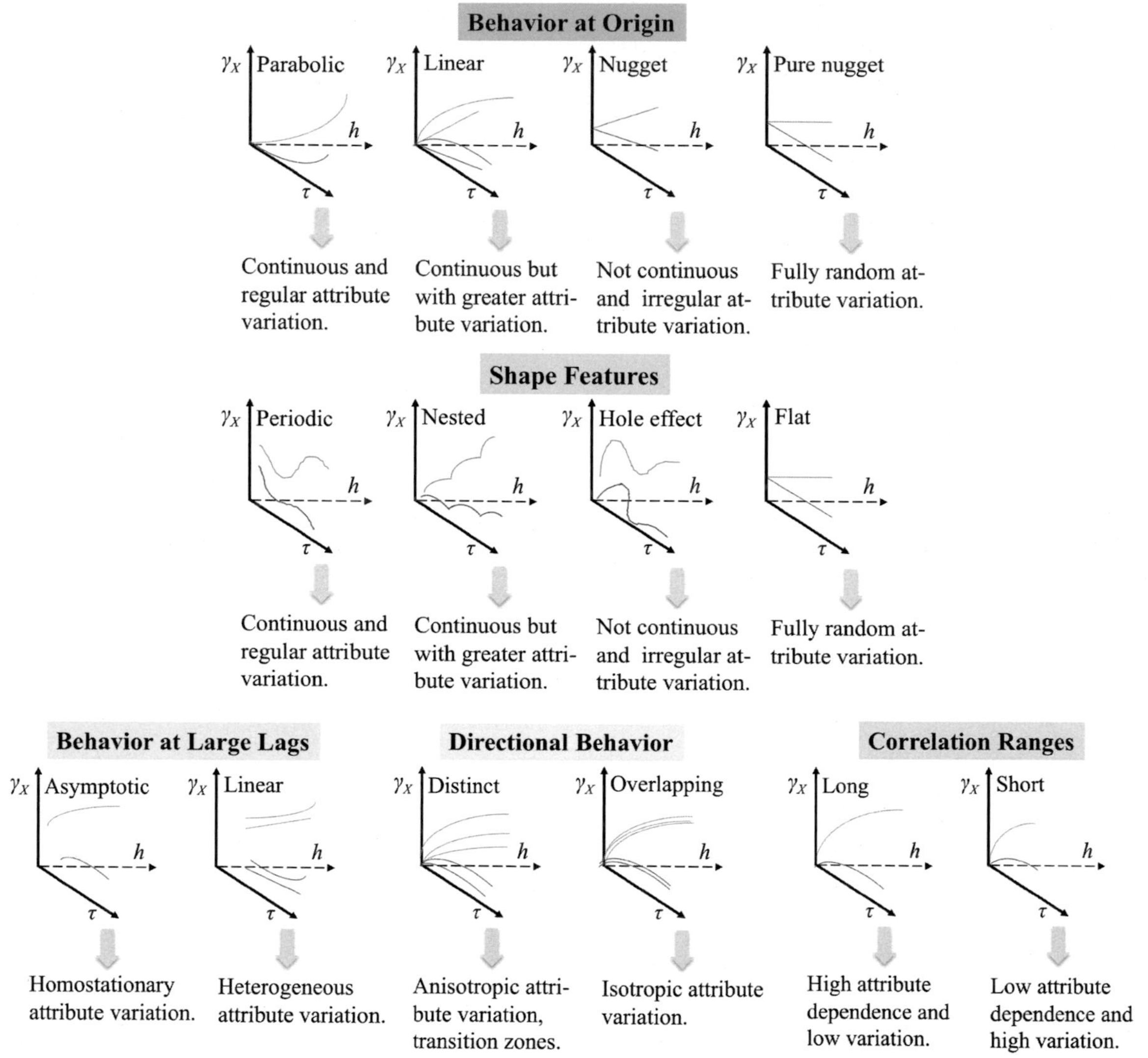

FIG. 3.13 Reading the variogram model.

gained is that in reality one should expect a rather nonstationary dynamic but spatially homogeneous sea surface variation. (b) The same variogram is closer to zero at short time separations than at short distances, which can be interpreted as indicating a smoother temporal than spatial sea surface variation.

- The log$PM_{2.5}$ concentration variogram of Fig. 3.7 depicts a considerably sharper increase in time than across space, which is potentially attributed to a smoother spatial than temporal log$PM_{2.5}$ variation. The asymptotic behavior at large h and τ (with a maximum variogram value of 0.528) indicates a homostationary variation. The space-time metric, $\Delta p = 0.01h + 4.94\tau$ (h in km and τ in days), seems to have a stronger temporal than spatial influence. There is a constant and slow increase in the variogram along space until a lag of approximately 140 km is reached. As they approach the 140 km distance, the variogram values fluctuate around a constant value (this is also true at zero and at longer separation times). On the other hand, there is a rapid increase of the variogram with time until $\tau \approx 4$ days. These features indicate a varying dependence between the concentrations in the domain defined by the 140 km and 4 days ranges.[ak] That the nugget $\delta_{0,t}(h)$ of the cross-sectional time variogram is smaller than the nugget effect $\delta_{0,s}(\tau)$ of the cross-sectional space variogram indicates a considerable smaller log$PM_{2.5}$ spatial variability between locations at the same time than temporal variability between times at the same location.

- The homicide variogram of Fig. 3.8 exhibits a very sharp slope at the origin, and a dampening periodic fluctuation in time, indicating a decreasing temporal periodicity of the homicide variation. The empirical law of Eq. (3.12) leads to the significant conclusion that, in this particular case, the homicide variogram is 2.3 times larger than the unemployment rate variogram, see Eq. (3.13). The inference drawn from this result is that the homicide variation is 230% larger than the unemployment variation.
- The rainfall variogram's shape at the origin, Figs. 3.9 and 3.10, is quadratic both across space and in time, implying rather smooth spatial and dynamic rainfall variations. The asymptotic variogram behavior at large distances and separations indicates a homostationary rainfall variation. All these conditions assure that a rather accurate space-time interpolation of the rainfall distribution is possible using the techniques described in Chapters 7–9.

3.5 Variography guidelines, tips, and rules of thumb

The variogram represents the investigator's quantitative understanding of an attribute's chronotopologic variability given the available site-specific evidence (datasets, charts) and relevant core knowledge (physical laws, scientific models). Being concerned with the systematic calculation and proper interpretation of the variogram in realistic situations, variography has significant effects in *CTDA* and beyond. *Inter alia*, variography plays a very important role in the development and implementation of high quality geostatistical estimation and mapping techniques (Chapter 8).

Example 3.23 Variographical information is critical in the production of a realistic grade and in lithologic modeling. It is apparent that an unrealistic variography parameter, like the nugget or the sill, can result in incorrect chronotopologic estimation error variances and, hence, it can make it difficult to obtain good site models; and, mistakes made in the interpretation of variogram ranges often lead to overestimation of resources.

The practical variography guidelines, tips, and rules of thumb listed in Table 3.8 could be useful to investigators and practitioners who need to develop a realistic variogram model of

[ak] As we will see in the following chapters on chronotopologic attribute estimation, only $PM_{2.5}$ sampling points within these ranges around an (unsampled) estimation point $\boldsymbol{p}_k$ are used. When the neighboring sampling points around the mapping grid point $\boldsymbol{p}_k$ reach a certain number, more sampling points at greater distances or time separations from $\boldsymbol{p}_k$ would have little influence on the estimated concentration value and error variance.

TABLE 3.8 Guidelines, tips, and rules of thumb used in empirical covariance and variogram computations.

Focus	Description
Physical consistency	Physical knowledge guides variogram computations: • Modeling assumptions should not conflict with attribute knowledge. • Parameter selection must make physical sense. • Domain choice[al] should be consistent with scientific information. Geostatistical modeling assumptions include: • Homostationarity (variogram is invariant under space-time translations). • Intrinsity (the variance of attribute increments between pairs of points depend on the space-time lags and are independent of point coordinates themselves).[am] • Normality (data obey a normal probability law). Modeling assumptions (homostationarity, normality, etc.) should be tested to ensure they are consistent with the physics of the case study of interest, or to determine whether assumption revision and/or data transformation[an] are necessary.
Data preprocessing	Before variography, if data exhibits systematic natural trends, it must be either removed by detrending (thus yielding homostationary residuals) or accounted for by nonstandard notions and techniques (generalized covariances, intrinsic Kriging, etc.). Skewed data distributions[ao] are transformed prior to variogram computing, because high-valued outliers have negative effects: • Yield poor parameter estimates. • Obscure actual attribute continuity and patterns. It is important to gain some understanding of the attribute variation scales. Irregularly distributed samples are visually analyzed to ensure an approximately uniform sample distribution for variography purposes. Samples of different sizes should be separated into different groups or different weights should be assigned to samples of different sizes to ensure the minimum support-effect. A popular way of dealing with cluster effects is to assign weights to the samples—other corrective measures should be attribute-dependent. Mixed data populations should be split into subsets with unique population parameters (means, variances, etc.), because variography using mixed populations can produce misleading results.
Variogram parameters	In approximate terms: • Nugget is the variogram value at near zero space-time lags (it is due to measurement errors or variation sources at lags less than the sampling intervals, etc.). • Sill is the variogram value of zero correlation at large lags where the analytical function becomes asymptotic.

[al] E.g., the R^{n+1} *vs.* $R^{n,1}$domains.

[am] In lay language, this means that the physical domain can be divided into subdomains with the same variogram.

[an] Using, e.g., a normal scores or an indicator transform, or a de-trending procedure (Section 3.1).

[ao] Many contaminant distributions, e.g., are positively skewed.

TABLE 3.8 Guidelines, tips, and rules of thumb used in empirical covariance and variogram computations—cont'd

Focus	Description
	• A correlation range is the critical lag below which correlation is positive and above which correlation is negative. Nugget and sill characterize the random aspect of the data, whereas the correlation range characterizes the space-time variability of the attribute at large lags. One should carefully distinguish between pure (or nugget-free) sill and total (or nugget-inclusive) sill. Sensitivity analysis of variogram parameters could help evaluate the variogram assumptions. The estimation of the nugget, sill, anisotropy, and major variability directions may be aided by generating omnidirectional and multidirectional empirical variograms.
Empirical variogram computation	Dynamic data at regular separations are often available for temporal variogram computation—but not spatial data at regular distance intervals. Spatial variograms are: • Computationally configured by lag magnitude, direction, and angular tolerance. • Typically calculated along the directions 0°, 30°, 60°, 90°, 120°, and 150°. Angular tolerance should: • Properly define the domain in which data are considered (e.g., data may be used that falls between given angles ±5 degrees). • Be minimized to describe the inherent anisotropy, if the data distribution is highly anisotropic. Multidirectional variograms are computed along principal variability directions defined on the basis of physical knowledge, and they may have different sets of parameters (nugget, sill, range, etc.) that can help detect regional anisotropies. In the case of limited data, the model parameters (nugget, sill, range, angular tolerance, etc.) should be inferred with the help of physical understanding of the phenomenon. Data outliers can cause erratic and unstable sill estimates. The total sill of the directional variogram should be less than or equal to the sample variance. The variogram values above the variance should not be considered, since above the variance line the sample behaves as a random variable. Ignoring distinct data populations may cause miscalculations in variogram estimation (e.g., larger nugget effects and shorter ranges than the actual ones). In human sciences (e.g., geography, public health and social sciences) only horizontal variograms are often computed, whereas in earth sciences (e.g., geology, hydrology, and geophysics) a variogram may need to be computed along the vertical direction too. Horizontal variograms: • May be not reliable because of sparse data. • Can give an indication of the strike orientation of the data. • Are important especially when directional continuities are not detected from physical information.

Continued

TABLE 3.8 Guidelines, tips, and rules of thumb used in empirical covariance and variogram computations—cont'd

Focus	Description
	The vertical variogram is: • Computed using all available lags in the downward direction. • Usually well-defined along the downward direction.
	Along any direction in space and in time the sampling interval should not exceed half the correlation range along the same direction.
	Hyperbolic (nonasymptotic) variogram shapes may indicate undetected data trends.
	Variograms with erratic shapes may indicate skewed datasets or datasets including extreme high or low values.
	Computations should be restricted to half the diagonal of the data extent, although this restriction can be affected by the domain shape.
	Rules of thumb concerning empirical variogram computation include: • If less than 50 data points are available, the empirical variogram may not be reliable and does not offer an accurate measure of chronotopologic dependence. • If 100–150 points are available, the computed variogram is considered reliable. • If more than 150 samples are used, one may have full confidence in the computed variogram values.
	It is often useful to generate omnidirectional experimental variograms that identify the sill and, if samples are well distributed in an equidimensional grid, estimate the nugget as well.
	Variogram roses may be constructed that readily identify anisotropy and major variability directions.
Variogram modeling	Variogram models constitute an effective means to represent data chronotopologic correlations in a convenient format through permissible analytical functions.
	Adequate variogram modeling is a critical factor in obtaining a physically meaningful variability characterization.
	Quantitative variogram modeling should be combined with qualitative knowledge,[ap] keeping in mind the basic scientific principles governing chronotopologic variability and heterogeneity.
	Since nature often exhibits complicated and confounding features (erraticness or lack of physical dependence, decreasing correlations with distance or separation, natural trends and cyclicity) that violate basic modeling assumptions, realistic variograms need to consist of elaborate combinations of standard variogram models.
	Selection of a variogram model involves a certain degree of subjectivity due to physical intuition, insight and experience, and, in many cases, it can be a source of uncertainty.
	In practice, a variogram model is typically chosen from predefined (permissible) classes of mathematical functions by fitting them to the empirical variogram values and computing the corresponding parameters (sill, nugget, range, etc.).

[ap] E.g., geologic reservoir aspects (such as environment, sequence stratigraphy, pore-space characteristics, reservoir thickness, porosity and permeability variation) can be used to generate information concerning major continuity directions, lateral extensions, and anisotropy indices.

TABLE 3.8 Guidelines, tips, and rules of thumb used in empirical covariance and variogram computations—cont'd

Focus	Description
	When new variogram models are constructed in terms of analytical functions, they must satisfy the variogram permissibility conditions or be obtained directly from physical laws.
	Given the occasional complexity of the variogram permissibility conditions, practitioners most of the time take advantage of some simple yet effective rules to construct new models, such as the sum of permissible models with positive coefficients.[aq]
Beyond variogram	The classical two-point variogram is not to be considered a panacea for chronotopologic variability characterization, i.e., there are cases where despite the investigator's best efforts the representations of attribute variability generated by the variogram do not look physically realistic. In such cases, alternative tools may be worth considering, like • The sysketogram and the contingogram functions. • The multipoint correlation functions (e.g., the trivariance measuring three-point correlation).

chronotopologic natural attribute distribution. This list also includes a number of potential variography pitfalls the investigators should be aware of. Due to its significance, some further comments are worth making concerning the guidelines, tips, and rules of thumb of Table 3.8, as follows.

An important early contemplation in variography is a careful review of the *physical characteristics* of the case study (geology, hydrology, ecology, oceanography, public health, etc., depending on the real-world situation).

> Variography results are highly dependent on both the investigator's **knowledge** of the physics of the phenomenon and the investigator's **practical experience**.

The variography quality also depends on the available public-domain literature, and, of course, on the investigator's acumen, insight, and industry experience. Any piece of physical information regarding attribute trends, anisotropy, and data noise should be taken into account by chronotopologic variography.

Example 3.24 Let us examine some real-world cases of geoscientific applications with diverse physical characteristics and representations.

One geological attribute per lithologic unit should be analyzed and modeled at a time and consistency with the available core knowledge be ensured.

- Bodies of subsurface contamination controlled by different structural units should also be treated separately for this purpose.
- Vein type mineralization should be treated separately from the disseminated type deposit, and data in the oxidized zone should be separated from the unoxidized zones for variography.
- Rigorous variography plays a key role in the consideration of the actual reservoir heterogeneity and geobodies variability, and in analyzing flow behavior via dynamic simulation. Well and log data may not be sufficient to support adequate variography (particularly along the horizontal directions), thus introducing considerable uncertainty in reservoir modeling and

[aq] For a discussion of such rules, see Christakos et al. (2017).

leading to oversimplification of geologic variations.
- In clastic reservoir modeling, different variogram models may represent distinct features of the facies and property distribution, hydrocarbon in place, recoverable resources, and production forecasts.
- A single unfolded coal seam bound by other lithological layers (say, sandstone/shale) and any planar structural elements (faults) should be considered as a homogenous geological unit.
- Poorly estimated variogram parameters (nugget, sill, range) can generate inaccurate variances leading to over- or under-estimation of hydrocarbon initially in place and recoverable resources and reserves.

Assuming that the right dataset has been collected, another important early step in variography is the careful *statistical analysis* of the dataset to make sure that the right data is acquired and to identify potential issues of concern. This includes the calculation of basic data statistics (means, ranges, standard deviations, and coefficients of variation), and the creation of cumulative frequency distribution plots, histograms, and scatterplots as necessary to gain an understanding of the data.

Example 3.25 Gaussian data transformation and atmospheric pollutant data trend removal should be usually performed prior to variography.

A variography trap occurs when *CTDA* is getting caught up in the "small" uncertainty that can be encompassed within the framework of a particular conceptualization and modeling approach, ignoring **bigger picture** uncertainties.

Situations like the previously mentioned can be very consequential, and include but are not limited to the following:

- Fundamentally different **conceptual models** of attribute variation or inter-dependency are valid than was originally anticipated.
- The occasional existence of a **larger-scale** chronotopology.
- Different **dominant** processes are at work (e.g., governing transport of a contaminant).

Example 3.26 Some typical cases in different scientific disciplines are as follows.
- Ecological conceptual models are thought of as hypotheses concerning the nature of ecological risks at contaminated sites. They differ from each other as regards the hypothesized sources of contaminants, routes of contaminant transport, contaminated media, exposure routes and endpoint receptors, each of which is associated with different levels of uncertainty.
- In geosciences, the site dataset is reviewed carefully, including specifically different sample lengths, well sizes, channel-sample, and drill-hole samples. For certain soil properties (e.g., permeability), there is usually more information available along the vertical than along the horizontal direction, which implies considerably reduced vertical than horizontal data uncertainty. Also, vertical trends and areal variations may exist, and the spacing of core data spacing can generate artifacts.
- In epidemiology, mortality data at the county level should be distinguished from those at the zip code or the state levels, since the two scales are linked to different kinds of uncertainty.

The spatio-dynamic distribution of attribute samples should be carefully examined from the different angles of data quality assurance (representability, accuracy, number, locations, instants) to determine if the current dataset is adequate for *CTDA* purposes. If this is not the case, further data collection and processing should be properly conducted to prepare for chronotopologic attribute estimation and mapping using the techniques to be discussed in Chapters 8 and 9.

Numerical **data adequacy** is a major issue in variography. Calculating the variogram model using a smaller number of samples than necessary can be a serious pitfall.

It is a fact of life that in many physical applications only sparse data is available. A typical reason that datasets may not be conducive to variogram analysis is that there are not enough sample points of a given attribute within a chronotopologic (geographo-dynamic) unit, spread uniformly to develop a good structure.

Example 3.27 This situation is potentially worrisome in *CTDA* studies. Specifically:

- In studies that use smaller sample sizes (e.g., soil cores, vegetation plots, etc.) and in short-term fieldwork where sample sizes of 40–60 data values are typical, the investigator is often confronted with the question: Should variograms even be attempted on such small sample sizes?
- In geosciences, sometimes drill-hole data is logged at unequal length intervals. If variograms are computed with samples of various sizes, the variogram may not represent the underlying structure of the attribute of interest. Samples of unequal size may not produce a realistic representative variogram because a larger sample may not provide information of equal weighting as the smaller samples do. Compositing the drill-hole data at equal length should reduce this support-effect. Ideally, drill-holes of unequal size should be treated separately, or their thickness should be weighted differently for variography analyses.
- Low-angle drill-hole samples and channel should be treated carefully to avoid unrealistic mathematical artifacts. Such samples should be de-clustered or should not be included (with other uniformly distributed samples) during variography.
- When well data is sparse, a careful interpretation of geologic reservoir aspects, such as environment, sequence stratigraphy, porosity, pore-space features, reservoir thickness, and permeability variations is needed to properly determine and facilitate quality control for the necessary inputs used in variogram modeling and interpretation.

Dataset *cleansing* should be part of the process, if required. Artifacts of data gathering practices should be filtered out so that the variogram represents the actual attribute variability.

Example 3.28 Data cleansing includes applying a cap and cutting assay values above the cap or throwing out high fliers or erroneous looking data. Composites of too small a length should be filtered out, as too many of them may create problems related to the support effect.

Variogram analysis of *mixed populations* can be misleading. Hence, these populations should be split into suitable subsets.

Example 3.29 In the study of regional poverty levels, one should not mix populations with different social features.

The chronotopologic distribution of the samples should be examined and care should be taken to reduce the impact of data *clustering*. To avoid the cluster effect in areas where samples are closely spaced, the samples should be de-clustered.

A useful variography step is to model and review the *omnidirectional* variogram to infer the sill of anisotropic variograms.

The sill of an **omnidirectional** variogram provides an approximate idea of the total sill of specific directional variograms.

Variogram values above the variance are not helpful in modeling. Depending on the type of data under consideration and the physical properties of the phenomenon (e.g., structural and lithological nature of the geology of the deposit), the variography may be as simple as a single omnidirectional variogram or as complex as multiple structures along each direction in a set of anisotropic variograms or a variogram rose.

As was emphasized earlier, variography can only implement models to play the role of the variogram that beyond being physically meaningful. They must also be mathematically *permissible*: a set of mathematical conditions must be satisfied by the candidate models.

Example 3.30 Only permissible variogram models should be used in the design of search (layering and vertical grid size) and parameter estimation for simulation techniques.

By way of a summary, then, variography distinguishes between three main stages with different objectives:

① In the 1st stage, **discrete** variogram values are calculated empirically.
② In the 2nd case, a **continuous** mathematical model is fitted to the empirical values.
③ In the 3rd stage, the model is **physically** understood and interpreted.

Well-performed variography can provide crucial information about the chronotopologic natural attribute distribution. Also, it should be noted that variogram modeling can be used in optimal sampling-design and to calculate key parameters of geostatistical estimation (Kriging, *BME* and other estimation techniques; see Chapters 8 and 9).

4 Chronotopologic block data analysis

Extensions of the standard *CTDA* operations have also been considered in applied sciences. These extensions are the result of real-world situations the study of which requires the modification of the standard *CTDA* concepts and tools, or even the development of new ones.

4.1 Basic concepts

In certain applications, the study issue is not point (chronolocation) data but rather *nonpoint* data, i.e., data with support consisting of a spatial domain (area or volume) and a time period. Then, the main study unit is as follows:

> A **chronoblock** unit $\boldsymbol{\mu}(\boldsymbol{p})$ consists of a spatial domain $\boldsymbol{v}$ centered at $\boldsymbol{s}$ and occurring at a specific time t.[ar]

In light of $\boldsymbol{\mu}(\boldsymbol{p})$, the standard *CTDA* is then extended to a *chronotopologic block data analysis* (*CTBDA*). A special case of $\boldsymbol{v}(\boldsymbol{p})$ is a *chronoareal* unit $\boldsymbol{a}(\boldsymbol{p})$ (two-dimensional block) consisting of a two-dimensional area $\boldsymbol{a}$ centered at $\boldsymbol{s}$ and occurring at a specific time t.

Example 4.1 Population scientists may be interested in the pattern of state-to-state migration within U.S. In this case the unit of observation is an individual state during a specified time period. The state is considered a representative areal unit, since a single observation is associated with the entire area enclosed by the spatial (i.e., the state) boundaries at a specified time. Other examples of representative areal units commonly encountered include: nations, counties, census tracts, and school districts, as well as natural regions such as watersheds, polluted areas, climate zones, and ore deposits.

Often scientists are concerned with the study of dynamic phenomena observed over a partitioning of a portion of a physical surface into mutually exclusive and nonoverlapping chronoareal units $\boldsymbol{a}(\boldsymbol{p})$. This conception naturally leads to the following notion (see Fig. 4.1A).

[ar] In practice, one is concerned with two- or three-dimensional blocks, i.e., $\boldsymbol{p} = (\boldsymbol{s};t) = (s_1, s_2;t)$ or $(s_1, s_2, s_3;t)$, in $\boldsymbol{a}(\boldsymbol{p})$.

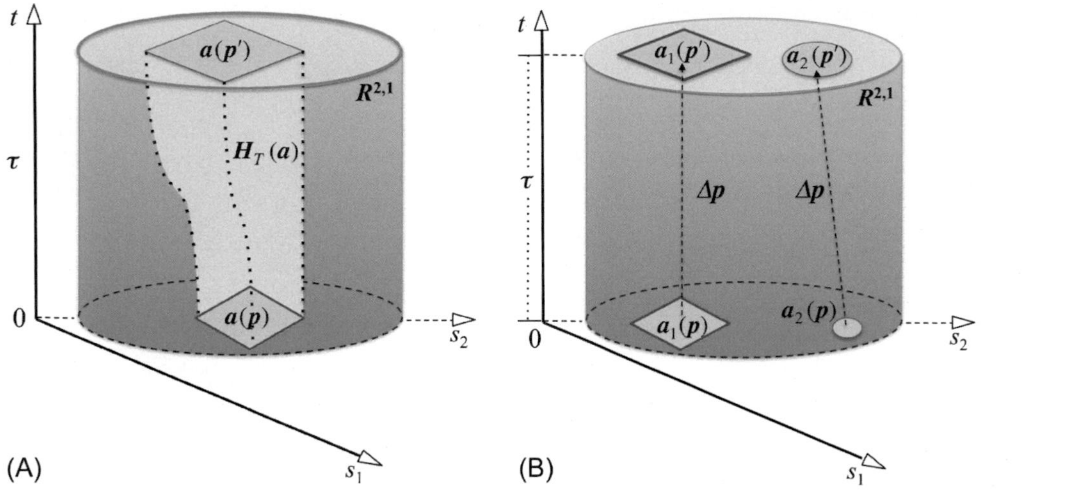

FIG. 4.1 (A) A chronoblock unit $\boldsymbol{a}(\boldsymbol{p})$ and its history $\boldsymbol{H}_T(\boldsymbol{a})$; and (B) an arrangement of different chronoblock units $\boldsymbol{a}_1(\boldsymbol{p})$ and $\boldsymbol{a}_2(\boldsymbol{p})$.

The **chronotopologic history** of the area a during the time period T is defined as

$$\boldsymbol{H}_T(\boldsymbol{a}) = \cup_{WL_p} \boldsymbol{a}(\boldsymbol{p}), \tag{4.1}$$

where $\boldsymbol{p} \in WL_p$,[as] i.e., $\boldsymbol{H}_T(\boldsymbol{a})$ is the union of all $\boldsymbol{a}$ along the WL_p during the period T. The **chronotopologic extent** of the history $\boldsymbol{H}_T(\boldsymbol{a})$ is usually represented as

$$|\boldsymbol{H}_T(\boldsymbol{a})| = \int_{WL_p} d\boldsymbol{p}|\boldsymbol{a}(\boldsymbol{p})| \tag{4.2}$$

(in space-time units), also denoted as simply $H_T(\boldsymbol{a})$ (nonbold H).

The readers may detect the similarity of the chronotopologic history with the worldtube notion (Section 2, Chapter 2). Moreover, the investigators are often interested in the relationship between a chronoblock unit $\boldsymbol{a}(\boldsymbol{p})$ centered at point $\boldsymbol{p}$ and its $\Delta\boldsymbol{p}$-translation $\boldsymbol{a}(\boldsymbol{p}') = \boldsymbol{a}(\boldsymbol{p} + \Delta\boldsymbol{p})$ centered at $\boldsymbol{p}' = \boldsymbol{p} + \Delta\boldsymbol{p}$, as is shown in Fig. 4.1B. Various *CTDA* operations are based on chronoblock units such as the following.

- Chronoblock **estimation** provides attribute estimates for unsampled chronoblock units and not for single points (chronolocations). The estimation error variance is generally lower in the chronoblock case than in the point case.

As the readers may anticipate, the above chronoblock estimation setting has many important *CTDA* applications (see, e.g., the chronoblock ordinary Kriging technique in Chapter 8).

4.2 Scale issues

There are many different types of analyses one can conduct on chronoblock data when several scales of observation are involved. In fact, the following motto seems to be widely accepted, with good reason:

Scale motto: It is the measurement or the observation scale that determines the phenomenon.

Moreover, the multileveled study of chronoblock data often needs to systematically take into consideration the following phenomenon.

[as] WL_p is the world line of point $\boldsymbol{p}$ (recall discussion in Chapter 2).

The **change of scale** (or **support**) effect involves the derivation of relationships between different scales (sizes and shapes of data).

Typically, the two main scale changes considered in practical situations are:

① **Up-scaling**, i.e., moving from a lower to a higher scale.
② **Down-scaling**, i.e., moving from a higher to a lower scale.

Example 4.2 Let us examine a few rather common-place public health situations involving the concept of scale.

- Table 4.1 presents the different scales of human exposure often used to classify key notions, such as space, time, age, exposure level, and health effects. The rigorous description of a natural attribute depends on the spatial and temporal scales at which the phenomenon is considered, while most attributes only exist as meaningful entities over limited scale ranges.
- In many epidemic studies data at a higher scale are available (say, county level or annual period), whereas the epidemiologist may be interested in data at a lower scale (say, local zip code level or monthly period), where a more meaningful human exposure analysis can be performed—this is a down-scaling process (see Fig. 4.2).

TABLE 4.1 Scale variations considered in human exposure studies.

Variable	Scale
Space	μm → km
Time	ms → years
Age	Young → old
Exposure	Micro → macro
Health effect	Local → global

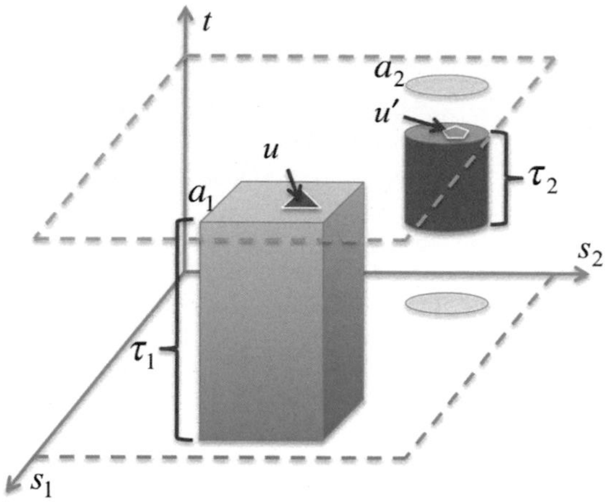

FIG. 4.2 An illustration of the down-scaling configuration for an epidemic field. The cube represents annually-documented disease data at the a_1 city level, the cylinder represents monthly disease data at the a_2 county level, the two circles represent daily disease data at the a_2 county level, and the triangle $\boldsymbol{u}$ and pentagon $\boldsymbol{u}'$ denote zip codes.

Next, we derive the basic relationships between chronolocation and chronoareal attribute data. Let $X(\boldsymbol{p})$ be a homostationary attribute represented by point data, with mean $\overline{X(\boldsymbol{p})} = \overline{X}$ and variogram $\gamma_X(\Delta\boldsymbol{p})$, where, as usual, $\boldsymbol{h}$ is the spatial distance between a pair of points $\boldsymbol{p}$ and $\boldsymbol{p}'$ and τ is the temporal separation between the points. Furthermore, let, respectively (Fig. 4.1A),[at]

$$\begin{aligned}\overline{\gamma}_X\left(a_p,\ a_p\right) &= \frac{1}{H_T^2(\boldsymbol{a})}\int_{H_T(\boldsymbol{a})}\int_{H_T(\boldsymbol{a})} d\boldsymbol{u}d\boldsymbol{u}'\gamma_X(\boldsymbol{u}-\boldsymbol{u}'),\\ \overline{\gamma}_X\left(a_p,\ a_{p'}\right) &= \frac{1}{H_T(\boldsymbol{a})H_T(\boldsymbol{a}')}\int_{H_T(\boldsymbol{a})}\int_{H_T(\boldsymbol{a}')} d\boldsymbol{u}d\boldsymbol{u}'\gamma_X(\boldsymbol{u}-\boldsymbol{u}'),\end{aligned} \quad (4.3a,b)$$

[at] For simplicity, $H_T(\boldsymbol{a})$ and $H_T(\boldsymbol{a}')$ denote $H_T(\boldsymbol{a}(\boldsymbol{p}))$ and $H_T(\boldsymbol{a}(\boldsymbol{p}'))$, respectively.

be the average variogram between all pairs of points within the same chronoarea $\boldsymbol{a_p}$ and the average variogram between all points of $\boldsymbol{a_p}$ and all points of a different chronoarea $\boldsymbol{a_{p'}}$. Investigators often want to relate these point statistics with the statistics of a chronoarea attribute, as follows.

The mean and variogram of the **chronoarea attribute** $X_a(\boldsymbol{p})$ defined as

$$X_a(\boldsymbol{p}) = \frac{1}{H_T(\boldsymbol{a})}\int_{H_T(\boldsymbol{a})} d\Delta\boldsymbol{p} X(\boldsymbol{p}+\Delta\boldsymbol{p}), \quad (4.4)$$

is computed by[au]

$$\overline{X_a(\boldsymbol{p})} = \frac{1}{H_T(\boldsymbol{a})}\int_{H_T(\boldsymbol{a})} d\Delta\boldsymbol{p}\overline{X(\boldsymbol{p}+\Delta\boldsymbol{p})} = \overline{X(\boldsymbol{p})} = \overline{X}, \quad (4.5)$$

$$\begin{aligned}\gamma_a(\Delta\boldsymbol{p}) &= \frac{1}{2}\overline{[X_a(\boldsymbol{p}) - X_a(\boldsymbol{p}')]^2} \\ &= \overline{\gamma}_X(\boldsymbol{a_p}, \boldsymbol{a_{p'}}) - \overline{\gamma}_X(\boldsymbol{a_p}, \boldsymbol{a_p}). \quad (4.6)\end{aligned}$$

One should keep in mind that Eqs. (4.4)–(4.6) are fundamental formulas that relate the variogram of the original attribute $X(\boldsymbol{p})$ of point data with the attribute $X_a(\boldsymbol{p})$ of the chronoarea data. This essentially means that if the chronotopologic statistics of $X(\boldsymbol{p})$ is known, the statistics of $X_a(\boldsymbol{p})$ can be readily calculated.

Example 4.3 In epidemiology, the $X_a(\boldsymbol{p})$ may denote mortality rate at the county level, and the $X(\boldsymbol{p})$ mortality rate at the zip code level.

As another illustration, the following example presents a case in which the block scale reduces to a point scale.

Example 4.4 Let $d(\boldsymbol{a_p})$ be the largest dimension of $\boldsymbol{a_p}$ in $R^{2,1}$ (e.g., in the special case that a_p is circular, $d(\boldsymbol{a_p})$ is its diameter). Assume that $d(\boldsymbol{a_p})$ is very small, say $d(\boldsymbol{a_p}) << |\boldsymbol{p} - \boldsymbol{p}'|$. Then, it can be shown that $\gamma_a(\Delta\boldsymbol{p}) \approx \gamma_X(\Delta\boldsymbol{p})$, i.e., the variogram at the chronoblock scale $(\boldsymbol{a_p})$ is approximately the same as that at the chronopoint scale $(\boldsymbol{p})$.

Another consequence of the change-of-scale effect is that using the same basic dataset, the *CTDA* outcomes depend on the size, shape, and chronotopologic arrangement of the block units.

Example 4.5 Fig 4.3A shows the distribution of temperature data (2048 data in total) in one city. Assume that the spatial coordinates of observation locations range from 1.5 to 9 with a $\delta s = 0.5$ interval and the temporal coordinate ranges from 1 to 8 with a $\delta t = 1$ interval. The shades in Fig. 4.3A indicate the levels of observed temperature values at the corresponding points (darker shades indicate higher temperatures). The mean, variance, and coefficient of variation of the observed temperatures in the original dataset are 0.5530, 0.5127, and 1.2947, respectively, as shown in Table 4.2. On the other hand, Table 4.2 also provides the descriptive statistics of the data subset for which the temperature values are equal or greater to 0. This is a data subset of smaller size than the original one, consisting of 1679 data in total, i.e., it includes 81.98% (0.8198) of the data values (i.e., those that satisfy the condition $X(\boldsymbol{p}) \geq 0$); the corresponding observed temperature mean, variance, and coefficient of variation are 0.6889, 0.5220, and 1.0489, respectively.

In certain circumstances, scientists may need to select chronoblock units to upscale a phenomenon from a lower to a higher scale (say, from the county to the state level, or from the daily to the monthly level).

Example 3.5(cont.): Consider *de novo* Fig. 4.3A. Investigators need to upscale temperature data from the county to the state level or from the daily level to the monthly level. For this purpose, four situations are considered:

[au] For simplicity, a_p and $a_{p'}$ denote $a(\boldsymbol{p})$ and $a(\boldsymbol{p}')$, respectively.

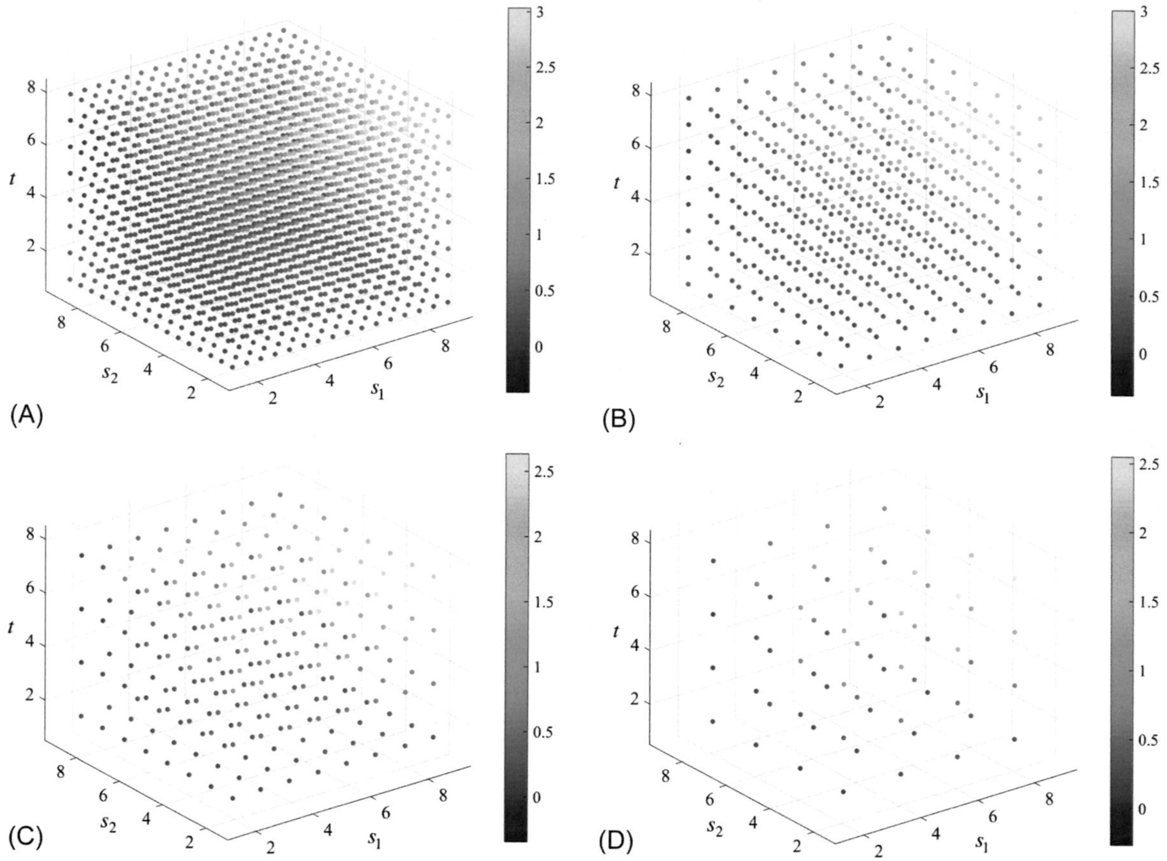

FIG. 4.3 (A) Original point chronotopologic temperature data (darker shades indicate higher temperatures); and chronotopologic areal units with various spatial and temporal scales, i.e., temperature upscaling was performed using (B) the spatial aggregation interval $\delta s = 1$; (C) the spatial and temporal aggregation intervals $\delta s = 1$ and $\delta t = 2$, respectively; and (D) the spatial and temporal intervals $\delta s = \delta t = 2$.

TABLE 4.2 Descriptive statistics of the four temperature situations shown in Fig. 4.3.

	$X(p)$				$X(p) \geq 0$			
	(a)	**(b)**	**(c)**	**(d)**	**(a)**	**(b)**	**(c)**	**(d)**
n	2048	512	256	64	1679	414	214	51
$\overline{x}$	0.5530	0.5399	0.5399	0.5102	0.6889	0.6843	0.6609	0.6597
σ_X^2	0.5127	0.5085	0.4904	0.4808	0.5220	0.5188	0.4967	0.4924
CV	1.2947	1.3206	1.2969	1.3593	1.0489	1.0525	1.0664	1.0638
Probability	1	1	1	1	0.8198	0.8086	0.8359	0.7969

(a) the original point dataset is shown in Fig. 4.3A (darker shades indicate higher temperatures);

(b) the upscaled attribute values in Fig. 4.3B obtained by aggregating the spatial data at each time instant using $\delta s = 1$ (i.e., the attribute values at points (1.5,1.5;1), (1.5,2;1), (2,1.5;1), and (2,2;1) were aggregated to get the averaged attribute value at point (1.75,1.75;1); the values at points (7.5,3.5;4), (7.5,4;4), (8,3.5;4), and (8,4;4) were aggregated to get the averaged attribute value at point (7.75,3.75;4), etc.;

(c) the upscaled attribute values in Fig. 4.3C obtained by aggregating the data using $\delta s = 1$ and $\delta t = 2$ (i.e., the values at points (1.5,1.5;1), (1.5,2;1), (2,1.5;1), (2,2;1), (1.5,1.5;2), (1.5,2;2), (2,1.5;2), and (2,2;2) were aggregated to get the averaged attribute value at point (1.75,1.75;1.5), etc.;

(d) the upscaled attribute values in Fig. 4.3D were obtained by aggregating data using $\delta s = 2$ and $\delta t = 2$ (i.e., values at points (1.5,1.5;1), (1.5,2;1), (1.5,2.5;1), (1.5,3;1), (2,1.5;1), (2,2;1), (2,2.5;1), (2,3;1), (2.5,1.5;1), (2.5,2;1), (2.5,2.5;1), (2.5,3;1), (3,1.5;1), (3,2;1), (3,2.5;1), (3,3;1), (1.5,1.5;2), (1.5,2;2), (1.5,2.5;2), (1.5,3;2), (2,1.5;2), (2,2;2), (2,2.5;2), (2,3;2), (2.5,1.5;2), (2.5,2;2), (2.5,2.5;2), (2.5,3;2), (3,1.5;2), (3,2;2), (3,2.5;2), and (3,3;2) were aggregated to get the averaged attribute value at point (2.25,2.25;1.5), etc.

Example 4.6 Fig. 4.4A–D shows a few cases of empirical variogram values (dots) and the fitted Gaussian variogram models obtained by using the corresponding data at various spatial and temporal scales. The four plots (A–D) are different from each other. Especially, the peak

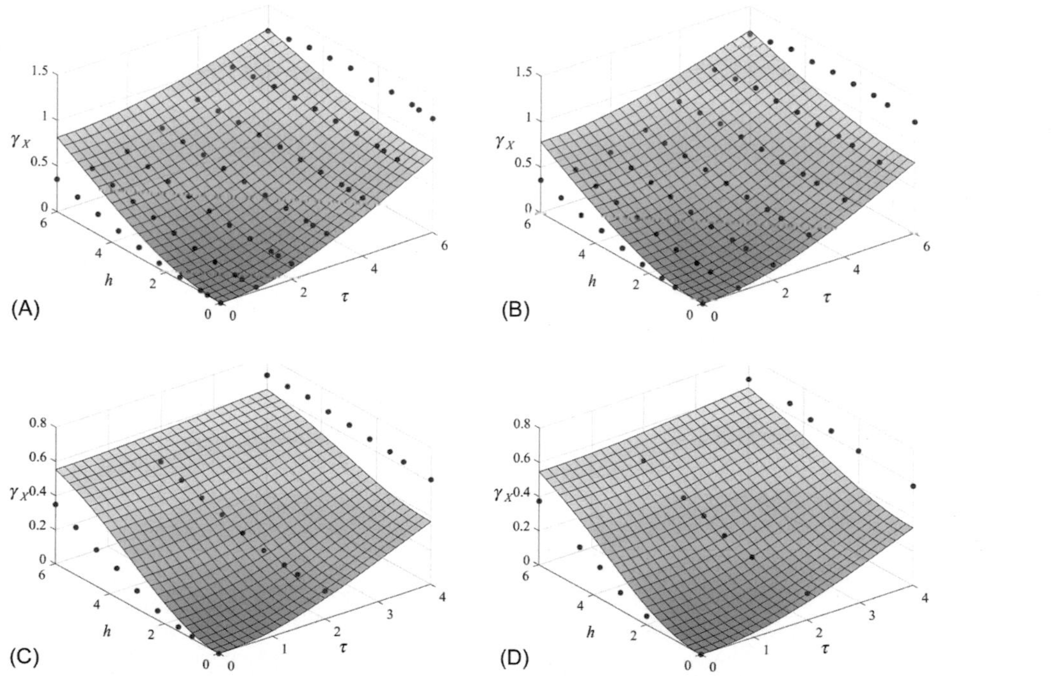

FIG. 4.4 Empirical variograms values and fitted Gaussian variogram models by using the corresponding data with various spatial and temporal scales as shown in Fig. 4.3A–D.

variogram values in Fig. 4.4C and D are lower than those in Fig. 4.4A and B. The sills of the four variograms in (A) through (D) are 1.6917, 1.8823, 0.6562, and 0.7125, respectively. After upscaling the original data, it was found that at certain spatial distances and temporal separations, there are no location pairs available, i.e., no empirical variogram values can be calculated at these distances and separations. Particularly, the number of dots in Fig. 4.4D is smaller than the corresponding numbers at Figs. 4.4A–C. One immediately notices the very different statistics of the block units of Fig. 4.4A–D as well as the distinct theoretical variogram shapes, which are essentially linking each choice of chronoblock unit with a different phenomenon.

Concluding this subsection, the take-home message of our earlier discussion could be stated as follows:

> Looking at the same phenomenon through data associated with varying observation scales, different **images of reality** are revealed.

Putting these various comments together we suggest that the motto about scale proposed at the beginning of this subsection turns out to be sound and valid.

5 Practice exercises

1. What is the difference between epistemic and nonepistemic randomness?
2. In the random field modeling context, are random field realizations constrained causally?
3. When thinking about a problem, should one take into account the observed realizations, the unobserved possible realizations, or both?
4. Are the random field realizations not necessarily equiprobable?

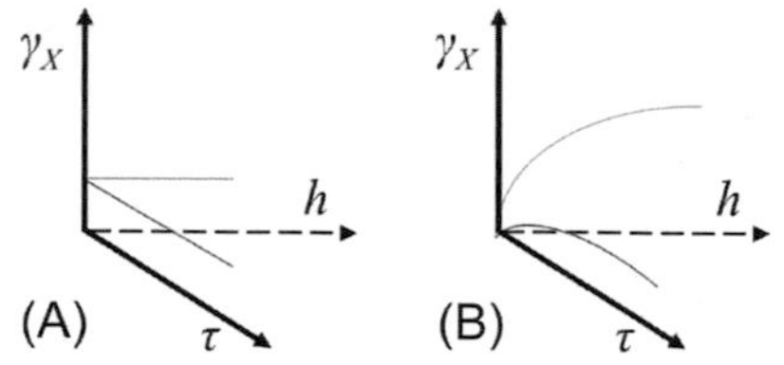

FIG. 5.1 Variogram shapes.

5. Which one of the two variograms in Fig. 5.1 suggests chronotopologic dependence (correlation) and which lack of dependence?
6. What is the significance of the variogram direction in the study of attribute variability? Why does the variogram change when we change the direction?
7. What is an empirical variogram and what does it measure?
8. How is attribute anisotropy represented in terms of the variogram?
9. What is chronotopologic homostationarity?
10. In your view, what are the main differences between classical statistics and geostatistics?
11. Is chronotopologic continuity consistent with the notion that small attribute values are in chronotopologic proximity to other small values, whereas high attribute values are close to other high values? Explain.
12. Associate each variogram shape $\gamma_X(h)$ in Fig. 5.2 with one of the following characterizations of the natural attribute $X(\boldsymbol{s})$:
    (i) Not correlated at any space distance or separation time.
    (ii) Correlated at any distance or separation.
    (iii) Correlated at a maximum distance or maximum separation.
13. Which of the following statements are valid?
    (a) Variogram range is a distance or separation beyond which there is no correlation.
    (b) Point pairs that are chronotopologically closer would have more similar attribute values and, thus, a smaller variogram value.

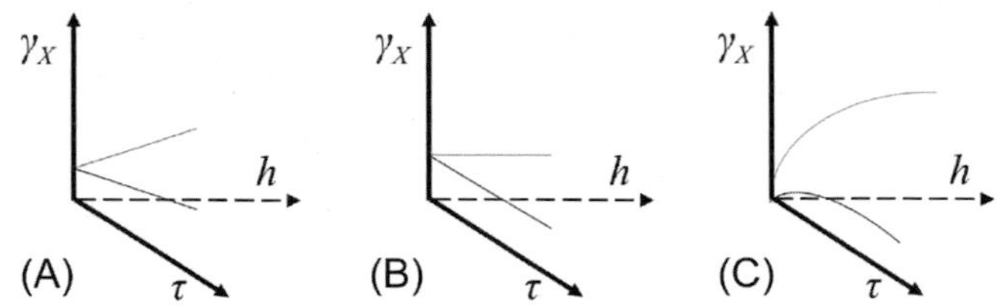

FIG. 5.2 Different variogram shapes.

(c) As point pairs get chronotopologically farther apart, the attribute values become more dissimilar, which results to higher variogram values.

14. Explain why the chronotopologic variogram changes with direction. What is this behavior called?

15. Consider the variogram shapes shown in Fig. 5.3 (at the origin and at large lags). Associate each one of these plots with the physical variation of the corresponding natural attribute distribution features:
   - **(a)** Completely erratic chronotopologic variation.
   - **(b)** Homostationary variation.
   - **(c)** Continuous and smooth variation.
   - **(d)** With periodic fluctuations (Tobler's law is violated).
   - **(e)** Noncontinuous and irregular chronotopologic variation.
   - **(f)** Superposition of various variation scales.
   - **(g)** Heterogeneous variation.

16. Let us recall the different interpretations of the covariance and variogram functions. Does the chronotopologic covariance consistently quantify the intuition that attribute values that are closer are more alike than values that are farther apart? On the other hand, does the chronotopologic variogram quantify the intuition that attribute values that are farther apart chronotopologically are more different than values that are closer? Explain.

17. Describe a few *S/TRF* models in your area of interest in terms of attribute, domain, and observed *vs.* unobserved realizations. Then, describe them in a mathematical setting.

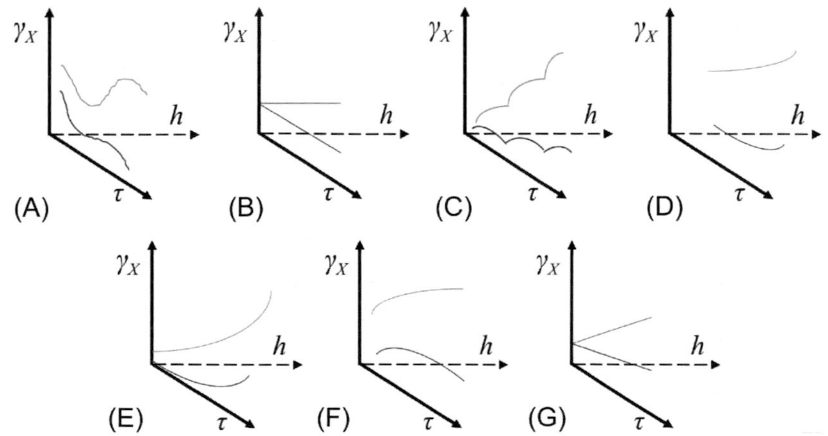

FIG. 5.3 Variogram plots.

18. Discuss the main difference between the assumptions of homogeneity, stationarity, homostationarity, and ergodicity.
19. Calculate the chronotopologic empirical covariance and variogram values for the dataset of Table 5.1 for selected distances h and separations τ. Use a software library of your choice to display the corresponding empirical covariance and variogram values. Lastly, choose among the known theoretical covariance and variogram models to fit to the empirical values.
20. Calculate the chronotopologic trend (mean function) of the dataset of Table 5.1. De-trend the original dataset and calculate the empirical covariance and variogram values of the de-trended dataset for selected distances h and separations τ. Then, use a software library of your choice to display the de-trended empirical covariance and variogram values. Lastly, choose among the known theoretical covariance and variogram models to fit to the calculated empirical values above.

TABLE 5.1 Chronotopologic dataset.

s_1	s_2	t	$X(\boldsymbol{p})$	s_1	s_2	t	$X(\boldsymbol{p})$
6.5	6.5	5	0.4858	6.5	6.5	7	1.1739
	7		0.4052	6.5	7		0.9387
	7.5		0.3467	6.5	7.5		0.7114
	8		0.2541	6.5	8		0.5422
	8.5		0.1649	6.5	8.5		0.3918
	9		0.0877	6.5	9		0.2870
7	6.5		0.5192	7	6.5		1.1642
	7		0.4362		7		0.9475
	7.5		0.3631		7.5		0.7315
	8		0.2801		8		0.5747
	8.5		0.1985		8.5		0.4220
	9		0.1140		9		0.3060
7.5	6.5		0.5169	7.5	6.5		1.1125
	7		0.4361		7		0.9456
	7.5		0.3531		7.5		0.7721
	8		0.2808		8		0.6075
	8.5		0.2080		8.5		0.4611
	9		0.1252		9		0.3324

TABLE 5.1 Chronotopologic dataset—cont'd

s_1	s_2	t	$X(p)$	s_1	s_2	t	$X(p)$
8	6.5		0.4962	8	6.5		1.0390
	7		0.4002		7		0.8928
	7.5		0.3220		7.5		0.7552
	8		0.2643		8		0.6043
	8.5		0.1999		8.5		0.4591
	9		0.1158		9		0.3383
8.5	6.5		0.4660	8.5	6.5		0.9638
	7		0.3628		7		0.8169
	7.5		0.2786		7.5		0.6939
	8		0.2419		8		0.5306
	8.5		0.1845		8.5		0.3939
	9		0.1028		9		0.2983
9	6.5		0.4265	9	6.5		0.8873
	7		0.3244		7		0.7400
	7.5		0.2507		7.5		0.6083
	8		0.2250		8		0.4614
	8.5		0.1651		8.5		0.3520
	9		0.0825		9		0.2753
6.5	6.5	6	0.8452	6.5	6.5	8	1.3654
	7		0.6870		7		1.1377
	7.5		0.5333		7.5		0.9114
	8		0.3873		8		0.7139
	8.5		0.2639		8.5		0.5401
	9		0.1618		9		0.3983
7	6.5		0.8597	7	6.5		1.2606
	7		0.7038		7		1.0244
	7.5		0.5492		7.5		0.8351
	8		0.4009		8		0.6979
	8.5		0.2800		8.5		0.5501
	9		0.1807		9		0.4096

Continued

TABLE 5.1 Chronotopologic dataset—cont'd

s_1	s_2	t	$X(p)$	s_1	s_2	t	$X(p)$
7.5	6.5		0.8457	7.5	6.5		1.1825
	7		0.6945		7		0.9520
	7.5		0.5389		7.5		0.8091
	8		0.3980		8		0.7157
	8.5		0.2838		8.5		0.5742
	9		0.1932		9		0.4232
8	6.5		0.8022	8	6.5		1.1564
	7		0.6544		7		0.9480
	7.5		0.5098		7.5		0.8277
	8		0.3773		8		0.7168
	8.5		0.2745		8.5		0.5964
	9		0.1917		9		0.4496
8.5	6.5		0.7532	8.5	6.5		1.1126
	7		0.6062		7		0.9326
	7.5		0.4710		7.5		0.8201
	8		0.3503		8		0.7121
	8.5		0.2549		8.5		0.5817
	9		0.1947		9		0.4374
9	6.5		0.6800	9	6.5		1.0664
	7		0.5459		7		0.8973
	7.5		0.4230		7.5		0.7804
	8		0.3128		8		0.6676
	8.5		0.2481		8.5		0.5374
	9		0.1980		9		0.4003

21. Use a Gaussian transformation technique to transform the dataset of Table 5.1. Calculate the empirical covariance and variogram values of the Gaussian-transformed dataset. Then, use a software library of your choice to display the Gaussian-transformed empirical covariance and variogram values. Lastly, choose among the known theoretical covariance and variogram models to fit to the calculated Gaussian-transformed empirical values above.

22. The standardized 2nd-, 3rd-, and 4th-order central moments are considered as the standard deviation, skewness, and kurtosis of a random variable. For the conversion from noncentral to central moments, the following binomial formula is used

$$\overline{(x-\overline{x})^n}=\sum_{k=0}^{n}\binom{n}{k}(-1)^{n-k}\overline{x^k}\,\overline{x}^{n-k}, \tag{5.1}$$

which is valid for a random variable or a random vector. If x is a random vector, the tensor product should be used. In the case of Gaussian *PDF*, Isserlis theorem can be used to decompose the higher-order moments into combinations of its 2nd-order moment. According to Isserlis' theorem, the mth-central moments can be estimated by the following rules: if $m=2n-1$, the central moment is zero; if $m=2n$, the central moment can be estimated by $\frac{(2n)!}{2^n n!}\sigma^{2,n}$, where n is a positive integer, and σ^2 is the variance. Therefore, based upon Isserlis' theorem, the computation of higher-order moments at a specific point can simply use Eq. (5.1). Using Eq. (5.1), express the central moments $\overline{(x-\overline{x})^2}$, $\overline{(x-\overline{x})^3}$ and $\overline{(x-\overline{x})^4}$.

23. Consider a homostationary random field $X(\boldsymbol{p})$. Show that the following relationships hold for the trivariance function:

(a) $\zeta_X(|\Delta\boldsymbol{p}|,|\Delta\boldsymbol{p}'|)=\zeta_X(|\Delta\boldsymbol{p}'|,|\Delta\boldsymbol{p}|)$
$=\zeta_X(|\Delta\boldsymbol{p}|-|\Delta\boldsymbol{p}'|,-|\Delta\boldsymbol{p}'|)$
$=\zeta_X(-|\Delta\boldsymbol{p}'|,|\Delta\boldsymbol{p}|-|\Delta\boldsymbol{p}'|)$
$=\zeta_X(|\Delta\boldsymbol{p}'|-|\Delta\boldsymbol{p}|,-|\Delta\boldsymbol{p}|)$
$=\zeta_X(-|\Delta\boldsymbol{p}|,|\Delta\boldsymbol{p}'|-|\Delta\boldsymbol{p}|)$.

(b) $\zeta_X(|\Delta\boldsymbol{p}|,|\Delta\boldsymbol{p}|)\neq\zeta_X(-|\Delta\boldsymbol{p}|,-|\Delta\boldsymbol{p}|)$,
(c) $\zeta_X(|\Delta\boldsymbol{p}|,-|\Delta\boldsymbol{p}|)=\zeta_X(-|\Delta\boldsymbol{p}|,|\Delta\boldsymbol{p}|)$.

Under what conditions do these relationships hold?

24. As was discussed in this chapter, the random field notion can be considered in both an intuitive and a mathematical setting. Consider the following assertions:

(a) The random field conception admits that the generator of reality is observable.
(b) One should take into account only the observed realizations.
(c) What the random field realizations have in common is that they satisfy (science- and logic-based) conditions believed to resemble the ones that prevail in reality realizations.
(d) The possible random field realizations are necessarily equiprobable.
(e) A key issue is whether some property of the phenomenon holds across all possible realizations or if it holds in a single realization.
(f) The term sample means that one sees only one (observed) realization among a collection of possible (but unobservable) ones.
(g) In the presence of natural laws, the random fields are constrained causally.

With which of the above assertions do you agree and with which do you not agree. Explain.

CHAPTER

6

Modern geostatistics

1 Toward a theory-driven *CTDA*

The classical (traditional) geostatistics discussed in Chapter 5 has been used for several decades in data analysis and modeling studies. Yet, the traditional concepts, language, and whole way of thinking seem in many cases inadequate considering the complexity of the models and the intensity and multiplicity of data sources emerging in natural sciences (physical equations, big data, etc.).

As a result, in many cases classical geostatistics cannot provide useful modeling tools to the rapidly advancing new scientific fields, data sources, and relevant *CTDA* applications. This inadequacy, naturally, led to new developments in the field:

Modern geostatistics is a major development in *CTDA* that adopted a **theory-driven** and **multisourced** perspective of knowledge gaining, processing, and interpretation.

Unlike other developments, modern geostatistics has not focused on purely technological advancements. Without neglecting the importance of new technologies, modern geostatisticians would find it more profitable in the long run to develop a sound theoretical background that rigorously accounts for the logic and science of the case, rather than to rely on collecting techniques and recipes via *Internet* searching. As a matter of fact, certain useful conclusions can be drawn based on the experience gained collectively over the years in many disciplines:

- Techniques and recipes, while temporarily fashionable, soon become obsolete, being of little **ontologic** and **epistemic** value,[a] neither improving in-depth understanding of the real phenomenon nor enhancing critical thinking.
- Many investigators admittedly do not always use **Internet** searching as rationally as they would like to think they do. Yet, it seems we live in an age where the ancient wisdom, *uti, non abuti*, is ignored.

[a] The term "ontologic" refers to the being of an entity (what is a tool, what is space, what is time, etc.), whereas the term "epistemic" refers to the act of acquiring knowledge.

https://doi.org/10.1016/B978-0-12-816341-2.00013-7

- The investigator's **beliefs** may differ from the epistemic underpinnings of the models used and the epistemic patterns of the data.

Threefold requirement

Therefore, modern geostatistics is concerned with those new developments in *CTDA* that satisfy the following three essential conditions:

① Involve an **epistemology** so that data analysis depends on *what* we know about the phenomenon as well as on *how* we know it.
② Be based on **integrative thinking** that offers a deeper theoretical understanding of the issues while taking into consideration an increasing body of knowledge and a rapidly enlarging database.
③ Not rely on quick fixes and techniques that are merely convenient but do not promote **intellectual development**.

Requirement ① distinguishes between what is the true nature of an attribute *vs.* how we come to understand it. Requirement ② implies that knowledge coming from theoretical models should be integrated with that contained in observational constraints. And, requirement ③ warns us that quick fixes and recipes discourage the real innovation that is needed to address complicated physical modeling and data intensive analysis.

1.1 Bayesian maximum entropy theory

A central position in the modern geostatistics perspective is occupied by a theory that, *ex constructione*, satisfies the scientific and logical inference conditions ①–③.

> The chronotopologic **Bayesian maximum entropy (*BME*)** theory studies a natural attribute by integrating both the relevant scientific body of information and the stochastic inference rules and logic standards of knowledge processing under conditions of *in-situ* uncertainty.

Otherwise said, the principal *BME* objective, around which all the other goals revolve, is multisourced knowledge integration. What, then, makes the *BME* point of view both interesting and challenging at the same time are its elemental knowledge integration features (Christakos, 1990, 2000; Christakos et al., 2002):

- It encourages the explicit determination of **physical *KB*** that are objectively available, and the development of **logically plausible rules** for knowledge processing and realistic map construction. This dual objective creates a body of information and a means of creative visualization.
- It recognizes that in *CTDA* reasoning the hypotheses made about a phenomenon are **epistemic**, in that the premises are known and empirically verifiable, in clear distinction to the assumptions of pure mathematical deduction that merely assume that propositions are true.
- Its methods can be placed in a **unified framework** demonstrating their importance to the development of modern geostatistics. *BME* theory is formulated in a way that preserves earlier geostatistics results, which are its limiting cases, and it also leads to novel and more general results not obtainable by traditional geostatistics.
- Its generalization power is further demonstrated by the fact that although the acronym "*BME*" indicates that its basic elements are the Bayesian and entropy notions, the underlying epistemology is not restricted by the acronym: the Bayesian concept can be replaced by **knowledge processing rules** based on stochastic logic, and information measures other than entropy may be considered (Section 3.3, later).

Because of features such as the above ones, *BME* theory introduces a fundamental insight into *CTDA* that plays a central role in its successful implementations in many scientific disciplines.

1.2 Spatiotemporal random field modeling

At the core of *BME* mathematics is the spatiotemporal random field (*S/TRF*) notion already introduced in Chapter 5. *S/TRF* modeling of natural attributes has led to considerable successes over the last several decades. As was argued in Section 2 of Chapter 5, the fundamental ideas underlying *S/TRF* modeling that have made it so successful for scientific analysis are: chronotopologic *continuum*, attribute *field*, and *complementarity*:

An *S/TRF* $X(\boldsymbol{p})$ is a collection of complementary field-realizations χ associated with the values of an attribute field at points $\boldsymbol{p}$ belonging to the chronotopologic continuum.

1.2.1 Chronotopologic point sets

In Table 1.1 a distinction is presented between subsets of points within the physical chronotopologic domain of interest, which will be useful in the modeling considerations of this and subsequent chapters.

The *S/TRF* theory plays a dominant role in *CTDA*. In light of the established notation of Table 1.1, the following description, which is rather a brief summary of the detailed mathematical *S/TRF* developments in Chapter 5, can contribute to this chapter's discussion:

S/TRF is a collection of correlated random variables $x_{map} = [x_{d1} \cdots x_{dn_d}\ x_{k1} \cdots x_{kn_k}]^T$ at points $\boldsymbol{p}_{map}$. The jth-realization at $\boldsymbol{p}_{map}$ is written as $\chi_{map}^{(j)} = [\chi_{d1}^{(j)} \cdots \chi_{dn_d}^{(j)}\ \chi_{k1}^{(j)} \cdots \chi_{kn_k}^{(j)}]^T$ ($j = 1, 2, \cdots$). In stochastic terms, $X(\boldsymbol{p})$ is fully described by the infinite sequence of *PDF* covering all continuum points, i.e., $f_X(\chi_{map})$.

Example 1.1 Suppose that a $PM_{2.5}$ study considers a set of four points, $\boldsymbol{p}_1, \boldsymbol{p}_2, \boldsymbol{p}_3$, and $\boldsymbol{p}_k$, in the space-time domain, see Fig. 1.1.[b] In this case, the above point set is denoted by the vector $\boldsymbol{p}_{map} = [\boldsymbol{p}_1, \boldsymbol{p}_2, \boldsymbol{p}_3, \boldsymbol{p}_k]^T$, where $\boldsymbol{p}_i = (s_{i1}, s_{i2}; t_i)$, $i = 1, 2, 3, k$, and $\chi_{map} = \{\chi_1, \chi_2, \chi_3, \chi_k\}$.

TABLE 1.1 Chronotopologic point sets to be considered in *CTDA*.

Entire point set	Vector $\boldsymbol{p}_{map} = [\boldsymbol{p}_1 \cdots \boldsymbol{p}_n]^T$, which includes the nodes of the chronotopologic mapping grid that covers the entire domain of interest
Hard data point subset	Vector $\boldsymbol{p}_h = [\boldsymbol{p}_{h1} \cdots \boldsymbol{p}_{hn_h}]^T \subset \boldsymbol{p}_{map}$, which includes the points where exact data is available
Soft data point subset	Vector $\boldsymbol{p}_s = [\boldsymbol{p}_{s1} \cdots \boldsymbol{p}_{sn_s}]^T \subset \boldsymbol{p}_{map}$, which includes the points where uncertain data is available
Data point subset	Vector $\boldsymbol{p}_d = [\boldsymbol{p}_{d1} \cdots \boldsymbol{p}_{dn_d}] = \boldsymbol{p}_h \cup \boldsymbol{p}_s$, i.e., the union of $\boldsymbol{p}_h$ and $\boldsymbol{p}_s$
Unsampled point subset	Vector $\boldsymbol{p}_k = [\boldsymbol{p}_{k1} \cdots \boldsymbol{p}_{kn_k}]^T$, which includes points of interest where, though, no empirical evidence exists[a]
Point subset link	Relationship holds between the set and the two main point subsets leading to the entire point set: $\boldsymbol{p}_{map} = \boldsymbol{p}_d \cup \boldsymbol{p}_k$

[a] *Attribute values may be generated at this subset of points, e.g., using data at points* $\boldsymbol{p}_d$ *and other knowledge sources.*

[b] The reason a distinction is made between the first three points and the fourth one will become clear later.

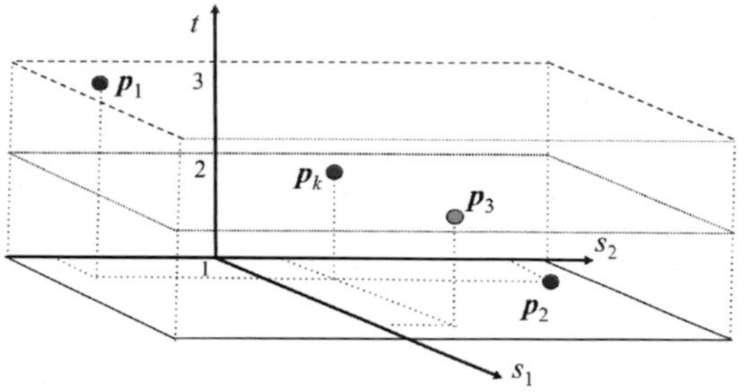

FIG. 1.1 Chronotopologic arrangement of three data points (p_1, p_2, p_3) and one estimation point (p_k). The corresponding coordinates are $p_1 = (10, -30;3)$, $p_2 = (10,50;1)$, $p_3 = (40,10;2)$ and $p_k = (10,10;2)$.

1.2.2 Physical and digital worlds: A synthetic reality perspective

Next, let us reflect on another intriguing *S/TRF* feature, which is particularly significant in view of continuing advances in computer and visualization technology.

- The distinction between the **real world** (that imposes physical constraints on natural phenomena) and the **digital world** is a current development created by powerful computers and used by *CTDA*.

The real world provides certain unique experiences (see, touch, smell, social collective experiences) and rich interactions that can hardly be replaced by digital artifacts and simulated environments. On the other hand, the flexibility of modern era digital tools has shown undeniable benefits that allow the partial overtaking of the real-world constraints.

In the middle of all this, *S/TRF* theory creates a bridge between the real and the digital worlds so that these two realms can complement each other.

Example 1.2 *S/TRF* realizations (simulations) $\chi_{map}^{(j)}$ of the phenomenon of interest can be generated in the digital world under a variety of possible scenarios in a way that would not be possible to do in real life.

Hence, another *S/TRF* perspective that is methodologically and instructionally significant in the modern geostatistics context is as follows:

> The *S/TRF* develops a **synthetic reality** in which digital information can be used in combination with its physical counterpart.

Although synthetic reality benefits from digital information anchored onto the physical world, its experiences still remain constrained by the physical environment. The *S/TRF* theory proposes a conceptual framework that can be progressively updated and improved in view of new evidence, and within which the investigator can travel between the physical and the virtual world. By providing augmented experiences, anchored in the physical reality, *S/TRF* modeling creates a digitally simulated environment where the investigators can benefit from both physical sensing and the flexibility of digital interaction. The result builds toward a unified way to look at the intersection between physical and digital realms combined on a single framework.

Despite the existence of certain challenges, which is a natural and even desirable feature of any active scientific field, chronotopologic geostatistics is quite well established in many dimensions.

2 Knowledge bases revisited

We already discussed the notion of *knowledge bases* (*KB*) in Section 2 of Chapter 3. In this chapter the *KB* notion is considered under the prism of modern geostatistics, the latter seen as a multidisciplinary affair involving *KB* from various sciences, not the province of pure a priori mathematics.

In this broad setting, the *total* knowledge used in real-world *CTDA* applications comes from a variety of sources. In principle, total knowledge includes all kinds of valid information available at a given moment and obtained by the competent scientist using effectively a scientific procedure. In this sense, the availability of physical knowledge is objective. Subjective bias enters when the scientist fails to use the appropriate procedures leading to the valid knowledge, even when such procedures are available. Accordingly, the following perspective seems to be rather crucial:

> *CTDA* investigators need to see beyond the purely formal notions and technical details and obtain a deeper understanding of the substantive issues related to **knowledge** and **knowing**.

Forms of knowledge and knowing

Along the above lines of thought, several distinctions can be made between different forms of knowledge and knowing, the most interesting of which are as follows:

① Between knowledge that is obtained by the **senses** and knowledge that is obtained by the **mind**.
② Between knowledge gained by the **direct experience** of the knower and knowledge gained through the **experience of others**.
③ Between knowledge relying on **common sense** and that derived from the different **scientific disciplines**.
④ Between the two primary bases of knowledge introduced in Chapter 3: the **core** or **general** knowledge base *G* and the **specificatory** or **case-specific** knowledge base *S*, such that $K = G \cup S$ is the total knowledge available.

Admittedly, each one of the above four distinctions has its own merit. For procedural reasons, we start by focusing on the distinction ④.

2.1 Core or general *KB* and the science-based *BME*

This is the first major kind of *KB* according to the earlier distinction, which can be described in more words as follows:

> **Core** or **general** ***G-KB*** denotes background knowledge and justified beliefs relative to the phenomenon of interest. The *G* may include physical laws, scientific theories and models, structured patterns and assumptions, analytic and synthetic statements, and previous experience with similar phenomena independent of any case-specific data.

Example 2.1 The statements "the chronotopologic distribution of the ozone concentration field is characterized by an exponential covariance model" and "the subsurface flow is governed by Darcy's law" are vague statements, since several ozone fields may share the same covariance model and various flow cases satisfy Darcy's law, so that both these statements constitute core or general knowledge.

The different kinds of *G-KB* above are offered a priori (i.e., before any case-specific evidence is obtained in the form of data), and *G* is considered "general" in the sense that it is vague enough to characterize a large class of fields or situations associated with the phenomenon of interest.

Remarkably, in many cases the consideration of theoretical *G-KB* may be needed for very practical reasons. In particular:

- The *G-KB* is necessary in real-world situations because of **unavailable evidence**: in many situations, such as the future occurrence of a phenomenon or the existence of inaccessible attribute domains, key evidence is not available at the current state.

2.1.1 G-KB semantics

From a semantic viewpoint, which can help an investigator understand the meaning of *KB* at a deeper level, the *G* may be divided into two major knowledge categories, namely:

Analytic statements, involving abstract quantities, symbols, and logical relations, with or without real counterparts; and **synthetic statements**, which include statements of fact, laws, and theories, and their truth or falsity is a matter of experience.

The distinction between analytic and synthetic statements depends on *why* the statements are true or false. An analytic statement is true by definition, so that its negation results in a contradiction or inconsistency. On the other hand, the truth or falsity of a synthetic statement can only be determined by relying on observation and experience, i.e., it cannot be determined by relying solely upon logic or examining the meaning of the words involved.

Example 2.2 The readers may appreciate the following instructive distinctions:

- Analytic statements without real counterparts, say, "if *A* or *B* and not *B*, then *A*," are justified on the basis of relations of concepts and theories.
- Other analytical statements have real counterparts, but their validity is a matter of definition and/or logic only, say, the statement "permeability values are nonnegative."
- On the other hand, a fact is a synthetic statement that has real counterparts, say, "the larger the hydraulic conductivity at a waste site is, the faster is the contaminant transport in the site," describes a fact.

2.1.2 Science-based BME

Models, laws, and theories are continuously developed in sciences in order to represent quantitatively the phenomenon of interest, with the twofold goal of understanding its underlying mechanisms and predicting its chronotopologic evolution. This twofold goal is an essential component of scientific inquiry. In view of these considerations, the following point should be stressed regarding the *BME* methodology.

A central objective of *BME* is the incorporation of all the essential **scientific constructs** (models, laws, and theories) into a rigorous *CTDA*.

Such an objective is motivated by certain decisive dilemmas that occasionally emerge in real-world *CTDA* but seem to be ignored. Some of these very practical dilemmas are described in the following example.

Example 2.3 Can we safely assume that we *know* what the word "mean value" denotes in a case study, or, to put it in a more methodologically suitable form, can we assume that we *agree* on what we mean by the word "mean value"? Is it the arithmetic average value of the available attribute data values or what is obtained by the averaged analytical solution of the physical attribute law? What if these two values diverge? Similar comments can be made for the covariance and variogram functions, as well.

Scientists explore nature for important chronotopologic patterns, which they then try to represent by models and organize into theories according to laws. Obviously models, laws, and theories are closely linked in scientific investigations. We examine each one of these three scientific notions in more detail.[c]

A **model** is a mental construct that possesses three properties: it is guided by the phenomenon of interest (*mapping* property), it only reflects a selection of the phenomenon's features (*reduction* property), and it is used in place of the phenomenon with a certain purpose in mind (*pragmatic* property).

[c] The notion of a "model" was formally introduced in Chapter 3.

Ex necessitate rei, one of the most important objectives of a model is to focus on the most significant or relevant features of the phenomenon while eliminating or suppressing minor details. This is a pragmatic requirement, since neither a single model can characterize a phenomenon completely, nor would such a model be desirable (its great complexity would make it too cumbersome to be useful). Beyond the above general modeling considerations, of particular interest is a modeling description favored in the *CTDA* context:

> A **model** essentially transforms an attribute value at a certain point in space and time toward a value at another point in space and time. This transformation is characterized by chronotopologic variation.

Naturally, a great variety of models can be constructed for any natural phenomenon, depending on the study objectives. Although this abundance seems overwhelming, it can be brought under control by distinguishing between different classes of models based on the following criteria:

- **Fully explanatory** models.
- **Partially explanatory** models (i.e., they sketch explanations that leave crucial details undetermined or hidden).
- **Conjecturing** models (i.e., they conjecture a "how-possibly" but not a "how-actually" explanation).
- Models that are merely **data summaries**, which, nevertheless, can be useful without explaining.

Ad rem, the classification criteria above pertain to different issues arising in connection with models, which raise important questions in *ontology* (what are the models), *epistemology* (how scientists learn and explain with models), and *semantics* (how models represent). Table 2.1 lists several model types together with their descriptions. Models of phenomena should be understood as models representing phenomena, and are valuable if they explain phenomena (explanatory, mechanistic) or provide satisfactory fits to the relevant data (predictive, empirical, statistical). The different model types in Table 2.1 are not mutually exclusive; instead, certain links exist between the various model types. This is the case, *verbi gratia*, of the boundary between mechanistic and empirical models, which is rather fuzzy, and there exists a wealth of interim types of models. Furthermore, some mechanistic models can be statistical, at the same time.

Each model type has its *pros* and *cons* and its implementation depends on the case at hand. A mechanistic model is, in principle, explanatory; since the model coefficients have natural (physical, biological, etc.) definitions, they may be measured independently of the dataset. Mechanistic models are more likely to work correctly when extrapolating beyond the observed conditions. Also, mechanistic models are preferred over empirical models as they are much more general and broader in application. On the other hand, fully explanatory mechanistic models are in many cases unfeasible. Phenomenologic, predictive, empirical, and statistical models often do not offer adequate phenomenon explanations. A phenomenologic model would have some theory in which the relevant phenomena are presupposed, but no theory that explains where these phenomena come from. Otherwise said, phenomenologic models describe the empirical relationship of phenomena to each other in a way that is consistent with fundamental theory, but is not directly derived from theory (not derived from first principles). It is widely considered that, all other things being equal, mechanistic models are more powerful than phenomenological (or statistical) models since they tell us about the underlying processes and driving patterns.

It is a matter of having enough data and constraints to be able to fit a nonexplanatory model

TABLE 2.1 Model types and their description.

Model characterization	Features
Fully explanatory	• Explanation of *why* a phenomenon is the way it is (e.g., it explains why a certain attribute at a specified location and time adopts a certain value) • Considered adequate to the extent that it describes the causal mechanisms that maintain, produce, or underlie the phenomenon under investigation • Primary goal is to test theoretical hypotheses, so its emphasis is on theoretically meaningful relationships between phenomenon parameters • It can be used in an *ex ante*[a] analysis and, thus, for prediction purposes
Partially explanatory	• Explanation needs to fit well to a sufficient portion of whatever is known about the phenomenon so that the model becomes useful • Substitute for full explanation, either because the full explanation is unavailable or because it is too cumbersome to be practical • Identify phenomenon parameters that have a scientifically meaningful and statistically significant relationship with an outcome of interest • Include fitting potentially theoretically important predictors, checking for statistical significance, evaluating effect sizes, and running diagnostics
Predictive	• The goal is to use the associations between phenomenon parameters to generate accurate predictions. • Variables that are used in a predictive model are based on association, not necessarily on scientific meaning. • There are cases when statistically significant variables will not be included in a predictive model (a statistically significant predictor that adds no predictive benefit may be excluded).
Mechanistic	• Relies on a hypothesized link between phenomenon parameters specified in terms of natural processes and mechanisms that govern the phenomenon • Can be fully or partially explanatory • Usually physically detailed (derived from first principles), and designed to be *causal* (i.e., it carries explanatory force to the extent that it reveals aspects of the causal structure of a mechanism) • Based on a mathematical description of the phenomenon (mechanical, chemical, biological, etc.)
Phenomenologic	• Only represents observable phenomenon properties and refrains from postulating hidden mechanisms • Describes *how* a phenomenon works (e.g., it does not explain why an attribute at a specified location and time adopts a certain value) • Relies on a hypothesized relationship between the dataset variables, where the relationship seeks only to best describe the data • Forgoes any attempt to explain why the phenomenon parameters interact the way they do, and simply attempts to describe a relationship, with the assumption that the relationship extends past the measured values • Can only be used in an *ex post* analysis and not in a predictive analysis
Empirical	• Is based on data (experimental, observational, survey, etc.) • Uses relatively simple approximations, like the 1st- and 2nd-order polynomials in regression analysis • May take the form of various "smoothing" functions • Need not have an underlying theory • Its emphasis could be to determine whether a relationship between phenomenon parameters is statistically significant
Statistical	• Usually serves as a phenomenologic model • Regression analysis is often used to create statistical models

[a] *Ex ante* means "before the event." When you're making a prediction, you're doing so *ex ante*. The opposite of *ex ante* is *ex post*, which means after the event.

(phenomenological, empirical, or statistical) to the observed phenomenon. Furthermore, since their primary goal is predictive accuracy, predictive models are created very differently than explanatory models. Being able to explain why a variable "fits" in the model is not of immediate concern for predictive models, which gives investigators the latitude to use predictive models that may not have any significant theoretical value. Statistics plays an important role in estimating unknown mechanistic or phenomenologic model coefficients using observed information.[d] *Regression* models are the standard form of statistical descriptions.

Example 2.4 A linear relation empirically established between a blood pressure drug and heart rate could be viewed as a statistical model. This model does not provide any insight concerning how the two variables are related biophysically. On the other hand, a mechanistic model of the phenomenon would involve a detailed consideration of the intermediate biophysical mechanisms (drug entering the system, binding to receptors, modulating levels of hormones, and acting as a signal to heart rate modulation, etc.). Naturally, this model would have more parameters, but it could accurately predict the phenomenon, initiate hypotheses, and be generalizable and interpretable.

If the predicted attribute is significant but only observable immediately before or at the time of the observed outcome, it cannot be used for predictions.

Example 2.5 Although water temperature is a very significant factor in determining whether a tropical storm turns into a hurricane, it is not a useful variable in a prediction model of the expected number of hurricanes during the upcoming season because it can only be measured immediately before an impending hurricane, which is too late for prediction purposes.

As it turns out, both theories and laws could be considered within the general modelling context.

A scientific **law** is the kind of model that provides a description, usually mathematical, of an observed phenomenon, which, though, does not necessarily explain why the phenomenon exists or what causes it.

Ut ita dicam, a scientific law is the summary of a pattern seen in nature, or a statement about an observed phenomenon or a unifying concept. The general laws of physics apply in a simple sense, but the underlying physical mechanisms are wildly different in the many regimes considered in the real world. Also, a law may not offer an adequate explanation of the phenomenon. The vast majority of the phenomena encountered in nature obey physical laws, which, in the case of classical physics, can be described in terms amenable to efficient *material causation*. This is why it is necessary that any sound *CTDA* approach should formally incorporate physical laws in a systematic and internally consistent manner.

Example 2.6 Newton's law of gravity or Mendel's law of independent assortment simply describe well-established observations about these phenomena, but not how or why they work.

A relevant warning is in order: Beyond the long established laws of nature, one should be aware of the case of *epiphenomenalism* that occurs when occasionally one has the illusion of deterministic cause-and-effect phenomena. That is, some relations may simply represent statistical correlations rather than deterministically established cause and effect physical laws (in the sense

[d] This is also known as the inverse problem.

that the elements underlying the causation are understood in purely mechanical terms). As is obvious from the above discussion, a scientific construct of interest is the theory.

A scientific **theory** is an in-depth explanation of the observed phenomenon.

In a sense, a scientific theory can be viewed as a sophisticated model possessing deep explanatory power. Although scientific laws and theories are supported by a large body of empirical data, accepted by the majority of scientists within that area of scientific study, and help to unify it, they are obviously not the same thing. In this respect, the theory is the reasoning behind a law, what laws to use, or how to apply a law.

Example 2.7 A law may be a representation, without explanatory power, of a relation that has been observed consistently in the past and it is expected to be so in the future. This is the case of Darcy's law of hydrology.[e] A theory is an intellectual construct that has explanatory power. This is the case of Navier-Stokes theory of the motion of viscous fluid substances. A law may be derived from a theory, say, Darcy's law is deduced from the Navier-Stokes theory by using an averaging process.

For *BME* purposes, the *G-KB* can be formalized in a way that includes a wide variety of scientific constructs (models, physical laws, and scientific theories) in the form of *algebraic* and *differential* equations that are realistic representations of reality. This is our next topic.

2.1.3 *Operational G-formalism*

In sciences, the datasets, concepts, and theories strongly depend on stochastic chronotopologic correlations and probabilistic dependencies. It is, therefore, useful to obtain a mathematical formulation of G that satisfies a set of conditions, such as the following:

❶ The G is expressed in terms of a series of suitable **functions** of the attribute realization χ_{map} at the point set $\boldsymbol{p}_{map}$ of interest, i.e.,

$$g_\alpha\left(\chi_{map}\right), \alpha=0,1,\ldots,N_c \tag{2.1}$$

(usually, by convention $g_0 = 1$). Depending on the in application, the explicit mathematical expression of Eq. (2.1) accounts for a variety of core knowledge sources including statistical moments, empirical relationships, and physical laws.

❷ Another condition that guides the choice of g_α's is that they are chosen so that the stochastic expectations $\overline{g_\alpha}$ are **calculable** at the point set $\boldsymbol{p}_{map}$ (which includes the sampled points and any unsampled points of interest).

❸ Lastly, the resulting equations must be **solvable** with respect to the unknown parameters of the g_α functions (these parameters are functions of $\boldsymbol{p}_{map}$).

The above three criteria can lead to a wide variety of g_α choices, depending on the G considered and the application of interest.

Example 2.8 These g_α functions, some of which will be discussed in more detail later, may include a wide variety of physically familiar two-point attribute arrangements expressing chronotopologic variation, such as

$$g_a\left(\chi_{map}\right)=\alpha_{ij}\chi^i(\boldsymbol{p}_i)\chi^j\left(\boldsymbol{p}_j\right),\ \alpha_{ij}e^{\beta_{ij}\chi(\boldsymbol{p}_i)\chi(\boldsymbol{p}_j)},$$

$$\alpha_{ij}\frac{\partial^{p+q}}{\partial s_i^p \partial t_j^q}\chi(\boldsymbol{p}_i)\chi\left(\boldsymbol{p}_j\right),$$

and combinations thereof, where the α_{ij} and β_{ij} are physical coefficients.

[e] The readers may notice that, in this sense, a law may also be seen as a summary representation of numerous facts without being explanatory, and it may also include nonobservable quantities.

An implication of the previously mentioned illustrative cases is that the selection of the g_α functions is both rich and predominantly application-dependent. In fact, one could think of g_α as a *physical* function that represents attribute features (in appropriate physical units) at a particular chronotopologic point (in the chosen coordinate system), rather than as a purely *mathematical* function that represents a particular functional relationship between a set of numbers.

Sets of g_α functions can be selected that represent attribute means, covariances, variograms, and higher-order statistics, which are expressed by the g_α **stochastic expectation.**

$$\overline{g_\alpha\left(\chi_{map}\right)} = \int d\chi_{map} g_\alpha\left(\chi_{map}\right) f_G\left(\chi_{map}, \boldsymbol{p}_{map}\right), \quad (2.2)$$

where $\alpha = 0, 1, \ldots, N_c$, and f_G is a PDF constructed on the basis of the *G-KB*.

The $\overline{g_\alpha}$ statistics are usually computed on the basis of core *KB* (physical laws, theoretical models) and/or specificatory *KB* (field data, experimental surveys). Table 2.2 (He and Kolovos, 2018) provided a g_α set derived from a few selected core knowledge constraints. In particular, the table lists the functions g_α when the mean, variance, and covariance functions of $X(\boldsymbol{p})$ are known at points $\boldsymbol{p}_i$ $(i = 1, \ldots, m, k)$.

2.1.4 Practical issues

As usual in such cases, certain practical issues emerge concerning the determination of the g_α functions and the corresponding $\overline{g}_\alpha$ averages.

Dictum factum, the following examples present a *step-by-step* process of obtaining the **functions** g_α and their **expectations** $\overline{g}_\alpha$ in practice based on Eq. (2.2).

Example 2.9 Assume that, in the $PM_{2.5}$ study of Example 1.1 of Section 1, the space distances are $h_{12} = 80$ km, $h_{13} = h_{23} = 50$ km, $h_{1k} = h_{2k} = 40$ km and $h_{3k} = 30$ km; and the time separations are $\tau_{12} = 2$ mos, $\tau_{13} = \tau_{1k} = \tau_{23} = \tau_{2k} = 1$ mos and $\tau_{3k} = 0$ mos.

Step 1: The *G-KB* consists of the $PM_{2.5}$ statistics models, i.e., the mean concentration model and the theoretical covariance model, respectively,

TABLE 2.2 Some g_α and $\overline{g}_\alpha$ functions expressing G-based constraints.

G-KB	α	g_α	$\overline{g}_\alpha$
Normalization	0	$g_0(\chi_{map}) = 1$	$\overline{g}_0 = 1$
Means $(m+1)$	1	$g_1(\chi_1) = \chi_1$	$\overline{g}_1 = \overline{X}_1$
	$\vdots$	$\vdots$	$\vdots$
	m	$g_m(\chi_m) = \chi_m$	$\overline{g}_m = \overline{X}_m$
	$m+1$	$g_{m+1}(\chi_k) = \chi_k$	$\overline{g}_{m+1} = \overline{X}_k$
Variances $(m+1)$	$m+2$	$g_{m+2}(\chi_1, \chi_1) = (\chi_1 - \overline{X}_1)^2$	$\overline{g}_{m+2} = \sigma_1^2$
	$\vdots$	$\vdots$	$\vdots$
	$2m+1$	$g_{2m+1}(\chi_m, \chi_m) = (\chi_m - \overline{X}_m)^2$	$\overline{g}_{2m+1} = \sigma_m^2$
	$2m+2$	$g_{2m+2}(\chi_k, \chi_k) = (\chi_k - \overline{X}_k)^2$	$\overline{g}_{2m+2} = \sigma_k^2$
Covariances $(\frac{m(m+1)}{2})$	$2m+3$	$g_{2m+3}(\chi_1, \chi_2) = (\chi_1 - \overline{X}_1)(\chi_2 - \overline{X}_2)$	$\overline{g}_{2m+3} = c_X(\boldsymbol{p}_1, \boldsymbol{p}_2)$
	$\vdots$	$\vdots$	$\vdots$
	$\frac{(m+4)(m+1)}{2}$	$g_{\frac{(m+4)(m+1)}{2}}(\chi_m, \chi_k) = (\chi_m - \overline{X}_m)(\chi_k - \overline{X}_k)$	$\overline{g}_{\frac{(m+4)(m+1)}{2}} = c_X(\boldsymbol{p}_n, \boldsymbol{p}_k)$

$$\overline{X}(\boldsymbol{p}) = \overline{X},$$
$$c_X(h,\tau) = e^{-\frac{h}{a_h}-\frac{\tau^2}{a_\tau^2}}, \qquad (2.3a,b)$$

where $\overline{X} = 65$ (μg/m^3), $\sigma_X^2 = 1$ ((μg/m^3)^2), $a_h = \frac{250}{3}$ (km), $a_\tau = \sqrt{3} = 1.73$ (mos),[f] and the covariance units are (μg/m^3)^2.

Step 2: The functions g_α associated with the *G-KB* are shown in Table 2.3, with $\alpha = 0, 1, \ldots, N_c$, where $m = 3$ is the number of data and $N_c = \frac{1}{2}(m+1)(m+4) = 14$.

Step 3: The corresponding expectations $\overline{g}_\alpha$ are also listed in Table 2.3. The means are listed in lines $\alpha = 1 - 4$ and the variances in lines $\alpha = 5 - 8$; the numerical covariance values in lines $\alpha = 9 - 14$ were obtained by inserting the distances ($h_{12} = 80$ km, $h_{13} = h_{23} = 50$ km, $h_{1k} = h_{2k} = 40$ km, $h_{3k} = 30$ km) and the separations ($\tau_{12} = 2$ mos, $\tau_{13} = \tau_{23} = \tau_{1k} = \tau_{2k} = 1$ mos, $\tau_{3k} = 0$ mos) into Eqs. (2.3a,b).

Obviously, higher-order statistical moments can be expressed by the g_α formalism without any theoretical difficulty. Next, we focus on the case where a body of knowledge is available in the form of empirical models formally expressed in terms of algebraic or differential equations.

TABLE 2.3 List of g_α and $\overline{g}_\alpha$ functions for Eqs. (2.3a,b).

α	g_α	$\overline{g}_\alpha$
0	1	1
1–4	χ_i ($i = 1,2,3,k$)	$\overline{X}_i = \overline{X(s_i,t_i)} = 65$
5–8	$(\chi_i - \overline{X})^2$ ($i = 1,2,3,k$)	$\sigma_i^2 = \sigma_X^2(s_i,t_i) = 1$
9	$(\chi_1 - \overline{X})(\chi_2 - \overline{X})$	$c_{12} = c_X(80,2) \approx 0.101$
10	$(\chi_1 - \overline{X})(\chi_3 - \overline{X})$	$c_{13} = c_X(50,1) \approx 0.393$
11	$(\chi_1 - \overline{X})(\chi_k - \overline{X})$	$c_{1k} = c_X(40,1) \approx 0.443$
12	$(\chi_2 - \overline{X})(\chi_3 - \overline{X})$	$c_{23} = c_X(50,1) \approx 0.393$
13	$(\chi_2 - \overline{X})(\chi_k - \overline{X})$	$c_{2k} = c_X(40,1) \approx 0.443$
14	$(\chi_3 - \overline{X})(\chi_k - \overline{X})$	$c_{3k} = c_X(40,0) \approx 0.698$

2.1.5 Empirical algebraic equation models

For many natural attributes, the g_α formalism can go beyond standard (direct) statistics calculations to include *G-KB* manifested in more involved quantitative forms, such as the following.

> Empirical relationships or models governing the chronotopologic change of a natural attribute $X(\boldsymbol{p})$ that are expressed in terms of **algebraic equations,**
>
> $$A[X(\boldsymbol{p}), w(\boldsymbol{p}), b_i] = 0, \qquad (2.4)$$
>
> where $w(\boldsymbol{p})$ and b_i ($i = 1, 2, \ldots, l$) are known attribute and model parameters, respectively, so that the stochastic Eq. (2.2) can be formulated at every point $\boldsymbol{p} \in \boldsymbol{p}_{map}$.

Example 2.9 (cont.): Consider *de novo* the arrangement of the $PM_{2.5}$ study of Example 2.9, where now we assume that the PM_{10} concentrations $X(\boldsymbol{p})$ are related to the available $PM_{2.5}$ measurements $w(\boldsymbol{p})$ by means of the empirical law

$$X(\boldsymbol{p}) - b_1 w(\boldsymbol{p}) - b_2 = 0, \qquad (2.5)$$

where $b_1 = 1.65$ and $b_2 = 3.3$ are experimental coefficients.

Step 1: Eq. (2.5) and related $PM_{2.5}$ statistics constitute the *G-KB*, in this case.

Step 2: The PM_{10} statistics models associated with Eq. (2.5) are

$$\overline{X} = 1.65\overline{w} + 3.3 = 110.55$$
$$= 1.65 \int d\omega_i \left(\omega_i + \frac{3.3}{1.65}\right) f_G(\omega_i, \boldsymbol{p}_i), \qquad (2.6a,b)$$

[f] To remind our readers, the spatial and temporal correlation ranges are $\varepsilon_s = 3a_h = 250$ (km) and $\varepsilon_t = \sqrt{3}a_\tau = 3$ (mos), respectively.

$$\sigma_X^2 = \overline{[X(\boldsymbol{p}_i) - 110.55]^2}$$
$$= 1.65^2\overline{[w(\boldsymbol{p}_i) - 65]^2} = 2.7225\sigma_w^2 \qquad (2.7a,b)$$
$$= 2.7225\int\int d\omega_i(\omega_i - 65)^2 f_G(\omega_i, \boldsymbol{p}_i),$$

$$c_X(\Delta\boldsymbol{p}) = \overline{X(\boldsymbol{p}_i)X(\boldsymbol{p}_j)} - \overline{X}^2 = 2.7225\overline{w(\boldsymbol{p}_i)w(\boldsymbol{p}_j)}$$
$$+ 5.445\left[\overline{w(\boldsymbol{p}_i)} + \overline{w(\boldsymbol{p}_j)}\right] + 10.89 - (1.65\overline{w} + 3.3)$$
$$= 2.7225\overline{w(\boldsymbol{p}_i)w(\boldsymbol{p}_j)} + 5.445(65 + 65) - 110.55^2$$
$$= 2.7225\overline{w(\boldsymbol{p}_i)w(\boldsymbol{p}_j)} - 11513.4525$$
$$= 2.7225\int\int d\omega_i d\omega_j\left[\omega_i\omega_j + \frac{5.445}{2.7225}(\omega_i + \omega_j)\right.$$
$$\left.+ \frac{10.89}{2.7225} - 110.55^2\right] f_G(\omega_i, \omega_j, \boldsymbol{p}_i, \boldsymbol{p}_j).$$
$$(2.8a\text{-}c)$$

And given that $\overline{w(\boldsymbol{p}_i)w(\boldsymbol{p}_j)} - c_w(h,\tau) + \overline{w}^2$ is known, Eq. (2.8b) gives.

$$c_X(\Delta\boldsymbol{p}) = 2.7225\left[e^{-\frac{3h}{250} - \frac{\tau^2}{3}} + 4225\right] - 11513.4525$$
$$= 2.7225\int\int d\omega_i d\omega_j\left[\omega_i\omega_j + \frac{5.445}{2.7225}(\omega_i + \omega_j)\right.$$
$$\left.+ \frac{10.89}{2.7225} - 110.55^2\right] f_G(\omega_i, \omega_j, \boldsymbol{p}_i, \boldsymbol{p}_j).$$
$$(2.9a,b)$$

Step 3: Using Eqs. (2.6a,b), (2.7a,b) the functions g_α of $X(\boldsymbol{p})$ and the corresponding expectations $\overline{g}_\alpha$ associated with the *G-KB* are shown in Table 2.4. In particular, the means are listed in lines $\alpha = 1$–4 and the variances in lines $\alpha = 5$–8; the numerical covariance values in lines $\alpha = 9$–14 were obtained by inserting the corresponding distances and time separations into Eq. (2.9a).

2.1.6 *Differential equation models*

Naturally, the g_α functions are adapted as understanding deepens and/or situations change. If the real-world situation happens to be more complicated, as is usually the case

TABLE 2.4 List of g_α functions and $\overline{g}_\alpha$ expectations for PM_{10} (X) of Eq. (2.5).

α	g_α	$\overline{g}_\alpha$
0	1	1
1–4	$1.65(\omega_i + \frac{3.3}{1.65})$ $(i = 1, 2, 3, k)$	110.55
5–8	$2.7225(\omega_i - 65)^2$ $(i = 1,2,3,k)$	2.7225
9	$2.7225\left[\omega_1\omega_2 + \frac{5.445}{2.7225}(\omega_1 + \omega_2) + \frac{10.89}{2.7225} - 110.55^2\right]$	0.275
10	$2.7225\left[\omega_1\omega_3 + \frac{5.445}{2.7225}(\omega_1 + \omega_3) + \frac{10.89}{2.7225} - 110.55^2\right]$	1.07
11	$2.7225\left[\omega_1\omega_k + \frac{5.445}{2.7225}(\omega_1 + \omega_k) + \frac{10.89}{2.7225} - 110.55^2\right]$	1.206
12	$2.7225\left[\omega_2\omega_3 + \frac{5.445}{2.7225}(\omega_2 + \omega_3) + \frac{10.89}{2.7225} - 110.55^2\right]$	1.07
13	$2.7225\left[\omega_2\omega_k + \frac{5.445}{2.7225}(\omega_2 + \omega_k) + \frac{10.89}{2.7225} - 110.55^2\right]$	1.206
14	$2.7225\left[\omega_3\omega_k + \frac{5.445}{2.7225}(\omega_3 + \omega_k) + \frac{10.89}{2.7225} - 110.55^2\right]$	1.90

in actu, the physical models are expressed in terms of more advanced mathematical functions:

> **Differential equations** representing the chronotopologic law of change of the natural attribute $X(\boldsymbol{p})$, like
>
> $$D[X(\boldsymbol{p}), w(\boldsymbol{p}), b_i] = 0, \qquad (2.10)$$
>
> and the associated *BIC*, at every point $\boldsymbol{p} \in \boldsymbol{p}_{map}$.

When one gives Eq. (2.10) its due weight, one realizes that it is methodologically appropriate to consider two distinct formulations of the *G-KB*. The first formulation is as follows:

Formulation ❶: Eq. (2.10) leads to *explicit* analytical g_α expressions of Eq. (2.2) at every point $\boldsymbol{p} \in \boldsymbol{p}_{map}$.

This formulation, when possible, provides a straightforward derivation of the g_α functions, as is illustrated in the following example.

Example 2.10 The following stochastic ordinary differential equation represents the unidimensional evolution of attribute X (fixed time t, which is suppressed for simplicity)

$$b_1 \frac{d}{ds} X(s) + X(s) = w(s), \tag{2.11}$$

where $X(0)=0$ is the boundary condition, $b_1 = 0.5$ is an empirical parameter and $w(s)$ is a Brownian random field with $\overline{w}=0$ and $c_w(s_i, s_j) = b_2 s_i$, where $b_2 = 0.2$ is an empirical parameter.

Step 1: Eq. (2.11) and related auxiliary conditions above constitute the *G-KB* in this case.

Step 2: The direct solution of Eq. (2.11) yields

$$\overline{X(s_i)} = 0 = \int d\chi_i \chi_i f_G(\chi_i, s_i), \tag{2.12a,b}$$

and

$$\begin{aligned} c_X(s_i, s_j) &= 0.2\left(e^{-2s_i} - 2s_i - 1\right)\left(1 - e^{-2s_j}\right) \\ &= \iint d\chi_i d\chi_j \chi_i \chi_j f_G\left(\chi_i, \chi_j, s_i, s_j\right), \end{aligned} \tag{2.13a,b}$$

where $i, j = 1, 2, 3, k$; for $i = j$, Eq. (2.13a,b) include the corresponding variances $\sigma_X^2(s_i)$.

Step 3: The functions g_α and their expectations $\overline{g}_\alpha$ corresponding to the *G-KB* are presented in Table 2.5.

Next we consider another possible formulation associated with Eq. (2.10), which can be more involved, yet it may offer a better alternative in certain applications.

Formulation ❷: Eq. (2.10) leads to expressions of the form

$$\iint d\chi_i d\chi_j g_\alpha D_\alpha [f_G] = 0, \tag{2.14}$$

where g_α ($\alpha = 0, 1, \ldots, N_c$) has an algebraic form associated with operator D_α, which includes derivatives of the attribute *PDF* with respect to the coordinates.

It turns out that, in the above formulation, while the statistics of the primary attribute $X(\boldsymbol{p})$ are implicit in Eq. (2.14), in many cases we do not need to solve for them. This fact has a special significance in the mathematical formulation of the *BME* mapping equations, to be discussed in Chapter 9. A few examples follow, which aim to clarify certain implementation aspects of Eq. (2.14).

TABLE 2.5 List of g_α and $\overline{g}_\alpha$ functions for X of Eq. (2.11).

α	g_α	$\overline{g}_\alpha$
0	1	1
1–4	χ_i ($i = 1,2,3,k$)	0
5–10	$\chi_i\chi_j$ ($j \geq i = 1,2,3$)	$0.2(e^{-2s_i} - 2s_i - 1)(1 - e^{-2s_j})$
11–14	$\chi_i\chi_k$ ($i = 1,2,3,k$)	$0.2(e^{-2s_i} - 2s_i - 1)(1 - e^{-2s_k})$

Example 2.11 The following stochastic ordinary differential equation represents the temporal evolution of attribute X (fixed location $\boldsymbol{s}$, which is suppressed for simplicity)

$$\frac{d}{dt} X(t) = bX(t), \tag{2.15}$$

where $b = 0.5$ is an empirical parameter, with random initial condition $X_0 \sim f_G(\chi, \boldsymbol{s}, 0) = e^{\sum_{\lambda=0}^{2} \mu_\lambda \chi^\lambda} = e^{-2113.42 + 65\chi - 0.5\chi^2}$, where $\overline{X_0} = 65$, $\sigma_0^2 = 1$, $\mu_0 = -\frac{1}{2}\left(\frac{\overline{X_0}^2}{\sigma_0^2} + \ln 2\pi\sigma_0^2\right) = 2113.42$, $\mu_1 = \frac{\overline{X_0}}{\sigma_0^2} = 65$, and $\mu_2 = -\frac{1}{2\sigma_0^2} = -0.5$.

Step 1: Eq. (2.15) and related auxiliary conditions above constitute the *G-KB*, in this case.

Step 2: The *G-KB* leads to the following equations expressed in the form of Eq. (2.14),

$$\begin{aligned} &2\lambda^{-1} \frac{d}{dt} \overline{X^\lambda(t)} = \overline{X^\lambda(t)}, \text{or} \\ &\int d\chi \chi^\lambda \left(2\lambda^{-1} \frac{d}{dt} - 1\right) f_G(\chi, t) = 0 \end{aligned} \tag{2.16a,b}$$

($\lambda = 1, 2$) for all times t; and

$$b^{-2}\frac{\partial^2}{\partial t_i \partial t_j} c_X\left(s, t_i, t_j\right) = c_X\left(t_i, t_j\right), \text{or}$$

$$\iint d\chi_i d\chi_j \left(\chi_i \chi_j - \overline{X}_i \overline{X}_j\right)\left(4\frac{\partial^2}{\partial t_i \partial t_j} - 1\right) f_G\left(\chi_i, \chi_j, t_i, t_j\right) = 0, \tag{2.17a,b}$$

where $i, j = 1, 2, 3, k$.

Step 3: The g_α functions and the D_α operators corresponding to the *G-KB* of $X(t)$ are shown in Table 2.6.

Example 2.12 The following stochastic advection-reaction partial differential equation describes polychlorinated biphenyls (*PCB*) concentrations along the Kalamazoo River (USA) introduced by the paper industry

$$\left(\frac{\partial}{\partial t} + q\frac{\partial}{\partial s}\right) X(\boldsymbol{p}) + \kappa X(\boldsymbol{p}) = 0, \tag{2.18}$$

where q (km/h) denotes the downstream velocity and κ the reaction rate ($1/h$). The boundary/initial condition $X_0 = X(\boldsymbol{p} = \boldsymbol{0})$ is a Gaussian random field with mean $\overline{X_0}$ and variance σ_0^2.

Step 1: Eq. (2.18) and related boundary/initial condition constitute the *G-KB* in this case.

Step 2: This *G-KB* leads to the following integrodifferential equations, again in the general form of Eq. (2.14) (Kolovos et al., 2002).

$$\begin{aligned}
&\iint d\chi_i d\chi_j f_G\left(\chi_i, \chi_j\right) = 1 \\
&\iint d\chi_i d\chi_j \chi_i \left(\frac{\partial}{\partial t_i} + q\frac{\partial}{\partial s_i} + \kappa\right) f_G\left(\chi_i, \chi_j\right) = 0 \\
&\iint d\chi_i d\chi_j \chi_i^2 \left(\frac{\partial}{\partial t_i} + q\frac{\partial}{\partial s_i} + 2\kappa\right) f_G\left(\chi_i, \chi_j\right) = 0 \\
&\iint d\chi_i d\chi_j \chi_j \left(\frac{\partial}{\partial t_j} + q\frac{\partial}{\partial s_j} + \kappa\right) f_G\left(\chi_i, \chi_j\right) = 0 \\
&\iint d\chi_i d\chi_j \chi_j^2 \left(\frac{\partial}{\partial t_j} + q\frac{\partial}{\partial s_j} + 2\kappa\right) f_G\left(\chi_i, \chi_j\right) = 0 \\
&\iint d\chi_i d\chi_j \chi_i \chi_j \left(\frac{\partial}{\partial t_i} + q\frac{\partial}{\partial s_i} + \kappa\right) f_G\left(\chi_i, \chi_j\right) = 0
\end{aligned} \tag{2.19a-f}$$

for all points $\boldsymbol{p}_i, \boldsymbol{p}_j$ of interest ($i, j = 1, \ldots, m, k$).

Step 3: The g_α functions and the D_α operators corresponding to the *G-KB* are presented in Table 2.7.

BME's remarkable ability to incorporate physical laws and empirical models of various types distinguishes it from most other geostatistical techniques. It is worth noticing that the formal complexity of the laws and models occurring in natural sciences varies considerably. In particular, *BME*'s incorporation of mechanistic models (in the form of differential equations, such as above) may be seen as an attempt to integrate explanatory variables in a

TABLE 2.6 The g_α functions and D_α operators for Eqs. (2.16b)–(2.17b).

α	g_α	D_α
0	1	1
1–4	χ_i ($i = 1,2,3,k$)	$2\frac{d}{dt} - 1$
5–8	χ_i^2 ($i = 1,2,3,k$)	$\frac{d}{dt} - 1$
9–11	$\chi_i\chi_j - \overline{X}_i\overline{X}_j$ ($j \geq i = 1,2,3$)	$4\frac{\partial^2}{\partial t_i \partial t_j} - 1$
12–14	$\chi_i\chi_k - \overline{X}_i\overline{X}_k$ ($i = 1,2,3$)	$4\frac{\partial^2}{\partial t_i \partial t_k} - 1$

TABLE 2.7 List of g_α functions and the D_α operators Eqs. (2.19a-c).

α	g_α	D_α
0	1	1
1 to (m + 1)	χ_i ($i = 1,\ldots,m,k$)	$\frac{\partial}{\partial t_i} + q\frac{\partial}{\partial s_i} + \kappa$
(m + 2) to (2 m + 2)	χ_i^2 ($i = 1,\ldots,m,k$)	$\frac{\partial}{\partial t_i} + q\frac{\partial}{\partial s_i} + 2\kappa$
(2 m + 3) to $\frac{(m+4)(m+1)}{2}$	$\chi_i\chi_j$ ($i \neq j = 1,\ldots,m, k$)	$\frac{\partial}{\partial t_i} + q\frac{\partial}{\partial s_j} + \kappa$

chronotopologic process. On the other hand, it is a matter of *de facto* convenience that a simpler model is often favored by those *CTDA* practitioners who consider that returns diminish quickly with increasing model complexity.

Putting the above considerations together, the investigators should carefully assess the relative evidential value of empirical *vs*. theoretical constructions in the light of two facts:

① The ever-present need to replicate basic empirical or experimental results, see the **specificatory** *KB* in the following subsection.
② The fact that a **theory** often uses advanced methods and usually covers a wide range of situations, whereas an **empirical** model looks for simple solutions and usually covers a limited range of situations.

We continue with a detailed study of case ①, i.e., the site-specific kinds of knowledge the investigators encounter in the real world.

2.2 Specificatory *KB*

As noted earlier, the second major kind of knowledge considered by *CTDA* is tailored to the specific features of the case study considered.

> The **specificatory** ***S-KB*** is targeted knowledge about the specific situation that may include both external or demonstrative evidence (actual measurements, perceptual or data of sense, etc.) and internal or inductive evidence (inferring one thing from another thing, empirical propositions, expertise with similar situations).

Admittedly, the main constituent of the *S-KB* is data, which constitutes not only the materials to be studied but, to a certain extent, the instruments with which investigators do their probing of Nature.

Example 2.13 For illustration, the following are two representative cases of specificatory evidence.

- The statement "the average value of the West Lyon field porosity data collected in West-Central Kansas is 11.3%" is a typical site-specific statement.
- A synthesis of specificatory historical data is discussed in Hunt et al. (2009), including annual runoff values obtained from the U.S. Geological Survey for gauging stations along Dry Creek and the Schubert University of California experimental field site, long-term precipitation records from nearby rain gauges obtained from the National Climatic Data Center, and annual evapotranspiration values obtained by independent measurement from atmospheric sensors deployed over an oak savannah ecosystem at the AmeriFlux Tonzi Ranch tower.

Naturally, several of the specificatory datasets considered in a real-world study are associated with attributes that are physically closely related to each other.

- The term **synergistic** datasets refers to distinct datasets, which, though, are related to each other physically.

The timely consideration of synergistic datasets in *CTDA* can have considerable benefits. The benefit of this kind of dataset is greater when one is dealing with accurate data, but it decreases as measurement error increases (the benefit will likely diminish for relatively large measurement errors).

Example 2.14 Synergistic weather forecasting involves the processing of surface pressure, thickness, and wind datasets together.

As it turns out, in the majority of *CTDA* applications those synergistic pieces of information consisting of hard and soft datasets to be described next play dominant roles. The issue here is to tie these two kinds of datasets together in theory, as they are tied together in practice.

2.2.1 *Hard and soft datasets*

For practical purposes, the following site-specific knowledge distinction is usually adopted in *CTDA* studies.

The dataset χ_d available as specificatory *S-KB* is formally expressed as

$$\chi_d = (\chi_h, \chi_s) = (\chi_1, \ldots, \chi_m), \quad (2.20)$$

where:

χ_h denotes **hard** data obtained from real-time observation devices with the probability of the attribute value in each datum being 1; and.

χ_s denotes **soft** data that include nonnegligible uncertainty (imprecision or vagueness) and may be expressed as

$$\begin{aligned} P_S[\Phi(x_s) \in I] &= F_S(I, \Phi) \\ &= \int_I d\Phi(\chi_s) f_S(\Phi(\chi_s)) \\ &= \int_I d\Phi(\chi_s) f_\Phi(\chi_s), \end{aligned} \quad (2.21)$$

where I is an interval of possible values, Φ is an empirical model, and f_S a PDF constructed on the basis of the *S*-KB.

Some examples of soft data are shown in Table 2.8. These soft data represent varying levels of understanding of uncertain observations leading to the direct calculation of the probabilities or to their indirect estimation from accumulated experience.

Both *KB* (general *G* and specificatory *S*) will be used in the *BME* theoretical construct leading to the chronotopologic map of the phenomenon (Chapter 9). These two *KB* must mesh in coherent interaction with the new information provided by the map in order to provide the investigator with an explanatory rationale for the phenomenon. Generally, when deriving the $g_\alpha(\chi_{map})$ and $f_S(\Phi)$ for a specific application, the investigator may want to look into the judgments of those who have the right expertise and experience and, in the process, establish some fruitful interactions.

TABLE 2.8 Examples of $\Phi, f_S(\Phi)$ and $F_S(I, \Phi)$.

Φ	$f_S(\Phi)$	$F_S(I, \Phi)$
χ_s	1	I^{-1}
χ_s	$f_S(\chi_s)$	$\int_I d\chi_s f_S(\chi_s)$

2.2.2 *The protean structure of knowledge*

Based on the above considerations, the readers would realize that the knowledge considered in *CTDA* is a rather protean notion, which, like *Proteus*,[g] can be represented in diverse ways (physical laws, empirical models, auxiliary information, hard and soft data), witnessing its richness. In the *CTDA* context, one knowledge representation can often effectively balance the lack of another representation. There seems no reason to doubt that important advances in *CTDA* will depend on adequately accounting for this protean diversity and take advantage of it.

3 Integrating lawful and dataful statistics

Sophisticated use of mathematics in *CTDA* should immerse the mathematical equations into the core and site-specific *KB* available, both at the modeling and the interpretation levels. When we give this view its due weight, we realize something crucial:

A mathematical model used in *CTDA* should be the best possible technically, and, also, it should be sound **epistemically**: (*a*) the processing of the *KB* objectively available is made by means of logically plausible rules, and (*b*) the updated knowledge is derived from coherent inferences.

[g] The mythical sea-god, who could take various shapes and forms.

Accordingly, in this section we find it appropriate to start with a discussion of certain further aspects of the epistemic paradigm underlying *BME* developments, before we present the formal analytical *BME* expressions and their implementation in the real world.

3.1 The epistemic paradigm: Cassandra and Pollyanna

Just as any other product of scientific reasoning and inquiry, *BME* relies on a sound *epistemic paradigm*, which introduces certain desiderata as regards the subsequent formal analysis. In particular (Christakos, 2000):

- *BME*'s claim is that understanding **epistemology** can enlighten considerably our mathematical investigations for the best *CTDA* approach possible.

This is a desideratum that admittedly leaves one with the following fundamental question: What should be the optimal *epistemic features* of a chronotopologic investigation?

From a broad philosophical standpoint, there could be more than one possible answer to this question, including the answer that it is in some cases unanswerable. But, where there is a *Cassandra* there is a *Pollyana*. Therefore, from a *CTDA* perspective, the epistemic *BME* paradigm is based on considerations that most philosophers of science would find acceptable, as follows:

- Along with the dominant tradition, the *BME* paradigm is regarded as an **evaluative** or **normative** paradigm, not a purely descriptive one, which distinguishes between three main stages of knowledge gaining, processing, and interpretation, as presented in Table 3.1.[h]

3.1.1 Main BME stages

As is described in Table 3.1, there are three distinct *BME* stages, each one with its own epistemic objectives and logical reasoning.

❶ *G*-based stage: There is an inverse relation between prior information and prior probability: the more vague and general a theory is, the more alternatives it includes (and it is, hence, more probable), but also the less informative it is.

Knowledge at the *G*-based stage comes epistemically though not necessarily genetically before observational experience. High prior information relative to *G* is sought, which is achieved in the stochastic setting by maximizing the expected value of the information measure, $\overline{\text{Info}_G}$, subject to *G*. A simple intuitive example of this idea is presented next (Christakos, 2000).

Example 3.1 Consider two competing weather forecasting theories. Weather forecasting theory *A* predicts that tomorrow it will either snow, or rain, or be cloudy (but not rain), or be sunny. Weather forecasting theory *B* predicts that tomorrow it will either rain or it will be cloudy (but not rain). Theory *A* is a very general theory that includes several possible alternatives. Hence, while it has a high probability $P[A]$ of turning out to be true, it is not a particularly informative theory, because it is incapable of discriminating among alternatives. Theory *B*, on the other hand, includes only two alternatives and, thus, the probability of being true is smaller, $P[B] < P[A]$. Since, however, it is capable of reducing the alternatives to only two, theory *B* is more informative than *A*.

❷ *S*-based stage: The *S-KB* is collected and organized in appropriate quantitative forms that can be explicitly incorporated into the *BME* formalism.

[h] Each stage is discussed in considerable detail in the following sections.

TABLE 3.1 Main *BME* stages.

Stage	Description	Formulation
Prior (*G*-based)	*BME* does not work in an intellectual vacuum, but it starts with a basic set of epistemic postulates in light of the *G*-KB	$\text{Info}_G = \log P_G^{-1}$ (3.1) $\left.\begin{array}{l}\max_{P_G} \overline{\text{Info}_G} \text{ subject to } G \\ \therefore\ P_G\end{array}\right\}$ (3.2)
Meta-prior or Pre-posterior (*S*-based)	The specificatory *KB* is gathered	*S* (Table 2.8 of Section 2)
Posterior (*K*-based)	Results from prior stage and knowledge from meta-prior stage are processed by means of inference rules to produce the attribute chronotopologic representation	$\left.\begin{array}{l}\text{Prob}_K[\chi_k] = \text{Prob}_G[\chi_k \mid \chi_d(S)] \\ K = G \cup S\end{array}\right\}$ (3.3)

The *S* knowledge has the characteristic that it is born pragmatically from experience and may change with use.

❸ *K*-based stage: This is a **synthesis** stage in which the *K*-based *posterior* probability is related to the *G*-based probability by means of a conditional probability knowledge-processing rule that draws inferences consistent with the available knowledge body and scientific reasoning.

This rule follows reason and evidence, even if, on occasion, it disagrees with a deep-stated belief system. In view of the stages ❶–❸, the following conclusion can be drawn.

> According to the *BME* perspective, the *G-KB* introduces **scientific consistency** into the *S-KB*, and the *S-KB* introduces **empirical conditioning** into *G-KB* in light of the specific site considered.

Interestingly, the above conclusion may be linked to the *verifiability* criterion used in sciences to constrain theoretical constructions by means of empirical evidence. With regard to material object claims, this criterion is known as "phenomenalism," and with regard to the theoretical claims of science, as "operationalism."

Whether a certain body of knowledge should be considered at the prior stage or at the pre-posterior stage is an issue of debate among science methodologists. Various possibilities may exist, depending on the situation, but one thing is for sure:

- We always have **prior** knowledge, but at different times we treat different knowledge as prior.

Example 3.2 A fascinating case of prior *vs.* pre-posterior knowledge was introduced by the discovery of non-Euclidean geometries. From a logical standpoint there is no a priori means of deciding which kind of geometry does in fact represent the relations between natural attributes in a particular situation. It is, thus, necessary to appeal to specificatory evidence (experimentation, etc.) to find out whether the question of geometry could be settled a posteriori.

There seems no reason to doubt that the prior-posterior knowledge distinction is an important issue that allows a certain amount of flexibility in the application of the *BME* approach. In the process, an important strand tying both the prior and posterior *KB* together is revealed.

3.2 Analytical expressions of the *BME* stages

Our discussion so far has provided a general conceptual framework of the main concepts and theses of knowledge processing. Particular analytical solutions can be obtained from the three

TABLE 3.2 Main *BME* stages where G and *S* are expressed by $\overline{g}_a$ and Φ; and $a = 0, 1, \ldots, N_c$.

Stage	Formulation
Prior (*G*-based)	$\max_{f_G}\left[\overline{\mathrm{Info}_G(\chi_{map})} = \overline{\log f_G^{-1}(\chi_{map})} = -\overline{\log f_G(\chi_{map})}\right]$ (3.4)
	subject to $\overline{g_a} = \int d\chi_{map} g_a f_G(\chi_{map})$ (3.5)
	$\therefore f_G(\chi_{map}) = e^{\sum_{\alpha=0}^{Nc}\mu_\alpha g}{}_\alpha = e^{\boldsymbol{\mu}^T \mathbf{g}}$ (3.6)
Meta-prior or Pre-posterior (*S*-based)	$\chi_d = \{\chi_h, \chi_s\}, \Phi, f_S$ (3.7)
Posterior[a] (*K*-based)	$f_K(\chi_k) = A^{-1}\int d\Phi f_S(\Phi) f_G(\chi_{map})$ (3.8) $A = \int d\Phi f_S(\Phi) f_G(\chi_h, \Phi)$ (3.9) $\chi_{map} = \{\chi_d, \chi_k\}$ (3.10)

[a] *A is a normalization parameter, independent of the points of interest* $\boldsymbol{p}_k$.

stages of Table 3.1 in Section 3 in terms of the corresponding *PDF* when the *G* and *S* are expressed in terms of, respectively, the $\overline{g}_a$ functions of Eq. (3.2) and the Φ function of Table 2.8 in Section 2. This is shown in Table 3.2, which essentially outlines a process whereby symbols serve as prompts for conceptual operations and reasoning mode as well as the recruitment of a body of background knowledge. At this point we would like to bring to the readers' attention the slightly different notation selected for convenience in Table 3.2 and the following developments: Subscripts *G*, *S*, and *K* denote the *KB* considered in the construction of the corresponding probability models. Herein, for simplicity the symbols χ_d, χ_h, and χ_s will be replaced with χ_d, χ_h, and χ_s, respectively. Similarly, the notation $\Phi(\chi_s)$, $g_a(\chi_{map})$, and $\boldsymbol{\mu}_a(\boldsymbol{p}_{map})$, which explicitly indicates that these quantities are functions of χ_s, χ_{map}, and $\boldsymbol{p}_{map}$, will be replaced with Φ, g_a, and $\boldsymbol{\mu}_a$, respectively, where the functional dependence is implicitly assumed.

Eqs. (3.4)–(3.10) introduce a general setting to evaluate the prediction uncertainty of the unsampled attribute values (χ_k) in view of the evidential attribute uncertainty (χ_s). When one gives this setting its due weight, one realizes that Eqs. (3.4)–(3.10) are internally linked: the solution of the optimization problem described by Eqs. (3.4)–(3.5) is given by Eq. (3.6) in terms of the *G*-based prior *PDF* $f_G(\chi_{map})$, where the associated Lagrange multipliers are functions of $\boldsymbol{p}_{map}$, i.e., $\boldsymbol{\mu}_a(\boldsymbol{p}_{map})$; and this solution is used in Eq. (3.8) to obtain the *K*-based posterior *PDF*. In particular, the coefficients μ_a ($a = 0, 1, \ldots, N_c$) in Table 3.2 are derived as the solutions of the system of Eqs. (3.4)–(3.5), where $f_G(\chi_{map})$ is given by Eq. (3.6).

The *BME* equations in Table 3.2 offer a concise and general description of the chronotopologic attribute phenomenon. Posterior information is defined as

$$\mathrm{Info}_K[\chi_k] = -\log f_K(\chi_k), \tag{3.11}$$

where there is an interesting intuitive implication of the last equation:

- The **observation** of χ_d lowers the number of possible realizations of χ_k.

The posterior stage of the *BME* approach introduces an important strand tying the core (prior) and the specificatory (preposterior) perspectives of the world together. Specifically, we consider the rather common case with

$$\begin{aligned}\chi_d &= \{x_h = \chi_h, x_s \sim f_S(\Phi)\},\\ \chi_{map} &= \{x_k \leq \chi_k, \chi_d\},\end{aligned} \tag{3.12a,b}$$

in which case the basic probabilistic expressions upon which the derivations at the posterior stage of the *BME* approach rely are

$$P_G\left[\chi_{map}\right] = \int_{-\infty}^{\chi_k} d\chi'_k \int d\Phi f_G(\chi'_k, \chi_h, \chi_s) f_S(\Phi), \tag{3.13}$$

$$P_G[\chi_d] = \int d\chi_s d\Phi f_G(\chi_h, \chi_s) f_S(\Phi) \equiv A, \tag{3.14}$$

and

$$P_K[x_k \leq \chi_k] = P_G[x_k \leq \chi_k | \chi_d] = \frac{P_G\left[\chi_{map}\right]}{P_G[\chi_d]}, \tag{3.15}$$

so that the combination of the above equations leads to Eq. (3.8) of Table 3.2.[i]

Just as certain rules tell the investigators how to measure distances or how to weigh objects, Eq. (3.15) reveals to them how to **update** their stochastic evaluation of a situation given new knowledge.

Particularly, Eq. (3.15) connects the new, *K*-based probability function at the posterior stage $P_K[x_k \leq \chi_k]$ with the old, *G*-based probability function $P_G[x_{map}]$. One observes that in terms of *K*, the probability $P_K[x_k \leq \chi_k]$ is a monadic function of χ_k (because its value depends on one thing—the occurrence of $x_k \leq \chi_k$). But in terms of *G*, the conditional probability $P_G[x_k \leq \chi_k | \chi_d]$ is a dyadic function of χ_k and χ_d (because its value depends on two related things—the occurrences of $x_k \leq \chi_k$ and χ_d). The *BME* $f_K(\chi_k)$ of Eq. (3.8) is functionally different from the direct formulation of the standard conditional (statistical) *PDF* defined in terms of the ratio of the joint *PDF* of x_k and x_d over the *PDF* of x_d.

Looking at special cases of a general mathematical framework often offers valuable insight into its intricacies.

Example 3.3 Three specific cases in Table 3.2 are worth further elaborating:

- When only hard data is considered, i.e., $\chi_d \equiv \chi_h$, the *BME* conditional probability $P_K[x_k \leq \chi_k]$ of Eq. (3.15) reduces to the classical conditional $P_G[x_k \leq \chi_k | \chi_h] = \frac{P_G[\chi_{map}]}{P_G[\chi_h]}$, which, in terms of the random field complementarity idea, may be given the following interpretation: If χ_d does indeed entail χ_k, then χ_{map} occurs in every random field *X*-realization in which χ_d occurs. But if χ_k is neither entailed by χ_d nor inconsistent with it, then χ_{map} occurs only in some of the possible *X*-realizations in which χ_d occurs (i.e., in those realizations in which χ_k also happen to occur). Therefore, one may take the ratio of the quantity of possible random field realizations in which χ_{map} occurs to the number of realizations in which χ_d occurs as determining the extent to which χ_d entails χ_k and, thus, defining the probability of χ_k given χ_d, namely, $P_G[x_k \leq \chi_k | \chi_d]$.
- Another commonly encountered case is when the hard data vector $\boldsymbol{x}_h = [x_{h1} \cdots x_{hn_h}]^{\mathrm{T}}$ at the point subset $\boldsymbol{p}_h$ is combined with the soft data vector $\boldsymbol{x}_s = [x_{s1} \cdots x_{sn_s}]^{\mathrm{T}}$ at the point subset $\boldsymbol{p}_s$ consisting of uncertain observations of the *PDF* form
$$f_S(\Phi) = [f_S(x_{s1}) \cdots f_S(x_{sn_s})]^{\mathrm{T}} \tag{3.16}$$
(Table 2.8 of Section 2). In this case, Eq. (3.8) yields the *BME* posterior *PDF* as (Christakos, 2000)
$$f_K(\chi_k) = A^{-1} \int_{-\infty}^{+\infty} d\chi_s f_G\left(\chi_{map}\right) f_S(\chi_s), \tag{3.17}$$

[i] More details concerning the derivation of these equations are presented later.

where $A = \int_{-\infty}^{+\infty} d\chi_s f_G(\chi_d) f_S(\chi_s)$, and the $d\Phi(\chi_s)$ and $f_S(\Phi(\chi_s))$ reduce to $d\chi_s$ and $f_S(\chi_s)$, respectively. The posterior *PDF* f_K fully characterizes the uncertainty of $x_k = [x_{k1} \cdots x_{kn_k}]^T$ at $p_k = [p_{k1} \cdots p_{kn_k}]^T$. Clearly, if no soft data χ_s is available at points p_s, i.e., $\chi_d \equiv \chi_h$, then $A = f_G(\chi_h)$, and Eq. (3.17) reduces to $f_K(\chi_k) = f_G(\chi_k | \chi_h)$.

- Another case of special interest is that in which one encounters observations that are uncertain yet valuable because they are available at the points p_k of interest themselves. In this case, Eq. (3.8) yields the *BME* posterior *PDF* as (Christakos et al., 2002)

$$f_K(\chi_k) = A^{-1} \int_{-\infty}^{+\infty} d\chi_s f_G\left(\chi_{map}\right) f_S(\chi_s) f_{S_k}(\chi_k), \quad (3.18)$$

where $f_{S_k}(\chi_k)$ denotes soft information at points p_k.

∴ As was expected, the $f_K(\chi_k)$ in the *BME* setting of Eqs. (3.17) and (3.18) are different than the standard conditional setting $f_G(\chi_k | \chi_d) = f_G^{-1}(\chi_d) f_G(\chi_{map})$.

3.2.1 A fusion of the empiricical and rational perspectives

As should be evident from the above considerations, the *BME* approach includes elements from both the *empiricist* and the *rationalist* traditions—just as scientific reasoning combines empiricist and rationalist approaches into an inseparable method. The *G*-stage is based on rationalism, since it involves conceptualization and logic, whereas the *S*-stage is based on empiricism since it relies on evidential support and verification. That is, one needs to start with a rational argument in *G* and then consider relevant observations in *S*.

Hence, in the *BME* context, rationalism and empiricism work together, since each one of them possesses elements essential to a workable solution. The rationalism of the *G*-stage recognizes the importance of experimentation and observation as the means by which reality penetrates one's understanding of a phenomenon, while the empiricism of the *S*-stage uses rational reasoning when observing, selecting, and processing data.

> Rationalism and empiricism can be integrated in *BME*, assuming that they are given their proper domains, in which **rationalism** produces structured theories and relational models and **empiricism** provides evidential support to these theories and models.

Indeed, without empirical knowledge (*S*), rational knowledge (*G*) may be of limited practical use, while without a rational system, empirical knowledge is unordered and of limited predictability value.

3.2.2 Logical and epistemic distinctions

In such a context, the distinction between prior and preposterior stages is not meant so much in a temporal sense. Rather the real meaning is that the boundary line between prior and posterior stages coincides with the boundary line between core *KB* and specificatory *KB*. Yet, these distinctions are of different nature.

> While the distinction between the two *KB* (*G vs. S*) is a matter of logic, the distinction between the two *BME* stages (prior *vs.* posterior) is an epistemic matter.

That is to say, the prior stage processes knowledge that has been characterized as general on the basis of some logical arguments, and the posterior stage involves knowledge, which—on the basis of the same logical arguments—was considered specificatory. Along these lines of thought, understanding the *BME* equations is not limited to connecting the symbols to natural attributes and being able to perform the operations in these equations, but an equally important task is to connect the mathematical operations in the *BME* equations to their physical meaning and to blend these equations with their implications in the real world.

3.2.3 *Analytical derivations*

For the interested readers, we briefly discuss some derivations and clarifications related to the *BME* equations of Table 3.2.

The *G-KB* is expressed in terms of the system of statistical Eqs. (3.5), also termed the *G-based constraints*. Since $g_0 = 1$, the *PDF* normalization constraint is a special case of Eq. (3.5) for $a = 0$, i.e., $\overline{g_0(\boldsymbol{x}_{map})} = 1 = \int d\chi_{map} f_G(\chi_{map})$. Using the Lagrange multipliers method (Rockafellar, 1993), the prior information maximization, Eq. (3.4), leads to the objective function.

$$\overline{\text{Info}_G(\boldsymbol{x}_{map})} = -\int d\chi_{map} f_G(\chi_{map}) \log f_G(\chi_{map}) + \sum_{a=0}^{N_C} \mu_a \left[\int d\chi_{map} f_G(\chi_{map}) g_a - \overline{g_a}\right], \tag{3.19}$$

where μ_a are unknown Lagrange multipliers. Due to the convexity of Eq. (3.19), the functional $\overline{\text{Info}_G(\boldsymbol{x}_{map})}$ is maximized with respect to f_G, i.e.,

$$\frac{d}{df_G}\overline{\text{Info}_G(\boldsymbol{x}_{map})} = -\log f_G(\chi_{map}) + \sum_{a=0}^{N_C} \mu_a g_a(\chi_{map}) = 0. \tag{3.20}$$

Based on Eq. (3.19), the *G*-based *PDF* is expressed as in Eqs. (3.6) and (3.9), where $A_0 = e^{-\mu_0}$ is a normalization constant. Eq. (3.6) is subject to the constraints of Eq. (3.5). In order to determine the Lagrange multipliers in Eq. (3.6), f_G is substituted into Eq. (3.5) to obtain the system

$$\overline{g_a(\boldsymbol{x}_{map})} = \int d\chi_{\text{map}} g_a(\chi_{map}) e^{\mu_0 + \sum_{\alpha=1}^{N_c} \mu_\alpha g_\alpha(\chi_{map})}, \tag{3.21}$$

$a = 0, 1, \cdots, N_C$. Eqs. (3.21) constitute a closed system that has a unique solution for μ_a (the number of unknowns $(N_C + 1)$ is the same with the number of equations). That is, the *PDF* is uniquely determined with respect to the *G-KB*. Eq. (3.15) yields[j]

$$f_K(\boldsymbol{x}_k) = F_G^{-1}(\chi_d) \frac{\partial}{\partial \chi_k} F_G(\chi_{map}), \tag{3.22}$$

where the $F_G(\chi_{map})$ and $F_G(\chi_d) = A$ are given by Eqs. (3.13) and (3.14), respectively. Lastly, since

$$\frac{\partial}{\partial \chi_k} F_G(\chi_{map}) = \frac{\partial}{\partial \chi_k} \int_{-\infty}^{\chi_k} d\chi'_k \int d\Phi f_S(\Phi) f_G(\chi'_k, \chi_d) = \int d\Phi f_S(\Phi) f_G(\chi_{map}), \tag{3.23}$$

Eq. (3.22) leads to the *K*-based *PDF* of Eq. (3.8) of Table 3.2.

3.2.4 More practical issues

Certain practical issues exist concerning the derivation of the prior and posterior *PDF* in chronotopologic environments. Naturally, the various computerized versions of the *BME* equations available take advantage of the computer's dual ability to store large amounts of various forms of knowledge and to process this knowledge in obedience to the logical procedures of the *BME* approach. In particular, to determine the parameters of the prior *PDF* f_G, one needs to solve the closed system of Eqs. (3.4)–(3.5).

Because this is a nonlinear system, its analytical solution is not generally feasible. Instead, a solution is commonly obtained using numerical algorithms to calculate the f_G parameters, including the following:

① **Multidimensional root-finding** algorithms (e.g., Newton-Raphson technique), which have been widely used in *BME* studies (Kolovos et al., 2002).

② **Maximum likelihood** algorithms, which view the Lagrange multiplier solution as an

[j] Notice the different subscripts, *K vs. G*, of the *PDF*s.

unconstrained optimization problem (Aitchison and Silvey, 1958).

③ **Iterative scaling** algorithms, including generalized iterative scaling (*GIS*, Darroch and Ratcliff, 1972) and improved iterative scaling (*IIS*, Della et al., 1995), which sequentially update each parameter until convergence is reached.

As usually happens in such cases, the selection of the most suitable numerical algorithm is application-dependent and a matter of experience with computational *BME* implementation.

The following examples present the *step-by-step* process of obtaining the *BME G*-based *PDF* $f_G(\chi_{map})$ and *K*-based *PDF* $f_K(\chi_k)$ *in actu*.

Example 3.4 This example revisits the $PM_{2.5}$ study of Example 1.1 of Section 1 and Example 2.9 of Section 2 above. The study involves the points $\boldsymbol{p}_1 = (10, -30; 3)$, $\boldsymbol{p}_2 = (10, 50; 1)$, $\boldsymbol{p}_3 = (40, 10; 2)$ and $\boldsymbol{p}_k = (10, 10; 2)$ in the space-time domain (Fig. 1.1). The corresponding space distances and time separations are $h_{12} = 80$ km, $h_{13} = h_{23} = 50$ km, $h_{1k} = h_{2k} = 40$ km, $h_{3k} = 30$ km; and $\tau_{12} = 2$ d, $\tau_{13} = \tau_{1k} = \tau_{23} = \tau_{2k} = 1$ d, $\tau_{3k} = 0$ d.

Step 1: The *G-KB* consists of the $PM_{2.5}$ mean and the covariance models of Eqs. (2.3a,b) of Section 2.

Step 2: The *S-KB* consists of the hard $PM_{2.5}$ data points $\boldsymbol{p}_1$ and $\boldsymbol{p}_2$ and the soft $PM_{2.5}$ datum point $\boldsymbol{p}_3$, i.e.,

$$\begin{aligned} \chi_1 &= 70, \\ \chi_2 &= 60, \\ \chi_3 &\sim U(62, 65), \end{aligned} \tag{3.24a-c}$$

in $\mu g/m^3$ units, where the χ_3 follows a uniform $PM_{2.5}$ probability distribution U with lower and upper limits 62 and 65 $\mu g/m^3$, respectively. No datum exists at point $\boldsymbol{p}_k$.

Step 3: The functions g_α and the associated expectations $\overline{g}_\alpha$ of the *G-KB* are listed in Table 3.3, $\alpha = 0, 1, \ldots, N_c = 14$, as in the previous examples.

Step 4: The 14 + 1 Lagrange multipliers μ_0, μ_1, μ_2, μ_3, μ_k, μ_{11}, μ_{22} μ_{33}, μ_{kk}, $\mu_{k1} = \mu_{1k}$, $\mu_{k2} = \mu_{2k}$, $\mu_{k3} = \mu_{3k}$, $\mu_{12} = \mu_{21}$, $\mu_{13} = \mu_{31}$, $\mu_{23} = \mu_{32}$, are solutions of a system of 15 equations with 15 unknowns:

$$\begin{aligned} &\overline{X}_i = 65 = A^{-1}\int d\chi_1 d\chi_2 d\chi_3 d\chi_k \chi_i e^{Q(\mu_i,\chi_i)}, i = 1,2,3,k, \\ &\sigma_i^2 = 1 = A^{-1}\int d\chi_1 d\chi_2 d\chi_3 d\chi_k (\chi_i - 65)^2 e^{Q(\mu_i,\chi_i)}, i = 1,2,3,k, \\ &c_{ij} = A^{-1}\int d\chi_1 d\chi_2 d\chi_3 d\chi_k (\chi_i - 65) \\ &\quad \times \left(\chi_j - 65\right) e^{Q(\mu_i,\chi_i)}, i \neq j = 1,2,3,k, \\ &\int d\chi_1 d\chi_2 d\chi_3 d\chi_k e^{Q(\mu_i,\chi_i)} = e^{-\mu_0} = A, \end{aligned} \tag{3.25a-d}$$

where

$$Q(\mu_i, \chi_i) = \sum\nolimits_{i=1,2}^{3,k} \mu_i \chi_i + \sum\nolimits_{i,j=1,2}^{3,k} \mu_{ij} (\chi_i - 65)\left(\chi_j - 65\right). \tag{3.25e}$$

In particular, the solution of Eqs. (3.25a-e) yields.

TABLE 3.3 List of g_α functions and $\overline{g}_\alpha$ expectations.

α	g_α	$\overline{g}_\alpha$
0	1	1
1–4	χ_i $(i = 1, 2, 3, k)$	$\overline{X}_i = 65$
5–8	$(\chi_i - \overline{X})^2$ $(i = 1, 2, 3, k)$	$\sigma_i^2 = 1$
9	$(\chi_1 - \overline{X})(\chi_2 - \overline{X})$	$c_{12} = c_X(80, 2) \approx 0.1009$
10	$(\chi_1 - \overline{X})(\chi_3 - \overline{X})$	$c_{13} = c_X(50, 1) \approx 0.3932$
11	$(\chi_1 - \overline{X})(\chi_k - \overline{X})$	$c_{1k} = c_X(40, 1) \approx 0.4434$
12	$(\chi_2 - \overline{X})(\chi_3 - \overline{X})$	$c_{23} = c_X(50, 1) \approx 0.3932$
13	$(\chi_2 - \overline{X})(\chi_k - \overline{X})$	$c_{2k} = c_X(40, 1) \approx 0.4434$
14	$(\chi_3 - \overline{X})(\chi_k - \overline{X})$	$c_{3k} = c_X(30, 0) \approx 0.6977$

$\mu_1 = \mu_2 = \mu_3 = \mu_k = 0$

$$\boldsymbol{\mu}_{ij} = \begin{bmatrix} \mu_{11} & \mu_{12} & \mu_{13} & \mu_{1k} \\ \mu_{21} & \mu_{22} & \mu_{23} & \mu_{2k} \\ \mu_{31} & \mu_{32} & \mu_{33} & \mu_{3k} \\ \mu_{k1} & \mu_{k2} & \mu_{k3} & \mu_{kk} \end{bmatrix} = -\frac{1}{2}\begin{bmatrix} \sigma_1^2 & c_{12} & c_{13} & c_{1k} \\ c_{21} & \sigma_2^2 & c_{23} & c_{2k} \\ c_{31} & c_{32} & \sigma_3^2 & c_{3k} \\ c_{k1} & c_{k2} & c_{k3} & \sigma_k^2 \end{bmatrix}$$
$$= \begin{bmatrix} -0.6394 & -0.0833 & 0.1269 & 0.2254 \\ -0.0833 & -0.6394 & 0.1269 & 0.2254 \\ 0.1269 & 0.1269 & -1.0188 & 0.6006 \\ 0.2254 & 0.2254 & 0.6006 & -1.1149 \end{bmatrix} \tag{3.26}$$

Step 5: The *G*-based *PDF* is

$$f_G(\chi_1, \chi_2, \chi_3, \chi_k) = e^{-29.8875 - 1.1149(\chi_k - 65)^2 + 0.6006(\chi_3 - 65)(\chi_k - 65) - 1.0188(\chi_3 - 65)^2}, \tag{3.27}$$

where from Eqs. (3.25e) and (3.26) we have

$$Q(\mu_i, \chi_i) = -29.8875 - 1.1149(\chi_k - 65)^2 + 0.6006(\chi_3 - 65)(\chi_k - 65) - 1.0188(\chi_3 - 65)^2. \tag{3.28}$$

Step 6: Lastly, the *S-KB* of Eq. (3.24a-c) is introduced so that the $K = G \cup S$-based *PDF* is

$$f_K(\chi_k) = \int_{62}^{65} d\chi_3 e^{-29.8875 - 1.1149(\chi_k - 65)^2 + 0.6006(\chi_3 - 65)(\chi_k - 65) - 1.0188(\chi_3 - 65)^2}$$
$$= \sqrt{\frac{\pi}{1.1088}} e^{-29.8875 + \left(\frac{0.3003^2}{1.0188} - 1.1149\right)(\chi_k - 65)^2} \left[\sqrt{\left(1 - e^{0.3003(\chi_k - 65)^2}\right)} - \sqrt{\left(1 - e^{3.3264 + 0.3003(\chi_k - 65)^2}\right)}\right], \tag{3.29}$$

where $A = 1$.

Example 3.5 The assumed space-time arrangement is as in the previous example.

Step 1: The *G-KB* this time consists of the $PM_{2.5}$ variogram model.

$$\gamma_X(\boldsymbol{\Delta p}) = 1 - e^{-\frac{3h}{250} - \frac{\tau^2}{3}} \tag{3.30}$$

in $(\mu g/m^3)^2$ units.

Step 2: The *S-KB* is the same as in the previous Example 3.4.

Step 3: The functions g_α and their expectations $\overline{g}_\alpha$ of the *G-KB* are shown in Table 3.4 ($\alpha = 0, 1, 2, 3$).

Step 4: The 3 + 1 Lagrange multipliers μ_α, i.e., $\mu_0, \mu_{1k}, \mu_{2k}, \mu_{3k}$, are solutions of the following system of 3 + 1 equations.

$$\gamma_{ik} = A^{-1}\int d\chi_1 d\chi_2 d\chi_3 d\chi_k \frac{1}{2}(\chi_i - \chi_K)^2 e^{Q(\mu_i, \chi_i)} \,(i = 1, 2, 3),$$
$$\int d\chi_1 d\chi_2 d\chi_3 d\chi_k e^{Q(\mu_i, \chi_i)} = e^{-\mu_0} = A, \tag{3.31a,b}$$

where

TABLE 3.4 List of g_α and $\overline{g}_\alpha$ functions.

α	g_α	$\overline{g}_\alpha$
0	1	1
1	$(\chi_1 - \chi_k)^2$	$\gamma_{1k}(40, 1) \approx 0.5566$
2	$(\chi_2 - \chi_k)^2$	$\gamma_{2k}(40, 1) \approx 0.5566$
3	$(\chi_3 - \chi_k)^2$	$\gamma_{3k}(30, 0) \approx 0.3023$

$$Q(\mu_i, \chi_i) = \frac{1}{2}\sum_{i=1}^{3} \mu_{ik}(\chi_i - \chi_k)^2. \quad (3.31c)$$

In particular, the solution of Eq. (3.31a-c) yields

$$\boldsymbol{\mu} = \begin{bmatrix} \mu_{1k} & 0 & 0 \\ 0 & \mu_{2k} & 0 \\ 0 & 0 & \mu_{3k} \end{bmatrix} = -\frac{1}{2}\begin{bmatrix} \gamma_{1k} & 0 & 0 \\ 0 & \gamma_{2k} & 0 \\ 0 & 0 & \gamma_{3k} \end{bmatrix}^{-1}$$

$$= \begin{bmatrix} -0.8983 & 0 & 0 \\ 0 & -0.8983 & 0 \\ 0 & 0 & -1.6540 \end{bmatrix} \quad (3.32)$$

Step 5: The *G*-based *PDF* is

$$f_G(\chi_1, \chi_2, \chi_3, \chi_k) = e^{-3817.775 + 117.779\chi_k - 0.8983\chi_k^2 - 0.827(\chi_3 - \chi_k)^2}, \quad (3.33)$$

where from Eqs. (3.31c) and (3.32) we have

$$Q(\mu_i, \chi_i) = \frac{1}{2}\left[\mu_{1k}(\chi_1 - \chi_k)^2 + \mu_{2k}(\chi_2 - \chi_k)^2 + \mu_{3k}(\chi_3 - \chi_k)^2\right]$$
$$= -3817.775 + 117.779\chi_k - 0.8983\chi_k^2 - 0.827(\chi_3 - \chi_k)^2. \quad (3.34)$$

Step 6: The *S-KB* of Eq. (3.24a-c) is introduced so that the $K = G \cup S$-based posterior *PDF* is

$$f_K(\chi_k) = \int_{62}^{65} d\chi_3 e^{-3817.775 + 117.779\chi_k - 0.8983\chi_k^2 - 0.827(\chi_3 - \chi_k)^2}$$
$$= \frac{1}{\sqrt{0.827}} e^{-3817.775 + 117.779\chi_k - 0.8983\chi_k^2}\left[\sqrt{\pi\left(1 - e^{-0.827(65-\chi_k)^2}\right)} - \sqrt{\pi\left(1 - e^{-0.827(62-\chi_k)^2}\right)}\right], \quad (3.35)$$

where $A = 1$.

Example 3.6 Consider Example 2.11 of Section 2. By substituting the *G*-based *PDF*.

$$f_G(\chi; t) = e^{\sum_{\lambda=0}^{2} \mu_\lambda(t)\chi^\lambda} \quad (3.36)$$

in the *BME* system of Eqs. (2.17a,b) of Section 2 we get

$$\int d\chi\chi^\lambda \left[\sum_{\beta=0}^{2} \frac{d}{dt}\mu_\beta(t)\chi^\beta\right] e^{\sum_{\eta=0}^{2} \mu_\eta(t)\chi^\eta}$$
$$= \lambda b \int d\chi\chi^\lambda e^{\sum_{\eta=0}^{2} \mu_\eta(t)\chi^\eta}, \quad (3.37)$$

$\lambda = 1, 2$, with the normalization equation $\int d\chi\, e^{\sum_{\eta=0}^{2}\mu_\eta(t)\chi^\eta} = 1$. The solution of the *BME* system of Eqs. (3.37) is $\frac{d}{dt}\frac{\mu_1}{\mu_2} = b\frac{\mu_1}{\mu_2}$ and $\frac{d}{dt}\frac{1}{\sqrt{\mu_2}} = b\frac{1}{\sqrt{\mu_2}}$, which can be also written as $\mu_1(t) = \mu_1(0)e^{bt}$ and $\mu_1(t) = \mu_2(0)e^{-2bt}$, where the initial conditions are related to the mean and variance of $X(t)$ at $t = 0$ by $\mu_1(0) = \frac{\overline{X(0)}}{\sigma_X^2(0)}$ and $\mu_2(0) = -\frac{1}{2\sigma_X^2(0)}$.

In light of our discussion so far, the following general conclusion is drawn that links the various parts together:

> In the *BME* setting, knowledge comes from a synthesis of concepts and experience that are processed in a stochastic sense, i.e., both **dataful** and **lawful** chronotopologic statistics are considered.

3.3 *BME* assets

As has been demonstrated in various parts of this book, as well as in the relevant literature, due to its versatility *BME* has certain important assets compared to the classical geostatistics approach (see He and Kolovos, 2018, and references therein):

① It satisfies sound **epistemic** ideals.
② It incorporates **physical knowledge** in a systematic manner.

③ As a result of asset ② (especially the inclusion of physical laws), *BME* exhibits a **global** attribute representation ability.
④ It restricts the shape of the attribute **probability law** on the basis of logical and physical considerations.
⑤ It accounts for chronotopologically **heterogeneous** data.
⑥ It provides several **novel** results that cannot be obtained by traditional methods.

For further illustration purposes, we briefly review a few selected *BME* assets compared to previous techniques, including some examples.

3.3.1 *The value of core knowledge*

Obviously, data analysis is, to a large extent, a data-driven process, i.e., it extracts information from the collected data. Therefore, the observed dataset will directly determine a study's results. However, a natural attribute usually obeys physical laws, principles, and relational mechanisms that constitute the attribute's core knowledge. i.e., the collected data themselves obey some inherent law, which, though, is not taken into account by classical geostatistics techniques.

> Accounting for **core** knowledge can offer critical information about essential aspects of the phenomenon, which is why, by its conception, the *BME* approach takes into consideration both core knowledge and site-specific empirical evidence.

Example 3.7 Several examples of core *KB* were discussed in the previous section. More real-world studies can be found in the *BME* literature (e.g., Kolovos et al., 2002; Lang and Christakos, 2018).

3.3.2 *The value of soft data*

To study a natural attribute, data from different sources are often collected at various scales of resolution, precision, and uncertainty. It is not easy to consider these data sources simultaneously and directly because they have different forms, and standard data analysis techniques usually require the same data form. *BME*, on the other hand, can handle different kinds of data, without necessarily making any restrictive assumptions, by properly expressing them as soft data. *Paucis verbis*:

> *BME* links different kinds of case-specific data in terms of the notion of **soft data**, which means different kinds of uncertain data are accounted for.

In the standard setting, sampling may consistently provide observations that occur frequently at specified locations or time periods, which are rather ignored to avoid repetitive patterns or misleadingly smooth behavior.

Example 3.8 To clarify certain issues concerning soft data, a few representative cases of this kind of data are discussed here.

- In *remote sensing*, the various sensors on satellites used to observe the Earth are regarded as a long distance camera, i.e., they take many photos of the earth at various locations and times. Given the unavoidable sun glint and clouds, the photos are not complete. Hence, scientists need to merge the photos taken by various sensors during different times, because the missing part of one photo taken by a certain sensor at a specific time may be found in an additional photo taken by another sensor at a different time. Also, photos taken by different sensors have different spatial resolutions, which make photo-merging a difficult matter. For this scope, Li et al. (2013) employed *BME* to merge sea surface temperature (*SST*) values from two sensors: Moderate Resolution Imaging Spectroradiometer (*MODIS*) with 4 km and 8 days spatial and temporal resolution, respectively, and Advanced Microwave Scanning Radiometer for Earth Observing System (*AMSR-E*) with 25 km and 1 day spatial and temporal resolution, respectively. Finer data can better reflect the details of spatial *SST* features. Using the soft data notion, data at coarse resolution was regarded as soft (*AMSR-E SST*), whereas data at fine resolution as hard (*MODIS SST*). An error model quantified the relationship between fine and coarse resolution (*MODIS SST* and *AMSR-E SST*, respectively), and the conditional *PDF* of *SST* at coarse resolution was obtained. Using the error model, the uncertainty caused by scale effects was effectively accounted for and the multi-resolution datasets were unified into a single scale. Then, *BME* was employed to handle the hard and soft data as site-specific knowledge for chronotopologic *SST* data analysis and mapping. More detailed information can be found in the cited literature.

- Christakos et al. (2005) studied the mortality and propagation of *Black Death* epidemic and generated 14th century *Black Death* chronotopologic maps by integrating diverse sources of information with various levels of uncertainty (such as surviving historical, civil, agricultural, health, and ecclesiastical sources).
- The histograms of daily CO and PM_{10} concentrations in Taiwan during April 2006–09 are plotted in Fig. 3.1. CO concentrations vary from 0.11 to 0.27 ppm and PM_{10} concentrations from 5 to 13 $\mu g/m^3$. It is unfair to use the mean, mode, minimum or maximum values of CO and PM_{10} to represent the pollution level during April. If we only have two PM_{10} samples with values, say, 5 and 6 $\mu g/m^3$, we cannot claim that PM_{10} in April ranges from 5 to 6 $\mu g/m^3$. A more realistic way to describe the pollution level is using a probability function derived from the weighted histogram of the daily observations (Yu and Wang, 2013), which is regarded as soft data for further *BME* analysis purposes.

3.3.3 *Auxiliary information*

Compared to the study domain, in the real world, the observed dataset is often sparse, which may result in inadequate or even inaccurate attribute description and inference. When investigators face this kind of problem, it is sometimes possible that although the number

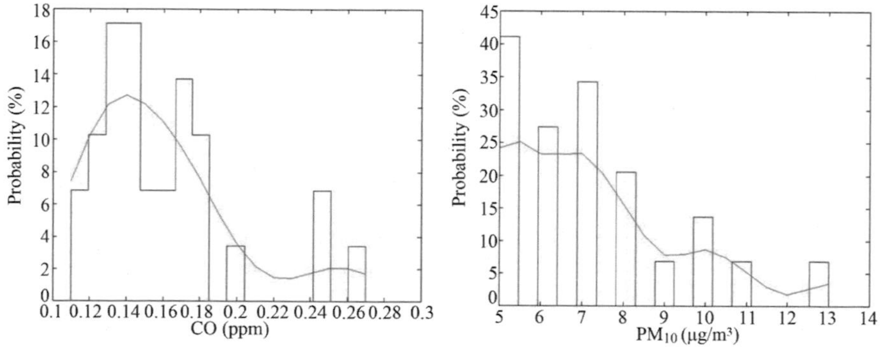

FIG. 3.1 Probabilistic form of the soft data in February for (A) CO and (B) PM_{10} during April 2006–09.

of attribute observations is limited, some relevant auxiliary variables are available across the study domain. In this case:

BME shows a significant ability in assimilating **auxiliary information**, thus essentially expanding considerably the attribute dataset.

Often, a quantitative model can be built to describe the relationship between the attribute and the associated auxiliary variables, in which case model estimates using these variables can serve as soft data. Let us examine a real case study.

Example 3.9 Fig. 3.2 shows the $PM_{2.5}$ monitoring stations in the Jing-Jin-Ji region of North China. The distribution of these stations is rather coarse compared to the entire study area. In order to improve the spatial coverage of $PM_{2.5}$ observations (hard data), some associated variables (aerosol optical depth, meteorological data, elevation, distance to the coastline, road network, land use) were collected to build an artificial neural network (*ANN*) characterizing the nonlinear relationship between $PM_{2.5}$ data and the other variables (He et al., 2018). Then, soft data points were considered to improve the spatial $PM_{2.5}$ coverage, i.e., the central part of the remote sensing (meteorologic) data, as shown in Fig. 3.2. A bootstrap method based on resampling was employed to quantify the uncertainty of the *ANN* outputs. In particular, a sufficiently large number (300) of *ANN* were trained by the bootstrap method, and at each point soft data was generated as a uniform distribution with lower and upper limits equal to the 95% confidence interval of the 300 *ANN* estimates, i.e., the *ANN* outputs served as soft data. By integrating soft with hard data, *BME* showed a better prediction accuracy and lower uncertainty compared to using only hard data.

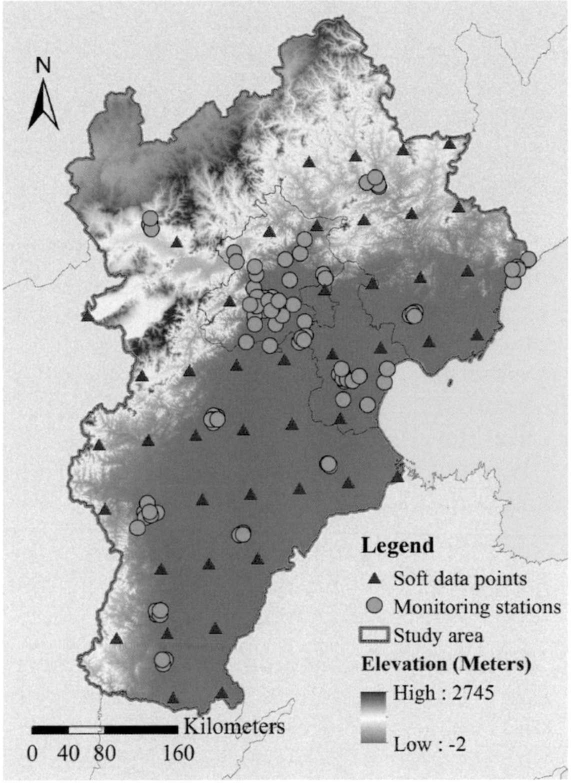

FIG. 3.2 Hard and soft $PM_{2.5}$ data locations in the Jing-Jin-Ji region in North China.

3.3.4 *Stochastic logic in CTDA*

It is not surprising, perhaps, that the statistical conditional of Eq. (3.3), $\text{Prob}_K[\chi_k] = \text{Prob}_G[\chi_k|\chi_d]$, is not the only way to derive a posterior mapping probability in modern geostatistics. Indeed, the *BME* approach can be generalized to include other kinds of updating mechanisms at the *K*-stage (Christakos et al., 2002).

Beyond conditional probability, other useful knowledge updating mechanisms can be expressed as the probability of **logical conditionals**, like the **implication** probability

$$P_G[\chi_d \to \chi_k] = P_G[\neg\chi_d \vee \chi_k] = P_G[\neg(\chi_d \wedge \neg\chi_k)], \quad (3.38)$$

and the **equivalence** probability

$$P_G[\chi_d \leftrightarrow \chi_k] = P_G[(\chi_d \wedge \chi_k) \vee (\neg\chi_d \wedge \neg\chi_k)] \quad (3.39)$$

of stochastic logic.

Eqs. (3.38)–(3.39) convey a stochastic deduction perspective. In particular, Eq. (3.38) expresses attribute value implication as the probability that

"either the data χ_d did not occur or the value χ_k occurred";

or of the argument

"it is not the case that χ_d occurred and the value χ_k did not occur."

- Eq. (3.38) is a probabilistic expression of the platitude that **valid**[k] arguments are necessarily truth preserving.

That is, the logical argument in the brackets, $\neg(\chi_d \wedge \neg\chi_k)$, is the definition of a valid or truth preserving argument necessarily: it is impossible for the premise χ_d to be true and the conclusion χ_k false simultaneously (alternatively, it must be the case that when χ_d is true, the χ_k is true).

Eq. (3.39) expresses attribute value equivalence probability as the probability that "either both values χ_d and χ_k occurred or both did not occur." The statistical formulation of Eq. (3.3) and the logical structure of Eqs. (3.38)–(3.39) reveal a deeper distinction, as follows:

- When **mathematical** logic is used in *CTDA*, because of the underlying deductive reasoning, if the premises are true the conclusions are always true. On the other hand, when **statistical inference** is used in *CTDA*, because of the underlying inductive reasoning, the premises may be true and the conclusions wrong.[l]

The logical *CTDA* conditionals above reveal that the relation of probability and deducibility involves the body of knowledge one actually has or the premises one actually accepts. In this respect, the statistical conditional probability $P_G[\chi_k|\chi_d]$ is not necessarily the probability of a logical conditional $P_G[\chi_d \to \chi_k]$ or $P_G[\chi_d \leftrightarrow \chi_k]$. In particular, the probability of the truth of the logical conditional "if data χ_d then estimate χ_k," $P_G[\chi_d \to \chi_k]$, is not necessarily equivalent to "estimate χ_k given data χ_d," $P_G[\chi_k|\chi_d]$. Indeed, the latter does not necessarily measure the probability of truth of $\chi_d \to \chi_k$ but rather its assertibility. In other words, $P_G[\chi_k|\chi_d]$ is the probability that χ_k occurs given that χ_d does, whether or not χ_d caused or gave rise to χ_k, whether or not there is any (probabilistic) subjunctive connection between χ_k and χ_d.

Despite their conceptual differences, Eq. (3.3) can be written in terms of the implication probability of Eq. (3.38) as[m]

$$P_G[\chi_k|\chi_d] = \frac{P_G[\chi_d \to \chi_k] - P_G[\neg\chi_d]}{P_G[\chi_d]}. \quad (3.40)$$

Eq. (3.40) links the conditional probability $P_G[\chi_k|\chi_d]$ and the implication probability $P_G[\chi_d \to \chi_k]$. Since $P_G[\chi_k|\chi_d] \neq P_G[\chi_d \to \chi_k]$, it implies that $(\chi_k|\chi_d) \neq (\chi_d \to \chi_k)$. Actually, some comparative properties are worth noting:

① It holds that $P_G[\chi_k|\chi_d] \leq P_G[\chi_d \to \chi_k]$, assuming that $P_G[\chi_d] \neq 0$; in which case the equality holds if and only if $P_G[\chi_d \to \chi_k] = 1$ (or, if and only if $P_G[\chi_d] = 1$ or $P_G[\chi_k|\chi_d] = 1$).

② The $P_G[\chi_k|\chi_d] = 1$ implies that $P_G[\chi_d \to \chi_k] = 1$ and vice versa assuming that $P_G[\chi_d] \neq 0$.

[k] The readers may recall that an argument is *valid* if and only if it is impossible for the premise to be true and the conclusion false, whereas an argument is *sound* if and only if it is both valid and its premise is actually true.

[l] This is why, e.g., $\chi_d \to \chi_k$ is also defined as $\neg(\chi_d \wedge \neg\chi_k)$.

[m] Similarly, Eq. (3.3) can be written in terms of the equivalence probability of Eq. (3.39), see Exercise 24.

③ If $P_G[\chi_d \to \chi_k] = 0$, then $P_G[\chi_k|\chi_d] = 0$, but $P_G[\chi_k|\chi_d] = 0$ implies only that $P_G[\chi_d \to \chi_k] < 1$.

④ Interestingly, $P_G[\chi_d \to \chi_k] = P_G[\neg\chi_k \to \neg\chi_d]$, but $P_G[\chi_k|\chi_d] \neq P_G[-\chi_d|\neg\chi_k]$.

The above stochastic logic considerations leave us with a number of questions, like the following:

What constraints do the implication and equivalence conditionals impose on *CTDA*?

How does the truth of χ_d guarantee the truth of χ_k in the logical conditionals?

Can there be more than one conclusion in a valid argument, like the implication conditional?

Occasione data, it should be stressed that stochastic logic is a promising avenue of *CTDA* research. The use of logic is as necessary as the use of perspective in painting—but only as a medium of expression, not as a criterion of reality. More details on the logical conditional probability and other kinds of stochastic logic conditionals and inferences are beyond the scope of this book, but they are discussed in detail in Christakos (2000, 2010) and Christakos et al. (2002). Certain related exercises are also included at the end of this chapter.

4 Rethinking chronotopologic dependence

As we saw in Chapter 5, in the context of random field correlation theory and classical geostatistics, the covariance function, and its mathematically equivalent yet on occasion more convenient tool in practice, the variogram function, have been widely used to characterize quantitatively the chronotopologic dependence (variation) of a natural attribute and generate attribute maps in earth and atmospheric sciences, ecology, natural resources, and public health (Monin and Yaglom, 1971; Bogaert, 1996; Castillo et al., 1996; Meyers et al., 1991; Olea, 1999; Hristopulos, 2002; Verfaillie et al., 2006; Ramos and Abreu, 2011; Liu et al., 2015; Glover et al. 2018).

4.1 Problems with tradition

Yet, despite its great successes in *CTDA* applications the covariance function (and, consequently, the variogram function) has some well-known limitations:

① **Indiscrimination**: the same covariance model may represent very different variation patterns.

② **Permissibility**: the covariance function must satisfy certain strict mathematical conditions that may be difficult to implement.

③ **Underevaluation**: direct covariance representation may seriously underestimate or even neglect strong physical correlations.

Discussions and specific examples concerning the limitations ①–③ can be found in Christakos (2010, 2017) and He et al. (2019a, b). As an illustration, consider the following example.

Example 4.1 Assume that the attribute distribution obeys an empirical law of the form $X(\boldsymbol{p}_j) = X^2(\boldsymbol{p}_i)$, $j > i$, where $X(\boldsymbol{p}_i)$ is a zero mean normal (Gaussian) *S/TRF*, $X(\boldsymbol{p}_i) \sim N(0, \sigma_X^2(\boldsymbol{p}_i))$. The *PDF* of $X(\boldsymbol{p}_j)$ is that of a chi-squared χ_1^2 distribution with one degree of freedom. Despite the fact that the attribute values at points $\boldsymbol{p}_i$ and $\boldsymbol{p}_j$ are functionally related via the empirical law, nevertheless, the covariance function is $c_X(\boldsymbol{p}_i, \boldsymbol{p}_j) = 0$, i.e., it neglects the presence of the physical dependence.

The above is a simple yet illustrative example of natural attribute modeling in which the mainstream covariance function offers an inadequate representation of the actual chronotopologic dependence. Moreover, several other cases are encountered in practice, where covariance and variogram turn out to be insufficient to model curvilinear or long-range continuous physical formations, thus leading to unrealistic models of the phenomenon.

Example 4.2 Unrealistic depositional elements are often simulated based on a covariance model that is an inadequate representation of long-range continuous geologic bodies.

Instead of postponing the matter *ad Kalendas Graecas*, there is a strong motivation to seek other measures of dependence that do not suffer from such drawbacks and may also perform better in applications. This is an area that is undergoing active development. Some possibilities have been considered that are discussed next.

4.2 Chronotopologic sysketogram and contingogram functions

The chronotopologic sysketogram and contingogram functions have been introduced as useful alternatives to the covariance-based (or variogram-based) characterization of the attribute variability, see Christakos (1991, 2010). In addition to not suffering the covariance and variogram drawbacks described in the previous section, the sysketogram and contingogram can also yield improved results as regards chronotopologic attribute estimation and mapping. Herein, unless otherwise stated, a random field $X(\boldsymbol{p})$ with zero mean and unit variance will be assumed, for simplicity of presentation.

4.2.1 The λ-ratio

Like other chronotopologic dependence measures, the so-called sysketogram and contingogram functions can be mathematically expressed in terms of the general stochastic expectation operator (Eq. 2.5a-b of Section 2, Chapter 5 with $m=2$),

$$\overline{S\left[X(\boldsymbol{p}_i), X\left(\boldsymbol{p}_j\right)\right]} = \int d\chi_i \int d\chi_j S\left[\chi_i, \chi_j\right] f_X\left(\chi_i, \chi_j\right), \tag{4.1a}$$

for different choices of the function $S[\chi_i, \chi_j]$. In particular, consider the ratio.

$$\lambda_{ij} = \frac{f_X\left(\boldsymbol{p}_i, \boldsymbol{p}_j\right)}{f_X(\boldsymbol{p}_i) f_X\left(\boldsymbol{p}_j\right)} = \frac{f_{X_{ij}}}{f_{X_i} f_{X_j}}, \tag{4.1b}$$

which compares *statistical dependence* represented by the bivariate *PDF* $f_{X_{ij}}$ *vs.* the *statistical independence* represented by the univariate *PDF* product $f_{X_i} f_{X_j}$. The chronotopologic attribute dependence is characterized in terms of the following functions (Christakos, 1991, 2010):

The **sysketogram** function is derived from Eq. (4.1a) by letting $S[\chi_i, \chi_j] = \log \lambda_{ij}$, i.e.,

$$\beta_X\left(\boldsymbol{p}_i, \boldsymbol{p}_j\right) = \overline{\log \lambda_{ij}}, \tag{4.2}$$

and the **contingogram** function is derived from Eq. (4.1a) by letting $S[\chi_i, \chi_j] = \lambda_{ij} - 1$, i.e.,

$$\psi_X\left(\boldsymbol{p}_i, \boldsymbol{p}_j\right) = \overline{\lambda_{ij}} - 1, \tag{4.3}$$

where the expectation operator $(\overline{\cdot})$ is with respect to $f_{X_{ij}}$.[n]

Example 4.3 As a simple illustration, if the univariate attribute *PDF* is $f_X = N(0,1)$ and the bivariate *PDF* is $f_{X_{ij}} = e^{X^2(\boldsymbol{p}_i) + X^2(\boldsymbol{p}_j)} f_{X_i} f_{X_j}$, the corresponding sysketogram and contingogram are $\beta_X(\boldsymbol{p}_i, \boldsymbol{p}_j) = \pi^{-1}$ and $\psi_X(\boldsymbol{p}_i, \boldsymbol{p}_j) = e^{\pi^{-1}} - 1$, respectively.

In principle, the investigator can start with any of the known bivariate *PDF* models, and then use Eqs. (4.2) and (4.3) to construct the corresponding sysketogram and contingogram model, respectively. The sysketogram and the contingogram are nonnegative functions, i.e., it is valid that

$$\beta_X, \psi_X \geq 0,$$

where the equality holds only in the case of independence, i.e., $f_{X_{ij}} = f_{X_i} f_{X_j}$ (an example of this limiting case is the uncorrelated normal *S/TRF*). Furthermore, using Schwartz's inequality, it can be shown that

$$\frac{c_X^2\left(\boldsymbol{p}_i, \boldsymbol{p}_j\right)}{\sigma_X^2(\boldsymbol{p}_i) \sigma_X^2\left(\boldsymbol{p}_j\right)} \leq \psi_X\left(\boldsymbol{p}_i, \boldsymbol{p}_j\right),$$

[n] I.e., $\overline{[\cdot]} = \overline{[\cdot]}\Big|_{f_{X_{ij}}} = \int\int d\chi_i d\chi_j f_{X_{ij}} \log[\cdot]$.

which provides a comparison of the corresponding c_X and ψ_X values characterizing the same physical variation. The equality holds in the case that $X(\boldsymbol{p})$ denotes an indicator random field, i.e., a field such that $X(\boldsymbol{p}) = 1$ (if $\chi \in I$; I is a real interval), and $= 0$ (otherwise).

The sysketogram Eq. (4.2) can also be written as the difference

$$\beta_X\left(\boldsymbol{p}_i, \boldsymbol{p}_j\right) = \overline{\log f_X^{-1}(\boldsymbol{p}_i)} - \int d\chi_j f_X\left(\boldsymbol{p}_j\right) \overline{\log f_X^{-1}\left(\boldsymbol{p}_i \mid \boldsymbol{p}_j\right)}, \quad (4.4)$$

where $f_X(\boldsymbol{p}_i|\boldsymbol{p}_j) = f_{X_{i|j}}$ denotes the conditional *PDF* of $X(\boldsymbol{p}_i)$ given $X(\boldsymbol{p}_j)$. The maximum sysketogram value is obtained when $\boldsymbol{p}_i \to \boldsymbol{p}_j$, and is equal to $\overline{\log f_X^{-1}(\boldsymbol{p})}$.

4.2.2 *Interpretation*

From a stochastic reasoning viewpoint, the $X(\boldsymbol{p})$ characterization provided by the sysketogram and the contingogram is considered cognitively general, *albeit* rather straightforward in interpretational terms:

- The sysketogram and contingogram functions can be interpreted as offering alternative measures of the **degree of departure** of the attribute dependence from chronotopologic independence.

In particular, the epistemic interpretation of Eq. (4.2) is as the information gain in knowing $f_{X_{ij}}$ compared to knowing $f_{X_i} f_{X_j}$ (the latter representing the statistical independence of $X(\boldsymbol{p}_i)$ and $X(\boldsymbol{p}_j)$), whereas the interpretation of Eq. (4.3) is as the difference between the statistical dependent case represented by $f_{X_{ij}}$ and the statistical independent case represented by $f_{X_i} f_{X_j}$ normalized by the latter. Similarly, the contingogram function offers a measure of the degree of $X(\boldsymbol{p})$'s departure from independence. *Paucis verbis*, the higher the value of ψ_X is, the more chronotopologically dependent the attribute distribution is (with profound consequences in attribute modeling, physical meaning, prediction, and mapping).

4.2.3 *Properties*

By their definition, the β_X and ψ_X have some notable properties that in many cases may favor their use in place of the mainstream covariance function c_X. These properties are listed in Table 4.1.[o]

Specifically, property P_1 implies that β_X and ψ_X contain more information about chronotopologic dependence than c_X; or, that β_X and ψ_X may uncover some dependence aspects of $X(\boldsymbol{p})$ that c_X cannot uncover. The usefulness of property P_2 is better appreciated in practice where the

TABLE 4.1 Properties of the sysketogram and contingogram functions.

Property	Description
P_1 Independence	The $c_X = 0$ implies lack of statistical correlations but not independence, whereas $\beta_X = 0$ or $\psi_X = 0$ implies independence
P_2 PDF-based	The c_X depends on both the *PDF* and the $X(\boldsymbol{p})$ values, whereas β_X and ψ_X depend only on the *PDF*
P_3 Absoluteness	The c_X is affected if the $X(\boldsymbol{p})$ is replaced by a one-to-one function of it, $\phi[X(\boldsymbol{p})]$, whereas β_X and ψ_X are not affected
P_4 Nonseparability	In the special case that c_X is space-time separable, the corresponding β_X and ψ_X are space-time nonseparable (i.e., the β_X and ψ_X offer a more general expression of space-time correlation than c_X)

[o] The same claim is valid as regards the mainstream variogram function γ_X.

discrete representations of β_X and ψ_X are used, in which case only the probability values are needed and not the numerical values of the corresponding attribute realizations. Property P_3 means that β_X and ψ_X are absolute rather than relative quantities, in the sense that the chronotopologic correlation defined by β_X and ψ_X are completely independent of the scale of measurement of $X(\boldsymbol{p})$. This property is very useful in physical applications in which the concepts of "scale of measurement" and "instruments window" play an important role. Similarly, when the attribute has random chronotopologic coordinates (e.g., distribution of aerosol particles), the β_X and ψ_X are independent of the coordinate system chosen. Property P_4 will be discussed in more detail.

Example 4.1(cont.): In the case of the empirical law of Example 4.1, it is valid that $c_X = 0$ but $\beta_X \neq 0$ and $\psi_X \neq 0$ (property P_1). I.e., unlike c_X, the β_X and ψ_X provide a physically meaningful characterization of the phenomenon in this case.

Example 4.4 The absoluteness property of β_X and ψ_X (property P_3) may bring to mind a fundamental result of modern physics, according to which only absolute quantities, say, quantities independent of the space-time coordinate system, can be used as essential components of a valid physical law.

Grosso modo, Eqs. (4.2) and (4.3) introduce, respectively, a logarithmic and an algebraic transformation of c_X so that the new dependence functions (β_X and ψ_X) do not suffer the limitations of the original dependence function (c_X) as described in $P_1 - P_4$.

4.2.4 Links to factoras and copulas

Another notable feature of the sysketogram and contingogram is their expressive *versatility*. Beyond Eqs. (4.2)–(4.3), other useful expressions of the sysketogram and contingogram are obtained in terms of the factoras and copulas *PDF* representation (Christakos, 1992, 2017). In particular, the ratio λ_{ij} in Eqs. (4.2)–(4.3) can be replaced by a factora or a copula function, directly leading to the following results.

The **factora**-based and the **copula**-based representations of the sysketogram and contingogram functions are, respectively,

$$\beta_X\left(\boldsymbol{p}_i, \boldsymbol{p}_j\right) = \log\overline{\theta_{X_{ij}}} = \overline{\log\varsigma_{X_{ij}}},$$

and

$$\psi_X\left(\boldsymbol{p}_i, \boldsymbol{p}_j\right) = \overline{\varsigma_{X_{ij}}} - 1,$$

where $\theta_{X_{ij}}$ and $\varsigma_{X_{ij}}$ are the corresponding factora and copula, respectively.[p]

Given that factoras and copulas can be generally expressed in terms of polynomials, the β_X and ψ_X can be written as.

$$\beta_X\left(\boldsymbol{p}_i, \boldsymbol{p}_j\right) = \log\sum\nolimits_{l=0}^{\infty}\theta_l\overline{\varpi_X^{(l)}(\boldsymbol{p}_i)\varpi_X^{(l)}\left(\boldsymbol{p}_j\right)},$$
$$\psi_X\left(\boldsymbol{p}_i, \boldsymbol{p}_j\right) = \sum\nolimits_{l=0}^{\infty}\theta_l\overline{\varpi_X^{(l)}(\boldsymbol{p}_i)\varpi_X^{(l)}\left(\boldsymbol{p}_j\right)} - 1, \tag{4.5a,b}$$

where $\varpi_X^{(l)}(\boldsymbol{p}_i)$ are complete sets of polynomials[q] of degree l in $X(\boldsymbol{p}_i)$ and $X(\boldsymbol{p}_j)$ that are orthonormal with respect to f_{X_i},[r] and θ_l are coefficients such

[p] Discussed in Christakos et al. (2011). Certain relationships between factoras and copulas were also examined in that work, which are useful in the context of the sysketogram and contingogram functions.

that $\theta_l\delta_{ll'} = \overline{\varpi_X^{(l)}(\boldsymbol{p}_i)\varpi_X^{(l')}(\boldsymbol{p}_j)}$.[s] In Eqs. (4.5a,b) lower-order statistics are linked to the first terms of the series, whereas higher-order statistics are linked to later terms.

Example 4.5 For illustration, let us examine two special cases of Eqs. (4.5a,b).

- Using 1st-order polynomial expansions, e.g., Eq. (4.5b) simply yields $\psi_X = \theta_1 c_X$, where the approximation error depends on θ_1.
- By letting $\varpi_X^{(l)}$ be Hermite polynomials $H_X^{a(l)}$, one obtains the contingogram model

$$\psi_X(\boldsymbol{p}_i, \boldsymbol{p}_j) = \sum\nolimits_{l=0}^{\infty} \rho_{ij}^{a(l)} \overline{H_X^{a(l)}(\boldsymbol{p}_i) H_X^{a(l)}(\boldsymbol{p}_j)} - 1,$$

where for $a(l) = l$ the $X(\boldsymbol{p})$ is normally distributed, and for $a(l) = 2l$ the $X(\boldsymbol{p})$ is not normally distributed.

- Eqs. (4.5a,b) have the significant trait that they can be used to derive **new** contingogram models directly from known factora and copula models.

4.2.5 *Important special cases*

At this point, let us consider the sysketogram and contingogram formulations in a few important cases of random field models that are encountered in physical applications.

❶ If $f_{X_{ij}}$ is a **bivariate normal** *PDF*, Eqs. (4.2) and (4.3) lead to the following expressions of the sysketogram and contingogram functions, respectively, in terms of the covariance function,

$$\begin{aligned} \beta_X(\boldsymbol{p}_i, \boldsymbol{p}_j) &= -\frac{1}{2}\log\left[1 - c_X^2(\boldsymbol{p}_i, \boldsymbol{p}_j)\right], \\ \psi_X(\boldsymbol{p}_i, \boldsymbol{p}_j) &= \frac{c_X^2(\boldsymbol{p}_i, \boldsymbol{p}_j)}{1 - c_X^2(\boldsymbol{p}_i, \boldsymbol{p}_j)}. \end{aligned} \quad (4.6a,b)$$

I.e., here the β_X and ψ_X are defined as the log transformation of the function $(1 - c_X^2)^{-\frac{1}{2}}$, with $0 \leq c_X^2 \leq 1$ (assuming a unit variance, for simplicity), and as the ratio of covariance functions, respectively. Eqs. (4.6a,b) are very useful in the construction of β_X and ψ_X models in practice (Section 5 below), starting from known covariance models.

- Even if the data are **not normally** distributed, a transformation into normally distributed values is usually possible, and then Eqs. (4.6a, b) can be applied onto the transformed data.

Furthermore, by inverting Eqs. (4.6a,b) we can express the covariance directly in terms of the sysketogram and the contingogram, i.e.,

$$c_X^2(\boldsymbol{p}_i, \boldsymbol{p}_j) = \begin{cases} 1 - e^{-2\beta_X(\boldsymbol{p}_i, \boldsymbol{p}_j)}, \\ \dfrac{\psi_X(\boldsymbol{p}_i, \boldsymbol{p}_j)}{1 + \psi_X(\boldsymbol{p}_i, \boldsymbol{p}_j)}. \end{cases} \quad (4.7a,b)$$

From the above equations, one can directly deduce that the sysketogram and contingogram are linked by

$$\begin{aligned} \beta_X(\boldsymbol{p}_i, \boldsymbol{p}_j) &= \frac{1}{2}\log\left[1 + \psi_X(\boldsymbol{p}_i, \boldsymbol{p}_j)\right], \\ \psi_X(\boldsymbol{p}_i, \boldsymbol{p}_j) &= e^{2\beta_X(\boldsymbol{p}_i, \boldsymbol{p}_j)} - 1, \end{aligned} \quad (4.8a,b)$$

which establish interesting links between β_X and ψ_X.

❷ In the case of **homostationarity**, the corresponding sysketogram and contingogram

[q] Hermite, Laguerre, etc.

[r] For example, $\varpi_X^{(0)}(\boldsymbol{p}_i) = 1$ and $\varpi_X^{(1)}(\boldsymbol{p}_i) = \frac{X(\boldsymbol{p}_i) - \overline{X(\boldsymbol{p}_i)}}{\sigma_X(\boldsymbol{p}_i)}$.

[s] For example, $\theta_0 = 1$ and $\theta_1 = \rho_{ij}$, the correlation coefficient.

functions obtained by Eqs. (4.6a,b) are, respectively, only functions of $\Delta p = p_i - p_j$; and independence practically occurs when $|\Delta p| \to \infty$.

Moreover, in the case of homostationarity closed form analytical expressions can be derived for the sysketogram and contingogram functions associated with specific covariance models, as in the following example.

Example 4.6 If $c_X(\Delta p) = e^{-(h+\tau)}$, where $h = |s_j - s_i|$ and $\tau = t_j - t_i$, Eqs. (4.6a,b) give

$$\beta_X(\Delta p) = (h+\tau) - \frac{1}{2}\log\left[e^{2(h+\tau)} - 1\right], \quad (4.9a)$$

and

$$\psi_X(\Delta_p) = \left[e^{2(h+\tau)} - 1\right]^{-1}, \quad (4.9b)$$

respectively.

- ❸ In the case of a space-time **separable** covariance, $c_X(\Delta p) = c_X(h)c_X(\tau)$, the corresponding sysketogram and contingogram obtained by Eqs. (4.6a,b) are

$$\begin{aligned} \beta_X(\Delta p) &= -\frac{1}{2}\log\left[1 - c_X^2(h)c_X^2(\tau)\right], \\ \psi_X(\Delta p) &= \frac{1}{c_X^{-2}(h)c_X^{-2}(\tau) - 1}, \end{aligned} \quad (4.10a,b)$$

respectively, which, though, are space-time **nonseparable** and, hence, they offer a more general representation of space-time correlation than the covariance.

Furthermore, Eqs. (4.10a,b) have an interesting implication as regards sysketogram and contingogram model construction:

New nonseparable, in general, $\beta_X(\Delta p)$ and $\psi_X(\Delta p)$ **models** can be derived directly from the purely spatial covariance models $c_X(h)$ of spatial statistics (geostatistics) and the purely temporal $c_X(\tau)$ covariance models of time series analysis.

For further visual investigation, using Eqs. (4.6a,b), the β_X and ψ_X are plotted in Fig. 4.1 as functions of c_X. We observe that the following inequalities are valid,

$$\beta_X \begin{cases} \leq c_X \text{ for } c_X \leq 0.917 \\ \geq c_X \text{ otherwise;} \end{cases} \quad (4.11a,b)$$

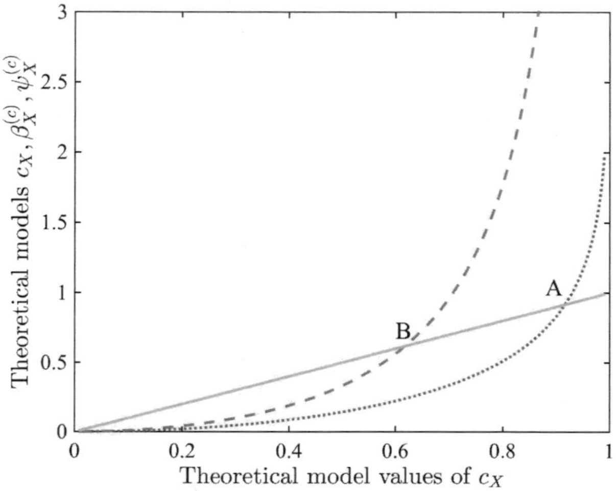

FIG. 4.1 Comparison of the theoretical models of c_X *(line)*, β_X *(dot line)* and ψ_X *(dash line)*. Covariance c_X values at points *A* and *B* are about 0.917 and 0.618, respectively.

and

$$\psi_X \begin{cases} \leq c_X \text{ for } c_X \leq 0.618 \\ \geq c_X \text{ otherwise.} \end{cases} \quad (4.11c,d)$$

Interestingly, Eqs. (4.11a) and (4.11c) determine a domain in the plot of Fig. 4.1 in which the β_X and ψ_X values lie below those of c_X, whereas Eqs. (4.11b) and (4.11d) indicate a domain in which the β_X and ψ_X values lie above those of c_X. In the plots of Fig. 4.1 the sysketogram and contingogram values may be viewed either as abbreviated (compressed) covariance values (in the parts of the plots determined by Eqs. (4.11a) and (4.11c)) or as magnified (stretched) covariance values (parts of the plot determined by Eqs. (4.11b) and (4.11d)). The β_X's compression ability is larger than that of the ψ_X's, whereas the ψ_X's magnification ability is larger than that of the β_X's. Two further conclusions are drawn: (*a*) When the maximum c_X value is larger than 0.917 (point *A* in Fig. 4.1), the β_X and ψ_X include both compressed and magnified covariance values; (*b*) At the β_X and ψ_X ranges of Fig. 4.1 corresponding to the intervals $c_X \in (0.917, 1)$ and $c_X \in (0.618, 1)$, respectively,[t] both β_X and ψ_X magnify the corresponding c_X values.

4.3 Construction techniques

Three different techniques for constructing sysketogram and contingogram models are suggested by the preceding analysis:

Technique ❶: Starting from any known **bivariate *PDF*** models and using Eqs. (4.2)–(4.3).

Technique ❷: Starting from any of the known **factora** and **copula** models and using different polynomial sets (Hermite, Laguerre, etc.) in Eqs. (4.5a,b).

Technique ❸: Directly from known **covariance** models using Eqs. (4.6a,b).

4.3.1 The direct technique

The idea of technique ❸ is particularly convenient in deriving useful sysketogram and contingogram models in practice:

> Fit to the empirical covariance values $\hat{c}_X$ a theoretical model c_X, and use Eqs. (4.6a,b) to obtain the corresponding sysketogram $\beta_X^{(c)}$ and contingogram $\psi_X^{(c)}$ **models**.

The superscript (*c*) is used to denote the direct links of the sysketogram and contingogram models with the covariance model via Eqs. (4.6a,b). As a matter of fact, this last technique is a particularly convenient way to derive sysketogram and contingogram models, for two practical reasons:

① Although Eqs. (4.6a,b) assume a bivariate normal *S/TRF*, nevertheless, the generated sysketogram and contingogram models can be used in many practical cases in which an **adequate fit** of the $\beta_X^{(c)}$ or $\psi_X^{(c)}$ to the data is possible.

② A **transformation** of nonnormal data into normally distributed values is usually possible.

The crux of the matter as regards the value of the direct technique is that it derives new $\beta_X^{(c)}$ and $\psi_X^{(c)}$ models from every known c_X model. Otherwise said, this is an approach that transforms a c_X model to $\beta_X^{(c)}$ and $\psi_X^{(c)}$ models but without retaining the limitations of c_X.

4.3.2 Analytical and visual comparisons

To obtain an analytical comparison of the features of c_X, $\beta_X^{(c)}$ and $\psi_X^{(c)}$, Table 4.2 presents a list of c_X models that are often used in *CTDA* applications with their corresponding $\beta_X^{(c)}$ and $\psi_X^{(c)}$ models. A visual comparison of these models is attempted in the following example.

[t] I.e., the ranges at which the β_X and ψ_X values are larger than the corresponding c_X values.

TABLE 4.2 Chronotopologic covariance models and the associated sysketogram and contingogram models.

Model no.	$c_X(\Delta \boldsymbol{p})$	$\beta_X^{(c)}(\Delta \boldsymbol{p})$	$\psi_X^{(c)}(\Delta \boldsymbol{p})$
1	$\frac{1}{(1+b\tau^2)^{1.5} e^{\frac{h^2}{2(1+b\tau^2)}}}\left[1-\frac{h^2}{2(1+b\tau^2)}\right]$ $b>0$	$-\frac{1}{2}\ln\left\{1-\frac{\left[1-\frac{h^2}{2(1+b\tau^2)}\right]^2}{(1+b\tau^2)^3 e^{\frac{h^2}{1+b\tau^2}}}\right\}$	$\frac{1-\frac{h^2}{1+b\tau^2}+\frac{h^4}{4(1+b\tau^2)^2}}{(1+b\tau^2)^3 e^{\frac{h^2}{1+b\tau^2}}+\frac{h^2}{1+b\tau^2}-\frac{h^4}{4(1+b\tau^2)^2}-1}$
2	$\frac{1}{(1+b\tau^{2c})^{0.5} e^{\frac{h^2}{(1+b\tau^{2c})}}}$ $b,\ c\in[0,1]$	$-\frac{1}{2}\ln\left[1-\frac{1}{(1+b\tau^{2c})e^{\frac{2h^2}{1+b\tau^{2c}}}}\right]$ $b>0$	$\frac{1}{(1+b\tau^{2c})e^{\frac{2h^2}{1+b\tau^{2c}}}-1}$
3	$e^{-\frac{\lvert h\pm\upsilon\tau\rvert}{\alpha}}$ $\alpha,\ \beta,\ \upsilon>0$	$-\frac{1}{2}\ln\left(1-e^{-\frac{2\lvert h\pm\upsilon\tau\rvert}{\alpha}}\right)$	$\frac{1}{e^{\frac{2\lvert h\pm\upsilon\tau\rvert}{\alpha}}-1}$
4	$e^{-\frac{(h\pm\upsilon\tau)^2}{\alpha^2}}$ $\alpha,\ \beta,\ \upsilon>0$	$-\frac{1}{2}\ln\left[1-e^{-\frac{2(h\pm\upsilon\tau)^2}{\alpha^2}}\right]$	$\frac{1}{e^{\frac{2(h\pm\upsilon\tau)^2}{\alpha^2}}-1}$
5	$ce^{-\left(\frac{bh}{\alpha}+\frac{\upsilon\tau}{\beta}\right)}$ $\alpha,\ \beta,\ b,\ c,\ \upsilon>0$	$-\frac{1}{2}\ln\left[1-c^2 e^{-2\left(\frac{bh}{\alpha}+\frac{\upsilon\tau}{\beta}\right)}\right]$	$\left(c^{-2}e^{\frac{2bh}{\alpha}+\frac{2\upsilon\tau}{\beta}}-1\right)^{-1}$
6	$\left[1+\frac{(h\pm\upsilon\tau)^2}{\beta^2}\right]^{0.5\nu} e^{-\frac{\lvert h\pm\upsilon\tau\rvert}{\alpha}}$ $\alpha,\ \beta,\ \nu>0$	$-\frac{1}{2}\ln\left\{1-\left[1+\frac{(h\pm\upsilon\tau)^2}{\beta^2}\right]^{\upsilon} e^{-\frac{2\lvert h\pm\upsilon\tau\rvert}{\alpha}}\right\}$	$\frac{\left[1+\frac{(h\pm\upsilon\tau)^2}{\beta^2}\right]^{\upsilon}}{e^{\frac{2\lvert h\pm\upsilon\tau\rvert}{\alpha}}-\left[1+\frac{(h\pm\upsilon\tau)^2}{\beta^2}\right]^{\upsilon}}$
7	$\left(1+\frac{\tau^2}{\nu^2}\right)^{-0.5\nu} e^{-\frac{h}{a}}$ $a,\ \nu>0$	$-\frac{1}{2}\ln\left[1-\left(1+\frac{\tau^2}{\nu^2}\right)^{-\nu} e^{-\frac{2h}{a}}\right]$	$\frac{1}{\left(1+\frac{\tau^2}{\nu^2}\right)^{\upsilon} e^{\frac{2h}{\alpha}}}$
8	$e^{-\left(\frac{h^2}{a^2}+\frac{\tau^2}{b^2}\right)^{0.5}}$ $a,\ b>0$	$-\frac{1}{2}\ln\left[1-e^{-2\left(\frac{h^2}{a^2}+\frac{\tau^2}{b^2}\right)^{0.5}}\right]$	$\frac{1}{e^{2\left(\frac{h^2}{\alpha^2}+\frac{\tau^2}{b^2}\right)^{0.5}}-1}$
9	$c\left(1+\frac{h^\lambda}{a^2}+\frac{\tau^\mu}{b^2}\right)^{-\nu}$ $\lambda,\ \mu\in(0,2];\ a,\ b,\ c,\ \nu>0$	$-\frac{1}{2}\ln\left[1-c^2\left(1+\frac{h^\lambda}{a^2}+\frac{\tau^\mu}{b^2}\right)^{-2\upsilon}\right]$	$\frac{c^2}{\left(1+\frac{h^\lambda}{a^2}+\frac{\tau^\mu}{b^2}\right)^{2\upsilon}-c^2}$

Example 4.7 Fig. 4.2 displays plots of some of the models of Table 4.2 on the basis of which certain observations can be made:

a. At short distances h and/or separations τ, contingogram values are larger than the corresponding covariance values, i.e., attribute variation in numerical terms is considerably larger when expressed by $\psi_X^{(c)}$ than by c_X.

b. At large h and/or τ, the $\psi_X^{(c)}$ tends to zero faster than c_X.

c. Unlike c_X, the $\psi_X^{(c)}$ can take values much higher than 1.

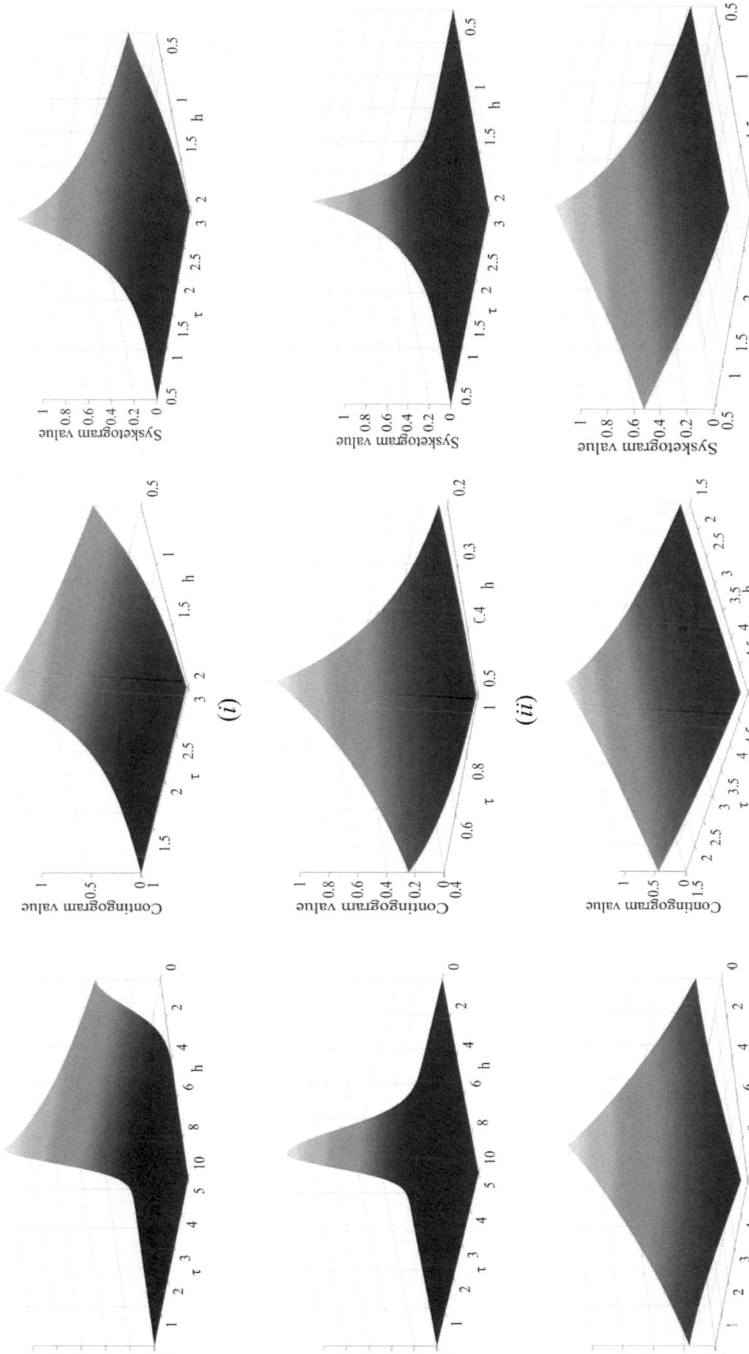

FIG. 4.2 Plots of covariance c_X *(left)* and associated contingogram $\psi_X^{(c)}$ *(middle)* and sysketogram $\beta_X^{(c)}$ *(right)* for models: *(i)* no.2 ($b = c = 0.5$), *(ii)* no.4 ($\alpha = \nu = 1$), and *(iii)* no.7 ($a = 10, \quad b = 5$).

∴ A noticeable implication of feature *a* concerns the implementation of chronotopologic estimation techniques, such as *Kriging* and *BME* (Chapters 8 and 9): For strongly chronotopologically dependent attributes, estimation techniques using data at short lags may generate more accurate attribute estimates in terms of $\psi_X^{(c)}$ than in terms of c_X models.[u]

Example 4.8 For further illustration, consider the covariance model No. 5 in Table 4.2[v] with $b = \upsilon = 3$, $c = 1$ (sill), $\alpha = \varepsilon_s = 554.3559$ (spatial correlation range) and $\beta = \varepsilon_t = 48.7061$) (temporal correlation range). The corresponding contingogram model is also shown in Table 4.2 (model No. 5). The value $c_X(h,\tau) = 0.618$ (point *A* in Fig. 4.1 occurs at the *h* and τ values satisfying the equation:

$$\varepsilon_t h + \varepsilon_s \tau = 48.7061h + 554.3559\tau = 4331.4838$$

Specifically, at $\tau = 0$ one gets $h \approx 88.931$, and at $h = 0$ one gets $\tau \approx 7.8135$. The space-time area under the line, say $U(h,\tau)$, is where $\psi_X(h,\tau) > c_X(h,\tau)$; while the area above the line, say $A(h,\tau)$, is where $\psi_X(h,\tau) < c_X(h,\tau)$. Thus, Fig. 4.2 gives a visual delimitation of the chronotopologic domain within which the contingogram is larger *vs*. smaller than the covariance (Fig. 4.3).

4.3.3 Sysketogram- and contingogram-based covariance models

Consider an ordinary covariance model c_X with coefficients c_0 (sill), ε_s (space range), and ε_t (time range), and let $\hat{c}_X$ denote the empirical covariance values available. Using Eqs. (4.6a,b) we obtain the $\hat{\beta}_X^{(\hat{c})}$ and $\hat{\psi}_X^{(\hat{c})}$ values from the empirical $\hat{c}_X$ values, and the analytical forms of the models $\beta_X^{(c)}$ and $\psi_X^{(c)}$ from the c_X model (at this stage, the specific values of c_0, ε_s and ε_t are unknown). The models $\beta_X^{(c)}$ and $\psi_X^{(c)}$ are fitted to the empirical values $\hat{\beta}_X^{(\hat{c})}$ and $\hat{\psi}_X^{(\hat{c})}$, and their coefficients $\{c_0^{(\beta)}, \varepsilon_s^{(\beta)}, \varepsilon_t^{(\beta)}\}$ and $\{c_0^{(\psi)}, \varepsilon_s^{(\psi)}, \varepsilon_t^{(\psi)}\}$ are estimated, accordingly. As a result:

> Using Eqs. (4.6a,b), the **sysketogram-based covariance** $c_X^{(\beta)}$ and the **contingogram-based covariance** $c_X^{(\psi)}$ models[w] are obtained from the $\beta_X^{(c)}$ and $\psi_X^{(c)}$ models, which are fitted to the original attribute dataset (Fig. 4.4).[x]

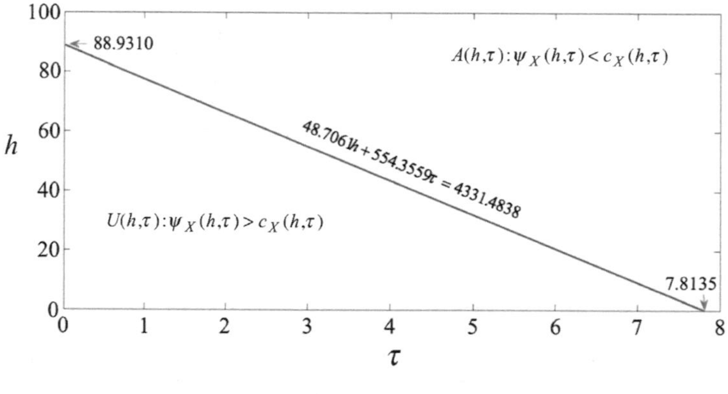

FIG. 4.3 Plot of the line $c_X(h,\tau) = 0.618 : 48.7061h + 554.3559\tau = 4331.4838$. The area $U(h,\tau)$ under the line is where $\psi_X > c_X$, whereas the area $A(h,\tau)$ above the line is where $\psi_X < c_X$.

[u] This situation is numerically demonstrated in the Western Pacific ocean salinity study to be discussed later.

[v] To be used later in the sea surface salinity study.

[w] Also called the sysketogram-based and contingogram-based covariance models.

[x] The superscripts β and ψ denote the direct link of the covariances to the sysketogram and contingogram functions via Eqs. (4.6a,b).

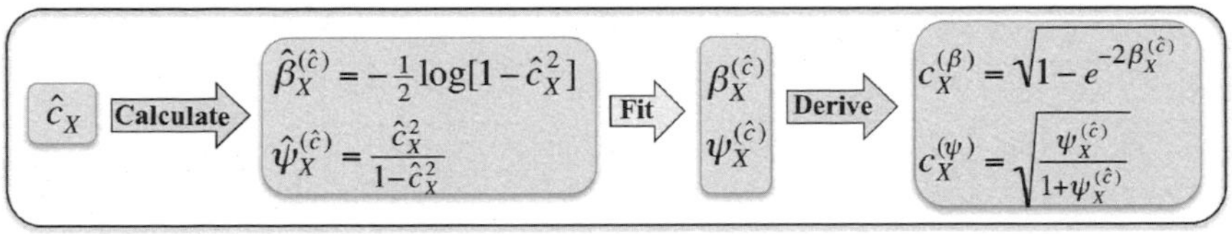

FIG. 4.4 Derivation of sysketogram-based covariance $c_X^{(\beta)}$ and contingogram-based covariance $c_X^{(\psi)}$ models.

One might wonder why one needs to go the extra distance and determine the sysketogram-based and the contingogram-based covariances. The answer is that the $c_X^{(\beta)}$ and $c_X^{(\psi)}$ are needed for technical reasons, since the sysketogram and the contingogram cannot directly serve as inputs to the existing estimation and mapping techniques. In addition, the $c_X^{(\beta)}$ and $c_X^{(\psi)}$ have certain interesting features: (*a*) the $c_X^{(\beta)}$ or $c_X^{(\psi)}$ models may offer better fits to the empirical covariance values than the c_X model, i.e., the new covariances $c_X^{(\beta)}$ or $c_X^{(\psi)}$ may offer better representations of empirical data than the ordinary covariance c_X.[y] (*b*) The $c_X^{(\beta)}$ and $c_X^{(\psi)}$ models may generate more accurate attribute estimates than the c_X model (this issue is discussed in Chapter 9). Both features *a* and *b* are illustrated in the following sections using simulated data and real-world measurements.

5 Applications

A posse ad esse, the theoretical developments in Section 4 can be tested with the help of a synthetic (simulation) experiment (Section 5.1) and a real-world case study (Section 5.2).

5.1 Synthetic experiment

The synthetic experiment is useful in testing new notions and techniques, since it allows the control of key parameters and conditions, which is not usually possible in a real-world study.

5.1.1 Synthetic dataset

A base of numerical data values $X(\boldsymbol{p}_i)$ was generated in a space-time domain consisting of 3610 ($19 \times 19 \times 10$) grid points $\boldsymbol{p}_i = (\boldsymbol{s}_i; t_i) = (s_{i1}, s_{i2}; t_i)$, $i = 1, 2, \ldots, 3610$, as described in Table 5.1. Next, the same process was implemented, and the mean, minimum, and maximum values, and the variance of the simulated dataset $X(\boldsymbol{p}_i)$ were found to be 1.2495, −0.3919, 3.1654, and 0.8836, respectively. Fig. 5.1 shows the visual similarity of the $Y(\boldsymbol{p}_i)$ and $X(\boldsymbol{p}_i)$ distributions (due to the fact that the simulated std. error deviations $\sigma_Y(\boldsymbol{p}_i)$ in Eq. (5.2) and the averaging process are rather small).

5.1.2 Computational issues and simulation results

The computational part of the synthetic experiment consisted of three main stages:

① 90% of the $X(\boldsymbol{p})$ database mentioned earlier was randomly selected to serve as the study's hard dataset, whereas the remaining data was used for validation purposes.
② The $c_X^{(\beta)}$ and $c_X^{(\psi)}$ models were computed following the procedure outlined in Table 5.2 and the available hard database.
③ By repeatedly applying the *BME* chronotopologic estimation technique[z] included in the *SEKS-GUI* software (Yu et al., 2007), each time using a different dependence function, i.e., one of the c_X, $c_X^{(\beta)}$ and $c_X^{(\psi)}$ models, the relative performance of

[y] Although the process leading to $c_X^{(\beta)}$ and $c_X^{(\psi)}$ models uses the same attribute database as c_X model, the $c_X^{(\beta)}$ and $c_X^{(\psi)}$ differ from the c_X model. This is because the $\beta_X^{(\hat{c})}$ and $\psi_X^{(\hat{c})}$ models fitted to the empirical $\hat{\beta}_X^{(\hat{c})}$ and $\hat{\psi}_X^{(\hat{c})}$ values, which were obtained from the empirical $\hat{c}_X$ values via Eqs. (4.6a,b), are not the same as the $\beta_X^{(c)}$ and $\psi_X^{(c)}$ models obtained directly from Eqs. (4.6a,b).

[z] This technique will be discussed in Chapter 9.

TABLE 5.1 Generation of the numerical database.

Step	Description	
1	A set of $Y(\boldsymbol{p}_i)$ values were generated at space-time grid nodes $\boldsymbol{p}_i$ (Fig 5.1A) by the sequential simulation technique (Yu et al., 2007) assuming a covariance model with weak spatial and temporal dependence $c_Y(\Delta\boldsymbol{p}) = 0.9\,c_{G,\,\varepsilon_s=10}(\boldsymbol{h})\,c_{G,\,\varepsilon_t=6}(\tau)$, where $\boldsymbol{h} = s_j - s_i$, $\tau = t_j - t_i$, and $c_{G,\,\varepsilon_s=10}(\boldsymbol{h})$ and $c_{G,\,\varepsilon_t=6}(\tau)$ denote, respectively, a spatial Gaussian component with range $\varepsilon_s = 10$ units and a temporal Gaussian component with range $\varepsilon_t = 6$ units	(5.1)
2	A noise term $v\sigma_Y(\boldsymbol{p}_i)$ was added to the $Y(\boldsymbol{p}_i)$ values above to obtain the new set of values $Z(\boldsymbol{p}_i) = Y(\boldsymbol{p}_i) + v\sigma_Y(\boldsymbol{p}_i)$, where v represents random values selected from a normal distribution and $\sigma_Y(\boldsymbol{p}_i)$ is the simulated std. error deviation of $Y(\boldsymbol{p}_i)$ at grid node $\boldsymbol{p}_i$ using the sequential simulation technique	(5.2)
3	The process of Eq. (5.2) was repeated 100 times, the corresponding $Z^{(j)}(\boldsymbol{p}_i)$ values ($j = 1, \ldots, 100$) were simulated at each node $\boldsymbol{p}_i$, and their arithmetic averages at all $\boldsymbol{p}_i$ ($i = 1, \ldots, 3610$) $X(\boldsymbol{p}_i) = \frac{1}{100}\sum_{j=1}^{100} Z^{(j)}(\boldsymbol{p}_i) = \overline{Z(\boldsymbol{p}_i)}$ were considered to be the simulated database available (see, three dimensional plot in Fig 5.1b)	(5.3)

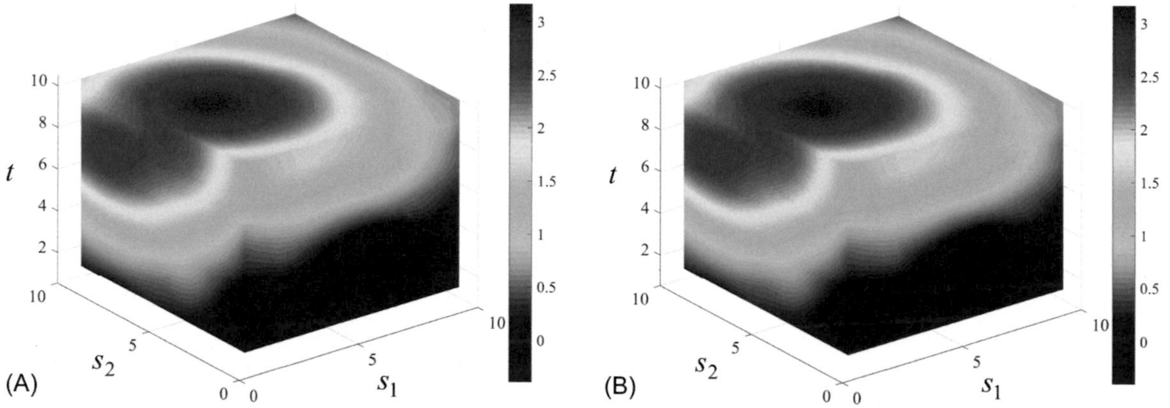

FIG. 5.1 (A) Simulated $Y = (\boldsymbol{p}_j)$ distribution in $R^{2,\,1}$; (B) simulated $X(\boldsymbol{p}_i)$ distribution in $R^{2,\,1}$.

these models—and by extension that of the $\beta_X^{(c)}$ and $\psi_X^{(c)}$ models—was assessed quantitatively in terms of the estimation errors they produced.

5.1.3 Fitting and accuracy performance

Using the procedure of Table 5.2, and assuming that in the present experiment the empirical $\hat{c}_X$ values range between 0 and 1,[aa] the original covariance, the sysketogram, the sysketogram-based covariance, the contingogram, and the contigogram-based covariance are all plotted in Figs. 5.2–5.4. Naturally, the original covariance model fits the original empirical covariance values better than the sysketogram-based and the contingogram-based covariances (Figs. 5.3B and 5.4B, respectively).

[aa] In fact, by using an n-score transformation method one can transfer the original dataset into one obeying the normal distribution.

TABLE 5.2 Computation of the space-time covariance and associated sysketogram and contingogram models.

Step	Description	
1	Empirical covariance values $\hat{c}_X$ are computed using the $X(\boldsymbol{p})$ database	
2	The theoretical covariance model $c_X(\Delta\boldsymbol{p}) = c_0 e^{-\left(\frac{3h^2}{\varepsilon_s^2}+\frac{3\tau^2}{\varepsilon_t^2}\right)}$ was selected to represent the space-time dependence of $X(\boldsymbol{p})$, where its coefficients ε_s, ε_t and c_0 denote the spatial correlation range, temporal correlation range and sill, respectively	(5.4)
3	The computed $\hat{c}_X$ values were used to derive empirical sysketogram $\hat{\beta}_X^{(\hat{c})}$ and contingogram $\hat{\psi}_X^{(\hat{c})}$ values in terms of Eqs. (4.6a) and (4.6b) of Section 4, using the convenient covariance separability property that applies in this case	
4	Theoretical $\beta_X^{(c)}$ and $\psi_X^{(c)}$ models are also derived from Eqs. (4.6a) and (4.6b) of Section 4, i.e., $\beta_X^{(c)}(\Delta\boldsymbol{p}) = -\frac{1}{2}\log\left[1 - c_0^2 e^{-2\left(\frac{3h^2}{\varepsilon_s^2}+\frac{3\tau^2}{\varepsilon_t^2}\right)}\right]$, $\psi_X^{(c)}(\Delta\boldsymbol{p}) = \left[c_0^2 e^{2\left(\frac{3h^2}{\varepsilon_s^2}+\frac{3\tau^2}{\varepsilon_t^2}\right)} - 1\right]^{-1}$.	(5.5) (5.6)
5	The c_X, $\beta_X^{(c)}$, and $\psi_X^{(c)}$ models are fitted to the empirical $\hat{c}_X$, $\hat{\beta}_X^{(\hat{c})}$, and $\hat{\psi}_X^{(\hat{c})}$ values, and the corresponding model coefficients ε_s, ε_t, and c_0 are calculated for each model[a]	
6	The sysketogram-based covariance $c_X^{(\beta)}$ and contingogram-based covariance $c_X^{(\psi)}$ models are derived from Eqs. (4.7a,b) of Section 4, i.e., $c_X^{(\beta)}(\Delta\boldsymbol{p}) = \left[1 - e^{-2\beta_X^{(c)}(h,\tau)}\right]^{\frac{1}{2}}$, $c_X^{(\psi)}(\Delta\boldsymbol{p}) = \frac{\left[\psi_X^{(c)}(h,\tau)\right]^{1/2}}{\left[1+\psi_X^{(c)}(h,\tau)\right]^{1/2}}$.	(5.7) (5.8)

[a] *The fitting process may use* MATLAB *software.*

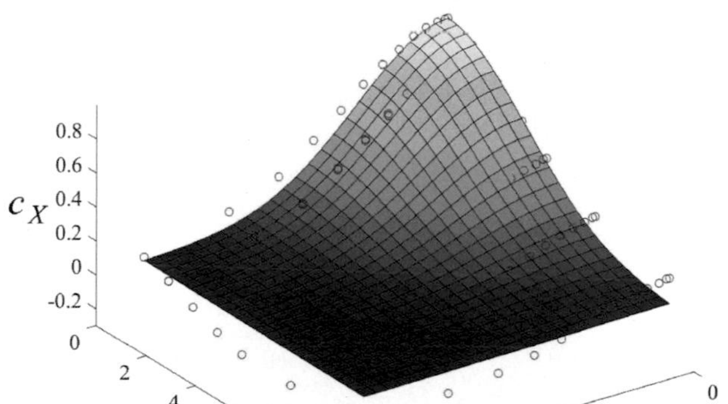

FIG. 5.2 Empirical $\hat{c}_X(\Delta\boldsymbol{p})$ values *(circles)* and fitted covariance model $\boldsymbol{c}_X(\Delta\boldsymbol{p})$ *(surface).*

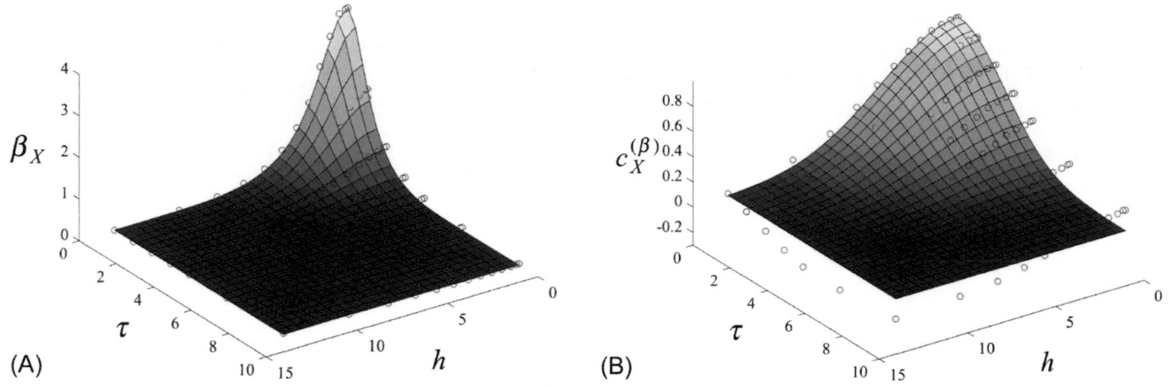

FIG. 5.3 (A) Empirical $\hat{\beta}_X^{(\hat{c})}(\boldsymbol{\Delta p})$ values *(circles)* and fitted model $\beta_X^{(c)}(\boldsymbol{\Delta p})$ *(surface)*; (B) Empirical $\hat{c}_X(\boldsymbol{\Delta p})$ values *(red circles)* and fitted model $c_X^{(\beta)}(\boldsymbol{\Delta p})$ *(surface)*.

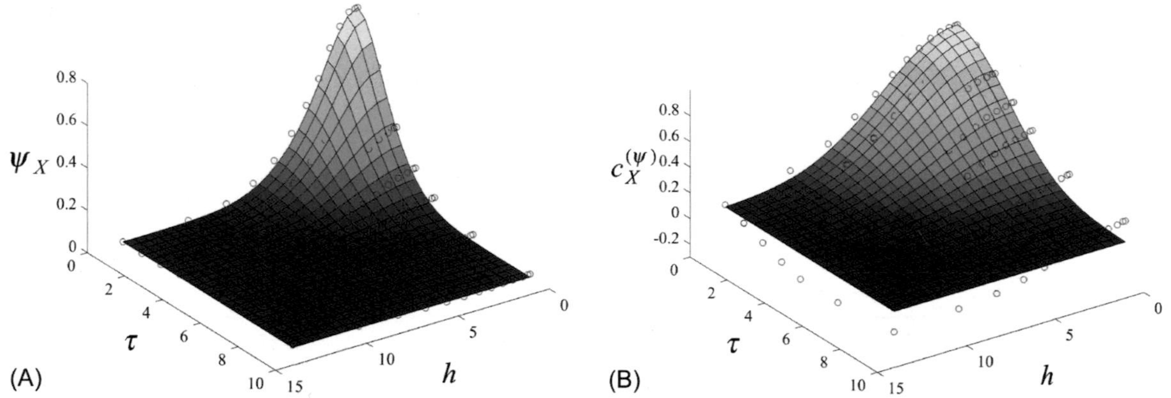

FIG. 5.4 (A) Empirical $\hat{\psi}_X^{(\hat{c})}(\boldsymbol{\Delta p})$ values *(circles)* and fitted model $\psi_X^{(c)}(\boldsymbol{\Delta p})$ *(surface)*; (B) Empirical $\hat{c}_X(\boldsymbol{\Delta p})$ values *(circles)* and fitted model $c_X^{(\psi)}(\boldsymbol{\Delta p})$ *(surface)*.

Let us have a look at the computational results of Figs. 5.2–5.4. The calculated coefficients of the c_X, $c_X^{(\beta)}$, and $c_X^{(\psi)}$ models are presented in Table 5.3. The spatial correlation range ($\varepsilon_s = 10.2854$) of $c_X^{(\beta)}$ is closest to the corresponding range ($\varepsilon_s = 10$) of the data simulation covariance of Eq. (5.4), whereas the temporal correlation range ($\varepsilon t = 6.0166$) of $c_X^{(\psi)}$ is closest to the corresponding range ($\varepsilon_t = 6$) of Eq. (5.4).

Since the $\hat{\beta}_X^{(\hat{c})}$ and $\hat{\psi}_X^{(\hat{c})}$ were calculated from Eqs. (4.6a,b) of Section 4, the $\hat{\beta}_X^{(\hat{c})}$ and $\hat{\psi}_X^{(\hat{c})}$ plots in Figs. 5.3A and 5.4A are functions of $\hat{c}_X$. Eqs. (4.11a) and (4.11c) of Section 4 determine a region in which the $\hat{\beta}_X^{(\hat{c})}$ and $\hat{\psi}_X^{(\hat{c})}$ values lie below

TABLE 5.3 Coefficients of the original *vs.* the sysketogram- and contingogram-derived covariance models.

Prediction based on:	ε_s	ε_t	c_0
c_X	9.6147	7.7754	0.8876
$c_X^{(\beta)}$	10.2854	6.6189	0.8876
$c_X^{(\psi)}$	10.6726	6.0166	0.8876

those of $\hat{c}_X$, whereas Eqs. (4.11b) and (4.11d) of Section 4 determine the region in which the $\hat{\beta}_X^{(\hat{c})}$ and $\hat{\psi}_X^{(\hat{c})}$ values lie above those of $\hat{c}_X$. In these plots the sysketogram and contingogram values may be interpreted either as "compressed" covariance values (parts of the plot determined by Eqs. (4.11a) and (4.11c) of Section 4) or as "stretched" covariance values (parts of the plot determined by Eqs. (4.11b) and (4.11d) of Section 4). Sysketogram expresses a higher level of covariance compression than contingogram (the lower the $\hat{c}_X$ value is, the greater is the extent of this compression, i.e., the lower are the $\hat{\beta}_X^{(\hat{c})}$ and $\hat{\psi}_X^{(\hat{c})}$ values). Although the transformation of the covariance model to the sysketogram and contingogram models leads to covariance compression at low $\hat{\beta}_X^{(\hat{c})}$ and $\hat{\psi}_X^{(\hat{c})}$ values (representing weak space-time dependence), it benefits $c_X^{(\beta)}$ and $c_X^{(\psi)}$ model-fitting because the focus is on high $\hat{\beta}_X^{(\hat{c})}$ and $\hat{\psi}_X^{(\hat{c})}$ value fitting at small space-time lags (Figs. 5.3A and 5.4A). I.e., a better fitting is achieved at high empirical $\hat{c}_X$ values by adjusting the coefficients of the $c_X^{(\beta)}$ and $c_X^{(\psi)}$ models (Figs. 5.3B and 5.4B). On the other hand, when the empirical covariance values belong to the ranges $\hat{c}_X \in (0.917, 1)$ and $\hat{c}_X \in (0.618, 1)$, the $\hat{\beta}_X^{(\hat{c})}$ and $\hat{\psi}_X^{(\hat{c})}$ values are larger than the corresponding $\hat{c}_X$ values. At these value ranges, the sysketogram and contingogram appear to "stretch" (or enlarge) the corresponding covariance values. Contingogram's stretching ability is larger than that of sysketogram's (Fig. 4.1 of Section 4). By stretching high $\hat{c}_X$ values, more detailed information is gained about $\hat{\beta}_X^{(\hat{c})}$ and $\hat{\psi}_X^{(\hat{c})}$, which results in better $c_X^{(\beta)}$ and $c_X^{(\psi)}$ model fitting.

At this point, and given that the *BME* chronotopologic attribute estimation technique will be introduced in Chapter 9, we wait until that chapter to test the estimation performance of the three covariance models (see Example 2.6, Section 2 of Chapter 9).

5.2 The sea surface salinity case study

The concern is the study of the chronotopologic *sea surface salinity* (*SSS*, in psu)[ab] in the Western Pacific Ocean (Maes et al., 2002; He et al., 2019a, b). To reiterate, this is a two-part case study: given that the *BME* chronotopologic attribute estimation technique will be introduced in Chapter 9, we wait until that chapter to present the generation of the *SSS* maps and assess their relative performance by comparing the map accuracy when the contingogram-based covariance model $c_X^{(\psi)}$ is used *vs.* when the ordinary covariance model c_X is used. Here, we limit ourselves to the presentation of the first part of the case study dealing with chronotopologic *SSS* data variability.

5.2.1 The SSS domain and the study motivation

The study region is between the latitudes 0° and 50° N and longitudes 100° and 150° E (it surrounds the eastern and southern sides of China and is adjacent to the Western Pacific), and the data time-span is from May 19th to 25th, 2015. Generally, *SSS* changes can affect coastal circulation and ecological environment. The study of *SSS* changes is of great significance for marine fisheries, marine engineering, and sustainable development, as well as the utilization of oceans (Kuang, 2009). Therefore, it is significant to improve the quality of *SSS* measurements and to deliver higher resolution *SSS* maps for the study region.

Although remote sensing techniques are widely used in ocean and marine sciences,[ac] using these techniques to estimate *SSS* values is difficult for two reasons:

[ab] Abbreviation for "practical salinity unit."

[ac] See, SMOS Team (2010); Melnichenko et al. (2014, 2016); Le Traon et al. (2015). Also, a review of remote sensing techniques for mapping *SSS* can be found in Klemas (2011).

① Satellite sensors lack sufficient resolution and coverage.

② Special attention is required when satellite-derived *SSS* data is used in areas with high *radio frequency interference* (*RFI*) and heavy rainfall close to the coastlines, where leakage of land signals may significantly affect the quality of data, and at high-latitude oceans where the L-band radiometer has poor sensitivity to *SSS* (Kao et al., 2018).

To deal with spatial resolution and coverage problems, a *CTDA* technique was used in terms of the contingogram-derived *SSS* covariance approach discussed in the previous section.

5.2.2 The SSS dataset and its statistical description

The *SSS* dataset was originated from satellite data provided by the Physical Oceanography Distributed Active Archive Center (*PODAAC*; Carr et al., 2006) and collected by Aquarius, which is a combination L-band radiometer and scatterometer designed to map the salinity field at the ocean surface from space, with a spatial resolution of 150 km and an accuracy of 0.2 psu globally (Lagerloef et al., 2008; Vine et al., 2010).

Considering the *RFI* problem, Level 3 data was chosen that has gridded and removed data significantly affected by *RFI*. Moreover, in order to validate the accuracy of the results, the final products were compared with measured data from Argo that can be downloaded from www.argo.org.cn. The Argo program consists of an *in-situ* global ocean array of over 3000 free-drifting profiling floats, providing near real-time salinity and temperature observations as a function of pressure in the upper 2 *km* of the ocean. The Argo contribution is very significant:

- **Continuous monitoring** of the upper ocean temperature, salinity, and velocity offers a rare opportunity to study the internal ocean structure, especially the long-term changes of marine environment factors. *SSS* precision measured by Argo is better than 0.01 psu, which is more accurate than the Aquarius mission goal and, therefore, provides *in-situ* reference for Aquarius data (Riser et al., 2008; Drucker and Riser, 2014).

As usual, let $X(\boldsymbol{p})$ denote *SSS* concentration at location $\boldsymbol{s} = (s_1, s_2)$, i.e., identified by the longitude (s_1) and latitude (s_2) pair in degrees, and at time t (denotes a specified day). The available dataset consists of 7626 locations that are shown as black dots in Fig 5.5A (the spatial data resolution is 1 degree of latitude/longitude and the spatial resolution of the generated *SSS* map is 0.25 degree of latitude/longitude). There are 207 Argo data in the study area during the period May 19th to 25th, 2015 (the Argo data locations are shown as black dots in Fig 5.5B). Although the spatial locations in the original dataset were presented in longitude and latitude (in degree units), spherical coordinates were used to calculate distances in subsequent calculations. The *SSS* dataset was subsequently de-trended and transformed to form a normal distribution.

5.2.3 Ordinary covariance- and contingogram-derived covariance

The plots of Fig. 5.6 show the empirical covariance $\hat{c}_X$ values of the *SSS* dataset together with the fitted ordinary space-time covariance model

$$c_X(\Delta\boldsymbol{p}) = e^{-3\left(\frac{h}{554.3559} + \frac{\tau}{48.7061}\right)}. \tag{5.9}$$

i.e., a nested model with two exponential components (model No. 5 of Table 4.2, Section 4 with $b = v = 3$, sill $c = 1\,\text{psu}^2$, spatial correlation range $\alpha = \varepsilon_s = 554.3559$ km, and temporal correlation range $\beta = \varepsilon_t = 48.7061$ days).[ad] The inverse ranges $\varepsilon_s^{-1} = 0.0018$ and $\varepsilon_t^{-1} = 0.0205$ determine

[ad] In view of the map scale in Fig. 5.1A, the spatial correlation range includes about 5 data locations along the EW and the NS directions. This means that, according to c_X, a maximum of 20 data locations could be used in estimation, see later Fig. 3.14 in Section 3 of Chapter 9.

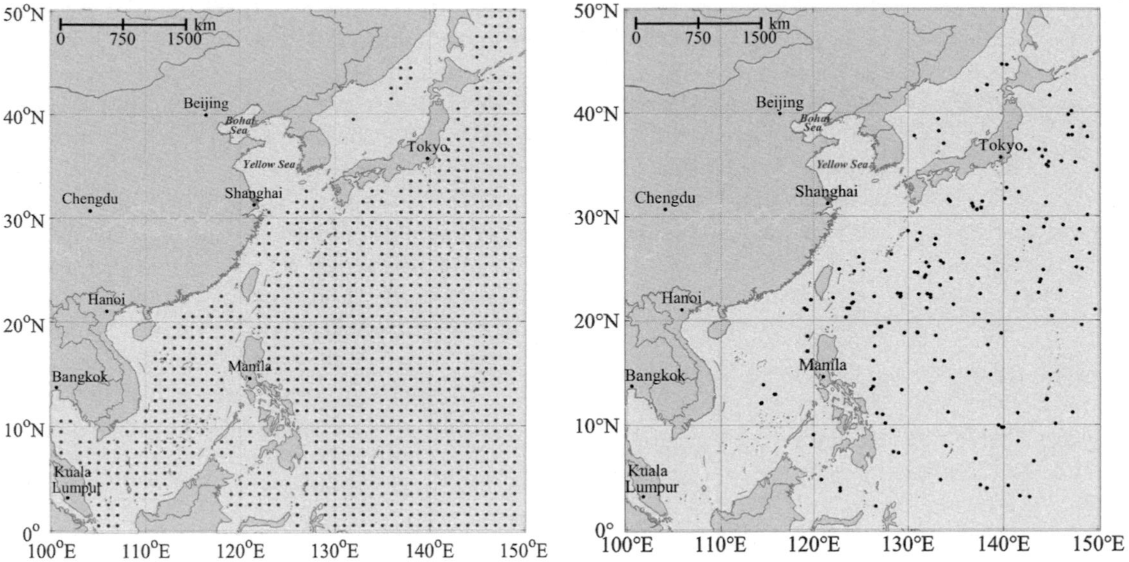

FIG. 5.5 (A) Map of the *SSS* data locations *(dots)* in the study area of Western Pacific Ocean; and (B) map of the Argo data locations *(dots)*.

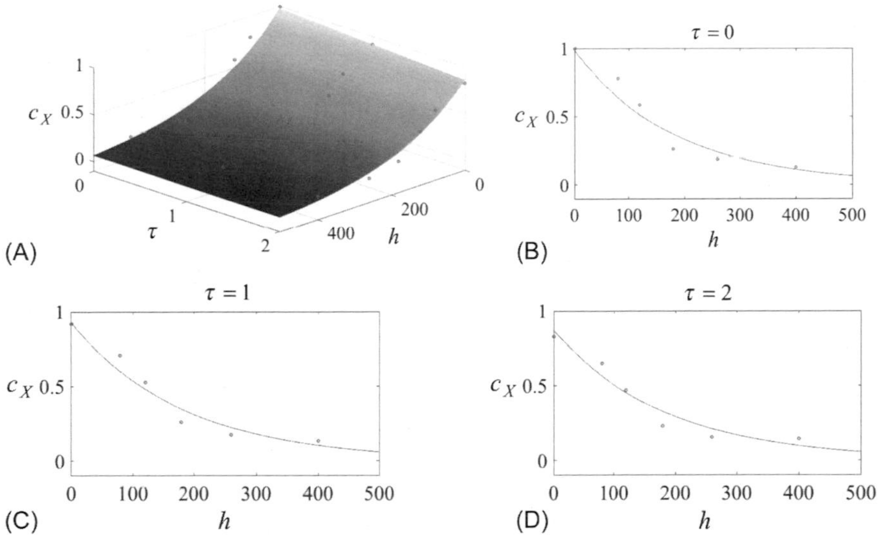

FIG. 5.6 (A) Empirical $\hat{c}_X(\boldsymbol{\Delta p})$ values *(circles)* and fitted model $c_X(h,\tau)$ *(surface)*; (B) spatial covariance cross-section at $\tau = 0$; (C) spatial covariance cross-section at $\tau = 1$; and (D) spatial covariance cross-section at $\tau = 2$.

the decay rate of the spatial and temporal *SSS* correlations, respectively.

Table 5.4 describes the approach leading to the contingogram-based covariance $c_X^{(\psi)}$, which, just as the ordinary covariance c_X, is computed using the same *SSS* database. Accordingly, Fig. 5.7 presents the computed empirical $\hat{c}_X^{(\hat{\psi})}$ covariance values together with a plot of the fitted covariance model[ae]

$$c_X^{(\psi)}(\Delta\boldsymbol{p}) = e^{-3\left(\frac{h}{806.3532}+\frac{\tau}{41.2011}\right)}, \tag{5.10}$$

where now $\varepsilon_s = 806.3532$ km and $\beta = \varepsilon_t = 41.2011$ days. Apparently, $c_X^{(\psi)}$ has a longer spatial correlation range but a shorter temporal range than c_X, i.e., the $c_X^{(\psi)}$ represents a slower decay rate of spatial *SSS* correlations and a faster decay rate of temporal *SSS* correlations than c_X. Also, by comparing Figs. 5.6 and 5.7 visually, it is easy to see that the $c_X^{(\psi)}$ shows a better fitting performance than the c_X, especially at small spatial distances and temporal separations.[af] This improvement may have benefitted from the previously noted contingogram trait that it may magnify (stretch) the empirical covariance values making the fit easier compared to the original empirical values.

Also, the numerical coefficients of these two models, c_X and $c_X^{(\psi)}$, and the *RMSE* values of the corresponding model fits to the empirical values are listed in Table 5.5. The $c_X^{(\psi)}$ model offers a better fit to the empirical values than the c_X model (see the corresponding *RMSE* values).

The transformation introduced by the contingogram function leads to a contingogram-based covariance notion $c_X^{(\psi)}$ that, in this case study, seems to offer a better representation of the empirical data than the standard covariance notion c_X does. This is another worth-noticing feature of the present contingogram-based *SSS* modeling. By way of a summary, then:

> Since many physical applications point to the need of developing novel **chronotopologic dependence** assessment tools, other than the mainstream covariance and variogram functions, the sysketogram and contingogram functions are interesting possibilities that possess certain theoretical and practical advantages over the mainstream tools.

TABLE 5.4 Computation of the space-time contingogram-based covariance model.

Step	*Description*
1	Calculated empirical covariance $\hat{c}_X$ values in Fig. 5.6A were used to generate empirical contingogram $\hat{\psi}_X^{(\hat{c})}$ values in terms of Eq. (4.6b) of Section 4 above
2	The form of the theoretical $\psi_X^{(c)}$ model was also derived from Eq. (4.6b) of Section 4.
3	The c_X and $\psi_X^{(c)}$ models were fitted to the empirical $\hat{c}_X$ and $\hat{\psi}_X^{(\hat{c})}$ values, and the corresponding model coefficients ε_s, ε_t and c_0 were computed for each model (the fitting process used the *MATLAB* software)
4	The contingogram covariance $c_X^{(\psi)}$ model was derived from $\psi_X^{(c)}$ using Eq. (4.6b) of Section 4

[ae] It is worth-noting that, in order to assure the consistency of the fitting method, both models c_X and $c_X^{(\psi)}$ are derived using the *Particle Swarm Optimization* algorithm (Poh et al., 2007; He et al., 2020a, b).

[af] Since we normally set the radius to search for neighboring points within a relatively small range, the more accurate model fit occurs in small spatial and temporal distances, in which case the estimates are more precise.

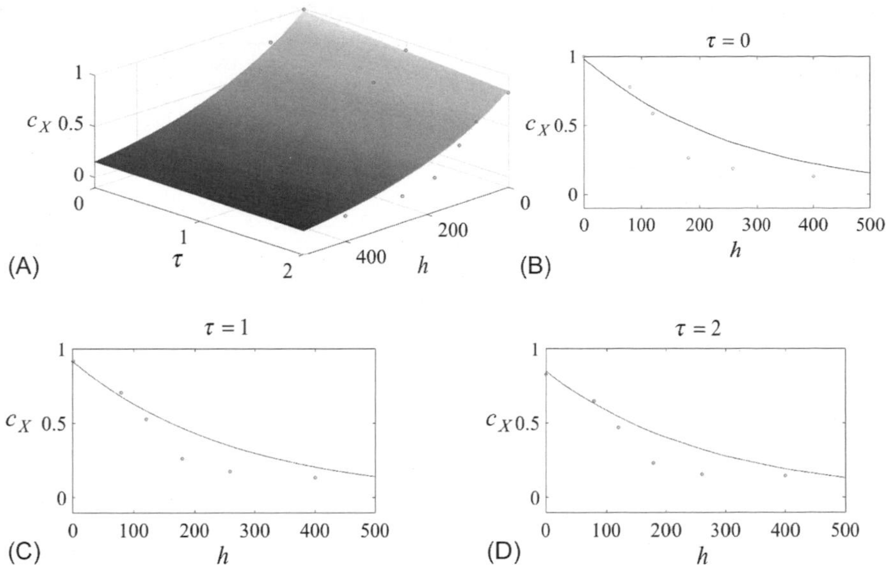

FIG. 5.7 (A) Empirical $\hat{c}_X(\Delta\boldsymbol{p})$ values (circles) and fitted model $c_X^{(\psi)}(\Delta\boldsymbol{p})$ (surface); (B) spatial covariance cross-section at $\tau = 0$; (C) spatial covariance cross-section at $\tau = 1$; and (D) spatial covariance cross-section at $\tau = 2$.

TABLE 5.5 Coefficients of the original covariance *vs.* the contingogram-derived covariance models.

Model	ε_s	ε_t	c_0	*RMSE*
c_X	554.3559	48.7061	0.9827	0.0647
$c_X^{(\psi)}$	806.3532	41.2011	0.9827	0.0603

After introducing the *BME* technique in Chapter 9, it will be used to generate the *SSS* chronotopologic maps and assess their relative performance by comparing the map accuracy when the contingogram-based covariance model $c_X^{(\psi)}$ is used *vs.* when the ordinary covariance model c_X is used. These tools could be of wide interest in earth, atmospheric, ocean and marine sciences, public health sciences, forecasting and mapping, objective analysis, geostatistics, and data assimilation.

6 Practice exercises

1. Which of the statements below describe hard data and which soft data?
 (a) Atmospheric temperature is currently 7°C.
 (b) Based on a forecasting model, the relative humidity will be 30% tomorrow.
 (c) Organic carbon concentration in soil sample measured by lab chemical analysis is 300 g/kg.
 (d) Historical records imply that $PM_{2.5}$ concentration in Haikou City, China on March 9, 2021 will be less than 5 μg/m^3.
 (e) Tiberius Menexes scored 94% on the final examination of the *CTDA* course.
 (f) Medical records include two cases: in one case the number of patients who suffered from flu in Beijing City during November 2015 was 500 without any missing

records, while in the other case the number of patients in Hangzhou City during November 2015 was 377 with missing records.

(g) Nitrate concentration mean and variance in a certain coastal Chinese region during the period January 1–25, 2019 were 4 mg/L and 0.9 $(\text{mg/L})^2$, respectively.

(h) The detection limit of a $PM_{2.5}$ instrument is 0.5 $\mu g/m^3$, and one gets several zero values.

2. Soil moisture is related to soil temperature. The soil moisture and the corresponding temperature were measured at 20 locations, Table 6.1. Generate soft data of soil moisture at 304.5, 305 and 308.8 K with Gaussian, uniform and interval probability distribution forms.
3. Consider the four hard data χ_i at points $p_i = (s_i, t_i)$, $i = 1, 2, 3, 4$, shown in Fig. 6.1. Theoretical models of the attribute means $\overline{X}_i$ and covariances c_{ij} are available. Describe the core *KB* of the case.
4. Find the g_α functions and D_α operators corresponding to the following stochastic ordinary differential equation describing a physical model

$$b_1 \frac{d}{ds} X(s) = X(s), \tag{6.1}$$

where b_1 is an empirical parameter.

5. Table 6.2 lists the $PM_{2.5}$-related variables and $PM_{2.5}$ concentrations in North China. Choose a model describing the relationship between AOD, PBLH, Wind, Elevation, Road length, Farm area and $PM_{2.5}$. Then, insert the values of Table 6.3 into your model to produce the corresponding soft $PM_{2.5}$ data that follow a Gaussian or a uniform probability law. Finally, combine the soft data (Table 6.3) and hard data (Table 6.4) to generate $PM_{2.5}$ estimates across the area during the first 4 days of October 2015 (the s_1 and s_2 coordinates of the area are shown in Tables 6.3 and 6.4) using the *BME* method. Given that the amount of data is rather large, the complete Tables 6.2–6.4 can be found on the website (https://www.dropbox.com/s/dr4fhodvrfmvzwv/Exercise%20of%20Ch%206.doc?dl=0).
6. Show that when $P_X[\chi_d] = 0$, the $P_X[\chi_k | \chi_d]$ is undefined, but the stochastic logic conditional $P_X[\chi_d \rightarrow \chi_k]$ can be defined, nevertheless.

TABLE 6.1 Dataset.

Temperature (K)	Soil moisture (%)
308.8695	10.4674
299.5238	25.2766
308.9605	9.8212
305.5883	17.2664
299.1705	27.2490
301.3420	23.0217
304.5626	20.0498
309.4901	10.6095
309.5787	10.2842
299.8914	28.2670
309.6471	8.8507
309.4860	11.4760
303.8245	20.8945
307.6034	12.7300
299.7026	26.8347
303.0611	18.7774
308.9888	8.7884
307.5065	13.6432
309.5139	11.5393
305.8689	18.3886

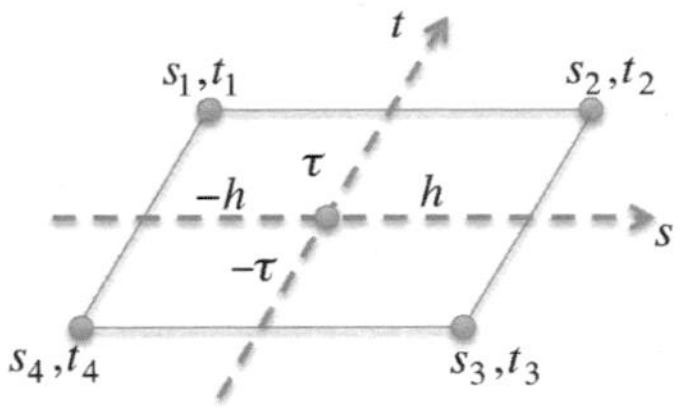

FIG. 6.1 Data distribution.

TABLE 6.2 $PM_{2.5}$-related variables and $PM_{2.5}$ concentrations.

No	s_1 coordinate	s_2 coordinate	*t*	AOD	PBLH	WIND	Elevation	Road	Farm area	$PM_{2.5}$
1	957,483.38	4,346,957.16	274	0.301	1011.94	6.98	52	2,130,885.63	0.00	14
2	950,970.50	4,346,207.47	274	0.156	968.19	7.27	57	2,126,104.03	0.00	13
3	952,713.12	4,340,610.35	274	0.338	984.61	7.09	52	1,842,910.07	1,622,838.94	16
4	955,127.93	4,352,709.97	274	0.286	994.51	7.07	47	1,915,623.80	13,072,869.24	13
5	961,052.99	4,348,279.71	274	0.183	1029.60	7.02	44	1,842,366.62	9,376,402.77	13
6	945,885.25	4,352,217.76	274	0.303	959.32	7.33	49	1,606,597.55	20,826,433.07	10
7	935,149.51	4,362,703.17	274	0.166	909.57	7.74	46	584,702.22	145,604,716.05	8
8	946,805.12	4,338,229.02	274	0.158	961.43	7.21	52	1,517,202.57	32,276,463.37	12
9	936,181.22	4,332,591.41	274	0.147	946.66	7.29	91	647,741.63	167,332,726.31	10
10	938,216.91	4,343,049.73	274	0.172	949.10	7.29	76	1,070,494.62	54,455,262.23	16

TABLE 6.3 Soft data.

No	s_1 coordinate	s_2 coordinate	*t*	AOD	PBLH	WIND	Elevation	Road	Farm area
1	1,029,214.03	4,647,675.32	274	0.048	784.42	12.80	1318	0.00	23,621,322.36
2	944,787.58	4,495,462.34	274	0.007	793.92	11.27	1016	20,615.90	12,171,292.05
3	1,098,357.86	4,514,756.18	274	0.032	1184.51	7.78	739	0.00	44,898,544.02
4	1,149,463.60	4,521,831.54	274	0.020	1208.54	8.26	766	0.00	128,023,960.86
5	901,206.75	4,419,590.81	274	0.063	721.13	9.96	1200	5087.97	3,516,151.04
6	952,915.49	4,425,426.94	274	0.056	863.18	8.49	727	59,254.70	2,163,785.25
7	1,004,586.46	4,431,588.48	274	0.030	989.72	6.33	270	105,362.00	115,582,195.65
8	1,056,217.62	4,438,075.17	274	0.049	1221.19	5.26	873	105,875.00	28,309,523.74
9	1,107,806.92	4,444,886.77	274	0.024	1297.23	5.47	726	35,695.30	63,020,245.52
10	1,159,352.32	4,452,022.99	274	0.019	1303.70	6.47	357	35,443.00	108,640,051.30

TABLE 6.4 Hard data.

No	s_1 coordinate	s_2 coordinate	t	$PM_{2.5}$	No	s_1 coordinate	s_2 coordinate	t	$PM_{2.5}$
1	957,483.38	4,346,957.16	274	14	195	826,942.40	4,007,374.70	275	23
2	957,205.12	4,342,036.89	274	13	196	957,483.38	4,346,957.16	276	26
3	950,970.50	4,346,207.47	274	13	197	957,205.12	4,342,036.89	276	26
4	952,713.12	4,340,610.35	274	16	198	950,970.50	4,346,207.47	276	28
5	955,127.93	4,352,709.97	274	13	199	952,713.12	4,340,610.35	276	28
6	961,052.99	4,348,279.71	274	13	200	955,127.93	4,352,709.97	276	27
7	945,885.25	4,352,217.76	274	10	201	961,052.99	4,348,279.71	276	27
8	935,149.51	4,362,703.17	274	8	202	945,885.25	4,352,217.76	276	28
9	939,018.82	4,353,141.26	274	6	203	935,149.51	4,362,703.17	276	23
10	946,805.12	4,338,229.02	274	12	204	946,805.12	4,338,229.02	276	28

7. Prove the following properties
   (a) $P_X[\neg\chi_k|\chi_d] = 1 - P_X[\chi_k|\chi_d]$;
   (b) $P_X[\chi_d] = P_X[\chi_k|\chi_d]P_X[\chi_d] + P_X[\neg\chi_k|\chi_d]P_X[\chi_d]$
   (c) $P_X[\chi_d \to \neg\chi_k] = 2 - P_X[\chi_d] - P_X[\chi_d \to \chi_k]$
   (d) $P_X[\chi_k|\chi_d] \neq P_X[\neg\chi_k|\neg\chi_d]$,
   (e) $P_X[\chi_d \to \chi_k] = P_X[\neg\chi_d \to \neg\chi_k]$,
   (f) $P_X[\chi_k|\chi_d] \leq P_X[\chi_d \to \chi_k]$, when $P_X[\chi_d] \neq 0$.

8. Show that $P_X[\chi_k|\chi_d] = 1$ implies that $P_X[\chi_d \to \chi_k] = 1$ and vice versa.

9. Prove the following assertions:
   (a) $P_X[\chi_k|\chi_d] = 1$ and $P_X[\chi_d] \neq 0$ imply that $P_X[\neg\chi_k|\chi_d] = 0$, regardless of the $P_X[\chi_d]$ value. If $P_X[\chi_d] = 0$, the $P_X[\chi_k|\chi_d]$ and $P_X[\neg\chi_k|\chi_d]$ are undefined, which makes sense intuitively. Why?
   (b) The *a* above holds if $X(\boldsymbol{p}_i)$ and $X(\boldsymbol{p}_j)$ are independent with $P_X[\chi_k] = 1$.
   (c) The *a* above does not hold if $X(\boldsymbol{p}_i)$ and $X(\boldsymbol{p}_j)$ are disjoint.

10. Show the following:
   (a) $P_X[\chi_d \to \chi_k] = 1$ and $P_X[\chi_d] = 1$ imply that $P_X[\chi_d \to \neg\chi_k] = 0$. But, if $P_X[\chi_d] = 0$, then $P_X[\chi_d \to \chi_k] = P_X[\chi_d \to \neg\chi_k] = 1$ always holds, which makes no sense intuitively. Why?
   (b) If $X(\boldsymbol{p}_i)$ and $X(\boldsymbol{p}_j)$ are independent, i.e., $P_X[\chi_k|\chi_d] = P_X[\chi_k]$, then $P_X[\chi_d \to \chi_k] = 1$ if $P_X[\chi_d] = 0$ or $P_X[\chi_k] = 1$.
   (c) Since $P_X[\chi_d] \neq 0$ by assumption of *a* above, one can only have $P_X[\chi_k] = 1$. Then, show that $X(\boldsymbol{p}_i)$ and $X(\boldsymbol{p}_j)$ are independent with $P_X[\chi_d] = P_X[\chi_k] = 1$.
   (d) If $X(\boldsymbol{p}_i)$ and $X(\boldsymbol{p}_j)$ are disjoint, then *a* above does not hold.

11. Repeat the calculations of Example 2.9(cont.) of Section 2 using a variogram model of your choice as part of the core *KB*.

12. In Example 2.9(cont.) of Section 2 show how the skewness and kurtosis measures can be expressed by the proposed g_α formalism. Use any numerical skewness and kurtosis values you find appropriate.

13. In what sense can an error be considered as information?
14. List some analytic and synthetic statements encountered in your field of expertise.
15. Derive a multipoint formulation of the equations of Table 3.2, Section 3.
16. Use the attribute dataset in Table 6.5 to calculate the empirical covariance, sysketogram, and contingogram values. Choose suitable models to fit to these values. Then, use the leave-one-out cross validation technique to test the performance of each model. Given that the amount of data is rather large, the complete Table 6.5 can be found in the website (https://www.dropbox.com/s/dr4fhodvrfmvzwv/Exercise%20of%20Ch%206.doc?dl=0).
17. Convert the following theoretical covariance models into sysketogram and contingogram models:

(a) $c_X(\Delta \boldsymbol{p}) = 2.4e^{-\frac{h^2}{8^2}}e^{-\tau^2}$

(b) $c_X(\Delta \boldsymbol{p}) = e^{-\frac{h^2}{3^2}}e^{-\frac{\tau}{3}}$

(c) $c_X(\Delta \boldsymbol{p}) = 2.4\left(1 - \frac{3h}{2} + \frac{h^3}{4}\right)e^{-\tau^2}$

(d) $c_X(h) = e^{-\frac{h^2}{3^2}}$

18. Describe the differences between the covariance, the sysketogram, and the contingogram functions. Discuss the merits of the sysketogram and contingogram compared to the classical covariance.
19. Derive Eqs. (4.6a,b) of Section 4 from Eqs. (4.2) and (4.3).
20. Based on the approach of the sea surface salinity case study discussed in this chapter, compute and interpret a chronotopologic attribute dataset in the field of your interest.
21. Describe conditions under which it is appropriate to use the sysketogram and contingogram functions rather than the classical covariance function.
22. Consider situations that the classical covariance cannot offer an inadequate representation of the actual chronotopologic dependence situation.

TABLE 6.5 Dataset.

s_1	s_2	t	X	s_1	s_2	t	X	s_1	s_2	t	X
1,480,997	4,458,841	1	47	1,353,941	4,322,969	3	99	1,154,884	4,339,393	5	166
1,472,582	4,468,904	1	47	1,302,487	4,402,917	3	99	1,161,574	4,340,171	5	169
1,091,088	4,323,697	1	51	1,346,150	4,373,789	3	107	1,007,277	4,345,440	5	174
935,338	4,140,374	1	55	1,353,065	4,294,132	3	110	1,016,565	4,198,685	5	185
1,408,718	4,458,456	1	58	1,163,981	4,414,801	3	112	1,100,370	4,112,158	5	190
1,320,434	4,303,084	1	59	1,354,169	4,291,400	3	113	1,144,570	4,092,021	5	193
1,516,325	4,435,803	1	67	1,231,788	4,341,547	3	115	964,095	4,279,696	5	194
1,009,922	4,262,892	1	71	1,123,981	4,400,626	3	116	1,129,094	4,112,649	5	195
1,368,258	4,465,443	1	74	1,237,623	4,340,333	3	120	1,031,332	4,294,803	5	196
1,052,407	4,311,959	1	75	1,161,574	4,340,171	3	125	1,062,032	4,311,048	5	197

23. As precipitation is the major input to water bodies, it is important to investigate any changes in rainfall in both time and space. Can the sysketogram and/or contingogram functions be useful in this respect, and how?
24. Prove that $P_G[\chi_k|\chi_d] \leq P_G[\chi_d \rightarrow \chi_k]$.
25. Express the statistical probability of Eq. (3.3) of Section 3 in terms of the equivalence probability of Eq. (3.39) of Section 3.
26. Show that if the attribute *PDF* is of the form $f_{X_{ij}} = e^{a[X^2(p_i)+X^2(p_j)]} f_{X_i} f_{X_j}$, where a is an empirical coefficient, the sysketogram and contingogram functions are given by

$$\beta_X(p_i, p_j) = a\left[\overline{X^2(p_i) + X^2(p_j)}\right] = \frac{a}{2\pi}\left[\overline{e^{-X^2(p_i)}} + \overline{e^{-X^2(p_j)}}\right],$$

$$\psi_X(p_i, p_j) = \overline{e^{a[X^2(p_i) + X^2(p_j)]}} - 1 = \overline{e^{\frac{a}{2\pi}\left[e^{-X^2(p_i)} + e^{-X^2(p_j)}\right]}} - 1,$$

respectively.

CHAPTER

7

Chronotopologic interpolation

1 Introduction

In the real-world, it is usually impossible to measure, observe or record a natural attribute value (e.g., atmospheric pollution, subsurface contamination, disease rate, soil moisture, crime rate) at all spatial locations and at every time instance of interest. This is the reason that one of the most important goals of *CTDA* is as follows:

> Develop techniques that assimilate the body of knowledge about a natural attribute, commonly in the form of a finite set of samples, to produce adequate **chronotopologic estimates** of the actual (but unobtainable) attribute values at a set of unsampled points. Attribute **maps** are visualizations of these numerical estimates together with the attribute data values over the chronotopologic domain of interest.

Following this description of the *CTDA* goals, some initial thoughts about the subject matter are as follows:

- Chronotopologic estimation and mapping must be **hybrid**, capable of considering space and time jointly but treating them differently. That is, a "homogenous" space-time (four-dimensional) representation may not be adequate in estimation, because time and space exhibit important differences.
- Many fields have been revolutionized by **technological advances** (mathematics by computer science, geometry by visualization, etc.). The increasing ubiquity of visualization technology used in mapping has made it possible to communicate knowledge and experience with more people and in more interesting ways.

Before proceeding further, an essential distinction of the main kinds of chronotopologic estimation should be described.

1.1 Chronotopologic estimation types

Consider an attribute represented by an *S/TRF* model,[a] $X(\boldsymbol{p})$, $\boldsymbol{p} = (\boldsymbol{s}, t) \in D \times T$, with D and T being the spatial area and time period, respectively, within which samples are available. "Estimation" is a general term. Yet, based on experience, there exist certain distinct

[a] The readers may want to review Section 2 of Chapter 5.

Quantitative Analysis and Modeling of Earth and Environmental Data
https://doi.org/10.1016/B978-0-12-816341-2.00014-9

manifestations of chronotopologic attribute estimation, as follows.

Interpolation: attribute estimates are sought at points within the sampling domain (*D*, *T*), i.e., the unsampled points $p_k = (s_k, t_k)$ at which attribute values are sought are such that $s_k \in D$ and $t_k \in T$.

Extrapolation: attribute estimates are sought at points $p_k = (s_k, t_k)$ outside the sampling area *D* but within the time period *T*, i.e., $s_k \notin D$ and $t_k \in T$.

Prediction: attribute estimates are sought at points within the sampling area *D* but outside the time period *T* (at future time instants), i.e., $s_k \in D$ and $t_k \notin T$ $(t_k > T)$.

Chronotopologic interpolation (*CTI*) is the most common, and easier to handle, manifestation of chronotopologic estimation. Both other manifestations (extrapolation and prediction) are more difficult to handle, and their meaningful resolution usually requires information beyond the limited set of available samples. Specifically, as has been emphasized in Christakos and Raghu (1996) and Christakos and Hristopulos (1998):

> Beyond numerical data, the **reliable** extrapolation and prediction of an attribute require the involvement of core knowledge (including physical laws and scientific theories; see also Section 2 of Chapter 2).

Ab hinc, the techniques covered in this chapter are primarily *CTI* techniques that are useful in physical and human studies. *CTI* is an essential feature of *TGIS*, and many *TGIS*-generated maps are the result of interpolation. Also, interpolation is frequently used as an aid in the decision-making process of many disciplines, such as resource exploration and health management.

Example 1.1 As was discussed in Section 1 of Chapter 5, physical phenomena leave distinct yet complementary signatures across space and in time. In earth and atmospheric sciences, separating the signature of the several sources in satellite and ground-based stations is carried out by means of chronotopologic interpolation. Interpolation techniques are used to generate a variety of surfaces like the digital terrain surface (using GNSS, LiDAR or conventional measurements), and the quasi-geoid conversion surface (using leveling observations).

Sine dubio, the list of interpolation techniques discussed in this chapter is not an exhaustive one. Additional estimation techniques (including extrapolation) are the topics of subsequent chapters.[b]

1.2 Linear CTI

In common with previous discussions about *S/TRF* theory, let $X(\boldsymbol{p})$ be an *S/TRF* representing an attribute that varies with point $\boldsymbol{p} = (\boldsymbol{s}, t)$, as shown in Fig. 1.1, where $\boldsymbol{s} = (s_1, s_2)$ denotes the spatial location and *t* denotes time (s_1 and s_2 are the spatial coordinates of the location). We recall that two chronotopologic lags are distinguished, depending on the type of data analysis considered, *S-TDA vs. STDA* (Table 2.1 in Section 2 of Chapter 2). Let $\hat{X}(\boldsymbol{p}_k)$ denote the sought interpolation at the point of interest $\boldsymbol{p}_k$, and $X(\boldsymbol{p}_i) = \chi_i$ denote the observed values at the sampled points $\boldsymbol{p}_i$ $(i = 1, 2, \ldots, m)$, Fig. 1.1. We start with the most widely used form of *CTI*.

> The general form of the **linear** *CTI* of $X(\boldsymbol{p})$ is expressed as the weighted average of the available sampled data.
>
> $$\hat{X}(\boldsymbol{p}_k) = \sum_{i=1}^{m} \lambda_i X(\boldsymbol{p}_i) \tag{1.1}$$
>
> where λ_i is the weight assigned to each sampled point, and *m* represents the number of sampled locations used.

[b] Incorporating salient forms of core knowledge in *CTI*, in addition to case-specific data, will be discussed in Chapters 9 and 10.

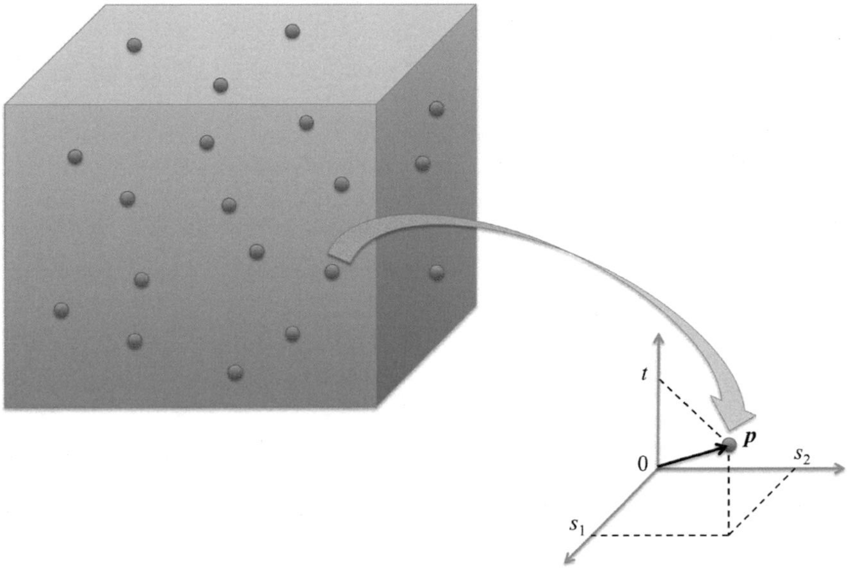

FIG. 1.1 Representation of chronotopologic point coordinates.

Example 1.2 Two commonly encountered interpolation applications are as follows.

- Several rainfall stations typically exist across a country, but between the stations, there is no recording of rainfall values. Nevertheless, one wants to derive rainfall estimates at unsampled locations too during specified time periods. An efficient way to do this is by using interpolation techniques for rainfall surface generation based on Eq. (1.1).
- Higher interpolation resolution is often required in atmospheric sciences including the use of satellite data affected by clouds and precipitation at scales where error evolution is highly nonlinear.

1.3 Classifications of *CTI* techniques

The *CTI* techniques can be further classified in a number of ways based on the relevant criteria chosen.

❶ *Chronopoint* (or simply *point*) *CTI vs. chronoblock CTI*.

In the point case, given a number of points with known coordinates and attribute values, new attribute values are interpolated at unsampled points. The interpolated grid points are often used as the data input to computer contouring algorithms. Once the grid of points has been determined, isolines (e.g., contours) can be threaded between them using *CTI*. Point to point interpolation is the most frequently performed type of *CTI* in *GIS*.

Example 1.3 This kind of *CTI* is used in applications involving weather station readings, spot heights, oil well readings, and porosity measurements.

In the chronoblock case (see, also, Section 3 of Chapter 8), given a set of data mapped on one set of source chronoblock, the attribute values are interpolated at an unsampled set of chronoblocks.

Example 1.4 Given population counts for census tracts, estimate populations for electoral districts.

❷ *Global CTI vs. local CTI*.

Global *CTI* uses a single interpolation function that applies across the entire domain of interest so that the interpolated values are globally interlinked (*exempli causa*, a change of an attribute value affects the entire interpolation map). Local *CTI* use functions that apply only within separate windows including subsets of points so that the interpolated values are locally interlinked (e.g., a change of an attribute value in a certain window affects the interpolation map only within the same window). As one might expect, global *CTI* generates smoother attribute maps than local *CTI*. Global *CTI* is appropriate when the attribute distribution is known to follow a specific pattern. Therefore, the distinction between global and local *CTI* is a matter of the domain considered and not a matter of different shapes of the interpolation function considered. Rather trivially, then, local becomes global when the window of the former covers the entire domain of interest.

❸ *Exact CTI vs. nonexact CTI.*

Exact *CTI* honors the attribute data points (i.e., at the sampling points *CTI* yields attribute values identical to the data), whereas non-exact *CTI* does not necessarily honor these points. *Kriging* techniques (Chapter 8) are exact *CTI*. Honoring data points is an important requirement in certain applications (assuming that there is no data noise), but not in others where the data may be soft or characterized by uncertainty (in which case, at the sampling points *CTI* yields estimates that differ from the data). Also, exact *CTI* may not be required when the processing of initial information is of the essence (see, e.g., the case of *Kalman filter* techniques, Christakos, 1992a).

❹ *Deterministic CTI vs. indeterministic CTI.*

Generally, deterministic *CTI* rely on the mechanistic perspective of the phenomenal world that assumes that the attribute data are error-free, and, the unknown attribute values can be interpolated with certainty. Indeterministic *CTI*, on the other hand, is based on the opposite viewpoint, i.e., the attribute data are characterized by *in-situ* uncertainty or noise, and the unknown attribute values can be interpolated only in statistical or probabilistic terms. Indeterministic *CTI* can be further classified as statistical and geostatistical.

The above distinctions involve different *modeling assumptions* about the underlying attribute variability. An understanding of the initial assumptions is the key to adequate interpolation. Based on such assumptions, this book studies three kinds of *CTI* techniques:

- **Deterministic** *CTI* techniques, like inverse distance or radial basis functions (to be reviewed in Section 2), which use simple functions to produce rather rough representations of the attribute distribution.
- **Statistical** techniques, like local sample means and regression (to be reviewed in Section 3), which basically fit different kinds of surfaces to the available dataset based on statistical criteria.
- **Geostatistical** techniques, like *Kriging* (Chapter 8) and *BME* (Chapter 9), which generate interpolations that account for both the attribute's chronotopologic correlation (dependence) structure and uncertain fluctuations, and they also provide an assessment of the associated interpolation accuracy.

A few brief comments are in order concerning the essential features of these *CTI* techniques. The deterministic techniques are based either on the degree of similarity or degree of smoothing in relation with neighboring data points. They are usually simple and easy to implement, but they are based on an over-simplistic choice of interpolation weights, they do not account for uncertainty sources and nonstationary dynamics, they ignore the presence of varying attribute patterns along different directions (anisotropy), and they do not provide any interpolation accuracy assessment. The statistical techniques use neighboring data averaging to interpolate uncertain values at unsampled points based on the degree of chronotopologic smoothing. They rely on restrictive assumptions, the data are assumed independent of each other, normally distributed and homogeneously invariant, and some of these techniques do not discriminate between samples at considerable distances from the interpolation point and those very close to it. Lastly, the geostatistical techniques estimate uncertain values at unsampled points based on an optimization criterion, *sensu lato* (technical or epistemic), and provide probabilistic assessments of interpolation quality based on chronotopologic correlation measures among data points. Naturally, they are more difficult to implement than the deterministic and the statistical techniques (more parameters need to be set, a system of equations needs to be solved, etc.), yet they are superior to them in almost any respect.

In the end, it is up to the investigators to decide what kind of technique best fits the objectives of the investigation and adequately satisfies the physical constraints of the situation. Whatever their choice may be, they should seriously consider the following reality check:

When selecting any of the above techniques, one should be aware of the **constraints** that the physical world imposes on the chronotopologic attribute estimates generated by the techniques.

These are application-dependent constraints, several of which are discussed throughout the book.

1.4 Mapping technology

The next *CTDA* stage is the visualization of the numerical interpolation results in an informative and creative manner.

In *CTDA* practice, numerical interpolation is followed by **mapping**, which is the visualization of the product of interpolation.

Particularly, in the graphical terms of Fig. 1.2A–C, the two relevant *CTDA* stages can be characterized as follows:

- **Computational** stage: With the help of an interpolation technique based on Eq. (1.1) the attribute samples of Fig. 1.2A are used to compute the set of interpolated attribute values on the chronotopologic regular grid of Fig. 1.2B.

- **Visualization** stage: With the help of visualization technology (Section 6, Chapter 1), the interpolated values of Fig. 1.2B are transformed into the chronotopologic map of Fig. 1.2C.

Mapping often involves a *base map*, i.e., a reference map that includes basic geographic features and is used to overlay the attribute data. A base map is available in different formats on the Internet (provided by governments or private organizations). The software used in the computational stage above would determine the appropriate base map format (i.e., depending on the mapping technique used, the base map may include user-defined quadrants and thematic boundaries).

Example 1.5 When assessing disease incidence, data aggregation may be required using a polygon grid consisting of zip codes, counties, or states.

1.4.1 The merging of two worlds

The readers may appreciate the fact that, unlike the physical world, the digital world, within which most of the interpolation techniques operate, is very flexible *albeit* often isolated from its physical counterpart. This being the case, it seems appropriate to follow a holistic kind of *CTDA* approach.

By complementing the **physical** and the **digital** worlds, digital information can be combined with its physical counterpart so that the

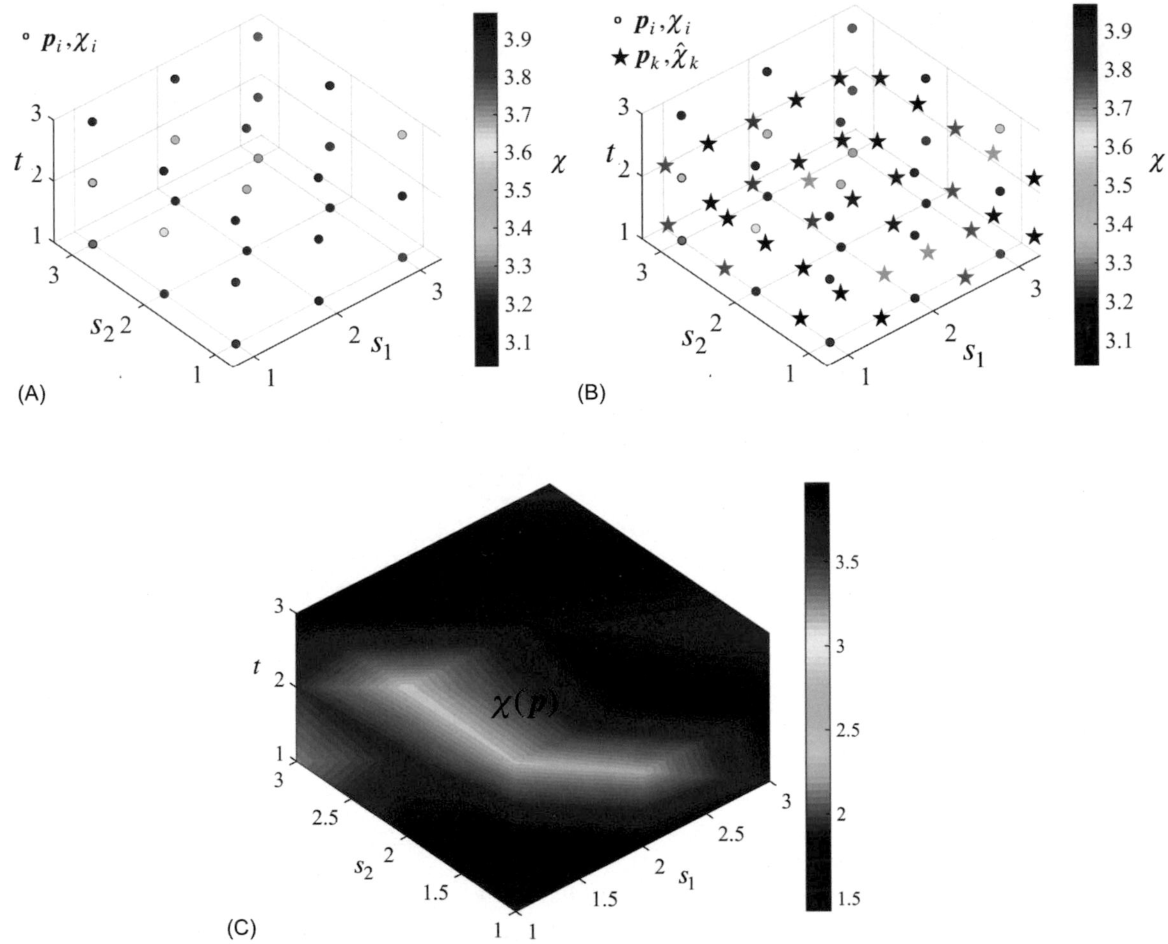

FIG. 1.2 The methodological process of chronotopologic mapping. (A) The 27 attribute data points (p_i, χ_i) are denoted by dots; (B) the attribute interpolation points ($p_k, \hat{\chi}_k$) are generated on a regular grid and denoted by asterisks, and (C) the resulting visualization in the form of a chronotopologic attribute map $\chi(p)$.

interpolation benefits from both the power of physical sensing and the flexibility of digital interaction.

Nowadays, there seems no reason to doubt that the lines between the physical and digital worlds are blurring at a fast pace, which leads to considerable changes occurring as new technologies affect the implementation of interpolation theories and techniques that already exist. Accordingly, *CTDA* interpolation software solutions become available that integrate the physical world with an increasingly detailed digital one, with connected devices constantly talking to one another worldwide. Although these software libraries do not actually involve any physical assets, rather they use digital technologies to facilitate their functioning, yet they operate in the real, physical world.

The following sections review some of the most commonly used interpolation techniques, according to the threefold classification discussed above.

2 Deterministic chronotopologic interpolation techniques

In this section, we review some of the most popular deterministic techniques. These techniques are characterized by their simplicity and easiness to implement, on the one hand, as well as by their considerable level of arbitrariness as regards the choice of the weights λ_i, on the other hand.

2.1 The inverse chronotopologic metric interpolation technique

In order to appreciate the meaning of interpolation, we will first consider a particularly simple *albeit* instructive technique of approximating attribute values at unsampled points.

Inverse chronotopologic metric (*ICTM*) interpolation produces λ_i-solutions of Eq. (1.1) of Section 1 above in which the weight given to each sample is inversely proportional to the metric $|\Delta \boldsymbol{p}_{ki}|$ between that sample point $\boldsymbol{p}_i$ and the interpolation point $\boldsymbol{p}_k$, i.e.,

$$\hat{X}(\boldsymbol{p}_k) = \frac{1}{A}\sum\nolimits_{i=1}^{m} \frac{1}{|\Delta \boldsymbol{p}_{ki}|} X(\boldsymbol{p}_i) \tag{2.1}$$

where $\lambda_i = \frac{1}{A|\Delta \boldsymbol{p}_{ki}|}$ and $A = \sum\nolimits_{i=1}^{m} \frac{1}{|\Delta \boldsymbol{p}_{ki}|}$ is a normalization parameter.

The *ICTM* basically assumes that each datum at point $\boldsymbol{p}_i$ has an influence on point $\boldsymbol{p}_k$ that diminishes with the increasing metric $|\Delta \boldsymbol{p}_{ki}|$ magnitude between the two points. Accordingly, the technique weights the data points that are chronotopologically closer to the interpolated point $\boldsymbol{p}_k$ more than those further away. When implemented in geographical studies, the *ICTM* technique expresses in quantitative terms the well-known *Tobler's* law (Section 2 of Chapter 1): things that are closer to each other affect each other to a larger degree than things that are farther apart. To obtain an attribute estimate at the unsampled point of interest $\boldsymbol{p}_k$, the *ICTM* uses sampling points $\boldsymbol{p}_i$ surrounding $\boldsymbol{p}_k$ in a way that the sample at $\boldsymbol{p}_i$ nearer to $\boldsymbol{p}_k$ will have greater influence (i.e., larger weight) on the interpolated value at $\boldsymbol{p}_k$ than samples that are farther away, which is why the term *"ICTM weighted"* (*ICTMW*) is sometimes used in the literature.

- A salient feature of the *ICTM* interpolation is that the expression of Eq. (2.1) depends in a rather fundamental manner on the form of the metric $|\Delta \boldsymbol{p}_{ki}|$ assumed, i.e., **bi-metric** (separate space-time metric) *vs.* **uni-metric** (unique spacetime metric).[c]

In practice, various metric forms can be considered, depending on the phenomenon. An illustrative example follows.

Example 2.1 In the *S-TDA* context, assuming a bi-metric, the weights λ_i in Eq. (2.1) may be viewed as the product of the distance $|\boldsymbol{h}_{ki}| = |\boldsymbol{s}_k - \boldsymbol{s}_i|$ (between the sample location $\boldsymbol{s}_i$ and the interpolation location $\boldsymbol{s}_k$) and the time separation $\tau_{ki} = |t_k - t_i|$ (between the sample timing t_i and the interpolation timing t_k), i.e.,

$$\lambda_i = \frac{1}{A|\Delta \boldsymbol{p}_{ki}|} = \frac{1}{A|\boldsymbol{h}_{ki}|\tau_{ki}} \tag{2.2a}$$

In the *STDA* setting, on the other hand, assuming a uni-metric, the weights λ_i in Eq. (2.1) may be viewed as inversely proportional to the composite (Pythagorean) metric, i.e.,

$$\lambda_i = \frac{1}{A|\Delta \boldsymbol{p}_{ki}|} = \frac{1}{A\sqrt{|\boldsymbol{h}_{ki}|^2 + \varepsilon\tau_{ki}^2}} \tag{2.2b}$$

where $\varepsilon = \left(\dfrac{\textit{spatial correlation range}}{\textit{temporal correlation range}}\right)^2$.

[c] At this point, the readers may find it useful to review Section 3 of Chapter 2.

ICTM usually produces attribute maps with gradual changes (although abrupt changes may result if a reduced number of data points is used locally). It is time for some numerical demonstrations of the notion of *ICTM* interpolation.

The following example presents a *step-by-step* **computational** process focusing on the implemention of the *ICTM* interpolation technique in practice.

Example 2.2 The department of meteorology monitored the amount of rainfall in a city using nine monitoring stations (Fig. 2.1). The spatial coordinates and time instances of the stations, s_i and t_i, respectively ($i = 1,\ldots,9$), and the corresponding amount of rainfall are presented in Table 2.1. The meteorology department wants to know the amount of rainfall at the location $s_k = s_{10}$ and $t_k = t_{10}$ in Table 2.1, where no monitoring occurs. Using the *ICTM* technique, the interpolated rainfall value at $s_k = s_{10}$ and $t_k = t_{10}$ is derived as follows.

Step 1: The distances between point $p_{10} = (s_{10,1},\ s_{10,2}, t_{10})$ and the other nine monitoring station points $p_i = (s_{i,1},\ s_{i,2}, t_i)\,(i = 1,\ldots,9)$ were calculated using the metric $|\Delta p_{10,i}| = \sqrt{|h_{10,i}|^2 + 10\tau_{10,i}^2}$, i.e.,

$$|\Delta p_{10,1}| = \sqrt{(s_{10,1} - s_{1,1})^2 + (s_{10,2} - s_{1,2})^2 + \varepsilon}$$

$$(t_{10} - t_1)^2 = \sqrt{(2-1)^2 + (1-1)^2 + 3\times(2-1)^2} = 2,$$

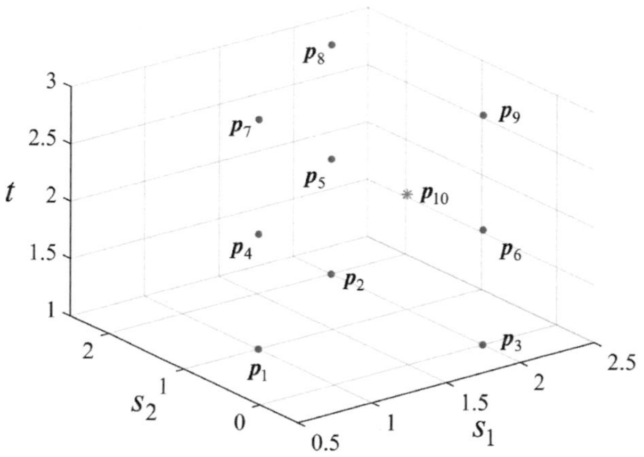

FIG. 2.1 The distribution of the nine meteorological monitoring points p_i, $i = 1, \ldots, 9$ (dots) and the unmonitored point p_{10} (asterisk).

TABLE 2.1 Information at the nine data stations i = 1–9 and the unsampled station k = 10.

No.	$s_{i,1}$	$s_{i,2}$	t	Rain (mm)	No.	$s_{i,1}$	$s_{i,2}$	t	Rain (mm)
1	1	1	1	12	6	2	0	2	18
2	2	2	1	13	7	1	1	3	15
3	2	0	1	15	8	2	2	3	18
4	1	1	2	14	9	2	0	3	17
5	2	2	2	16	10	2	1	2	?

$$|\Delta p_{10,2}| = \sqrt{(s_{10,1} - s_{2,1})^2 + (s_{10,2} - s_{2,2})^2 + \varepsilon(t_{10} - t_2)^2}$$
$$= \sqrt{(2-2)^2 + (1-2)^2 + 3 \times (2-1)^2} = 2,$$

$$|\Delta p_{10,3}| = \sqrt{(s_{10,1} - s_{3,1})^2 + (s_{10,2} - s_{3,2})^2 + \varepsilon(t_{10} - t_3)^2}$$
$$= \sqrt{(2-2)^2 + (1-0)^2 + 3 \times (2-1)^2} = 2,$$

$$|\Delta p_{10,4}| = \sqrt{(s_{10,1} - s_{4,1})^2 + (s_{10,2} - s_{4,2})^2 + \varepsilon(t_{10} - t_4)^2}$$
$$= \sqrt{(2-1)^2 + (1-1)^2 + 3 \times (2-2)^2} = 1,$$

$$|\Delta p_{10,5}| = \sqrt{(s_{10,1} - s_{5,1})^2 + (s_{10,2} - s_{5,2})^2 + \varepsilon(t_{10} - t_5)^2}$$
$$= \sqrt{(2-2)^2 + (1-2)^2 + 3 \times (2-2)^2} = 1,$$

$$|\Delta p_{10,6}| = \sqrt{(s_{10,1} - s_{6,1})^2 + (s_{10,2} - s_{6,2})^2 + \varepsilon(t_{10} - t_6)^2}$$
$$= \sqrt{(2-2)^2 + (1-0)^2 + 3 \times (2-2)^2} = 1,$$

$$|\Delta p_{10,7}| = \sqrt{(s_{10,1} - s_{7,1})^2 + (s_{10,2} - s_{7,2})^2 + \varepsilon(t_{10} - t_7)^2}$$
$$= \sqrt{(2-1)^2 + (1-1)^2 + 3 \times (2-3)^2} = 2,$$

$$|\Delta p_{10,8}| = \sqrt{(s_{10,1} - s_{8,1})^2 + (s_{10,2} - s_{8,2})^2 + \varepsilon(t_{10} - t_8)^2}$$
$$= \sqrt{(2-2)^2 + (1-2)^2 + 3 \times (2-3)^2} = 2,$$

$$|\Delta p_{10,9}| = \sqrt{(s_{10,1} - s_{9,1})^2 + (s_{10,2} - s_{9,2})^2 + \varepsilon(t_{10} - t_9)^2}$$
$$= \sqrt{(2-2)^2 + (1-0)^2 + 3 \times (2-3)^2} = 2.$$

Step 2: The normalization parameter A was found to be

$$A = \sum_{i=1}^{9} \frac{1}{|\Delta p_{10,i}|}$$
$$= \frac{1}{2} + \frac{1}{2} + \frac{1}{2} + 1 + 1 + 1 + \frac{1}{2} + \frac{1}{2} + \frac{1}{2} = 6.$$

Step 3: Lastly, the desired interpolation value of rainfall at location 10 was calculated as

$$\hat{X}(p_{10}) = \frac{1}{A} \sum_{i=1}^{9} \frac{1}{|\Delta p_{10,i}|} X(s_i)$$
$$= \frac{1}{6}\left(\frac{12}{2} + \frac{13}{2} + \frac{15}{2} + 14 + 16 + 18 + \frac{15}{2} + \frac{18}{2} + \frac{17}{2}\right)$$
$$= 15.5.$$

2.1.1 Pros and cons

The use of the *ICTM* technique relies on the assumption that the data values decrease in interpolation influence linearly as a function of $|\Delta p_{ki}|^{-1}$. This is, admittedly, a rather restrictive assumption. Furthermore, the *ICTM* does not account for variational anisotropy, it does not necessarily honor local (high or low) attribute values but rather looks at a moving average of nearby data points, and it tends to estimate local trends. Yet, the *ICTM* has been among the earliest computer-implemented interpolation techniques, mainly due to the fact that it is easily programmed. It is also easily generalizable in a rather straightforward way.

2.2 The generalized ICTM technique

An anticipated and interesting extension of Eq. (2.1) with considerably higher flexibility in practice is obtained as follows.

The **generalized inverse chronotopologic metric (*GICTM*)** interpolation

$$\hat{X}(p_k) = \sum_{i=1}^{m} \lambda_i X(p_i), \tag{2.3}$$

where $\lambda_i = \frac{1}{A|\Delta p_{ki}|^{\rho}}$, $A = \sum_{i=1}^{m} \frac{1}{|\Delta p_{ki}|^{\rho}}$, and the power ρ can assume any positive real value (e.g., $\rho = 0.5,\ 1,\ 2,\ 3$), depending on the application.

Example 2.3 Possible metrics to be used in Eq. (2.3) include the bi-metric

$$|\Delta p_{ki}|^{\rho} = |h_{ki}|^{\rho_1} \tau_{ki}^{\rho_2},$$

where $\rho = \rho_1 + \rho_2$, and the uni-metric

$$|\Delta p_{ki}|^{\rho} = \left(|h_{ki}|^2 + \varepsilon \tau_{ki}^2\right)^{\frac{\rho}{2}}.$$

Curves of $|h_{ki}|^{\rho_1}$ for different ρ_1 values are shown in Fig. 2.2A. Similar plots can be drawn for $\tau_{ki}^{\rho_2}$ assuming different ρ_2 values. For a fixed ρ_2, the larger the ρ_1 is, the faster the $|h_{ki}|^{\rho_1}$ increases, and the faster the corresponding weight $\lambda_i \propto |h_{ki}|^{-\rho_1}$ declines. Similarly, for a fixed ρ_1, the larger ρ_2 is, the faster the $\tau_{ki}^{\rho_2}$ increases, and the faster the corresponding weight $\lambda_i \propto \tau_{ki}^{-\rho_2}$ declines. On the other hand,

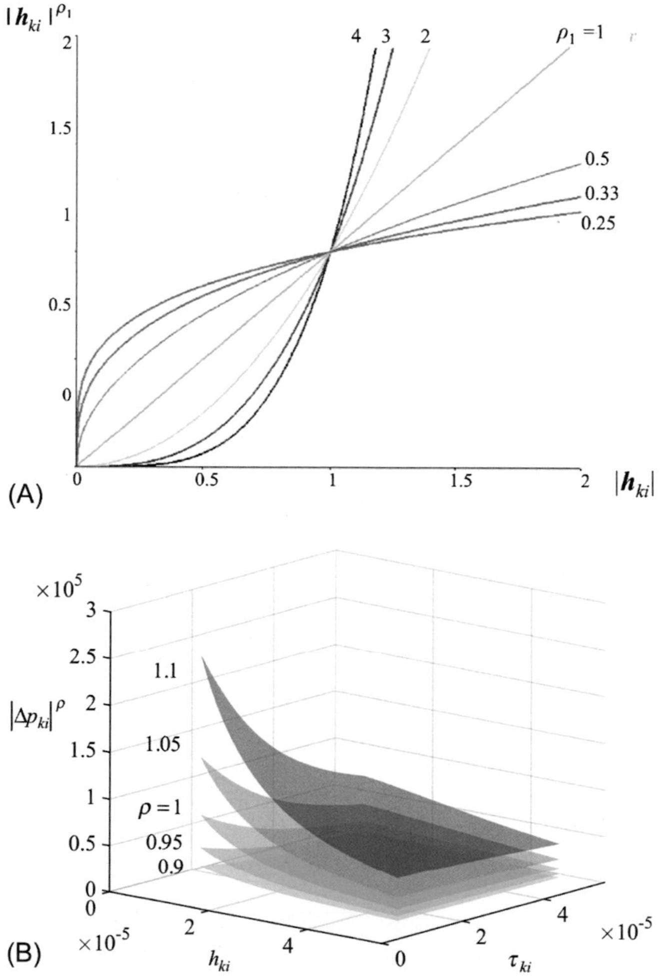

FIG. 2.2 (A) Curves of $|h_{ki}|^{\rho_1}$ *vs.* $|h_{ki}|$ of the separable metric for various values of ρ_1, and (B) surfaces of $|\Delta p_{ki}|^{\rho}$ *vs.* $|h_{ki}|$ and $|\tau_{ki}|$ of the nonseparable metric for various values of ρ.

surfaces of the uni-metric $|\Delta p_{ki}|^{\rho}$ are shown in Fig. 2.2B. The larger the ρ is, the faster the weight declines with the nonseparable metric.

Example 2.4 Just as with the *ICTM* technique, in geographical sciences the *GICTM* techniques are basically considered as "Toblerian" interpolators: they assume that each sample point has a local influence that diminishes with the metric $|\Delta p|$ (to a power of ρ), and assign higher weights to sample points closer to the interpolation point than to those farther away. In this respect, *GICTM* may be seen as a *generalized Toblerian interpolator.*

Ut aequum, the next goal of our discussion is the numerical demonstration of the *GICTM* interpolation.

The following example presents the *step-by-step* **computational** process of implementing the *GICTM* technique in practice.

Example 2.2(cont.): The solution of Eq. (2.3) for $\rho = 2$ is a special case of *GICTM* that assumes that the weight given to each sample is inversely proportional to the squared metric $|\Delta p_{10,i}|^2$ between that sample points p_i ($i = 1, \ldots, 9$) and the interpolation points p_{10}, i.e.,

$$\hat{X}(p_{10}) = \frac{1}{A}\sum\nolimits_{i=1}^{9} \frac{1}{\left|\Delta p_{10,i}\right|^2} X(p_i) \tag{2.4}$$

where $\lambda_i = \frac{1}{A|\Delta p_{10,i}|^2}$, and $A = \sum\nolimits_{i=1}^{9} \frac{1}{|\Delta p_{10,i}|^2}$. This is also known as the *inverse squared chronotopologic metric (ISCTM)* technique. Using the *ISCTM* technique, the interpolated rainfall value at point in Example 2.2 can be derived as follows.

Step 1: The distances between location p_{10} and the other nine locations of the monitoring stations were calculated as in Example 2.2 above.

Step 2: The normalization parameter A was found to be

$$A = \sum\nolimits_{i=1}^{9} \frac{1}{\left|\Delta p_{10,i}\right|^2} = \frac{1}{2^2} + \frac{1}{2^2} + \frac{1}{2^2} + 1 + 1 + 1 + \frac{1}{2^2} + \frac{1}{2^2} + \frac{1}{2^2} = 4.5$$

Step 3: The *ISCTM* interpolation value of rainfall at location 10 was calculated as

$$\hat{X}(p_{10}) = \frac{1}{A}\sum\nolimits_{i=1}^{9} \frac{1}{\left|\Delta p_{10,i}\right|^2} X(s_i) = \frac{1}{4.5}\left(\frac{12}{4} + \frac{13}{4} + \frac{15}{4} + 14 + 16 + 18 + \frac{15}{4} + \frac{18}{4} + \frac{17}{4}\right) = 15.67$$

One notices that this value is slightly higher than the *ICTM* value (15.5) of Example 2.2, as was expected.

To reiterate: a basic assumption of this sort of interpolator is that the attribute values at sampled points closer to the unsampled interpolation point are more similar to it than the values of farther away points, although this similarity is expressed in different extents by different ρ values. The weights λ_i diminish as the distance h_{ki} increases, especially when the value of the power ρ increases, so nearby samples have a heavier weight and exert more influence on interpolation, and the resultant interpolation is local.

2.2.1 *Power selection*

Since how fast the λ_i value decreases depends on the value of ρ, the main factor affecting *GICTM* interpolations is the power ρ. In this respect, the following is an obvious limitation of the *GICTM* techniques.

- Generally, there is not a rigorous way of knowing a priori which form of the weight λ_i or which **power** of ρ is best for the phenomenon of interest.

A few more comments are in order concerning power selection in real-world *GICTM* applications. A useful criterion for choosing the value of the power ρ is the minimization of an interpolation *error indicator*, like the mean squared interpolation error or the mean absolute interpolation error. It is expected that the smoothness of the interpolated attribute surface increases as ρ increases (in this sense, the interpolations may turn out to be less satisfactory when, say, $\rho = 1$ or 2 than when $\rho = 4$). *GICTM* is sometimes referred to as chronotopologic moving average ($\rho = 0$), linear interpolation when ($\rho = 1$), and weighted moving average ($\rho \neq 1$).

Example 2.5 A simulated point arrangement with the associated attribute data values is shown in Fig. 2.3A, with the uni-metric $|\Delta p_{ki}|^{\rho} = \left(|h_{ki}|^2 + \tau_{ki}^2\right)^{-\frac{\rho}{2}}$), $i = 1, \ldots, 36$. Fig. 2.3B–D presents the attribute interpolation maps

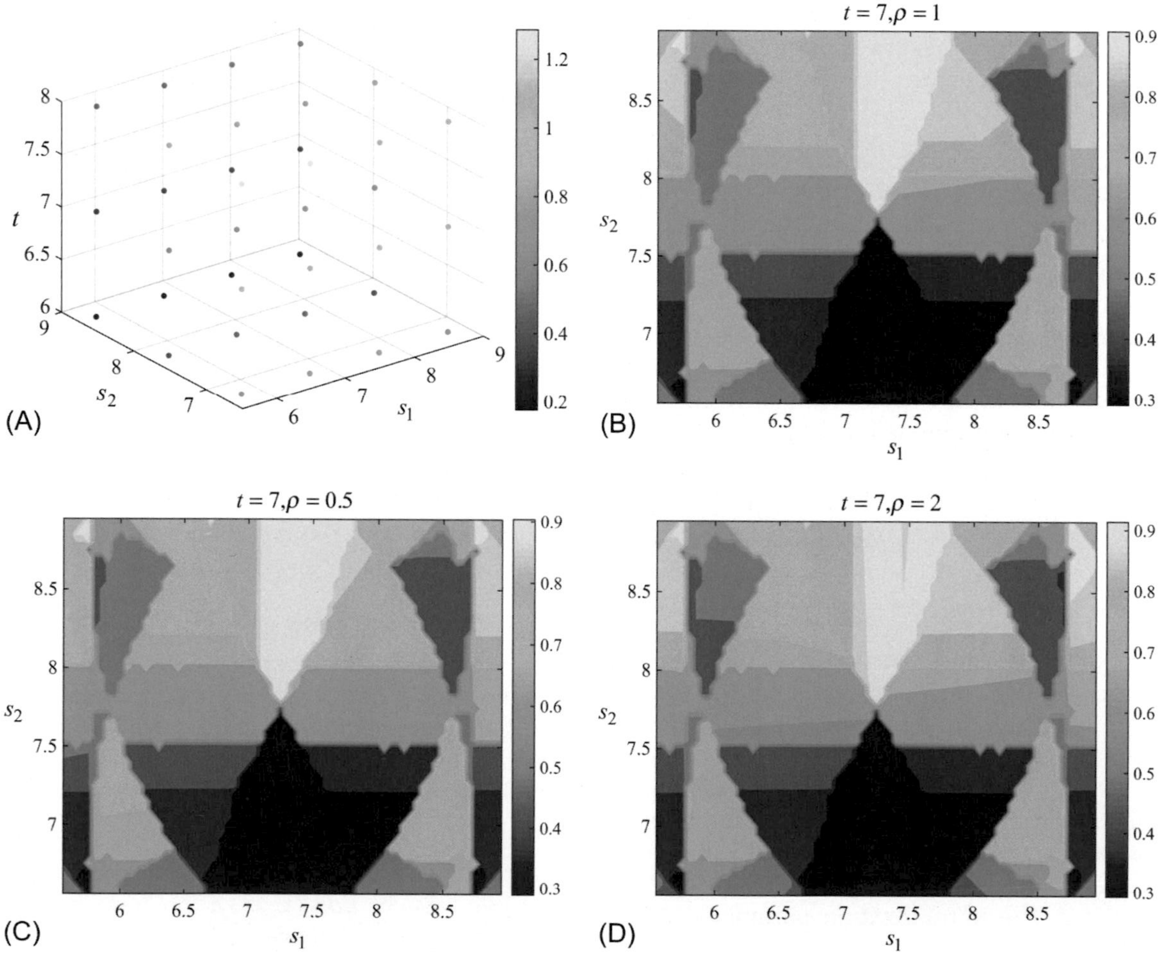

FIG. 2.3 Based on the same simulated dataset shown in (A), maps using the *GICTM* technique with different ρ values were obtained in (B)–(D).

at time $t = 7$ generated by *GICTM* with $\rho = 0.5$, 1, and 2. The mapped attribute values increase with increasing ρ values. Some differences are also observed in the spatial attribute patterns. As was expected, these results are metric-dependent.

2.2.2 Interpolation neighborhood

Precisely because the *GICTM* interpolation technique assumes that attribute values at points that are close to each other tend to be more alike than those that are not, as the points get farther away the measured values will have little relationship with the attribute value at the interpolation point. Accordingly, in order to speed up computation one may only use points that are sufficiently close to the interpolation point. It is common practice, then, to limit the number of data that are used when interpolating the unknown attribute value at a point by specifying a search subdomain around this point.

The **interpolation neighborhood** is a chronotopologic subdomain that ideally includes neighboring points that are the strongest correlated to each other and to the interpolation point.

The specified shape of the neighborhood restricts *ad rem* how far and when to look for the measured values to be used in interpolation. Other neighborhood parameters may restrict the locations and times that will be used within that shape.

Example 2.6 Fig. 2.4 presents a cylindrical neighborhood in $R^{2,1}$, which is the *GICTM* domain (methodologically, these are *S-TDA* kinds of techniques, Chapter 1). In $R^{2,1}$, the neighboring points $(\boldsymbol{s}_i, t_i)$ are considered such that the bi-metric $(|\boldsymbol{h}_{ki}|, \tau_{ki})$ satisfies the constraints $h_{ki} = |\boldsymbol{s}_k - \boldsymbol{s}_i| \leq r$ and $\tau_{ki} = |t_k - t_i| \leq \tau$ $(i = 1, ..., 5)$, where $2r$ and 2τ are the diameter and height of the cylinder centered

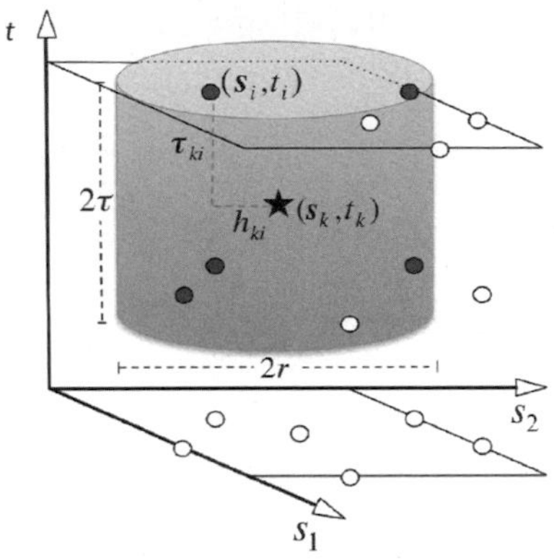

FIG. 2.4 Cylindrical neighborhood defined by $h_{ki} \leq r$ and $\tau_{ki} \leq \tau$, and containing five samples (solid circles) that are used to interpolate the unknown value at the interpolation point (asterisk). Sample points outside the cylinder (blank circles) are too far from the interpolation point so their weights are essentially zero.

at $\boldsymbol{p}_k = (\boldsymbol{s}_k, t_k)$, respectively. There are 5 data points (blue dots) within the space-time neighborhood.

2.2.3 Pros and cons

Obvious advantages of the *GICTM* techniques are that they are easy to understand and easy to automate. The *GICTM* performance could improve by adjusting the power ρ and the chronotopologic search radius, or by incorporating a physical model. The disadvantages of the *GICTM* techniques can be rather serious, depending on the physical application considered, including that the *GICTM* techniques determine the weights by a straightforward yet arbitrary recipe, they are susceptible to adverse effects of clustered samples in high-valued areas, they do not account for lawful dependencies, the resulting interpolations may not honor the sample values (which is counter-intuitive), and they do not offer any measure of interpolation error. Generally, these techniques may not produce realistic results if the data point configuration is sparse or irregular.

2.3 The chronotopologic natural neighbor interpolation technique

This technique is a geometric chronotopologic interpolator that differs from the inverse metric weighted techniques discussed above in the manner the interpolation weights are conceived and computed.

The **chronotopologic natural neighbor** (***CTNN***) interpolation technique computes the weights λ_i of Eq. (1.1) of Section 1 using natural neighborhood regions generated around each point $\boldsymbol{p}_i$ of the dataset (the size and shape of these regions would depend on the geometry of the data point distribution).[d]

To achieve its goals, *CTNN* employs the so-called *Voronoi* scheme. Given a set of points $S = \{\boldsymbol{p}_1, \ldots, \boldsymbol{p}_m\}$ in the domain $R^{n,1}$, the set of all points closer to a specified point, say, $\boldsymbol{p}_1 \in S$, than to any other point in S constitutes the interior of a (sometimes unbounded) convex polytope called the *Voronoi* cell for $\boldsymbol{p}_1$. To construct the *Voronoi* cell for point $\boldsymbol{p}_1$, one can follow the procedure in Table 2.2. According to this procedure, given a set of data points $\boldsymbol{p}_i$, the natural neighbor coordinates associated with $\boldsymbol{p}_i$ are defined from the *Voronoi* diagram of $\boldsymbol{p}_i$, and the potential *Voronoi* cell of the interpolation point $\boldsymbol{p}_k$ "steals" part of the existing data point cells.

More specifically, let $V(\boldsymbol{p}_k)$ denote the volume of the potential *Voronoi* cell of $\boldsymbol{p}_k$ and $V_i(\boldsymbol{p}_k)$ denote the volume of the subcell that would be "stolen" from the $\boldsymbol{p}_i$ cell by the $\boldsymbol{p}_k$ cell. The natural neighbor weight of $\boldsymbol{p}_k$ with respect to the data point $\boldsymbol{p}_i$ is

$$\lambda_i(\boldsymbol{p}_k) = \frac{V_i(\boldsymbol{p}_k)}{V(\boldsymbol{p}_k)}, \tag{2.5}$$

which defines the weight of the datum at $\boldsymbol{p}_i$ on the interpolation at $\boldsymbol{p}_k$, according to Eq. (1.1) of Section 1 above. It is note-worthy that given a set of data points, there is only one *Voronoi* diagram that can be generated, i.e., neighborhood assignment is unique. Moreover, the layout of the *Voronoi* cells ensures that for each data point neighbors in all directions will be selected.

Example 2.7 In Fig. 2.5 the *CTNN* weight, $\lambda_{left}(\boldsymbol{p}_k)$, is calculated as the ratio of the magnitude $V_{left}(\boldsymbol{p}_k) = |EFCE'F'C'|$ of the chronotopologic

TABLE 2.2 Construction of *Voronoi* cell for point $\boldsymbol{p}_1$.

Step	Description
1	For each point $\boldsymbol{p}_i \in S$ consider the set of points in $R^{n,1}$ closer to $\boldsymbol{p}_1$ than to any other $\boldsymbol{p}_i \in S$, $i \neq 1$ (as described above)
2	Determine the *Vonoroi* cell for $\boldsymbol{p}_1$ as the intersection of all these sets
3	The set of such polytopes tessellates the entire $R^{n,1}$ domain, and defines the *Voronoi* diagram corresponding to the set S

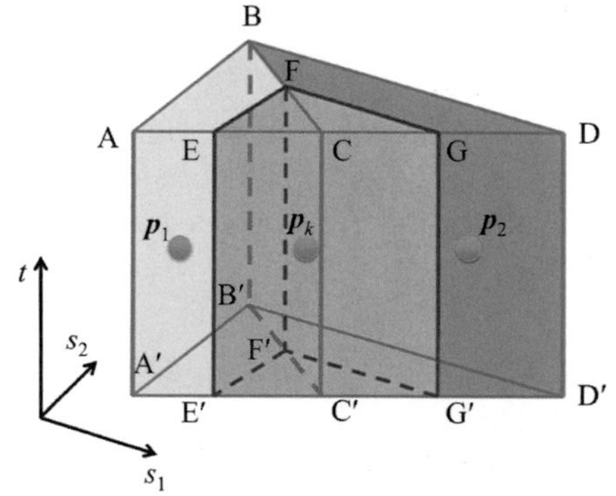

FIG. 2.5 *Voronoi* cells of the *CTNN* interpolation technique.

[d] In other words, *CTNN* searches for the closest samples to the interpolation point and assigns weights to them based on proportionate subdomains.

triangular cube <EFCE′F′C′ > over the magnitude $V(p_k) = |EFGE'F'G'|$ of the larger volume <EFGE′F′G′>. Clearly, $V_{left}(p_k) + V_{right}(p_k) = V(p_k)$.

2.3.1 Pros and cons

The *CTNN* technique seems to be particularly effective for dealing with a variety of data themes exhibiting clustered distributions. Among other *CTNN* features worth noticing are that it provides good slope continuity, and it yields smooth and visually appealing results in the case of unstructured data commonly encountered in geophysical applications. According to Sibson (1980), the idea of a neighbor is formalized in a natural way and made quantitative, and the properties of the technique depend on a geometric identity relating this neighborliness measure to a point. Among the drawbacks of *CTNN* interpolation is that it does not account for the physics of the interpolated attribute and its chronotopologic variability. Moreover, the technique can create artifacts when the data are sparse. While the interpolation surface that *CTNN* generates is continuous throughout its domain, the slope is not. There are places where it is not adequate to ensure a smooth transition of slope (e.g., first-derivative continuity). Slope discontinuities may occur at data points.

2.4 Concluding remarks

There are a few final comments worth making about the deterministic interpolation techniques. Beyond the ones discussed above, there are several other deterministic techniques (admittedly, most of them are predominantly used in the purely spatial domain). Some of these techniques are briefly reviewed next.

- The **spline** *CTI* technique generates attribute values using a function that minimizes the overall attribute surface curvature, thus producing a smooth surface that passes through the data points.

Its disadvantages include that the data are often noisy, in which case the perfect data fit provided by spline can lead to interpolation errors. Also, it may exhibit different minimum and maximum values than the data, and the spline functions are sensitive to outliers due to the inclusion of the original data values at the sample points since it operates best for gently varying surfaces.

- The **radial basis function** *CTI* technique involves linear combinations of terms based on a single univariate function.

The radial basis function is a special case of splines. It is an exact *CTI* technique (it honors the data) that works well when the phenomenon is smoothly varying. Its drawbacks include the fact that its interpolations can be above the maximum and below the minimum data value, and it is inadequate when the data are noisy or when large changes in the attribute values occur within short distances.

- The **trend** *CTI* technique fits a smooth polynomial function surface to the attribute data points that changes gradually and captures coarse-scale data features.

The trend surface is a global *CTI* that assumes that the general trend of the interpolated attribute map is independent of possible random errors at the sampling points. Among its drawbacks is that it usually captures broad space-time attribute features, high-order polynomials generate numerical issues, it is highly sensitive to outliers and fitting errors, and it does not provide an interpolation error assessment. It may be useful to distinguish between defining and accompanying properties of the term "deterministic interpolation" above.

- The **Fourier series** *CTI* technique approximates the surface by overlaying a series of sine and cosine waves.

This is a global interpolator that is most appropriate when the attribute datasets exhibit a periodic behavior (e.g., chronotopologic variations of sea surface heights).

Chronotopologic data continuity and Tobler's law are defining properties of deterministic interpolation. On the other hand, total attribute surface curvature minimization is an accompanying property of deterministic interpolation, since it holds for some techniques (e.g., spline) but not for others (e.g., *ICTM* or *NNCT*).

By way of a summary, the deterministic *CTI* techniques of this section are characterized by their relative simplicity and ease of implementation which are features appreciated by those who prefer to handle the interpolation problem as arbitrarily as *Alexander the Great* dealt with the *Gordian knot*. Yet, more often than not this is not an adequate approach for the complex phenomena the investigators encounter in nature. And, even if they accept the "simplicity" of the Gordian knot perspective as the guide to scientific investigation, there are still certain serious reservations to be noted, some of which were described above.

3 Statistical chronotopologic interpolation techniques

The development of this group of techniques involves a paradigm shift from a deterministic way of thinking, which was the case of the techniques of the previous section, to a statistical way of thinking. Many of these techniques to be discussed in this section are extensions of well-known data-driven techniques in a chronotopologic domain. Several of these techniques were originally developed in a space-free setting, and then considered in a space-dependent context (particularly in geographical sciences).

3.1 The chronotopologic sample mean technique

We start with the simplest form of statistical interpolation. It is considered a rather "naive"[e] technique, but it is sometimes mentioned in the literature because it has historically been the starting point for more advanced statistical interpolation techniques.[f]

- The **chronotopologic sample mean** (***CTSM***) interpolation technique uses the simplest λ_i-solution of Eq. (1.1) of Section 1 that leads to the elementary interpolator expressed as the statistical average

$$\hat{X}(\boldsymbol{p}_k) = \frac{1}{m}\sum\nolimits_{i=1}^{m} X(\boldsymbol{p}_i), \tag{3.1}$$

  where $m = m_s m_t$ is the total number of points (consisting of, say, m_s locations, and m_t time instances).

Obviously, Eq. (3.1) weights all neighboring samples equally. The resulting interpolations are usually quite smooth, which in some cases may be good for contouring. Eq. (3.1) is also known as the *chronotopologic unweighted moving average* (*CTUMA*) technique, because it does not discriminate between a sample that is at considerable distance from the interpolation point and one which is very close to it (which is counter-intuitive, hence, the characterization "naive" technique).

Example 3.1 Using the local sample mean technique of Eq. (3.1), we can derive an interpolation value of rainfall at location 10 (Example 2.2 above), $\hat{X}(\boldsymbol{p}_{10}) = \frac{1}{9}\sum_{i=1}^{9}$

$$X(\boldsymbol{p}_i) = \frac{12+13+15+14+16+18+15+18+17}{9} = 15.33 \text{ mm}.$$

This value is smaller than both the *ICTM* and the *ISCTM* values (15.5 and 15.67, respectively, see Example 2.2 of Section 2 earlier).

[e] Generally, the term "naive" refers to convenient and useful assumptions that may not be valid or that are weak approximations of reality.

[f] It is also a last resort kind of a tool when no other more sophisticated means are available, and, then, one would try the naive tool as an initial guess without putting much thought in.

3.1.1 *Pros and cons*

The rather obvious advantages of *CTSM* are that it is easy to understand, fairly easy to perform manually, and very easy to automate. In practice, it can be useful in local interpolation applications in which the distances (separations) between the interpolation point and the data points do not vary considerably and the attribute variability can be assumed locally isotropic. However, serious disadvantages of *CTSM* are that it fails to account for spatial distance and time separation to nearby samples, i.e., it sacrifices the dynamic local details of the phenomenon for global smoothness, it is very susceptible to adverse effects of clustered samples in high-valued areas, it does not account for physical space-time structure (distributional dependency, correlation, anisotropy), and the resulting interpolations may not honor the sample values.

3.2 The chronotopologic regression interpolation techniques

The following technique relies on classical statistical regression, properly extended in the chronotopologic domain.

Chronotopologic regression (***CTR***) interpolation assumes that the data are independent of each other, normally distributed and homogeneous in variance, and then explores a possible functional relationship between the *primary* attribute[g] $X(\boldsymbol{p})$ and a set of relevant *explanatory* attributes[h] $Y_i(\boldsymbol{p})$ that are easy to measure.

The information provided by the explanatory attributes is sometimes called *secondary* information. Customarily, the primary attribute occupies the ordinate axis, and the explanatory attributes occupy the abscissa axis. Yet, in many studies, which variable goes into which axis turns out to be an arbitrary decision, potentially leading to different results.

Example 3.2 In a rather typical real-world case, the primary attribute may be the chronotopologic pollutant concentration and the explanatory attributes the geographical coordinates and topographic elevation.

In view of the above consideration, *CTR* is a technique for interpolating the value of a dependent attribute $X(\boldsymbol{p})$, based on the value of the independent attributes $Y_i(\boldsymbol{p})$. At this point, for instructional reasons we focus on the case where there is only one independent variable $Y(\boldsymbol{p})$. This is called *simple* regression (as opposed to *multiple* regression, which handles two or more independent variables).

- A widely considered ***CTR* model** links the dependent and independent variables as (see, also, Fig. 3.1)

$$X(\boldsymbol{p}) = f[Y(\boldsymbol{p});\ \boldsymbol{\theta}] + \varepsilon(\boldsymbol{p}) \tag{3.2}$$

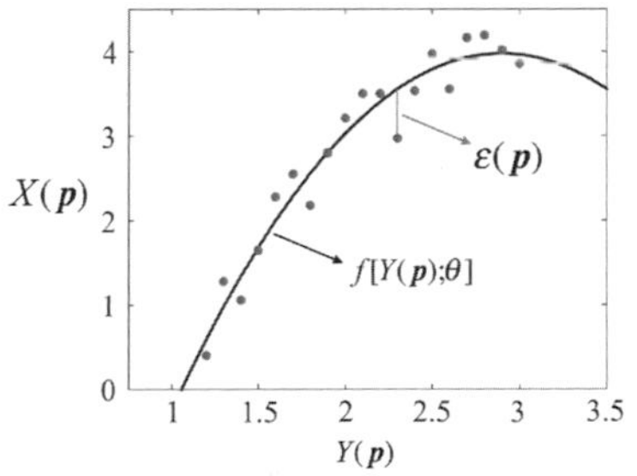

FIG. 3.1 General setup of the *CTR* technique.

[g] Also termed a regressed or dependent variable.

[h] Also termed independent variables, ancillary variables or regressors.

where $X(\boldsymbol{p})$ and $Y(\boldsymbol{p})$ are the primary attribute and the explanatory attributes, respectively, $\boldsymbol{\theta}$ is a vector of unknown parameters, f is a continuous function that is judiciously selected by the user, and $\varepsilon(\boldsymbol{p})$ is the residual or error term.

The function f can be determined by combining the existing attribute observations with a thorough understanding of the physical links between the primary and the secondary variables. Causality or the physics of the process may help in resolving the *indetermination* issue that may arise (i.e., in principle, there may exist an infinite number of curves or surfaces that satisfy a given dataset). Having selected f, the parameters $\boldsymbol{\theta}$ are computed by minimizing the total error. Then, Eq. (3.2) can be used to interpolate the $X(\boldsymbol{p})$ attribute at points $\boldsymbol{p} = \boldsymbol{p}_k$, to obtain $\hat{X}(\boldsymbol{p}_k)$.

3.2.1 *Linear case*

For all practical purposes, the most common version of the *CTR* interpolator of Eq. (3.2) is the one introduced below.

The **chronotopologic linear regression (*CTLR*)** interpolation technique is obtained from Eq. (3.2) when f is a linear function, i.e.,

$$X(\boldsymbol{p}) = \theta_0 + \theta_1 Y(\boldsymbol{p}) + \varepsilon(\boldsymbol{p}) \tag{3.3}$$

where θ_0 is a constant, and θ_1 is the regression coefficient.

In practice, given a set of observations, the $X(\boldsymbol{p}_k)$ value is directly approximated by the *CTLR* of Eq. (3.3) as

$$\hat{X}(\boldsymbol{p}_k) = \hat{\theta}_0 + \hat{\theta}_1 Y(\boldsymbol{p}_k) \tag{3.4}$$

where $\hat{\theta}_0$ and $\hat{\theta}_1$ are estimates of θ_0 and θ_1, respectively, obtained by fitting Eq. (3.4) to the available observations. In this respect, the *CTLR* technique would seek the line (say, least squares regression line) that best represents observations in a bivariate dataset.

Routinely, one uses a computational tool, among the many software packages available (e.g., *R*, *SAS*, *Matlab*, *SPSS*), to compute the coefficients $\hat{\theta}_0$ and $\hat{\theta}_1$. The X and Y values are entered into the software program, and the tool solves for each parameter.

Example 3.3 For illustration purposes, one can use the following analytical expressions:

$$\begin{aligned} \hat{\theta}_1 &= \rho_{XY}\frac{\sigma_X}{\sigma_Y} = \frac{\sum_{i=1}^{N}\left[\chi_i - \overline{X}\right]\left[\psi_i - \overline{Y}\right]}{\sum_{i=1}^{N}\left[\psi_i - \overline{Y}\right]^2}, \\ \hat{\theta}_0 &= \overline{X} - \hat{\theta}_1 \overline{Y}, \end{aligned} \tag{3.5a-b}$$

where ρ_{XY} is the XY correlation coefficient, σ_X and σ_Y denote the standard deviations of X and Y, χ_i and ψ_i denote realizations of X and Y, and $\overline{X}$ and $\overline{Y}$ denote the mean values of X and Y, respectively.

A measure of how well the *CTLR* technique of Eq. (3.4) explains and interpolates unknown $X(\boldsymbol{p})$ values is the so-called *coefficient of determination*,

$$R^2 = 1 - \frac{SSE}{SSD} \tag{3.6}$$

where $SSD = \sum_i \left[\chi_i - \overline{X}\right]^2$ measures the deviations of the observations $X(\boldsymbol{p}_i) = \chi_i$ from their mean $\overline{X}$ and $SSE = \sum_{i=1}^{N} \left[\chi_i - \hat{\chi}_i\right]^2$ measures the deviations of observations from their interpolated values generated by Eq. (3.4) at the observation points.

The coefficient R^2 is a key component of the *CTLR* technique with the following properties.

① The higher the R^2 (the smaller the *SSE*) is, the more reliable the interpolations $\hat{X}(\boldsymbol{p}_k) = \hat{\chi}_k$ obtained from the technique.
② R^2 is interpreted as the proportion of the variance in the dependent variable that is predictable from the independent variable.
③ $R^2 = 0$ means that the dependent attribute $X(\boldsymbol{p}_k)$ cannot be predicted from the independent attribute $Y(\boldsymbol{p}_k)$. $R^2 = 1$ means the dependent variable can be predicted

without error from the independent variable.

④ The value $R^2 \in (0,1)$ indicates the extent to which $X(\boldsymbol{p}_k)$ is predictable (e.g., $R^2 = 0.20$ means that 20% of the variance in $X(\boldsymbol{p}_k)$ is predictable from $Y(\boldsymbol{p}_k)$). Otherwise stated the R^2 value is indicative of the level of explained variability in the *CTLR* technique of Eq. (3.4), i.e., it is used as a guideline to measure the accuracy of the technique.

⑤ Under certain conditions, an R^2-value may be simply calculated as the square of the correlation coefficient between the original and interpolated data values. In this case, the R^2 value is not directly a measure of how good the interpolated values are, but rather a measure of how good an interpolator might be constructed from these values.

Next, we will investigate the numerical implementation of the above formulas and assertions.

The following example presents a *step-by-step* **computational** process focusing on the implementation of the *CTLR* technique in practice.

Example 3.4 In order to explore quantitatively their relationship, temperature (*Tem*, *Y*), and chlorophyll (*Chl*, *X*) were measured at various space-time points, see Table 3.1. Then, a step-by-step *CTLR* technique was employed to characterize such a relationship, as follows:

TABLE 3.1 Temperature (Tem) and chlorophyll (Chl) measurements at various points; the chlorophyll estimates (Chl_est) are also listed.

No.	Longitude	Latitude	*T*	Tem	Chl	Chl_est
1	−68.0564	43.8479	773	13.5950	1.1287	1.2441
2	−68.0562	43.5782	773	15.9750	0.7903	0.8095
3	−68.0562	43.6091	773	15.2050	0.8929	0.9501
4	−68.0555	43.7241	773	15.2750	0.8577	0.9373
5	−68.0548	43.6508	773	15.2650	0.8605	0.9392
6	−68.0543	43.7968	773	14.9650	1.0251	0.9939
7	−68.0542	43.5267	773	16.0050	0.7194	0.8040
8	−68.0542	43.7657	773	14.5850	1.0655	1.0633
9	−68.0535	43.6924	773	15.4250	0.9427	0.9099
10	−68.0530	43.8383	773	13.9400	1.1474	1.1811
11	−68.0528	43.5684	773	16.0250	0.7901	0.8004
12	−68.0528	43.5994	773	15.4150	0.8557	0.9118
13	−68.0521	43.7144	773	15.3250	0.9374	0.9282
14	−68.0517	43.8797	773	13.9200	1.3032	1.1848
15	−68.0514	43.6410	773	15.3950	0.8795	0.9154

Continued

TABLE 3.1 Temperature (Tem) and chlorophyll (Chl) measurements at various points; the chlorophyll estimates (Chl_est) are also listed—cont'd

No.	Longitude	Latitude	*T*	Tem	Chl	Chl_est
16	−68.0509	43.7871	773	14.6100	1.0496	1.0588
17	−68.0508	43.5170	773	16.1800	0.7100	0.7721
18	−68.0507	43.7560	773	14.7100	0.9904	1.0405
19	−68.0501	43.6827	773	15.1450	0.9096	0.9611
20	−68.0496	43.8286	773	13.8050	1.1701	1.2058
21	−68.0494	43.5587	773	16.1350	0.7681	0.7803
22	−68.0486	43.7047	773	15.3300	1.0098	0.9273
23	−68.0483	43.8700	773	13.6950	1.1696	1.2259
24	−68.0480	43.6004	773	15.5350	0.8097	0.8898
25	−68.0480	43.6313	773	15.2200	0.8400	0.9474
26	−68.0473	43.5072	773	16.1800	0.6907	0.7721
27	−67.7301	43.3912	776	17.6000	0.6413	0.5127
28	−67.7301	43.9494	776	15.9750	0.7623	0.8095
29	−67.7300	43.6241	776	16.4200	0.7931	0.7282
30	−67.7300	43.2791	776	16.6450	0.8630	0.6871
31	−67.7299	43.9078	776	16.0950	0.7899	0.7876
32	−67.7296	44.0704	776	13.9500	1.3365	1.1793
33	−67.7295	43.4614	776	16.9600	0.7288	0.6296
34	−67.7295	44.1119	776	13.4700	1.3041	1.2670
35	−67.7294	43.3492	776	17.5000	0.6127	0.5310
36	−67.7291	43.2285	776	16.5750	0.7196	0.6999
37	−67.7286	43.8273	776	15.9250	0.7676	0.8186
38	−67.7285	43.6647	776	16.2450	0.8638	0.7602
39	−67.7283	43.7770	776	15.2400	0.9734	0.9437
40	−67.7280	43.5021	776	16.8550	0.6993	0.6488
41	−67.7280	43.3814	776	17.6100	0.6053	0.5109
42	−67.7280	43.9396	776	16.0750	0.8304	0.7912
43	−67.7280	43.6143	776	16.6450	0.7971	0.6871
44	−67.7278	43.8979	776	15.9300	0.8164	0.8177
45	−67.7278	43.7352	776	15.2500	1.1479	0.9419

TABLE 3.1 Temperature (Tem) and chlorophyll (Chl) measurements at various points; the chlorophyll estimates (Chl_est) are also listed—cont'd

No.	Longitude	Latitude	*T*	Tem	Chl	Chl_est
46	−67.7276	44.0605	776	14.5250	1.1981	1.0743
47	−67.7274	43.4516	776	16.9650	0.6679	0.6287
48	−67.7274	44.1021	776	13.7300	1.3114	1.2195
49	−67.7274	43.3394	776	17.3100	0.6269	0.5657
50	−67.7272	44.0103	776	15.8700	0.7766	0.8287
51	−67.7264	43.6549	776	16.3150	0.8193	0.7474
52	−67.7262	43.7671	776	14.8650	1.0901	1.0122
53	−67.7260	43.4923	776	16.7700	0.6914	0.6643
54	−67.7260	43.3716	776	17.6750	0.6022	0.499
55	−67.7259	43.9297	776	16.1600	0.8186	0.7757
56	−67.7259	43.6045	776	16.3550	0.8223	0.7401
57	−67.7258	43.8880	776	16.0350	0.7940	0.7985
58	−67.7257	43.7254	776	15.3950	1.0473	0.9154
59	−67.7255	44.0506	776	15.0050	1.0279	0.9866
60	−67.7254	43.4418	776	16.7700	0.6489	0.6643
61	−67.7253	43.3297	776	17.2550	0.6788	0.5757
62	−67.7251	44.0004	776	16.0000	0.7380	0.8049
63	−67.8914	43.3381	777	16.6300	0.5071	0.6899
64	−67.8909	43.3202	777	16.8550	0.5460	0.6488
65	−67.8820	43.3383	777	16.5550	0.5513	0.7036
66	−67.8816	43.3203	777	16.9750	0.5984	0.6269
67	−67.8811	43.3024	777	17.2550	0.5912	0.5757
68	−67.8386	43.4262	777	16.4150	0.5917	0.7291
69	−67.8349	43.4293	777	16.4000	0.5874	0.7319
70	−67.8322	43.3212	777	16.8050	0.5912	0.6579
71	−67.8316	43.3031	777	16.9850	0.6129	0.6251
72	−67.8311	43.2851	777	16.8200	0.6311	0.6552
73	−67.8228	43.3216	777	17.0350	0.6252	0.6159
74	−67.8224	43.3036	777	16.7350	0.6306	0.6707
75	−67.8220	43.2856	777	17.0250	0.6300	0.6177

Continued

TABLE 3.1 Temperature (Tem) and chlorophyll (Chl) measurements at various points; the chlorophyll estimates (Chl_est) are also listed—cont'd

No.	Longitude	Latitude	*T*	Tem	Chl	Chl_est
76	−67.8215	43.2675	777	16.6700	0.6859	0.6826
77	−67.8173	43.2707	777	16.9600	0.6746	0.6296
78	−67.7783	43.3910	777	16.6550	0.6208	0.6853
79	−67.7778	43.3729	777	16.9200	0.6390	0.6369
80	−67.7748	43.3944	777	16.6900	0.5778	0.6789
81	−67.7744	43.3764	777	16.6500	0.5950	0.6862
82	−67.7739	43.3583	777	16.8450	0.6008	0.6506
83	−67.7735	43.3402	777	16.8750	0.5895	0.6451
84	−67.7730	43.3221	777	16.9100	0.6083	0.6387
85	−67.7725	43.3041	777	16.8650	0.6012	0.647
86	−67.7720	43.2860	777	16.5650	0.6837	0.7018
87	−67.7644	43.3590	777	17.0350	0.6426	0.6159
88	−67.7640	43.3409	777	16.8350	0.6195	0.6524
89	−67.7636	43.3229	777	16.7700	0.6395	0.6643
90	−67.7632	43.3048	777	16.9100	0.6307	0.6387
91	−67.7628	43.2867	777	16.8750	0.6691	0.6451
92	−67.7623	43.2686	777	16.7600	0.6964	0.6661
93	−67.7619	43.2505	777	16.5450	0.7177	0.7054
94	−67.7582	43.2721	777	16.6050	0.6885	0.6944
95	−67.7578	43.2540	777	16.3850	0.6963	0.7346
96	−67.7574	43.2359	777	16.2250	0.7227	0.7638

Step 1: To obtain the linear equation, the average value of measured temperature Y and the corresponding chlorophyll X were calculated:

$$\overline{Y}=\frac{1}{96}\sum_{i=1}^{96}\psi_i$$
$$=\frac{1}{96}(13.5950+15.9750+\cdots+16.2250)$$
$$=16.0715$$

$$\overline{X}=\frac{1}{96}\sum_{i=1}^{96}\chi_i$$
$$=\frac{1}{96}(1.1287+0.7903+\cdots+0.7227)=0.7919$$

Step 2: The coefficients $\hat{\theta}_0$ and $\hat{\theta}_1$ were computed from Eqs. (3.5a-b) as

$$\hat{\theta}_1=\frac{\sum_{i=1}^{96}\left[\chi_i-\overline{X}\right]\left[\psi_i-\overline{Y}\right]}{\sum_{i=1}^{96}\left[\psi_i-\overline{Y}\right]^2}=-0.1826,$$

$$\hat{\theta}_0 = \overline{X} - \hat{\theta}_1 \overline{Y} = 0.7918 + 0.1826 \times 16.0715 = 3.7268$$

where

$$\sum_{i=1}^{96} [x_i - \overline{X}][\psi_i - \overline{Y}]$$

$$= (1.1287 - 0.7919)(13.5950 - 16.0715) + (0.7903 - 0.7919)(15.9750 - 16.0715) + \ldots + (0.7227 - 0.7919)(16.2250 - 16.0715),$$

$$\sum_{i=1}^{96} [\psi_i - \overline{Y}]^2 = (13.5950 - 16.0715)^2 + (15.9750 - 16.0715)^2 + \ldots + (16.2250 - 16.0715)^2.$$

Step 3: The linear equation between the relative chlorophyll interpolation ($\hat{X}$) and the measured temperature Y is

$$\hat{X} = 3.7268 - 0.1826Y;$$

see Fig. 3.2. Introducing the Y values into the linear equation, the interpolated values of relative air density $\hat{X} = \hat{\chi}_i$ were obtained as 1.2441, 0.8095, 0.9501, ... and 0.7683, see the *Chl_est* in Table 3.1.

Step 4: Finally, the coefficient of determination was calculated from Eq. (3.6) as

$$R^2 = 1 - \frac{SSE}{SS_{XX}} = \frac{\sum_{i=1}^{96} [\chi_i - \hat{\chi}_i]^2}{\sum_{i=1}^{96} [\chi_i - \overline{X}_i]^2}$$

$$= 1 - \frac{(1.1287 - 1.2441)^2 + \ldots + (0.7227 - 0.7638)^2}{(1.1287 - 0.7919)^2 + \ldots + (0.7227 - 0.7919)^2}$$

$$= 0.8594$$

∴ It can be, thus, concluded that 85.94% of the variance in relative chlorophyll is predictable from temperature.

3.2.2 *Pros and cons*

It is worth stressing that the *CTLR* technique is appropriate when the following, rather restrictive, conditions are satisfied in the real-world:

① The primary attribute $X(\boldsymbol{p})$ has a **linear** relationship with the secondary attribute $Y(\boldsymbol{p})$. To check this, one should make sure that the XY-scatterplot (like that in Fig. 3.1) is linear and that the residual $\varepsilon(\boldsymbol{p})$ variation shows a random pattern.

② For each $Y(\boldsymbol{p})$ value, the **probability law** of $X(\boldsymbol{p})$ has the same standard deviation.

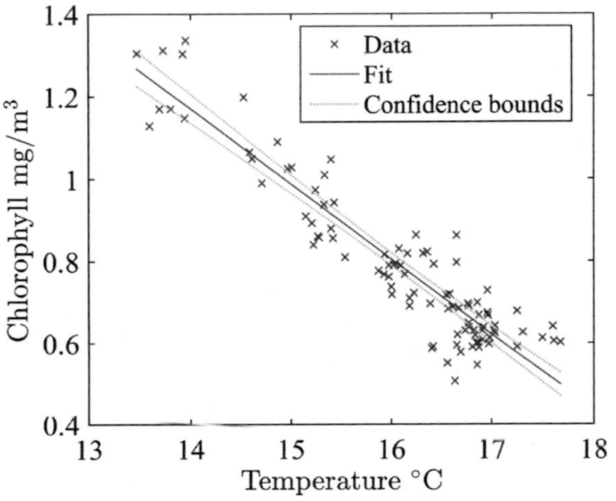

FIG. 3.2 Linear regression between temperature and chlorophyll.

When this condition is satisfied, the variability of the residuals will be relatively constant across all values of $Y(\boldsymbol{p})$, which is easily checked on a residual plot.

③ For any given value of $Y(\boldsymbol{p})$, the $X(\boldsymbol{p})$ values are **independent**, as indicated by a random pattern on the residual plot.

④ The $X(\boldsymbol{p})$ values are roughly **normally** distributed (i.e., symmetric and unimodal). A histogram will show the shape of the probability distribution—a little skewness is acceptable if the sample size is large.

It should be kept in mind that when one uses a regression equation, one should not use values of the independent attribute that are outside the range of values used to create the equation. This can lead to the extrapolation issues and produce unreasonable attribute estimates.

3.3 Concluding remarks

There are several other statistical interpolation techniques that are not described in sufficient detail in this book. Among them, the following techniques are worth noticing in the *CTDA* setting considered in this book.

- The **chronotopologic classification** (***CTCl***) interpolation technique uses easily accessible soft information to divide the domain of interest into a number of representative units and provides the interpolated value of attribute $X(\boldsymbol{p}_k)$ as

$$\hat{X}(\boldsymbol{p}_k) = \overline{X} + \alpha_l + \varepsilon(\boldsymbol{p}_k) \quad (3.7)$$

where l is the unit within which $\boldsymbol{p}_k$ belongs, $\overline{X}$ is the attribute mean over the region of interest, α_l is the deviation between $\overline{X}$ and the attribute mean within unit l, $\overline{X}_l$, and $\varepsilon(\boldsymbol{p}_k)$ is the residual (pooled within-unit) error.

In the description above, *CTCl* is based on the choice of the unit l using soft information that is representative of the local attribute variation and is characterized by the statistical moments (mean, variance) of the attribute measured at points within the representative unit (Li et al., 2013).

Example 3.5 Easily accessible soft information includes soil types, vegetation types, and administrative domains. For numerical illustration, assume that the domain of interest has been divided into three subdomains ($L = 3$) with attribute means $\overline{X}_1 = 2.5$, $\overline{X}_2 = 3.5$ and $\overline{X}_3 = 5.5$, whereas the mean over the entire domain is $\overline{X} = 4$. The interpolation point lies within subdomain $l = 2$, and $\varepsilon(\boldsymbol{p}_k) = 0.1$. Then, $\alpha_l = \overline{X} - \overline{X}_2 = 0.5$, and according to Eq. (3.7), $\hat{X}(\boldsymbol{p}_k) = 4 + 0.5 + 0.1 = 4.6$.

As it turns out, the *CTCl* interpolation of Eq. (3.7) can be computed from the *CTLR* solution by specifying the prime attribute as a response variable and the soft information as an explanatory factor with k classes. Accordingly, the *CTCl* shares the same assumptions as *CTLR* interpolation. For various reasons, this is not a particularly popular *CTDA* technique.

- The **chronotopologic trend surface** (***CTTS***) interpolation technique is essentially a special case of *CTLR* interpolation that only uses geographic coordinates to predict the values of the primary attribute.

CTTS separates the data into regional trends and local variations; it shares the same assumptions as *CTLR* interpolation.

- The **chronotopologic splines** (***CTSp***) interpolation technique consists of polynomials of degrees $\boldsymbol{\nu} = (\nu_1, \nu_2, \ldots, \nu_n)$ in space and μ in time expressed as (Christakos, 2017)

$$\eta_{\nu,\mu}(\boldsymbol{s}, t) = \sum_{\alpha=1}^{N_n(\nu,\mu)} c_\alpha \eta_\alpha(\boldsymbol{s}, t) \quad (3.8)$$

where $c_{\rho,\zeta}$ are estimated coefficients, $\eta_\alpha(\boldsymbol{s}, t) = \boldsymbol{p}^{(\rho,\zeta)} = \boldsymbol{s}^{\rho} t^{\zeta} = \Pi_{i=1}^{n} s_i^{\rho_i} t^{\zeta}$, $\boldsymbol{\rho} = (\rho_1, \rho_2, \ldots, \rho_n)$, $\rho = |\boldsymbol{\rho}| = \sum_{i=1}^{n} \rho_i$, $\nu = |\boldsymbol{\nu}| = \sum_{i=1}^{n} \nu_i$, and $N_n(\nu, \mu) = (\mu + 1) \sum_{\rho=0}^{\nu} \frac{(\rho+n-1)!}{\rho!(n-1)!}$ is the number of monomials that depends on the dimension n and the orders ν, μ.

The *CTSp* interpolation is a local rather than a global technique. The polynomials describe pieces of a chronotopologic surface fitted together[i] so that they are linked smoothly. The places where the pieces join are called knots. The choice of knots is arbitrary, yet, it may have a dramatic impact on interpolation. For degrees $\nu = 1, 2,$ or 3, a spline is called linear, quadratic or cubic, respectively, in space; and similarly, for $\mu = 1, 2,$ or 3, a spline is called linear, quadratic or cubic, respectively, in time (typically, splines of degree 3 are termed *cubic* splines).

- The **local chronotopologic trend surface** (***L-CTTS***) interpolation technique fits a polynomial surface at each interpolation point using nearby samples.

There are two *L-CTTS* versions. The first is a local polynomial regression fitting. The second is a *bilinear* or *bicubic* spline that was developed to implement bivariate interpolation onto a grid for irregularly distributed point data. Both the *CTSp* and *L-CTTS* techniques are rather unable to choose the smoothness level.

4 Practice exercises

1. What kinds of chronotopologic estimation problems can *CTDA* deal with?
2. Describe the defining and accompanying properties of the interpolation techniques presented in this chapter.
3. Do you agree or not that continuity involves the concept that small values of an attribute are in close chronotopologic proximity to other small values, while high values are close to other high values? Explain.
4. In Fig. 4.1 we have the chronotopologic arrangement of eight data points p_1, p_2, p_3, p_4, p_1', p_2', p_3', p_4' and one interpolation point (denoted as p_0). The distances and

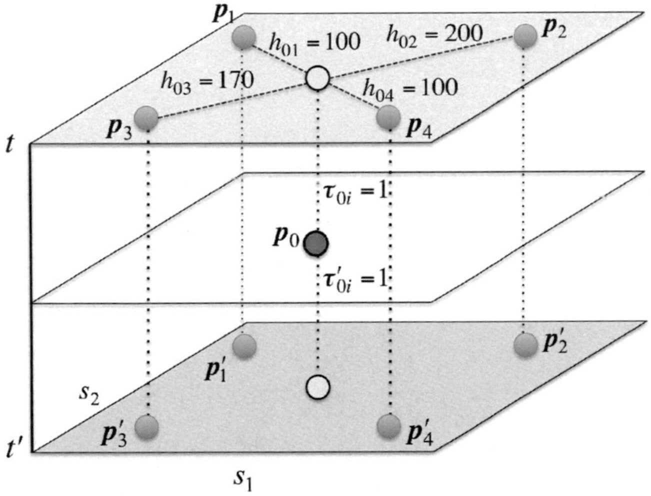

FIG. 4.1 Data configuration and distances between samples and the estimated point.

[i] I.e., each piece is associated with a small number of data points exactly.

separations between the data and the interpolation point are also shown in Fig. 4.1, and the metric considered is of the non-separable form $|\Delta p_{ki}|^{\rho} = \left(|h_{ki}|^2 + \varepsilon\tau_{ki}^2\right)^{\frac{\rho}{2}}$ with $\varepsilon = 20$. Also, the values at p_1, p_2, p_3, p_4, p_1', p_2', p_3', p_4' are 40, 42, 43, 41, 45, 43, 43, 42, respectively. Using the *GICTM* technique derive interpolations at point p_0 for $\rho = 1$ and 5.

5. Use the same dataset shown in exercise 4 above assuming $\varepsilon = 10$ and $\varepsilon = 30$. Describe the role of the metric parameter ε in chronotopologic interpolation.
6. Think of a few real-world cases that render the assumptions of the following techniques so unrealistic as to be perniciously misleading: (a) *ICTM*, (b) *CTLR*, (c) *CTSM*, and (d) *CTCl*.
7. Fig. 4.2 shows 15 sampling points and their chronotopologic coordinates with the corresponding attribute values listed in Table 4.1. Using the *GICTM* technique ($\rho = 1$ and 2) interpolate the attribute value at $p_0 = (s_{0,1}, s_{0,2}; t) = (2.25, 3; 2)$.
8. Following Example 3.4 of Section 3, use the *CTLR* technique to explore the relationship between sea surface temperature and sea surface chlorophyll (Table 4.2).

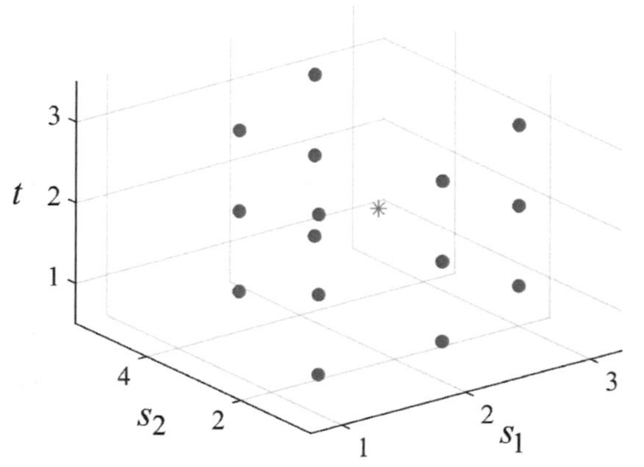

FIG. 4.2 Sampling points (dots) and estimation point (asterisk).

TABLE 4.1 Dataset.

	s_1	s_2	$t = 1$	$t = 2$	$t = 3$
A	1	1	2.01	2.16	2.51
B	2	1	2.52	2.44	2.66
C	3	2	2.43	2.70	2.51
D	2.5	5	2.67	2.65	2.75
E	1.5	4	2.15	2.40	2.16

TABLE 4.2 Dataset.

Longitude	Latitude	*T*	Tem	Chl
−67.7197	43.5032	776	16.8455	0.6828
−67.7197	43.3825	776	17.7505	0.5936
−67.7196	43.9406	776	16.2355	0.8100
−67.7196	43.6154	776	16.4305	0.8137
−67.7195	43.8989	776	16.1105	0.7854
−67.7194	43.7363	776	15.4705	1.0387
−67.7192	44.0615	776	15.0805	1.0193
−67.7191	43.4527	776	16.8455	0.6403
−67.7190	43.3406	776	17.3305	0.6702
−67.7188	44.0113	776	16.0755	0.7294
−67.8851	43.3490	777	16.7055	0.4985
−67.8846	43.3311	777	16.9305	0.5374
−67.8757	43.3492	777	16.6305	0.5427
−67.8753	43.3312	777	17.0505	0.5898
−67.8748	43.3133	777	17.3305	0.5826
−67.8323	43.4371	777	16.4905	0.5831
−67.8286	43.4402	777	16.4755	0.5788
−67.8259	43.3321	777	16.8805	0.5826
−67.8253	43.3140	777	17.0605	0.6043
−67.8248	43.2960	777	16.8955	0.6225
−67.8165	43.3325	777	17.1105	0.6166

CHAPTER

8

Chronotopologic krigology

1 The emergence of geostatistical Kriging

As was discussed in Chapter 7, in most applications the available measurements and observations usually provide information on a finite data point basis. The goal is to extend this information to a much larger field basis. No model, however sophisticated, will ever tell us exactly what an attribute's value will be at every point that has not been observed. As far as *CTDA* is concerned, the answer to this problem can only be in the form of attribute interpolations that allow a certain level of real-world uncertainty. In this respect, the interpolation requirement can be clearly stated:

> A good **chronotopologic interpolation** technique should handle the available data under conditions of uncertainty in a realistic and efficient way as to get the most out of them.

Geostatistical chronotopologic interpolation, then, emerged naturally as a result of the profound inadequacies of previous interpolation techniques, some of which were discussed in the preceding Chapter 7.[a]

Example 1.1 In the *ICTM* interpolation techniques of Section 2 of Chapter 7, weight assignment at a sampling point relies on the sample's distance from the interpolation location and/or its time separation from the interpolation instant, i.e., they focus solely on the geometrical placement and temporal arrangement of the samples relative to the interpolation point. Many of these techniques suffer from certain drawbacks: they are independent of direction, do not account for the physical characteristics of the attribute (such as the physical content of sample values, high *vs.* low pollutant regions, isotropic *vs.* anisotropic disease incidence variation, smooth *vs.* erratic ecological pattern), and ignore potential uncertainty sources. Such techniques do not consider, say, the kind of mineral interpolated in a deposit, the category of the chemical contaminating the groundwater, or the type of disease that became an epidemic.

Alter apars, the geostatistical interpolation techniques do not suffer the drawbacks of the

[a] As the readers recall, limitations of previous techniques include considering space and time in isolation, and not seeking the rigorous incorporation of *in-situ* uncertainties.

Quantitative Analysis and Modeling of Earth and Environmental Data
https://doi.org/10.1016/B978-0-12-816341-2.00003-4

interpolation techniques mentioned in the above example, and they have been widely used in applied sciences. For these two reasons, the geostatistical interpolation is presented in considerable detail in this chapter. An important lesson geostatisticians (and *CTDA* investigators, in general) learn from experience is as follows:

- When a good attribute variogram or covariance model is available, and the underlying modeling assumptions are satisfied by the attribute data, a group of geostatistical interpolation techniques collectively known as **Kriging** can provide accurate attribute interpolations.

In literary terms, unlike *Isaiah* who was long on outrage but short on facts, the Kriging users try to soft-pedal any concerns and conform as best as possible to the line of hard facts (attribute data, in this case).

1.1 Kriging logic

Before delving into the technical details of Kriging in its various forms, let us first look briefly at the underlying fundamental reasoning.

The basic **logic** of Kriging is sequential, i.e., it possesses three distinct components forming a logical sequence:

$$Input \rightarrow Generator \rightarrow Output,$$

where

- the *input* basically consists of the available attribute samples, and the chronotopologic variability model (variogram, covariance) and the search parameters (radius, range, etc.);
- the *generator* consists of statistical optimization criteria (minimum mean squared interpolation error, unbiasedness); and
- the *output* consists of chronotopologic attribute interpolations and associated interpolation errors generated at unsampled points in terms of a weighted sample average.

Although the sequential logic process above does not constitute what one would call complicated reasoning, certain issues may emerge due to the facts that its components are closely interdependent and should be internally consistent (these are by no means trivial issues, since, as we will see in Chapter 9, it can lead to serious *circulus in probando* problems, like self-referentiality). Several types of Kriging have been proposed based on the above logical process, which are of varying levels of complexity. The Kriging type that applies in a particular study depends on both an *ontic* factor (natural attribute's properties) and an *epistemic* factor (investigator's modeling assumptions). In addition, the Kriging logic components have their own technical characteristics and implementation challenges, some of which will be discussed in this chapter.

1.2 Technical characteristics

Technically, Kriging can be seen as a *statistical regression* approach used as a way to handle *uncertainty* at points of interest that differ from the sampling points, thus making a radical departure from alternative approaches like deterministic *ICTM* interpolation (Chapter 7).

As usual, let $\hat{X}(\boldsymbol{p}_k)$ be the interpolated attribute value at an unsampled point $\boldsymbol{p}_k$ and $X(\boldsymbol{p}_i)$ be the measured attribute values at points $\boldsymbol{p}_i$, $i = 1, \ldots, m$. A common feature of most Kriging techniques is that the generated $\hat{X}(\boldsymbol{p}_k)$ values satisfy two statistical criteria (or conditions, depending on how one looks at it):

- **Unbiasedness**: on the average $\hat{X}(\boldsymbol{p}_k)$ is equal to $X(\boldsymbol{p}_k)$, i.e.,

$$\overline{X(\boldsymbol{p}_k)} = \overline{\hat{X}(\boldsymbol{p}_k)} \qquad (1.1)$$

- **Accuracy maximization**: the mean squared chronotopologic interpolation error, $e_X^2(\boldsymbol{p}_k)$, is minimized so that

$$\min\left\{ e_X^2(\boldsymbol{p}_k) = \overline{\left[\hat{X}(\boldsymbol{p}_k) - X(\boldsymbol{p}_k)\right]^2} \right\} \qquad (1.2)$$

Eqs. (1.1) and (1.2) are the two basic Kriging interpolation criteria.

1.3 Generic linear Kriging interpolator

The starting point of several popular Kriging techniques is a simple yet powerful stochastic formulation of the problem that satisfies the criteria introduced by Eqs. (1.1) and (1.2), as follows[b]:

The **generic linear** interpolator at the core of the mainstream Kriging techniques has the form

$$\hat{X}(\boldsymbol{p}_k) = \sum_{i=1}^{m} \lambda_i X(\boldsymbol{p}_i) \tag{1.3}$$

where λ_i are weights assessing the relative contribution of each datum $X(\boldsymbol{p}_i)$ to the interpolated $\hat{X}(\boldsymbol{p}_k)$ value. The associated **generic interpolation error** has the form

$$e_X^2(\boldsymbol{p}_k) = \sigma_X^2(\boldsymbol{p}_k) + \sum_{i=1}^{m} \sum_{j=1}^{m} \lambda_i \lambda_j c_X(\boldsymbol{p}_i, \boldsymbol{p}_j) - 2\sum_{i=1}^{m} \lambda_i c_X(\boldsymbol{p}_i, \boldsymbol{p}_k) \tag{1.4}$$

where $\sigma_X^2(\boldsymbol{p}_k)$ is the interpolation variance at point $\boldsymbol{p}_k$, $c_X(\boldsymbol{p}_i, \boldsymbol{p}_j)$ and $c_X(\boldsymbol{p}_i, \boldsymbol{p}_k)$ are the covariances between the point pairs $\boldsymbol{p}_i$, $\boldsymbol{p}_j$ and $\boldsymbol{p}_i$, $\boldsymbol{p}_k$, respectively.

Eqs. (1.3) and (1.4) provide a formal description of the linear Kriging interpolator, and as such it is a strictly mathematical description written in *computationable* terms. In this description, the main issue is obviously the computation of the weights λ_i so that the criteria of Eqs. (1.1) and (1.2) are satisfied in view of the linear interpolator form of Eq. (1.3) and the interpolation error form of Eq. (1.4).

1.4 The Kriging decalogue

There are a few general comments that should be made *prima facie* regarding the generic Kriging Eqs. (1.3) and (1.4), which are presented in Table 1.1. In addition to the comments ①–⑩ of Table 1.1, one more point is worth stressing as regards possible Kriging interpretations, as follows.

- The difference in the quality of the **interpretations** of Eqs. (1.3) and (1.4) proposed by different investigators should be understood in the context of the body of knowledge that the investigators activate in connection with the equations.

Example 1.2 When the body of knowledge includes the physical mean, Eqs. (1.3) and (1.4) are reduced into a particular set of Kriging equations, which distinguish them from other possible sets of Kriging equations (see the case of the space-time simple Kriging interpolator in Section 2.2 below).

In the following sections, the covariance and variogram will be both used in order to familiarize the readers with the implementation of these tools in attribute interpolation.

1.5 Kriging neighborhood

Another important concern is how much data will support how much interpolation. This is, actually, a classical problem of induction and statistical inference. To put it in more words:

The issue here concerns the number m of data points around $\boldsymbol{p}_k$ to be used in Eq. (1.3), which is a matter that clearly depends on the kind of Kriging **neighborhood** selected.

Example 1.3 Two representative cases are considered in Fig. 1.1 that illustrate some aspects of the chronotopologic neighborhood issue.

- Fig 1.1A presents a spherical neighborhood in the R^{2+1} domain, where the neighboring points $\boldsymbol{p}_i$ are considered such that $|\Delta\boldsymbol{p}_{ki}| \leq \varpi$, the ϖ

[b] Herein, the equivalent notations for point $\boldsymbol{p}_i$ *vs.* $(\boldsymbol{s}_i; t_i)$, space-time lag $\Delta\boldsymbol{p}_{ij}$ *vs.* $(\boldsymbol{h}_{ij}, \tau_{ij})$, and attribute $X(\boldsymbol{p}_i)$ *vs.* $X(\boldsymbol{s}_i; t_i)$ will be used interchangeably, depending on the context.

TABLE 1.1 The Kriging decalogue.

No.	Description
①	Eq. (1.3) is a **conditional** rather than a categorical interpolation device: if the data $X(\boldsymbol{p}_i)$ are valid, then the interpolation $\hat{X}(\boldsymbol{p}_k)$ will be valid
②	Eq. (1.4) holds in theory for any covariance c_X form. Yet, in practice certain **assumptions** are usually made concerning the c_X shape (homostationary, isostationary, etc.)
③	Eq. (1.3) is an optimal interpolator in the sense of Eq. (1.2) only if $X(\boldsymbol{p})$ is represented by a **Gaussian** random field, whereas for a non-Gaussian $X(\boldsymbol{p})$ the linear interpolator is nonoptimal, in general
④	Starting from Eqs. (1.3) and (1.4), several **types** of linear Kriging have been suggested. Each type determines a different set of weights λ_i assigned to the data $X(\boldsymbol{p}_i)$ and, hence, a different attribute interpolation $\hat{X}(\boldsymbol{p}_k)$ and associated interpolation error $\overline{e_X^2(\boldsymbol{p}_k)}$ are obtained
⑤	Although the interpolator of Eq. (1.3) is of the linear form that was also used by the interpolators of Chapter 7, the computation of the interpolation weights in Chapter 7 was *ad hoc*, whilst the Kriging weights λ_i are computed based on a data-driven weight function satisfying rigorous statistical **criteria**: $\overline{e_X^2(\boldsymbol{p}_k)}$ minimization and, on occasion, $\hat{X}(\boldsymbol{p}_k)$ unbiasedness
⑥	Eq. (1.3) generates interpolations at unsampled points using samples at nearby points, yet, these samples may contain nonnegligible **errors** (measurement noise, operator error, model inaccuracy). Then, Eqs. (1.3) and (1.4) are combined with an extra equation describing these errors (Section 2 below)
⑦	An expression analogous to Eq. (1.3) holds in terms of the **variogram** (see following section). Covariance and variogram represent the chronotopologic attribute variation pattern (Chapter 5) and allow insight used by Kriging to generate interpolations that are mathematically rigorous and physically meaningful
⑧	Since Kriging interpolation directly depends on the covariance (or variogram) model, a matter of great significance is model **quality**, in the sense of how closely a model must approximate the actual covariance (or variogram) to make Kriging a more adequate interpolator compared to its competitors
⑨	When the attribute is known to obey a **physical law**, say, in the form of a partial differential equation (Chapter 10), the implications for the Kriging interpolator should be considered
⑩	The **interpretation** of Eqs. (1.3) and (1.4) is not fixed but determined from the specific *CTDA* context

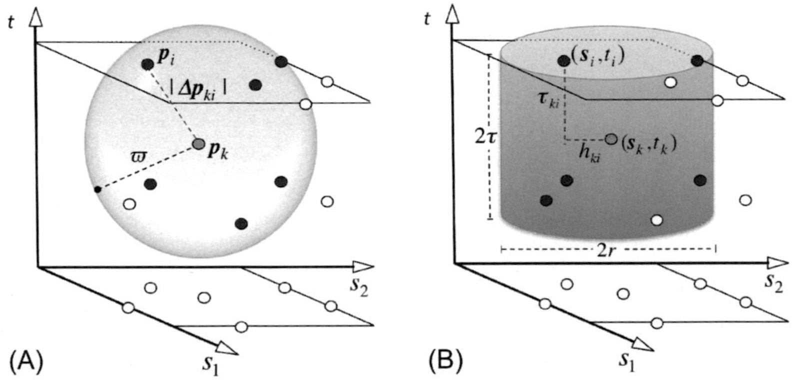

FIG. 1.1 Types of Kriging neighborhoods: (A) spherical neighborhood in R^{2+1} with $|\Delta p_{ki}| \leq |\Delta \boldsymbol{p}|$; and (B) cylindrical neighborhood in $R^{2,1}$ with $h_{ki} \leq r$ and $\tau_{ki} \leq \tau$. The blue and blank dots indicate, respectively, data included and not included in the neighborhoods.

is the radius of the space-time sphere centered at $\boldsymbol{p}_k$, and the metric may assume one of the forms discussed in Section 3 of Chapter 2, say, $|\Delta \boldsymbol{p}_{ki}| = \sqrt{h_1^2 + h_2^2 + \varepsilon\tau^2}$ (Example 3.5 of Section 3 of Chapter 2). There are $m = 6$ data points (blue dots) within the spacetime neighborhood (blank dots represent data points that are not included in the neighborhood).
- Fig 1.1B presents a cylindrical neighborhood in the $R^{2,1}$ domain, where the neighboring points $(\boldsymbol{s}_i; t_i)$ are considered such that $|\Delta \boldsymbol{p}_{ki}| = (h_{ki}, \tau_{ki})$, the $h_{ki} = |\boldsymbol{s}_k - \boldsymbol{s}_i| \leq r$ and $\tau_{ki} = |t_k - t_i| \leq \tau$, where $2r$ and 2τ are the radius and height of the cylinder, respectively, centered at $\boldsymbol{p}_k$. There are $m = 5$ data points (blue dots) within the space-time neighborhood.

A point worth noticing is that in many cases the separated space-time $R^{n,1}$ domain characterization will provide satisfactory Kriging results, whereas in some other circumstances it will be necessary to think of a unified spacetime R^{n+1} domain characterization (both characterizations were examined in detail in Section 2 of Chapter 2). Although the theoretical development of Kriging in this chapter includes both characterizations, in the majority of applications in practice the users of the Kriging techniques seem to adopt the separated space-time characterization of the physical domain, in which case these techniques are usually referred to as *space-time Kriging interpolators*, where a metric with both a space-time separate or a composite form may be used (Chapter 2).

1.6 A review of Kriging techniques

As we saw earlier, a common feature of most Kriging techniques is that the generated interpolated values satisfy two statistical conditions (or constraints, depending on how one looks at the matter): they produce unbiased and interpolation error variance minimizing attribute values at the unsampled points. Beyond these conditions, additional assumptions are occasionally introduced whilst some others are relaxed, depending on the physical situation and the study objectives, as follows:

1. Popular Kriging techniques assume that the attribute distribution obeys the Gaussian (normal) probability law and it is *homostationary*,[c] and that the interpolator has a linear form such as Eq. (1.3).[d] These techniques include **Space-time Ordinary Kriging** (*STOK*,[e] Section 2.1 below), which is used when the mean is unknown, and **Space-time Simple Kriging** (*STSK*, Section 2.2), which is used when the mean is known.
2. By relaxing the homostationarity distribution assumption, the resulting estimation technique is **Space-time Intrinsic Kriging** (*STInK*), which can study *heterogeneous*[f] attributes with varying means. This type of Kriging is not discussed further in this book; instead, the interested readers are referred to Christakos and Hristopulos (1998), and references therein.
3. When the interpolator linearity assumption is relaxed, the resulting *nonlinear* estimation techniques are the **Space-time Disjunctive**

[c] Homostationarity means a constant mean over the domain and variogram depending only on the space-time lag (Chapter 5).

[d] See also comment 3 in Table 1.1 earlier.

[e] Using *STOK* as a starting point, more complex types of Kriging have been developed that relax one or more of the restrictive assumptions.

[f] Heterogeneity means a mean varying over the domain of interest and a variogram depending on the point coordinates and not merely on space-time lag (Chapter 5).

Kriging (*STDK*) and the **Space-time Indicator Kriging** (*STIK*), which are based on nonlinear transformations of attributes.

❹ Some useful *recursive* formulations of Kriging estimators have also been suggested in the literature. In this context, the **Kalman** filter was introduced in geostatistics by Christakos (1985), and further space-time versions of the Kriging techniques have been discussed in Christakos and Hristopulos (1998).

Occasione data, the geostatistics technique that makes the fewer restrictive assumptions and allows the incorporation of the widest range of knowledge bases is based on the theory of Bayesian Maximum Entropy (*BME*), which will be the topic of the subsequent Chapter 9.

1.7 Kriging implementation

Methodologically, the implementation of space-time Kriging involves four main steps, as outlined in Table 1.2.

As with the class of interpolation techniques considered in Chapter 7, one may distinguish between *defining* and *accompanying* properties of the Kriging class of techniques.

Example 1.4 Error variance minimization is a defining property of the class, whereas linearity is an accompanying property, since it holds for class members like *STOK*, *STSK*, and *STInK* but not for other members like *STDK*. Homostationarity is an accompanying property of Kriging, since it holds for *STOK* and *STSK* but not for *STInK*.

1.8 Kriging visualization

To reiterate: Kriging was developed to interpolate the attribute values at unsampled points of interest, since investigators want to know the values of a natural attribute over the entire study domain. Yet this is not the end of the story:

> After the chronotopologic attribute interpolation has been carried out, the next important step is the **visualization** of the attribute interpolations in terms of an informative attribute **map**. In the visualization context, attribute mapping is not the same as form, it is a way of seeing form.

Developing the map grid requires certain decisions based on experience and judgment, such as how large each grid cell needs to be and how to distribute the cells over the domain of interest. One possible approach is as follows: a

TABLE 1.2 Kriging implementation steps.

Step	Description
①	Carefully review the available knowledge bases as regards quality, uncertainty, reliability, relevancy, etc. (recall discussion in Section 2 of Chapter 3).
②	Consider the chronotopology of the situation (Chapter 2) and decide which domain representation to assume (R^{n+1} or $R^{n,1}$), and which metric $\lvert \Delta \boldsymbol{p} \rvert$ (separable or composite) best fits the physics of the application.[a]
③	Choose the modeling assumptions (homogeneity, normality, etc., Chapter 5) that are physically justified.
④	In the light of steps ①–③ and previous experience, make a judicious choice of the most suitable Kriging technique.

[a] *The readers may be reminded that it is mostly application-dependent how the weight of the time dimension should be considered in combination with that of the space dimension in a chronotopologic metric.*

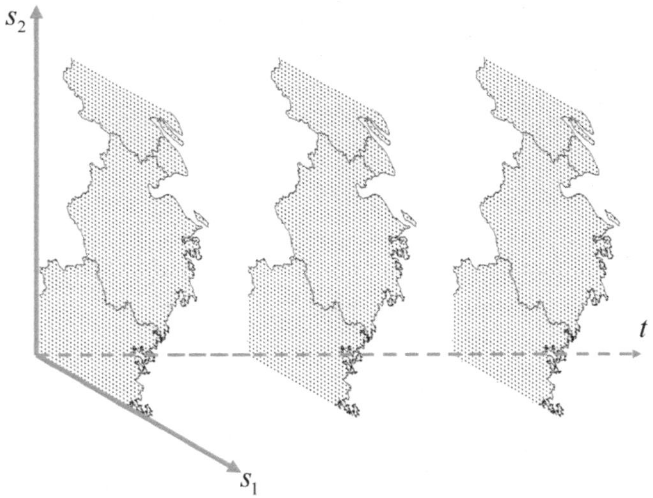

FIG. 1.2 Time-dependent fine grid node arrangement along the coastal area of China.

fine-mapping grid is set up across the entire study region at various time instances; then, the mean attribute value is calculated within each grid cell, and all of these means are combined to compute a global average attribute value for each year. Several computer software libraries can be used for visualization purposes in practice, such as *ArcGIS*, *R*, and *Matlab* (Chapter 1).

Example 1.5 Grids with chronotopologic resolution $10\text{ km} \times 10\text{ km} \times 1\text{ day}$ are set in the coastal area of China; see a zoom-in scale in Fig. 1.2. The attribute values at each geographical grid and at various instances will be predicted by Kriging techniques, and finally a map of the attribute is obtained (see, Fig. 2.5 in Section 2 below).

Next, two major classifications of the Kriging techniques are reviewed that can transform data into useful information, while adequately accounting for the specifics of the real-world application.

2 1st Kriging classification

Before proceeding with case-dependent Kriging classification matters, we would like to draw the readers' attention to the following general recommendation:

> It is a crucial matter that an investigator fully understands the main **properties** and **capabilities** of each Kriging interpolation technique before applying it in a real-world study.

Needless to say, the above recommendation addresses a crucial matter. Certain studies have estimated that selecting an unsuitable technique could lead to interpolation errors up to ±50% (e.g., Dominy et al., 2002). We start with the first, and perhaps most fundamental, classification of linear space-time Kriging techniques.

2.1 Space-time ordinary kriging (*STOK*)

This is, perhaps, the most widely used type of Kriging. Its basic properties are listed in Table 2.1—left column. It may not be surprising that there are certain properties that *STOK* shares with other Kriging types (like *STSK*, Section 2.2).

Methodologically, rather than choosing the interpolation weights λ_i in an *ad hoc* manner or according to arbitrary rules (as is the case, e.g., with the *ICTM* technique), *STOK* instead takes a goal-oriented approach, i.e., the weights λ_i

TABLE 2.1 *STOK vs. STSK.*

STOK	*STSK*
Attribute values follow the Gaussian probability law (if not, an appropriate data transformation is applied)	
Attribute distribution is homostationary	
Attribute mean is constant but unknown	Attribute mean is constant and known $\overline{X}(\boldsymbol{p}) = \overline{X}$
Variogram depends only on chronotopologic lags between point pairs and not on the points themselves	
Variogram function is calculated reliably from available information	
	$X(\boldsymbol{p}) = \overline{X} + \varepsilon(\boldsymbol{p})$ $\varepsilon(\boldsymbol{p})$ known at data points (since $\overline{X}$ is known)

satisfy the two criteria introduced by Eqs. (1.1) and (1.2) of Section 1 earlier.[g]

In practice, the *STOK* can account for local attribute mean fluctuations by limiting the validity of the homostationarity assumption to properly chosen space-time varying neighborhoods (say, as in Example 1.3 of Section 1). The implication is that the attribute mean may vary throughout the original domain (although it remains unknown), and only strictly within each selected neighborhood the mean is assumed constant, say $\overline{X}$ (although the value of $\overline{X}$ may vary from one neighborhood to the next).

2.1.1 STOK equations

The intuitively appealing properties expressed mathematically by Eqs. (1.1) and (1.2) of Section 1 lead to the following *STOK* system of equations:

$$\sum\nolimits_{i=1}^{m} \lambda_i \gamma_X\left(\Delta \boldsymbol{p}_{ij}\right) + \mu = \gamma_X\left(\Delta \boldsymbol{p}_{kj}\right), \; j = 1, \ldots, m,$$
$$\sum\nolimits_{i=1}^{m} \lambda_i = 1. \tag{2.1a, b}$$

I.e., the weights assigned to the input data are such that the interpolation error is kept to a minimum and their sum is adjusted to 1. Moreover, as hinted earlier, the following *STOK* feature is worth noting.

> By serving as the input to the *STOK* equations the **variogram** γ_X represents certain aspects of attribute's space-time distribution that have a very significant impact on the generated interpolations.

The system of Eqs. (2.1a,b) is solved with respect to λ_i and μ ($i = 1, \ldots, m$). Each weight λ_i depends on the *shape* of the variogram model, and on the chronotopologic *lags* between the data points $\boldsymbol{p}_i = (\boldsymbol{s}_i; t_i)$ and the interpolated point $\boldsymbol{p}_k = (\boldsymbol{s}_k; t_k)$, as well as between the data points $\boldsymbol{p}_i$ themselves. Lastly, the λ_i and μ solutions of Eqs. (2.1a, b) are substituted into the following linear expressions:

$$\hat{X}\left(\boldsymbol{p}_k\right) = \sum\nolimits_{i=1}^{m} \lambda_i X\left(\boldsymbol{p}_i\right),$$
$$e_X^2\left(\boldsymbol{p}_k\right) = \sum\nolimits_{i=1}^{m} \lambda_i \gamma_X\left(\Delta \boldsymbol{p}_{ki}\right) + \mu, \tag{2.2a, b}$$

which provide the desired *STOK* interpolated attribute value and the associated interpolation error variance at every point $\boldsymbol{p}_k$.

The *STOK* outputs, i.e., $\hat{X}\left(\boldsymbol{p}_k\right)$ and $e_X^2(\boldsymbol{p}_k)$, are:

(a) statistically optimal and

[g] In this sense, *STOK* is a *best linear unbiased interpolation* (*BLUI*) technique.

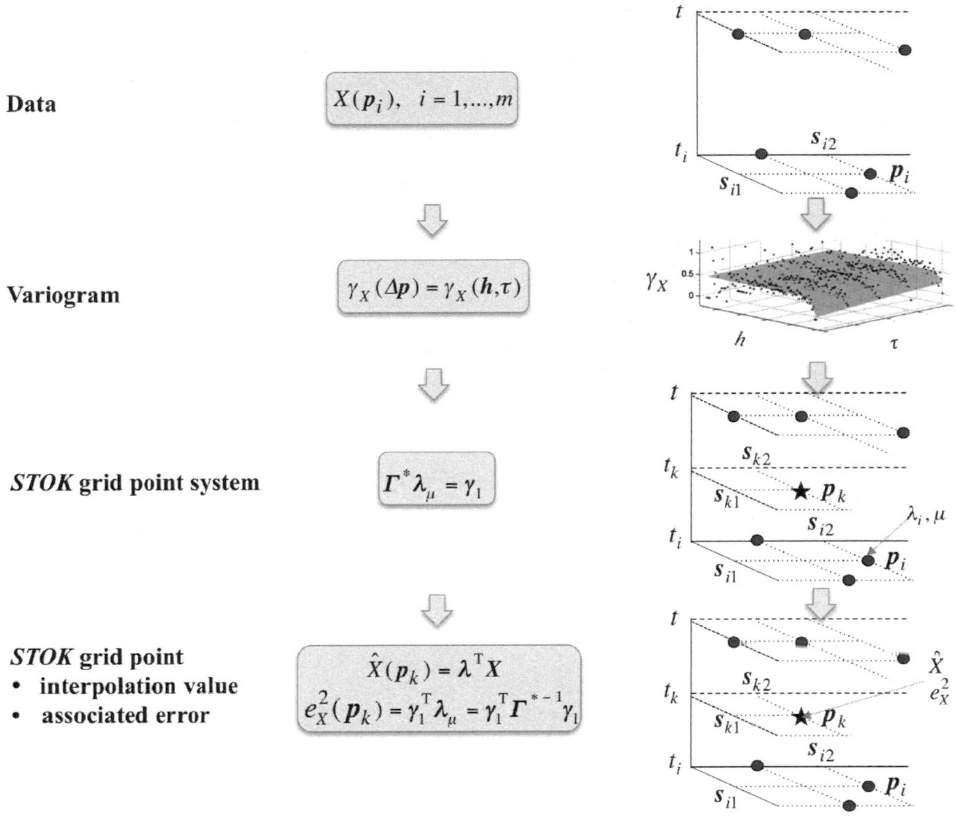

FIG. 2.1 A graphical outline of the *STOK* technique.

(b) constrained to the neighborhood **data** and the **variogram** model.

For instructional purposes, and not only, the *STOK* attribute interpolation process is graphically outlined in Fig. 2.1. The *STOK* equations are rewritten in a matrix form suitable for computational purposes, where

$$
\begin{aligned}
X &= [X(\boldsymbol{p}_1)\cdots X(\boldsymbol{p}_N)]^{\mathrm{T}},\\
\boldsymbol{\lambda} &= [\lambda_1\cdots\lambda_m]^{\mathrm{T}},\\
\boldsymbol{\lambda}_\mu &= [\boldsymbol{\lambda}\,\mu]^{\mathrm{T}},\\
\boldsymbol{\gamma} &= [\gamma_X(\Delta\boldsymbol{p}_{k1})\ldots\gamma_X(\Delta\boldsymbol{p}_{km})]^{\mathrm{T}},\\
\boldsymbol{\gamma}_1 &= [\boldsymbol{\gamma}\,\mathbf{1}]^{\mathrm{T}},
\end{aligned}
\tag{2.3a-e}
$$

$$
\boldsymbol{\Gamma} = \begin{bmatrix} \gamma_X(\mathbf{0}) & \gamma_X(\Delta\boldsymbol{p}_{12}) & \cdots & \gamma_X(\Delta\boldsymbol{p}_{1m}) \\ \gamma_X(\Delta\boldsymbol{p}_{21}) & \gamma_X(\mathbf{0}) & \ldots & \gamma_X(\Delta\boldsymbol{p}_{2m}) \\ \vdots & \vdots & & \vdots \\ \gamma_X(\Delta\boldsymbol{p}_{m1}) & \gamma_X(\Delta\boldsymbol{p}_{m2}) & \ldots & \gamma_X(\mathbf{0}) \end{bmatrix},
\quad
\boldsymbol{\Gamma}^* = \begin{bmatrix} \boldsymbol{\Gamma} & \mathbf{1} \\ \mathbf{1}^{\mathrm{T}} & 0 \end{bmatrix},
\tag{2.4a, b}
$$

and $\mathbf{1}^{\mathrm{T}} = [1\ldots1]$.

In practice, one is usually dealing with a large number of interpolation points $\boldsymbol{p}_k$ (often on the order of thousands), which are the nodes of a properly selected space-time grid that covers the domain of interest. *Pro rata*, this implies that an equally large number of *STOK* systems have to be solved. The solution, of course, involves fast computational techniques and suitable software libraries (see discussion later).

2.1.2 Analytical STOK solutions

As noted earlier, in the majority of practical applications the *STOK* system of Eqs. (2.1a, b) is solved numerically with the help of a computer software library. Yet, despite the convenient common practice to favor computational approaches, it is always instructive to examine the possibility of deriving some closed-form solutions. There is a very good motivation for such an effort:

> **Analytical** solutions can provide valuable insight into the physical meaning of the solutions that cannot be obtained by numerical approaches.

Arguably, the mathematical solution of the linear *STOK* system is basically a matrix inversion operation. In particular, the solution is about the inversion of the variogram $\boldsymbol{\Gamma}$ matrix with dimensions $(m+1) \times (m+1)$, see Fig. 2.1.

- The **inversion** of the $\boldsymbol{\Gamma}^*$ matrix can be expressed on the basis of its decomposition into submatrices, as follows,

$$\boldsymbol{\Gamma}^{*-1} = \begin{bmatrix} \boldsymbol{\Gamma}_1 & \boldsymbol{\Gamma}_2 \\ \boldsymbol{\Gamma}_3 & \boldsymbol{\Gamma}_4 \end{bmatrix}^{-1}$$
$$= \begin{bmatrix} (\boldsymbol{\Gamma}_1 - \boldsymbol{\Gamma}_2\boldsymbol{\Gamma}_4^{-1}\boldsymbol{\Gamma}_3)^{-1} & -(\boldsymbol{\Gamma}_1 - \boldsymbol{\Gamma}_2\boldsymbol{\Gamma}_4^{-1}\boldsymbol{\Gamma}_3)^{-1}\boldsymbol{\Gamma}_2\boldsymbol{\Gamma}_4^{-1} \\ -\boldsymbol{\Gamma}_4^{-1}\boldsymbol{\Gamma}_3(\boldsymbol{\Gamma}_1 - \boldsymbol{\Gamma}_2\boldsymbol{\Gamma}_4^{-1}\boldsymbol{\Gamma}_3)^{-1}\boldsymbol{\Gamma}_4^{-1} + \boldsymbol{\Gamma}_4^{-1}\boldsymbol{\Gamma}_3(\boldsymbol{\Gamma}_1 - \boldsymbol{\Gamma}_2\boldsymbol{\Gamma}_4^{-1}\boldsymbol{\Gamma}_3)^{-1}\boldsymbol{\Gamma}_2\boldsymbol{\Gamma}_4^{-1} \end{bmatrix}, \quad (2.5a, b)$$

where $\boldsymbol{\Gamma}_1$, $\boldsymbol{\Gamma}_2$, $\boldsymbol{\Gamma}_3$ and $\boldsymbol{\Gamma}_4$ are submatrices of arbitrary size, $\boldsymbol{\Gamma}_1$ and $\boldsymbol{\Gamma}_4$ must be square (to be invertible), and $\boldsymbol{\Gamma}_4$ and $\boldsymbol{\Gamma}_1 - \boldsymbol{\Gamma}_2\boldsymbol{\Gamma}_4^{-1}\boldsymbol{\Gamma}_3$ must be nonsingular.

This matrix inversion technique is particularly useful if $\boldsymbol{\Gamma}_4$ is diagonal and $\boldsymbol{\Gamma}_1 - \boldsymbol{\Gamma}_2\boldsymbol{\Gamma}_4^{-1}\boldsymbol{\Gamma}_3$ is a small matrix.[h] The technique could be used in a sequential manner to invert $\boldsymbol{\Gamma}$ matrices of increasing dimensionality. A few instructive examples follow.

Example 2.1 We examine two implementations of Eq. (2.5a, b), leading to the solution of the *STOK* systems and the generation of the corresponding interpolations and associated statistical errors.

Case ❶

Consider the arrangement of two data points $\boldsymbol{p}_1$ and $\boldsymbol{p}_2$ with one interpolation point $\boldsymbol{p}_k$. Let $\Delta\boldsymbol{p}_{ij} = (\boldsymbol{h}_{ij}, \tau_{ij})$ $(i, j = k, 1, 2)$.

Step 1: The Kriging system is written in this case as

$$\boldsymbol{\Gamma}^*\boldsymbol{\lambda}_\mu = \boldsymbol{\Gamma}_1\boldsymbol{\lambda}_\mu = \begin{bmatrix} \gamma_X(0) & \gamma_X(\Delta\boldsymbol{p}_{12}) & 1 \\ \gamma_X(\Delta\boldsymbol{p}_{21}) & \gamma_X(0) & 1 \\ 1 & 1 & 0 \end{bmatrix} \begin{bmatrix} \lambda_1 \\ \lambda_2 \\ \mu \end{bmatrix}$$
$$= \begin{bmatrix} 0 & \gamma_{12} & 1 \\ \gamma_{21} & 0 & 1 \\ 1 & 1 & 0 \end{bmatrix} \begin{bmatrix} \lambda_1 \\ \lambda_2 \\ \mu \end{bmatrix} = \boldsymbol{\gamma}_1 \quad (2.6)$$

where we simply set $\boldsymbol{\Gamma}^* = \boldsymbol{\Gamma}_1$ in Eq. (2.5a, b), since the inverse of a 3×3 matrix is known.[i-]
Step 2: The solution of Eq. (2.6) is

[h] Notice that these are the only matrices requiring inversion.

[i] Recall that if $\boldsymbol{A} = \begin{bmatrix} a & b & c \\ d & e & f \\ g & h & k \end{bmatrix}$, then $\boldsymbol{A}^{-1} = \frac{1}{a(ek-fh)-b(kd-fg)+c(dh-eg)} \begin{bmatrix} ek-fh & ch-bk & bf-ce \\ fg-dk & ak-cg & cd-af \\ dh-eg & gb-ah & ae-bd \end{bmatrix}$.

$$\boldsymbol{\lambda}_\mu = \begin{bmatrix} \lambda_1 \\ \lambda_2 \\ \mu \end{bmatrix} = \boldsymbol{\Gamma}_1^{-1}\boldsymbol{\gamma}_1$$
$$= \begin{bmatrix} -\frac{1}{2\gamma_{12}} & \frac{1}{2\gamma_{12}} & \frac{1}{2} \\ \frac{1}{2\gamma_{12}} & -\frac{1}{2\gamma_{12}} & \frac{1}{2} \\ \frac{1}{2} & \frac{1}{2} & -\frac{\gamma_{12}}{2} \end{bmatrix} \begin{bmatrix} \gamma_{k1} \\ \gamma_{k2} \\ 1 \end{bmatrix}$$
$$= \begin{bmatrix} \frac{\gamma_{k2}-\gamma_{k1}}{2\gamma_{12}} + \frac{1}{2} \\ \frac{\gamma_{k1}-\gamma_{k2}}{2\gamma_{12}} + \frac{1}{2} \\ \frac{\gamma_{k1}+\gamma_{k2}-\gamma_{12}}{2} \end{bmatrix} \tag{2.7}$$

where $\gamma_{k1} = \gamma_X(\boldsymbol{\Delta p}_{k1})$, $\gamma_{k2} = \gamma_X(\boldsymbol{\Delta p}_{k2})$, and $\gamma_{12} = \gamma_{21}$.

Step 3: Then, the attribute interpolation and the associated interpolation error are given by

$$\hat{X}(\boldsymbol{p}_k) = \boldsymbol{\lambda}^T X = \lambda_1 X(\boldsymbol{p}_1) + \lambda_2 X(\boldsymbol{p}_2)$$
$$= \left[\frac{1}{2} - \frac{\gamma_{k1}-\gamma_{k2}}{2\gamma_{12}}\right] X(\boldsymbol{p}_1) + \left[\frac{1}{2} + \frac{\gamma_{k1}-\gamma_{k2}}{2\gamma_{12}}\right] X(\boldsymbol{p}_2),$$
$$e_X^2(\boldsymbol{p}_k) = \boldsymbol{\gamma}_1^T \boldsymbol{\lambda}_\mu = [\gamma_{k1}\ \gamma_{k2}\ 1] \begin{bmatrix} \frac{\gamma_{k2}-\gamma_{k1}}{2\gamma_{12}} + \frac{1}{2} \\ \frac{\gamma_{k1}-\gamma_{k2}}{2\gamma_{12}} + \frac{1}{2} \\ \frac{\gamma_{k1}+\gamma_{k2}-\gamma_{12}}{2} \end{bmatrix}$$
$$= \gamma_{k1} + \gamma_{k2} - \frac{(\gamma_{k1}-\gamma_{k2})^2}{2\gamma_{12}} - \frac{\gamma_{12}}{2}. \tag{2.8a, b}$$

Step 4: Note that if $\gamma_{k1} > \gamma_{k2}$, then $\lambda_1 < \lambda_2$; whereas if $\gamma_{k1} < \gamma_{k2}$, then $\lambda_1 > \lambda_2$. In the space-time symmetric case, $\boldsymbol{h}_{k1} = \boldsymbol{h}_{k2}$ and $\tau_{k1} = \tau_{k2}$, so that $\gamma_{k1} = \gamma_{k2}$; the last set of equations reduces to

$$\hat{X}(\boldsymbol{p}_k) = \boldsymbol{\lambda}^T \boldsymbol{X} = \lambda_1 X(\boldsymbol{p}_1) + \lambda_2 X(\boldsymbol{p}_2) = \frac{1}{2}[X(\boldsymbol{p}_1) + X(\boldsymbol{p}_2)] \tag{2.9a}$$

$$e_X^2(\boldsymbol{p}_k) = \boldsymbol{\gamma}_1^T \boldsymbol{\lambda}_\mu = [\gamma_{k1}\ \gamma_{k2}\ 1] \begin{bmatrix} \frac{1}{2} \\ \frac{1}{2} \\ \frac{\gamma_{k1}+\gamma_{k2}-\gamma_{12}}{2} \end{bmatrix} = 2\gamma_{k1} - \frac{\gamma_{12}}{2} \tag{2.9b}$$

- Due to **symmetry**, the interpolated value is independent of the variogram, whereas the interpolation error variance increases with increasing γ_{k1} and γ_{k2} values or with decreasing γ_{12} values.

Case ❷

Next, three data points $\boldsymbol{p}_1$, $\boldsymbol{p}_2$, and $\boldsymbol{p}_3$ are considered with one interpolation point $\boldsymbol{p}_k$.

Step 1: The Kriging system is

$$\boldsymbol{\Gamma}^* \boldsymbol{\lambda}_\mu = \begin{bmatrix} \gamma_X(\boldsymbol{0}) & \gamma_X(\boldsymbol{\Delta p}_{12}) & \gamma_X(\boldsymbol{\Delta p}_{13}) & 1 \\ \gamma_X(\boldsymbol{\Delta p}_{21}) & \gamma_X(\boldsymbol{0}) & \gamma_X(\boldsymbol{\Delta p}_{23}) & 1 \\ \gamma_X(\boldsymbol{\Delta p}_{31}) & \gamma_X(\boldsymbol{\Delta p}_{31}) & \gamma_X(\boldsymbol{0}) & 1 \\ 1 & 1 & 1 & 0 \end{bmatrix} \begin{bmatrix} \lambda_1 \\ \lambda_2 \\ \lambda_3 \\ \mu \end{bmatrix}$$
$$= \begin{bmatrix} 0 & \gamma_{12} & \gamma_{13} & 1 \\ \gamma_{21} & 0 & \gamma_{23} & 1 \\ \gamma_{31} & \gamma_{32} & 0 & 1 \\ 1 & 1 & 1 & 0 \end{bmatrix} \begin{bmatrix} \lambda_1 \\ \lambda_2 \\ \lambda_3 \\ \mu \end{bmatrix} = \begin{bmatrix} \boldsymbol{\Gamma}_1 & \boldsymbol{\Gamma}_2 \\ \boldsymbol{\Gamma}_3 & \boldsymbol{\Gamma}_4 \end{bmatrix} \begin{bmatrix} \boldsymbol{\lambda} \\ \mu \end{bmatrix}$$
$$= \boldsymbol{\gamma}_1 \tag{2.10}$$

where

$$\boldsymbol{\Gamma}_1 = \begin{bmatrix} 0 & \gamma_{12} \\ \gamma_{21} & 0 \end{bmatrix}, \quad \boldsymbol{\Gamma}_2 = \begin{bmatrix} \gamma_{13} & 1 \\ \gamma_{23} & 1 \end{bmatrix},$$
$$\boldsymbol{\Gamma}_3 = \begin{bmatrix} \gamma_{31} & \gamma_{32} \\ 1 & 1 \end{bmatrix}, \quad \boldsymbol{\Gamma}_4 = \begin{bmatrix} 0 & 1 \\ 1 & 0 \end{bmatrix},$$
$$\boldsymbol{\gamma}_1 = [\gamma_{k1}\ \gamma_{k2}\ \gamma_{k3}\ 1]^T, \quad \boldsymbol{\lambda} = [\lambda_1\ \ \lambda_2\ \ \lambda_3]^T.$$

Step 2: From Eq. (2.5a, b), the solution is

$$\boldsymbol{\lambda}_\mu = \begin{bmatrix} \lambda_1 \\ \lambda_2 \\ \lambda_3 \\ \mu \end{bmatrix} = \boldsymbol{\Gamma}^{*-1}\boldsymbol{\gamma}_1 = \begin{bmatrix} (\boldsymbol{\Gamma}_1 - \boldsymbol{\Gamma}_2\boldsymbol{\Gamma}_4^{-1}\boldsymbol{\Gamma}_3)^{-1} & -(\boldsymbol{\Gamma}_1 - \boldsymbol{\Gamma}_2\boldsymbol{\Gamma}_4^{-1}\boldsymbol{\Gamma}_3)^{-1}\boldsymbol{\Gamma}_2\boldsymbol{\Gamma}_4^{-1} \\ -\boldsymbol{\Gamma}_4^{-1}\boldsymbol{\Gamma}_3(\boldsymbol{\Gamma}_1 - \boldsymbol{\Gamma}_2\boldsymbol{\Gamma}_4^{-1}\boldsymbol{\Gamma}_3)^{-1} & \boldsymbol{\Gamma}_4^{-1} + \boldsymbol{\Gamma}_4^{-1}\boldsymbol{\Gamma}_3(\boldsymbol{\Gamma}_1 - \boldsymbol{\Gamma}_2\boldsymbol{\Gamma}_4^{-1}\boldsymbol{\Gamma}_3)^{-1}\boldsymbol{\Gamma}_2\boldsymbol{\Gamma}_4^{-1} \end{bmatrix} \begin{bmatrix} \gamma_{k1} \\ \gamma_{k2} \\ \gamma_{k3} \\ 1 \end{bmatrix} \tag{2.11}$$

where[j]

$$\boldsymbol{\Gamma}_1^{-1} = -\frac{1}{\gamma_{12}^2}\begin{bmatrix} 0 & -\gamma_{12} \\ -\gamma_{21} & 0 \end{bmatrix},\ \boldsymbol{\Gamma}_2^{-2} = \frac{1}{\gamma_{13}-\gamma_{23}}\begin{bmatrix} 1 & -1 \\ -\gamma_{23} & \gamma_{13} \end{bmatrix},\ \boldsymbol{\Gamma}_3^{-1} = \frac{1}{\gamma_{31}-\gamma_{32}}\begin{bmatrix} 1 & -\gamma_{32} \\ -1 & \gamma_{31} \end{bmatrix},$$

$$\boldsymbol{\Gamma}_4^{-1} = -\begin{bmatrix} 0 & -1 \\ -1 & 0 \end{bmatrix} = \begin{bmatrix} 0 & 1 \\ 1 & 0 \end{bmatrix},$$

$$\left(\boldsymbol{\Gamma}_1 - \boldsymbol{\Gamma}_2\boldsymbol{\Gamma}_4^{-1}\boldsymbol{\Gamma}_3\right)^{-1} = \frac{1}{\gamma_{12}(\gamma_{12}-2\gamma_{23}-2\gamma_{13})+(\gamma_{13}-\gamma_{23})^2}\begin{bmatrix} 2\gamma_{23} & \gamma_{12}-\gamma_{13}-\gamma_{23} \\ \gamma_{12}-\gamma_{13}-\gamma_{23} & 2\gamma_{13} \end{bmatrix},$$

$$-\left(\boldsymbol{\Gamma}_1 - \boldsymbol{\Gamma}_2\boldsymbol{\Gamma}_4^{-1}\boldsymbol{\Gamma}_3\right)^{-1}\boldsymbol{\Gamma}_2\boldsymbol{\Gamma}_4^{-1} = \frac{1}{\gamma_{12}(\gamma_{12}-\gamma_{23}-\gamma_{13})-\gamma_{12}(\gamma_{13}+\gamma_{23})+(\gamma_{13}-\gamma_{23})^2}\times\begin{bmatrix} \gamma_{13}-\gamma_{12}-\gamma_{23} & \gamma_{23}(\gamma_{23}-\gamma_{12}-\gamma_{13}) \\ \gamma_{23}-\gamma_{12}-\gamma_{13} & \gamma_{13}(\gamma_{13}-\gamma_{12}-\gamma_{23}) \end{bmatrix},$$

$$-\boldsymbol{\Gamma}_4^{-1}\boldsymbol{\Gamma}_3\left(\boldsymbol{\Gamma}_1 - \boldsymbol{\Gamma}_2\boldsymbol{\Gamma}_4^{-1}\boldsymbol{\Gamma}_3\right)^{-1} = \frac{1}{\gamma_{12}(\gamma_{12}-2\gamma_{23}-2\gamma_{13})+(\gamma_{13}-\gamma_{23})^2}\times\begin{bmatrix} \gamma_{13}-\gamma_{12}-\gamma_{23} & \gamma_{23}-\gamma_{12}-\gamma_{13} \\ \gamma_{23}(\gamma_{23}-\gamma_{12}-\gamma_{13}) & \gamma_{13}(\gamma_{13}-\gamma_{12}-\gamma_{23}) \end{bmatrix},$$

$$\boldsymbol{\Gamma}_4^{-1} + \boldsymbol{\Gamma}_4^{-1}\boldsymbol{\Gamma}_3\left(\boldsymbol{\Gamma}_1 - \boldsymbol{\Gamma}_2\boldsymbol{\Gamma}_4^{-1}\boldsymbol{\Gamma}_3\right)^{-1}\boldsymbol{\Gamma}_2\boldsymbol{\Gamma}_4^{-1} = \frac{1}{\gamma_{12}(\gamma_{12}-2\gamma_{23}-2\gamma_{13})+(\gamma_{13}-\gamma_{23})^2}\times\begin{bmatrix} 2\gamma_{12} & \gamma_{12}(\gamma_{12}-\gamma_{13}-\gamma_{23}) \\ \gamma_{12}(\gamma_{12}-\gamma_{13}-\gamma_{23}) & 2\gamma_{12}\gamma_{13}\gamma_{23} \end{bmatrix}.$$

Hence,

$$\boldsymbol{\lambda}_\mu = \begin{bmatrix} \lambda_1 \\ \lambda_2 \\ \lambda_3 \\ \mu \end{bmatrix} = \frac{1}{\gamma_{12}(\gamma_{12}-2\gamma_{23}-2\gamma_{13})+(\gamma_{13}-\gamma_{23})^2}$$
$$\times\begin{bmatrix} 2\gamma_{23}\gamma_{k1}+(\gamma_{12}-\gamma_{13}-\gamma_{23})\gamma_{k2}+(\gamma_{13}-\gamma_{12}-\gamma_{23})\gamma_{k3}+\gamma_{23}(\gamma_{23}-\gamma_{12}-\gamma_{13}) \\ (\gamma_{12}-\gamma_{13}-\gamma_{23})\gamma_{k1}+2\gamma_{13}\gamma_{k2}+(\gamma_{23}-\gamma_{12}-\gamma_{13})\gamma_{k3}+\gamma_{13}(\gamma_{13}-\gamma_{12}-\gamma_{23}) \\ (\gamma_{13}-\gamma_{12}-\gamma_{23})\gamma_{k1}+(\gamma_{23}-\gamma_{12}-\gamma_{13})\gamma_{k2}+2\gamma_{12}\gamma_{k3}+\gamma_{12}(\gamma_{12}-\gamma_{13}-\gamma_{23}) \\ (\gamma_{23}-\gamma_{12}-\gamma_{13})\gamma_{23}\gamma_{k1}+(\gamma_{13}-\gamma_{12}-\gamma_{23})\gamma_{13}\gamma_{k2}+(\gamma_{12}-\gamma_{13}-\gamma_{23})\gamma_{12}\gamma_{k3}+2\gamma_{12}\gamma_{13}\gamma_{23} \end{bmatrix} \tag{2.12}$$

Step 3: Then, the attribute interpolation and the associated error are given by

$$\hat{X}(\boldsymbol{p}_k) = \boldsymbol{\lambda}^T\mathbf{X} = [\lambda_1\ \ \lambda_2\ \ \lambda_3]\begin{bmatrix} X(\boldsymbol{p}_1) \\ X(\boldsymbol{p}_2) \\ X(\boldsymbol{p}_3) \end{bmatrix} = \lambda_1 X(\boldsymbol{p}_1) + \lambda_2 X(\boldsymbol{p}_2) + \lambda_3 X(\boldsymbol{p}_3)$$
$$= \frac{1}{\gamma_{12}(\gamma_{12}-2\gamma_{23}-2\gamma_{13})+(\gamma_{13}-\gamma_{23})^2}$$
$$\times\{[2\gamma_{23}\gamma_{k1}+(\gamma_{12}-\gamma_{13}-\gamma_{23})\gamma_{k2}+(\gamma_{13}-\gamma_{12}-\gamma_{23})\gamma_{k3}+\gamma_{23}(\gamma_{23}-\gamma_{12}-\gamma_{13})]X(\boldsymbol{p}_1)$$
$$+[(\gamma_{12}-\gamma_{13}-\gamma_{23})\gamma_{k1}+2\gamma_{13}\gamma_{k2}+(\gamma_{23}-\gamma_{12}-\gamma_{13})\gamma_{k3}+\gamma_{13}(\gamma_{13}-\gamma_{12}-\gamma_{23})]X(\boldsymbol{p}_2)$$
$$+[(\gamma_{13}-\gamma_{12}-\gamma_{23})\gamma_{k1}+(\gamma_{23}-\gamma_{12}-\gamma_{13})\gamma_{k2}+2\gamma_{12}\gamma_{k3}+\gamma_{12}(\gamma_{12}-\gamma_{13}-\gamma_{23})]X(\boldsymbol{p}_3)\}, \tag{2.13a}$$

[j] Recall that if $\boldsymbol{A} = \begin{bmatrix} a\ b \\ c\ d \end{bmatrix}$, then $\boldsymbol{A}^{-1} = \frac{1}{ad-bc}\begin{bmatrix} d-b \\ -c\ a \end{bmatrix}$

$$e_X^2(\boldsymbol{p}_k) = \boldsymbol{\gamma}_1^T \boldsymbol{\lambda}_\mu = [\gamma_{k1} \ \ \gamma_{k2} \ \ \gamma_{k3} \ \ 1] \begin{bmatrix} \lambda_1 \\ \lambda_2 \\ \lambda_3 \\ \mu \end{bmatrix} = \frac{1}{\gamma_{12}(\gamma_{12} - 2\gamma_{23} - 2\gamma_{13}) + (\gamma_{13} - \gamma_{23})^2}$$
$$\times \{[2\gamma_{23}\gamma_{k1} + (\gamma_{12} - \gamma_{13} - \gamma_{23})\gamma_{k2} + (\gamma_{13} - \gamma_{12} - \gamma_{23})\gamma_{k3} + \gamma_{23}(\gamma_{23} - \gamma_{12} - \gamma_{13})]\gamma_{k1}$$
$$+ [(\gamma_{12} - \gamma_{13} - \gamma_{23})\gamma_{k1} + 2\gamma_{13}\gamma_{k2} + (\gamma_{23} - \gamma_{12} - \gamma_{13})\gamma_{k3} + \gamma_{13}(\gamma_{13} - \gamma_{12} - \gamma_{23})]\gamma_{k2}$$
$$+ [(\gamma_{13} - \gamma_{12} - \gamma_{23})\gamma_{k1} + (\gamma_{23} - \gamma_{12} - \gamma_{13})\gamma_{k2} + 2\gamma_{12}\gamma_{k3} + \gamma_{12}(\gamma_{12} - \gamma_{13} - \gamma_{23})]\gamma_{k3}$$
$$+ (\gamma_{23} - \gamma_{12} - \gamma_{13})\gamma_{23}\gamma_{k1} + (\gamma_{13} - \gamma_{12} - \gamma_{23})\gamma_{13}\gamma_{k2} + (\gamma_{12} - \gamma_{13} - \gamma_{23})\gamma_{12}\gamma_{k3} + 2\gamma_{12}\gamma_{13}\gamma_{23}\}. \quad (2.13b)$$

The following numerical example uses the analytical results of Example 2.1 to examine the uneven effect of space-time *symmetry* on the interpolated attribute value and interpolation error.

Example 2.2 In Fig. 2.2, an attribute interpolation value is sought at point $(\boldsymbol{s}_k; t_k)$ using two neighboring data points $(\boldsymbol{s}_1; t_1)$ and $(\boldsymbol{s}_2; t_2)$, where the attribute values are, respectively, $X(\boldsymbol{p}_1) = 10$ and $X(\boldsymbol{p}_2) = 8$ (attribute units). The corresponding spatial distances and temporal separations are $h_{k1} = |\boldsymbol{s}_k - \boldsymbol{s}_1| = |\boldsymbol{s}_k - \boldsymbol{s}_2| = h_{k2} = 1$, $h_{12} = |\boldsymbol{s}_1 - \boldsymbol{s}_2| = \sqrt{2} \approx 1.41$ (space units), and $\tau_{12} = |t_1 - t_2| = 4$ (time units). Two illustrative cases are examined next.

Case ❶

The first variogram model is assumed to be of the combined exponential-Gaussian form

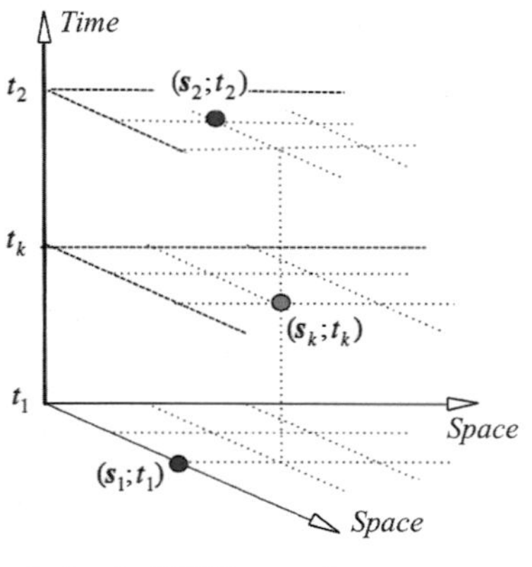

FIG. 2.2 A simple *STOK* example.

$$\gamma_X(h, \tau) = \left(1 - e^{-h}\right)\left(1 - e^{-\tau^2}\right)$$

in (attribute units)^2. Hence, $\gamma_{11} = \gamma_X(h_{11}, \tau_{11}) = 0$ and $\gamma_{22} = \gamma_X(h_{22}, \tau_{22}) = 0$. Because of space-time symmetry, $h_{k1} = h_{k2}$ and $\tau_{k1} = \tau_{k2}$, so that it is valid that $\gamma_{k1} = \gamma_X(h_{k1}, \tau_{k1}) = \gamma_X(h_{k2}, \tau_{k2}) = \gamma_{k2} = (1 - e^{-1})(1 - e^{-2^2}) = 0.6205$, and $\gamma_{12} = \gamma_X(h_{12}, \tau_{12}) = \gamma_X(h_{21}, \tau_{21}) = \gamma_{21} = \left(1 - e^{-\sqrt{2}}\right)\left(1 - e^{-4^2}\right) = 0.7569$.

Using the analytical expressions derived in Example 2.1 above, the interpolated attribute value and the associated interpolation error are obtained as

$$\hat{X}(\boldsymbol{p}_k) = \left[\frac{1}{2} - \frac{\gamma_{k1} - \gamma_{k2}}{2\gamma_{12}}\right] X(\boldsymbol{p}_1) + \left[\frac{1}{2} + \frac{\gamma_{k1} - \gamma_{k2}}{2\gamma_{12}}\right] \times$$
$$X(\boldsymbol{p}_2) = \frac{1}{2}[X(\boldsymbol{p}_1) + X(\boldsymbol{p}_2)] = \frac{1}{2}[10 + 8] = 9,$$

$$e_X^2(\boldsymbol{p}_k) = \gamma_{k1} + \gamma_{k2} - \frac{(\gamma_{k1} - \gamma_{k2})^2}{2\gamma_{12}} - \frac{\gamma_{12}}{2}$$

$$= \gamma_{k1} + \gamma_{k2} - \frac{\gamma_{12}}{2} = 0.6205 + 0.6205 - \frac{0.7569}{2} = 0.8626$$

in attribute units and (attribute units)^2, respectively, where $\mu = \frac{\gamma_{k1} + \gamma_{k2} - \gamma_{12}}{2} = 0.2421$ (i.e., the μ-introduced uncertainty due to the unknown attribute mean increases the *STOK* error variance). The analysis leads to the following conclusion:

- Due to space-time **symmetry**, the weights λ_1 and λ_2 and the corresponding interpolated value $\hat{X}(\boldsymbol{p}_k)$ do not depend on the variogram shape. Only the multiplier μ and the interpolation error $e_X^2(\boldsymbol{p}_k)$ do, in this case.

Case ❷

A different variogram model is assumed next, say, of the space-time linear form

$$\gamma_X(h,\tau) = h\tau,$$

where $\gamma_{11} = \gamma_{22} = \gamma_{k1} = \gamma_{k2} = 0$ and $\gamma_{12} = 4\sqrt{2} = 5.6568$. It is found that

$$\hat{X}(\boldsymbol{p}_k) = \left[\frac{1}{2} - \frac{\gamma_{k1}-\gamma_{k2}}{2\gamma_{12}}\right] X(\boldsymbol{p}_1) + \left[\frac{1}{2} + \frac{\gamma_{k1}-\gamma_{k2}}{2\gamma_{12}}\right] \times$$
$$X(\boldsymbol{p}_2) = \frac{1}{2}[X(\boldsymbol{p}_1) + X(\boldsymbol{p}_2)] = \frac{1}{2}[10+8] = 9,$$
$$e_X^2(\boldsymbol{p}_k) = \gamma_{k1} + \gamma_{k2} - \frac{(\gamma_{k1}-\gamma_{k2})^2}{2\gamma_{12}} - \frac{\gamma_{12}}{2}$$
$$= \gamma_{k1} + \gamma_{k2} - \frac{\gamma_{12}}{2} = 2 + 2 - \frac{5.6568}{2} = 1.1716.$$

This interpolation error is higher than the one obtained earlier, i.e., 1.1716 *vs.* 0.8626. By way of a summary:

- Although the variogram models above differ, the same interpolated attribute value is obtained in cases ❶ and ❷, but with a higher error variance in the case ❷, which may be due to the sharper slope of the second variogram at the origin.[k]

2.1.3 STOK interpretation

The *STOK* equations provide a description of the solution of the interpolation problem by employing an appropriate mathematical formalism. Being able to meaningfully interpret these equations, undoubtedly takes particular knowledge and skill. This is not a trivial effort, yet it is a rewarding one, since the adequate interpretation of the *STOK* equations can improve the understanding of the relation between the computed equation parameters and how they can affect the results, as well as their particular physical meaning. This line of thought leads to the following conclusion.

> The intuitive appeal of the *STOK* formulation is that it expresses the interpolated attribute value $\hat{X}(\boldsymbol{p}_k)$ as a combination of nearby samples, $X(\boldsymbol{p}_i)$, weighted by some measure (λ_i) of their **similarity** to the interpolated value, which is measured statistically in terms of variogram values between point pairs in the domain of interest.

Along these lines, the *STOK* formulation introduced earlier has some other properties with noticeable physical significance, as highlighted in Table 2.2. In many real-world studies, these properties enable *STOK* to take into account several aspects of chronotopologic distribution that make scientific sense and can generate realistic and accurate attribute interpolations that are testable. Based on the properties of Table 2.2, additional application-dependent interpretations can be derived. Some illustrations are given in the next example.

Example 2.3 Let us consider properties ① and ③.

- An implication of property ① in Table 2.2 is that a data point arrangement achieves better domain coverage and generates more information if it consists of evenly distributed points than if it consists of clustered subsets of points.
- A common issue is the interpretation of the variogram nugget (Section 3 of Chapter 5) in relation to the exact interpolation aspect of property ③. This property implies that if the nugget is interpreted as a measurement error, Kriging interpolation would want to smooth over, whereas if the nugget is interpreted as fine-scale (microscale) variability, Kriging interpolations at the data points should coincide with the data values.

[k] It holds that $\left.\frac{\partial}{\partial h}(1-e^{-h})\left(1-e^{-\tau^2}\right)\right|_{h=0} = 1-e^{-\tau^2}$ *vs.* $\left.\frac{\partial}{\partial h}h\tau\right|_{h=0} = \tau$, and $\left.\frac{\partial}{\partial \tau}(1-e^{-h})\left(1-e^{-\tau^2}\right)\right|_{\tau=0} = 0$ *vs.* $\left.\frac{\partial}{\partial \tau}h\tau\right|_{\tau=0} = h.$

TABLE 2.2 *STOK* properties and their description.

Property	Description
① Data point arrangement	• Accounts for relative data locations and instances • All data points are not granted the same importance, and data quality may vary from point to point in the arrangement
② Variation structure	• Differs among attributes, depending on the underlying attribute mechanisms and the natural laws they obey
③ Interpolated value	• Is unbiased, meaning that, on the average, the interpolated attribute mean is equal to the actual mean • Depends on "data-interpolation" and "data-data" point lags • Is an exact interpolator (it restores the actual attribute value at the data points, assuming noiseless data) • Usually relies more on data points close to the interpolation point than on distant ones • Remains unchanged if variogram is multiplied by a constant factor
④ Interpolation error variance	• Depends on the "data-interpolation" and "data-data" point lags • Also depends on the data point configuration • Is low in the data vicinity and gets larger with increasing "data-interpolation" lags • Does not explicitly depend on the data values • Is multiplied by a constant factor, if the variogram is multiplied by the same factor

According to Eq. (2.2b), both the variogram $\gamma_X(\Delta \boldsymbol{p}_{ki})$ and the multiplier μ contribute to the interpolation error $e_X^2(\boldsymbol{p}_k)$ by introducing uncertainties that can be interpreted as being the result of interpolation-data variability and mean indeterminacy. Another important *STOK* feature materializes in terms of the explicit analytical solution of the interpolation problem. As was noted earlier, an analytical *STOK* solution possesses an interpretation power that can reveal essential interpolation properties, something that could not be possible to do in terms of a routine numerical *STOK* solution. *Inter alia*, an analytical *STOK* solution enables the understanding of the relation between the physical quantities in the solution and how it may change under certain conditions. The following example is illustrative in this respect.

Example 2.4 When a practitioner routinely uses a computer software library, say *SEKS-GUI*, to implement the *STOK* technique, the available dataset is inserted into the computer software, which then automatically solves the *STOK* system of Eqs. (2.1a, b) to directly generate numerical values for the interpolator of Eq. (2.1a) and the associated interpolation error of Eq. (2.2b). These are just numbers that do not provide any further physical insight into the inter-workings of the *STOK* technique. On the contrary, the analytical *STOK* solutions can provide considerable insight into the particular interpolator in terms of their explicit expressions of the corresponding variogram functions. For illustration, let us revisit the analytical *STOK* Eqs. (2.13a, b) of Example 2.1 above, and consider the rather common case in practice that interpolation involves moving local neighborhoods with relative short distances between the interpolated and the data points. In this case, it is sufficient to focus on the linear part of the variogram close to the origin, $ah\tau$, $h = |\boldsymbol{h}|$. A realistic concern could be the anticipated change of the interpolated attribute value if instead of the appropriate coordinate system *CS*, a convenient system *CS′* is used that underestimates the actual geographical distance h between pairs of points by ε%, on the average. As it turns out,[1] some valuable insight is gained by the analytical approach that cannot be obtained by the corresponding numerical *STOK* solution, as follows.

[1] The detailed analytical proof is left as an exercise for the readers, see Exercise 10 at the end of this chapter.

- The interpolated value is **independent** of the coordinate system, whereas the associated interpolation error changes by a fixed (predetermined) amount $(1-\varepsilon)$. This happens because of two factors: the relativeness of the *STOK* weights expressed in terms of variogram ratios and the variogram linearity assumed.

The take-home message of the above example is that many important answers to interpolation problems can be obtained using analytical thinking, without the need to plug in numbers and carry out long arithmetic calculations with their unavoidable approximations. According to the routine computational thinking, on the other hand, the same answer should not result from the structure of the *STOK* equations but, instead, it should be the result of numerical calculations—usually implying that the investigator could not have reached the answers without doing the calculations.[m]

2.1.4 STOK implementation—A case study

There are certain practical issues associated with the realistic and effective implementation of the *STOK* technique, which point up some of the procedural and computational issues that investigators encounter. Undoubtedly, this is the case with the implementation of any mathematical results in the real world environment, and *STOK* is not an exception to the general rule.

- The following instructive real-world case study presents a typical step-by-step **implementation** of the *STOK* technique in practice.

Example 2.5 Fig 2.3A displays the arrangement of $PM_{2.5}$ monitoring stations and the corresponding $PM_{2.5}$ concentrations ($\mu g/m^3$) along the *coastal region of China* at randomly selected days during the period October 1–31, 2015. We seek to predict the $PM_{2.5}$ concentration at point $\boldsymbol{p}_k$, indicated by the symbol star ($\star$) in Fig 2.3B, with coordinates $s_k = (952.516, 4351.753)$ (in units of km) at time t_k=October 16, 2015. It is noteworthy that the coordinates of the original monitoring stations and of the interpolation point were longitudes and latitudes. Since the earth can be regarded as a sphere, the longitudes and latitudes cannot be used to calculate the real distance between two points. Thus, all points were projected on a plane by Albert projection, and the units of the coordinates changed from degrees to meters. Then, *STOK* implementation involves the following steps:

Step 1: To find the neighboring points to be used in the *STOK* interpolation of $PM_{2.5}$ concentration at point $\boldsymbol{p}_k$, the space-time searching radii were determined. A time searching radius of 1 day covering the period October 15–17 and a spatial searching radius of 7 km were selected so that three monitoring stations were included, see circle in Fig 2.3B. The longitudes and latitudes of the neighboring points are listed in Table 2.3.

Step 2: Using the data processing techniques described in Chapter 4, the available $PM_{2.5}$ samples were used to compute the empirical variogram values shown in Fig 2.4A. By means of the modeling techniques also described in Chapter 4, a variogram model consisting of exponential spatial and temporal components, i.e.,

$$\gamma_X(h,\tau) = 1887.102\left\{2 - 0.972e^{-\frac{h}{510.189}} - e^{-\frac{\tau}{3.892}} - \left[0.028 + 0.972\left(1 - e^{-\frac{h}{510.189}}\right)\right]\left(1 - e^{-\frac{\tau}{3.892}}\right)\right\},$$

was subsequently fitted to the empirical variogram values (Fig. 2.4B). The fitted model shows that a nugget exists, i.e., the variogram value is not zero at $h=\tau=0$. The variogram model quantifies the space-time variation among $PM_{2.5}$ samples and enables $PM_{2.5}$ interpolation at any point along the coastal China and surrounding regions using *STOK*.

Step 3: The distances and separations between pairs of interpolation and data points and between the data points themselves were calculated and displayed in Table 2.4.[n]

[m] The hidden message here is that analytical solutions, when possible, have the advantage of being cheap (they do not require expensive equipments and high computational costs).

[n] According to Table 2.3, some points have the same spatial coordinates but different temporal instances.

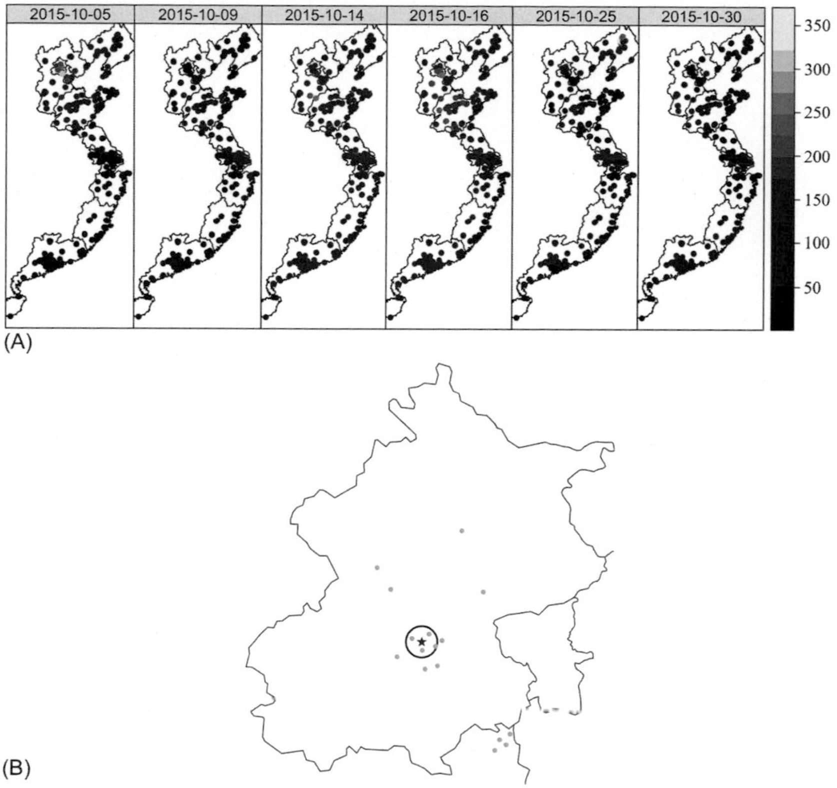

FIG. 2.3 (A) Distribution of monitoring stations along coastal China with daily $PM_{2.5}$ concentrations ($\mu g/m^3$) during October 5–30, 2015. Dots denote monitoring stations, the color bar represents $PM_{2.5}$ concentrations. (B) Zoom in map (located in the highest $PM_{2.5}$ region in north of the coastal area) showing the estimation point $\boldsymbol{s}_k$, the star is the $PM_{2.5}$ interpolation location, and the circle indicates the searching neighborhood.

TABLE 2.3 Neighbor points $(\boldsymbol{s}_i; t_i) = (s_{i1}, s_{i2}; t_i)$ used in *STOK* interpolation at $(\boldsymbol{s}_k; t_k) = (s_{k1}, s_{k2}; t_k)$.

Point i	s_{i1} (km)	s_{i2} (km)	Time t_i (day)	$PM_{2.5}$ concentration $X(\boldsymbol{s}_i; t_i)$ ($\mu g/m^3$)
1	958.602	4349.724	2015-10-15	124
2	952.633	4347.933	2015-10-15	108
3	948.139	4353.203	2015-10-15	104
4	955.689	4355.173	2015-10-15	121
5	958.602	4349.724	2015-10-16	243
6	952.633	4347.933	2015-10-16	210
7	948.139	4353.203	2015-10-16	187
8	955.689	4355.173	2015-10-16	221
9	958.602	4349.724	2015-10-17	325
10	952.633	4347.933	2015-10-17	305
11	948.139	4353.203	2015-10-17	293
12	955.689	4355.173	2015-10-17	315
k	952.516	4351.753	2015-10-16	???

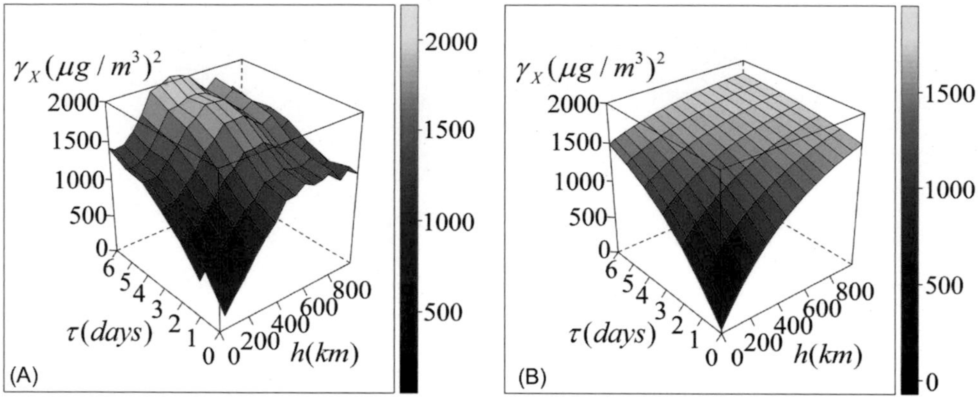

FIG. 2.4 (A) Empirical variogram values and (B) the fitted variogram model.

TABLE 2.4 Distances and separations (h, τ) between pairs of points in Table 2.3.

Point	1	2	3	4	5	6	7	8	9	10	11	12	*k*
1	(0,0)	(6.23,0)	(11.03,0)	(6.18,0)	(0,1)	(6.23,1)	(11.03,1)	(6.18,1)	(0,2)	(6.23,2)	(11.03,2)	(6.18,2)	(6.42,1)
2	(6.23,0)	(0,0)	(6.93,0)	(7.86,0)	(6.23,1)	(0,1)	(6.93,1)	(7.86,1)	(6.23,2)	(0,2)	(6.93,2)	(7.86,2)	(3.82,1)
3	(11.03,0)	(6.93,0)	(0,0)	(7.80,0)	(11.03,1)	(6.93,1)	(0,1)	(7.80,1)	(11.03,2)	(6.93,2)	(0,2)	(7.80,2)	(4.61,1)
4	(6.18,0)	(7.86,0)	(7.80,0)	(0,0)	(6.18,1)	(7.86,1)	(7.80,1)	(0,1)	(6.18,2)	(7.86,2)	(7.80,2)	(0,2)	(4.67,1)
5	(0,1)	(6.23,1)	(11.03,1)	(6.18,1)	(0,0)	(6.23,0)	(11.03,0)	(6.18,0)	(0,1)	(6.23,1)	(11.03,1)	(6.18,1)	(6.42,0)
6	(6.23,1)	(0,1)	(6.93,1)	(7.86,1)	(6.23,0)	(0,0)	(6.93,0)	(7.86,0)	(6.23,1)	(0,1)	(6.93,1)	(7.86,1)	(3.82,0)
7	(11.03,1)	(6.93,1)	(0,1)	(7.80,1)	(11.03,0)	(6.93,0)	(0,0)	(7.80,0)	(11.03,1)	(6.93,1)	(0,1)	(7.80,1)	(4.61,0)
8	(6.18,1)	(7.86,1)	(7.80,1)	(0,1)	(6.18,0)	(7.86,0)	(7.80,0)	(0,0)	(6.18,1)	(7.86,1)	(7.80,1)	(0,1)	(4.67,0)
9	(0,2)	(6.23,2)	(11.03,2)	(6.18,2)	(0,1)	(6.23,1)	(11.03,1)	(6.18,1)	(0,0)	(6.23,0)	(11.03,0)	(6.18,0)	(6.42,1)
10	(6.23,2)	(0,2)	(6.93,2)	(7.86,2)	(6.23,1)	(0,1)	(6.93,1)	(7.86,1)	(6.23,0)	(0,0)	(6.93,0)	(7.86,0)	(3.82,1)
11	(11.03,2)	(6.93,2)	(0,2)	(7.80,2)	(11.03,1)	(6.93,1)	(0,1)	(7.80,1)	(11.03,0)	(6.93,0)	(0,0)	(7.80,0)	(4.61,1)
12	(6.18,2)	(7.86,2)	(7.80,2)	(0,2)	(6.18,1)	(7.86,1)	(7.80,1)	(0,1)	(6.18,0)	(7.86,0)	(7.80,0)	(0,0)	(4.67,1)
k	(6.42,1)	(3.82,1)	(4.61,1)	(4.67,1)	(6.42,0)	(3.82,0)	(4.61,0)	(4.67,0)	(6.42,1)	(3.82,1)	(4.61,1)	(4.67,1)	(0,0)

For illustration, the distance and separation between data points 1 and 4 are 68.414 km and 1 day, respectively (these are the numbers shown in row 5-column 2 or row 2-column 5 of Table 2.4). By introducing the distance-separation pair into the variogram model obtained in *Step* 2, the corresponding variogram values are obtained, as shown in Table 2.5.

Step 4: In this case, six neighbor points were used to predict the $PM_{2.5}$ concentration at location $s_k = (952.516, 4351.753)$ on the day t_k= October 16, 2015. Accordingly, the *STOK* system of equations was set up as below,

TABLE 2.5 Variogram values of the paired points.

Point	1	2	3	4	5	6	7	8	9	10	11	12	k
1	52.84	75.11	92.05	74.92	468.45	485.67	498.78	485.53	789.89	803.21	813.35	803.10	486.18
2	75.11	52.84	77.57	80.88	485.67	468.45	487.58	490.14	803.21	789.89	804.69	806.67	479.04
3	92.05	77.57	52.84	80.68	498.78	487.58	468.45	489.98	813.35	804.69	789.89	806.55	481.21
4	74.92	80.88	80.68	52.84	485.53	490.14	489.98	468.45	803.10	806.67	806.55	789.89	481.37
5	468.45	485.67	498.78	485.53	52.84	75.11	92.05	74.92	468.45	485.67	498.78	485.53	75.76
6	485.67	468.45	487.58	490.14	75.11	52.84	77.57	80.88	485.67	468.45	487.58	490.14	66.53
7	498.78	487.58	468.45	489.98	92.05	77.57	52.84	80.68	498.78	487.58	468.45	489.98	69.34
8	485.53	490.14	489.98	468.45	74.92	80.88	80.68	52.84	485.53	490.14	489.98	468.45	69.54
9	789.89	803.21	813.35	803.10	468.45	485.67	498.78	485.53	52.84	75.11	92.05	74.92	486.18
10	803.21	789.89	804.69	806.67	485.67	468.45	487.53	490.14	75.11	52.84	77.57	80.88	479.04
11	813.35	804.69	789.89	806.55	498.78	487.58	468.45	489.98	92.05	77.57	52.84	80.68	481.21
12	803.10	806.67	806.55	789.89	485.53	490.14	489.98	468.45	74.92	80.88	80.68	52.84	481.37
k	486.18	479.04	481.21	481.37	75.76	66.53	69.34	69.54	486.18	479.04	481.21	481.37	52.84

$$
\begin{aligned}
&\lambda_1\gamma_X(h_{1,1},\tau_{1,1})+\lambda_2\gamma_X(h_{2,1},\tau_{2,1})+\cdots+\lambda_{12}\gamma_X(h_{12,1},\tau_{12,1})+\mu=\gamma_X(h_{k,1},\tau_{k,1})\\
&\lambda_1\gamma_X(h_{1,2},\tau_{1,2})+\lambda_2\gamma_X(h_{2,2},\tau_{2,2})+\cdots+\lambda_{12}\gamma_X(h_{12,2},\tau_{12,2})+\mu=\gamma_X(h_{k,2},\tau_{k,2})\\
&\lambda_1\gamma_X(h_{1,3},\tau_{13})+\lambda_2\gamma_X(h_{2,3},\tau_{2,3})+\cdots+\lambda_{12}\gamma_X(h_{12,3},\tau_{12,3})+\mu=\gamma_X(h_{k,3},\tau_{k,3})\\
&\vdots\\
&\lambda_1\gamma_X(h_{1,11},\tau_{1,11})+\lambda_2\gamma_X(h_{2,11},\tau_{2,11})+\cdots+\lambda_{12}\gamma_X(h_{12,11},\tau_{12,11})+\mu=\gamma_X(h_{k,11},\tau_{k,11})\\
&\lambda_1\gamma_X(h_{1,12},\tau_{1,12})+\lambda_2\gamma_X(h_{2,12},\tau_{2,12})+\cdots+\lambda_{12}\gamma_X(h_{12,12},\tau_{12,12})+\mu=\gamma_X(h_{k,12},\tau_{k,12})\\
&\lambda_1+\lambda_2+\cdots\lambda_{12}=1.
\end{aligned}
$$

Subsequently, using the variogram values shown in Table 2.5, the *STOK* system becomes

$$
\begin{aligned}
&52.84\lambda_1+75.11\lambda_2+92.05\lambda_3+\cdots+803.10\lambda_{12}+\mu=486.18\\
&75.11\lambda_1+52.84\lambda_2+77.57\lambda_3+\cdots+806.07\lambda_{12}+\mu=479.04\\
&92.05\lambda_1+77.57\lambda_2+52.84\lambda_3+\cdots+806.55\lambda_{12}+\mu=481.21\\
&\vdots\\
&813.35\lambda_1+804.69\lambda_2+789.89\lambda_3+\cdots+80.68\lambda_{12}+\mu=481.21\\
&803.10\lambda_1+806.67\lambda_2+806.55\lambda_3+\cdots+52.84\lambda_{12}+\mu=481.37\\
&\lambda_1+\lambda_2+\lambda_3+\cdots+\lambda_{11}+\lambda_{12}=1.
\end{aligned}
$$

The above system can also be written in a convenient matrix form as

$$
\begin{bmatrix}
52.84 & 75.11 & 92.05 & \cdots & 803.10 & 1\\
75.11 & 52.84 & 77.57 & \cdots & 806.07 & 1\\
92.05 & 77.57 & 52.84 & \cdots & 806.55 & 1\\
\vdots & \vdots & \vdots & & \vdots & 1\\
813.35 & 804.69 & 789.89 & \cdots & 80.68 & 1\\
803.10 & 806.67 & 806.55 & \cdots & 52.84 & 1\\
1 & 1 & 1 & \cdots & 1 & 0
\end{bmatrix}
\begin{bmatrix}\lambda_1\\ \lambda_2\\ \lambda_3\\ \vdots\\ \lambda_{11}\\ \lambda_{12}\\ \mu\end{bmatrix}
=\begin{bmatrix}486.18\\ 479.04\\ 481.21\\ \vdots\\ 481.21\\ 481.37\\ 1\end{bmatrix}
$$

Step 5: Solving the *STOK* system, the weights were obtained as

$$
\begin{aligned}
&[\lambda_1\ \lambda_2\ \lambda_3\ \lambda_4\ \lambda_5\ \lambda_6\ \lambda_7\ \lambda_8\ \lambda_9\ \lambda_{10}\ \lambda_{11}\ \lambda_{21}\ \mu]^{\mathrm{T}}\\
&\quad=[-2.3\times10^{-4}-9.78\times10^{-5}-2.54\times10^{-4}\\
&\qquad-1.25\times\ 10^{-4}\,6.02\times10^{-2}\,3.638\times10^{-1}\,2.891\\
&\qquad\times10^{-1}\,2.882\times10^{-1}-2.3\times10^{-4}\\
&\qquad-9.78\times10^{-5}-2.54\times10^{-4}\\
&\qquad-1.25\times10^{-4}-2.2727]^{\mathrm{T}}.
\end{aligned}
$$

Step 6: According to Eqs. (2.2a, b), the interpolated $PM_{2.5}$ concentration and the corresponding error variance were calculated as

$$
\begin{aligned}
&\hat{X}(\boldsymbol{p}_k)=\sum\nolimits_{i=1}^{12}\lambda_iX(\boldsymbol{p}_i)\\
&=-2.3\times10^{-4}\times124-9.78\times10^{-5}\\
&\quad\times108+\cdots-1.25\times10^{-4}\times315-2.2727=208.5066,\\
&e_X(\boldsymbol{p}_k)=\left[\sum\nolimits_{i=1}^{12}\lambda_i\gamma_X(h_{ki},\tau_{ki})+\mu\right]^{\frac{1}{2}}\\
&\quad=[-2.3\times10^{-4}\times486.18-9.78\times10^{-5}\\
&\quad\times479.04+\cdots-1.25\times10^{-4}\\
&\quad\times481.37-2.2727]^{\frac{1}{2}}=8.1181,
\end{aligned}
$$

respectively (both in μg/m^3). The interpolation error e_X is rather low compared to the interpolated $PM_{2.5}$ value (about 3.9%). For further illustration, the interpolated $PM_{2.5}$ values over the region of interest during October 16, 2015, are presented in the form of a $PM_{2.5}$ map in Fig. 2.5. Some comments are in order:

- The error e_X is directly related to λ_i, μ and γ_X. The e_X's first part is a weighted γ_X mean. Variogram values are high, causing high interpolation errors, because the fitted global model γ_X is valid over the entire coastal

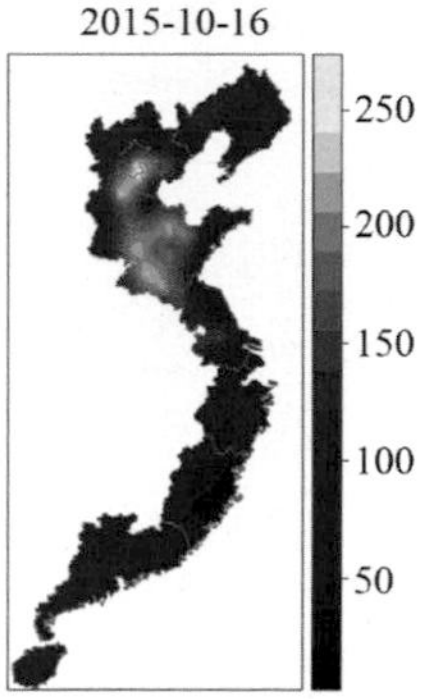

FIG. 2.5 The $PM_{2.5}$ concentration map during October 16, 2015.

China region, i.e., it accounts for the overall $PM_{2.5}$ variability. The e_X could be reduced if a local γ_X model was used that accounted for smaller-scale $PM_{2.5}$ variability, not captured by the global model.

- Negative weights λ_i occur when data near the interpolation points screen outlying data or when clustered points are separated by very small lags (much smaller than the variogram ranges). When applied to high data values, negative λ_i may produce artifacts and meaningless interpolations. Then, negative λ_i must be corrected (by removing clustered data, selecting suitable variogram models). In this study, the exponential variogram shape did not contribute to any such problems.

Step 7: The interpretation of such maps is crucial in understanding the phenomenon. Map interpretation relies on intuition and previous experience, is helped by insight into the physical order lying behind the visual appearance, but it can also benefit considerably from the interpretation of the corresponding formal *STOK* equations.

Last but not least, the readers of this book do not need to be reminded of the following basic methodological rule:

- The interpretation of any chronotopologic attribute maps should take potential subsequent **actions** into consideration.

This is because *STOK* interpolation basically takes place in the realm of knowledge, that is, in the world of contemplation, and seeks to adopt a scientific attitude toward the modeling assumptions of the case. On the other hand, the use of the interpolation maps takes place in the realm of worldly action and seeks to perform a certain act.[o]

2.1.5 Effect of data noise

Often the available attribute dataset is uncertain, i.e., the measurements or observations contain nonnegligible errors (e.g., see Section 2 of Chapter 1). In such cases, the Kriging techniques can be properly modified to account for data noise.

Example 2.6 For illustration, assume that in the presence of data noise expressed by a measurement model of the form

$$Y(\boldsymbol{p}_i) = X(\boldsymbol{p}_i) + U(\boldsymbol{p}_i),$$

$i = 1, \ldots, m$, where $X(\boldsymbol{p}_i)$ is the actual attribute value, $Y(\boldsymbol{p}_i)$ is the measured value, and $U(\boldsymbol{p}_i)$ is the measurement error, which in *CTDA* studies is commonly considered a white noise (*wn*) with zero mean and variance σ_U^2. Then, as is discussed in Exercise 11, the *STOK* system must be modified by subtracting the term σ_U^2 ($j = 1, \ldots, m$) on the left side of Eq. (2.1a).

Example 2.6 discussed a rather straightforward modification of the *STOK* system in the presence of a rather simple (linear, additive)

[o] Our readers may recall (Chapter VII) the metaphorical comparison of *Socrates* as the seeker after truth and *Alexander* as the man of action.

measurement model. Naturally, analytical and computational complications may emerge when more involved forms of measurement models need to be considered.

2.2 Space-time simple kriging (*STSK*)

STSK is an interpolation technique similar to *STOK*, with a few differences in their assumptions. In particular, the unknown mean assumption of *STOK* is replaced with the following ones:

- The attribute mean is **constant** and **known**, $\overline{X}(\boldsymbol{p}) = \overline{X}$.
- The **decomposition**, $X(\boldsymbol{p}) = \overline{X} + \varepsilon(\boldsymbol{p})$ is used, where the $\varepsilon(\boldsymbol{p})$ is known at the data points, since $\overline{X}(\boldsymbol{p}) = \overline{X}$ is known.

Just as was the case with *STOK*, the *STSK* too can be modified to allow for the presence of data noise.

2.2.1 *STSK equations*

As a result of the above considerations, the corresponding *STSK* system of equations is as follows,

$$\sum_{i=1}^{m} \lambda_i \gamma_X\left(\Delta \boldsymbol{p}_{ij}\right) + c_0\left[\sum_{i=1}^{m} \lambda_i - 1\right] = \gamma_X\left(\Delta \boldsymbol{p}_{kj}\right), \quad j = 1, \dots, m \tag{2.14}$$

where c_0 is the variance of $X(\boldsymbol{p})$. The system of Eq. (2.14) must be solved for λ_i ($i = 1, \dots, m$). As was the case with *STOK*, the *STSK* weights λ_i are based not only on the lags between the data points and the interpolation point, but also on the overall space-time arrangement of the data points. To incorporate this arrangement in the λ_i values, the attribute variability must be quantified in terms of the variogram function. Eqs. (2.14) can be also written in a matrix form, see Exercise 13 at the end of this chapter.

In view of Eq. (2.14), the corresponding *STSK* interpolation and associated error variance are given by

$$\hat{X}(\boldsymbol{p}_k) = \overline{X} + \sum_{i=1}^{m} \lambda_i \left[X(\boldsymbol{p}_i) - \overline{X}\right] = \overline{X} + \sum_{i=1}^{m} \lambda_i X(\boldsymbol{p}_i),$$
$$e_X^2(\boldsymbol{p}_k) = c_0\left[1 - \sum_{i=1}^{m} \lambda_i\right] + \sum_{i=1}^{m} \lambda_i \gamma_X(\Delta \boldsymbol{p}_{ki}), \tag{2.15a, b}$$

respectively. Hence, on the basis of a finite dataset the *STSK* generates attribute interpolations on a regular mapping grid. The *STSK* interpolated values are affected by the data mean, which was not the case with *STOK*.

The *STSK* process is graphically outlined in Fig. 2.6. The *STSK* is rewritten in a matrix form suitable for computational purposes, where $\boldsymbol{\lambda} = [\lambda_1 \cdots \lambda_m]^{\mathrm{T}}$, $\boldsymbol{\gamma} = [\gamma_X(\Delta \boldsymbol{p}_{k1}) \dots \gamma_X(\Delta \boldsymbol{p}_{km})]^{\mathrm{T}}$, and $\mathbf{1}^{\mathrm{T}} = [1 \dots 1]$.

Example 2.7 Irregularly spaced weather stations in a region are used to create surfaces of temperature or air pressure in the space-time domain; and satellites are increasingly using multi-spectral and hyper-spectral images that represent climate attributes such as the above.

2.2.2 *STSK interpretation*

Regarding its interpretation, the *STSK* technique shares some of the relevant *STOK* properties, with some occasional modifications. In particular:

① The significance of the **known mean** assumption is that there exists reliable prior knowledge concerning the attribute (e.g., the mean is available in the form of a space-time trend derived from a physical law or empirical model).

② *STSK* is an **unbiased** interpolator, since the assumption of constant mean and Eq. (1.1) Section 1 imply that $\overline{\hat{X}(\boldsymbol{p}_k)} = \overline{X}$.[p] On the other hand, for *STSK* it is generally valid that

[p] As noted earlier, the meaning of unbiasedness is that when interpolating an attribute at several points, some of them are above the actual values and some below, so that, on average, the difference between the interpolated attribute values and the actual values should be zero.

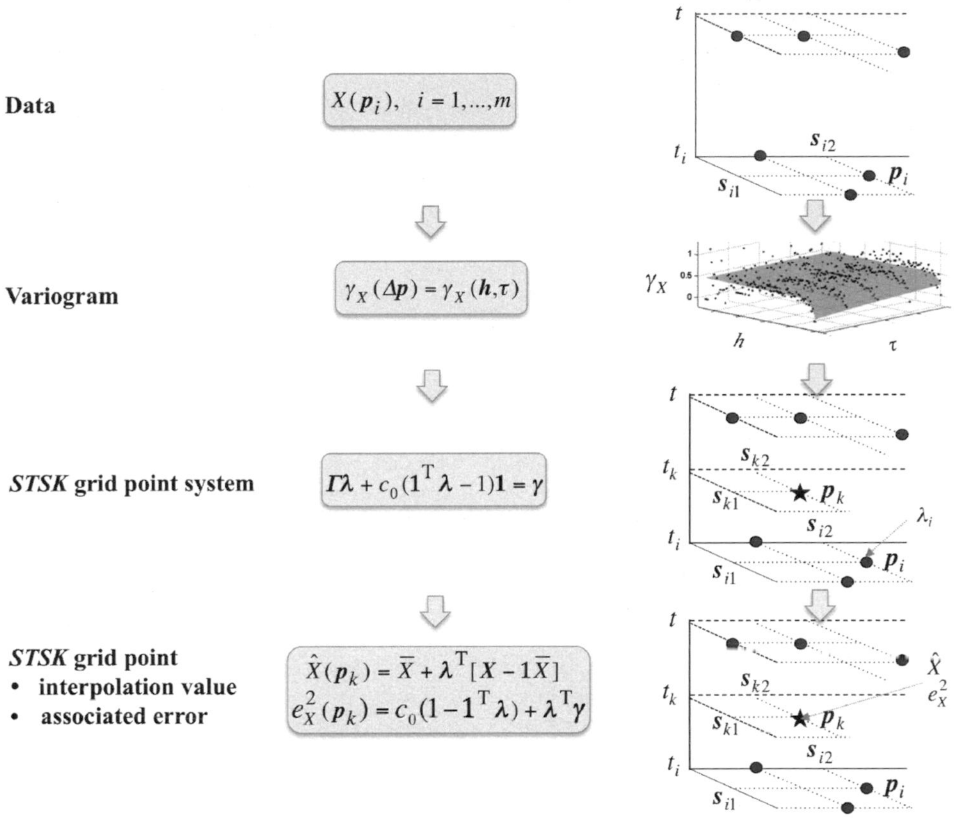

FIG. 2.6 A graphical outline of the *STSK* technique.

$$1 - \sum_{i=1}^{m} \lambda_i \neq 0$$

—compare with Eq. (2.1b). In fact, the greater the value of $1 - \sum_{i=1}^{m}\lambda_i$, the more the interpolator is drawn toward the attribute mean.

The *STSK* assumption of a known constant mean $\overline{X}$ is sometimes unrealistic *in-situ*. Instead, there are applications in which a physical model governing the attribute distribution is available that explicitly determines a space- and time-dependent trend $\overline{X}(\boldsymbol{p})$. Then, one can take the difference between that model and the observations, i.e., the residuals $e(\boldsymbol{p}) = X(\boldsymbol{p}) - \overline{X}(\boldsymbol{p})$, and use *STSK* on the residuals, given that the *e*-trend is zero, $\overline{e}(\boldsymbol{p}) = 0$. Since the mean is assumed known, $e(\boldsymbol{p})$ is also known at the observation points. On the other hand, in *STOK* the mean and residuals must be estimated separately.

③ In the special case that the interpolation point is also a data point, i.e., $\boldsymbol{p}_k = \boldsymbol{p}_i$, it holds that

$$\hat{X}(\boldsymbol{p}_k) = X(\boldsymbol{p}_i), \quad e_X^2(\boldsymbol{p}_k) = \gamma_X(\Delta\boldsymbol{p}_{ii}) = 0. \qquad (2.16a, b)$$

Therefore, like *STOK*, the *STSK* too is an **exact** interpolator.

A brief comparison of the two interpolation techniques is presented in Table 2.6 (also, recall Table 2.1).

An additional interpretation issue is as follows: how a technique is presented by an

TABLE 2.6 A further comparison of *STOK vs. STSK*.

Issue	*STOK*	*STSK*
Sum of *STOK* weights:	Is equal to 1, which is why the technique involves a Lagrange multiplier μ	Is not equal to 1, since the technique does not involve any μ
Interpolated value:	Does not depend on attribute mean	Depends on attribute mean
Interpolation error:	Is influenced by μ	Does not involve any μ

equation matters in determining the resulting conceptualizations, i.e., changes in the symbolic structure or arrangement may cause changes in conceptualization. This realization can have a direct effect on *STOK* interpretation:

- A difference in the arrangement of the symbols in the interpolation equations could imply a different **organization** of the body of knowledge, which could, in turn, lead to different interpretations of the situation.

This is obviously valid in the case of the *STSK* interpolators as is illustrated in the next example.

Example 2.8 In terms of covariance functions, the *STSK* system of Eq. (2.14) above can be written in two mathematically equivalent forms, namely,

$$c = C\lambda$$

and

$$\lambda = C^{-1}c,$$

where c is the data-interpolation covariance vector, C is the data covariance matrix, and λ is the weight vector. Although their mathematical formulations are equivalent, the conceptualizations of the two Kriging equations could be different. Specifically, the 2nd equation focuses on defining the weight vector as the product of the inverse data covariance matrix and the data-interpolation covariance vector, in which case the equal sign is understood as connecting two identical entities. The 1st equation, on the other hand, encourages the consideration of a causal relationship between physical correlations: the data correlations (expressed by C) and the resulting data-interpolation correlations (expressed by c), in which case the equal sign may carry the very different meaning of connecting effect to cause.

2.2.3 *More practical issues*

Just as with *STOK*, when implementing the *STSK* technique one encounters a number of practical issues that need to be carefully considered. One potential issue is the accurate solution of large *STOK* linear systems of equations that may emerge in an application, and another issue concerns the cost-effective implementation of the interpolation technique, which relies on the appropriate choice of a suitable computer software for the case study of interest.

- The following example presents the *step-by-step process* of **implementing** the *STSK* technique in **practice**.

Example 2.9 Revisiting Example 2.5, but here the *STSK* interpolation technique is used, instead. The *STSK* system is

$$\begin{aligned}
&\lambda_1\gamma_X(h_{1,1},\tau_{1,1}) + \lambda_2\gamma_X(h_{2,1},\tau_{2,1}) + \cdots \\
&\quad +\lambda_{12}\gamma_X(h_{12,1},\tau_{12,1}) = \gamma_X(h_{k,1},\tau_{k,1}) \\
&\lambda_1\gamma_X(h_{1,2},\tau_{1,2}) + \lambda_2\gamma_X(h_{2,2},\tau_{2,2}) + \cdots \\
&\quad +\lambda_{12}\gamma_X(h_{12,2},\tau_{12,2}) = \gamma_X(h_{k,2},\tau_{k,2}) \\
&\lambda_1\gamma_X(h_{1,3},\tau_{13}) + \lambda_2\gamma_X(h_{2,3},\tau_{2,3}) + \cdots \\
&\quad +\lambda_{12}\gamma_X(h_{12,3},\tau_{12,3}) = \gamma_X(h_{k,3},\tau_{k,3}) \qquad (2.17)\\
&\vdots \\
&\lambda_1\gamma_X(h_{1,11},\tau_{1,11}) + \lambda_2\gamma_X(h_{2,11},\tau_{2,11}) + \cdots \\
&\quad +\lambda_{12}\gamma_X(h_{12,11},\tau_{12,11}) = \gamma_X(h_{k,11},\tau_{k,11}) \\
&\lambda_1\gamma_X(h_{1,12},\tau_{1,12}) + \lambda_2\gamma_X(h_{2,12},\tau_{2,12}) + \cdots \\
&\quad +\lambda_{12}\gamma_X(h_{12,12},\tau_{12,12}) = \gamma_X(h_{k,12},\tau_{k,12})
\end{aligned}$$

Considering that each equation in Eq. (2.14) has the same component $c_0[\sum_{i=1}^{m}\lambda_i - 1]$, we can delete it as was done in Eq. (2.17). Then introducing the particular variogram values, Eq. (2.17) becomes

$$\begin{bmatrix} 52.84 & 75.11 & 92.05 & \cdots & 803.10 \\ 75.11 & 52.84 & 77.57 & \cdots & 806.07 \\ 92.05 & 77.57 & 52.84 & \cdots & 806.55 \\ \vdots & \vdots & \vdots & \ddots & \vdots \\ 813.35 & 804.69 & 789.89 & \cdots & 80.68 \\ 803.10 & 806.67 & 806.55 & \cdots & 52.84 \end{bmatrix} \begin{bmatrix} \lambda_1 \\ \lambda_2 \\ \lambda_3 \\ \vdots \\ \lambda_{11} \\ \lambda_{12} \end{bmatrix} = \begin{bmatrix} 486.18 \\ 479.04 \\ 481.21 \\ \vdots \\ 481.21 \\ 481.37 \end{bmatrix}$$

By solving the last system of equations, the weights are obtained

$$\begin{aligned} &[\lambda_1\ \lambda_2\ \lambda_3\ \lambda_4\ \lambda_5\ \lambda_6\ \lambda_7\ \lambda_8\ \lambda_9\ \lambda_{10}\ \lambda_{11}\ \lambda_{12}\ \mu]^{\mathrm{T}} \\ &= [-9.8\times 10^{-4} - 4.17\times 10^{-4} - 1.1\times 10^{-3} \\ &\quad -5.3\times 10^{-4}\, 6.01\times 10^{-2}\, 3.638\times 10^{-1} \\ &\quad 2.889\times 10^{-1}\, 2.881\times 10^{-1} - 9.8\times 10^{-4} \\ &\quad -4.17\times 10^{-4} - 1.1\times 10^{-3} - 5.3\times 10^{-4}]^{\mathrm{T}}. \end{aligned}$$

The mean and variance of $X(\boldsymbol{s}_i; t_i)$ can be calculated as

$$\begin{aligned} \overline{X} &= \frac{1}{12}\sum\nolimits_{i=1}^{12} X(\boldsymbol{p}_i) \\ &= \frac{1}{12}(124+108+104+121+\cdots+315) = 213, \\ c_0 &= \frac{1}{12}\sum\nolimits_{i=1}^{12}\left[X(\boldsymbol{p}_i) - \overline{X}\right]^2 \\ &= \frac{1}{12}\left[(124-213)^2 + (108-213)^2 + (104-213)^2 \right. \\ &\quad \left. + (121-213)^2 + \cdots + (315-213)^2\right] = 6562.67. \end{aligned}$$

The *STSK*-interpolated $PM_{2.5}$ concentration and the corresponding error variance can be calculated numerically as

$$\begin{aligned} \hat{X}(\boldsymbol{p}_k) &= \overline{X} + \sum\nolimits_{i=1}^{12}\lambda_i\left[X(\boldsymbol{p}_i) - \overline{X}\right] \\ &= 213 - 9.80e-4\times(124-213) - 4.17e-4 \\ &\quad \times(108-213) - 0.0011\times(104-213) + \cdots \\ &\quad - 5.30e-4\times(315-213) \\ &= 208.5066, \end{aligned}$$

$$\begin{aligned} e_X(\boldsymbol{p}_k) &= \left[c_0\left[1-\sum\nolimits_{i=1}^{12}\lambda_i\right] + \sum\nolimits_{i=1}^{12}\lambda_i\gamma_X(h_{ki}, \tau_{ki})\right]^{\frac{1}{2}} \\ &= [6562.67\times(1+9.80e-4+4.17e \\ &\quad -4+\cdots+5.30e-4) - 9.80e \\ &\quad -4\times 486.18 - 4.17e-4\times 479.04 \\ &\quad +\cdots - 5.30e-4\times 481.37]^{\frac{1}{2}} = 22.6944, \end{aligned}$$

respectively.

Lastly, as is the case with any chronotopologic interpolation technique, the attribute interpolation values computed by *STSK* are also assigned to a properly selected grid, depending on the application. The grid size needs to match the sampling density and scale at which the attributes of interest occur. The investigators can always try to produce maps by using the most detailed grid size that the interpolators allow them. Then, they can slowly test how the interpolation accuracy changes with coarser grid sizes and finally select a grid size that allows maximum detail, while being computationally effective.

2.2.4 *Violation of Tobler's law*

The *STSK* technique, as well as the *STOK* technique and other types of Kriging, can lead to more violations of Tobler's law of geography. Two of them are described next.

① We start with the so-called **screen effect**, i.e., samples that are further away from the interpolation points may have a greater influence on the interpolated attribute values than points much closer to the interpolation points.

To clarify how violation ① could happen in practice consider the following case: With the screen effect, the influence of an observational point, say $\boldsymbol{p}_i$, will be reduced by the addition of one or more observations, say $\boldsymbol{p}_{i'}$ ($i' = 1, 2, \ldots$), at the intermediate points between point $\boldsymbol{p}_i$ and the interpolation point, say $\boldsymbol{p}_k$. Then, the influence on interpolation at $\boldsymbol{p}_k$ of another observational point, say $\boldsymbol{p}_j$, can be larger than that of $\boldsymbol{p}_i$ even if $\boldsymbol{p}_j$ is located at a larger distance from $\boldsymbol{p}_k$

than $\boldsymbol{p}_i$. As the screen effect can make the influence of distant observational points (like the points $\boldsymbol{p}_{i'}$ in the above case) negligible, the use of a sampling subset in Kriging is a safe practice compared with other weighting methods.

② With the **declustering** property, several sample points close together will have collectively the weight of a single sample point located near the centroid of the cluster.

Screen effect and declustering are two interesting Kriging features that are due to the fact that Kriging weights depend not only on the space distances and time separations between sample and attribute interpolation points but also on the distances and separations among sample points themselves.

2.3 Space-time Indicator Kriging (*STIK*)

The *STIK* is based on the introduction of an *I*-binary random field, which takes the values 0 and 1, corresponding, respectively, to the presence or absence of an event, an object, or the occurrence of a specified attribute value.

Example 2.10 Here we see a few representative cases of *I* implementation.

- In *meteorology*, the value of *I* is 1 if a weather event (say, rainfall) occurs, and 0 if it doesn't occur.
- In *ecology*, a datum may provide information on whether or not a point belongs to a biodiverse domain, in which case the binary random field *I* indicates what is the case.

For *CTDA* purposes involving numerical attribute values, the following *I*-binary transformation is used that links the original random field representing the attribute of interest with the *I*-binary random field:

The original continuous *S/TRF* $X(\boldsymbol{p})$ is transformed into an **indicator *S/TRF*** by considering a specific threshold ζ, i.e.,

$$I(\boldsymbol{p};\zeta) = \begin{cases} 1 & \text{if } X(\boldsymbol{p}) \,\}\text{A3}\,\zeta \\ 0 & \text{otherwise} \end{cases} = \begin{cases} 1 & \text{with } P[X(\boldsymbol{p}) \,\}\text{A3}\,\zeta] \\ 0 & \text{with } P[X(\boldsymbol{p}) > \zeta] \end{cases} \quad (2.18\text{a, b})$$

for each fixed nonnegative number (threshold) ζ. The $I(\boldsymbol{p};\zeta)$ is also known as **characteristic** *S/TRF*.

Formally, then, the indicator *S/TRF* is a function of two arguments: $X(\boldsymbol{p})$ and ζ. Eqs. (2.1a, b) determine the indicator *S/TRF* in terms of the random attribute realizations χ and their probabilities P. One of the useful properties of this is that $\overline{I(\boldsymbol{p};\zeta)} = P[X(\boldsymbol{p}) \leq \zeta]$, which has the following useful interpretation:

- The space-time distribution of the mean indicator $\overline{I(\boldsymbol{p};\zeta)}$ values represents the distribution of the attribute **probabilities** $P[X(\boldsymbol{p}) \leq \zeta]$.

Example 2.11 Let us examine the $PM_{2.5}$ data collected from the coastal region of China presented in Table 2.7 and set the threshold $\zeta = 20\mu g/m^3$; then the new binary random field is obtained as is shown in Table 2.7.

Next, the chronotopologic variability or dependence structure can be assessed as follows. Assume that the threshold is set as $\zeta = \zeta_0$, and let all attribute data $X(\boldsymbol{p}_i) = X(\boldsymbol{s}_i;t_i)$ be transformed to a binary form $I(\boldsymbol{p}_i;\zeta_0)$ following Eqs. (2.18a, b).

TABLE 2.7 Indicator values.

i	s_{i1}	s_{i2}	t_i	$X(\boldsymbol{s}_i;t_i)$	$I(\boldsymbol{s}_i;t_i;\zeta)$
1	1,370,291.27	2,703,808.15	2015-10-15	26	0
2	1,399,266.96	2,765,782.56	2015-10-15	32	0
3	1,370,291.27	2,703,808.15	2015-10-16	19	1
4	1,399,266.96	2,765,782.56	2015-10-16	21	0
5	1,370,291.27	2,703,808.15	2015-10-17	18	1
6	1,399,266.96	2,765,782.56	2015-10-17	21	0

Using the empirical covariance and variogram computation formulas introduced in Chapter 4, the corresponding indicator covariance and variogram values can be found as.

$$\hat{c}_I\left(\boldsymbol{p}_i, \boldsymbol{p}_j; \zeta_0\right) = \frac{1}{N}\sum_{i=1}^{N}\left[I(\boldsymbol{p}_i; \zeta_0) - \overline{I(\boldsymbol{p}; \zeta_0)}\right] \times \left[I\left(\boldsymbol{p}_j; \zeta_0\right) - \overline{I(\boldsymbol{p}; \zeta_0)}\right],$$

$$\hat{\gamma}_I\left(\boldsymbol{p}_i, \boldsymbol{p}_j; \zeta_0\right) = \frac{1}{2N}\sum_{i=1}^{N}\left[I(\boldsymbol{p}_i; \zeta_0) - I\left(\boldsymbol{p}_j; \zeta_0\right)\right]^2, \qquad (2.19\text{a, b})$$

respectively, where $\overline{I(\boldsymbol{p}; \zeta_0)}$ is the average value of all considered realizations of the binary random field $I(\boldsymbol{p};\zeta_0)$. As usual, theoretical covariance (c_I) and variogram (γ_I) models can be subsequently fitted to the empirical values calculated by Eqs. (2.19a, b).

2.3.1 STIK equations

In order to interpolate the indicator value at the point of interest $\boldsymbol{p}_k = (\boldsymbol{s}_k; t_k)$, we can also set up a *STIK* system of equations (similar to the *STOK* system).

$$\sum_{i=1}^{m}\lambda_i\gamma_I\left(\Delta\boldsymbol{p}_{ij}; \zeta_0\right) + \mu = \gamma_I\left(\Delta\boldsymbol{p}_{kj}; \zeta_0\right), \ j = 1, \ldots, m,$$
$$\sum_{i=1}^{m}\lambda_i = 1. \qquad (2.20\text{a, b})$$

Usually, in practice the $\Delta\boldsymbol{p}_{ij}$ and $\Delta\boldsymbol{p}_{kj}$ are replaced by $(\boldsymbol{h}_{ij}, \tau_{ij})$ and $(\boldsymbol{h}_{kj}, \tau_{kj})$, respectively. Then, Eqs. (2.20a, b) can be solved for the weights λ_i, and similarly the indicator interpolation value and the corresponding indicator interpolation error variance can be calculated as

$$\hat{I}(\boldsymbol{p}_k; \zeta_0) = \sum_{i=1}^{m}\lambda_i I(\boldsymbol{p}_i; \zeta_0),$$
$$e_I^2(\boldsymbol{p}_k; \zeta_0) = \sum_{i=1}^{m}\lambda_i\gamma_I(\Delta\boldsymbol{p}_{ki}; \zeta_0) + \mu, \qquad (2.21\text{a, b})$$

respectively. Alternatively, the *STIK* system can also be set up similarly to the *STSK* one, as below:

$$\sum_{i=1}^{m}\lambda_i\gamma_I\left(\Delta\boldsymbol{p}_{ij}; \zeta_0\right) + c_0\left[\sum_{i=1}^{m}\lambda_i - 1\right] = \gamma_I\left(\Delta\boldsymbol{p}_{kj}; \zeta_0\right), \ j = 1, \ldots, m \qquad (2.22)$$

and the indicator interpolated value and error variance can be calculated as

$$\hat{I}(\boldsymbol{p}_k; \zeta_0) = \overline{I(\boldsymbol{p}_i; \zeta_0)} + \sum_{i=1}^{m}\lambda_i\left[I(\boldsymbol{p}_i; \zeta_0) - \overline{I(\boldsymbol{p}_i; \zeta_0)}\right],$$
$$e_I^2(\boldsymbol{p}_k; \zeta_0) = c_0\left[1 - \sum_{i=1}^{m}\lambda_i\right] + \sum_{i=1}^{m}\lambda_i\gamma_I(\Delta\boldsymbol{p}_{ki}; \zeta_0), \qquad (2.23\text{a, b})$$

where $\overline{I\left(\boldsymbol{p}_i; \zeta_0\right)}$ is the average value of the considered indicator realizations. An example can illustrate some matters of the preceding analysis.

Example 2.11(cont.): We continue looking into the same $PM_{2.5}$ dataset used before (Example 2.11), but this time we examine two different threshold situations.

Step 1: The two thresholds, $\zeta_1 = 20$ and $\zeta_2 = 60\,\mu g/m^3$, were chosen to quantitatively assess the difference between the two situations.

Step 2: The $X(\boldsymbol{p}_i) = X(\boldsymbol{s}_i; t_i)$ values were transformed into the indicator values $I(\boldsymbol{s}_i; t_i; \zeta_1)$ and $I(\boldsymbol{s}_i; t_i; \zeta_2)$ of the two thresholds using Eq. (2.18).

Step 3: The empirical variogram values $\hat{\gamma}_{I,\zeta_i}$ and the corresponding fitted theoretical variogram models γ_{I,ζ_i} $(i = 1, 2)$ for the two thresholds are shown in Fig. 2.7. The analytical expressions of the theoretical variogram models include spatial γ_{IS,ζ_i} and temporal components γ_{IT,ζ_i}, as shown in Table 2.8. As one can observe, the sill of the fitted variogram model γ_{I,ζ_2} associated with the threshold $\zeta_2 = 60\,\mu g/m^3$ is larger than that of the variogram model γ_{I,ζ_1} associated with the threshold $\zeta_1 = 20\,\mu g/m^3$.

Step 4: Using the above variogram models, the generated *STIK* maps are plotted in Fig. 2.8. Interpretationally, the two maps depict the distribution of the probabilities that the $PM_{2.5}$ data are less or equal to the two thresholds, 20 and 60 $\mu g/m^3$. Specifically, the following conclusion is drawn:

- In the region and period of interest, there exist two areas of interest that deserve the investigator's attention: (a) a small area with a high probability of $PM_{2.5}$ concentrations being less than or equal to 20 $\mu g/m^3$, and (b) a large area with a high probability of $PM_{2.5}$ concentrations being less than or

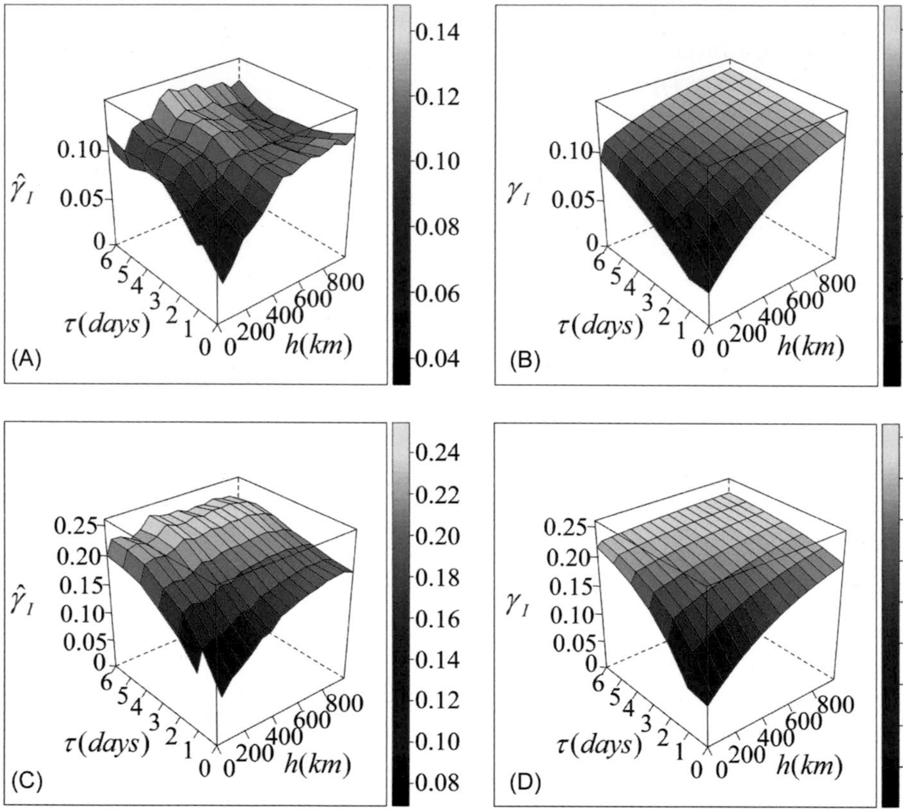

FIG. 2.7 (A), (C) Empirical variogram values, and (B)–(D) the fitted theoretical variogram models of $I(\mathbf{s}_i; t_i, \zeta_1)$ and $I(\mathbf{s}_i; t_i, \zeta_2)$, respectively.

equal to 60 μg/m^3, especially in the south part of the coastal region.

2.3.2 *Pros and cons*

The original motivation behind the development of the *STIK* technique was the handling of *non-Gaussian* data. Through the years *STIK* has been used successfully in several attribute data analysis and modeling applications. Yet, *STIK* also has certain drawbacks (theoretical and practical) that limit its applicability. These include (i) its neglect of cross-dependencies among different thresholds, and (ii) its self-contradiction due to its representation of non-Gaussian attribute distributions solely in terms of low-order statistics.

Because of these drawbacks, when dealing with non-Gaussian data many investigators increasingly favor the implementation of the Bayesian maximum entropy (*BME*) technique (see Chapters 6 and 9).

3 Second Kriging classification: point, chronoblock and functional

So far, point Kriging techniques have been considered, i.e., interpolation techniques that were concerned with point samples. The physical meaning of the term "point sample" is that its size is much smaller than the spatial distances and time separation considered in *CTDA*, and

TABLE 2.8 Analytic forms of fitted theoretical variogram models for two threshold values.

Threshold ζ_i, $i = 1,2$	Fitted theoretical variogram model	Model components
20	$\gamma_{I,\zeta_1} = 0.1487(\gamma_{IS\zeta_1} + \gamma_{IT\zeta_1} - \gamma_{IS\zeta_1}\gamma_{IT\zeta_1})$	$\gamma_{IS,\zeta_1} = 0.2276 + 0.7724\left(1 - e^{-\frac{h}{599.939}}\right)$ $\gamma_{IT,\zeta_1} = 0.1045 + 0.8955\left(1 - e^{-\frac{\tau}{6.95}}\right)$
60	$\gamma_{I,\zeta_2} = 0.2377(\gamma_{IS_2} + \gamma_{IT_2} - \gamma_{IS_2}\gamma_{IT_2})$	$\gamma_{IS,\zeta_2} = 0.3025 + 0.6975\left(1 - e^{-\frac{h}{600.004}}\right)$ $\gamma_{IT,\zeta_2} = 1 - e^{-\frac{\tau}{2.469}}$

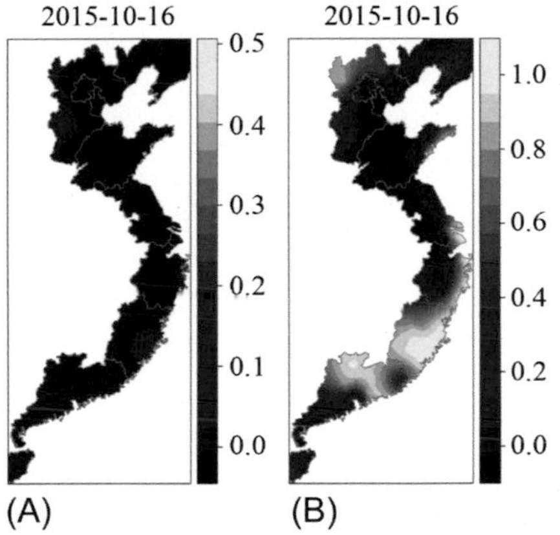

FIG. 2.8 The indicator maps using the two thresholds, (*A*) $\zeta_1 = 20\,\mu g/m^3$ and (*B*) $\zeta_2 = 60\,\mu g/m^3$.

attribute interpolations are at unsampled points. The Kriging techniques below, instead, are concerned with cases of *nonpoint* interpolation, i.e., attribute interpolations are sought at unsampled space-time blocks or attribute functionals are observed.

3.1 Chronoblock Kriging (CBK)

The notion of a *chronoblock* refers to a domain with specified (finite) dimensions across space and time, which are determined by the needs of the real-world application. In the commonly encountered $R^{2,1}$ and $R^{3,1}$ domains, a chronoblock is also termed a *chronoarea* and a *chronovolume*, respectively. *CBK* is, then, interested in interpolations of linear attribute averages inside a certain chronoblock support that is intermediate in size between the observation support and the sampling support.

In the ***CBK*** setting, the quantity of interest may be expressed mathematically in the integral form

$$X_a(s_k; t_k) = \left|a_{(s_k;\, t_k)}\right|^{-1} \int_{a(s_k;\, t_k)} du\, d\upsilon\, X(s_k - u; t_k - \upsilon) \quad (3.1)$$

where $a_{(s_k;t_k)}$ is the space-time domain centered at the interpolation point of interest $\boldsymbol{p}_k = (s_k; t_k)$.

On the basis of a set of data at points $\boldsymbol{p}_j = (s_j; t_j)$, the investigators want to generate attribute $X_a(\boldsymbol{s}_k; t_k)$ interpolations, $\hat{X}_a(\boldsymbol{s}_k; t_k)$, at $a_{(s_k;t_k)}$. Interestingly, the types of point Kriging considered in previous sections can be seen as limiting cases of a chronotopologically averaged Kriging formulation to be introduced here. Not surprisingly, then, the *Chronoblock Ordinary Kriging* (*CBOK*) is the most common form of *CBK*.

3.1.1 CBOK equations

In view of Eq. (3.1), the *CBOK* system is a linear set of equations,

$$\sum_{i=1}^{m} \lambda_i \gamma_X\left(\Delta \boldsymbol{p}_{ij}\right) + \mu = \overline{\gamma}_X\left(\boldsymbol{p}_j, a_{\boldsymbol{p}_k}\right), \ j = 1, \ldots, m,$$
$$\sum_{i=1}^{m} \lambda_i = 1, \quad (3.2a, b)$$

where the new entity is $\overline{\gamma}_X\left(\boldsymbol{p}_j, a_{\boldsymbol{p}_k}\right)$, i.e., the average variogram between the data points $\boldsymbol{p}_j$ and all the points within the domain a_{p_k}. As usual, this system must be solved for λ_i ($i = 1, \ldots, m$). As was the case with *STOK* and *STSK*, the *CBOK* weights λ_i are based not only on the metric Δp_{ki} between the data points and the interpolation point, but also on the overall arrangement of the data points, as well as the chronogeometry of a_{p_k}. The *CBOK* attribute interpolation and the associated interpolation error variance are as follows,

$$\hat{X}_a(p_k) = \boldsymbol{\lambda}^T \boldsymbol{\chi}_d = \sum_{i=1}^{m} \lambda_i X(p_i),$$
$$e^2_{X_a}(p_k) = \sum_{i=1}^{m} \lambda_i \bar{\gamma}_X(p_i, a_{p_k}) + \bar{\gamma}_X(a_{p_k}, a_{p_k}) + \mu, \quad (3.3a, b)$$

where $\bar{\gamma}_X(a_{p_k}, a_{p_k})$ is the average variogram between all pairs of points within the domain a_{p_k}.

The purpose of *CBOK* is to produce direct chronoblock averages from point measurements, not by averaging point interpolations. The main peculiarity of *CBOK* is the use of supports of two different sizes for the observations and the interpolator, and, hence, variograms (or covariances) between two different support sizes.

3.2 Space-time functional Kriging (*STFK*)

This kind of Kriging is of interest when evidence in the form of indirect attribute observations is available or the evidence consists of attribute functions over a space-time domain.

Example 3.1 Typical physical functionals include topographic slope, rainfall precipitation curves recorded at a set of weather stations (instruments return a long record of observation over time), infiltration water, pollutant concentration not exceeding an environmental threshold, and contaminant attributes at various scales and quantitative links between the results obtained at each one of these scales.

Mathematically, a functional over a space-time domain Λ centered at $\boldsymbol{p}_k$ may be generally expressed as (Christakos, 2000)

$$X_\Lambda(\boldsymbol{p}_k) = \Phi[X(\boldsymbol{p}), \Lambda] \quad (3.4)$$

where the form of the functional Φ may depend on the physics of the problem, and the $X(\boldsymbol{p})$ is usually called a "point" natural attribute.

Linear and nonlinear functionals Φ have been considered, including the chronoblock attribute (Section 3.1 above), an attribute gradient, an attribute curve, or an attribute exceedance indicator. In light of Eq. (3.4), physically the size of a point sample is much smaller than the size of Λ.

Example 3.2 Two illustrative cases of the functional are as follows.

- Eq. (3.1) is a special case of Eq. (3.4) when
$$\Lambda = a,$$
$$\Phi[\cdot] = |a|^{-1} \int_a d\boldsymbol{u} dv[\cdot]. \quad (3.5a, b)$$
- Another interesting case is when
$$\Lambda = a,$$
$$\Phi[\cdot] = |a|^{-1} \int_a d\boldsymbol{u} dv \theta(\boldsymbol{s}_k - \boldsymbol{u}; t_k - v)[\cdot], \quad (3.6a, b)$$
where θ is the density of receptors in the neighborhood of a, and $X(\boldsymbol{s};t)$ is a specified health effect.
- ***STFK*** is used when the functional Φ of the attribute $X(\boldsymbol{s};t)$ is observed instead of the attribute itself, and an approximation $\hat{\Phi}$ of Φ needs to be found.

Particularly, the approximation $\hat{X}_\Lambda(\boldsymbol{p}_k) = \hat{\Phi}(\boldsymbol{\chi}_d)$, is such that the conditions are satisfied, $\overline{\hat{\Phi}(\boldsymbol{\chi}_d) - X_\Lambda(\boldsymbol{p}_k)} = 0$ and $\min_{\hat{\Phi}} \int d\boldsymbol{\chi}_d \int d\chi_k [\hat{\Phi}(\boldsymbol{\chi}_d) - \chi_k]^2 f_X(\boldsymbol{\chi}_d, \chi_k)$, or.

$$\overline{\hat{X}_\Lambda(\boldsymbol{p}_k)} = \overline{X_\Lambda(\boldsymbol{p}_k)},$$
$$\int d\chi_k \frac{\partial}{\partial \hat{\Phi}} [\hat{\Phi}(\boldsymbol{\chi}_d) - \chi_k]^2 f_X(\boldsymbol{\chi}_d, \chi_k) = 0, \quad (3.7a, b)$$

where $\hat{\Phi}$ may be linear or nonlinear. In the linear case, $\hat{\Phi}(\boldsymbol{\chi}_d) = \boldsymbol{\lambda}^T \boldsymbol{\chi}_d$, the corresponding *STFK* equations are essentially modifications of Eqs. (3.2a, b) and (3.3a, b) depending on the particular form of Φ. A more detailed discussion of the theory of functional random fields can be found in Christakos et al. (2017).

4 Mapping accuracy indicators and cross-validation tests

Naturally, before the investigators use the final attribute map generated by one of the interpolation techniques discussed in this book in their risk assessment, decision-making, etc. studies, they should have some idea of how well the selected technique interpolates the attribute values at the unsampled points of interest. This is, of course, an application-dependent issue.

4.1 Mapping accuracy indicators

Beyond the Kriging error e_X^2 defined earlier, several numerical indices can be used to quantify attribute mapping accuracy based on a certain kind of a comparison of the interpolated values *vs.* the observed values. The most obvious and commonly used comparison scheme is established as follows.

Accuracy indicators (*AcI*) are expressed as functions of the interpolation error defined as the difference between the interpolated attribute value $\hat{\chi} = \hat{X}(\boldsymbol{p})$ and the observed value $\chi = X(\boldsymbol{p})$ at point $\boldsymbol{p}$, i.e.,

$$AcI(\boldsymbol{p}) = f(\hat{\chi} - \chi) \tag{4.1}$$

where the form of the function f depends on the accuracy conception assumed.

A list of *AcI* of the kind suggested by Eq. (4.1) are presented in Table 4.1. Let us have a closer look at them:

TABLE 4.1 A list of mapping *AcI*.

Index	Mathematical expression
Coefficient of correlation (r)	$\frac{\sum_{i=1}^{m}(\chi_i - \bar{\chi}_i)(\hat{\chi}_i - \bar{\hat{\chi}}_i)}{\sqrt{\sum_{i=1}^{m}(\chi_i - \bar{\chi}_i)^2}\sqrt{\sum_{i=1}^{m}(\hat{\chi}_i - \bar{\hat{\chi}}_i)^2}}$
Determination of coefficient (R^2)	$1 - \frac{SS_{res}}{SS_{tot}} = 1 - \frac{\sum_{i=1}^{m}(\chi_i - \hat{\chi}_i)^2}{\sum_{i=1}^{m}(\chi_i - \bar{\chi}_i)^2}$
Mean error (ME)	$\frac{1}{m}\sum_{i=1}^{m}(\hat{\chi}_i - \chi_i)$
Mean absolute error (MAE)	$\frac{1}{m}\sum_{i=1}^{m}\lvert\hat{\chi}_i - \chi_i\rvert$
Mean absolute relative error ($MARE$)	$\frac{1}{m}\sum_{i=1}^{m}\lvert\frac{\hat{\chi}_i - \chi_i}{\chi_i}\rvert$
Root mean squared error ($RMSE$)	$\sqrt{\frac{1}{m}\sum_{i=1}^{m}(\chi_i - \hat{\chi}_i)^2}$

- The ***r* index** quantifies the strength of the correlation between a set of interpolated values $\hat{\chi}_i$ and the observations χ_i ($i = 1, \ldots, m$, m is the number of observations). The $\bar{\chi}_i$ and $\bar{\hat{\chi}}_i$ are the arithmetic means of χ_i and $\hat{\chi}_i$, respectively. The r varies between 0 and 1, and a higher r value indicates a stronger correlation.
- The $\boldsymbol{R^2}$ **index** is the proportion of the arithmetic interpolation variance that is predictable from the observed values. The SS_{res} and SS_{tot} denote the residual sum of the squared differences between $\hat{\chi}_i$ and χ_i and the total sum of the squared deviations of χ_i from $\bar{\chi}_i$, respectively.
- The ***ME* index** detects a straightforward bias expressed in terms of the differences between $\hat{\chi}_i$ and χ_i.
- The ***MAE* index** quantifies the absolute deviation between $\hat{\chi}_i$ and χ_i.
- The ***MARE* index** is a relative indicator of the absolute deviation between $\hat{\chi}_i$ and χ_i normalized by χ_i.
- The ***RMSE* index** is the arithmetic standard deviation of the squared residuals $\hat{\chi}_i - \chi_i$.

Example 4.1 In ocean sciences, remote sensing techniques (i.e., satellite sensors) are usually implemented to observe sea surface chlorophyll-a concentrations. By comparing remote sensing interpolations and *in-situ* observations, the performance of the remote sensing technique is evaluated. Table 4.2 lists a set of interpolated and observed chlorophyll-a concentrations. The observation and interpolated means are,

TABLE 4.2 Interpolated and observed values.

No.	Interpolated values $\hat{\chi}_i$	Observed values χ_i	No.	Interpolated values $\hat{\chi}_i$	Observed values χ_i
1	6.44	6.83	11	5.70	6.09
2	4.48	4.87	12	0.53	0.14
3	7.56	7.95	13	3.02	3.41
4	6.69	7.08	14	9.82	10.21
5	7.53	7.92	15	7.70	10.19
6	6.02	6.41	16	3.50	3.89
7	1.15	0.76	17	1.91	2.30
8	1.83	1.44	18	1.07	1.46
9	8.55	8.93	19	1.27	0.88
10	0.63	0.24	20	3.19	3.58

$$\bar{\chi}_i = \frac{1}{20}\sum_{i=1}^{20}\chi_i = \frac{1}{20}(6.83 + 4.87 + \cdots + 3.58) = 4.73 \text{ and } \bar{\hat{\chi}}_i = \frac{1}{20}\sum_{i=1}^{20}\hat{\chi}_i = \frac{1}{20}(6.44 + 4.48 + \cdots + 3.19) = 4.43.$$

Then, the six accuracy indices are calculated as.

$$r = \frac{\sum_{i=1}^{20}(\chi_i - \bar{\chi}_i)(\hat{\chi}_i - \bar{\hat{\chi}}_i)}{\sqrt{\sum_{i=1}^{20}(\chi_i - \bar{\chi}_i)^2}\sqrt{\sum_{i=1}^{20}(\hat{\chi}_i - \bar{\hat{\chi}}_i)^2}} = \frac{(6.83 - 4.73)(6.44 - 4.43) + \cdots(3.58 - 4.73)(3.19 - 4.43)}{\sqrt{(6.83 - 4.73)^2 + \cdots(3.58 - 4.73)^2}\sqrt{(6.44 - 4.43)^2 + \cdots(3.19 - 4.43)^2}} = 0.9885,$$

which is very close to 1 (excellent accuracy);

$$R^2 = 1 - \frac{\sum_{i=1}^{20}(\chi_i - \hat{\chi}_i)^2}{\sum_{i=1}^{20}(\chi_i - \bar{\chi}_i)^2} = 1 - \frac{(6.83 - 6.44)^2 + \cdots + (3.58 - 3.19)^2}{(6.83 - 4.73)^2 + \cdots + (3.58 - 4.73)^2} = 0.9579.$$

I.e., about 95.8% of the interpolation variance is predictable from data;

$$ME = \frac{1}{20}\sum_{i=1}^{20}(\hat{\chi}_i - \chi_i) = \frac{1}{20}[(6.44 - 6.83) + \cdots + (3.19 - 3.58)] = -0.2995,$$

i.e., the interpolation mean $\bar{\hat{\chi}}_i$ underestimates the data mean $\bar{\chi}_i$ by only about 2%;

$$MAE = \frac{1}{20}\sum_{i=1}^{20}|\hat{\chi}_i - \chi_i| = \frac{1}{20}[|6.44 - 6.83| + \cdots + |3.19 - 3.58|] = 0.4945,$$

which is the mean absolute deviation between interpolation and data;

$$MARE = \frac{1}{20}\sum_{i=1}^{20}\left|\frac{\hat{\chi}_i - \chi_i}{\chi_i}\right| = \frac{1}{20}\left[\left|\frac{6.44 - 6.83}{6.83}\right| + \cdots + \left|\frac{3.19 - 3.58}{3.58}\right|\right] = 0.3569,$$

which is the normalized mean absolute deviation between interpolation and data; and

$$RMSE = \sqrt{\frac{1}{20}\sum_{i=1}^{20}(\hat{\chi}_i - \chi_i)^2}$$
$$= \sqrt{\frac{1}{20}\left[(6.83 - 6.44)^2 + \cdots + (3.58 - 3.19)^2\right]}$$
$$= 0.6739,$$

which is the arithmetic standard deviation of the squared deviations between interpolation and data.

In light of the above list of *AcI*, ultimately the next step would be to perform an exhaustive sampling and assess the quality of interpolation by comparing the new data with the collocated attribute estimates produced by the original data. According to this reasoning, the best interpolator should be the one with the best accuracy statistic on the basis of a previously agreed criterion. As conclusive as this approach may be, in practice its implementation is impossible to prohibitively expensive and, also, it seems to defeat the purpose of interpolation. Yet, there is hope in the form of cross validation.

4.2 Cross validation

Fortunately, the interpolation *vs*. observation statistics can serve as diagnostics, so to speak, that indicate whether the interpolator and its associated parameter values are sound and reasonable. In particular, beyond the impractical *AcI* computation based on exhaustive sampling, a similar computation using partial or limited sampling has considerable merits.

Cross-validation compares the data values at a selected number of points with the attribute interpolations at the same points obtained without including these data in the interpolation process.

4.2.1 *"Leave-one-out" and other strategies*

Using the cross-validation notion above, the interpolated *vs*. the observed attribute values are compared, thus gaining useful information about the quality (accuracy, informativeness) of the interpolation techniques considered. The above is also known as the "leave-one-out" cross-validation method, i.e., one point is removed from the dataset at each time. In practice, it has been shown that cross-validation is a valuable alternative to the impossibility of exhaustive sampling that helps investigators make informed decisions as to which interpolation technique offers the best attribute interpolations across space and time. Generally, comparisons or remedial actions based on cross-validation are more direct and conclusive if the collection of cross-validation interpolation errors is reduced to some key statistics.

Example 4.2 By dividing each cross-validation error D_i by the standard *STOK* error deviation $e_X(\boldsymbol{p}_i)$, the investigator can compare the magnitudes of both errors. If the average of the squared standardized cross-validation errors is not far from one, i.e., $\frac{1}{m}\sum_{i=1}^{m}\frac{D_i^2}{\sigma_X^2(\boldsymbol{p}_i)} \approx 1$, then the actual interpolation error is equal, on average, to the *STOK* interpolation error.

The five steps of a cross-validation procedure based on the "leave-one-out" strategy are listed in Table 4.3. An implementation of this strategy is examined in the following example.

TABLE 4.3 The "leave-one-out" cross-validation procedure.

Step	Description
①	Use the complete set of available data to compute the covariance or variogram model
②	Remove each datum $\chi_i = X(\boldsymbol{p}_i)$, one at a time, and derive an interpolation $\hat{\chi}_i = \hat{X}(\boldsymbol{p}_i)$ of χ_i
③	The interpolation $\hat{\chi}_i$ and the datum χ_i at the omitted point $\boldsymbol{p}_i$ are compared in terms of the cross-validation error difference $D_i = \hat{\chi}_i - \chi_i$
④	This procedure is repeated for a second data point, and so on
⑤	After a sufficient number of repetitions of steps ①–④, if the average of the cross-validation errors D_i is $\overline{D} = \frac{1}{m}\sum_{i=1}^{m} D_i \approx 0$, there is no systematic interpolation bias. A negative (positive) $\overline{D}$ indicates systematic over-interpolation (under-interpolation)

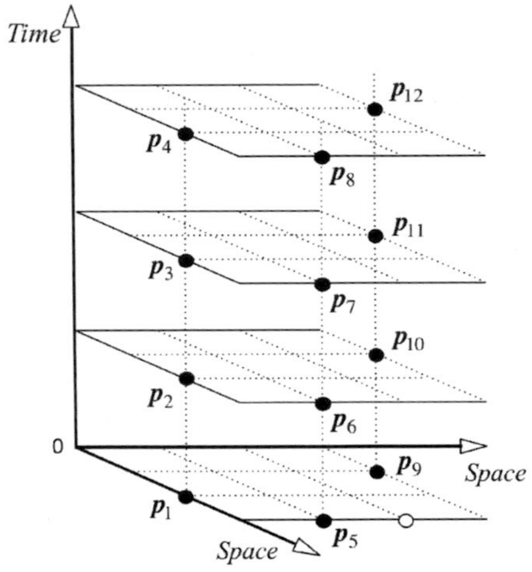

FIG. 4.1 Chronotopologic data arrangement serving to illustrate the implementation of the various cross-validation techniques.

Example 4.3 Consider the chronotopologic data arrangement consisting of the twelve data points shown in Fig. 4.1. The implementation of the "leave-one-out" cross-validation technique involves the following steps:

(i) datum at p_1 is temporarily removed, interpolated using the remaining data, and put back into place;

(ii) datum at p_2 is temporarily removed, interpolated, and put back in place;

…; and

(iii) datum at p_{12} is temporarily removed, interpolated, and put back in place.

In addition to the "leave-one-out" cross-validation above, there are other worth-mentioning cross-validation methods that implement different removing-point strategies. Some of them are described, step-by-step, in Table 4.4. Each one of these different strategies has its *pros* and *cons*. Naturally, the choice among them is rather a matter of investigator experience and preference.

Example 4.3(cont.): Illustrations of the cross-validation techniques of Table 4.4 are presented with reference to Fig. 4.1:

- The implementation of the *10-fold cross-validation* technique consists of the following steps: (i) the

TABLE 4.4 Removing-point strategies.

Technique	Description
10-fold cross-validation	① As in the "leave-one-out" cross-validation technique (Table 4.3). ② Dataset randomly divided into 10 folds, each fold having approximately 10% of the data points. ③ Removing each fold, one at a time, and interpolating at the datum point. ④ The process is repeated 10 times until all folds are interpolated once. ⑤ As in the leave-one-out cross-validation technique (Table 4.3)
Location-based cross-validation (data are derived from many stationary monitoring stations at various instances)	① As in the leave-one-out cross-validation technique (Table 4.3). ② Dataset is divided into several groups in terms of location, i.e., the number of groups is equal to the number of locations. ③ The removing-point strategy consists of removing data points, one at a time, and deriving data interpolations. ④ Process is repeated until the data at all locations are interpolated once. ⑤ As in the leave-one-out cross-validation technique (Table 4.3).
Time-based cross-validation (data are derived during many instances at various locations)	① As in the leave-one-out cross-validation technique (Table 4.3). ② Dataset is divided into several groups in terms of time instances (i.e., number of groups = number of time instances). ③ The strategy consists of removing, in turn, all spatial data at each time instance, and interpolating at these data points. ④ The process is repeated until the spatial data at all time instances are interpolated once. ⑤ As in the leave-one-out cross-validation technique (Table 4.3).

dataset is divided into 10 folds, p_1, $\{p_2,p_3\}$, $\{p_4,p_5\}$, p_6, p_7, ..., p_{11} and p_{12}; (ii) the datum at p_1 is temporarily removed, interpolated, and put back; (iii) data at $\{p_2,p_3\}$ are temporarily removed, interpolated, and put back; ...; and (xi) datum at p_{12} is temporarily removed, interpolated, and put back.

- The implementation of the *location-based cross-validation* technique consists of the following steps: (i) data at $p_1 - p_4$ are temporarily removed, interpolated, and put back; (ii) data at $p_5 - p_8$ are temporarily removed, interpolated, and put back; and (iii) data at $p_9 - p_{12}$ are temporarily removed, interpolated, and put back.
- The implementation of the *time-based cross-validation* technique consists of the following steps: (i) data at p_1, p_5, p_9 are temporarily removed, interpolated, and put back; (ii) data at p_2, p_6, p_{10} are temporarily removed, interpolated, and put back; (iii) data at p_3, p_7, p_{11} are temporarily removed, interpolated, and put back; and (iv) data at p_4, p_8, p_{12} are temporarily removed, interpolated, and put back.

Example 4.4 A numerical cross-validation illustration is shown in Table 4.5. A set of 7627 data points are available (due to space limitations, Table 4.5 only shows 273 cross-validation

TABLE 4.5 A numerical cross-validation example.

s_1	s_2	t	$X(s;t)$	$\hat{X}(s;t)$	$\hat{e}(s;t)$	s_1	s_2	t	$X(s;t)$	$\hat{X}(s;t)$	$\hat{e}(s;t)$
101	7.5	1	32.255	32.843	0.588	109	7.5	2	33.523	33.557	0.034
101	7.5	2	32.368	32.881	0.513	109	7.5	3	33.523	33.559	0.036
101	7.5	3	32.155	32.858	0.703	109	8.5	1	32.425	33.208	0.783
101	8.5	1	31.737	32.218	0.481	109	8.5	2	32.898	33.510	0.612
101	8.5	2	31.891	32.435	0.544	109	8.5	3	32.898	33.523	0.625
101	8.5	3	31.846	32.250	0.404	109	9.5	1	33.276	33.481	0.205
101	9.5	1	31.537	32.458	0.921	109	9.5	2	33.420	33.600	0.180
101	9.5	2	31.762	32.254	0.492	109	9.5	3	33.420	33.609	0.189
101	9.5	3	31.687	32.180	0.493	110	3.5	2	34.133	33.946	−0.187
101	10.5	1	31.985	32.772	0.787	110	3.5	3	34.133	33.946	−0.187
101	10.5	2	32.078	32.570	0.492	110	4.5	1	33.807	33.804	−0.003
101	10.5	3	32.052	32.583	0.531	110	4.5	2	33.915	33.892	−0.023
102	10.5	1	31.944	32.852	0.908	110	4.5	3	33.915	33.920	0.005
102	10.5	2	32.043	32.858	0.815	110	5.5	1	33.731	33.829	0.098
102	10.5	3	32.043	32.880	0.837	110	5.5	2	33.707	33.844	0.137
103	6.5	1	32.894	32.967	0.073	110	5.5	3	33.707	33.878	0.171
103	6.5	2	32.894	32.978	0.084	110	6.5	1	33.745	33.682	−0.063
103	6.5	3	32.894	33.089	0.195	110	6.5	2	33.787	33.809	0.022

Continued

TABLE 4.5 A numerical cross-validation example—cont'd

s_1	s_2	t	$X(s;t)$	$\hat{X}(s;t)$	$\hat{e}(s;t)$	s_1	s_2	t	$X(s;t)$	$\hat{X}(s;t)$	$\hat{e}(s;t)$
103	7.5	1	32.871	32.905	0.034	110	6.5	3	33.787	33.771	−0.016
103	7.5	2	32.871	32.962	0.091	110	7.5	1	33.377	33.617	0.240
103	7.5	3	32.874	32.927	0.053	110	7.5	2	33.723	33.711	−0.012
103	8.5	1	32.974	32.971	−0.003	110	7.5	3	33.723	33.711	−0.012
103	8.5	2	32.974	32.929	−0.045	110	8.5	1	33.233	33.538	0.305
103	8.5	3	32.971	32.897	−0.074	110	8.5	2	33.609	33.668	0.059
104	3.5	1	33.068	33.258	0.190	110	8.5	3	33.609	33.674	0.065
104	3.5	2	33.068	33.225	0.157	110	9.5	1	33.756	33.697	−0.059
104	3.5	3	33.068	33.222	0.154	110	9.5	2	33.837	33.803	−0.034
104	4.5	1	33.190	32.992	−0.198	110	9.5	3	33.837	33.818	−0.019
104	4.5	2	33.190	32.975	−0.215	110	10.5	1	34.048	33.816	−0.232
104	4.5	3	33.190	32.975	−0.215	110	10.5	2	34.102	33.844	−0.258
104	5.5	1	33.017	32.939	−0.078	110	10.5	3	34.102	33.836	−0.266
104	5.5	2	33.017	32.945	−0.072	111	4.5	1	33.643	33.812	0.169
104	5.5	3	33.017	32.944	−0.073	111	4.5	2	33.686	33.862	0.176
104	6.5	1	32.783	32.967	0.184	111	4.5	3	33.686	33.883	0.197
104	6.5	2	32.783	32.953	0.170	111	7.5	1	33.963	33.836	−0.127
104	6.5	3	32.783	32.967	0.184	111	7.5	2	33.963	33.918	−0.045
104	7.5	1	33.046	33.095	0.049	111	7.5	3	33.963	33.916	−0.047
104	7.5	2	33.046	33.061	0.015	111	8.5	1	33.834	33.832	−0.002
104	7.5	3	33.046	33.066	0.020	111	8.5	2	33.834	33.906	0.072
104	8.5	1	33.158	33.261	0.103	111	8.5	3	33.834	33.906	0.072
104	8.5	2	33.158	33.161	0.003	111	9.5	1	33.957	33.865	−0.092
104	8.5	3	33.158	33.201	0.043	111	9.5	2	33.957	33.869	−0.088
105	0.5	1	33.217	33.261	0.044	111	9.5	3	33.957	33.883	−0.074
105	0.5	2	33.217	33.297	0.080	111	10.5	1	34.003	33.874	−0.129
105	0.5	3	33.217	33.265	0.048	111	10.5	2	34.003	33.893	−0.110
105	1.5	1	33.278	33.326	0.048	111	10.5	3	34.003	33.894	−0.109
105	1.5	2	33.278	33.321	0.043	111	11.5	1	33.772	33.934	0.162
105	1.5	3	33.278	33.322	0.044	111	11.5	2	33.772	33.941	0.169
105	2.5	1	33.325	33.263	−0.062	111	11.5	3	33.772	33.949	0.177

TABLE 4.5 A numerical cross-validation example—cont'd

s_1	s_2	t	$X(s;t)$	$\hat{X}(s;t)$	$\hat{e}(s;t)$	s_1	s_2	t	$X(s;t)$	$\hat{X}(s;t)$	$\hat{e}(s;t)$
105	2.5	2	33.325	33.288	−0.037	111	12.5	1	33.831	33.883	0.052
105	2.5	3	33.325	33.268	−0.057	111	12.5	2	33.831	33.862	0.031
105	3.5	1	33.028	33.253	0.225	111	12.5	3	33.831	33.857	0.026
105	3.5	2	33.028	33.182	0.154	111	13.5	1	33.837	33.920	0.083
105	3.5	3	33.028	33.219	0.191	111	13.5	2	33.837	33.918	0.081
105	4.5	1	32.719	33.180	0.461	111	13.5	3	33.837	33.863	0.026
105	4.5	2	32.719	33.174	0.455	111	14.5	1	33.746	33.960	0.214
105	4.5	3	32.719	33.180	0.461	111	14.5	2	33.746	33.956	0.210
105	5.5	1	32.773	33.291	0.518	111	14.5	3	33.746	33.922	0.176
105	5.5	2	32.773	33.257	0.484	111	16.5	1	34.045	33.929	−0.116
105	5.5	3	32.773	33.258	0.485	111	16.5	2	34.045	33.919	−0.126
105	6.5	1	33.075	33.286	0.211	111	16.5	3	34.045	33.924	−0.121
105	6.5	2	33.075	33.158	0.083	112	4.5	1	33.757	33.927	0.170
105	6.5	3	33.075	33.202	0.127	112	4.5	2	33.757	33.994	0.237
105	7.5	1	33.316	33.031	−0.285	112	4.5	3	33.757	33.998	0.241
105	7.5	2	33.316	33.027	−0.289	112	5.5	1	34.491	33.995	−0.496
105	7.5	3	33.316	33.027	−0.289	112	5.5	2	34.491	34.023	−0.468
106	0.5	1	32.994	33.177	0.183	112	5.5	3	34.491	34.017	−0.474
106	0.5	2	32.994	33.167	0.173	112	6.5	1	34.475	34.246	−0.229
106	0.5	3	32.994	33.191	0.197	112	6.5	2	34.475	34.255	−0.220
106	1.5	1	33.284	33.235	−0.049	112	6.5	3	34.475	34.270	−0.205
106	1.5	2	33.284	33.223	−0.061	112	7.5	1	34.122	34.174	0.052
106	1.5	3	33.284	33.260	−0.024	112	7.5	2	34.122	34.120	−0.002
106	2.5	1	33.406	33.271	−0.135	112	7.5	3	34.122	34.167	0.045
106	2.5	2	33.406	33.277	−0.129	112	8.5	1	33.979	33.981	0.002
106	2.5	3	33.406	33.263	−0.143	112	8.5	2	33.979	33.989	0.010
106	4.5	1	33.408	33.400	−0.008	112	8.5	3	33.979	33.989	0.010
106	4.5	2	33.408	33.261	−0.147	112	9.5	1	33.796	33.945	0.149
106	4.5	3	33.408	33.367	−0.041	112	9.5	2	33.796	33.945	0.149
106	5.5	1	33.688	33.422	−0.266	112	9.5	3	33.796	33.930	0.134

Continued

TABLE 4.5 A numerical cross-validation example—cont'd

s_1	s_2	t	$X(s;t)$	$\hat{X}(s;t)$	$\hat{e}(s;t)$	s_1	s_2	t	$X(s;t)$	$\hat{X}(s;t)$	$\hat{e}(s;t)$
106	5.5	2	33.688	33.436	−0.252	112	10.5	1	33.743	33.912	0.169
106	5.5	3	33.688	33.440	−0.248	112	10.5	2	33.743	33.900	0.157
106	6.5	1	33.469	33.413	−0.056	112	10.5	3	33.743	33.892	0.149
106	6.5	2	33.469	33.472	0.003	112	11.5	1	33.900	33.889	−0.011
106	6.5	3	33.469	33.423	−0.046	112	11.5	2	33.900	33.883	−0.017
106	7.5	1	32.683	33.374	0.691	112	11.5	3	33.900	33.881	−0.019
106	7.5	2	32.683	33.453	0.770	112	12.5	1	34.001	33.906	−0.095
106	7.5	3	32.683	33.459	0.776	112	12.5	2	34.001	33.933	−0.068
107	0.5	1	32.814	33.230	0.416	112	12.5	3	34.001	33.930	−0.071
107	0.5	2	32.814	33.266	0.452	112	13.5	1	34.009	33.884	−0.125
107	0.5	3	32.814	33.262	0.448	112	13.5	2	34.009	33.898	−0.111
107	1.5	1	33.154	33.202	0.048	112	13.5	3	34.009	33.900	−0.109
107	1.5	2	33.154	33.227	0.073	112	14.5	1	33.931	33.872	−0.059
107	1.5	3	33.154	33.198	0.044	112	14.5	2	33.931	33.883	−0.048
107	2.5	1	33.176	33.400	0.224	112	14.5	3	33.931	33.879	−0.052
107	2.5	2	33.176	33.435	0.259	112	15.5	1	34.064	33.859	−0.205
107	2.5	3	33.176	33.476	0.300	112	15.5	2	34.064	33.821	−0.243
107	4.5	1	33.392	33.566	0.174	112	15.5	3	34.064	33.849	−0.215
107	4.5	2	33.373	33.612	0.239	112	16.5	1	33.925	33.894	−0.031
107	4.5	3	33.373	33.585	0.212	112	16.5	2	33.925	33.853	−0.072
107	5.5	1	33.773	33.614	−0.159	112	16.5	3	33.925	33.896	−0.029
107	5.5	2	33.810	33.641	−0.169	112	17.5	1	33.914	33.826	−0.088
107	5.5	3	33.810	33.555	−0.255	112	17.5	2	33.914	33.819	−0.095
107	6.5	1	33.740	33.532	−0.208	112	17.5	3	33.914	33.838	−0.076
107	6.5	2	33.813	33.569	−0.244	113	6.5	1	34.397	34.285	−0.112
107	6.5	3	33.813	33.464	−0.349	113	6.5	2	34.397	34.344	−0.053
107	7.5	1	33.361	33.362	0.001	113	6.5	3	34.397	34.339	−0.058
107	7.5	2	33.489	33.481	−0.008	113	7.5	1	34.288	34.264	−0.024
107	7.5	3	33.489	33.477	−0.012	113	7.5	2	34.288	34.234	−0.054
107	8.5	1	33.524	33.297	−0.227	113	7.5	3	34.288	34.235	−0.053
107	8.5	2	33.776	33.336	−0.440	113	8.5	1	34.118	34.084	−0.034

TABLE 4.5 A numerical cross-validation example—cont'd

s_1	s_2	t	$X(s;t)$	$\hat{X}(s;t)$	$\hat{e}(s;t)$	s_1	s_2	t	$X(s;t)$	$\hat{X}(s;t)$	$\hat{e}(s;t)$
107	8.5	3	33.776	33.319	−0.457	113	8.5	2	34.118	34.094	−0.024
108	4.5	2	33.677	33.887	0.210	113	8.5	3	34.118	34.090	−0.028
108	4.5	3	33.677	33.783	0.106	113	9.5	1	34.008	33.930	−0.078
108	5.5	1	33.766	33.754	−0.012	113	9.5	2	34.008	33.917	−0.091
108	5.5	2	33.775	33.784	0.009	113	9.5	3	34.008	33.904	−0.104
108	5.5	3	33.775	33.677	−0.098	113	10.5	1	33.876	33.915	0.039
108	6.5	1	33.577	33.634	0.057	113	10.5	2	33.876	33.869	−0.007
108	6.5	2	33.704	33.679	−0.025	113	10.5	3	33.876	33.828	−0.048
108	6.5	3	33.704	33.598	−0.106	113	11.5	1	33.964	33.903	−0.061
108	7.5	1	33.161	33.417	0.256	113	11.5	2	33.964	33.901	−0.063
108	7.5	2	33.313	33.604	0.291	113	11.5	3	33.964	33.901	−0.063
108	7.5	3	33.313	33.582	0.269	113	12.5	1	33.983	33.911	−0.072
108	9.5	2	33.430	33.575	0.145	113	12.5	2	33.983	33.915	−0.068
108	9.5	3	33.430	33.638	0.208	113	12.5	3	33.983	33.945	−0.038
109	3.5	1	34.031	33.716	−0.315	113	13.5	1	33.857	33.912	0.055
109	3.5	2	34.310	33.807	−0.503	113	13.5	2	33.857	33.930	0.073
109	3.5	3	34.310	33.908	−0.402	113	13.5	3	33.857	33.883	0.026
109	4.5	1	34.005	33.918	−0.087	113	14.5	1	33.760	33.870	0.110
109	4.5	2	34.180	33.820	−0.360	113	14.5	2	33.760	33.923	0.163
109	4.5	3	34.180	33.924	−0.256	113	14.5	3	33.760	33.879	0.119
109	5.5	1	34.018	33.796	−0.222	113	15.5	1	33.726	33.835	0.109
109	5.5	2	33.809	33.765	−0.044	113	15.5	2	33.726	33.875	0.149
109	5.5	3	33.809	33.841	0.032	113	15.5	3	33.726	33.836	0.110
109	6.5	1	33.681	33.660	−0.021	113	16.5	1	33.541	33.798	0.257
109	6.5	2	33.810	33.660	−0.150	113	16.5	2	33.541	33.803	0.262
109	6.5	3	33.810	33.649	−0.161	113	16.5	3	33.541	33.799	0.258
109	7.5	1	33.025	33.303	0.278						

results). The "$X(s;t)$" column lists the measured value at each point, the "$\hat{X}(\boldsymbol{s};t)$" column gives the interpolated value at the same point when we remove it from the input (i.e., use the other 7626 points to interpolate), the "$\hat{e}(\boldsymbol{s};t)$" column is simply the difference between the measured and interpolated values. Because either over- or under-interpolation occurs at each point, the error may be either positive or negative. Averaging the errors is not very useful if one wants to see the overall error –one will end up with a value that is essentially zero due to these alternating positive and negative values. Thus, in order to assess the extent of interpolation error, one can square each term, and then take the average of these squared errors. This average is the *MSE* indicator, and the square root of *MSE* is the *RMSE* indicator (defined in Table 4.1).

4.2.2 *Cross-validation guidelines*

Generally, the cross-validation *guidelines* to follow are based on diagnostic statistics and account for the following considerations:

① The ideal *CTDA* method should result in no errors and a correlation coefficient of 1 between measurements and interpolations—this is rarely the case in practice, of course.

② Large individual errors may be the effect of blunders or outliers. Data preparation and coding (locations, time instances, etc.) must be checked carefully.

③ A slope different from 45° in the regression line between measurement and interpolations denotes conditional bias.

④ A highly desirable interpolator is one generating errors D_i that are chronotopologically uncorrelated (i.e., resulting in a pure nugget variogram).

⑤ If the ratio of the interpolation error over the standard error deviation, $\frac{D_i}{e_X(\boldsymbol{p}_i)}$, follows a standard normal distribution, one cannot discard the possibility that the errors are multi-normally distributed.

Generally, cross-validation is a useful tool for testing various kinds of Kriging results in most practical applications.

A simple yet effective *step-by-step process* of **testing** interpolation in practice can be established by means of cross-validation.

This systematic process can be implemented with efficiency and its outcomes are easily interpretable. An instructive real-world numerical $PM_{2.5}$ case study is discussed next.

Example 4.5 Let us look into a recent $PM_{2.5}$ real-world study (Yang et al., 2018). The authors selected one of the most populated and developed areas in China, namely southern Jiangsu Province (Fig. 4.2).

Step 1: The $PM_{2.5}$ data was collected at 53 monitoring stations at 5 cities during the period January 1—December 31, 2014 (totally, 365 days), with the space-time points being described as

$$\boldsymbol{p}_i = (\boldsymbol{s};t)_i, \quad i = 1,\ldots,m = 53 \times 365 = 19{,}345. \tag{4.2a}$$

Recall that an alternative way is as

$$(\boldsymbol{s}_j; t_{j'}), \quad j = 1,\ldots,53, \; j' = 1,\ldots,365 \tag{4.2b}$$

Then, the entire $PM_{2.5}$ dataset can be described as

$$\boldsymbol{X} = \left[PM_{2.5}(\boldsymbol{p}_1)\ldots\; PM_{2.5}(\boldsymbol{p}_{19345})\right]^T \tag{4.3}$$

Concerning descriptive statistics (Fig. 4.3), the mean $PM_{2.5}$ concentration was 65.63 $\mu g/m^3$ and the coefficient of variation (*CV*) was 0.57, indicating a medium level variability of the monitored $PM_{2.5}$ data ($0.1 < CV < 1$). The daily $PM_{2.5}$ concentrations averaged over all monitoring sites available in the study region during the year 2014 are plotted in Fig. 4.4, exhibiting a seasonal pattern with elevated concentrations during spring and winter due to seasonal fluctuations of both the emission and meteorological conditions.

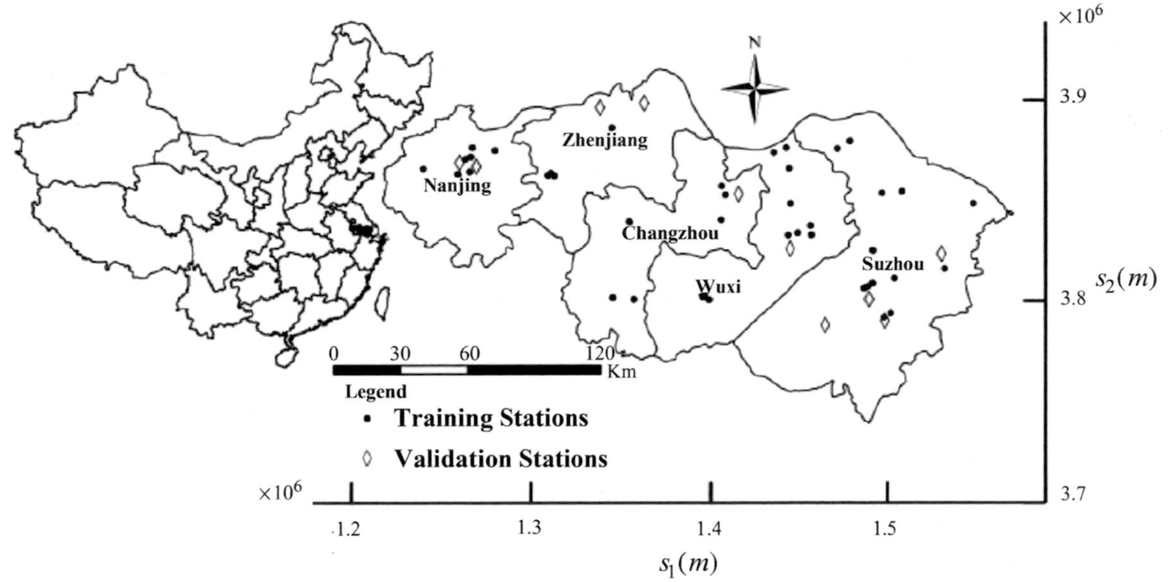

FIG. 4.2 Geographical location of southern Jiangsu Province (China) and the spatial locations of $PM_{2.5}$ monitoring stations.

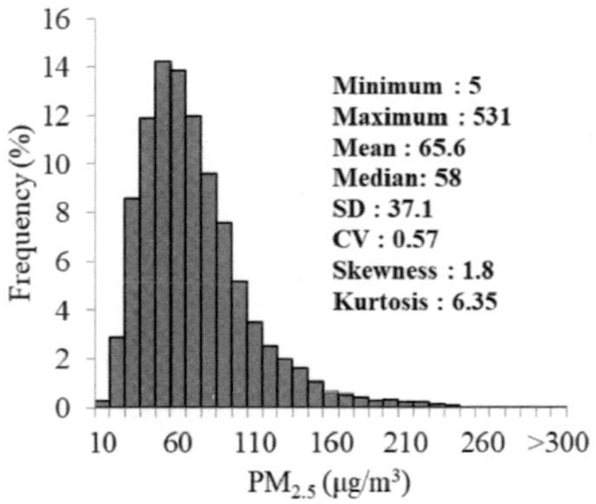

FIG. 4.3 Descriptive statistics and histogram of the $PM_{2.5}$ concentrations (μg/m^3) in southern Jiangsu during 2014 (*SD*, standard deviation; *CV*, coefficient of variation).

Step 2: In order to implement *STOK* in $PM_{2.5}$ mapping, the empirical variogram values were calculated by.

$$\hat{\gamma}_{PM_{2.5}}(\boldsymbol{h};\tau)=\frac{1}{2N(h,\tau)}\sum_{i=1}^{N(h,\tau)}[PM_{2.5}(s_i;t_i)-PM_{2.5}(s_i+\boldsymbol{h};t_i+\tau)]^2 \quad (4.4a)$$

where $\boldsymbol{h}=\boldsymbol{s}_i-\boldsymbol{s}_j$, $\tau=t_i-t_j$ and $i=j=1, \ldots, 53\times 365$, and a theoretical variogram model.

$$\gamma_{PM_{2.5}}(\boldsymbol{h},\tau)=0.53-0.51\left[1+0.29\left(10^{-5}\boldsymbol{h}+4.94\tau\right)^2\right]^{1.12} e^{-0.61\,|10^{-5}\boldsymbol{h}+4.94\tau|} \quad (4.4b)$$

was fitted to the empirical variogram values, see Fig. 4.5 (the dots represent the empirical variogram values, while the shadow surface represents the fitted theoretical variogram model). The contribution of the time dimension in a chronotopologic metric is mostly *application-dependent*. In this case study, then the metric $\boldsymbol{\Delta p}=10^{-5}\boldsymbol{h}+4.94\tau$ was selected so that the model was written as

$$\gamma_{PM_{2.5}}(\boldsymbol{\Delta p})=0.53-0.51\left[1+0.29\boldsymbol{\Delta p}^2\right]^{1.12} e^{-0.61\,|\boldsymbol{\Delta p}|} \quad (4.4c)$$

Step 3: To interpolate the $PM_{2.5}$ concentration at the point of interest $\boldsymbol{p}_k$, the *STOK* system can be set up as, $\boldsymbol{\Gamma}^*\boldsymbol{\lambda}_\mu=\boldsymbol{\gamma}_1$, or

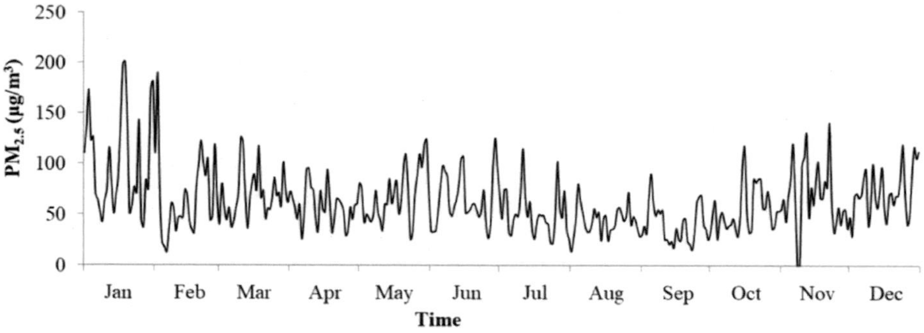

FIG. 4.4 Daily variations of monitored $PM_{2.5}$ mean concentrations ($\mu g/m^3$) in southern Jiangsu during 2014.

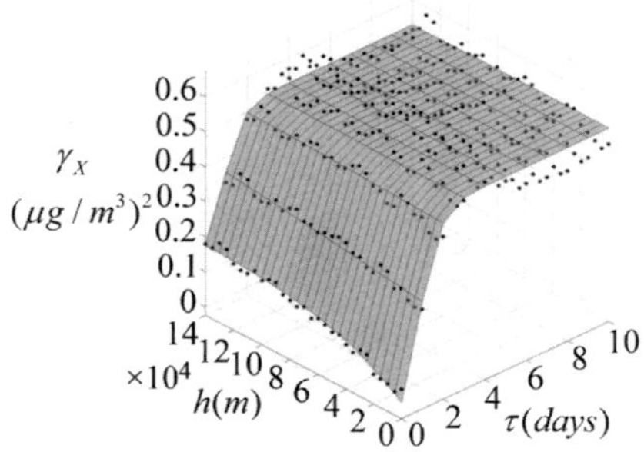

FIG. 4.5 Empirical variogram values (dots) and the fitted theoretical variogram model (shadow surface).

$$\begin{bmatrix} \gamma_{PM_{2.5}}(\boldsymbol{p}_1, \boldsymbol{p}_1) & \gamma_{PM_{2.5}}(\boldsymbol{p}_1, \boldsymbol{p}_2) & \cdots & \gamma_{PM_{2.5}}(\boldsymbol{p}_1, \boldsymbol{p}_{19345}) & 1 \\ \gamma_{PM_{2.5}}(\boldsymbol{p}_2, \boldsymbol{p}_1) & \gamma_{PM_{2.5}}(\boldsymbol{p}_2, \boldsymbol{p}_2) & \cdots & \gamma_{PM_{2.5}}(\boldsymbol{p}_2, \boldsymbol{p}_{19345}) & 1 \\ \vdots & \vdots & \cdots & \vdots & \vdots \\ \gamma_{PM_{2.5}}(\boldsymbol{p}_{19345}, \boldsymbol{p}_1) & \gamma_{PM_{2.5}}(\boldsymbol{p}_{19345}, \boldsymbol{p}_2) & \cdots & \gamma_{PM_{2.5}}(\boldsymbol{p}_{19345}, \boldsymbol{p}_{19345}) & 1 \\ 1 & 1 & \cdots & 1 & 0 \end{bmatrix} \begin{bmatrix} \lambda_1 \\ \lambda_2 \\ \vdots \\ \lambda_{19345} \\ \mu \end{bmatrix} = \begin{bmatrix} \gamma_{PM_{2.5}}(\boldsymbol{p}_1, \boldsymbol{p}_k) \\ \gamma_{PM_{2.5}}(\boldsymbol{p}_2, \boldsymbol{p}_k) \\ \vdots \\ \gamma_{PM_{2.5}}(\boldsymbol{p}_{19345}, \boldsymbol{p}_k) \\ 1 \end{bmatrix} \quad (4.5)$$

The large matrix above (19346×19346) makes the solution of the system of Eq. (4.5) for the weights λ_i and multiplier μ at all data points computationally time-consuming, and, moreover, the large number of points included in the matrix are redundant for interpolation purposes because of the screen effects.

Step 4: In view of the above considerations, a moving window was used to search for space-time point neighborhoods (Fig. 4.6). In this case study, the spatial and temporal searching radii ($|\boldsymbol{h}|$ and τ) were set equal to 140 km and 4 days, respectively. The space-time neighborhoods (points) within these radii were finally used for *STOK* interpolation, in which case the system of Eq. (4.5) was rewritten as

$$
\begin{bmatrix}
\gamma_{PM_{2.5}}(p_1,p_1) & \gamma_{PM_{2.5}}(p_1,p_2) & \cdots & \gamma_{PM_{2.5}}\left(p_1,p_{m_k}\right) & 1 \\
\gamma_{PM_{2.5}}(p_2,p_1) & \gamma_{PM_{2.5}}(p_2,p_2) & \cdots & \gamma_{PM_{2.5}}\left(p_2,p_{m_k}\right) & 1 \\
\vdots & \vdots & \cdots & \vdots & \vdots \\
\gamma_{PM_{2.5}}\left(p_{m_k},p_1\right) & \gamma_{PM_{2.5}}\left(p_{m_k},p_2\right) & \cdots & \gamma_{PM_{2.5}}\left(p_{m_k},p_{m_k}\right) & 1 \\
1 & 1 & \cdots & 1 & 0
\end{bmatrix}
\begin{bmatrix} \lambda_1 \\ \lambda_2 \\ \vdots \\ \lambda_m \\ \mu \end{bmatrix}
=
\begin{bmatrix} \gamma_{PM_{2.5}}(p_1,p_k) \\ \gamma_{PM_{2.5}}(p_2,p_k) \\ \vdots \\ \gamma_{PM_{2.5}}(p_m,p_k) \\ 1 \end{bmatrix}
\tag{4.6}
$$

where the number of space-time neighborhoods m_k is much less than 19,345, i.e., $m_k \ll 19{,}345$. The new *STOK* system of Eq. (4.6) was further used for mapping and cross-validation purposes.

Step 5: In order to test the performance of *STOK* in $PM_{2.5}$ interpolation by means of cross-validation, the stations were sorted/labeled according to their s_2 values (Fig. 4.2); and, for every group of 5 stations one was sequentially selected as the validation station, so that in the end a total number of 3611 samples from 10 stations (almost 20% of all monitoring stations) served as the validation dataset. This can be regarded as a variant location-based cross-validation technique. The pairs of interpolated-observed $PM_{2.5}$ concentrations of the validation set are shown in Fig. 4.7. The Pearson correlation coefficient r (0.918), the *ME* ($-0.89\ \mu g/m^3$), and the *RMSE* ($15.18\ \mu g/m^3$) were used to evaluate the *STOK* interpolation performance. The intercept and slope values of the regression line were 10.41 and 0.858, respectively. Since the slope is sufficiently close to 1, it implied that in the validation set the interpolated concentrations were very close to the observed $PM_{2.5}$ concentrations. Hence, the three accuracy indicators clearly demonstrated the high accuracy of the *STOK* technique in the mapping of $PM_{2.5}$ concentrations in the southern Jiangsu Province during 2014.

Step 6: Using the same *STOK* system of Eq. (4.6), the $PM_{2.5}$ concentrations were interpolated using a space-time resolution of $2\,km \times 2\,km \times 1\,day$, see Fig. 4.8A. The annual mean concentration and the corresponding coefficient of variation are shown in Fig. 4.9A and C depicting increasing concentration trends from east to west in the study region, thus indicating that meteorological factors had a significant effect on the $PM_{2.5}$ concentration levels, because the high wind speeds, relative humidity, abundant rainfall, etc. of the eastern coastal areas

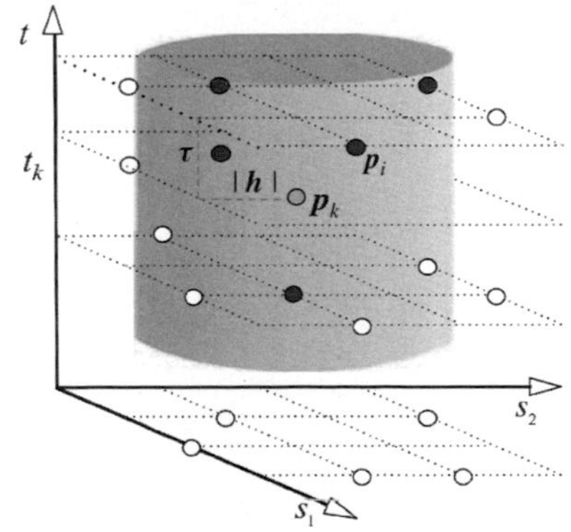

FIG. 4.6 Moving window employed to search for neighborhoods (space-time points p_i) around an interpolation point p_k; $|h|$ is the spatial searching radius and τ is the temporal searching radius.

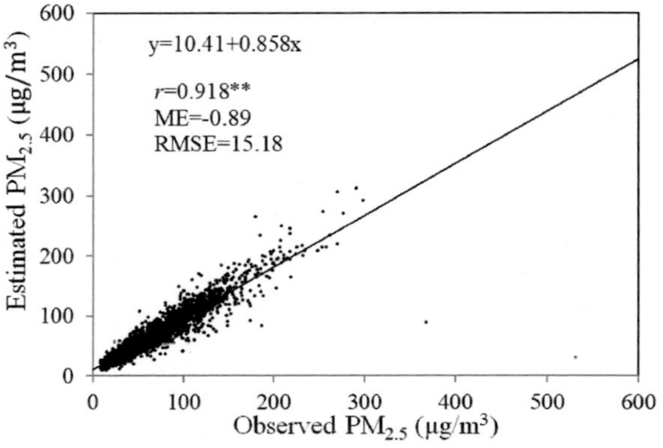

FIG. 4.7 Cross-validation results for the *STOK* technique.

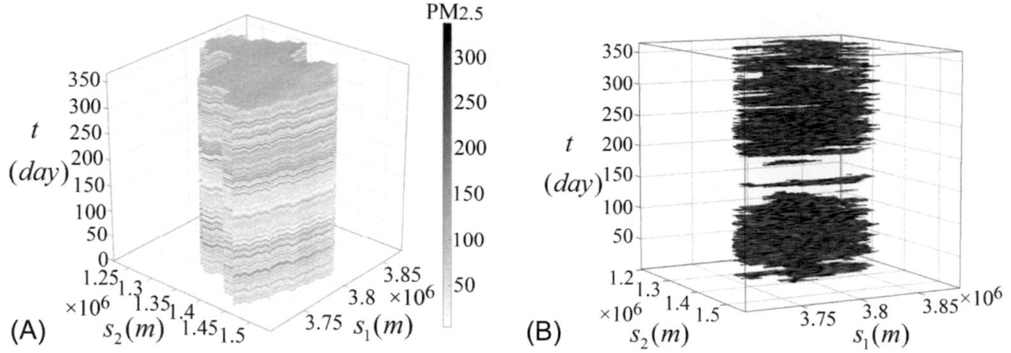

FIG. 4.8 (A) Plot of the space-time map of predicted $PM_{2.5}$ concentrations generated by the *STOK* technique. (B) Polluted space-time domain identified by the *STOK* technique.

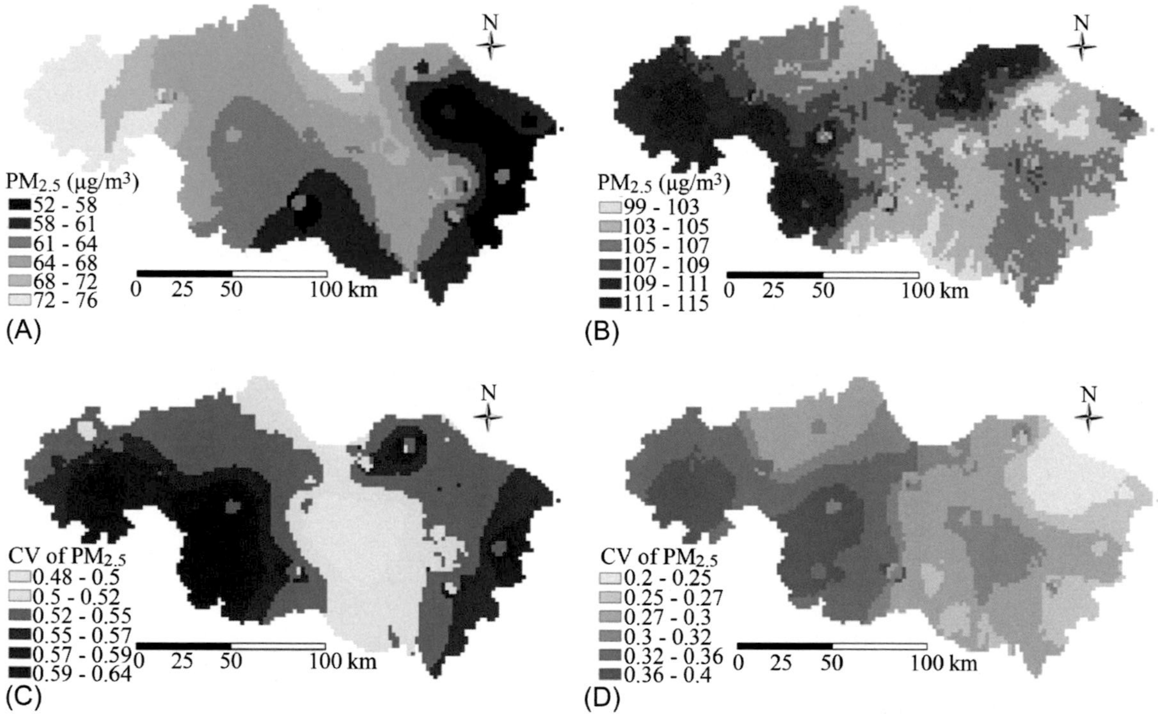

FIG. 4.9 (A) Geographical distribution of annual mean $PM_{2.5}$ concentrations ($\mu g/m^3$), and (C) corresponding *CV*; (B) geographical distribution of the average $PM_{2.5}$ concentration during the days when concentration exceeded the threshold value, and (D) corresponding *CV*.

create favorable conditions for the diffusion of $PM_{2.5}$ concentrations in the study region. According to the China National Ambient Air Quality Standard, the $PM_{2.5}$ ambient concentration was divided into six intervals corresponding to six levels, see Table 4.6. Fig. 4.8B presents the polluted space-time domain in southern Jiangsu (in fact, as is shown in Fig. 4.10 later, about 29.3% of the chronotopologic domain was $PM_{2.5}$ polluted). Fig. 4.9B shows the spatial

TABLE 4.6 Grading standards of $PM_{2.5}$ ambient concentration.

Level of air quality	Mean $PM_{2.5}$ value within 24 h ($\mu g/m^3$)	Type of air quality
Level 1	0–35	Excellent
Level 2	35–75	Good
Level 3	75–115	Light pollution
Level 4	115–150	Moderate pollution
Level 5	150–250	Heavy pollution
Level 6	>250	Serious pollution

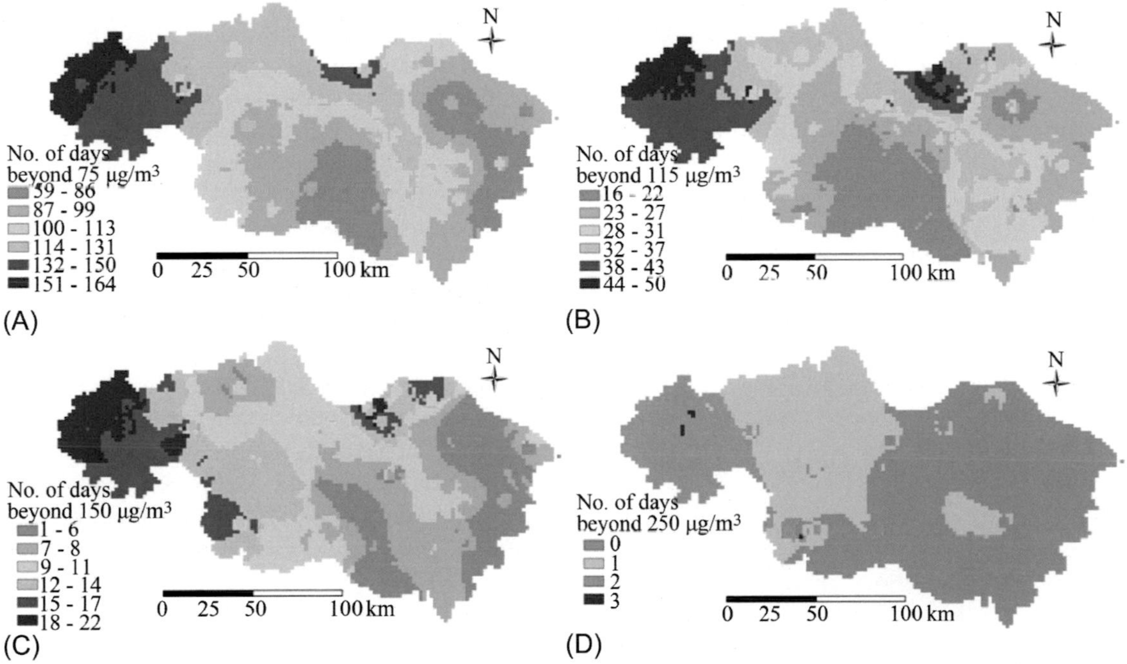

FIG. 4.10 Spatial distribution of the number of days of 2014 during which $PM_{2.5}$ concentrations exceeded (*A*) 75, (*B*) 115, (*C*) 150, and (*D*) 250 $\mu g/m^3$.

distribution of average $PM_{2.5}$ concentration during the days when it is greater than a threshold value (75 $\mu g/m^3$) and the corresponding *CV* map is shown in Fig. 4.9D, which shows a similar pattern compared to Fig. 4.9A and C.

Step 7: As is shown in Figs 4.10A–D, the number of days during 2014 with $PM_{2.5}$ concentrations exceeding the thresholds of 75, 115, 150, and 250 $\mu g/m^3$ varied from 59 to 164, 16 to 50, 1 to 22, and 0 to 3 days, respectively, in the entire study area. The plots in Figs. 4.10A–C showed a similar spatial pattern as the distribution of the annual mean $PM_{2.5}$ concentrations (Fig 4.9A). This apparently happens because the higher the annual mean, the more days the $PM_{2.5}$ concentration exceeds the corresponding threshold. The number of days with $PM_{2.5}$ pollution $>75\ \mu g/m^3$ were more than 151 in the western part of the study area, and it was heavily polluted (according to the classification of Table 4.6) during more than 18 of these days. Nanjing, the provincial capital of Jiangsu, experienced 3 days of serious pollution (see, Fig. 4.10D). On the other hand, as is clear in Fig. 4.10A–D, the eastern part of the study area experienced relatively good $PM_{2.5}$ air quality.

Step 8: Based on Fig. 4.11, in the time direction (day sequence) the $PM_{2.5}$ concentrations exhibited a similar temporal pattern from west to east, characterized by 4 peaks and 3 troughs. The interpretation of this pattern is as follows. The first and last peaks appeared during winter and were caused by local meteorologic conditions, such as the high frequency of the inversion layer, lower rainfalls (most of the winter days were cloudy and only about 13% of these days had light rain), and human activities (e.g., heating energy consumption). After the first peak, the $PM_{2.5}$ concentrations began to gradually decrease, affected by the strong winds during spring (Fig 4.12A), the gradual increase of the surface temperature, and the gradual vegetation enrichment (this is when the first trough occurs). Then, the rainy season arrived (called the plum rainy season) and the accumulated pollutants in the region could not spread due to a subtropical high that resulted in hazy weather (the second peak). During summer, the prevailing wind direction, mainly from an eastern marine air mass (Fig 4.12B), with good cleaning performance and strong atmospheric wet deposition, contributed to the wet deposition and dilution of air pollutants (second trough). The autumn harvest season that followed was the peak straw-burning period, when a large amount of smoke was produced by open burning. Adverse meteorologic conditions (small wind forces etc., Fig 4.12A) and the various pollutant sources led to a $PM_{2.5}$ concentration increase in the area (third peak). After the autumn harvest, the local pollution sources gradually decreased, resulting in a small $PM_{2.5}$ trough (the third one). Mean $PM_{2.5}$ concentrations showed a significant decreasing trend from west to east (similar to that of Fig 4.9A).

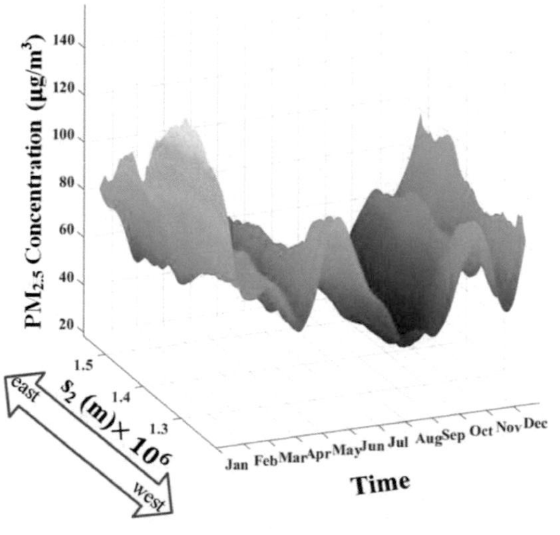

FIG. 4.11 Three-dimensional plot of the temporal trend of $PM_{2.5}$ concentrations during 2014 by the Dynamic Harmonic Regression method.

4.2.3 *Pros and cons*

Cross-validation is often a fast and inexpensive way of indirectly testing anything about Kriging interpolation or about the data. However, one must implement it with caution keeping in mind the following:

1. Cross-validation does not indicate whether an observation, interpolation, parameter, or assumption is inadequate.
2. Complex interdependencies among errors have precluded finding their distribution,

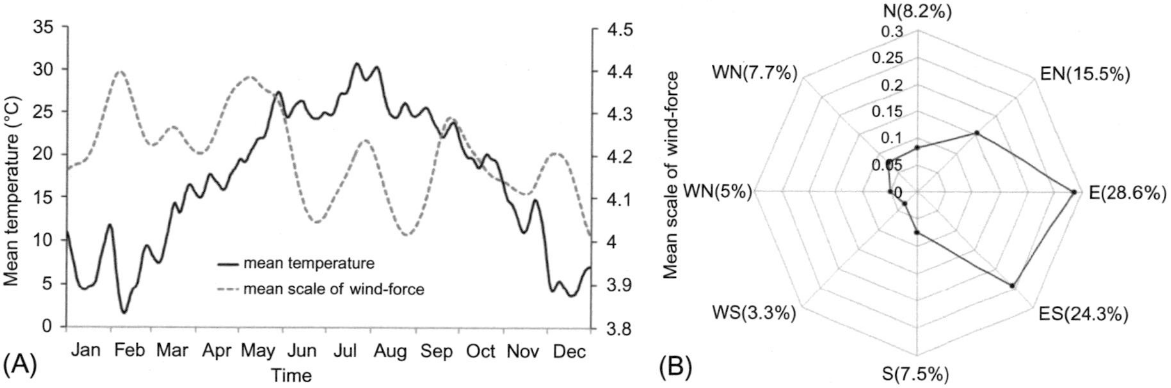

FIG. 4.12 (A) Plot of the mean temperature and mean scale of wind-force along with day sequence, and (B) rose plot of wind direction in the summer of 2014. The temperature, wind-force, and wind direction data were collected from the Jiangsu Meteorological Bureau. The daily data were recorded for each city. The daily "wind-force" data were calculated by averaging "wind-force" values from the five cities.

thus hindering a rigorous analysis or testing of the results.

3. Some investigators argue that, in a sense, cross-validation somehow "cheats" by using all data to estimate the variogram model; but, after completing cross-validation, some data points may be set aside as unusual, thus requiring the variogram model to be refit.

Nonetheless, cross-validation remains a useful tool primarily to dispel blunders and assist in drawing honest conclusions by comparison, instead of making arbitrary decisions or assumptions.

5 Applied krigology: benefits and concerns

CTDA considers both deterministic and stochastic techniques of attribute modeling and interpolation. Chapter 7 presented some well-known groups of deterministic interpolators that use mathematical functions to calculate attribute values at unknown points based on the degree of similarity in relation with neighboring data points. Geostatistical Kriging interpolation, which is the focus of this chapter, is based on a different conception of interpolation and its objectives.

> **Krigology** is a *CTDA* term that refers to the theory and application of Kriging interpolation by combining probabilistic and statistical techniques to generate attribute values at all points within the domain of interest and to assess the interpolation quality in terms of a measure of the associated interpolation accuracy based on the attribute dependency pattern among data points.

For a variety of reasons, Kriging interpolators have been very popular in applied sciences. Some of them are investigated next.

5.1 Benefits of Kriging interpolation

Early on in this chapter, we considered the question "why Krige". The quick answer was then given that when a good variogram model is available, Kriging can provide attribute interpolations best representing the distribution of the attribute data.

In practical applications, the decision on whether or not to use Kriging will depend upon various factors such as the chronotopologic data distribution and structure, the physics of the attribute, post-Kriging analyses, etc. Accordingly, when using a fundamental *CTDA* tool, like Kriging, one should duly appreciate its potential advantages as well as drawbacks compared to other techniques.

Example 5.1 If a considerable number of data exist on a near-perfect grid pattern, it is likely that Kriging will produce comparable attribute interpolations to other techniques, but it can reduce considerably or eliminate the problematic *bulls-eye effect* (often observed in inverse distance square interpolation, Chapter 7). If the samples are clustered, then a *declustering* technique should be used before variogram modeling.

Further key benefits of the Kriging interpolation techniques over other techniques are as follows:

① **Flexibility**: Krigology allows considerable flexibility in the selection of the interpolation technique that is most appropriate for the particular natural attribute distribution.

Over the years, the family of Kriging techniques has provided useful and versatile solutions to the interpolation problem. Commonly used linear Kriging techniques, like *STOK* and *STSK*, were discussed above. Among the best nonlinear techniques are *BME* (Chapter 9), as well as disjunctive Kriging and lognormal Kriging for highly skewed distributions (Christakos, 1992a).

Example 5.2 The geology of a specific mineralization offers valuable insight concerning the choice of the most suitable Kriging technique for the case.

② **Accuracy assessment**: Beyond the optimal (in a statistical sense) attribute interpolations, Kriging techniques also provide attribute interpolation errors in terms of the Kriging error variance or standard deviation.

While the Kriging interpolation values are physical, the Kriging interpolation error is conceptual. Yet, the latter's contribution in all kinds of risk analysis studies is essential.

Example 5.3 *STOK* is a local interpolator for the unknown attribute values, which is optimal in the sense that it minimizes the statistical interpolation error variance and is unbiased.

③ **Confidence intervals**: Kriging error variances or standard deviations can be used to derive confidence intervals of attribute interpolations.

Example 5.4 Assuming that the interpolation error is *normally* distributed, the actual attribute value at the unsampled point of interest $\boldsymbol{p}_k$ belongs to the interval $\left[\hat{X}(\boldsymbol{p}_k) - 1.96\sigma_X(\boldsymbol{p}_k),\ \hat{X}(\boldsymbol{p}_0) + 1.96\sigma_X(\boldsymbol{p}_k)\right]$ with 95% probability.

5.2 Concerns with Kriging interpolation

Certain aspects of Kriging interpolation need special attention in order to understand the underlying methodology. Accordingly, we investigate here some potential concerns with Kriging as a motivation to continue addressing them and, on occasion, to potentially search for more advanced chronotopologic interpolators:

① **Negative weights**: As noted earlier, due to clustering some samples may receive negative weights resulting in negative interpolated attribute values. Similar results may occur if a highly structured variogram is used, ignoring inherent randomness (nugget effect).

Such undesirable results may be avoided or corrected if a more realistic variogram is used and/or the data is de-clustered.

② **Search parameters**: Small search radii may result in very conservative interpolations as compared to very large search parameters, which, on the other hand, may cause over-smoothing.

Search parameters play an important role in Kriging, which is why they should be designed with consideration of the chronotopologic data structure.

③ **Covariance and variogram parameters**: Over-estimated nuggets lead to over-smoothed Kriging maps; long correlation ranges lead to larger Kriging neighborhoods than short ones; the stronger the dependency of the attribute, the more accurate the estimation, etc.

These parameters can impact the Kriging results and the subsequent decisions to varying degrees.

Example 5.5 In mining, an overestimated nugget effect can lead to financial losses by ore misclassification, whereas an underestimated nugget effect leads to overly selective (too continuous) Kriging estimates causing ore blocks incorrectly being sent to the waste dump and waste blocks being sent to the process plant (Morgan, 2011).

④ **Minimum mean square error techniques**: Belonging to this kind of technique, Kriging generates realistic interpolations when the average sampling interval is one order of magnitude smaller than the minor axis of the variogram range. As sampling becomes sparser, the Kriging maps may exhibit an increasing smoothing, which may not be appropriate.

The degree of severity of the minimum mean squared error effects above on Kriging interpolation will depend on the real-world study.

⑤ **Chronotopologic variation assumptions**: Mainstream Kriging interpolation should be used only if the underlying assumptions (homostationarity, normality, linearity, etc.) are met. If the data distribution does not meet these assumptions, other, non-standard Kriging techniques (e.g., intrinsic and disjunctive Krigings) could be helpful.

While theoretically well-developed, these nonstandard Kriging techniques have not been sufficiently tested in practice (e.g., standard software is not yet available in the public domain).

⑥ **Data availability**: A basic requirement of using a Kriging technique is having a sufficient number of appropriately distributed data points (samples).

Yet, there is no definite criterion for the number of samples to be used in chronotopologic Kriging; instead, *ad hoc* choices are made depending on the particular study. Data *clustering* (proximity) is also a concern as regards the adequate number of samples.

- In sum, an important factor to consider when choosing a specific Kriging technique is the technical features-physical characteristics **association**, i.e., what kind of features each Kriging technique has and how compatible these features are with the physical attribute characteristics.

Example 5.6 *STOK* should be used when the attribute mean is constant but not known. When the attribute mean is known, *STSK* should be used instead. *STIK* should be used when non-Gaussian probability data distributions are considered.

6 Practice exercises

1. An investigator predicts that if the *BIC* (boundary/initial conditions) of a physical phenomenon are thus and so, if such and such events (which may well be within control) do not intervene, then the outcome (probably and approximately and within the limits of error) will be so and so. Is the investigator's prediction categorical or conditional, and why?
2. Is the mechanism of Kriging interpolation categorical or conditional, and why?
3. In light of the data of Example 2.5 of Section 2 here we assume that the fitted theoretical model is $\gamma_X(h,\tau) = 100\left(1 - 0.972e^{-\left(\frac{h}{510.189} + \frac{\tau}{3.892}\right)}\right)$.

Calculate the *STOK* interpolated value and variance at point p_k in Table 2.3 of Section 2.

4. Choose some less space-time points near the neighbor points shown in Table 2.3 of Section 2 and calculate the *STOK* interpolation value and the corresponding interpolation error. Compare them with the values shown in Example 2.5 of Section 2 and suggest any empirical law that could be accounted for by *STOK*. In doing so, keep in mind the suggestion of *Periander of Corinth* concerning the importance of established laws: "Use food that is fresh but laws that are old."

5. Compare the results in Exercise 4 with Example 2.9 of Section 2, and suggest any empirical law that could be accounted for by *STOK* and *STIK*.

6. Which of the following statements concerning the *STOK* weights are valid?

(a) They depend on the variogram model.

(b) They depend on the spatial and temporal distance of sample points to the estimation point.

(c) They depend on the data values.

(d) They depend on the space-time relationships among samples around the estimation point.

7. Which of the following statements are valid?

(a) The steeper the variogram curve near the origin, the more influence the closest neighbors have on attribute's space-time interpolation. As a result, the interpolation maps will be less smooth.

(b) An interpolation technique that produces an attribute value identical to the attribute value at a sampled point is known as a deterministic interpolator.

(c) An interpolation technique that produces an attribute value identical to the attribute value at a sampled point is known as an exact interpolator.

(d) An interpolation is unbiased when some of the interpolated values are above the actual values and some below, so that, on average, the difference between the interpolations and the actual values is zero.

(e) In geostatistical Kriging, the sum of the interpolation weights is always one.

(f) Toblerianism in Kriging implies that value pairs that are closer to each other should have a smaller measurement difference than those farther away from one another. The extent that this assumption is valid can be detected in the empirical variogram.

8. Using Eq. (1.1) of Section 1, prove Eq. (1.2) of Section 1.

9. Table 6.1 shows a dataset with the space-time coordinates and the corresponding attribute values. Compute the empirical variogram values by choosing appropriate spatial and temporal lags, and use a simple theoretical variogram model to fit to the empirical variogram. Then, using these results implement the leave-one-out cross-validation process and the six indicators (Table 4.1 of Section 4) to evaluate the accuracy of the *STOK* and *STSK* techniques. Given that the number of data is rather large, the complete Table 6.1 can be found on the website (https://www.dropbox.com/scl/fi/879542m39762mu2a5ybp8/Exercise-of-Ch-8.doc?dl=0&rlkey=4q7qp8p16vj5q3mjaudrzrfse)

10. In the analytical *STOK* Eqs. (2.13a, b) of Example 2.1 of Section 2, assume that interpolation involves a linear space-time variogram $\gamma_X(h,\tau) = ah\tau$, $h = |\boldsymbol{h}|$. How much will the actual interpolated attribute value and the associated error change if instead of the actual coordinate system *CS*, a convenient system *CS'* is used that underestimates the actual distance h between pairs of points by $\varepsilon\%$, on the average.

11. Assume that in the presence of data noise, the measurement model has the form $Y(\boldsymbol{p}_i) = X(\boldsymbol{p}_i) + U(\boldsymbol{p}_i)$, where $i = 1, \ldots, m$, $X(\boldsymbol{p}_i)$ is the actual attribute value, $Y(\boldsymbol{p}_i)$ is the measured value, and $U(\boldsymbol{p}_i)$ is the measurement error which in *CTDA* studies

TABLE 6.1 Dataset of Exercise 9.

s_1	s_2	t	X	s_1	s_2	t	X	s_1	s_2	t	X
1,480,997	4,458,841	1	47	1,353,941	4,322,969	3	99	1,154,884	4,339,393	5	166
1,472,582	4,468,904	1	47	1,302,487	4,402,917	3	99	1,161,574	4,340,171	5	169
1,091,088	4,323,697	1	51	1,346,150	4,373,789	3	107	1,007,277	4,345,440	5	174
935,338	4,140,374	1	55	1,353,065	4,294,132	3	110	1,016,565	4,198,685	5	185
1,408,718	4,458,456	1	58	1,163,981	4,414,801	3	112	1,100,370	4,112,158	5	190
1,320,434	4,303,084	1	59	1,354,169	4,291,400	3	113	1,144,570	4,092,021	5	193
1,516,325	4,435,803	1	67	1,231,788	4,341,547	3	115	964,095	4,279,696	5	194
1,009,922	4,262,892	1	71	1,123,981	4,400,626	3	116	1,129,094	4,112,649	5	195
1,368,258	4,465,443	1	74	1,237,623	4,340,333	3	120	1,031,332	4,294,803	5	196
1,052,407	4,311,959	1	75	1,161,574	4,340,171	3	125	1,062,032	4,311,048	5	197
…											

is commonly considered a white noise with zero mean and variance σ_U^2. Show that in this case the *STOK* system must be modified by subtracting the term σ_U^2 ($j = 1, \ldots, m$) on the left side of Eq. (2.1a) of Section 2.

12. As was shown in the text, the *STOK* system of equations can be written in matrix form as

$$\begin{aligned}\boldsymbol{\Gamma\lambda} + \mathbf{1}\mu &= \boldsymbol{\gamma},\\ \mathbf{1}^{\mathrm{T}}\boldsymbol{\lambda} &= 1,\end{aligned}$$

where $\mathbf{1}^{\mathrm{T}} = [1 \ldots 1]$, $\boldsymbol{\lambda} = [\lambda_1 \cdots \lambda_m]^{\mathrm{T}}$, $\boldsymbol{\gamma} = [\gamma_X(\Delta\boldsymbol{p}_{k1}) \ldots \gamma_X(\Delta\boldsymbol{p}_{km})]^{\mathrm{T}}$, and

$$\boldsymbol{\Gamma} = \begin{bmatrix} \gamma_X(\mathbf{0}) & \gamma_X(\Delta\boldsymbol{p}_{12}) \cdots & \gamma_X(\Delta\boldsymbol{p}_{1m}) \\ \gamma_X(\Delta\boldsymbol{p}_{21}) & \gamma_X(\Delta\boldsymbol{p}_{22}) \ldots & \gamma_X(\Delta\boldsymbol{p}_{2m}) \\ \vdots & \vdots & \vdots \\ \gamma_X(\Delta\boldsymbol{p}_{m1}) & \gamma_X(\Delta\boldsymbol{p}_{m2}) \ldots & \gamma_X(\mathbf{0}) \end{bmatrix}.$$

Show that the *STOK* system can also be written as (Fig. 2.1 of Section 2)

$\boldsymbol{\Gamma}^*\boldsymbol{\lambda}_\mu = \boldsymbol{\gamma}_1$, where $\boldsymbol{\lambda}_\mu = [\boldsymbol{\lambda}\,\mu]^{\mathrm{T}} = [\lambda_1 \cdots \lambda_m\,\mu]^{\mathrm{T}}$, $\boldsymbol{\gamma}_1 = [\boldsymbol{\gamma}\,1]^{\mathrm{T}} = [\gamma_X(\Delta\boldsymbol{p}_{k1}) \ldots \gamma_X(\Delta\boldsymbol{p}_{km})\;1]^{\mathrm{T}}$, and

$$\boldsymbol{\Gamma}^* = \begin{bmatrix} \boldsymbol{\Gamma} & \mathbf{1} \\ \mathbf{1}^{\mathrm{T}} & 0 \end{bmatrix} = \begin{bmatrix} \gamma_X(\mathbf{0}) & \gamma_X(\Delta\boldsymbol{p}_{12}) \cdots & \gamma_X(\Delta\boldsymbol{p}_{1m}) & 1 \\ \gamma_X(\Delta\boldsymbol{p}_{21}) & \gamma_X(\mathbf{0}) & \ldots \gamma_X(\Delta\boldsymbol{p}_{2m}) & 1 \\ \vdots & \vdots & \vdots & \vdots \\ \gamma_X(\Delta\boldsymbol{p}_{m1}) & \gamma_X(\Delta\boldsymbol{p}_{m2}) \ldots & \gamma_X(\mathbf{0}) & 1 \\ 1 & 1 \;\ldots & 1 & 0 \end{bmatrix}.$$

13. The *STSK* system of equations can be written in matrix form as

$$\boldsymbol{\Gamma\lambda} + c_0(\mathbf{1}^{\mathrm{T}}\boldsymbol{\lambda} - 1)\mathbf{1} = \boldsymbol{\gamma}$$

Can the *STSK* system above also be written in the more concise form $\boldsymbol{\Gamma}_?\boldsymbol{\lambda}_? = \boldsymbol{\gamma}_?$, and what are the detailed forms of $\boldsymbol{\Gamma}_?$, $\boldsymbol{\lambda}_?$, and $\boldsymbol{\lambda}_?$ in this case?

14. Consider Example 2.1 of Section 2. (a) Show that when the variograms between the data points $\boldsymbol{p}_i$ and the interpolation point $\boldsymbol{p}_k$ are such that $\gamma_{k1} = \gamma_{k2} = \gamma_{k3} = \gamma_k$, the λ_i weights ($i = 1,2,3$) and the interpolation $\hat{X}(\boldsymbol{p}_k)$ do not depend on γ_k, whereas the multiplier μ and the interpolation error $e_X^2(\boldsymbol{p}_k)$ depend on γ_k; (b) when in addition it holds that $\gamma_{12} = \gamma_{23} = \gamma_{13} = \gamma$, find the corresponding

λ_i weights $(i = 1,2,3)$, multiplier μ, interpolation $\hat{X}(\boldsymbol{p}_k)$ and interpolation error $e^2_X(\boldsymbol{p}_k)$.

15. If the nugget is interpreted as fine-scale (microscale) variability, what is the effect of slight point changes on the corresponding changes in *STOK* interpolations, and why?

16. Examine the following situations:

(a) The arrangement of Fig. 6.1 consists of the data points $\boldsymbol{p}_1 = (\boldsymbol{s}_1; t_1)$, $\boldsymbol{p}_2 = (\boldsymbol{s}_2; t_2)$, $\boldsymbol{p}_3 = (\boldsymbol{s}_3; t_3)$ and the interpolation point $\boldsymbol{p}_k = (\boldsymbol{s}_k; t_k)$. Let $\gamma_X(\boldsymbol{h}, \tau) = \gamma_S(\boldsymbol{h}) + \gamma_T(\tau)$ be a space-time separable variogram model. Show that the *STOK* interpolation at point $\boldsymbol{p}_k$ depends only on the datum at point $\boldsymbol{p}_2$.

(b) More generally, if a space-time separable variogram model $\gamma_X(\boldsymbol{h}, \tau) = \gamma_S(\boldsymbol{h}) + \gamma_T(\tau)$ is assumed and the same set of locations are sampled at every instance considered, does the *STOK* interpolator at the point of interest depend only on data at the same instance?

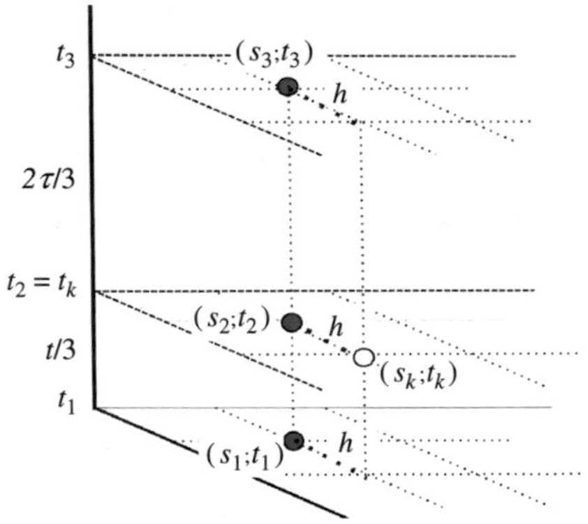

FIG. 6.1 Arrangement with data points $\boldsymbol{p}_1 = (\boldsymbol{s}_1;\ t_1)$, $\boldsymbol{p}_2 = (\boldsymbol{s}_2;\ t_2)$, $\boldsymbol{p}_3 = (\boldsymbol{s}_3;\ t_3)$ and interpolation point $\boldsymbol{p}_k = (\boldsymbol{s}_k; t_k)$.

17. Consider the chronotopologic arrangement of Fig. 6.1 and let $\gamma_X(\boldsymbol{h}, \tau) = c_S(0)\gamma_T(\tau) + \gamma_S(\boldsymbol{h})$ $c_T(0) - \gamma_S(\boldsymbol{h})\gamma_T(\tau)$ be a space-time separable variogram model. Derive analytically the *STOK* interpolation at point $\boldsymbol{p}_k = (\boldsymbol{s}_k; t_k)$.

18. Let $e^2_X(\boldsymbol{p})$ be the STSK error variance of the $X(\boldsymbol{p})$ attribute interpolation. Show that $e^2_X(\boldsymbol{p}) \leq \sigma^2_X(\boldsymbol{p})$.

CHAPTER

9

Chronotopologic BME estimation

1 Epistemic underpinnings

Just as any other product of scientific reasoning, the chronotopologic estimation[a] of natural attributes and the subsequent mapping ought to be based on a sound epistemic[b] framework or paradigm. Accordingly, the adoption of an epistemic framework of chronotopologic estimation involves the following postulate.

Understanding the underlying **epistemology** can enlighten considerably one's mathematical investigations for the best (formally and substantively) chronotopologic estimation approach possible.

The epistemic perspective is a necessary complement to technical advances that constitute the vast majority of *CTDA* developments nowadays. The following epistemic characteristics of *BME*-based chronotopologic estimation are worth discerning[c]:

① Epistemically important qualities are those that can be determined **before the event**, i.e., before the attribute represented by the map actually takes place.

② A number of issues in statistical inference theory can be illuminated by the *BME* perspective: it is not only the technical inferences that are of concern to the investigator, but also **experiential**

[a] This chapter refers to attribute estimation rather than mere interpolation (as was the case of the previous Chapter 8), since, by incorporating core knowledge in the form of physical laws and scientific theories, *BME* can go beyond mere interpolation into the domains of extrapolation and prediction (recall that the term "estimation" includes interpolation and extrapolation or prediction, as was discussed in Section 4 of Chapter 1).

[b] The following point should be stressed regarding the meaning of the term "epistemic" in the *CTDA* setting: viewed as an epistemic process in the realm of contemplation, chronotopologic estimation adopts a certain cognitive attitude toward data and assumptions leading to the generated map, whereas viewed as an ordinary process in the realm of worldly action, chronotopologic estimation generates a map without the need of contemplation (i.e., generating a map as one would if one knew that the data and assumptions were surely true).

[c] The relevant *BME* theory was discussed in Chapter 6.

https://doi.org/10.1016/B978-0-12-816341-2.00015-0

inferences (i.e., inferences concerning the investigator's ordinary experience).

③ To transform **facts** into useful information takes particular knowledge and skill. Metaphorically speaking, facts would be seen as ventriloquists' dummies: sitting on a real expert's knee, they may be made to utter words of substance; otherwise they say nothing important or they talk nonsense.

④ Chronotopologic attribute estimation constitutes a **multidisciplinary** endeavor involving various sciences, not the province of pure a priori statistics. In this setting, new routes open when one type of data crosses another.

Putting these features together, attribute estimates and maps are sought that are the result of a process that is a priori *informative* (carrying as much information as possible), *well-supported* by evidence, and *posteriori cogent* (having a high posterior probability of being correct, rather than certain correctness, which is an aspect that cannot be guaranteed before the event). Due to the natural variations and uncertainties involved in the description of such natural attributes, both requirements of informativeness and cogency involve *conditional* probabilities relative to the different *KB* considered at each stage. Otherwise said, this double epistemic goal of *BME* would be summarized by the following narrative:

- *BME* aims at prior **information maximization** given *G* as well as posterior **probability maximization** given *S*.

It is with these two objectives, disparate as they may seem in some respects, that this chapter is chiefly concerned. The upshot of this twofold perspective is that at the prior stage attribute representations are sought that are maximally informative relative to *G*. Among these representations, we subsequently seek the ones that are highly probable, but we want them to achieve this posterior stage probability on the basis of *S* and not in terms of *G* alone (as was done at the prior stage).

The above paradigm is based on considerations that most philosophers of science will find acceptable. Along with the dominant tradition, we regard it as an evaluative or normative paradigm, not a purely descriptive one. This is a chronotopologic estimation paradigm that distinguishes between the three main stages of knowledge-gaining, processing, and interpretation presented in Chapter 6 (each stage will be discussed below), in which case the following correspondence should be made:

- In the present mapping setting, the **point vector** $\boldsymbol{p}_k = [p_{k1} \cdots p_{kn_k}]^T$ in Table 1.1 (Section 1 of Chapter 6) specifically corresponds to points where chronotopologic attribute estimates are sought.

Ad rem, the epistemic arguments described above directly impose quantitative constraints on the chronotopologic attribute estimation techniques considered. This issue is discussed next.

2 Mathematical developments

Before proceeding further, it is *ad sum* worth formulating a chronotopologic estimation problem of wide interest in natural sciences[d]:

> ***BME* estimation problem**: Generate chronotopologic estimates $\hat{X}(\boldsymbol{p})$ of the unknown attribute $X(\boldsymbol{p})$ values at points of interest, where $X(\boldsymbol{p})$ is associated with a set of core *KB* functions $g_\alpha(\chi_{map})$ as in Eqs. (2.1) and (2.2) of Section 2 of Chapter 6 and a set of specificatory *KB* $\chi_d = (\chi_h, \chi_s)$ as in Eqs. (2.20) and (2.21) of Section 2 of Chapter 6.

[d] The readers may find it useful to recall the notation in Table 7.1, Section 7 of Chapter 1.

It is noteworthy that the above formulation encompasses single-point and multipoint considerations. In particular:

① **Single-point** analysis deals with an estimate $\hat{\chi}_k$ of the actual attribute value χ_k at a single point $\boldsymbol{p}_k$.
② **Multipoint** analysis deals with the simultaneous derivation of estimates $\hat{\chi}_{k1}$, ..., $\hat{\chi}_{km_k}$ of the actual attribute values χ_{k1}, ..., χ_{km_k} at points $\boldsymbol{p}_{k1}, \ldots, \boldsymbol{p}_{km_k}$, where $m_k \leq n_k$.[e]

Although theoretical *BME* (Chapter 6) was set up in a way that readily includes the multipoint case, yet, for practical reasons single-point estimation is considered here. In most applications, the mapping points lie on the nodes of a chronotopologic grid.

2.1 *BME* chronotopologic estimation methodology

Let us now examine the epistemic ideals of the previous section in light of the estimation problem outlined above. The readers should keep in mind that among the main objectives of *BME* are to organize the investigator's thinking about the phenomenon of interest (*epistemic* component) and to predict reality as accurately as possible given the available objective knowledge (*ontic* component). Next, we consider the discussion in Section 3 of Chapter 6, in the present chronotopologic mapping setting.

2.1.1 Prior stage: Core knowledge processing

As we saw in Section 3 of Chapter 6, at the *prior* stage, the probability function considered is relative to the core knowledge *G*, i.e., in the mapping case,

$$P_G\left[\boldsymbol{\chi}_{map}\right] = p \in [0, 1], \tag{2.1}$$

which reads that "the probability of the map $\boldsymbol{\chi}_{map} = (\boldsymbol{\chi}_d, \boldsymbol{\chi}_k)$ given *G* is *p*." Another way of describing the meaning of Eq. (2.1) is by stating that, at this stage, the probability judgments about $\boldsymbol{\chi}_{map}$ are relative to the available core knowledge *G*.

❶ The prior mapping probability of Eq. (2.1) is constructed based on the available **core knowledge** ***G***, i.e., it is part of the *BME* epistemic process.

Example 2.1 A well-known situation of core knowledge processing is unconditional simulation producing various field realizations on the basis of its mean and covariance. While useful in the broad characterization of chronotopologic variability, an unconditional simulation may be of limited value for mapping purposes.

One may find it intriguing that Eq. (2.1) can be also interpreted in terms of the *complementarity* idea (Section 1 of Chapter 6): given *G*, the probability function in Eq. (2.1) measures the relative number of random fields realizations in which $\boldsymbol{\chi}_{map}$ occurs over all possible realizations. The operational features of this *BME* stage are, though, fundamentally different than the ones considered in the classical geostatistics literature (including most of the developments in Chapter 8).

As has been discussed earlier (Section 3.1 of Chapter 6), the following inverse relationship holds between the notions of information and probability:

- The prior mapping stage assumes an **inverse relationship** between the information provided by the map and the probability of map occurrence: the more informative a map is, the less probable it is to occur.

[e] While in single-point analysis individual estimates are obtained at points $\boldsymbol{p}_{k1}, \ldots, \boldsymbol{p}_{kn_k}$, one point at a time, in multipoint analysis joint estimates are obtained simultaneously at points $\boldsymbol{p}_{k1}, \ldots, \boldsymbol{p}_{km_k}$.

This relationship actually describes a standard epistemic rule: the more vague and general a theory is, the more alternatives it includes (it is, thus, more probable), but also the less informative it is. Conversely, the more alternatives a theory excludes, the more informative (less probable) it is. To put it succinctly:

The more a theory **forbids**, the more it tells us.

In quantitative terms, the information may be expressed as (Table 3.1 of Section 3 of Chapter 6) $Info_G[\chi_{map}] = P_G^{-1}[\chi_{map}]$. For technical reasons, given that probabilities can be very small, it is often convenient to work with logarithms so that the information is expressed as (see Eq. (3.1) of Section 3, Chapter 6),

$$Info_G\left[\chi_{map}\right] = -\log P_G\left[\chi_{map}\right]. \tag{2.2}$$

Because the information we seek to maximize at the prior stage is conditioned to the available G, this stage is also known as the *core knowledge processing* stage.

2.1.2 Metaprior or preposterior stage: Evidential support

At this stage, we collect and organize specificatory knowledge S in appropriate quantitative forms that can be explicitly incorporated into the *BME* formulation. In the estimation (mapping) context, the *S-KB* often refers primarily to the dataset χ_d, which is also denoted as $\chi_d(S)$.[f]

❷ The quality and quantity of the hard and soft data collected at the metaprior or preposterior mapping stage is a matter of experimental investigation producing **evidential support** rather than part of the epistemic process of the prior stage.

As we discussed in Sections 2 and 3 of Chapter 6, in real-world applications an investigator may be dealing with a variety of possible *KB*, which could make the analysis at the metaprior stage, not a trivial task.

2.1.3 Posterior stage: Total knowledge processing

At this stage, the new probability function is relative to both the core knowledge G of the prior stage and the specification knowledge S of the preposterior stage, i.e.,

$$P_K[\chi_k] = p' \in [0, 1], \tag{2.3}$$

which means that the probability of estimates χ_k in light of total knowledge $K = G \cup S$ is p',[g] i.e., Eq. (2.3) offers a measure of the credibility or assertibility of this statement.

❸ The posterior probability of Eq. (2.3) is constructed based on the **total knowledge *K***.

Expressions like Eq. (2.3) assert reasoning relations between knowledge K and the attribute estimates χ_k. In statistical terms: given K, the probability of Eq. (2.3) provides a measure of the relative number of random attribute realizations in which χ_k occurs over all possible attribute realizations.

Example 2.2 A well-known case of specification knowledge processing in geostatistics is a conditional simulation, which—by incorporating a set of data—is of greater predictive value than unconditional simulation.

The probability functions of Eqs. (2.1) and (2.3) assume a connection between mapping and the available knowledge at each stage. In other words:

- The probabilities considered in stages 1 and 3 are epistemic (objective and relational) based on **theoretical** conceptions (prior), supported

[f] The symbol S in parentheses is occasionally used to indicate that the dataset χ_d is part of the *S-KB* considered at the metaprior (or preposterior) stage.

[g] The probability value p' of Eq. (2.3) is generally different than the value p of Eq. (2.1).

by **perceptual** data (stage 2) and related to **inductive** evidence (posterior).

2.1.4 Knowledge synthesis

The three-stage analysis above has left us with a final issue to be considered within the epistemic framework of modern geostatistics, namely:

> How should we **synthesize** knowledge from the prior and metaprior chronotopologic estimation stages all the way to the posterior estimation stage?

There are various ways to do this. A particularly efficient way is by means of the knowledge processing rule obtained by the following approach. At the posterior (*K*-based) stage it holds that

$$P_K[\chi_d(S)] = 1, \quad (2.4)$$

which expresses the obvious fact that given that $\chi_d(S)$ occurred, its probability is one. On the other hand, at the prior (*G*-based) stage, it is usually valid that $P_G[\chi_d(S)] < 1$. Furthermore, the following relation holds between conditional probabilities:

$$P_K[\chi_k|\chi_d(S)] = P_G[\chi_k|\chi_d(S)], \quad (2.5)$$

which simply means that the evidential impact of $\chi_d(S)$ on the chronotopologic attribute values χ_k is already fully assessed in assigning the *G-conditional* probability (as indicated by the conditioning *S-KB* in $\chi_d(S)$) and, hence, the fact that $\chi_d(S)$ actually occurred is no reason to change this assessment. In other words, Eq. (2.5) implies that the probability we assign to the attribute values χ_k *assuming* that $\chi_d(S)$ will turn out to be true is equal to the probability we would assign to χ_k on learning that $\chi_d(S)$ *indeed* turned out to be true. In light of Eq. (2.4), and by definition of the conditional probability, it is also valid that

$$P_K[\chi_k|\chi_d(S)] = \frac{P_K[\chi_k \wedge \chi_d(S)]}{P_K[\chi_d(S)]} = P_K[\chi_k] \quad (2.6)$$

(since $\chi_d(S)$ is contained in the *K-KB*). Taking into consideration Eqs. (2.5) and (2.6) yields the *Bayesian conditionalization* principle (see, also, Eq. (3.3) of Section 3, Chapter 6):

$$P_K[\chi_k] = P_G[\chi_k|\chi_d(S)] = \frac{P_G\left[\chi_{\text{map}}(S)\right]}{P_G[\chi_d(S)]} \quad (2.7)$$

(notice the different subscripts referring to the appropriate *KB*), which is essentially Eq. (3.15) of Section 3 of Chapter 6 considered in the present mapping context. This leads to the following update of the earlier assertion ❸:

❸′ The posterior mapping probability of Eq. (2.3) is related to the prior mapping probability of Eq. (2.1) by means of the relationship of Eq. (2.7), which is the **knowledge synthesis** rule sought.

An instructive analogy may be appropriate here: just as certain rules tell us how to measure distances or how to weigh objects, Eq. (2.7) tells us how to update one's evaluation of a situation given new knowledge. Lastly, while we seek posterior attribute estimates that are highly probable, we nevertheless want them to achieve this probability on the basis of specificatory knowledge and not in terms of core knowledge alone:

- This **cogency** requirement seeks the maximization of probability $P_K[\chi_k]$ with respect to χ_k, thus generating a posteriori maximally probable attribute estimates $\hat{\chi}_k$.

Given that the χ_{map} of the prior stage probability $P_G\left[\chi_{map}\right]$, Eq. (2.1), includes the attribute values at both the data points and the estimation points and the χ_k of the posterior stage probability $P_K[\chi_k]$, Eq. (2.3), refers only to the estimation points, Eq. (2.7) relates this new estimation probability $P_K[\chi_k]$with the prior map probability $P_G\left[\chi_{map}\right]$.

2.1.5 Posterior probability interpretation

In terms of the random field complementarity idea (Section 2 of Chapter 5), Eq. (2.7) may be given the following interpretation: If $\chi_d(S)$ does indeed entail χ_k, then $\chi_{map}(S)$ occurs in every attribute realization in which $\chi_d(S)$ occurs. But if χ_k is neither entailed by $\chi_d(S)$ nor inconsistent with it, then $\chi_{map}(S)$ occurs only in some of the possible attribute realizations in which $\chi_d(S)$ occurs (i.e., those in which χ_k also happen to occur). Therefore, one may take the ratio of the quantity of possible random field realizations in which $\chi_{map}(S)$ occurs to the quantity of realizations in which $\chi_d(S)$ occurs as determining the extent to which $\chi_d(S)$ entails χ_k and, thus, defining the probability of χ_k given $\chi_d(S)$, $P_G[\chi_k|\chi_d(S)]$. The following argument can then be made:

- From a scientific reasoning viewpoint the aim of the mapping paradigm above is to **constrain induction**.

As a matter of fact, this is the chief purpose of the *G*-constraints on information maximization at the prior mapping stage as well as the *S*-constraints on probability maximization at the posterior estimation stage. In this way, we can avoid generating innumerable fruitless maps in our search for useful generalizations.

Example 2.3 The past decade has seen the emergence of an increasing number of geomagnetic studies seeking to integrate dynamical models (*G-KB*) and geophysical observations (*S-KB*) to constrain the state of the Earth's core. Also, valuable findings have been the result of numerical simulations that integrate primitive equations.

2.1.6 *Self-reference issues*

At this point, an interesting observation to be made is that conventional attribute interpolation techniques based on statistical regression (Chapter 8) appear to be logically circular, in the sense that they start by using data to calculate the experimental covariance or variogram values, and then return to the same data to calculate attribute interpolations at unsampled points. This is known in logic as the *self-reference* problem.

Metaphorically speaking, it is like a rider on a merry-go-round ultimately coming back to the starting point, no matter how many valuable lessons the rider may have learned while swinging around the circuit. *BME*'s answer to the circularity problem is to incorporate physical laws at the prior stage and to assimilate data at the posterior stage. As a matter of fact, it can be argued that in most cases in which the physical laws are not known or considered, chronotopologic regression is essentially an educated interpolation.

Our discussion so far provided a general conceptual presentation of knowledge processing rules and standards in a chronotopologic estimation setting. It was necessary that some key *BME* ideas be made clear, and, also, it was important to make clear the methodological purpose why these ideas needed to be made clear. The derivation of an analytically tractable mathematical formulation of the *BME* conceptual rules and standards that is efficiently implementable in real-world applications is the concern of the following sections.

2.2 The chronotopologic *BME* estimation equations

The salient point of the previous section was that the proposed epistemic paradigm places modern geostatistics in a pragmatic framework, the central theses of which are as follows:

A chronotopologic mapping approach should be **context-dependent** (being guided by the *KB* objectively available), satisfy **logically plausible rules**, and always be **relevant** to the goals of the specific study.

2.2.1 *Formalization of the BME estimation*

Following the approach of Eq. (2.7), the solution of the chronotopologic estimation (and mapping) problem described in the previous section is readily obtained from the results presented in Table 3.2 of Section 3 of Chapter 6. In particular, for the present purposes the equations of Table 3.2 of Section 3 of Chapter 6 can be conveniently summarized as

$$\int d\chi \left(g(\chi_{map}) - \overline{g}\right) e^{\mu \cdot g(\chi_{map})} = 0,$$
$$\int d\Phi(\chi_s) f_S(\Phi(\chi_s)) f_G(\chi_{map}) = A f_K(\chi_k), \quad (2.8a, b)$$

where the coefficients μ_α ($\alpha = 0, \ldots, N_C$) of the vector $\boldsymbol{\mu}$ express quantitatively the importance of each *G-KB* function g_α in the estimation context.

Eq. (2.8a) can be solved with respect to the coefficients μ_α, which are then substituted in Eq. (2.8b) to obtain the attribute *PDF* $f_K(\chi_k)$ at all points $\boldsymbol{p}_k$ of interest. Metaphorically speaking, the *BME* passes the "Equations on a napkin" test based on the view that a mathematical theory is not worth its name unless it can be summarized on a napkin.

2.2.2 The BME types of estimators

Given that $f_K(\chi_k)$ has been obtained from the solution of the *BME* Eq. (2.8a, b) above, several kinds of chronotopologic estimates $\hat{\chi}_k$ can be derived. In particular:

- The *BME* **mode** estimate $\hat{\chi}_{k,mode}$: It is such that

$$f_K(\hat{\chi}_{k,mode}) = \max_{\chi_k} f_K(\chi_k), \quad (2.9a)$$

and it is computed as the solution of the equation $\frac{\partial}{\partial \chi_k} f_K(\chi_k)\Big|_{\chi_k = \hat{\chi}_{k,mode}} = 0$,[h] which can be written as

$$\int d\Phi(\chi_{soft}) f_S(\Phi(\chi_{soft})) \frac{\partial}{\partial \chi_k} e^{\sum_{\alpha=1}^{N_c} \mu_\alpha g_\alpha(\chi_{data}, \chi_k)}\Big|_{\chi_k = \hat{\chi}_{k,mode}} = 0. \quad (2.9b)$$

- *BME* **mean** estimate $\hat{\chi}_{k,mean}$: It is defined as

$$\hat{\chi}_{k,mean} = \overline{X(\boldsymbol{p}_k) | \chi_d} = \int d\chi_k f_K(\chi_k) \chi_k \quad (2.10)$$

- *BME* **median** estimate $\hat{\chi}_{k,med}$: It is defined as

$$\hat{\chi}_{k,med} = F_G^{-1}(0.5, \boldsymbol{p}_k) : \int_{\chi_k \leq \hat{\chi}_{k,med}} d\chi_k f_K(\chi_k) = \tfrac{1}{2} \quad (2.11)$$

- *BME* **percentile *p*** estimate $\hat{\chi}_{k,p}$: It is derived as the solution of

$$\int_{-\infty}^{\hat{\chi}_{k,p}} d\chi_k f_K(\chi_k) = p \quad (2.12)$$

Eqs. (2.9)–(2.12) convey an epistemic point of view as well as a body of knowledge. *Id est BME* attribute estimates above have a scientific warrant, and therefore epistemic justification. The choice of the most suitable *BME* estimate is application-dependent and relies on the successful coordination of three factors:

① Satisfactory **track record** of attribute estimation in similar situations.
② **Theoretical background** promoting its use against other possible estimates.
③ **Explanatory rationale** provided by the attribute estimate.

In addition to the attribute estimates above, the *BME* theory provides various assessments of the accuracy of these estimates, as follows.

2.2.3 The BME estimation uncertainty

Reporting the uncertainty of scientific conclusions is one of the hallmarks of *BME* thinking, which follows the narrative:

> It is a sensical approach that investigators should take only the **risks** they can measure.

In the case of chronotopologic estimation, these risks can be measured in terms of suitable indicators of the estimation uncertainty.

[h] Eq. (2.9a) is obtained from the probability maximization criterion $\frac{\partial}{\partial \chi_k} f_K(\chi_k)\Big|_{\chi_k = \hat{\chi}_{k,mode}} = 0$.

Generally, the definition of such indicators depends on the shape of the posterior *PDF* (symmetric *vs.* asymmetric, single maximum *vs.* multiple maxima, etc.). In particular, the estimation error can be computed in more than one way, as follows:

- **Symmetric** attribute *PDF*:

$$e_X(\boldsymbol{p}_k) = \left\{\overline{\left[X(\boldsymbol{p}_k) - \hat{X}(\boldsymbol{p}_k)\right]^2}\right\}^{\frac{1}{2}}. \quad (2.13)$$

- **Gaussian** attribute *PDF*:

$$e_X(\boldsymbol{p}_k) \approx \left[-\frac{d^2}{d\chi_k^2}\log f_K(\chi_k)\bigg|_{\chi_k=\hat{\chi}_k}\right]^{-\frac{1}{2}}. \quad (2.14)$$

- **Asymmetric** *PDF*:

$$e_X(\boldsymbol{p}_k) = SD[f_K(\chi_k)], \quad (2.15)$$

where, as usual, *SD* denotes standard deviation.

An uncertainty indicator can be defined in a straightforward manner when the *PDF* has a single maximum; if this is not the case, the attribute estimation uncertainty should be considered separately for each maximum. Joint confidence sets of the *BME* estimates, in particular, can be derived from the posterior *PDF*. A single-point confidence set is the familiar confidence interval (Christakos and Li, 1998). A multipoint confidence set requires a multidimensional plot; e.g., a contour map in the case of two points. A confidence set takes into consideration dependencies between all variables, as expressed by the posterior *PDF*. In some cases, however, the posterior *PDF* does not have a simple shape, which can make theoretical analysis difficult. In such cases, a numerical analysis may be the proper choice.

- A *BME* **index set** may be defined as

$$\Lambda_\beta = \left\{\chi_k:\ 1 - \frac{f_K(\chi_k)}{f_K(\hat{\chi}_k)} \le \beta\right\}, \quad (2.16)$$

where $0 \le \beta \le 1$. If the posterior *PDF* is Gaussian or can be approximated by a Gaussian *PDF*, numerically efficient expressions are available for Λ_β (Choi et al., 1998).[i]

We conclude with an interpretation of the uncertainty analysis above in the spirit of knowledge theory:

> Epistemically, Eqs. (2.13)–(2.16) may be seen as quantitative expressions of the investigator's **conscious ignorance**.

2.2.4 *Generalization power*

Due to the *BME*'s versatility, its basic equations possess significant generalization power on the basis of theory and data. As a consequence, the following essential *BME* features are of considerable interest in the *CTDA* context (Christakos, 1998a, b, 2000):

① Many of the existing statistical interpolation techniques (Kriging, regression) can be derived as **special cases** of the chronotopologic *BME* estimation theory.

② *BME* is a **nonlinear** chronotopologic estimator, in general.

③ It incorporates **multisourced knowledge** bases (core and site-specific).

④ It can process **higher-order statistics** associated with non-Gaussian data distributions, like skewness and trivariogram functions (Christakos, 2017).

⑤ It has **extrapolation** (or **prediction**) abilities due to its incorporation of physical laws and other forms of core knowledge.

⑥ It handles **functional**, **vector,** and **multipoint** chronotopologic estimation cases.

⑦ It is **computationally** efficient, particularly in chronotopologic estimation cases where closed-form analytical expressions are available.

[i] Several other indicators of the mapping uncertainty can be derived from the basic *BME* equations, but are not discussed here due to space limitations. The interested readers are suggested to consult the relevant literature on the subject.

Feature ① is one of the essential traits of a novel theory that successfully replaces old ones (a new theory should preserve earlier ones that remain valid under limiting conditions). Features ② and ③ are in contrast to mainstream Kriging interpolators (Chapter 8), which are, in principle, restricted to a linear combination of hard data and do not provide a systematic mechanism to incorporate soft data and prior information. Although linearity is a potentially serious drawback of Kriging, among the euphoria that its simplicity generates, it is often overlooked. Feature ④ is useful in practice when the initial non-Gaussian data distributions cannot be transformed into a Gaussian one, in which case the data distribution needs to be represented in terms of higher-order statistics. *Ad pondus omnium*, feature ⑤ is very important in chronotopologic attribute estimation requiring the involvement of scientific principles and theories. Feature ⑥ refers to phenomena where the focus

is not a single attribute but a set of related attributes, a function of an attribute, or the joint chronotopologic dependency of attribute values at various points. Feature ⑦ is particularly valid in chronotopologic estimation cases where closed-form analytical expressions are available.

Noticeably, the *BME* generalization may occur at several levels of data availability that extend from abundance to near absence, in which case due caution is required. At the abundance level, careful generalization can take place, whereas at the near absence level a generalization is considerably riskier and may take on the nature of a hypothesis.

2.2.5 A BME estimation outline

As with the Kriging techniques in Chapter 8 (Figs. 2.1 and 2.6 of Section 2), the *BME* process can be graphically outlined as in Fig. 2.1, which is, essentially, a graphical representation of the preceding discussion.

G-KB $\quad g(\chi_{map})$

S-KB $\quad \chi_h$ and χ_s

BME grid point system $\quad \int d\chi \, (g(\chi_{map}) - \bar{g}) e^{\mu \cdot g(\chi_{map})} = 0$

BME grid point
- **PDF**
- **estimated value**
- **associated error**

$$f_K(\chi_k) = A^{-1} \int d\Phi(\chi_s) f_S(\Phi(\chi_s)) f_G(\chi_{map})$$
$$\hat{X}(p_k) = \hat{\chi}_{k,mean}, \; \hat{\chi}_{k,mod}, \; \hat{\chi}_{k,med}, \; \hat{\chi}_{k,p}$$
$$e_X(p_k) = SD[f_K(\chi_k)]$$

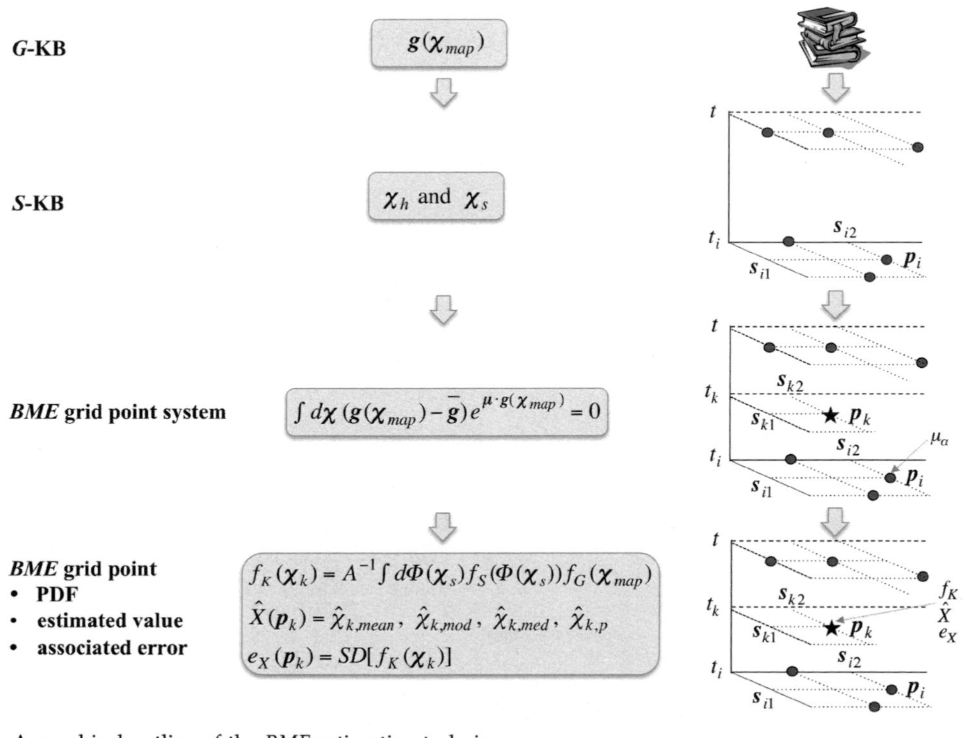

FIG. 2.1 A graphical outline of the *BME* estimation technique.

In view of the discussion so far, for the formal *BME* estimation approach to be of scientific interest, it should possess two key characteristics:

❶ It should be based on a **logical** process.
❷ It should be written in **computational** terms that link formal representation with natural realization.

We have already discussed the significance of the first characteristic. The second characteristic provides sufficient motivation for addressing certain practical issues concerning the computational *BME* implementation.

Example 2.4 Consider the arrangement of Example 3.4, Section 3 of Chapter 6, where now $\boldsymbol{p}_k$ is an estimation point in the chronotopologic domain.

Steps 1–6: As in Example 3.4, Section 3 of Chapter 6.

Steps 7: The *BME* mode estimate $\hat{\chi}_{k,\,\mathrm{mod}}$ at point $\boldsymbol{p}_k$ is the solution of the equation

$$\frac{d}{d\chi_k} f_K(\chi_k)\Big|_{\chi_k=\hat{\chi}_{k,\,mod}} = 0,$$

where $f_K(\chi_k)$ is given by Eq. (3.29), Section 3 of Chapter 6, or

$$\int_{62}^{65} d\chi_3 e^{-29.8875-1.1149(\chi_k-65)^2+0.6006(\chi_3-65)(\chi_k-65)-1.0188(\chi_3-65)^2} \frac{d\left[-29.8875-1.1149(\chi_k-65)^2+0.6006(\chi_3-65)(\chi_k-65)-1.0188(\chi_3-65)^2\right]}{d\chi_k}\Bigg|_{\chi_k=\hat{\chi}_{k,\,mod}}$$

$$= \int_{62}^{65} d\chi_3 \left[0.6006(\chi_3-65) - 2\times 1.1149\left(\hat{\chi}_{k,\,\mathrm{mod}}-65\right)\right] e^{Q(\mu_i,\chi_i)}$$

$$= -2.2298\left(\hat{\chi}_{k,\,\mathrm{mod}}-65\right)$$

$$-\frac{0.6006}{2\sqrt{1.0188\pi}}\left[\frac{e^{-\frac{0.3607}{4\times1.0188}\left[\left(\hat{\chi}_{k,\,\mathrm{mod}}-65\right)\right]^2} - e^{-1.0188\left[-3-\frac{0.6006}{2\times1.0188}\left(\hat{\chi}_{k,\,\mathrm{mod}}-65\right)\right]^2}}{\sqrt{e^{-1.0188\left[-3-\frac{0.6006}{2\times1.0188}\left(\hat{\chi}_{k,\,\mathrm{mod}}-65\right)\right]^2} - e^{-1.0188\left[-\frac{0.6006}{2\times1.0188}\left(\hat{\chi}_{k,\,\mathrm{mod}}-65\right)\right]^2}}} + \frac{\left(\hat{\chi}_{k,\,\mathrm{mod}}-65\right)}{\sqrt{1.0188}}\right] = 0,$$

which gives the space-time estimate $\hat{\chi}_{k,mod} = 65.1099\ \mu\mathrm{g/m}^3$.

Accordingly, although the above theoretical developments serve as a guide to what constitutes a *BME* estimation perspective, there are still certain practical issues to be addressed. This is because in practice the investigator operates in the intersection between the physical world and the digital world, the former constraining the flexibility of the latter. *A posse ad esse*, some of these computational issues will be investigated next.

The following numerical examples present the *step-by-step* process of computing the ***BME* mode** attribute estimate and the associated estimation error in practice.

Example 2.5 Consider the arrangement of Example 3.5, Section 3 of Chapter 6, where now $\boldsymbol{p}_k$ is an estimation point in the space-time domain.

Steps 1–6: As in Example 3.5, Section 3 of Chapter 6.

Steps 7: The *BME* mode estimate $\hat{\chi}_{k,\,\mathrm{mod}}$ at point $\boldsymbol{p}_k$ is the solution of the equation

$$\frac{d}{d\chi_k} f_K(\chi_k)\Big|_{\chi_k=\hat{\chi}_{k,\,mod}} = 0,$$

where $f_K(\chi_k)$ is given by Eq. (3.35), Section 3 of Chapter 6, or

$$\int_{62}^{65} d\chi_3 e^{-3817.775+117.779\chi_k-0.8983\chi_k^2-0.827(\chi_3-\chi_k)^2} \left.\frac{d\left[-3817.775+117.779\chi_k-0.8983\chi_k^2-0.827(\chi_3-\chi_k)^2\right]}{d\chi_k}\right|_{\chi_k=\hat{\chi}_{k,mod}}$$

$$=\int_{62}^{65} d\chi_3 e^{Q(\mu_i,\chi_i)}\left[117.779-2\times0.8983\hat{\chi}_{k,mod}+2\times0.827(\chi_3-\hat{\chi}_{k,mod})\right]$$

$$=117.779-2\times0.8983\hat{\chi}_{k,mod}-\sqrt{\frac{0.827}{\pi}}\frac{e^{-0.827(65-\hat{\chi}_{k,mod})^2}-e^{-0.827(62-\hat{\chi}_{k,mod})^2}}{\sqrt{1-e^{-0.827(65-\hat{\chi}_{k,mod})^2}}-\sqrt{1-e^{-0.827(62-\hat{\chi}_{k,mod})^2}}}=0,$$

which gives the space-time estimate $\hat{\chi}_{k,mod}=65.3818\ \mu g/m^3$.

The next example examines the *BME* estimation results obtained using different theoretical covariance conceptions in the context of the *G-KB*.

Example 2.6 The synthetic experimental setup of Section 5, Chapter 6 is revisited. The relative estimation performance of the three kinds of covariance conceptions previously discussed (namely, the *ordinary* covariance c_X, the *sysketogram-based* covariance $c_X^{(\beta)}$, and the *contingogram-based* covariance $c_X^{(\psi)}$) is presented numerically in Table 2.1. Attribute estimates generated by *BME* using the $c_X^{(\psi)}$ model showed the best accuracy performance in both *MAE* and *RMSE* terms, followed by the $c_X^{(\beta)}$ and the c_X models. The improved performance of the $c_X^{(\beta)}$ and $c_X^{(\psi)}$ models is explained as follows.[j] Sysketogram and contingogram values may be interpreted either as "compressed" (reduced) or as "stretched" (enlarged) covariance values (sysketogram indicates a higher level of covariance compression than contingogram), which benefits $c_X^{(\beta)}$ and $c_X^{(\psi)}$ model fitting. This situation further influences estimation because it is the high covariance values (indicating strong dependency at short space and time lags) that are usually accounted for by the estimation process. Hence, its performance in terms of the sysketogram and contingogram is better than that in terms of the standard covariance, in this case. On the other hand, the sysketogram and contingogram appear to "stretch" the corresponding empirical covariance values ranging as $\hat{c}_X \in (0.917, 1)$ and $\hat{c}_X \in (0.618, 1)$, respectively. The stretching results in a better fitting of the $c_X^{(\beta)}$ and $c_X^{(\psi)}$ models, as well as an improved c_X fitting at small space and time lags (associated with the high dependency c_X part) that leads to more accurate estimation as mentioned earlier (the stronger the dependency of the attribute, the more accurate the estimation). This is, perhaps, the reason that using the contingogram implies better estimation performance than using the sysketogram, in this case.

TABLE 2.1 Estimation accuracy using the ordinary, the sysketogram-based and the contingogram-based covariance models.

Prediction based on:	*MAE* (×10^{-3})	*RMSE* (×10^{-3})
c_X	6.6702	12.4794
$c_X^{(\beta)}$	6.6439	12.4483
$c_X^{(\psi)}$	6.6342	12.4382

The preceding examples point up some problems that an investigator may encounter when implementing *BME* in complicated real-world studies:

① Chief among these problems is the **integral computation**.

In the above examples, the integral calculation was analytically possible, yet, in most

[j] The readers may recall the relevant discussion in Section 5 of Chapter 6.

realistic situations, it can only be done numerically.

② Another problem is the adequate **formalization** of soft data sources, the level of difficulty in doing so depending on the particular case study.

The above implementation issues will be revisited in other parts of this book (the interested readers may also consult the relevant *BME* literature).

TABLE 3.1 A summary of *BME* mapping formulas used in the case studies.

***BME* estimator and associated error**	**Formula**
Mean	$\hat{\chi}_{k,mean} = \int d\chi_k \chi_k f_K(\chi_k)$
Mode	$\hat{\chi}_{k,\text{mode}} = \max_{\chi_k} f_K(\chi_k)$
Percentile (quantile)	$\hat{\chi}_k = \chi_k^p : \int_{-\infty}^{\chi_k^p} d\chi_k f_K(\chi_k) = p$
Error variance	$\sigma_X^2 = \int d\chi_k (\chi_k - \hat{\chi}_{k,\text{mean}})^2 f_K(\chi_k)$

3 An overview of real world *BME* case studies

Let us recall, once more, that the main goal of many real-world case studies is to estimate the values of natural attributes at unsampled points within the chronotopologic domain of interest. In doing so the following objective is central.

Investigators seek a better understanding of the phenomenon of interest with the help of **maps** of its chronotopologic attribute distributions.

As we saw in previous sections, according to the *BME* approach, following the derivation of the posterior *PDF* at the mapping points, these *PDF* are used to calculate the expected attribute values (mean, mode, and percentiles) at these points, together with the associated estimation errors (see Table 3.1 for a convenient summary).

A few selected case studies follow, which illuminate a number of *BME* implementation issues in the real world and, also, they demonstrate the considerable versatility of the *BME* approach.

3.1 A study of climate-*HFRS* associations

Potential associations between climatic conditions and *hemorrhagic fever with renal syndrome* (*HFRS*) are of serious environmental health concern. The investigation of the chronotopologic features of such associations was the subject of the *CTDA* study by He et al. (2018).

3.1.1 The *HFRS* dataset

A recently completed dataset was originally used by He et al. (2018) consisting of monthly *HFRS* cases in 127 Eastern China cities during January 2005–December 2016. The dataset was collected by the China Information System for Disease Control and Prevention (*CISDCP*). These cities are distributed in 19 provinces, autonomous regions, and metropolitan areas in Eastern China with a total area of approximately 2,820,000 km^2, see Fig. 3.1. This figure shows that the northeastern and western parts of the study region have a high number of *HFRS* infection cases.[k]

3.1.2 Methodology and result

The *multivariate El Niño southern oscillation index* (*MEI*)[l] is regarded as a proxy describing global climate dynamics quantitatively, especially in terms of worldwide temperature and precipitation levels. An outline of the method used in this case study is given in Fig. 3.2. The

[k] The quantitative analysis in this work was performed in terms of the number of *HFRS* cases.

[l] The *MEI* is computed in terms of six oceanic and meteorologic variables: sea level pressure, zonal and meridional components of surface wind, sea surface temperature, surface air temperature, and total cloudiness fraction of sky. As regards *MEI* interpretation, large positive *MEI* values indicate the presence of El Niño phenomena, whereas large negative *MEI* values indicate the presence of La Nina phenomena (Wolter and Timlin, 2011).

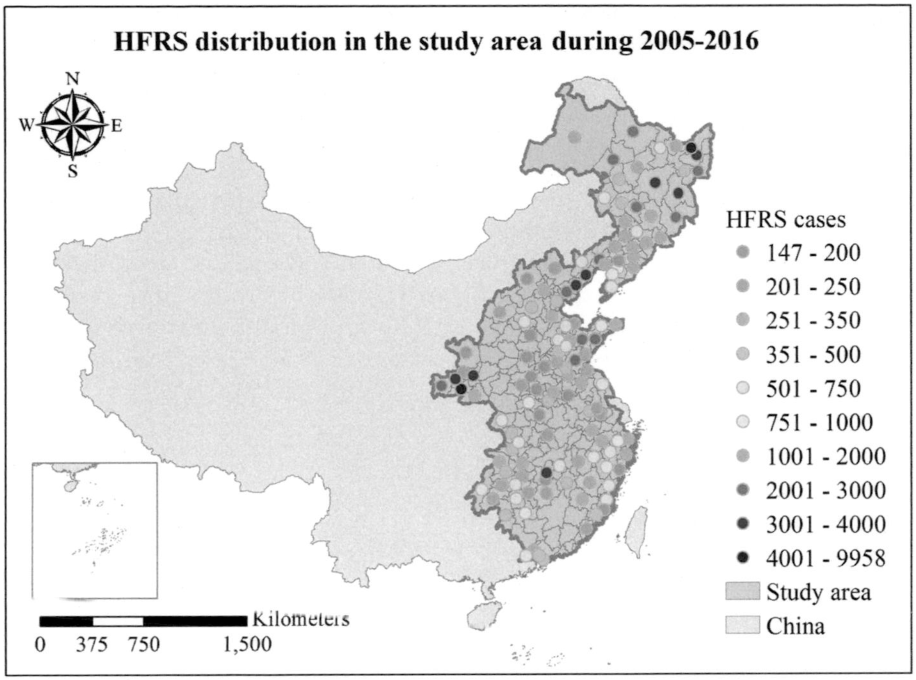

FIG. 3.1 Distribution of *HFRS* cases in the Eastern China study region during the period 2005–16.

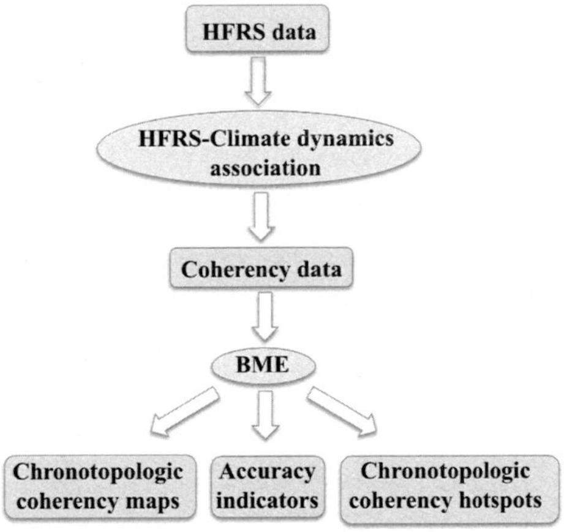

FIG. 3.2 An outline of the Eastern China *HFRS* case study methodology.

corresponding computational steps are as follows:

Step 1: As climate has been associated with *HFRS* spread and outbreaks, the *MEI* was used to assess the association between *HFRS* cases at each Chinese city in terms of *wavelet coherency analysis*. The chronotopologic coherency variation (strength of *MEI-HFRS* association) in conditions of *in-situ* uncertainty is mathematically represented by the random field $C(\boldsymbol{p})$, where

$\boldsymbol{p} = (\boldsymbol{s};t)$ denotes the geographical coordinates $\boldsymbol{s} = (s_1, s_2)$ and time instant t, so that

$$\text{Chronotopologic domain } \boldsymbol{p}: \quad \begin{cases} \boldsymbol{s} = (s_1, s_2) \in \text{Eastern China region} \\ t \in 2005-2016 \text{ time period.} \end{cases} \tag{3.1}$$

The internal covariation between two time series was expressed quantitatively in terms of the synchronicity strength of the trends of the two series. This implied that the larger the coherency value, the stronger the association between climate dynamics and *HFRS* infections. At each city a separate coherency time series was generated to characterize the temporal coherency variability, thus producing coherency data that were distributed across space and time.

Step 2: The *S-KB* consisted of the coherency $C(\boldsymbol{p})$ dataset described above. The calculated empirical coherency covariance values shown in Fig. 3.3 imply that the coherency between *HFRS* and global climate dynamics is spatially dependent and temporally sustained. The

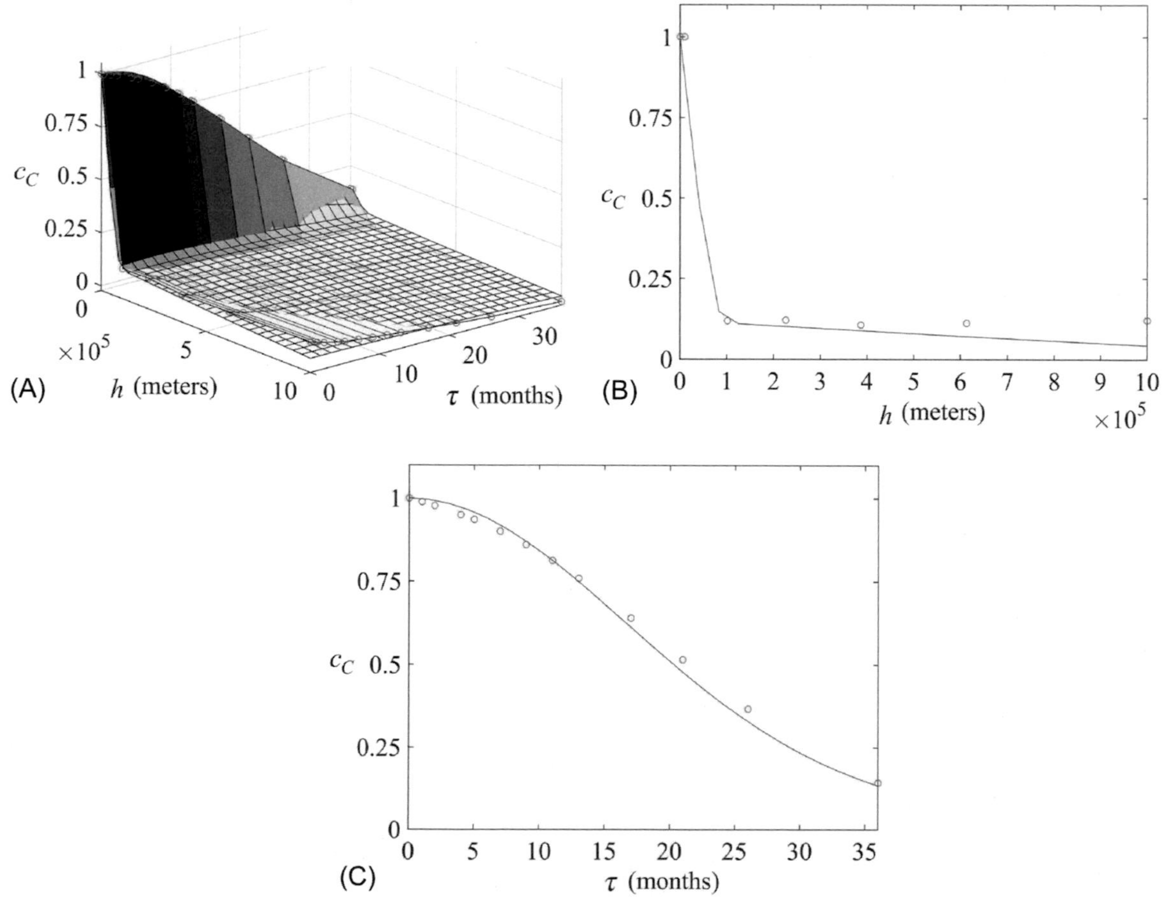

FIG. 3.3 (A) Empirical space-time coherency covariance values (dots) and fitted theoretical model (surface). (B) Marginal empirical spatial covariance values (dots) and fitted model (continuous line). (C) Marginal empirical temporal covariance values (dots) and fitted model (continuous line).

G-KB consisted of the theoretical coherency mean $\overline{C}(\boldsymbol{p})$ and the theoretical covariance model representing space-time coherency variation,

$$c_C(|\boldsymbol{h}|,\tau)=0.88\left[1-\frac{3|\boldsymbol{h}|}{2\times10^5}+\frac{1}{2}\left(\frac{|\boldsymbol{h}|}{10^5}\right)^3\right]e^{-\left(\frac{\sqrt{3}\tau}{40}\right)^2}+0.12\left[1-\frac{3|\boldsymbol{h}|}{4.4\times10^6}+\frac{1}{2}\left(\frac{|\boldsymbol{h}|}{2.2\times10^6}\right)^3\right]e^{-\left(\frac{\sqrt{3}\tau}{72}\right)^2}, \quad (3.2)$$

where $|\boldsymbol{h}|$ is the spatial distance (in m), τ is temporal separation (in mos), and the model parameters were estimated by fitting them to the empirical covariance values (Fig. 3.3).

Step 3: The *BME* theory was used in this study to estimate the coherency $C(\boldsymbol{p})$ as a function of the points $\boldsymbol{p}$ in the domain of Eq. (3.1).[m] As $C(\boldsymbol{p})$ is mathematically described by its *PDF* $f_C(\boldsymbol{p};\chi)$, where χ are possible coherency realizations at $\boldsymbol{p}$, *BME* constructed the $f_C(\boldsymbol{p};\chi)$ by integrating the *G-KB* and the *S-KB* described in Step 2. The basic set of *BME* Eq. (2.8a, b) of Section 2 above, expressed in terms of $C(\boldsymbol{p})$, was solved to obtain the coherency values at unsampled points on a 10 km × 10 km × 1 month space-time grid covering the domain of Eq. (3.1). Specifically, Fig. 1.3 in Section 1 of Chapter 3 presented a few selected months of the above *BME*-generated coherency maps that offer a detailed picture of the strength of the *HFRS* cases-climate dynamics association in Eastern China for each month of 2012. Furthermore, to avoid the edge effects of wavelet coherency analysis during the first and the last year of the study period, the 120 maps of the period 2006–15 were used to explore further the climate-*HFRS* association pattern across Eastern China. Based on the calculated coherency values, the region was divided vertically into four parts (south, middle-south, middle-north, and north parts) suggesting that local characteristics can affect the association (Fig. 3.4A): the south and middle-north parts exhibited the highest and the lowest coherency values, respectively, whereas the other two parts showed medium values; a few low coherency values were observed in the south, middle-south, and north parts of the study region.

Step 4: Using a 10-fold cross-validation technique, the *BME*–based $C(\boldsymbol{p})$ estimation performance was subsequently evaluated in terms of three accuracy indicators: the $MAE=0.00271$, the $RMSE=0.0139$, and the $R^2=0.991$ of the simple linear regression model relating observed and predicted values. These results confirmed that *BME* generates accurate coherency estimates.

Step 5: Given that climate exerts nonpoint impacts on *HFRS* infections, the coherency values between *HFRS* and climate dynamics are expected to exhibit local cluster characteristics. Hot spot analysis can efficiently assess high- *vs.* low-valued coherency clusters across space, i.e., it can identify strong *vs.* weak associations between global climate dynamics and *HFRS*-infected areas (a hot spot means high coherency values at a set of neighboring locations; a low spot means the opposite, i.e., low coherency values at a set of neighboring locations). Specifically:

∴ The geographical heterogeneity of coherency in the study region shown in Fig. 3.4A is weak, implying that both high and low coherency values are clustered geographically.

∴ Obvious clusters were observed in the map of Fig. 3.4B (notice the geographic distribution of hot and cold spots corresponding to the global climate-*HFRS* association map of Fig. 3.4A). The distribution of spots in Fig. 3.4A delimits geographically high- and low-valued clusters,

[m] In particular, the *SEKS-GUI* software library was selected to produce these space-time coherency maps.

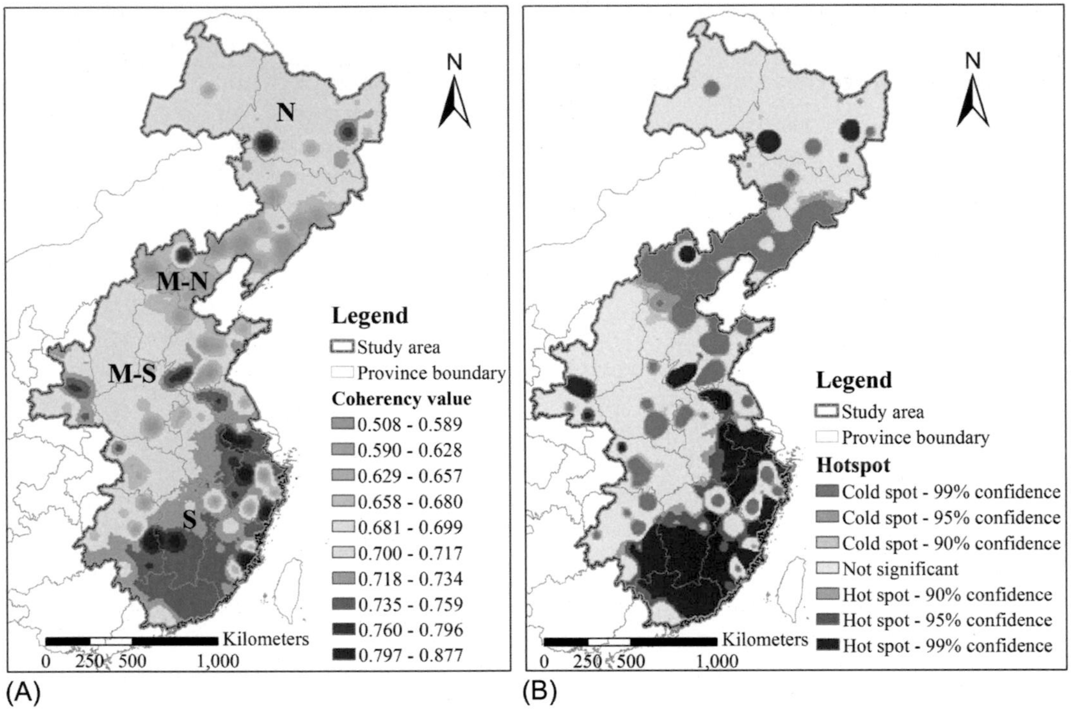

FIG. 3.4 Association between *HFRS* cases and global climate dynamics. (A) Map of the association strength, and (B) map of the association hot spots. The N, M-N, M-S, and S represent the north, middle-north, middle-south, and south parts of the study region, respectively.

whereas Fig. 3.4B makes it easier to detect visually the boundaries of hot and cold spots. ∴ Most clusters with high association coherency (hot spots) were located in the south part of the study region, whereas most clusters with low coherency (cold spots) were located in the middle-north part. Certain hot spots were close to each other, and the same is true for some of the cold spots. As a result, larger spatially continuous hot spot areas as well as a larger cold spot area occurred, particularly, in the south and middle-north parts of Fig. 3.4B. The other two parts (middle-south and north) did not exhibit any noticeable distributions of continuous nonpoint hot/cold spots. Yet, a number of individual point hot/cold spots were still found at certain places.

3.2 A class-dependent study of chronotopologic *HFRS* distributions

The objective of this case study was the investigation of *HFRS* incidence distributions with considerable chronotopologic heterogeneity features, which motivated the consideration of *BME* as a data class-dependent approach, in this case. This new *BME* perspective had some interesting methodological consequences that were discussed in He et al. (2019b).

3.2.1 The HFRS dataset

During the period January 2005–December 2013, China's Information System for Disease Control and Prevention (*CISDCP*) recorded *HFRS* cases at 130 counties of the Heilongjiang Province, covering an area of approximately

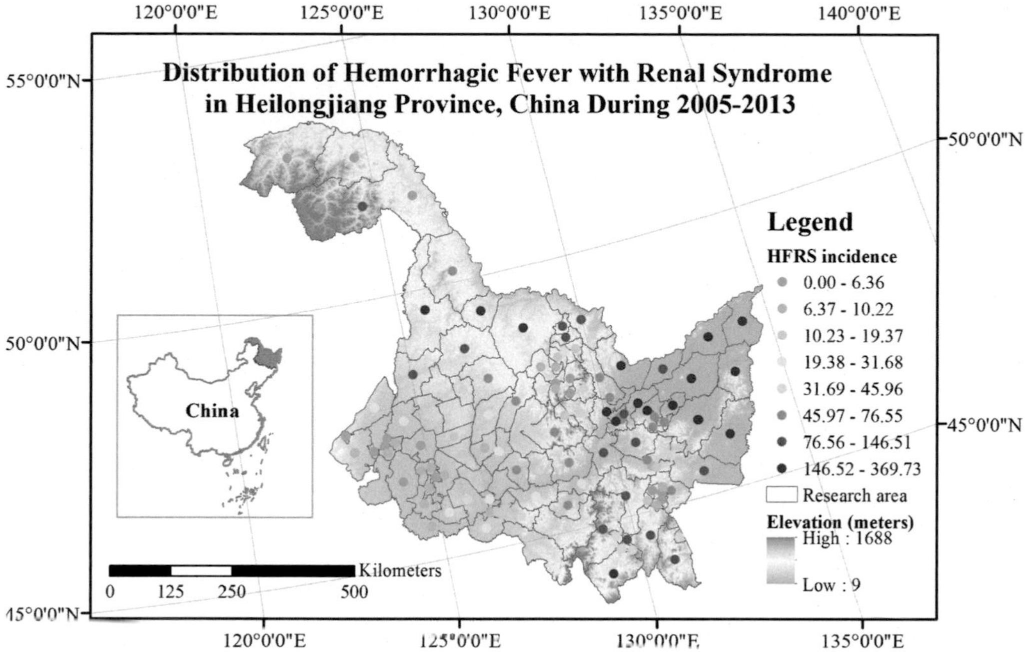

FIG. 3.5 Population-standardized *HFRS* incidences for the period 2005–13 (i.e. total incidences/10^5 capita), in the 130 counties of the Heilongjiang Province in China (dots correspond to county centroids).

473 thousand km^2, and a population of 38.35 million. The monthly *HFRS* case data was population-standardized using the corresponding demographic data obtained from the National Bureau of Statistics of China. Fig. 3.5 shows the geographical distribution of *HFRS* population-standardized incidences during the entire study period (i.e., January 2005–December 2013) in the 130 counties of the Heilongjiang Province. High *HFRS* chronotopologic variability was observed during the period 2005–13 (skewness = 6.265 and kurtosis = 64.007).[n]

3.2.2 HFRS modeling and methodology

The following modeling decisions were made by He et al. (2019b) in order to transform the *HFRS* data into useful knowledge for decision-making purposes:

① The chronotopologic distribution of population-standardized *HFRS* incidences was mathematically represented as an *S/TRF*, $X(\boldsymbol{p})$, where $\boldsymbol{p} = (\boldsymbol{s};t)$ denotes the geographical coordinates $\boldsymbol{s} = (s_1, s_2)$ and time instant t, so that, in this case,

$$\text{Chronotopologic domain} \quad \boldsymbol{p}: \begin{cases} \boldsymbol{s} = (s_1, s_2) \in \text{Heilongjiang (China) Province} \\ t \in 2005-2013 \text{ time period.} \end{cases} \tag{3.3}$$

② Selected *HFRS* incidence interval classes at points $\boldsymbol{p} = (\boldsymbol{s};t)$ and $\boldsymbol{p}' = (\boldsymbol{s}';t')$ of the chronotopologic domain of Eq. (3.3) were denoted as

$$I_m = [\zeta_l^m, \zeta_u^m] \text{ and } I_n = [\zeta_l^n, \zeta_u^n], \tag{3.4a, b}$$

respectively, where the subscripts m, $n = 1$, 2, 3, 4 are the class identification numbers,

[n] The maps in the current study were all produced using the ArcGIS 10.2 software.

and the subscripts l and u are the lower and upper limit, respectively, of each class interval. These *class-based* (*categorical*) incidences were denoted as $X(\boldsymbol{p}) \in I_m$, i.e., the *HFRS* incidence at point $\boldsymbol{p}$ belongs to the interval class I_m, and $X(\boldsymbol{p}') \in I_n$, which means that $X(\boldsymbol{p}') \in I_n$, i.e., the incidence at point $\boldsymbol{p}'$ belongs to the class I_n.

③ Two notions were used to describe quantitatively the *HFRS* pattern across space and time: the *global size* of each incidence class, and the *chronotopologic arrangement* of the different incidence classes relative to each other. Individual *HFRS* incidence classes may visually appear to occupy mutually exclusive "patches" of various sizes within the chronotopologic domain of Eq. (3.3) (e.g., any pair of classes I_m and I_n may have or may not have common boundaries). Also, the patches may spread uniformly throughout the domain of interest, or they may appear to be elongated along a particular direction, in which case the *HFRS* pattern was characterized as anisotropic.

The modeling decision above leads to a modification of the *BME* setting that is appropriate for the *HFRS* incidence study, as follows:

A **class-dependent** ***BME*** **(cd-*BME*)** was implemented in the *HFRS* case study that separately analyzes *HFRS* data classes that are defined in terms of incidence percentiles (Table 3.2).

An outline of the *HFRS* case study method is given in Fig. 3.6, which involved the following steps:

Step 1: Data classification and processing sought to represent the chronotopologic *HFRS* data amenable for *BME*-based analysis. Categorical incidence analysis was based on the four incidence interval classes, I_m, $m = 1, 2, 3, 4$, mentioned earlier.

Step 2: The distribution connectivity of incidence classes was determined by their chronotopologic correlation structure, which made the latter a key notion of the quantitative *HFRS* study.

Step 3: The *cd-BME* produced *HFRS* incidence maps by assimilating the relevant *G-KB* (*HFRS* mean and covariance functions) and *S-KB* (log-transformed *HFRS* data $Y(\boldsymbol{p})$) in the domain of Eq. (3.3).

Step 4: The *cd-BME* mapping results were subsequently superimposed and back-transformed to obtain the final *HFRS* incidence maps, which provided information about the actual form of the chronotopologic *HFRS* spread.

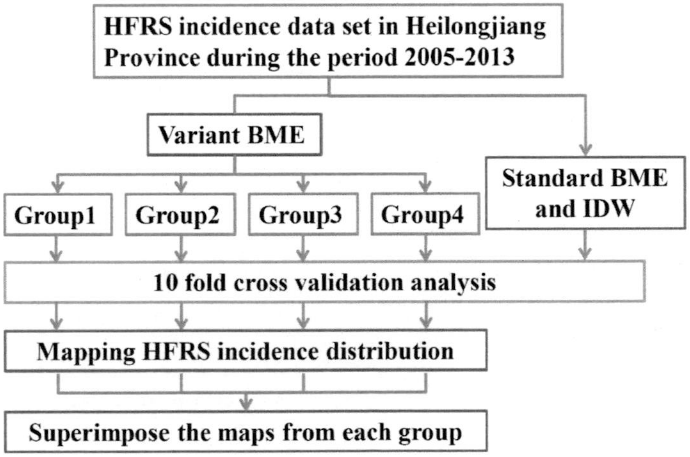

FIG. 3.6 An outline of the Heilongjiang *HFRS* case study methodology.

Step 5: For comparison purposes, the *standard BME* approach and the inverse chronotopologic metric interpolator (*ICTM*, Chapter 7) were also employed to analyze the original dataset without class-decomposition. *BME*-based computational data analyses used the software library *SEKS-GUI* (Yu et al., 2007), whereas the *ICTM* technique was implemented using the *R* software.

Step 6: To evaluate the performance of the different approaches, a 10-fold cross-validation analysis was conducted that involved the *RMSE*, the *MAE*, and the R^2 accuracy indicators.

Next, the above six steps are described in more detail.

3.2.3 Dataset classification and processing

Noticeably, in this case study the *HFRS* data classification and processing involved the following prime activities:

① The *HFRS* $X(\boldsymbol{p})$ data was log-transformed by means of the equation

$$Y(\boldsymbol{p}) = \log_{10}(X(\boldsymbol{p}) + 1), \quad (3.5)$$

which has the advantage that it accounts for 0 values (these are noteworthy as they may offer useful clues regarding, e.g., the absence of disease or individual immunity).

② As noted earlier, the *HFRS* dataset was categorized into four classes in terms of percentiles as shown in Table 3.2. Considering that 58% of the original data consisted of 0 values, the remaining 42% of the data was divided into four classes by percentile so that each of them includes 10.5% of the data, and, subsequently, all 0s were added to the 1st class, i.e., 0%–68.5%, 68.5%–79%, 79%–89.5% and 89.5%–100%.

③ Due to the high *HFRS* data variability exhibiting a heavy-tailed distribution, the intraclass data variability was reduced significantly by classifying the data in terms of incidence percentiles. The particular incidence classes were selected because they allowed a sufficient number of space-time points in each class for mapping purposes. Within the I_1, I_2, I_3, and I_4 classes (Table 3.2) there existed, respectively, 9617 points with 130 overlapping locations, 1470 points with 90 overlapping locations, 1478 points with 101 overlapping locations, and 1475 points with 96 overlapping locations. Hence, there existed at least 90 locations in each incidence class (which can be regarded as a spatial data coverage condition for improved mapping accuracy).[o]

TABLE 3.2 The four *HFRS* incidence classes and their descriptive statistics.

		Original dataset $X(p)$		Log-transformed dataset $Y(p)$	
Class (I_m) No.	**Percentile**	**Lower limit (ζ_l^m)**	**Upper limit (ζ_u^m)**	**Lower limit**	**Upper limit**
1 (I_1)	0–68.5	0 (ζ_l^1)	0.3576 (ζ_u^1)	0	0.1328
2 (I_2)	68.5–79	0.3577 (ζ_l^2)	0.6830 (ζ_u^2)	0.1329	0.2261
3 (I_3)	79–89.5	0.6831 (ζ_l^3)	1.5277 (ζ_u^3)	0.2262	0.4027
4 (I_4)	89.5–100	1.5278 (ζ_l^4)	26.4884 (ζ_u^4)	0.4032	1.4439

Note: Units of min and max of original data are cases/10^5 capita.

[o] Based on empirical considerations, rigorous *HFRS* mapping requires that for each estimation (mapping grid) point a certain number of space-time data points should exist around it that belong to its category.

④ Two kinds of *HFRS* spread patterns were considered for comparison purposes: *outward HFRS* spread linking a specific incidence class to the entire set of classes, and *inward HFRS* spread that is concerned with incidence transition from the entire set of classes to a specific class.

3.2.4 Chronotopologic correlation of HFRS incidence

Ex necessitate rei, the assessment of the *HFRS* incidence chronotopologic correlation is a key component of the *HFRS* study. Monthly geographical correlations of the original *HFRS* incidence $X(\boldsymbol{p})$ were calculated (108 covariances, in total) within the domain of Eq. (3.3). The following interpretations of the chronotopologic correlation results were made:

① The temporal variation of the *HFRS* mean was smooth and periodic, constantly fluctuating around the 0.25 incidence value (Fig. 3.7A).

② The covariance sill values explained the incidence variance in the Heilongjiang Province during the same month, exhibiting noticeable peaks at 12-mos periods, the size of which was reduced with time (Fig. 3.7B). These high peaks implied the presence of large uncertainty (or the presence of outbreaks at specific locations) in the *HFRS* variation during the peak times.

③ The correlation ranges determine the *HFRS* domain of influence, showing a rough variation with time, which, like the *HFRS* mean variation, also exhibits a periodic character (Fig. 3.7C). The maximum and minimum values of the correlation range during the period 2005–13 are 21,700 m and 897,650 m, respectively.

④ The functional shapes of the *HFRS* mean, covariance sill, and range were remarkably similar (e.g., in Fig. 3.8 large (small) *HFRS* mean values are directly linked to long (short) correlation ranges). Interpretationally, the two distinct peaks of the *HFRS* mean plot during June and November, which coincide with the corresponding peaks of the mean sill plot, detected the *HFRS* outbreaks that occurred during these months.

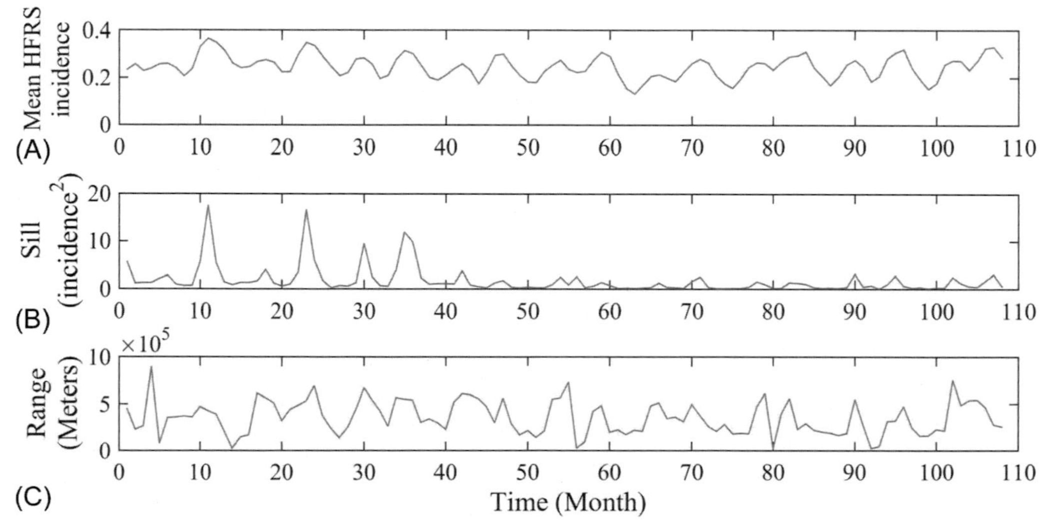

FIG. 3.7 Monthly variation of (A) the *HFRS* mean, (B) covariance sill, and (C) range during 2005–13.

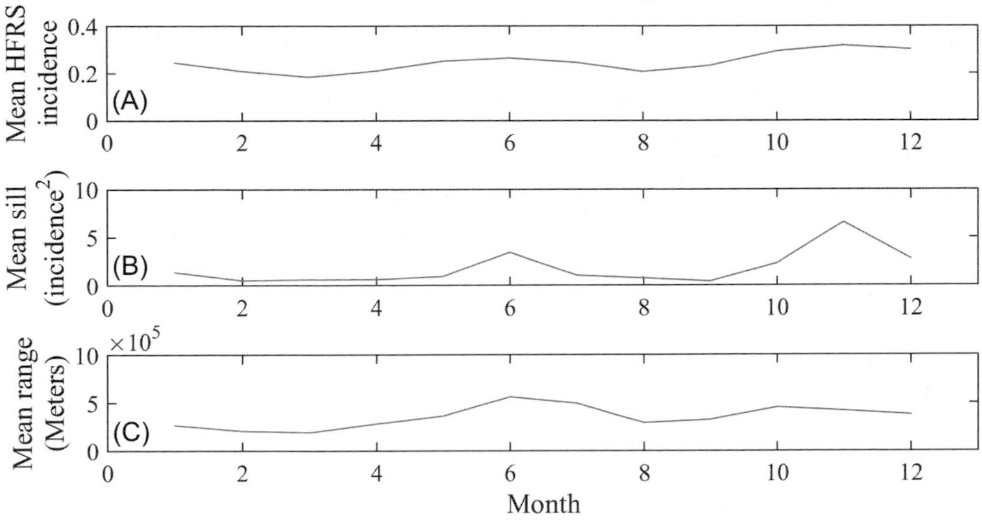

FIG. 3.8 Plots of (A) the *HFRS* mean and covariance parameters, i.e., (B) sill and (C) range for the same month (January–December) averaged over the corresponding months of the period 2005–13.

Subsequently, the empirical covariance values and the fitted theoretical models of the log-transform *HFRS* incidences $Y(\boldsymbol{p})$ (Fig. 3.9) were derived separately for each interval class of Table 3.2. It is noteworthy that the chronotopologic dependence ranges of the *HFRS* incidences vary with the incidence level (Table 3.3). All theoretical covariance models, $c_X(\Delta\boldsymbol{p}) = c_X(|\boldsymbol{h}|, \tau)$, were space-time separable. Yet, the spatial part of the covariance model of class I_4 (corresponding to the highest incidence level) was more complicated than the other three, further indicating the different space-time variation patterns of class I_4. Some key variation features of the empirical covariance values and the corresponding theoretical models are observed in Fig. 3.9:

① For all *HFRS* incidence classes considered, the nonzero slopes of the spatial covariance components at the origin indicated that the log-transformed incidences exhibited intense localized variations. With the exception of the incidence class 1, the slopes of the temporal *HFRS* covariance components at the origin were zero, implying that temporal incidence variation is much smoother than spatial incidence variation.

② Visual inspection of the temporal and spatial covariance plots for different incidence classes within the *HFRS* domain of Eq. (3.3) reveals that the chronotopologic structure of incidence distributions embodied in the shapes of the covariance functions depends on the incidence level, with dependencies in the time being stronger than those in space.

③ The spatial and temporal covariance lags beyond which *HFRS* incidence dependencies are negligible (referred to as spatial and temporal correlation ranges, respectively) vary among the different incidence classes selected.

④ The spatial and temporal covariance components of the four *HFRS* incidence classes were approximately zero-valued at large h and τ lags, indicating that the log-transformed incidence is homostationary. The selected theoretical models (Table 3.3) provided good fits to the empirical covariance values, thus validating the

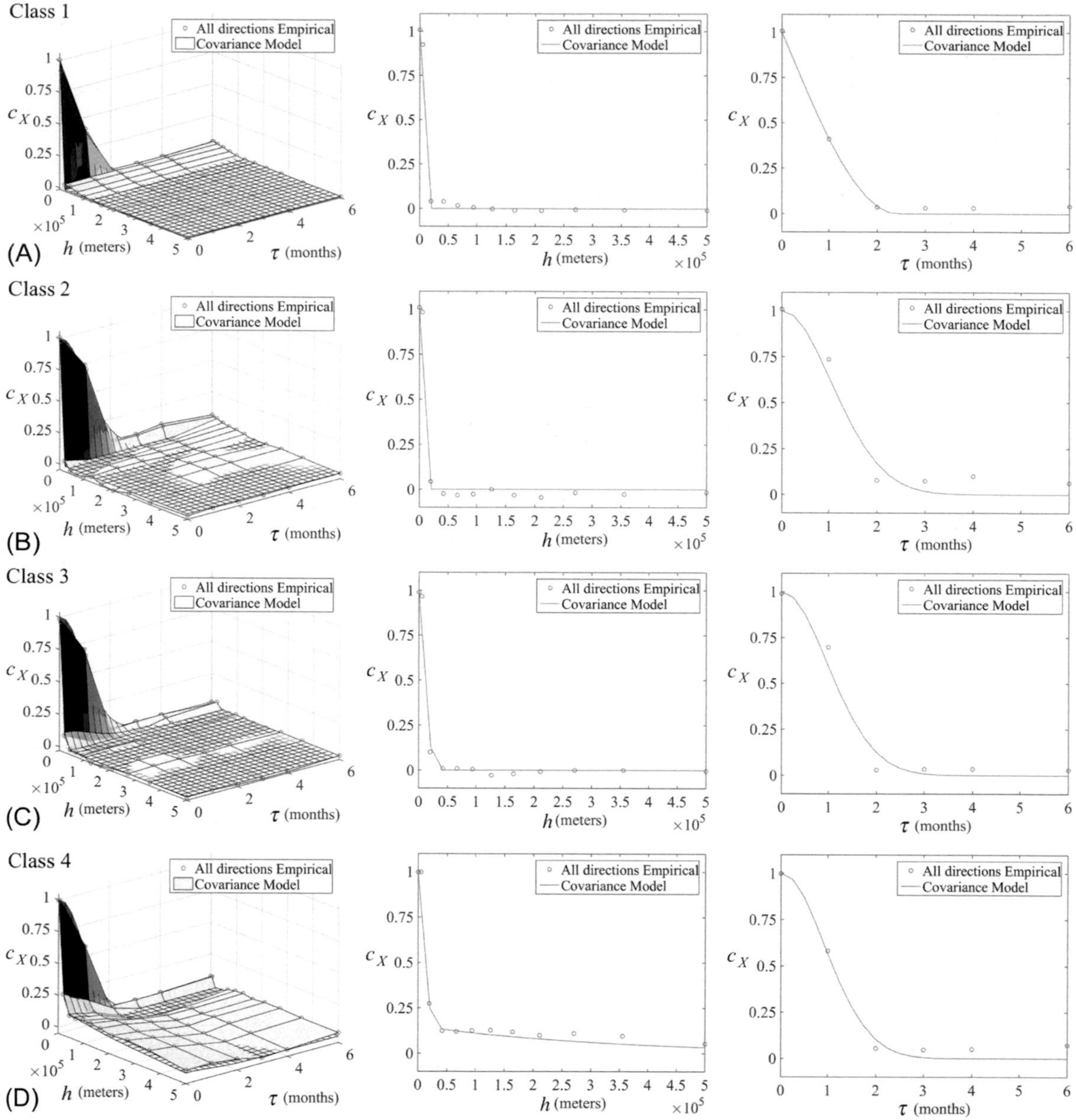

FIG. 3.9 Plots of the empirical covariance values (circles) and fitted theoretical models (surface and line in 3D, and 2D plots, respectively) of the log-transformed *HFRS* incidences (county-level) for each incidence class.

TABLE 3.3 Spatial and temporal dependency ranges and fitted theoretical covariance models of the four *HFRS* incidence classes.

Class No.	Spatial ranges (km)	Temporal ranges (months)	Fitted covariance model $c_X(\lvert h \rvert, \tau)$
1	10 (*Sph* model)	2.4 (*Sph* model)	*Sph*(10000)*Sph*(2.4)
2	20 (*Sph* model)	2.6 (*Gau* model)	*Sph*(20000)*Gau*(2.6)
3	30 (*Sph* model)	2.4 (*Gau* model)	*Sph*(30000)*Gau*(2.4)
4	30 (*Sph* model) 1000 (*Exp* model)	2.3 (*Gau* model)	0.85*Sph*(30000)*Gau*(2.3) + 0.15*Exp*(1000000)*Gau*(2.3)

Note: *Sph*, *Gau*, and *Exp* denote the standard Spherical, Gaussian, and Exponential models, respectively. These models are readily available in the relevant software libraries, like the *SEKS-GUI* used in this work.

adequate theoretical representation of the actual *HFRS* variation in the domain of Eq. (3.3).

3.2.5 Accuracy performance of CD-BME mapping

Tenfold cross-validation demonstrated that *cd-BME* outperforms the standard (class-independent) *BME* and *ICTM* techniques in terms of the *RMSE*, *MAE*, and R^2 values calculated using the entire *HFRS* incidence dataset (Table 3.4). At each individual *HFRS* interval class, the performance of both the *cd* and the standard *BME* techniques worsened with increasing incidence level (from class I_1 to I_4), but in all cases the *cd-BME* performance remained superior (Table 3.5). In terms of the *MAE* indicator, the accuracy improvement

TABLE 3.4 Accuracy performance of class-dependent, standard *BME* and *ICTM* implementations in *HFRS* incidence estimation.

Methods	*RMSE*	*MAE*	R^2
Class-dependent *BME*	0.74	0.19	0.72
Standard *BME*	1.11	0.43	0.44
ICTM	1.18	0.53	0.34

Note: The *cd-BME* results were derived by combining the 10-fold cross-validation values obtained for each class. Units of *RMSE* and *MAE* are cases/10^5 capita.

TABLE 3.5 Performance of the three *HFRS* incidence estimation techniques for each class.

	cd-BME		*BME*		*ICTM*	
Class no.	*RMSE*	*MAE*	*RMSE*	*MAE*	*RMSE*	*MAE*
1	0.10	0.04 (65.16%)	0.30	0.11	0.42	0.28
2	0.09	0.07 (82.13%)	0.45	0.37	0.38	0.28
3	0.24	0.19 (72.32%)	0.79	0.67	0.64	0.55
4	2.26	1.36 (42.18%)	3.22	2.34	3.39	2.43

Note: The standard *BME* and *ICTM* results were obtained by dividing the 10-fold cross-validation result from the entire *HFRS* incidence dataset into four classes according to the divide process of *cd-BME*. The improvement of *cd-BME* compared to *BME* is shown in brackets.

gained by *cd-BME* compared to standard *BME* is 65.16%, 82.13%, 72.32%, and 42.18% for the four incidence classes considered.

3.2.6 Chronotopologic mapping of the HFRS incidence

A total of 108 chronotopologic *HFRS* distribution maps with resolution 5 km × 5 km × 1 month were obtained for each interval class (i.e., $108 \times 4 = 432$ maps were generated, in total), and subsequently superimposed to produce the final *HFRS* distribution maps. For illustration, Fig. 3.10 shows the selected 2008 *HFRS* incidence map. The study yielded certain findings: *HFRS* incidences in the domain of Eq. (3.3) exhibited two noticeable peaks during June and November. The first peak triggers the rapid spread of *HFRS* cases during the August–November period toward the western part of Heilongjiang Province and also toward some counties in the eastern part of the province. After November, the *HFRS* incidences decrease in some counties, and the number of counties suffering high levels of *HFRS* infections also decreases. The western part of the Heilongjiang Province exhibits a lower number of *HFRS* incidences than the eastern part.

An informative visualization of *HFRS* spread over Heilongjiang during 2008 is given by the categorical chronotopologic maps in terms of the four different incidence interval classes defined earlier and shown in Fig. 3.11. Specifically, the maps of Fig. 3.11 present to scale the monthly distribution of "patches" with incidence classes I_2, I_3, and I_4 amidst class I_1. An apparent feature of Fig. 3.11 is that a few parts of the Heilongjiang region are dominated by small incidence patches, whereas in some other places the *HFRS* patches seem to be large. These incidence patches are mutually exclusive, bounded, and they have varying sizes. Certain *HFRS* patches seem to be distinctly elongated toward some preferred direction, indicating the presence of anisotropy. Sometimes the *HFRS* patches are clearly separated from one another, some other times they share common boundaries, and yet some other times they are completely surrounded by another class (e.g., in various monthly maps the classes I_2, I_3, and I_4 occur as patches within the class I_1). The proportions of regional cover by each incidence class show the dominance of class I_1 within which all the other classes are embedded, so that class I_1 acts like a background class.

3.2.7 Mean HFRS patch distances

This case study also calculated the mean distances across patches during each month of the period 2005–13 using the centers of each incidence patch, with one exception: since the area of class I_1 is the background domain that did not consist of patches in Fig. 3.11, a center cannot be defined for class I_1 in Fig. 3.11, and, hence, mean distances could not be calculated for this class.

Mean distance plots across patches during each month of 2005–13 are shown in Fig. 3.12 (e.g., the mean distance between classes I_2 and I_3 during March 2008 is 258.3 km). These mean distances were interpreted as effective ranges beyond which the transition probabilities remained essentially constant. As we see in Fig. 3.12, the temporal mean distance variations across patches are stationary, i.e., they fluctuate around constant values: 272 km (mean distance across I_2 patches), 304 km (across $I_2 - I_3$ patches), 340 km (across $I_2 - I_4$ patches), 255 km (across I_3 patches), 273 km (across $I_3 - I_4$ patches), and 273 km (across I_4 patches). The longest mean distance during 2005–13 occurred across $I_2 - I_4$ patches and the shortest across I_2 patches.

Interestingly, the reciprocals of the mean distances between incidence patches can also serve as the intensity parameters of the four classes. As the above numerical results demonstrated, there are small differences among the intensity parameters of the four *HFRS* classes, with the relatively most intense being the I_3 patches (4×10^{-4}km^{-1}), followed by the I_2, I_4 and

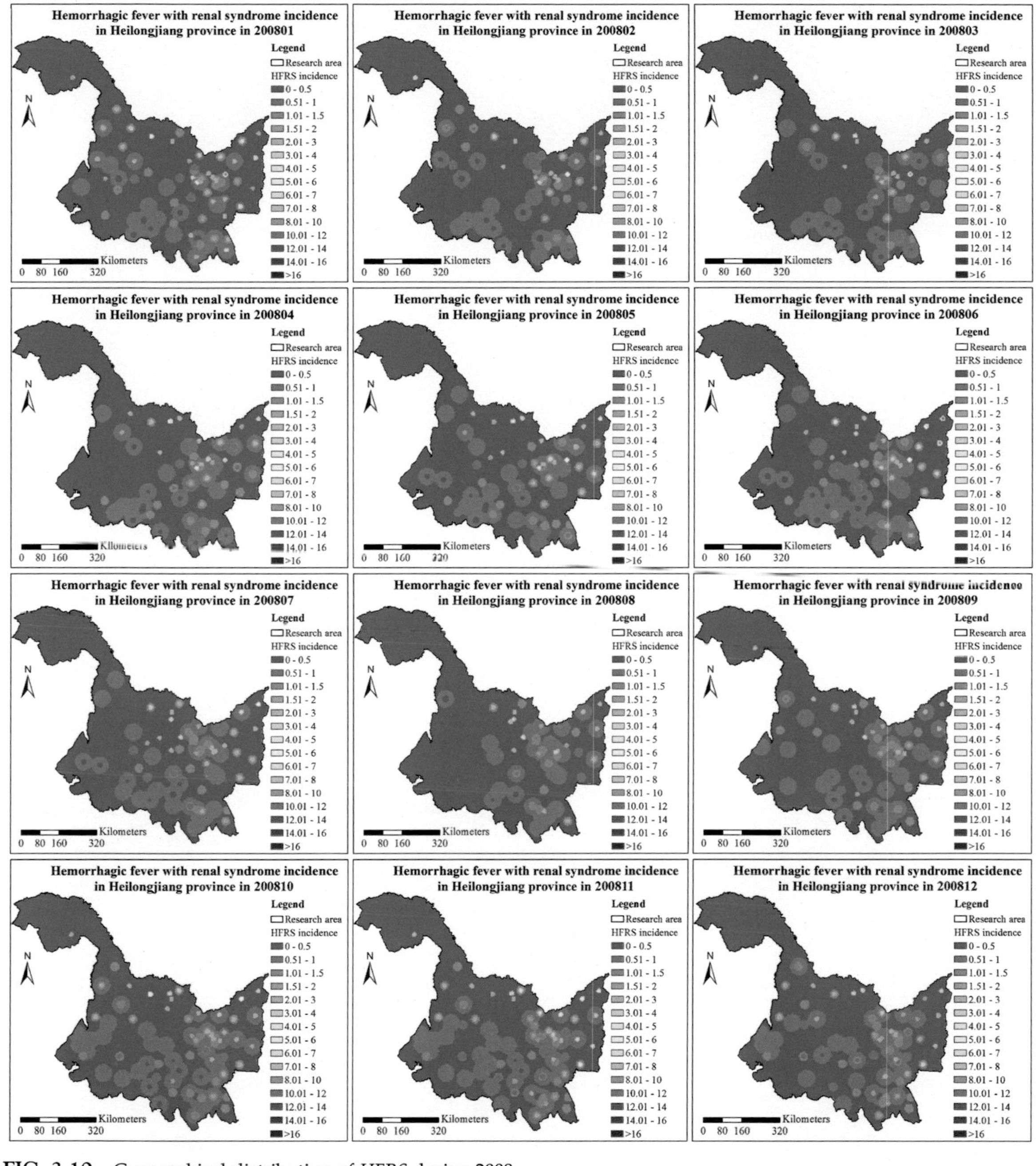

FIG. 3.10 Geographical distribution of *HFRS* during 2008.

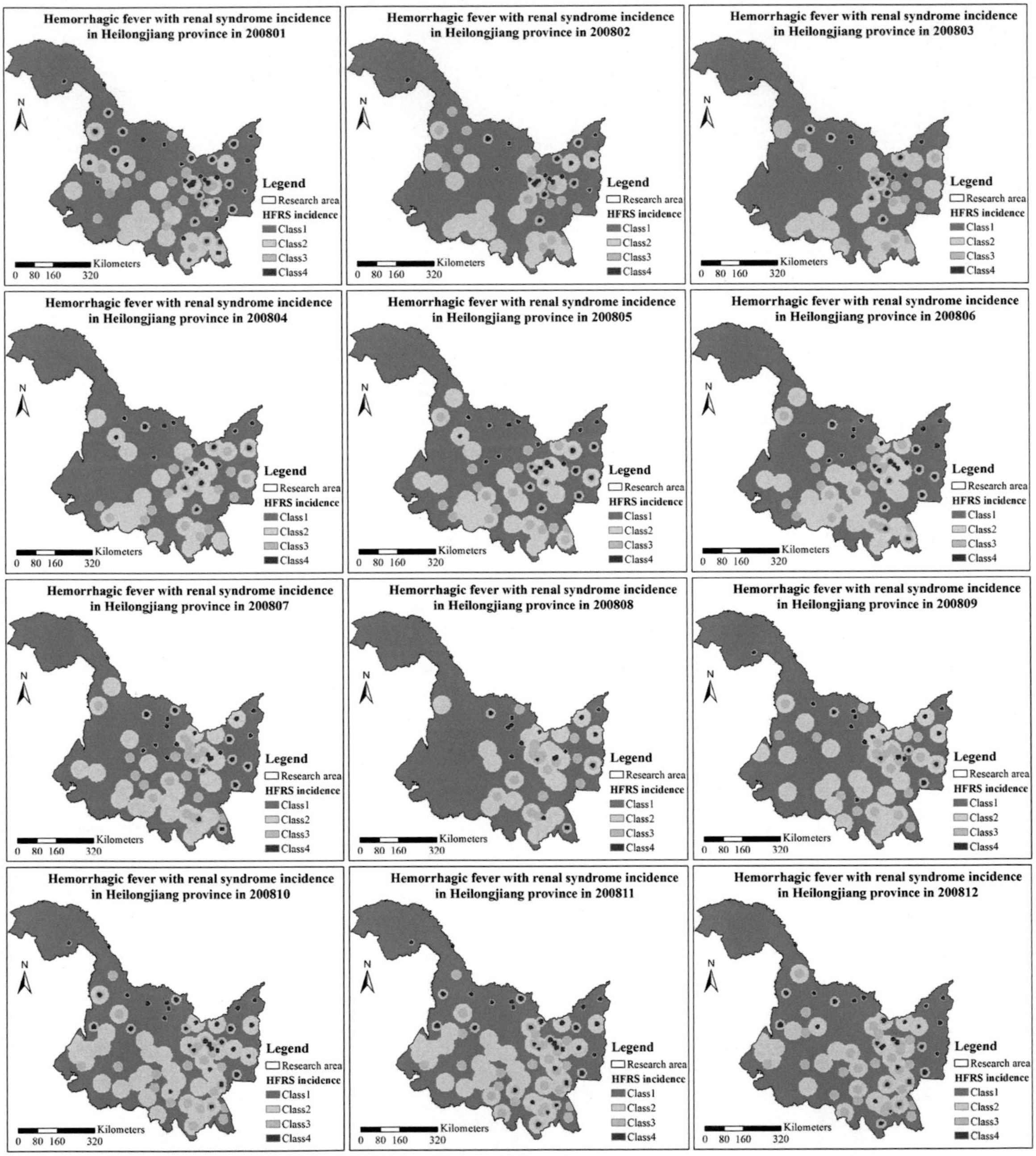

FIG. 3.11 *HFRS* incidence maps of Heilongjiang Province during 2008 in terms of the incidence classes.

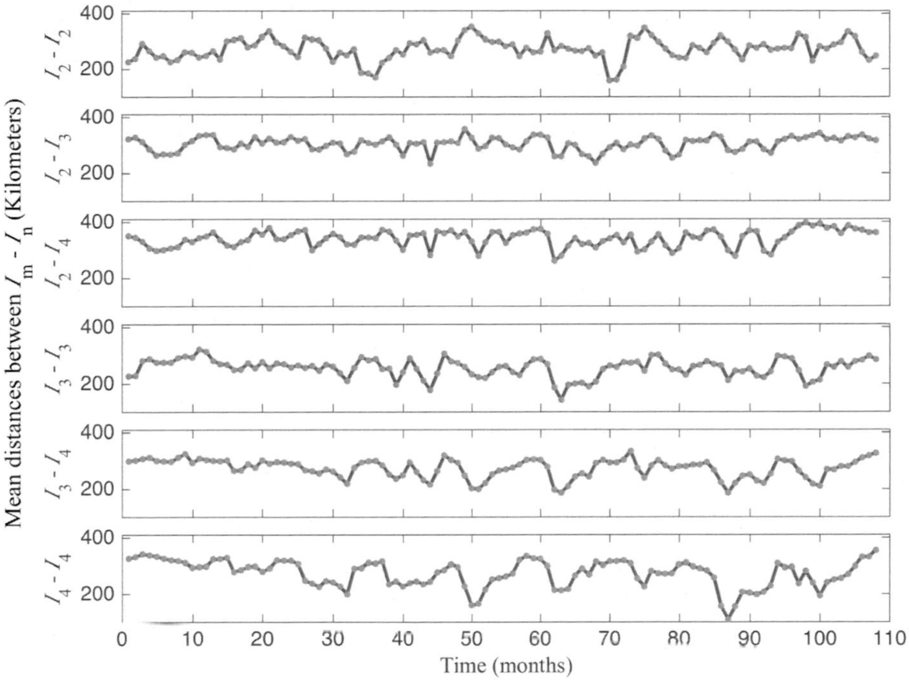

FIG. 3.12 Plots of mean distances across *HFRS* incidence patches during 2005–13.

$I_3 - I_4$ patches ($3.6 \times 10^{-4}\text{km}^{-1}$), and the least intense being the $I_2 - I_3$ ($3.3 \times 10^{-4}\text{km}^{-1}$) and $I_2 - I_4$ ($2.9 \times 10^{-4}\text{km}^{-1}$) patches. The mean coefficients of variation of the *HFRS* incidence during 2005–13 were 57.6532, 0.0232, 0.0824, and 0.2997 for the I_1, I_2, I_3, and I_4 class, respectively.

3.3 The sea surface salinity case study

The *sea surface salinity* (*SSS*) case study initially presented in Section 5.2 of Chapter 6 is revisited here. Following the work of He et al. (2020a), here we discuss how the *BME* approach was used to generate *SSS* maps and assess their relative performance by comparing the map accuracy when the contingogram-based covariance model $c_X^{(\psi)}$ was used *vs*. the accuracy when the ordinary covariance model c_X was implemented. An outline of the *SSS* study method is given in Fig. 3.13.

3.3.1 *SSS maps and their interpretation*

The *BME* chronotopologic estimation technique was implemented sequentially, first using the $c_X^{(\psi)}$ model and then the c_X model to obtain *SSS* estimates across space during the period May 19th–25th of 2015.[P] The resulting series of *SSS* maps are plotted in Fig. 3.14A–N. By interpreting these maps in a physical oceanography context, an improved understanding was gained concerning important Western Pacific Ocean salinity aspects, like the ways seawater

[P] This technique was used in the Western Pacific ocean study because it does not suffer the limitations of other interpolation techniques (Chapters 7–9), like the generalized inverse distance (does account for *SSS* anisotropy across space and relies on the ad hoc characterization of the salinity dependency pattern), or splines (they over-smooth the sea surface, the change becomes underwhelming, and they tend to retain smaller features of salinity distribution).

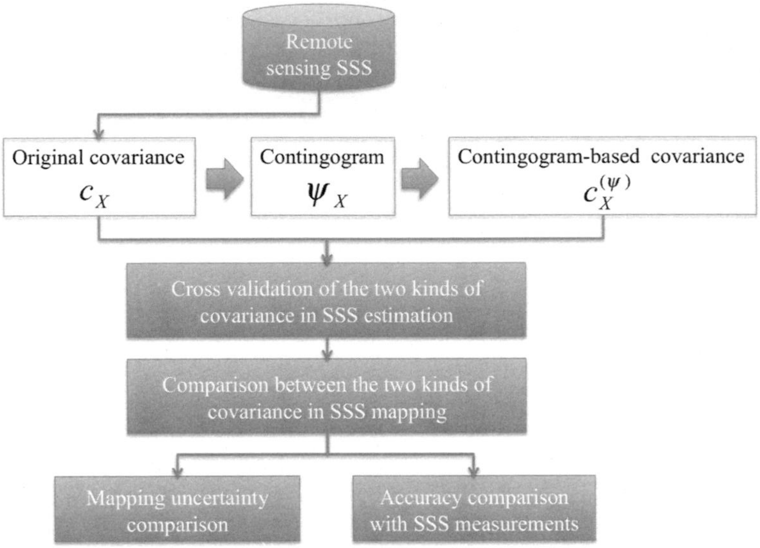

FIG. 3.13 An outline of the *SSS* study methodology.

salinity varied over time and across space, during which periods of time the highest variations occurred, what was the form of the spatial and temporal salinity trends, and what could be causing the composite space and time changes. Some specific results of the *BME* analysis by He et al. (2020a) are described next:

① A visual inspection of Fig. 3.14A–N revealed that, as was expected, the generated *SSS* maps were physically consistent with *SSS* empirical laws. Overall, the *SSS* distribution revealed in these figures tends to be zonal, and the salinity of ocean water is higher than that of coastal water under the runoff effect.

② The saltiest water was found at mid-latitudes, between 20°N and 30°N, which was physically linked to high evaporation and relatively low rainfall at mid-latitudes.

③ On the other hand, less salty waters were found near the equator where greater rainfall occurs and at high latitudes where melted sea ice freshens the surface.

④ Subarctic low salinity water meets subtropical high salinity water at around 42°N, where an obvious salinity gradient forms a "front."

⑤ The seafront near Hokkaido was caused by the junction between the Oyashio and Kuroshio Currents. The confluence of these currents disturbs and overflows seawater, bringing nutrients from below to the surface. This process makes plankton flourish, providing rich bait for fish and, thus, producing the well-known Hokkaido Fishery.

3.3.2 *SSS* mapping performance of the $c_X^{(\psi)}$ and c_X models

Some interesting observations were made by comparing the two different sets of maps, i.e., Fig. 3.14A–G *vs*. Fig. 3.14H–N. The *SSS* maps lead to the conclusion that there existed certain areas at which the $c_X^{(\psi)}$-based mapping generated different salinity estimates than the c_X-based mapping. In particular:

① At the Bohai Sea and the north part of the Yellow Sea, the c_X-based mapping could not produce *SSS* estimates during the first

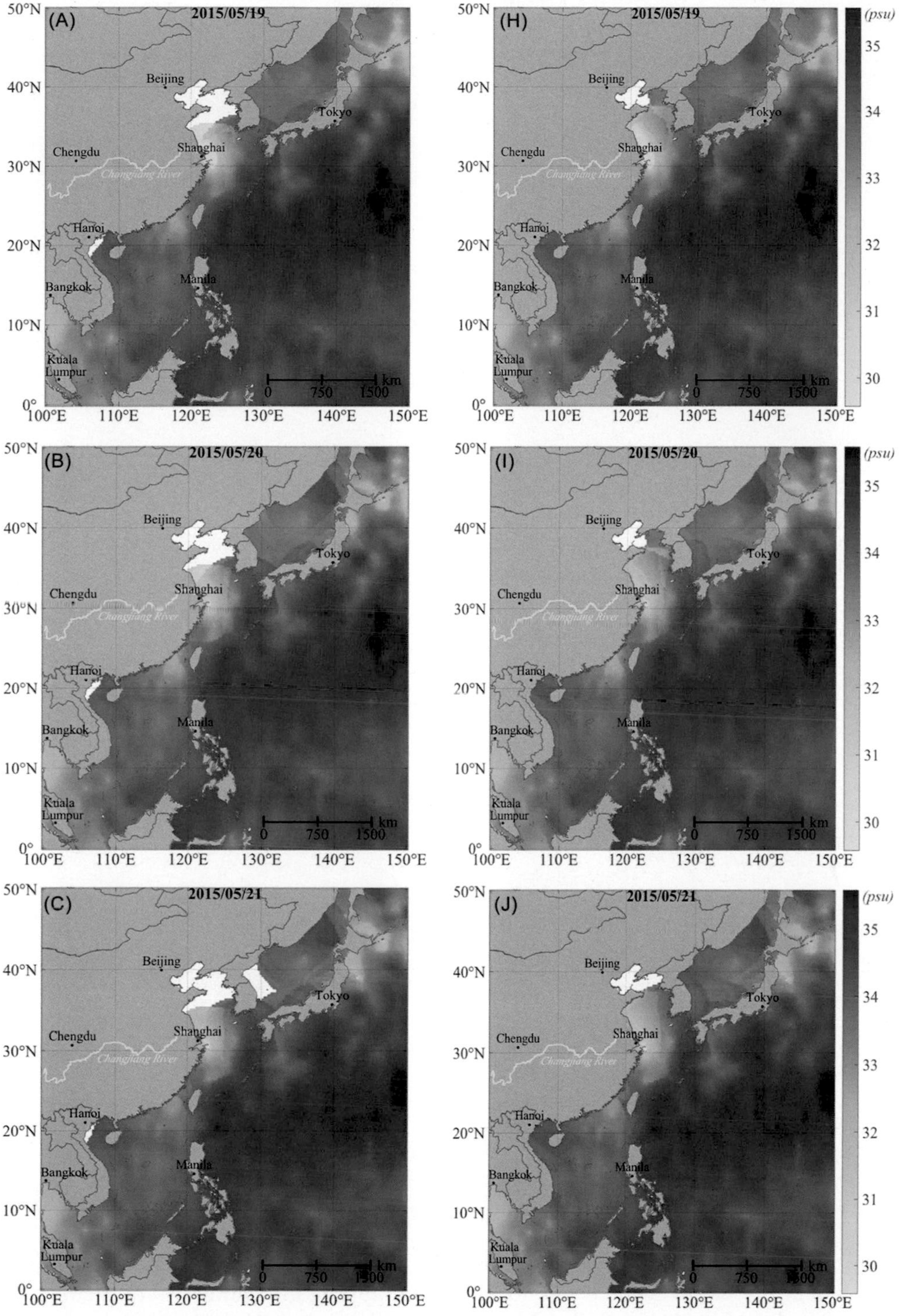

FIG. 3.14 *SSS* maps during May 19th–25th, 2015 obtained: (A–G) based on the $c_X^{(\psi)}$ model, and (H–N) based on the c_X model (no salinity estimates were generated in the white areas due to the lack of nearby data).

(Continued)

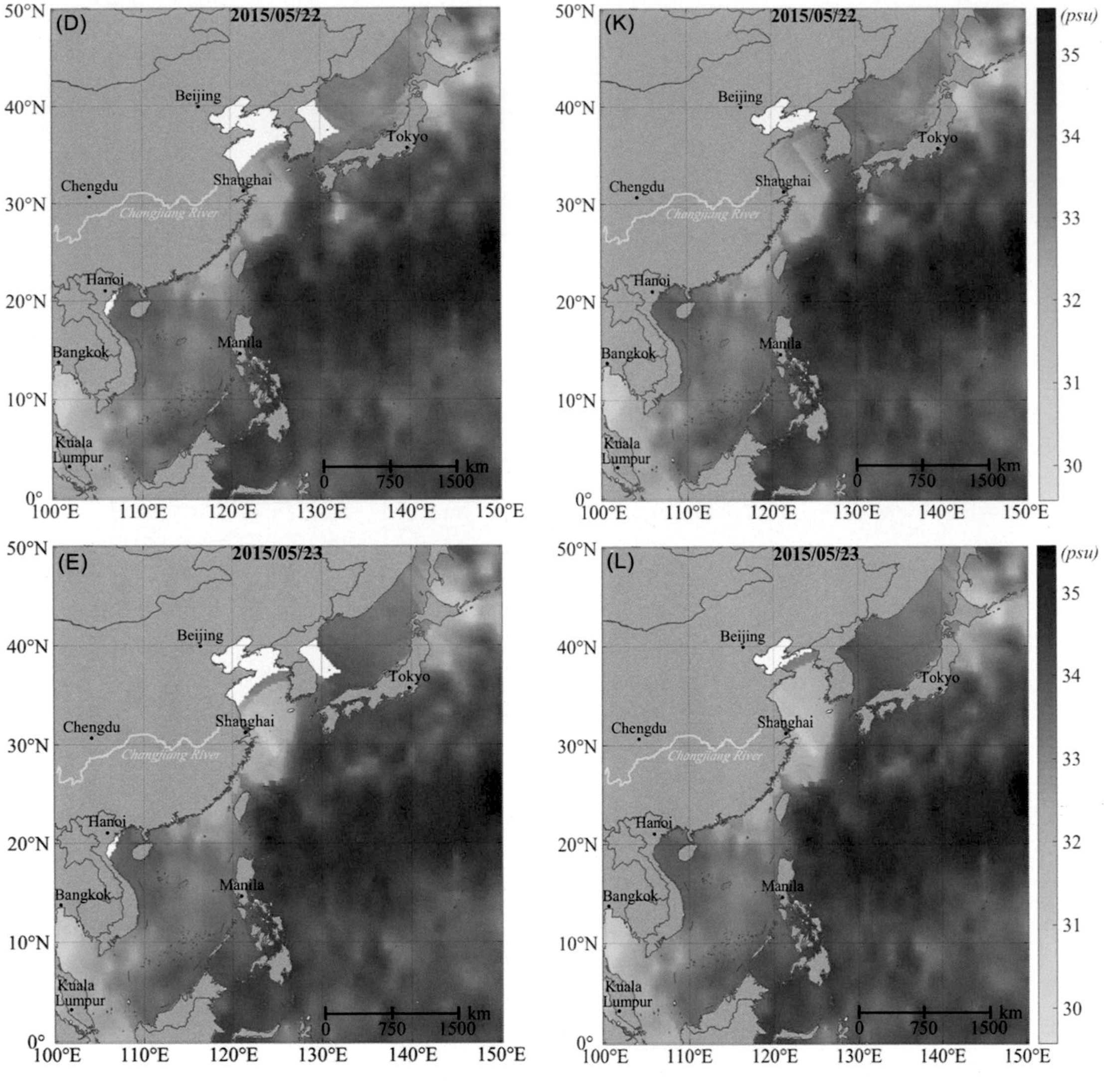

FIG. 3.14, Cont'd

day. The no-estimate area expanded continuously toward the central part of the Yellow Sea and a part of the Japanese Sea (this is confirmed by a visual inspection of Fig. 3.14A–G).

② Although the $c_X^{(\psi)}$-based mapping has some missing data at the Bohai Sea and the north part of the Yellow Sea, it showed a better coverage consistency (this is confirmed in Fig. 3.14H–N).

③ Quantitatively, the adequacy of the spatial *SSS* coverage is assessed by the *coverage ratio* defined as the number of *SSS* grid nodes with data over the total number of mapping grid nodes. In particular, Table 3.6 summarizes the coverage ratios of Aquarius

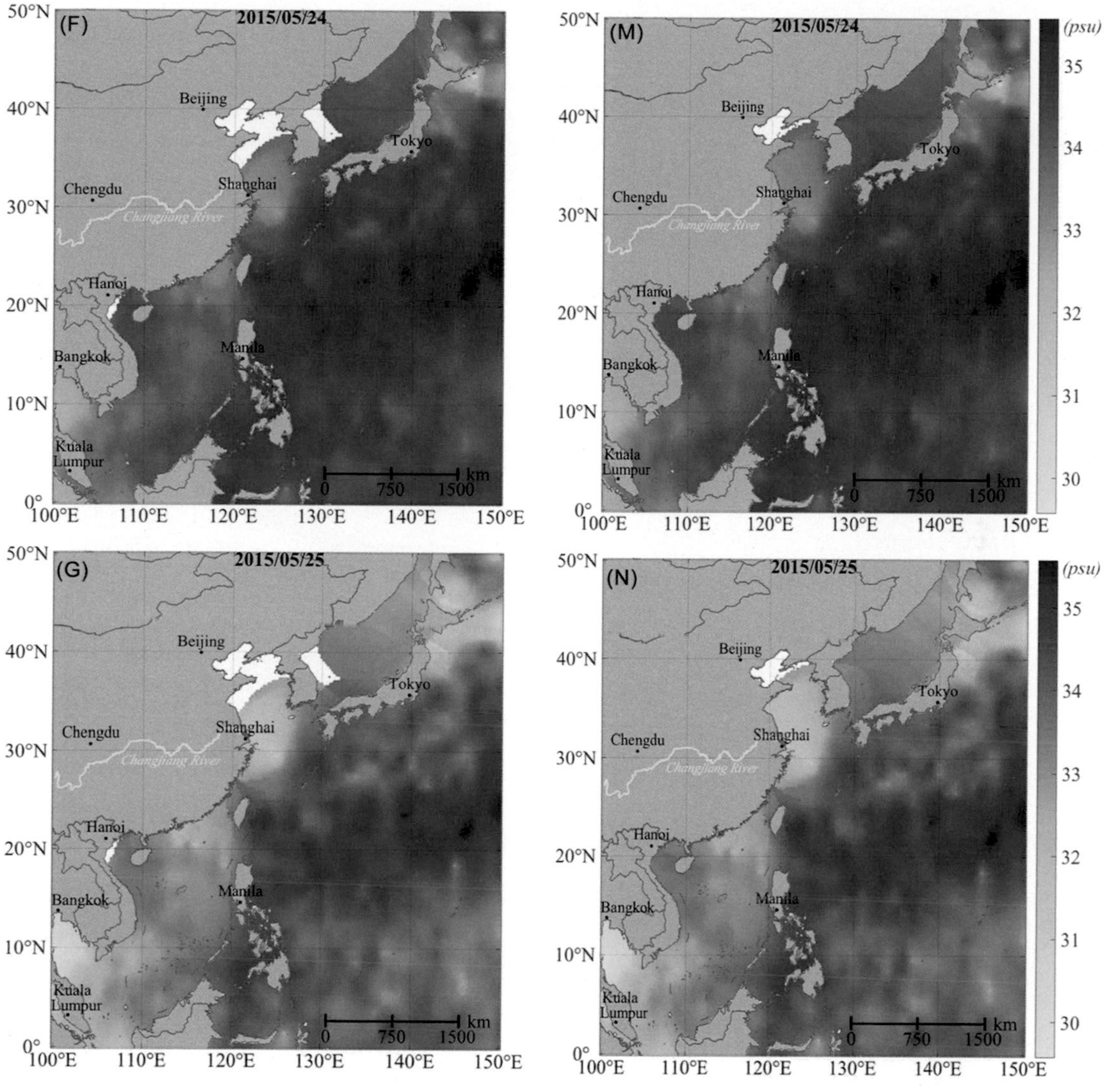

FIG. 3.14, Cont'd

TABLE 3.6 Coverage ratios (%) of Aquarius *SSS* data, c_X-based *SSS* estimates and $c_X^{(\psi)}$-based *SSS* estimates.

Data	DAY1	DAY2	DAY3	DAY4	DAY5	DAY6	DAY7
Aquarius *SSS* data	4.3819	4.3939	4.3099	4.3699	4.3699	4.3659	4.3019
c_X-based estimates	98.1289	98.1289	97.5732	97.3253	97.7531	97.7531	97.7531
$c_X^{(\psi)}$-based estimates	99.3283	99.3283	99.0884	99.0764	99.3483	99.3483	99.3483

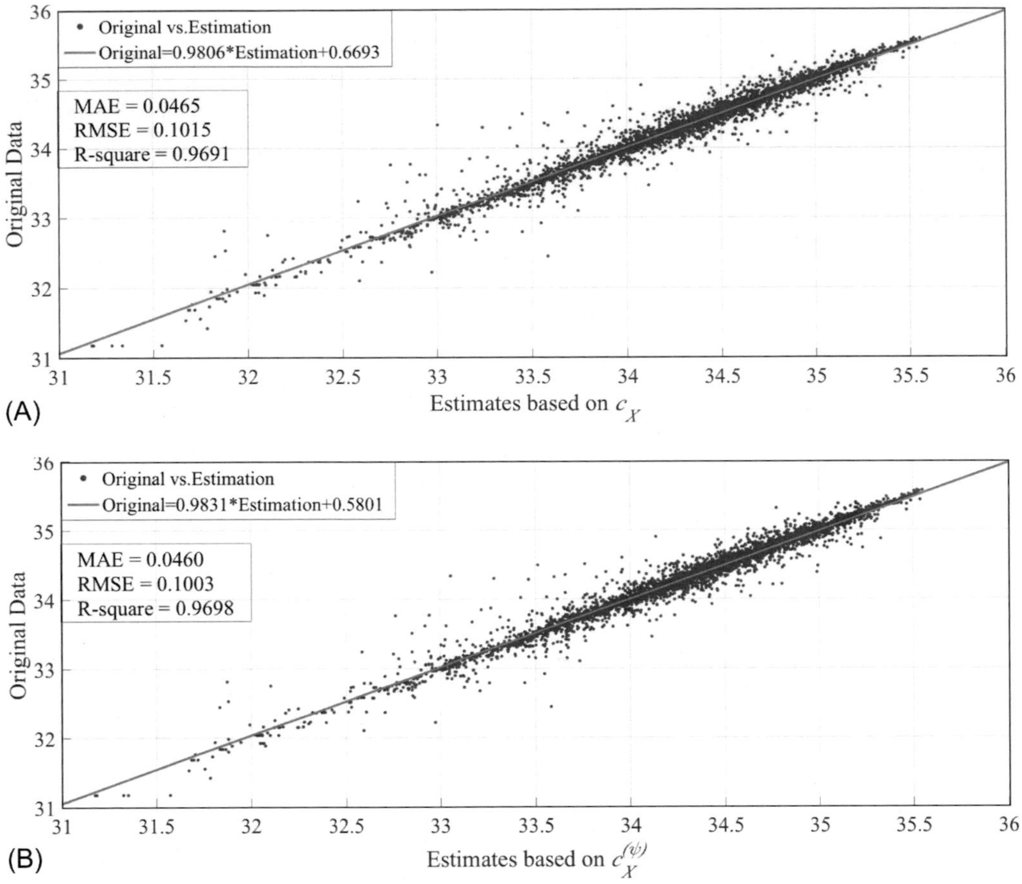

FIG. 3.15 Cross-validation of actual data at the validation points *vs.* estimates obtained at the same points by (A) the $c_X^{(\psi)}(h,\tau)$, and (B) the $c_X(h,\tau)$ covariance models.

SSS data, *SSS* estimates based on c_X, and *SSS* estimates based on $c_X^{(\psi)}$ in the study region during May 19th–25th of 2015 (the Aquarius coverage ratio in Table 3.6 fluctuates between 4.3019% and 4.3939%, yet the coverage ratios of the c_X-based and $c_X^{(\psi)}$-based *SSS* estimates are approximately 97.77% and 99.27%, respectively, which remain almost constant in the study region and period). Thus, the c_X-based and the $c_X^{(\psi)}$-based estimates, especially the latter, have much higher coverage ratios than the Aquarius *SSS* data, and they also show better stability.

Paucis verbis, despite the rather limited amount of *SSS* data, some interesting conclusions were drawn from the overall *CTDA*, in this case.

3.3.3 *Comparative SSS mapping accuracy*

In this case study, the chronotopologic estimation (mapping) performance of the $c_X^{(\psi)}$ and c_X models was evaluated in several ways:

① As is the usual practice, the mapping accuracy was assessed in terms of the R^2, *MAE,* and *RMSE* values of the *SSS* estimates that the $c_X^{(\psi)}$ and c_X models produced (Table 3.7). The *SSS* estimates obtained by adopting the $c_X^{(\psi)}$ model were more accurate than those derived by adopting the c_X model.[q] The theoretical contingogram model offers a better fit to the empirical *SSS* values at short spatial distances and small temporal separations, which implies a better contingogram-based *SSS* estimation performance compared to the original covariance-based estimation performance.

② The cross-validation plots in Fig. 3.15A and B (actual remote sensing *SSS* data at the validation points *vs.* remote sensing *SSS* estimates obtained at the same points by the two kinds of covariance models) show that the cross-validation performance of *SSS* estimation in terms of the $c_X^{(\psi)}$ model is better than in terms of the c_X model.

③ The *SSS* estimates were also compared with field *SSS* measurements. As the grid node locations were not completely consistent with the Argo data, in order to facilitate comparison the *SSS* values at the locations of the Argo data were sought. The *MAE* and R^2 values of the $c_X^{(\psi)}$-based estimation were 0.2006 psu and 0.6755, respectively, and for the c_X-based estimation they were 0.2007 psu and 0.6754, respectively, showing that *SSS* was only slightly more accurately estimated by the $c_X^{(\psi)}$ model.

TABLE 3.7 Space-time *SSS* estimation accuracy using the original and the contingogram-based covariance models.

Estimation based on	*MAE*	*RMSE*	R^2
$c_X^{(\psi)}$	0.0460	0.1003	0.9698
c_X	0.0465	0.1015	0.9691

[q] Only neighboring points were used in *SSS* estimation.

④ A series of *SSS* estimation error maps, $e_X^{(\psi)}(\boldsymbol{s};t)$ and $e_X^{(c)}(\boldsymbol{s};t)$, were plotted that were associated with the $c_X^{(\psi)}$ and c_X models, respectively. The *SSS* estimation error was particularly high along the coastal area, where hard data values were sparse or even nonexistent. Naturally, at hard data points the *SSS* estimation error was 0. As the distance from these points increased, so did the error. Fig. 3.16 presents maps of the differences $e_X^{(c)}(\boldsymbol{s};t) - e_X^{(\psi)}(\boldsymbol{s};t)$ between the *SSS* estimation errors at each point. These differences indicated that the $c_X^{(\psi)}$ model leads to better *SSS* estimates with lower estimation errors than the c_X model. Most of the area in Fig. 3.16 is covered with positive values, i.e., $e_X^{(c)}(\boldsymbol{s};t) - e_X^{(\psi)}(\boldsymbol{s};t) > 0$ (in particular, the area in lighter shade indicates $e_X^{(c)}(\boldsymbol{s};t) - e_X^{(\psi)}(\boldsymbol{s};t) > 0.01$ psu).

☞ As will be discussed in the next chapter, the chronotopologic *BME* estimation theory can offer an interesting approach for solving partial differential equations representing **natural laws** under conditions of uncertainty.

4 Practice exercises

1. Write a brief essay comparing, in the context of chronotopologic estimation, the specialized world-view that takes contemplation as fundamental *vs.* the specialized worldview that takes action as fundamental.
2. What is meant by the term "conscious ignorance."
3. Assume that we have four hard data χ_i at points $\boldsymbol{p}_i = (\boldsymbol{s}_i; t_i)$, $i = 1, 2, 3, 4$, shown in Fig. 4.1 and the attribute covariances c_{ij} are also available. List the core knowledge of the case and estimate the attribute value $\boldsymbol{p}_k = (\boldsymbol{s}_k; t_k)$.

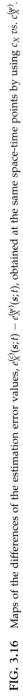

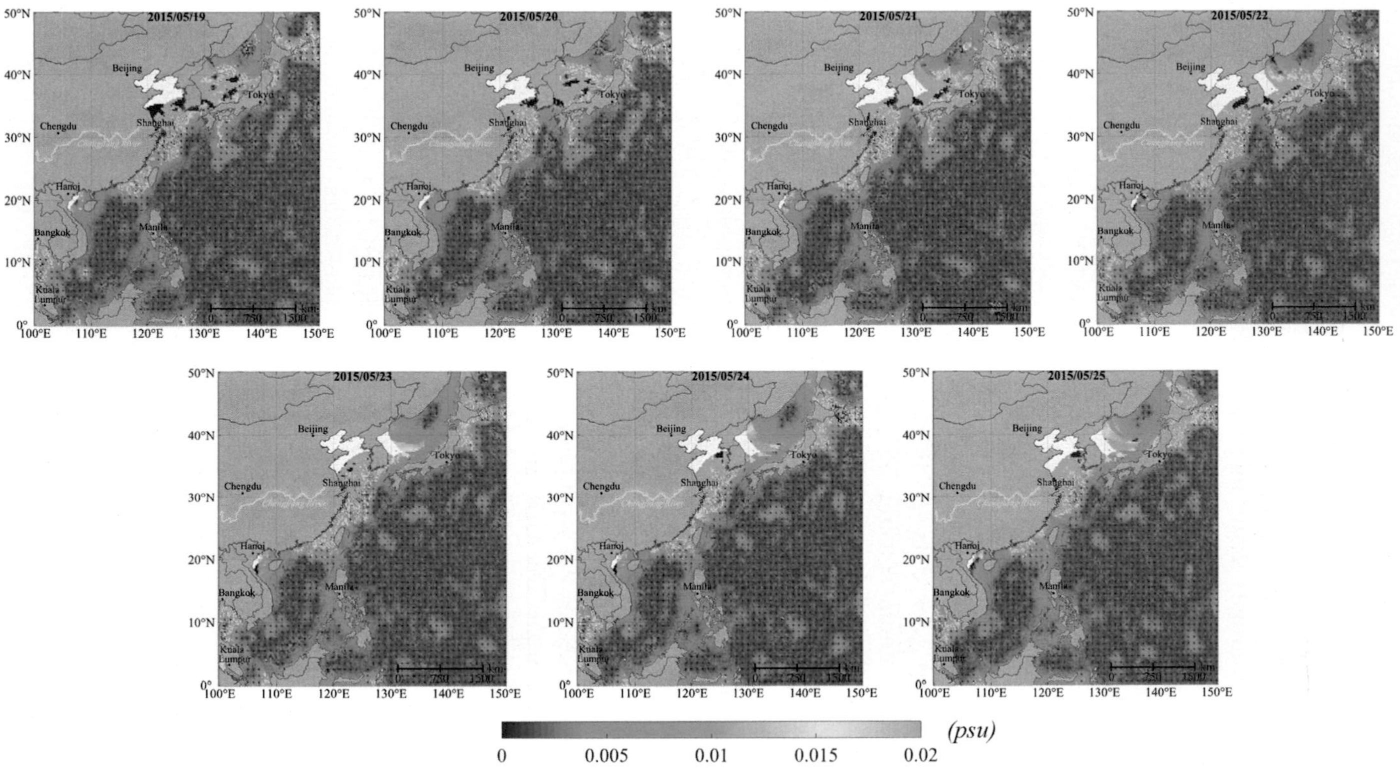

FIG. 3.16 Maps of the differences of the estimation error values, $e_X^{(c)}(\boldsymbol{s};t) - e_X^{(\psi)}(\boldsymbol{s};t)$, obtained at the same space-time points by using c_X *vs.* $c_X^{(\psi)}$.

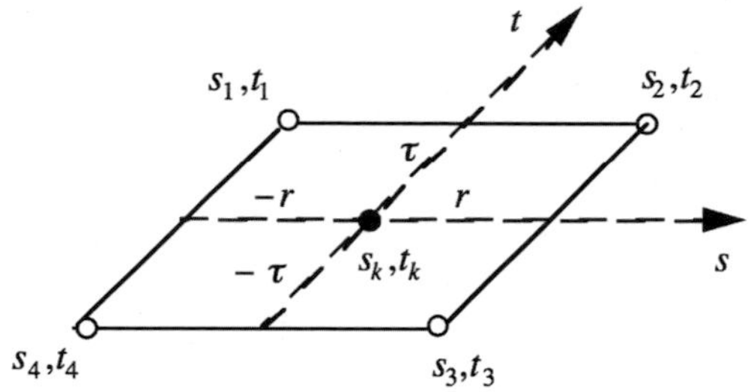

FIG. 4.1 Data distribution.

4. Calculate the mean, mode, and error variance of the attribute $X(\boldsymbol{p})$ given a *PDF* $f_X(\chi)$ of the form (a) $\frac{1}{\sqrt{2\pi}}e^{-(\chi-2)^2}$, (b) $\frac{1}{\sqrt{2\pi}}e^{-(\chi-2)^2}+\frac{1}{\sqrt{4\pi}}e^{-\frac{(\chi-2)^2}{4}}$, (c) $\frac{1}{3x\sqrt{2\pi}}e^{-\frac{(\ln\chi-2)^2}{6}}$ $(\chi\in(-\infty,+\infty))$.

5. Three data points (including two hard data and one soft datum) and one estimation point is shown in Fig. 1.1 of Section 1 in Chapter 6. The general knowledge consists of the humidity (unit: %) mean value and the theoretical covariance model, i.e., $\overline{X}=70, c_X(h,\tau)=e^{-\frac{3h}{200}-\frac{\tau^2}{3}}$, and the site-specific knowledge consists of the hard humidity data points $\boldsymbol{p}_1$ and $\boldsymbol{p}_2$ and the soft humidity datum point $\boldsymbol{p}_3$, i.e., $\chi_1=67$, $\chi_2=72$, $\chi_3\sim U(66,71.5)$, where the χ_3 follows a uniform humidity probability distribution U with lower and upper limits 66% and 71.5%, respectively. No datum exists at point $\boldsymbol{p}_k$. Calculate the *BME* mode estimate $\hat{\chi}_k$ at a point $\boldsymbol{p}_k$.

6. Table 4.1 shows a dataset representing a record (hard data) of rainfall. Assume there is a soft

TABLE 4.1 Dataset.

s_1	s_2	t	$X(p)$	s_1	s_2	t	$X(p)$
7	1	6	1.51	8.5	2	8	2.62
7	1.5	6	1.56	8.5	2.5	8	2.55
7	2	6	1.58	8.5	3	8	2.46
7	2.5	6	1.59	9	1	8	2.45
7	3	6	1.60	9	1.5	8	2.47
7.5	1	6	1.49	9	2	8	2.45
7.5	1.5	6	1.51	9	2.5	8	2.39
7.5	2	6	1.51	9	3	8	2.31
7.5	2.5	6	1.52	9.5	1	8	2.25
7.5	3	6	1.52	9.5	1.5	8	2.28
8	1	6	1.45	9.5	2	8	2.25
8	1.5	6	1.45	9.5	2.5	8	2.18

Continued

TABLE 4.1 Dataset—cont'd

s_1	s_2	t	$X(p)$	s_1	s_2	t	$X(p)$
8	2	6	1.44	9.5	3	8	2.10
8	2.5	6	1.43	10	1	8	2.03
8	3	6	1.45	10	1.5	8	2.06
8.5	1	6	1.38	10	2	8	2.03
8.5	1.5	6	1.37	10	2.5	8	1.95
8.5	2	6	1.36	10	3	8	1.88
8.5	2.5	6	1.35	7	1	9	3.65
8.5	3	6	1.35	7	1.5	9	3.63
9	1	6	1.29	7	2	9	3.54
9	1.5	6	1.28	7	2.5	9	3.39
9	2	6	1.28	7	3	9	3.21
9	2.5	6	1.27	7.5	1	9	3.57
9	3	6	1.24	7.5	1.5	9	3.55
9.5	1	6	1.19	7.5	2.5	9	3.33
9.5	1.5	6	1.18	7.5	3	9	3.16
9.5	2	6	1.19	8	1	9	3.42
9.5	2.5	6	1.19	8	1.5	9	3.40
9.5	3	6	1.16	8	2	9	3.34
10	1	6	1.10	8	2.5	9	3.23
10	1.5	6	1.09	8	3	9	3.07
10	2	6	1.11	8.5	1	9	3.23
10	2.5	6	1.11	8.5	1.5	9	3.22
10	3	6	1.08	8.5	2	9	3.17
7	1	7	2.19	8.5	2.5	9	3.08
7	1.5	7	2.24	8.5	3	9	2.94
7	2	7	2.26	9	1	9	3.00
7	2.5	7	2.26	9	1.5	9	3.00
7	3	7	2.22	9	2	9	2.97
7.5	1	7	2.15	9	2.5	9	2.89
7.5	1.5	7	2.20	9	3	9	2.78
7.5	2	7	2.22	9.5	1	9	2.76

TABLE 4.1 Dataset—cont'd

s_1	s_2	t	**X(p)**	s_1	s_2	t	**X(p)**
7.5	2.5	7	2.21	9.5	1.5	9	2.76
7.5	3	7	2.16	9.5	2	9	2.74
8	1	7	2.08	9.5	2.5	9	2.68
8	1.5	7	2.13	9.5	3	9	2.58
8	2	7	2.14	10	1	9	2.50
8	2.5	7	2.13	10	1.5	9	2.51
8	3	7	2.08	10	2	9	2.49
8.5	1	7	2.00	10	2.5	9	2.43
8.5	1.5	7	2.02	10	3	9	2.35
8.5	2	7	2.02	7	1	10	3.95
8.5	2.5	7	2.01	7	1.5	10	3.90
8.5	3	7	1.96	7	2	10	3.81
9	1	7	1.89	7	2.5	10	3.65
9	1.5	7	1.89	7	3	10	3.43
9	2	7	1.88	7.5	1	10	3.83
9	2.5	7	1.85	7.5	1.5	10	3.79
9	3	7	1.82	7.5	2	10	3.71
9.5	1	7	1.75	7.5	2.5	10	3.58
9.5	1.5	7	1.74	7.5	3	10	3.38
9.5	2	7	1.72	8	1	10	3.68
9.5	2.5	7	1.71	8	1.5	10	3.64
9.5	3	7	1.69	8	2	10	3.57
10	1	7	1.58	8	2.5	10	3.47
10	1.5	7	1.57	8	3	10	3.30
10	2	7	1.56	8.5	1	10	3.49
10	2.5	7	1.57	8.5	1.5	10	3.46
10	3	7	1.55	8.5	2	10	3.40
7	1	8	3.07	8.5	2.5	10	3.31
7	1.5	8	3.02	8.5	3	10	3.17
7	2	8	2.94	9	1	10	3.29

Continued

TABLE 4.1 Dataset—cont'd

s_1	s_2	t	$X(p)$	s_1	s_2	t	$X(p)$
7	2.5	8	2.84	9	1.5	10	3.26
7	3	8	2.74	9	2	10	3.20
7.5	1	8	2.99	9	2.5	10	3.12
7.5	1.5	8	2.94	9	3	10	2.99
7.5	2	8	2.86	9.5	1	10	3.05
7.5	2.5	8	2.75	9.5	1.5	10	3.03
7.5	3	8	2.65	9.5	2	10	2.98
8	1	8	2.85	9.5	2.5	10	2.89
8	1.5	8	2.82	9.5	3	10	2.78
8	2	8	2.75	10	1	10	2.80
8	2.5	8	2.66	10	1.5	10	2.78
8	3	8	2.56	10	2	10	2.73
8.5	1	8	2.65	10	2.5	10	2.65
8.5	1.5	8	2.65	10	3	10	2.54

data at point $p_3 = (7.3,2.1;9)$ with uniform distribution $U(3.3,3.5)$. Try to get the empirical covariance and use some theoretical covariance model to fit it, and then calculate the *BME* posterior probability density function at point $p = (7.5,2;9)$ by searching two hard neighbor data and one soft neighbor data. Then calculate the *BME* mean, mode, and error variance according to the posterior probability density function. The electronic version of Table 4.1 can be found on the website (https://www.dropbox.com/scl/fi/n0ghlggcknj5nqluzvdw3/Exercise-of-Ch-9.doc?dl=0&rlkey=9t6515ptz1i9kvts7gwyme561).

7. Assuming the soft data in the previous exercise follows a uniform distribution $U(3.34,3.48)$, use the same calculated covariance and hard data to calculate the *BME* mean, mode, and error variance according to the posterior probability density function. Then, compare the results with the previous ones, and give some comments on the quality of soft data.
8. Consider a space-time dataset of your interests and use the *BME* technique for mapping purpose by following the cases in Section 3.
9. When facing a practical decision, one may need to decide between one of two approaches:
   a. The active approach: One must use the data one has instead of waiting to get more data or better information before acting. To decline to act on the grounds that the available information is imperfect may not be an option open to one.[r]

[r] In fact, one may be wrong about the law of falling bodies. But it is irrational for a man who has leaped from the top of a skyscraper to depend on the falsity of that law to save him from death.

b. The contemplative approach: When the available data are not considered satisfactory for the problem at hand, one should seek additional and/or better information before one acts. To act in a rush could be irresponsible and even dangerous. What are the pros and cons of the above alternative approaches, in your view?

CHAPTER

10

Studying physical laws

1 The important role of physical *PDE* in *CTDA*

As we saw in previous chapters, the realistic study of natural phenomena encounters two main *KB*: the *core* (*G*) *KB* and the *site-specific* (*S*) *KB*. The physical law is a prime component of *G-KB*. No reader of this book, then, needs to be reminded that classical (macro-scale) physical laws rely heavily on the cause and effect principle (they can be described in terms amenable to efficient material causation). The realization of the importance of *lex naturae* has two significant implications in scientific *CTDA*. Indeed, as it has been repeatedly emphasized throughout the book:

❶ Given the universal acceptance among scientists of the **lawfulness** of natural phenomena that is thrust upon them by the reality of consistently observed facts, it would be unforgivable for any *CTDA* study that claims to be scientific to ignore its significance.

❷ Given the **uncertainty** sources characterizing real-world phenomena, these physical laws should be presented in a stochastic form that rigorously accounts for these sources.

Investigators in all scientific disciplines continuously construct, test, compare and interpret physical laws and scientific theories that *CTDA* cannot neglect. In many cases, the analytical formulation of physical laws and models (theoretical and empirical or phenomenological) has the important advantage that offers valuable insight about the phenomenon of interest (in terms of its interpretation power, the physical meaning of the solutions, the law or model parameters and how they relate with the phenomenon attributes), which is not possible in terms of numerical formulations.

Example 1.1 He et al. (2020b) proposed a modified *susceptible-exposed-infected-removed model* (*SEIR*) to study the spread of COVID-19 in various regions of the world: three Chinese regions (Zhejiang, Guangdong, Xinjiang), South Korea, Italy and Iran (see, also, Example 2.1 of Section 2 in Chapter 3). The corresponding analytical equations were as follows:

Quantitative Analysis and Modeling of Earth and Environmental Data
https://doi.org/10.1016/B978-0-12-816341-2.00008-3

$$\frac{d}{dt}S(t) = -\frac{\beta_1 I(t)S(t)}{N} - \frac{\beta_2 E(t)S(t)}{N},$$
$$\frac{d}{dt}E(t) = \frac{\beta_1 I(t)S(t)}{N} + \frac{\beta_2 E(t)S(t)}{N} - \alpha E(t),$$
$$\frac{d}{dt}I(t) = \alpha E(t) - \gamma I(t) - \lambda I(t),$$
$$\frac{d}{dt}C(t) = \gamma(t)I(t),$$
$$\frac{d}{dt}D(t) = \lambda(t)I(t),$$

(1.1a-e)

where N is the regional population size, $S(t)$, $E(t)$, $I(t)$, $C(t)$ and $D(t)$ are, respectively, the number of susceptible, exposed, infected, cured and dead individuals at time t. The model parameters $\beta_1(t)$ and $\beta_2(t)$ denote the rate of disease transmission when an individual comes in contact with infected and exposed individuals, respectively; the α is the probability that the exposed individuals become infected; and the parameters $\gamma(t)$ and $\lambda(t)$ are the disease cure rate and death rate, respectively. Based on the analytical formulation of Eqs. (1.1a–e) further insight was gained regarding the COVID-19 spread:

① The infection contact rate $q(t) = \frac{\beta_1(t)}{N(\gamma(t)+\lambda(t))}$ expresses the fraction of population that comes into contact with an infected individual during the infection period.

② The exposure rate included one part that was proportional to the contacts between susceptible and infected individuals (assessed by β_1), and another part that was proportional to the contacts between susceptible and exposed individuals (assessed by β_2). Then, useful assessments were made possible: since infected individuals in China were immediately sent to hospitals for isolation and medical treatment, it was set $\beta_1 < \beta_2$; based on COVID-19 literature, $\beta_2 = 5\beta_1$, $\alpha = \frac{1}{7}$; cured and dead individuals were removed at the rate $\gamma(t) + \lambda(t)$, where at $t = 0$, $\gamma(0) = \lambda(0) = 0$.

③ The reproductive ratio $R_0(t) = q(t)S_0$ was defined (S_0 is initial number of susceptible cases) that represented the number of secondary infections at any time t caused by an initial primary infection. Then, the epidemic was considered under control at that time t_{ECT} (epidemic control time, *ECT*) after which the $R_0(t)$ was consistently smaller than $L_0(t) = \frac{I(t)}{I(t)+5E(t)}$.

④ Most importantly, simulated values of the model parameters in Eqs. (1.1a–e) (β_1, γ and λ) were obtained in the regions of interest by assuming alternative disease control measures (working assumption), and, then, these model parameter values were used to compare the effectiveness of these measures. In this way, He et al. (2020b) showed that certain disease control measures were particularly effective in flattening and shrinking the COVID-19 case curve, which could effectively reduce the severity of the disease and mitigate medical burden.

Physical laws and theories can have a variety of mathematical forms. The most common form is the topic of the remainder of this chapter.

1.1 Stochastic partial differential equations

In light of the main considerations ❶ and ❷ above, a relevant observation can be made that is of central importance in many *CTDA* applications, as follows (see, also, Chapter 6).

> A prime component of the *G-KB* is formalized in terms of **stochastic partial differential equations** (***SPDE***) that represent physical laws governing the phenomena of interest or theories explaining the underlying processes in conditions of uncertainty.

The readers are reminded that, generally speaking, an *SPDE* is a physical *PDE* under conditions of uncertainty. Specifically, *SPDE* may mathematically represent different kinds of physical laws based on mechanistic and phenomenologic models (see Table 1.3, Section 1 of Chapter 3). A list of physical laws and the corresponding *SPDE* can be found in Christakos (2017). For illustration, we consider the following example.

Example 1.2 Stochasticity is a crucial component of mechanistic models, like the ones described here. Furthermore:

- Statistical thermodynamics provides several stochastic mechanistic models in the form of *SPDE*.
- A mechanistic model described by an *SPDE* is the sound propagation in random media (e.g., atmosphere).
- Pharmacokinetic/pharmacodynamics (*PK/PD*) models are cases of mechanistic modeling in which a system of differential equations describes absorption/elimination processes of chemical compounds in the body. These equations may contain random parameters, random observation errors and random dynamic (temporal) variability of processes in the subject body.

The *S-KB*, on the other hand, consists of hard and soft data (uncertain evidence) relevant to the phenomenon of interest,[a] as well as other kinds of secondary or auxiliary information, including the boundary and initial conditions (*BIC*) of the physical *SPDE*. As we also saw in Section 1 of Chapter 6, the following statements are valid:

- While attribute interpolation draws heavily on *S-KB*, **reliable extrapolation** or **prediction** requires the involvement of *G-KB* beyond numerical data, including physical laws in the form of *SPDE*. Accordingly, the *S-KB* is a necessary but not sufficient component of a scientific investigation, and only when combined with the *G-KB* the outcome can become a sufficient component too.

The *G-KB* and *S-KB*, *ut aequum*, constitute the corpus of knowledge that a rigorous and meaningful *CTDA* has to take into consideration.

Example 1.3 The continuous refinement of sophisticated theoretical and numerical models involving large numbers of equations and computer simulations (*G-KB*) against high-quality data from distributed sensor networks (*S-KB*) forms the basis for understanding how the ocean functions, making accurate predictions and potentially managing global ocean, which is perhaps the most powerful climate modulating system on the planet.

To slightly formalize matters, *ad usum*, let $X(\boldsymbol{p})$ be an attribute with chronotopology $\boldsymbol{p}$ obeying an *SPDE*. Then, the following problem is the focus of this chapter:

- Generate a **solution** of this *SPDE* (considered the main component of the *G-KB*) in the light of the available *S-KB*.

Stochasticity is an essential part of the solution to this problem, whereas considerable random measurement and observational errors may also be present. There are different ways to solve an *SPDE*. Some of the most important are discussed next.

1.2 A *BME* perspective

BME theory, viewed from a new angle in this chapter, can offer an interesting solution to the *SPDE* problem above. Generally, one can distinguish between three mainstream groups of techniques used to solve an *SPDE* (Mercier, 1985; Evensen, 1994; Anderson and Anderson, 1999; Robinson and Lermusiaux, 2002; Wang and Shen, 2010; Emmanouil et al., 2012; Lebreton et al., 2012):

[a] We recall that the term "hard data" denotes a measurement or observation that, for all practical purposes, is considered error-free, and, hence, it can be expressed in terms of a unique numerical value. On the other hand, the term "soft data" refers to a measurement or observation that is characterized by a degree of uncertainty, and, hence, it can be expressed in terms of an interval, a probability distribution, an empirical chart etc., depending on the *in-situ* situation.

① The **direct numerical** solution in terms of X realizations, usually generated via *Monte Carlo* simulation.

② The **statistical assimilation** solution in terms of X realizations, which combines statistical techniques (Kalman filters, etc.) and *S-KB* assimilation models.

③ The **standard stochastic** solution in terms of the *PDF* $f_X \equiv f_G$, with $G \equiv SPDE$.

In addition to the three mainstream *SPDE* solution approaches above, a 4th group based on the *BME* perspective has been proposed:

④ The ***BME*** solution is obtained in terms of the *PDF* $f_X \equiv f_K$, where $K = G \cup S$.

A comparative outline of the methodologies underlying the four groups of techniques above is presented in Fig. 1.1. According to this outline, techniques belonging to group ① do not account for *S-KB* and do not provide a full stochastic characterization of $X(\boldsymbol{p})$ in terms of its *PDF*; techniques belonging to group ② account for *S-KB* but still do not provide a full stochastic characterization; techniques belonging to group ③ provide a full stochastic characterization in terms of f_G but do not account for the *S-KB*; whereas the techniques of group ④ introduce a synthesis of the *G-KB* with the *S-KB* based on *BME* theory (the resulting total *KB* is denoted as $K = G \cup S$) and derive the integrated *PDF* $f_K = f_{G \cup S}$ at each spacetime point of interest $\boldsymbol{p}$ (Serre and Christakos, 1999; Kolovos et al., 2002; Lang and Christakos, 2018).

Hence, the *BME* approach in group ④ introduces a certain organization of the complete body of knowledge that does not have the apparent drawbacks of techniques ①, ② and ③. On the contrary, *BME* is a particularly flexible approach.

BME connects the mathematical operations to the total body of knowledge *K*, and generates both (a) a stochastically **complete** *SPDE* solution in terms of the ***PDF*** f_K, and (b) several **partial** *SPDE* solutions in terms of individual *X* **realizations** and $\overline{S[X]}$ **expectation operators**,[b] depending on the objectives of the study.

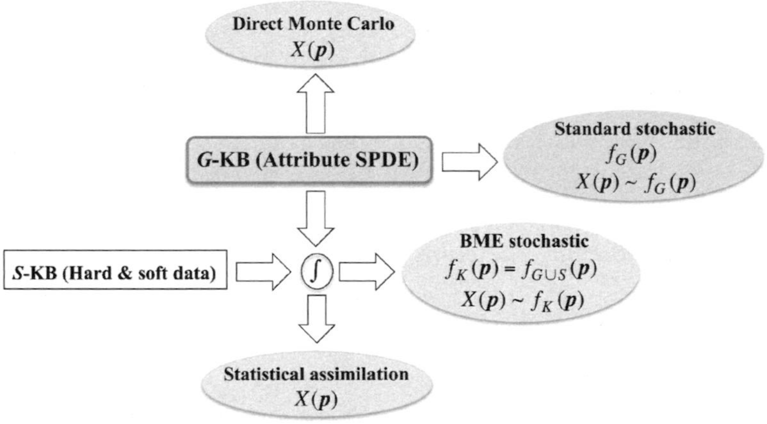

FIG. 1.1 A methodological comparison of the direct Monte Carlo, the statistical assimilation, the standard stochastic, and the *BME* stochastic techniques of ocean attribute *PDE* solution. The symbol " $\int$ " means that the *G*- and *S-KB* are integrated by *BME* to produce the corresponding *PDF*.

[b] *S* denotes a general expectation operator (which includes the mean, mode and median solutions), see Eqs. (2.5a-b), Section 2 of Chapter 5.

In case (b) above, the *SPDE*, which transforms an attribute at a certain point in space and time towards another point in space and time, is in principle characterized by *chronotopologic correlation* expressed mathematically by the operator $S[X]$. Then, by connecting the mathematical operations to the total body of knowledge, *BME* enhances them with physical meaning and, thus, a potentially deeper insight is gained into the phenomenon. The *BME* perspective concerning what constitutes a physical *PDE* solution, as outlined above, is presented in more detail in the following sections of this chapter.

2 BME solution of a physical law

In sciences, one may encounter several kinds of *SPDE* representations of physical laws in conditions of *in-situ* uncertainty (Miller et al., 1999; Lermusiaux et al., 2006; Kharif et al., 2009; Vitanov et al., 2013). Given the importance of such laws in scientific investigations, since the beginning of its development, *BME* theory has focused on the incorporation of physical laws in statistical *CTDA* with applications involving the prediction and mapping of earth, atmospheric, ocean and health phenomena (e.g., Christakos and Raghu, 1996; Serre and Christakos, 1999; Kolovos et al., 2002; Lang and Christakos, 2018). These works paved the way for applications of the *BME* approach to many kinds of physical phenomena, though it certainly requires a closer look at issues associated with the emergence of big data (e.g., rapid changes as captured with modern data may not be reproduced yet with primitive equations, which may need to be updated, and multisourced data errors must be accounted for; see, also, Chapters 12 and 13).

2.1 Deriving the *PDF* of the *SPDE*

Let us recall some relevant notation from previous chapters. The $\chi = (\chi_d, \chi_k)$ denotes the attribute $X(\boldsymbol{p})$ realization through all chronotopologic points $\boldsymbol{p} = (\boldsymbol{p}_1, \ldots, \boldsymbol{p}_m)$ covering the domain of interest; the χ_h and χ_s denote hard and soft data, respectively, so that the total dataset is denoted as $\chi_d = (\chi_h, \chi_s)$; and χ_k are the unsampled attribute values (to be estimated by the *BME* solution). The prior *PDF* f_G of the attribute physical law represented by the *SPDE* is given by Eq. (3.6), Section 3 of Chapter 6, i.e.,

$$f_G(\chi) = e^{\boldsymbol{\mu}^T \boldsymbol{g}} \tag{2.1}$$

where, as usual, the vector $\boldsymbol{g}$ consists of functions expressing mathematically the knowledge about $X(\boldsymbol{p})$ obtained from the *SPDE*, and the vector $\boldsymbol{\mu}$ consists of coefficients the values of which are obtained by solving the integral Eq. (3.5), Section 3 of Chapter 6.

Example 2.1 A set of $\boldsymbol{g}$ functions with the corresponding expectations $\overline{\boldsymbol{g}}$ was listed in Table 2.2, Section 2 of Chapter 6. Many other examples of $\boldsymbol{g}$ functions can be found in the relevant *BME* literature. Note that in this case, the statistics associated with the g-functions are the attribute means $\overline{X(\boldsymbol{p}_i)} = \overline{X}_i$ at the point $\boldsymbol{p}_i$ and the covariance functions $\overline{(X(\boldsymbol{p}_i) - \overline{X}_i)(X(\boldsymbol{p}_{i'}) - \overline{X}_{i'})} = c_{ii'}$ between the points $\boldsymbol{p}_i$ and $\boldsymbol{p}_{i'}$, $i, i' = 1, \ldots, m, k$.

2.2 *SPDE*-derived chronotopologic statistics

As a matter of fact, there are two approaches to calculate mean and covariance functions from the *SPDE*:

1. Using **analytical** techniques, i.e., in some cases the chronotopologic means and covariances of the attribute can be derived directly from the *SPDE* and associated *BIC* by means of closed form expressions (Christakos, 2017).
2. Using **numerical** techniques (e.g., Monte Carlo simulation, Graham et al., 1996), several possible attribute realizations $\chi_i^{(j)} = X^{(j)}(\boldsymbol{p}_i)$ can be generated that satisfy

the *SPDE* and associated *BIC*, where $\chi_i^{(j)}$ denotes the jth realization at point $\boldsymbol{p}_i$. Attribute means and covariances can be calculated from these realizations as

$$\overline{X}_i = \frac{1}{M}\sum\nolimits_{j=1}^{M} \chi_i^{(j)}, \tag{2.2}$$

$$c_{ii'} = \frac{1}{M}\sum\nolimits_{j=1}^{M} \chi_i^{(j)}\chi_{i'}^{(j)} - \overline{X}_i\overline{X}_{i'}, \tag{2.3}$$

where M denotes the number of numerical simulations considered.

For illustration purposes, we start with a presentation of a step-by-step study of the analytical approach ❶ starting with a physical law from environmental sciences.

Example 2.2 Let us explore the chronotopologic distribution of ocean pollution governed by the nondispersive point pollutant transport law on the sea surface (Bracco et al., 2009). In particular, we consider the dynamic (time-varying) unidimensional case of this law represented by the advection-reaction law of sea surface pollution (Lang and Christakos, 2018),

$$\left(\frac{\partial}{\partial t} + \nu\frac{\partial}{\partial s} + \kappa\right)X(\boldsymbol{p}) = 0, \tag{2.4}$$

where $X(\boldsymbol{p})$ is the pollutant value at the space-time point $\boldsymbol{p}$ determined by its space coordinate s and time t, ν is a constant flow velocity and κ is the reaction rate. Below, we implement the step-by-step process leading to the *BME* solution of Eq. (2.4).

Step 1: To gain physical insight into the pollution situation, the behavior of Eq. (2.4) is studied using certain physically justified assumptions. Due to *in-situ* uncertainties, the *BIC* $X_0 = X(\boldsymbol{p}_0)$ is assumed to follow a normal (Gaussian) *PDF* with known mean $\overline{X}_0$ and variance σ_0^2, i.e., $X_0 \sim N(\overline{X}_0, \sigma_0^2)$. It should be stressed that understanding Eq. (2.4) is not just about mastering the relevant mathematical operations.

- Understanding the *SPDE* of Eq. (2.4) is about making **connections** of the mathematical operations to essential ocean pollution knowledge. Understanding how the mathematics describing ocean pollution connects to insights into how the physical system behaves can also provide a safety net against possible errors in the relevant computations.

Step 2: Under the above modeling assumptions and knowledge organization perspective, the solution of the physical law represented by the *SPDE* of Eq. (2.4) is expressed by

$$X(\boldsymbol{p}) = X_0 e^{(\varepsilon v - \kappa)t - \varepsilon s}, \tag{2.5}$$

where ε denotes the inverse of the advection process length, and follows a Normal law with known mean $\overline{\varepsilon}$ and variance σ_ε^2, i.e., $\varepsilon \sim N(\overline{\varepsilon}, \sigma_\varepsilon^2)$, and the X_0 and ε are statistically independent. At this point a significant observation can be made:

- The *SPDE* solution of Eq. (2.5) is directly expressed in terms of the **physical parameters** ν, κ, ε.

Step 3: The means and covariances can be derived analytically from the physical law of Eq. (2.4) and associated *BIC* by means of

$$\overline{X}_i = \overline{X}_0 e^{(vt_i - s_i)\left[\overline{\varepsilon} + \frac{1}{2}(vt_i - s_i)\sigma_\varepsilon^2\right] - \kappa t_i}, \tag{2.6}$$

$$c_{ii'} = \left(\sigma_0^2 + \overline{X}_0\right) e^{\overline{\varepsilon}[v(t_i + t_{i'}) - (s_i + s_{i'})] + \frac{1}{2}[v(t_i + t_{i'}) - (s_i + s_{i'})]^2 \sigma_\varepsilon^2 - \kappa(t_i + t_{i'})} - \overline{X}_i\overline{X}_{i'}. \tag{2.7}$$

Eqs. (2.6) and (2.7) show that the distribution of the ocean pollutant is chronotopologically heterogeneous, i.e., spatially nonhomogeneous/temporally nonstationary, being characterized by varying spatial drifts and temporal trends.

Step 4: The ocean pollutant *SPDE* solution associated with the realization solution of Eq. (2.5) and subject to the statistics of Table 2.1, Section 2 of Chapter 6 is given by Eq. (2.1) above, in particular,

$$f_G(\chi) = A^{-1} e^{\sum_{i=1}^m \mu_i \chi_i + \sum_{i,i'=1}^m \mu_{ii'}(\chi_i - \overline{X}_i)(\chi_{i'} - \overline{X}_{i'})}, \quad (2.8)$$

where $A = -\left[2(2\pi)^m |\boldsymbol{\mu}|^{-1}\right]^{\frac{1}{2}}$, and the standard Lagrange multipliers forming matrix $\boldsymbol{\mu}$ are $\mu_i = 0$, and $\mu_{ii'} = -\frac{1}{2}c_{ii'}^{-1}$, where $c_{ii'}^{-1}$ (i, $i' = 1, \ldots, m$) is the inverse of the covariance $c_{ii'}$. Given that Eq. (2.8) is a function of $\overline{X}_i$ and $c_{ii'}$, it can be directly expressed in terms of the physical parameters ν, κ and ε and their statistics (via Eqs. (2.6) and (2.7)). Therefore, an updated and, perhaps, even more significant observation is made:

- An advantage of the probability solution of Eq. (2.8) over the realization solution of Eq. (2.5) is that the former, in addition to the law parameters ν, κ and ε, also accounts for the **uncertainty** statistics $\overline{X}_0$, $\overline{\varepsilon}$, $\sigma_{X_0}{}^2$ and σ_ε^2 (otherwise said, it helps us understand how the law parameters and *in-situ* uncertainties affect the solution).

By way of example summary: given the means $\overline{X}_i$ and covariances $c_{ii'}$ of ocean pollutant concentrations (calculated either numerically by Eqs. (2.2)–(2.3) or analytically by Eqs. (2.6)–(2.7)), we can compute the Lagrange multipliers μ_i and $\mu_{ii'}$, and then substitute them into Eq. (2.8) to obtain $f_G(\chi)$, which is the *PDF* of the physical law of Eq. (2.4). In fact, $f_G(\chi)$ is the *PDF* obtained by standard (direct) techniques of solving the *SPDE* representing physical laws, like that of Eq. (2.4).

We continue with an illustration of the numerical approach ❷ above. In the process, the readers could find it instructive to consider some key differences between approaches ❶ and ❷.

Example 2.2(cont.) Table 2.1 below displays selected values of the *SPDE* parameters of Eq. (2.4), the *BIC* chosen, and the space-time dimensions of the ocean pollution domain.

Step 1: A Monte Carlo simulation technique was used to directly generate possible solutions $\chi_i^{(j)}$ of the ocean pollution *SPDE* of Eq. (2.4) subject to the associated *BIC*. A resolution of $N = 21 \times 21$ nodes across space and time was assumed (the spatial and temporal units were 0.5 km and 0.5 h, respectively, covering the 10 km × 10 h study domain). For illustration, the chronotopologic pollutant concentration distribution in Fig. 2.1A is one possible simulation, $\chi^{(1)}(\boldsymbol{p})$, of the solution $X(\boldsymbol{p})$.

TABLE 2.1 *SPDE* parameters, *BIC* and domain size of simulation experiment.

Parameters, *BIC* and domain size	Values
Current velocity ν	1 km/h
Reaction rate κ	0.2 h^{-1}
BIC concentration mean $\overline{X}_0$	20 ppm
BIC concentration variance $\sigma_{X_0}{}^2$	1 ppm^2
Noise mean $\overline{\varepsilon}$	0.1 km^{-1}
Noise variance σ_ε^2	0.005 km^{-2}
Spatial domain size *S*	10 km
Temporal domain size *T*	10 h

Step 2: Using Eq. (2.2), the numerical mean ocean pollutant distribution is plotted in Fig. 2.1B that was obtained by averaging $M = 10{,}000$ simulations $\chi^{(j)}(\boldsymbol{p})$, $j = 1, \ldots, 10{,}000$. The mean shows a downward trend across space and time. Compared to the ocean pollutant mean, the ocean pollutant realizations exhibit significant fluctuations. Furthermore, for comparison purposes, the analytical mean pollutant concentration of Eq. (2.5) is plotted in Fig. 2.2A, and the percentage errors between the analytical mean pollutant plot (Fig. 2.2A) and the numerical mean pollutant plot (Fig. 2.1B) are displayed in Fig. 2.2B.

- The maximum errors are less than 1%, which demonstrate the very good **accuracy** of the numerical ocean pollutant simulation results, in this case.

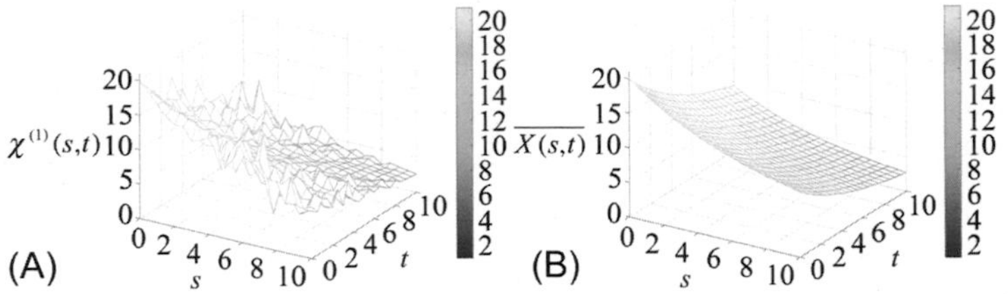

FIG. 2.1 (A) A simulation $\chi^{(1)}(s,t)$ of the stochastic solution $X(s,t)$ of the ocean pollution Eq. (2.4) subject to associated *BIC*. (B) Mean pollutant distribution $\overline{X(s,t)}$ found by averaging the simulations $\chi^{(j)}(s,t)$, $j=1, \ldots, 10{,}000$..

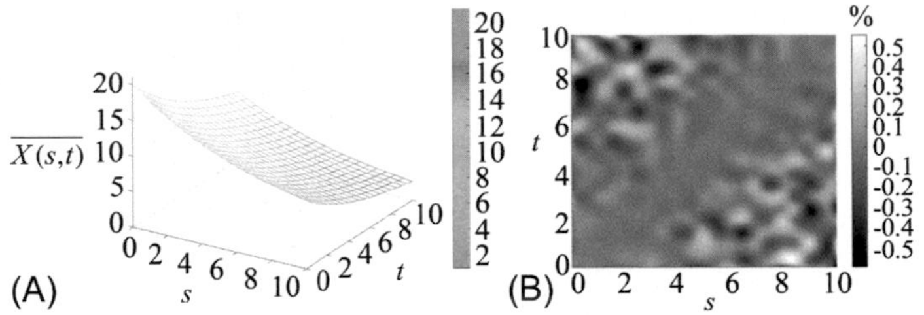

FIG. 2.2 (A) Analytical mean space-time pollutant distribution $\overline{X(s,t)}$. (B) Percentage error between numerical and analytical mean values across space-time.

Step 3: Based on Eqs. (2.8)–(2.9a,b), the *PDF* f_G associated with the ocean pollutant *SPDE* of Eq. (2.4) were derived at each point of the 10 km × 10 h study domain and plotted in Fig. 2.3. All f_G plots are symmetric with varying peaks, meaning that the probabilities of the generated solutions vary from one point to another and they also depend on the values of the physical parameters ν, κ and ε. These plots will be also useful in subsequent calculations and comparisons.

2.3 Assimilating specificatory knowledge (*S-KB*)

In practice, the case-specific *S-KB* is collected using different methods, and it may include hard data (i.e., noise-free observations) and/or soft information (e.g., measurements or auxiliary variables containing measurement noise or human operator errors). As mentioned earlier, the hard data at points $\boldsymbol{p}_h$ are denoted by the vector $\boldsymbol{\chi}_h$, and the soft data at points $\boldsymbol{p}_s$ by

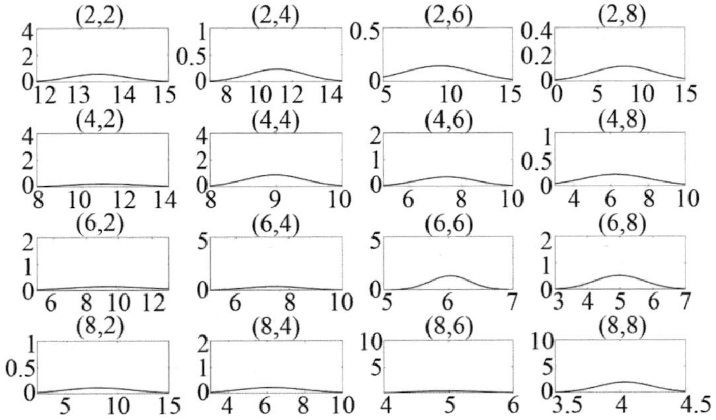

FIG. 2.3 Plots of the *PDF* f_G obtained from the ocean pollutant *SPDE* of Eq. (2.4) at the space-time points (2,2), (2,4), …, (8,8).

the vector χ_s. Soft data χ_s are usually expressed in the form of probability distributions $f_S(\chi_s)$ of pollutant concentrations (this includes intervals of concentration values and empirical charts), i.e.,

$$\chi_s \sim f_S(\chi_s); \tag{2.9a}$$

whereas the hard data can be formally expressed as

$$\chi_h \sim \delta_S(\chi_h), \tag{2.9b}$$

so that $\chi_h = \int d\chi \delta_S(\chi - \chi_h)\chi$, where δ_S is the delta function. As usual, the entire data vector can be written as $\chi_d = (\chi_h, \chi_s)$, and the corresponding set of data points as $p_d = (p_h, p_s)$. The vector of points p in Eq. (2.1) includes the vector of the data points p_d and the vector of the unsampled points p_k with unknown pollutant concentrations χ_k. The χ_k concentrations need to be predicted across space and time in terms of the *BME* solution of the attribute *SPDE* (considered as the *G KB*) subject to the available $S - KB$ (i.e., hard χ_k and soft data $f_S(\chi_s)$), so that $\chi = (\chi_h, \chi_s, \chi_k)$.

Example 2.2(cont.) To demonstrate the case-specific data assimilation in to the *BME* solution of the advection-reaction *SPDE*, we proceed as follows:

Step 1: In the case of the sea surface pollution law of Eq. (2.4) above we assume that the *S-KB* consists of a hard datum $\chi_h = 14.5$ ppm at point $p_C = (s_C, t_C) = (2\,\text{km},\ 1\,\text{h})$, and soft data at points p_A and p_B with coordinates $(s_A, t_A) = (6\,\text{km},\ 9\,\text{h})$ and $(s_B, t_B) = (5\,\text{km},\ 0\,\text{h})$, respectively. In real practice, a seawater quality monitoring system may be used to gauge pollutant concentrations or the concentrations may be obtained from other sources, like remote sensing images by means of empirical functions.

Step 2: For comparison purposes, at each soft datum point three different kinds of data *PDF* $f_S(\chi_s)$ were considered that shared the same numerical mode value (this choice was made for practical reasons, since in reality the *PDF* mode is often easier to observe than other statistical measures of central tendency):

(a) Uniform probability distributions with range [3.5 ppm, 5 ppm] at point p_A and range [9 ppm, 12 ppm] at point p_B (Fig. 2.4A).

(b) Gaussian distributions with mean 5.5 ppm and standard deviation 0.5 ppm at p_A (Fig. 2.4B), and mean 8 ppm and standard deviation 1 ppm at p_B (Fig. 2.4B).

(c) Triangular probability distributions with interval range [4, 5.5] ppm and mode 5 ppm at p_A, and range [10.5, 13] ppm and mode 11 ppm at p_B (Fig. 2.4C).

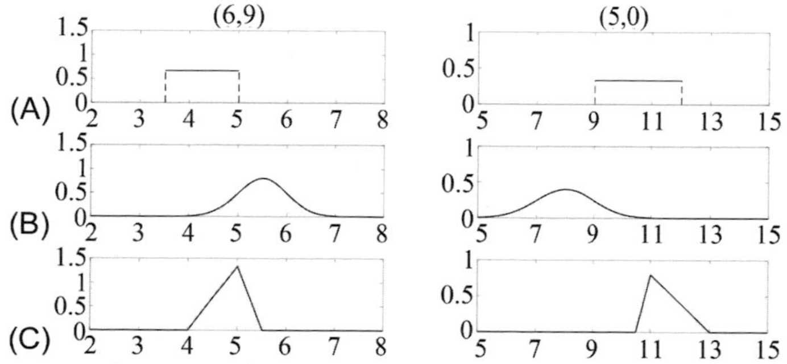

FIG. 2.4 The soft data *PDF* $f_S(\chi_s)$ at points $p_A = (6\,\text{km}, 9\,\text{h})$ and $p_B = (5\,\text{km}, 0\,\text{h})$ assuming (A) uniform, (B) Gaussian and (C) triangular data probability shapes.

2.4 Updating the physical *SPDF* in light of *S-KB*

In light of the *S-KB* described above, the updated pollutant *PDF* $f_K(\chi_k)$ of attribute solutions at the unsampled points χ_k is obtained by means of the integral Eq. (3.8), Section 3 of Chapter 6. Given that the *S-KB* consists of hard data χ_h and soft data $f_S(\chi_s)$ (Fig. 2.4A–C), Eq. (2.10) becomes

$$f_K(\chi_k) = A^{-1}\int d\chi_s f_S(\chi_s) f_G(\chi_d, \chi_k), \qquad (2.10)$$

where the total *S-KB* has been taken into consideration in constructing the $f_K(\chi_k)$, and $A = \int d\chi_s f_S(\chi_s) f_G(\chi_d)$ is a normalization constant. In other words, Eq. (2.10) provides the *posterior PDF* $f_K(\chi_k)$ that is an update of $f_G(\chi)$ after the *S-KB* has been also taken into account, i.e., the subscript $K = (G, S)$ denotes the total *KB* that includes both *G* and *S*.

For illustration, in the special case that $f_S(\chi_s)$ is a uniform *PDF* over the interval *I* of possible pollutant concentration values, Eq. (2.10) reduces to

$$f_K(\chi_k) = A^{-1}\int_I d\chi_s f_G(\chi),$$

where $A = \int_I d\chi_s f_G(\chi_d)$. Given $f_K(\chi_k)$, various attribute solutions can be derived at each point $\boldsymbol{p}_k$ in terms of the pollutant *mean*, the *mode* and the *median* given by Eqs. (2.9)–(2.11), Section 2 of Chapter 9.

- The choice of one of the **alternative** *BME* solutions above should depend on the particular conditions of the *in-situ* ocean pollution application under consideration.

Example 2.2(cont.) In the case of the sea surface pollution of Example 2, by considering the hard datum at the point $\boldsymbol{p}_C$ and the soft data shown in Figs 2.4A–C, Eq. (2.10) was used to calculate the updated pollutant concentration PDF at each one of the 21 × 21 nodes of the chronotopologic domain of interest.

Step 1: In particular, Eq. (2.10) can be expressed in the following computationally efficient formulation:

$$f_K(\chi_k) = f_G(\chi_k|\chi_d) = A^{-1}\int d\chi_s f_S(\chi_s) f_G(\chi_s, \chi_k|\chi_h) f_G(\chi_h), \qquad (2.11)$$

where $f_G(\chi_k|\chi_d)$ is a conditional Gaussian law with mean vector $\overline{\mathbf{X}_k|\mathbf{X}_d} = \overline{\mathbf{X}_k} + C_{X_kX_d}C^{-1}_{X_hX_h}(\chi_h - \overline{\mathbf{X}}_h)$ and covariance matrix $C_{X_k|X_d} = C_{X_kX_k} - C_{X_kX_d} C_{X_hX_h}{}^{-1}C_{X_dX_h}$, and $A = \int d\chi_s f_S(\chi_s) f_G(\chi_s|\chi_h) f_G(\chi_h)$ is a constant.

Step 2: Figs. 2.5–2.7 display the *PDF* $f_K(\chi_k)$ of ocean pollutant concentrations obtained by Eq. (2.11) (although the *PDF* were obtained at all

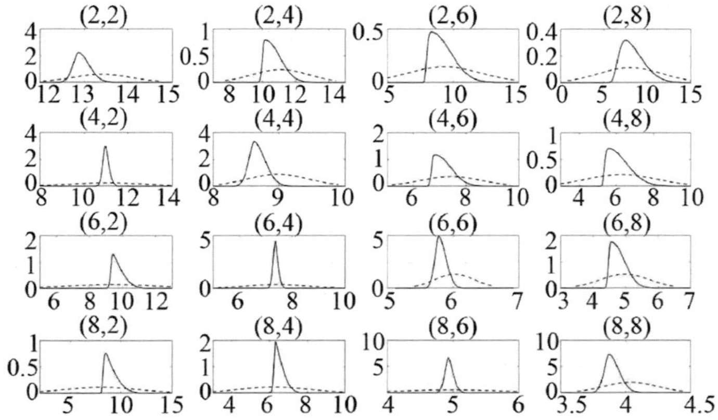

FIG. 2.5 *K*-based *PDF* f_K (solid lines) at the space-time points (2 km,2 h), (2 km,4 h), …, (8 km,8 h) for uniformly distributed soft data (Fig. 2.4A); the *G*-based *PDF* f_G (dash lines) are also shown for comparison.

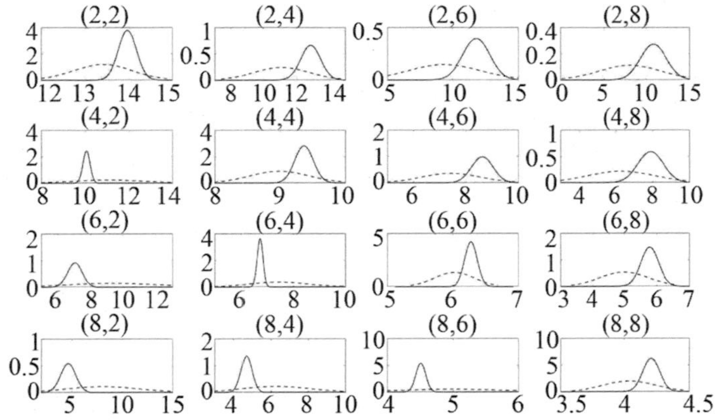

FIG. 2.6 *K*-based *PDF* f_K (solid lines) at the space-time points (2 km,2 h), (2 km,4 h), …, (8 km,8 h) for normally distributed soft data (Fig. 2.4B); the *G*-based *PDF* f_G (dash lines) are also shown for comparison.

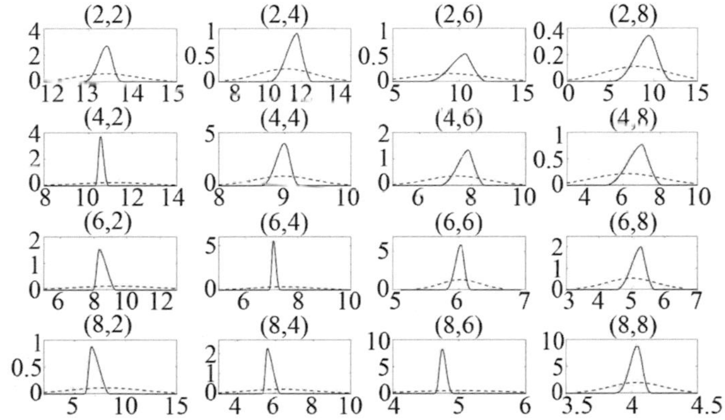

FIG. 2.7 *K*-based *PDF* f_K (solid lines) at the space-time points (2 km,2 h), (2 km,4 h), …, (8 km,8 h) for triangularly distributed soft data (Fig. 2.4C); the *G*-based *PDF* f_G (dash lines) are also shown for comparison.

21 × 21 space-time nodes, due to obvious space limitations, the *PDF* at selected space-time nodes (2km, 2h), (2km, 4h), …, (8km, 8h) are shown). The plots in these figures clearly depict the differences between the *G*-based *PDF* (Fig. 2.3, also shown in dashed lines in Figs. 2.5–2.7) *vs.* the *K*-based *PDF*. These differences are physically attributed to the effect of the soft data that f_K incorporates. The shapes of f_K are asymmetric and more involved than the smooth and symmetric shapes of f_G, the f_K shapes change from point to point to a much larger degree than the f_G shapes, and the widths of f_G are considerably longer (implying higher uncertainty) than those of f_K at all points considered. Hence, the following observation can be made:

- The standard solution of the ocean pollution law of Eq. (2.4) is more **uncertain** than the *BME* solution and, accordingly, its ocean pollution predictions are less accurate than those obtained by the *BME* solution (see, also, further results in this respect below).

Step 3: Given the *PDF* f_K of pollutant concentrations across space and time, we can also

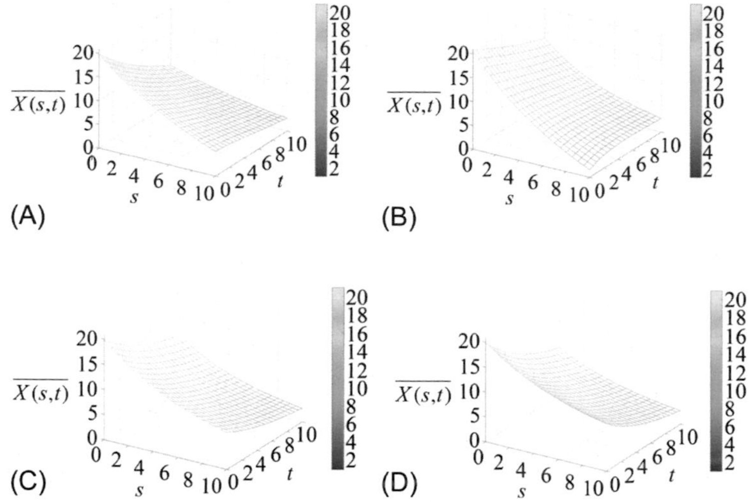

FIG. 2.8 The mean chronotopologic distributions $\overline{X(s,t)}$ calculated from f_K using soft data with (A) uniformly, (B) normally and (C) triangularly shaped distributions; (D) the mean chronotopologic distribution calculated from f_G (notice that this is the same plot as in Fig. 2.2A).

calculate several ocean pollution solutions. For illustration, in Fig. 2.8A–C we plot the chronotopologic distributions of the ocean pollutant mean obtained by the f_K for the three kinds of soft data of Fig. 2.4A–C. The chronotopologic distribution of the ocean pollutant mean obtained by the *PDF* f_G of Eq. (2.8) is also plotted in Fig. 2.8D, for comparison purposes. There are certain visually detectable differences between the four ocean pollution prediction maps in Fig. 2.8A–D, which further demonstrate the considerable effects of the differences among the soft data on ocean pollution assessment.

Step 4: Moreover, the accuracy of each one of these ocean pollution maps will be calculated in Section 2.5 below in terms of the distributions of the pollutant prediction error standard deviations. This is an important *CTDA* aspect, since the error is valuable information about the phenomenon. *Inter alia*, error assessment can play a critical role in pollution risk analysis that describes uncertain scenarios wherein chosen actions yield a range of possible outcomes that are quantified by distinct statistical features in the prediction error distribution.

2.5 Accuracy comparison of the standard stochastic solution *vs.* the *BME* stochastic solution

After the corresponding *PDF* (f_G and f_K) have been calculated across space and time, the uncertainty (or the corresponding accuracy) of the chronotopologic ocean pollutant concentration predictions of the simulation experiment above can be calculated in terms of the prediction error standard deviation at each prediction point $(\boldsymbol{s}_k, t_k)$, i.e.,

$$e_X(\boldsymbol{s}_k, t_k) = \left[\int d\chi_k (\chi_k - \overline{x}_k)^2 f(\chi_k)\right]^{\frac{1}{2}}, \qquad (2.12)$$

where $k = 1, \ldots, 21 \times 21$, and $f = f_G$ or $f = f_K$.

Example 2.2(cont.) As was anticipated earlier, some more steps must be included in the study:

Step 4(cont.): The distributions of the ocean prediction error standard deviation $e_X(\boldsymbol{s}_k, t_k)$ are plotted in Fig. 2.9A–D. It is obvious that the $e_X(\boldsymbol{s}_k, t_k)$ values of ocean pollutant predictions for $f = f_K$ are considerably smaller than those for $f = f_G$. In particular, the averaged prediction error standard deviations in Fig. 2.9A–D are: 0.575 ppm (map of Fig. 2.9A), 0.680 ppm (map

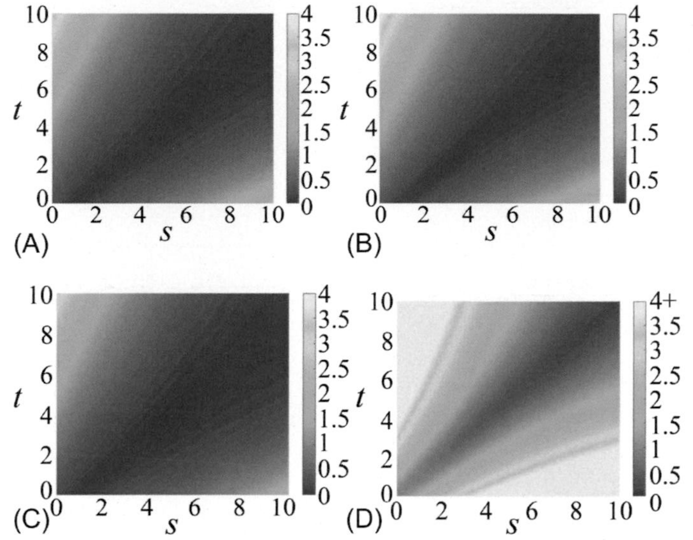

FIG. 2.9 Distribution of the prediction error $e_X(s_k, t_k)$ using f_K. The corresponding soft data was a (A) uniform, (B) normal and (C) triangular distribution. And, (D) distribution of $e_X(s_k, t_k)$ using f_G.

of Fig. 2.9B), 0.498 ppm (map of Fig. 2.9C), and 2.265 ppm (map of Fig. 2.9D). Furthermore, the maps of Fig. 2.9A–D display a certain similarity with the error image of Fig. 2.2B.

Step 5: These results lead to the following noticeable conclusions:

① By being able to calculate f_K at every point of the domain, the *BME* approach outperforms the standard stochastic technique that can only calculate f_G.

② The fact that the triangular soft data leads to the most accurate pollutant concentration solutions may indicate that this kind of soft data is the most appropriate case-specific information for the ocean pollution study considered.

Lastly, Fig. 2.10 presents an outline of the *BME* approach to the solution of the advection-reaction law of ocean pollutant concentration (*G-KB*) considered in the simulation study. The *BME* approach successfully assimilated hard data and three kinds of soft data: uniformly, normally and triangularly distributed observations (*S-KB*). The solution is expressed in terms of the full pollutant concentration $f_K(\chi_k)$ at each point in the chronotopologic domain of interest.

3 *BME* solution of an epidemic law

Disease diffusion can vary significantly from place to place and from time to time for a number of reasons, including heterogeneity of the hosts and pathogens, physical and social environments, and interactions across space and time. Moreover, uncertainties linked to population movement and records of infected individuals can increase the difficulty of understanding the spread of an infectious disease. The so-called *Susceptible-Infected-Recovered* (*SIR*; e.g., Allen and Burgin, 2000) models are widely implemented to represent the disease evolution of populations over time.

Angulo et al. (2013) proposed a realistic *space-time* extension of the purely temporal *SIR* model, i.e., a metapopulation model, in the context of *BME* theory, which led to a powerful synthesis of the two, jointly called the ***BME-SIR*** approach.

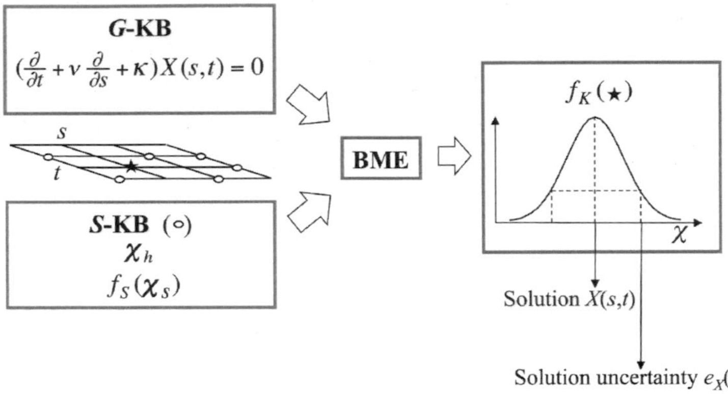

FIG. 2.10 An outline of the implementation of the *BME* approach in the ocean pollution study.

In the *BME* context, then, *SIR* constitutes a key component of the *G-KB*. The distribution of the fraction of infected population is represented as the random field $X(\boldsymbol{p})$. Similarly, $Y(\boldsymbol{p})$ and $Z(\boldsymbol{p})=Z(\boldsymbol{s},t)$ are random fields representing the distributions, respectively, of the fraction of the population that is susceptible to become infected and the fraction of the population that has recovered and is immune. Basic relationships between $X(\boldsymbol{p})$, $Y(\boldsymbol{p})$ and $Z(\boldsymbol{p})$ exist as

$$\begin{aligned} &X(\boldsymbol{p})+Y(\boldsymbol{p})+Z(\boldsymbol{p})=1,\\ &Y(\boldsymbol{s};0)=1-X(\boldsymbol{s};0),\\ &Z(\boldsymbol{s};0)=0, \end{aligned} \tag{3.1a-c}$$

where $X(\boldsymbol{s},0)$, $Y(\boldsymbol{s},0)$, and $Z(\boldsymbol{s},0)$ denote the initial conditions (*IC*) of the corresponding population fractions. Then, the modeling of the combined space-time distributions $X(\boldsymbol{p})$, $Y(\boldsymbol{p})$ and $Z(\boldsymbol{p})$ is described by the following *SIR* model in matrix form:

$$\begin{aligned} &\boldsymbol{X}(t+1|\,t)=\boldsymbol{A}\boldsymbol{X}(t)+\boldsymbol{W}_t\\ &\boldsymbol{P}(t+1|\,t)=\widehat{\boldsymbol{A}}\boldsymbol{P}(t|\,t)\widehat{\boldsymbol{A}}^T+\boldsymbol{Q}_t\\ &\boldsymbol{P}(t+1|\,t+1)=\left\{I-\boldsymbol{P}(t+1|\,t)H^T\left[H\boldsymbol{P}(t+1|\,t)H^T\right]^{-1}H\right\}\boldsymbol{P}(t+1|\,t), \end{aligned} \tag{3.2a-c}$$

where $\boldsymbol{X}(t)$ and $\boldsymbol{X}(t+1\,|\,t)$ are vectors containing the current and predicted states of infected disease counts and the (immune and infection) rates of the *SIR* model, $\boldsymbol{I}$ is the unit matrix, and the matrices $\boldsymbol{A}$ and $\widehat{\boldsymbol{A}}$ are the transition and Jacobian matrices characterizing the dynamics of the *SIR* model. In particular, the elements of $\boldsymbol{A}$ are functions of the population fraction $q_{s,t}$ that resides at the domain $(\boldsymbol{s},t)$, the population portion $k_{s,\,t}$ that resides at $(\boldsymbol{s},t)$ and does not migrate, the rate $a_{s,\,t}$ that an infected individual at the domain recovers and becomes immune, the corresponding rate of infection transmission $b_{s,t}$ during an encounter of one infected and one susceptible individual, and a function $\varphi_{s,t}$ with a smooth shape similar to that of the infected population covariance function (note that $a_{s,t}$ and $b_{s,t}$ allow one to include information about regional topography and local climatic conditions). The vector $\boldsymbol{W}_t$ models the dynamic uncertainty of infected states across space, which cannot be represented by *SIR* modeling, and is characterized by the covariance matrix $\boldsymbol{Q}_t$. The observation matrix $\boldsymbol{H}$ contains only 0 s and 1 s indicating data presence across space. $\boldsymbol{P}(t)$, $\boldsymbol{P}(t+1\,|\,t)$ and $\boldsymbol{P}(t+1\,|\,t+1)$ are the current, predicted and updated state covariances (for technical details, see Angulo et al., 2013).

Eqs. (3.2a-c), describing the stochastic properties of disease dynamics across space, are solved using the *BME* approach above, which, at the same time, accounts for the *S-KB* (including the uncertain infected observations and the initial conditions of the transmission and recovery rates). This approach was implemented in the

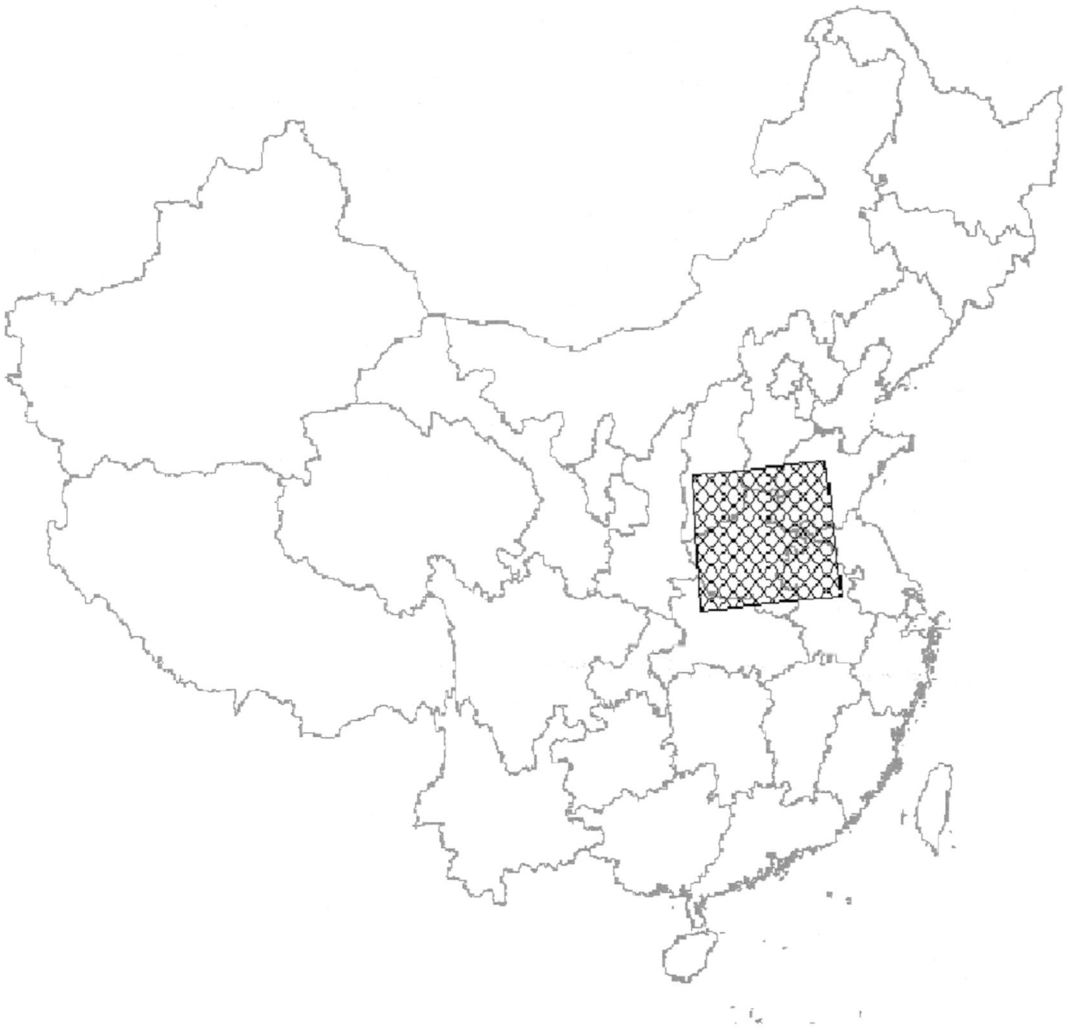

FIG. 3.1 The study region and its location in China.

case study described in the following real-world example.

Example 3.1 Angulo et al. (2013) used the *BME* method to solve the *SIR* equations of the *hand-foot-mouth disease* (*HFMD*) at a region in China with relatively high disease incidence. The study focus was the disease evolution at 145 counties that extend between 111°E to 118°E, and 32°N to 37°N (Fig. 3.1). The data were weekly-aggregated *HFMD* rates (infected cases per 10,000 people) over a period of 20 weeks that spanned from September 27–October 3, 2008 ($t = 1$ week) to February 7–13, 2009 ($t = 20$ weeks). The main steps of the *BME* approach used in this study were as follows:

Step 1: The *SIR* model of Eqs. (3.2a-c), which belongs to the *G-KB* of the case study, was used to model the *HFMD* dynamics across space. The parameters and initial conditions of the *SIR* model were computed next.

Step 2: The initial spatial spread of infecteds, $X(\mathbf{s}, t = 1)$, consisted of observed cases, no recovered individuals were assumed at $t = 1$, and the remaining population consisted of susceptibles. One-week disease duration was considered; relocation occurred sparsely during the 20-week study period that was accounted for by a Gaussian kernel function with 0.1 bandwidth (97.55% of the population on average did not relocate during the study period). The recovery and transmission rates had initial values $a = 0.1$, $b = 0.4$ and variances $\sigma_a^2 = 0.05$, $\sigma_b^2 = 0.1$. The initial covariance $\boldsymbol{P}(t = 1 | t = 1)$ for $X(\boldsymbol{s}, t = 1)$ was computed and a model was fitted with a 0.07 nugget effect and a spherical component with a 0.07 sill and a 3° spatial range.

Step 3: All observations belonging to the *S-KB* were assumed to be uncertain, and each rate was randomly sampled from a uniform distribution that is 1 unit wide. Observed incidence rates that were reported to be exactly 0 were represented by soft uniform distributions with rates between [0,1]. The soft interval width selection was a conservative estimate on the basis of the recorded national average rates for *HFMD* (3.69 in 2008 and 8.68 in 2009 based on the corresponding population sizes and the reported *HFMD* cases).

Step 4: On the basis of the above inputs, space-time distributions were obtained of the *HFMD* infected $X(\boldsymbol{p})$, susceptible $Y(\boldsymbol{p})$, and recovered $Z(\boldsymbol{p})$ population fractions throughout the 20-week period. At each consecutive time instance, the *SIR* model parameters a and b came progressively closer to the values that best interpreted the *HFMD* dataset. This process is also guided by updating the model with new data at every time step. Fig. 3.2 illustrates how the *SIR* parameter values estimated from the *HFMD*

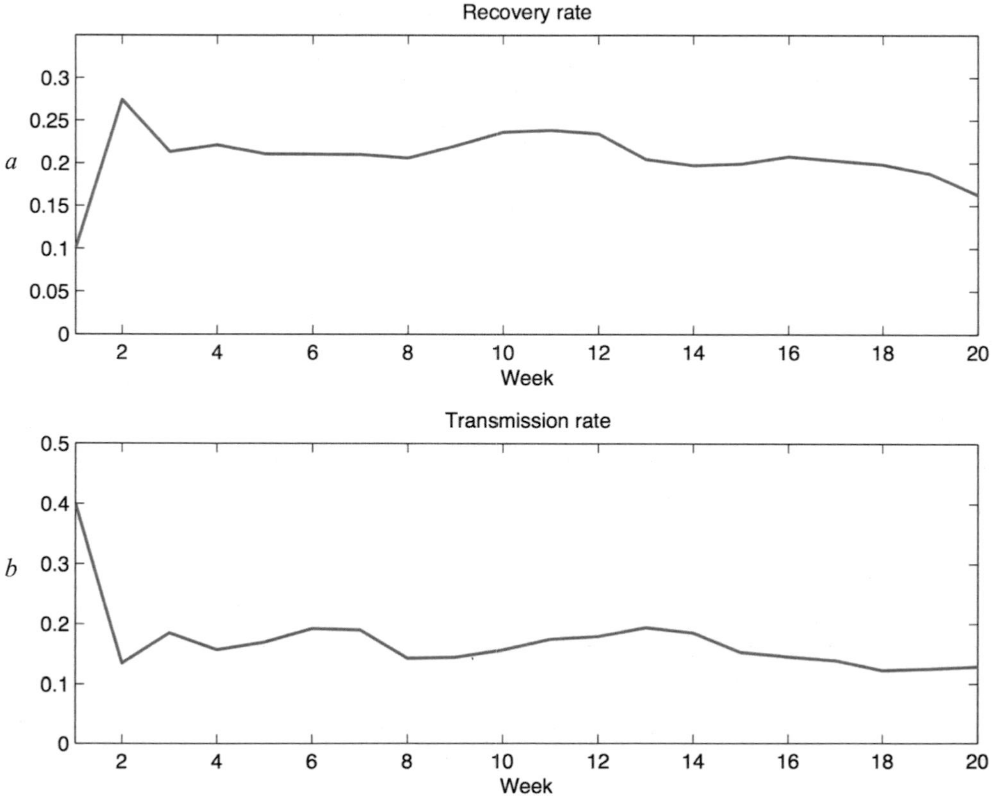

FIG. 3.2 Estimated transmission and recovery rates of the *HFMD* study.

data reached equilibrium. The resulting *BME-SIR* model predicts an approximate mean transmission rate $b \approx 0.17$, and an approximate mean recovery rate $a \approx 0.21$. Despite the arbitrary initial a and b values, relatively accurate parameter estimates were reached rather fast (within about 2–4 weeks).

Step 5: Maps of the *BME-SIR* solution for the infected distribution means are produced for each of the 20 weeks of this study. Fig. 3.3 shows these means within the region of interest at selected weekly instances, and Fig. 3.4 illustrates the corresponding solution error at these instances. The solution error throughout the study was found to range between 0.0067 and 0.2884. These values reflect that the *SIR* solutions using *BME* also account for *HFMD* observation uncertainty. In sum, this real-world case study indicates that the *BME* solution of the *SIR* equations can provide an informative overview of the disease evolution. Also, this application showed how *BME-SIR* can be effectively used to compute the disease spread based on highly uncertain data, without any distributional assumptions.

- The ***BME-SIR*** solution can assimilate both theoretical disease diffusion dynamics and the uncertain chronotopologic disease data.

As a result, the characteristics of disease evolution can be revealed over time, even in cases when the disease data are highly uncertain. By way of a summary, the chronotopologic *BME-SIR* approach was shown to have certain attractive features in this case study:

① It represented the population dynamics of infectious diseases within and across localities.

② It took into consideration the composite space and time variation of disease features.

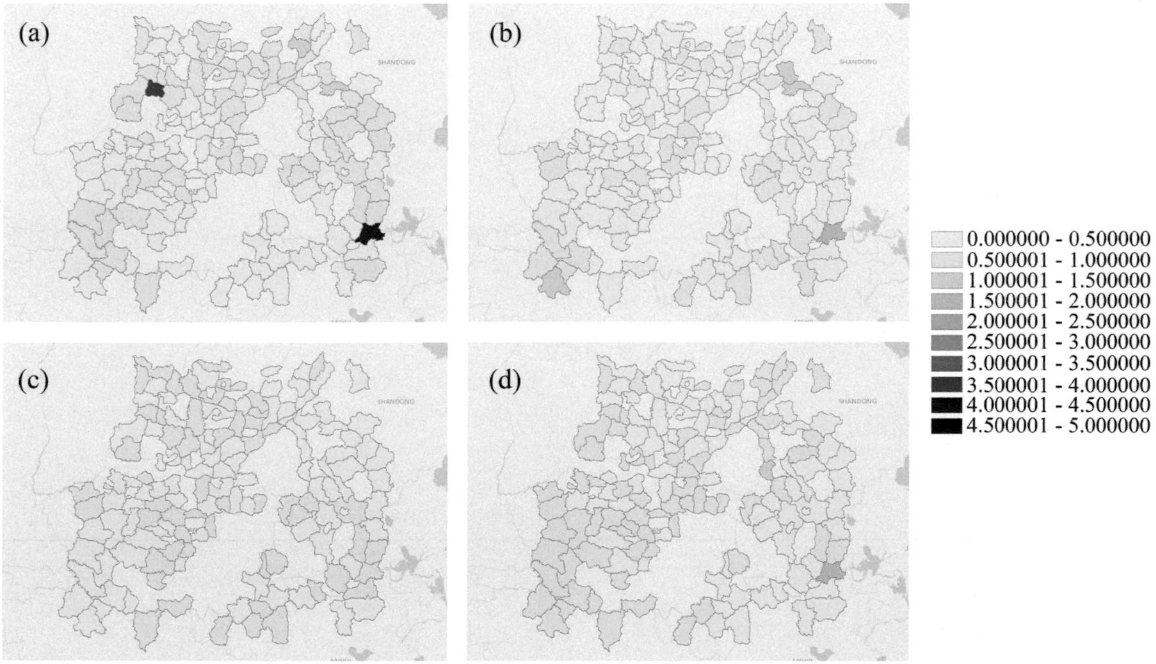

FIG. 3.3 *BME-SIR* population *HFMD* incidences (cases per 10,000 people) in the study region for four selected week instances: (A) $t = 5$, (B) $t = 10$, (C) $t = 15$, and (D) $t = 20$.

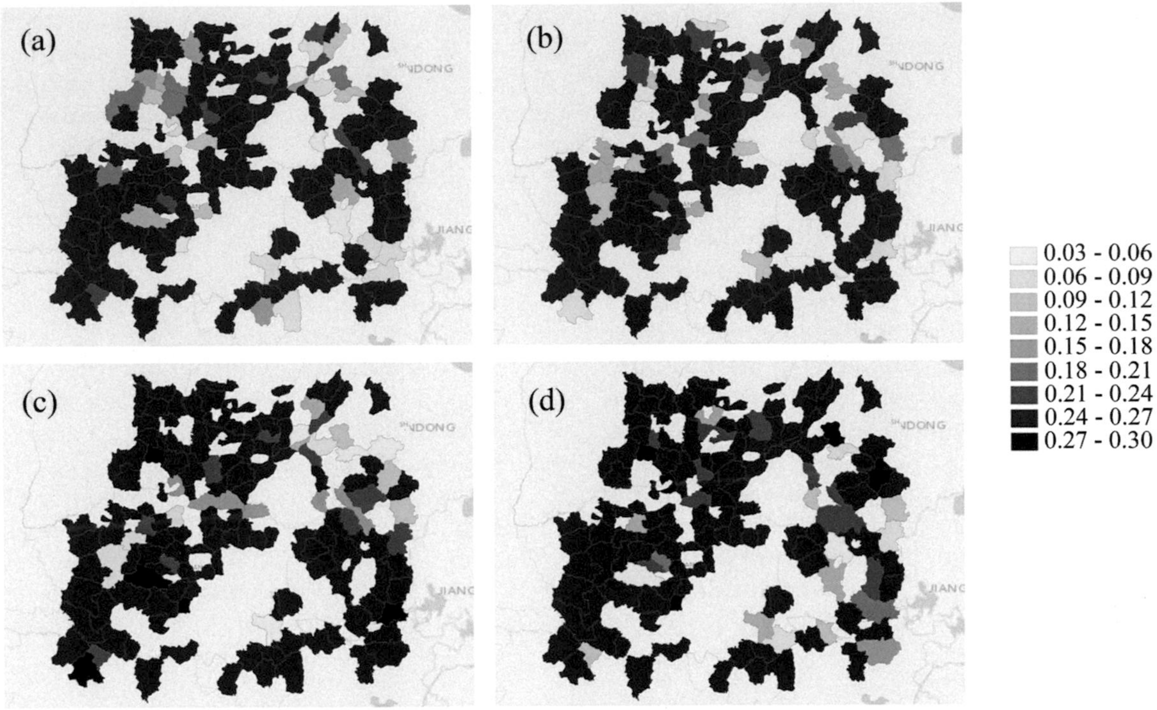

FIG. 3.4 *BME-SIR* standard error for the *HFMD* incidences shown in Fig. 3.3.

③ It accounted for observation uncertainties (e.g., in the records of infected individuals).
④ In addition to the *SIR* disease dynamics, it integrated different sources of knowledge (e.g., hard and soft disease data together with epidemic models and physical laws).
⑤ It updated the chronotopologic model parameters in real time.

4 Comparing core and specificatory probabilities

This section discusses certain further potential improvements offered by the *BME* solution of an *SPDE* representing a physical law:

Although at the *G*-stage of the *BME* approach the same *SPDE* solution may be derived as that derived by the standard stochastic approach, improved results are obtained at the subsequent ***K*-stage** that cannot be obtained by the standard stochastic solution techniques.

The above claim is essentially twofold, and it will be demonstrated accordingly in this section.

4.1 Generalization and realism

We start by focusing on two characteristics of the *BME* approach compared to the mainstream ones.

The *BME* solution is more **general**, in the sense that it reproduces the classical stochastic solution strictly under the same conditions, but it can also go one step beyond and become more **realistic** in the sense that it accounts for important site-specific information.

For illustration purposes, a numerical example is examined next.

Example 4.1 A natural attribute $X(t)$ obeys the following stochastic physical law,

$$\frac{d}{dt}X(t)=bX(t), \tag{4.1}$$

where $b=1$ (for simplicity) is an empirical parameter, and the physical law's *IC* is $X_0=X(t=0)\sim f_X(\chi_0)\equiv N(\overline{X_0}=1,\sigma_0=1)$.

Step 1: Using the *standard stochastic* approach (Fig. 1.1 of Section 1), the *PDF* associated with Eq. (4.1) is the Gaussian

$$\begin{aligned} f_G(\chi,t) &= f_G(\chi_0=\chi e^{-bt})e^{-bt} \\ &= \frac{1}{\sqrt{2\pi}\,\sigma_0}e^{-\sigma_0^{-2}\left(\frac{1}{2}\overline{X_0}^2-\overline{X_0}\,e^{-bt}\chi+\frac{1}{2}e^{-2bt}\chi^2\right)-bt} \\ &= \frac{1}{\sqrt{2\pi}}e^{-\frac{1}{2}+e^{-t}\chi-\frac{1}{2}e^{-2t}\chi^2-t}. \end{aligned} \tag{4.2}$$

As was shown in Christakos (2017) and elsewhere, the same result is obtained using *BME*.

Step 2: To perform some numerical simulations, using Eqs. (4.1) and (4.2) it is found that at time $t=1$ the *PDF* is

$$f_G(\chi,1)=\frac{1}{\sqrt{2\pi}}e^{-\frac{3}{2}+e^{-1}\chi-\frac{1}{2}e^{-2}\chi^2}. \tag{4.3}$$

Step 3: A sample is considered at $t=1$ that follows a uniform distribution, $f_S(t=1)\sim U\left(0,\frac{1}{\sqrt{2\pi}}\right)$.

Step 4: Given the sample, *BME* can be again employed to calculate the *K-PDF* as

$$\begin{aligned} f_K(\chi,1) &= A^{-1}f_G(\chi,1)f_S(\chi,1)=\frac{f_G(\chi,1)f_S(\chi,1)}{\int d\chi f_G(\chi,1)f_S(\chi,1)} \\ &= \frac{e^{-1.5+e^{-1}\chi-0.5e^{-2}\chi^2}}{\int_0^{+\infty}d\chi e^{-1.5+e^{-1}\chi-0.5e^{-2}\chi^2}} \\ &= \frac{e^{-1.5+e^{-1}\chi-0.5e^{-2}\chi^2}}{2.1089}. \end{aligned} \tag{4.4}$$

Hence, when comparing the standard *vs.* the *BME* solutions to Eq. (4.1) above, the interesting conclusion is drawn:

- Initially, both the standard and the *BME* solutions derive the same *G-PDF*, f_G. But, *BME* can go one step further and assimilate specificatory empirical evidence *S*, thus obtaining an **augmented** *K-PDF*, f_K, that accounts for a richer knowledge body than the standard stochastic approach.

4.2 The *S-KB* effect on solution uncertainty

Next, we examine further the effect of specificatory empirical evidence (*S-KB*) on the final *BME* solution at the *K*-stage.

The incorporation of ***S-KB*** can improve considerably the standard stochastic solution of a physical law by reducing the number of possible attribute realizations and the uncertainty of the *SPDE* solution.

For illustration purposes, a numerical example is examined next.

Example 4.2 Consider again Eq. (4.1), where now $X_0=1$ is the law's *IC*, and b is a random coefficient, e.g., $b\sim U[0,\ 1]$.

Step 1: Solving Eq. (4.1) under these conditions gives

$$X(t)=X_0e^{bt}=e^{bt}. \tag{4.5}$$

For illustration, given that $b\sim U[0,1]$, the standard stochastic solution of Eq. (4.5) at $t=2$ follows the uniform *PDF*,

$$X(2)\sim U(2.3052, 2.7257), \tag{4.6}$$

i.e., $f_G(\chi,2)=\frac{1}{2.7257-2.3052}=2.7381$. In *BME* terms, this is considered the *G*-based *PDF* at $t=2$. Moreover, by uniformly varying the values of b between 0 and 1, several possible $X(t)$ realizations of Eq. (4.5) are obtained, see curves in Fig. 4.1A. The large number of realizations implies a considerable attribute *uncertainty*. In fact, only information about the attribute trend is obtained from Fig. 4.1A (i.e., no other interesting conclusion can be drawn from this figure).

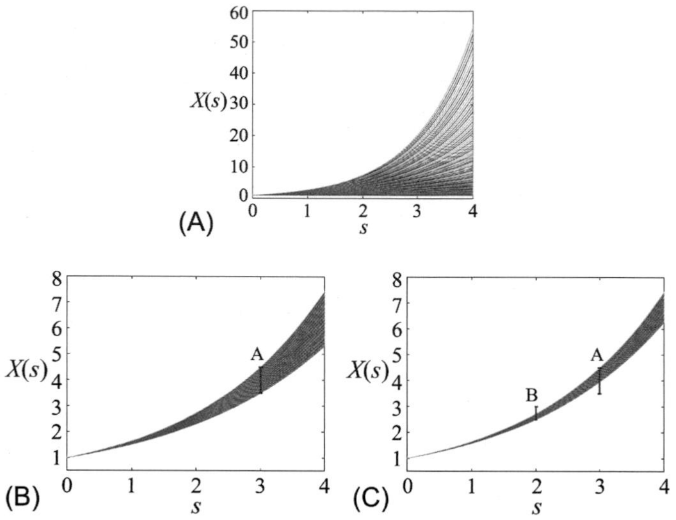

FIG. 4.1 Changes of possible X-realizations by assimilating site-specific data in to the core knowledge. (A) Only core knowledge is considered; (B) One site-specific datum is added; (C) Two site-specific data are added.

Step 2: Therefore, a sampling strategy is needed for a better understanding of the attribute. For this purpose, assume that using the appropriate equipment a sample A (Fig. 4.1B) is obtained at time at $t = 3$ that is found to obey a uniform *PDF*,

$$X(3) \sim U(3.5, 4.5), \tag{4.7}$$

i.e., $f_S(\chi, 3) = \frac{1}{4.5-3.5} = 1$. As a result of sampling, the set of curves in Fig. 4.1A is reduced to the much smaller set of curves of Fig. 4.1B, where b ranges from 0.4176 to 0.5014. Due to the much smaller range of b values, Fig. 4.1B offers a less uncertain representation of the X attribute's variation than Fig. 4.1A.

Step 3: Yet, there are still many possibilities (attribute realizations) in Fig. 4.1B that honor the sample A and simultaneously satisfy the physical law of Eq. (4.1). Therefore, it is decided that more sampling is needed to further reduce uncertainty, i.e., to reduce the number of possible curves (realizations) in Fig. 4.1B.

Step 4: Next, one actually collects a sample B at $t = 2$ (Fig. 4.1C), and it turns out that it follows a uniform probability *PDF*,

$$X(2) \sim U(2.5, 3), \tag{4.8}$$

i.e., $f_S(\chi, 2) = \frac{1}{3-2.5} = 2$. By introducing sample B, the number of curves reduces even further, see Fig. 4.1C, i.e., an even less uncertain representation of the phenomenon is obtained compared to step 2 above. Indeed, the width of the *PDF* in Eq. (4.8) is half of that of the *PDF* in Eq. (4.7), and the possible b values in Eq. (4.8) range from 0.4587 to 0.5014 *vs.* 0.4176 to 0.5014 in Eq. (4.7).

- Hence, when comparing the standard *vs.* the *BME* solutions to the physical law above, the **uncertainty** of the *BME* solution of the physical law is considerably smaller than that of its standard solutions.

Step 5: *BME* theory can also be used to explicitly compute the posterior $f_K(\chi,2)$ in terms of the prior $f_G(\chi,2)$ and the site-specific knowledge $f_S(\chi,2)$, as follows,[c]

[c] Notice that the integration range [2.5, 2.7257] is the intersection of the uniform intervals of Eqs. (4.7) and (4.8).

$$f_K(\chi,2)=A^{-1}f_G(\chi,2)f_S(\chi,2)=\frac{f_G(\chi,2)f_S(\chi,2)}{\int d\chi f_G(\chi,2)f_S(\chi,2)}$$
$$=\frac{2.7381\times 2}{\int_{2.5}^{2.7257}\int d\chi 2.7381\times 2}=4.43066. \tag{4.9}$$

The corresponding *BME* expected attribute solution and its error standard deviation are

$$\hat{X}_K(2)=\int d\chi\chi f_K(\chi,2)=\int_{2.5}^{2.7257} d\chi\chi 4.43066$$
$$=2.6128, \tag{4.10a}$$

and

$$\sigma_K(2)=\sqrt{\int d\chi\left[\chi-\hat{X}_K(2)\right]^2 f_K(\chi,2)}$$
$$=\sqrt{\int_{2.5}^{2.7257} d\chi(\chi-2.6128)^2 4.43066}$$
$$=0.06515, \tag{4.10b}$$

respectively. These results can be compared with the corresponding results obtained by the standard stochastic approach, i.e.,

$$\hat{X}_G(2)=\int d\chi\chi f_G(\chi,2)=\int_{2.3052}^{2.7257} d\chi\chi 2.7381$$
$$=2.8962, \tag{4.11a}$$

and

$$\sigma_G(2)=\sqrt{\int d\chi\left[\chi-\hat{X}_G(2)\right]^2 f_K(\chi,2)}$$
$$=\sqrt{\int_{2.5}^{2.7257} d\chi(\chi-2.8962)^2 2.7381}=0.22856, \tag{4.11b}$$

- Clearly, the *BME* solution has a 3.5 times smaller **statistical error** than the standard stochastic solution, i.e., the former is much more accurate than the latter.

Based on the results discussed above, we would like to conclude the discussion of the present chapter with the following suggestion:

> Combining **site-specific** with **core** knowledge (physical law in the above case) can help one gain a deeper understanding of the phenomenon. In many cases, the more site-specific knowledge is included, the more improved the understanding and the more realistic the representation of the phenomenon is.

As a matter of fact, in many real-world applications neglecting to incorporate site-specific information into the *SPDE* solution may render the solution so unrealistic as to be perniciously misleading. The *BME* approach to the *SPDE* solution can successfully address this important problem, which is one of the reasons that the approach is currently undergoing very active development.

5 Practice exercises

1. Calculate the mean value $\overline{X(\boldsymbol{p})}$ and covariance $c_X(\boldsymbol{\Delta p})$ of the following empirical laws:

(a) $X(\boldsymbol{p})=w(\boldsymbol{p})+e^{-5t}$,

(b) $X(\boldsymbol{p})=0.13w(\boldsymbol{p})+3.01$,

(c) $X(\boldsymbol{p})=1.52w(\boldsymbol{p})$,

(d) $X(\boldsymbol{p})=3.87w(\boldsymbol{p})^2+1.45$,

where $w(\boldsymbol{p})$ has a known mean $\overline{w}$ and covariance $c_w(\boldsymbol{\Delta p})$.

2. If $\overline{X}=10$ and $\sigma_X^2=2$, what are the values of σ_{-X}^2, σ_{10X+3}^2, and $\sigma_{\frac{2}{3}X+\frac{4}{3}}^2$?

3. Consider the physical law

$$\left[a\frac{d}{ds}+b\frac{d}{dt}+1\right]X(\boldsymbol{p})=w(\boldsymbol{p}),$$

where a and b are known parameters, $X(\boldsymbol{p})$ is an unknown attribute with $X(\mathbf{0})=0$, and $w(\boldsymbol{p})$ is an observable attribute with $\overline{w(\boldsymbol{p})}=0$

and $c_w(\Delta \boldsymbol{p}) = mh + n\tau$ (m, n are known parameters).

Calculate the $c_X(\Delta \boldsymbol{p})$ in terms of the known parameters a, b, m and n.

4. Revisit Example 2.2 of Section 2 and derive analytically:
  (a) the mean and covariance of Eqs. (2.6) and (2.7), and
  (b) the *PDF* of Eq. (2.8).

5. Starting with Eq. (2.10) of Section 2, derive the *K-PDF* of Eq. (4.4) of Section 4.

6. Revisit Example 2.2 of Section 2, and derive analytically Eq. (2.2).

7. Consider the two-dimensional extension of the advection-reaction *SPDE* of Section 2 governing contaminant concentration $X(\boldsymbol{p})$,

$$\left[\frac{\partial}{\partial t} + \boldsymbol{\nu} \cdot \nabla - \nabla \cdot \boldsymbol{a} \nabla + \beta\right] X(\boldsymbol{p}) = \eta(\boldsymbol{p}),$$

where $\boldsymbol{\nu}$ is a velocity vector, $\boldsymbol{a}$ is a diffusion matrix, $\beta > 0$ is a dumping coefficient, and $\eta(\boldsymbol{p})$ is a forcing or source-sink term. The advection term $\boldsymbol{\nu} \cdot \nabla X$ accounts for transport effects, the term $\nabla \cdot \boldsymbol{a} \nabla X$ represents diffusion, and the term βX denotes damping effects. Making all the necessary assumptions of your choice concerning initial and boundary conditions and *SPDE* parameters $\boldsymbol{\nu}$, $\boldsymbol{a}$, β and $\eta(\boldsymbol{p})$ (e.g., $\eta(\boldsymbol{p})$ is often represented by a Gaussian random field), solve the above *SPDE* in terms of $X(\boldsymbol{p})$ realizations using the *BME* approach.

8. Starting with the *PDF* of the form (a) $\frac{1}{\sqrt{2\pi}}e^{-(\chi-2)^2}$, (b) $\frac{1}{\sqrt{2\pi}}e^{-(\chi-2)^2} + \frac{1}{\sqrt{4\pi}}e^{-\frac{(\chi-2)^2}{4}}$, and (c) $\frac{1}{3x\sqrt{2\pi}}e^{-\frac{(\ln\chi-2)^2}{6}}$, derive the corresponding analytical prediction error expressions of Eq. (2.12) of Section 2.

9. Under what conditions is probability considered to be fundamentally an epistemic concept?

10. Are the relations of probability and deducibility the ones involving the body of knowledge we actually have or the premises we actually accept, and why?

11. In the presence of natural laws, does random field modeling impose causal constraints on the attribute of interest and why?

CHAPTER

11

CTDA by dimensionality reduction

1 The motivation

As was stressed in previous chapters, chronotopologic mapping can improve our understanding of the *in-situ* distribution of a natural attribute and offer valuable information for risk assessment and management purposes. It is, then, appropriate to start the discussion in this chapter by focusing on the motivation behind another chronotopologic attribute interpolation and map construction approach. This motivation is different than that behind all previous techniques. From our discussion so far a first key conclusion has been drawn:

❶ It is a matter of physical insight that **combined** space and time attribute dependencies and interactions be considered in real-world *CTDA* applications, rather than focusing separately on isolated space and time dependencies.

However, as the chronotopologic variability of the natural attribute can be very complex (e.g., due to the presence of nonlinearities), the data-based computational calculation of a space-time covariance is practically inefficient for big datasets (Wikle and Cressie, 1999). This is true for different kinds of models, as is illustrated next.

Example 1.1 In *CTDA* applications, separable or nonseparable space-time covariance models are fitted to the corresponding empirical covariance values.[a] In the separable models, attribute dependencies are usually represented by the sum or the product of purely spatial and purely temporal covariance components (Rouhani and Hall, 1989), which can make the model parameter estimation process easy and fast. On the other hand, the nonseparable group of covariance or variogram models, including the product-sum, the metric, and the sum-metric models (De Cesare et al., 2001), provide a better interpretation of the space-time attribute structure. However, the complicated form makes it difficult to specify the parameters in nonseparable covariance or variogram models.

Moreover, attribute interpolation at unsampled points requires the determination of an adequate *chronotopologic metric* between the unsampled points and selected neighboring sampling points.

[a] The same is true, of course, in terms of variogram models.

https://doi.org/10.1016/B978-0-12-816341-2.00002-2

❷ Technically, either a uni- or a bi-metric is used in *CTDA* (recall Section 3 of Chapter 2). In both cases, the associated **computational cost** of neighboring sample searching could be considerably high.

Example 1.2 A commonly used uni-metric is the $|\Delta \boldsymbol{p}|_P = \sqrt{h_1^2 + h_2^2 + \varepsilon\tau^2}$, where ε is a ratio determined by expert judgment (Example 3.5, Section 3 of Chapter 2). This metric involving both spatial and temporal components makes neighboring sample searching a much more computationally costly procedure than searching using solely the spatial component.

Naturally, the difficulties ❶ and ❷ above intensify in the case of a *big dataset*. Therefore, it is of great importance to develop attribute interpolation techniques with the dual objective of reducing the computational cost (especially when dealing with big datasets) while accounting for physical space-time attribute dependencies. In an effort to address these issues, Xie et al. (2001) considered certain potentially computationally cost-effective schemes, including the following:

- Keeping the number of data points as small as possible.
- Calculating covariance functions as fast as possible.
- Calculating realistic covariance functions using as few data-pairs as possible.

Unfortunately, it was found that these schemes lead to unrealistic results in practice. In other words, none of the above three schemes are practically useful when confronted with a case-specific dataset and a given software library. For these reasons, this chapter examines a very different perspective towards addressing the difficulties ❶ and ❷ above.

2 The space-time projection (*STP*) method

2.1 Introduction

Christakos et al. (2017) proposed a new technique to confront previous attribute interpolation difficulties in practice by reducing the dimensionality of the original domain:

> The **Space-Time Projection** (*STP*) technique lessens computational effort, and at the same time, new theoretical insight is gained, by transferring the original explicitly space-time interpolation problem into a spatial interpolation problem that is implicitly time-dependent.

Essentially, the *STP* technique seeks to address in practice the issues discussed in the previous section by projecting the study of the attribute of interest in a convenient domain that possesses certain attractive features:

① Reduced dimensionality.
② Improved representation of space-time variation.
③ Higher interpolation accuracy.
④ Lower computational cost.

2.1.1 The machine learning link

Moving from a domain of higher dimensionality to one of lower dimensionality is a conceptually appealing idea in *CTDA* studies, which, in many cases, can also be proven to be practically useful. Another potentially appealing idea is that which views *STP* in a wider context:

- In a sense, *STP* may be viewed as a dimensionality reduction algorithm of **machine learning** (Chapter 12), i.e., an algorithm that seeks to make the chronotopologic attribute interpolation problem manageable by reducing its dimensionality.

Particularly, the core element of the *STP* technique is the realization that the use of random fields of reduced dimensionality can have considerable modeling benefits (theoretical and computational) in the study and interpolation of attribute distributions in conditions of *in-situ* uncertainty.

2.1.2 The threefold notion

Methodologically, the *STP* is based on the threefold idea of "Transform-Solve-Backtransform," which involves three stages:

1. **Coordinate transformation**: *STP* reduces the higher-dimensionality dataset of the space-time domain $R^{n,1}$ to the lower-dimensionality dataset of the spatial domain R^n—a reduction with considerable modeling and computational advantages.
2. **Solution**: It solves the attribute interpolation problem using only the transformed (projected) dataset in the reduced dimensionality pseudo-spatial domain R^n.[b]
3. **Backtransformation**: It brings the results back to the original $R^{n,1}$ domain.

The readers may notice that the study of an attribute in the $R^{n,1}$ domain can be complicated not only by the fact that one more dimension is present, but by the additional fact that this dimension is time, which is physically different than the spatial dimensions. The *STP* idea, then, is to temporarily "compress" the time information at the transformation stage, solve the attribute mapping problem in the much simpler domain of the temporally compressed data, and then release the compressed time data information at the backtransformation (final) stage. In this setting, the compressed domain (R^n) is a transitional stage whose purpose is to simplify the data analysis. These steps are discussed in the following subsections (more details can be found in Christakos et al., 2017, Christakos, 2017).

2.2 Dimensionality reduction transformation

According to Christakos et al. (2017), a coordinate transformation that fits the above dimensionality reduction requirement can be developed in terms of an appropriate change of space-time coordinates, as follows:

For any point $\boldsymbol{p} = (\boldsymbol{s}, t)$, a **dimensionality reduction transformation** (***DRT***) is generally defined as the change of coordinates

$$\boldsymbol{p} \in R^{n,1} \mapsto \widehat{\boldsymbol{s}} = \varpi(\boldsymbol{p}) \in R^n, \tag{2.1a}$$

where $\mapsto$ denotes transformation and ϖ is a suitable function of the original $\boldsymbol{p}$ coordinates such that an attribute $X(\boldsymbol{p})$ is represented as

$$X(\boldsymbol{p}) = \widehat{X}\left(\widehat{\boldsymbol{s}}\right), \tag{2.1b}$$

where $\widehat{X}\left(\widehat{\boldsymbol{s}}\right)$ is the projection of $X(\boldsymbol{p})$ on the R^n domain.

Apparently, a key *DRT* aspect is the term "suitable function," that is, the function ϖ must be selected so that $\varpi(\boldsymbol{p}) \in R^n$ and the computations involving $\widehat{X}(\widehat{\boldsymbol{s}})$ are considerably simpler than those involving the original $X(\boldsymbol{p})$. In fact, there is a real need for *DRT*, because it can allow for shortcuts in the expensive modeling of complex real-world studies, it could improve the understanding of the underlying physical mechanisms, and since it relies on a smaller number of parameters, it is ideal for synthetic approaches (see, e.g., the synthetic *STP-BME* approach in Chapter 12).

2.2.1 Linear DRT

Below we focus on a simple yet particularly fruitful selection of the function $\varpi(\boldsymbol{p}) = \boldsymbol{s} - \boldsymbol{v}t$, which leads to a *DRT* that plays a prime role in the development of the *STP* technique.

[b] The term "pseudo" is used here since, as we will see later, the time-dependency is implicitly considered in this domain.

- Starting with the original attribute $X(\boldsymbol{p}) = X(s,t)$, to any point $\boldsymbol{p} \in R^{n,1}$ apply the **linear coordinate transformation**

$$\boldsymbol{p} \in R^{n,1} \mapsto \widehat{\boldsymbol{s}} = \boldsymbol{s} - \boldsymbol{v}t \in R^n \quad (2.2a)$$

to define the transformed attribute $\widehat{X}\left(\widehat{\boldsymbol{s}}\right)$, where for each $\boldsymbol{p}$ the transformation vector multiplier $\boldsymbol{v} = (v_1 \dots v_n) \in R^n$ is chosen so that the equality

$$X(\boldsymbol{p}) = X(\boldsymbol{s} - \boldsymbol{v}t, 0) = \widehat{X}\left(\widehat{\boldsymbol{s}}\right) \quad (2.2b)$$

holds.

A few comments are in order. The $R^{n,1}$ is an explicitly space-time domain which, for reference purposes, is termed the *higher dimensionality* (*HD*) domain of the original attribute $X(\boldsymbol{p})$; whereas R^n is considered in a space domain (with implicit time-dependence) termed the *lower dimensionality* (*LD*) domain of the transformed (projected) $\widehat{X}(\widehat{\boldsymbol{s}})$. This means that while the time information is explicit in terms of $\boldsymbol{p} \in R^{n,1}$, it is compressed in terms of $\widehat{\boldsymbol{s}} \in R^n$. For Eq. (2.2b) to hold for every different point $\boldsymbol{p}$, the $\boldsymbol{v}$ must be, in general, a function of s and t. Eq. (2.2b) establishes an inner relationship between the original and the projected attributes, the domain of the latter attribute having one dimension less than that of the former. The *HD* domain points will be projected onto the *LD* domain points by means of Eq. (2.2a), which matches the two sets of points one-to-one.

In real-world studies, one usually encounters a three-dimensional dataset, i.e., the attribute domain has two spatial dimensions and one temporal dimension, i.e., the points

$$\boldsymbol{p} = (s_1, s_2; t) \in R^{2,1} \quad (2.3a)$$

of the original space-time domain, see Fig. 2.1A, are transferred (projected) onto the points

$$\widehat{\boldsymbol{s}} = \left(\widehat{s}_1, \widehat{s}_2\right) = (s_1 - v_1 t, s_2 - v_2 t) \quad (2.3b)$$

in a purely spatial domain, see Fig. 2.1B. *Ut aequum*, most of the case studies investigated in this chapter deal with datasets distributed in the $R^{2,1}$ domain. The quantitative determination of the *DRT* in Eq. (2.3b) requires the calculation of the vector $\boldsymbol{v}$ – how this can be done is discussed in the following subsection.

In certain applications, it may be physically meaningful to view $\boldsymbol{v}$ as a *velocity* vector, in which case $\widehat{X}(\widehat{\boldsymbol{s}})$ is an attribute "traveling" along the $\boldsymbol{v}$-direction at a distance $|\boldsymbol{v}|\,t$, where $v = |\boldsymbol{v}|$ is the corresponding speed.

Example 2.1 In this modeling setting, it is worth discussing a few illustrative cases regarding possible $\boldsymbol{v}$ interpretations.

- In *physical sciences* Eq. (2.2b) has been viewed as a modeling assumption introducing the so-called *frozen* turbulence model (e.g., Conan et al., 2000), which is a different conception than that introduced by the *DRT* above.
- In *public health sciences*, if a high disease incidence region is detected in the study moving from an urban to a rural area, it implies that the high disease incidence region travels along the v-direction at a distance

$$|\boldsymbol{v}|t = vt = \left[v_1^2 + v_2^2\right]^{\frac{1}{2}} t,$$

without significant change. In this way, each point of the original disease incidence $X(\boldsymbol{s},t)$ distribution with coordinates $(s_1, s_2; t)$ can be "projected" onto a point of the traveling distribution $\widehat{X}(\widehat{\boldsymbol{s}})$ with coordinates

$$\widehat{\boldsymbol{s}} = \left(\widehat{s}_1, \widehat{s}_2\right) = (s_1 - vt, s_2 - vt).$$

- As an illustration, using *DRT* of Eq. (2.2a), the distribution of a set of data points before and after projection is shown in Fig. 2.2A and B, respectively.

2.2.2 Space-time statistics reduction

Starting from Eqs. (2.2a)–(2.2b) and following the implementation of the *DRT*, the following result is obtained:

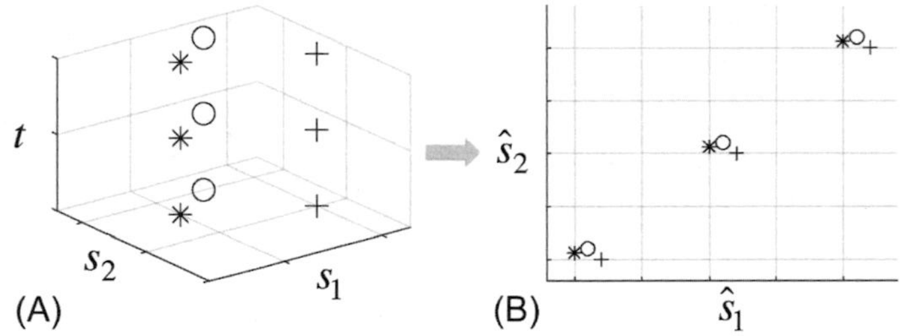

FIG. 2.1 The projection process: (A) original points $\boldsymbol{p} = (s_1, s_2, t)$ in the $R^{2,1}$ domain, and (B) transferred (projected) points $\widehat{\boldsymbol{s}} = \left(\widehat{s}_1, \widehat{s}_2\right) = (s_1 - v_1 t, s_2 - v_2 t)$ in the R^2 domain.

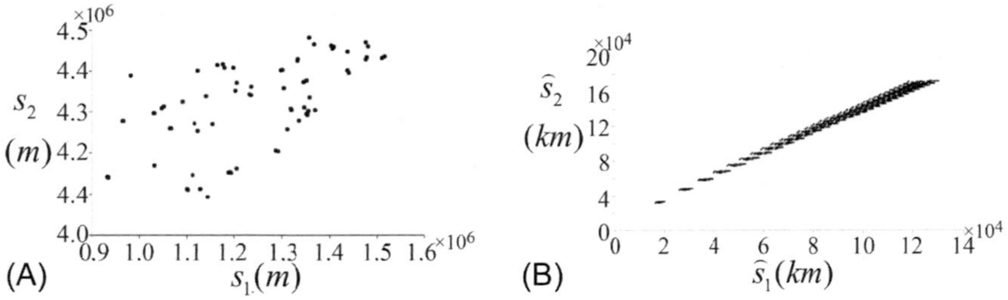

FIG. 2.2 Distribution of data points in (A) the *HD* domain, (s_1, s_2) at time t, and (B) the *LD* domain, $\left(\widehat{s}_1, \widehat{s}_2\right) = (s_1 - vt, s_2 - vt)$.

The **statistics** (attribute mean, covariance, and variogram functions) in the original *HD* domain $R^{n,1}$ are related to the projected ones on the *LD* domain R^n as

$$\overline{X(\boldsymbol{p})} = \overline{\widehat{X}\left(\widehat{\boldsymbol{s}}\right)}, \tag{2.4a}$$

$$c_X(\Delta \boldsymbol{p}) = c_X(\boldsymbol{h} - \boldsymbol{v}\tau, 0) = c_{\widehat{X}}\left(\widehat{\boldsymbol{h}}\right), \tag{2.4b}$$

$$\gamma_X(\Delta \boldsymbol{p}) = \gamma_X(\boldsymbol{h} - \boldsymbol{v}\tau, 0) = \gamma_{\widehat{X}}\left(\widehat{\boldsymbol{h}}\right), \tag{2.4c}$$

where

$$(\Delta \boldsymbol{p}) \in R^{n,1}, \tag{2.5a}$$

$$\widehat{\boldsymbol{h}} = (\boldsymbol{h} - \boldsymbol{v}\tau) \in R^n. \tag{2.5b}$$

Methodologically, the *DRT* introduced by Eqs. (2.2a)–(2.2b) and Eqs. (2.4a)–(2.4c) links two perspectives of attribute representation, one in the $R^{n,1}$ domain and one in the R^n domain. While these two perspectives are different, they communicate with its other through the attribute transformation. This is an element that makes *DRT* methodologically attractive. For example, we can mathematically transfer attribute modeling and mapping into a more convenient domain and then properly link the results in this domain to the attribute modeling and mapping in the original domain. Another direct consequence of the above *DRT* results in the attribute interpolation context is as follows:

- Since Eqs. (2.4a)–(2.4c) are derived directly from Eqs. (2.2a)–(2.2b), the **vector** $\boldsymbol{v}$ that satisfies Eqs. (2.2a)–(2.2b) will also satisfy Eqs. (2.4a)–(2.4c).

As will be shown below, an immediate implication of the above result is that, since explicit expressions are usually available for the covariance and the variogram models, it may be preferable to determine $\boldsymbol{v}$ based on Eqs. (2.4a)–(2.4c), and then use it to compute the attribute coordinates in Eqs. (2.2a)–(2.2b). Moreover, the covariance or variogram interrelations in Eqs. (2.4b) and (2.4c) may be associated with different kinds of attribute distributions. The observed interrelations need to be quantified for various purposes, including testing for the presence of significant space-time attribute correlations, assessing the attribute dependence ranges, selecting theoretical models (covariance or variogram) and physical laws (e.g., partial differential equations, *PDE*) to summarize the observed attribute characteristics.

2.3 Multiplier determination

A key step of the implementation of the *SPT* technique is the determination of the vector multiplier $\boldsymbol{v}$. There are several ways to achieve this goal, and in this subsection, we will present two of them.

2.3.1 *Analytical $\boldsymbol{v}$-expressions*

A simple way to determine $\boldsymbol{v}$ is as follows: According to Eq. (2.4b) or (2.4c), an inner relationship (inter-dependence) among the three arguments, $\boldsymbol{h}$, τ and $\boldsymbol{v}$, can be expressed as $V(\boldsymbol{h},\tau,\boldsymbol{v}) = 0$, where the form of the function V depends on that of the attribute covariance or variogram model considered.

A useful in practice form of V is obtained as

$$V: \begin{cases} c_X(\Delta\boldsymbol{p}) - c_X(\boldsymbol{h}-\boldsymbol{v}\tau, 0) = 0, \\ \gamma_X(\Delta\boldsymbol{p}) - \gamma_X(\boldsymbol{h}-\boldsymbol{v}\tau, 0) = 0, \end{cases} \quad (2.6a,b)$$

where the traveling vector $\boldsymbol{v}$ can be computed in terms of the spatial and temporal correlation ranges (ε_s and ε_t, respectively) of the attribute covariance or variogram function by solving Eq. (2.4b) or Eq. (2.4c), depending on the case.

The distances in the *HD* and *LD* domains are related by $\widehat{\boldsymbol{h}} = \boldsymbol{h} - v\tau$, where each pair of $(\boldsymbol{h},\tau)$-values is related to a unique pair of $\left(v, \widehat{\boldsymbol{h}}\right)$-values. Hence, after $\boldsymbol{v}$ has been computed, the projected distance $\widehat{\boldsymbol{h}}$ can be calculated too.

The following real-world examples present the *step-by-step* computation of the **multiplier $\boldsymbol{v}$** in practice.

Example 2.2 $PM_{2.5}$ concentrations were collected at 96 national air quality monitoring stations in Shandong Province (China) during the period January 1–31, 2014 (Christakos et al., 2018). Then, the coordinates $\widehat{s} = \left(\widehat{s}_1, \widehat{s}_2\right) = (s_1 - vt, s_2 - vt)$ were computed in R^2, as follows.

Step 1: Fig. 2.3 displays the empirical variogram values of the $PM_{2.5}$ distribution, and the fitted theoretical variogram model

$$\gamma_X(\Delta\boldsymbol{p}) = 192.4^2 \times \left[1 - \frac{1}{(8.0215\tau^{0.8276}+1)^{0.0355}} e^{-\frac{0.0019|\boldsymbol{h}|^{0.2562}}{(8.0215\tau^2+1)^{0.0639}}}\right] \quad (2.7)$$

in the original domain $R^{2,1}$ with arguments $(|\boldsymbol{h}|, \tau) \in R^{1,1}$.

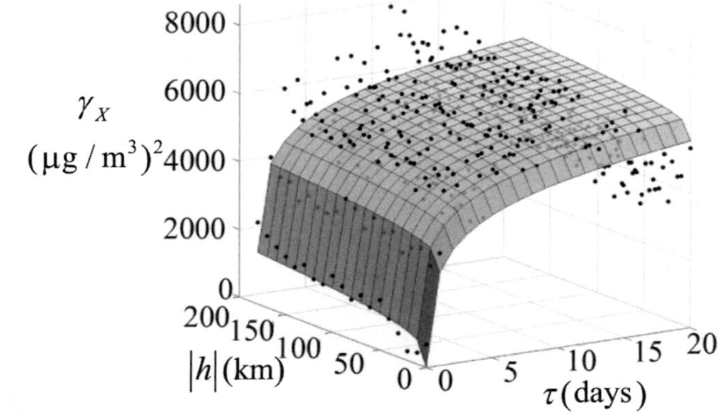

FIG. 2.3 Empirical variogram values (black dots) and theoretical variogram model γ_X (continuous surface) of the original space-time $PM_{2.5}$ distribution.

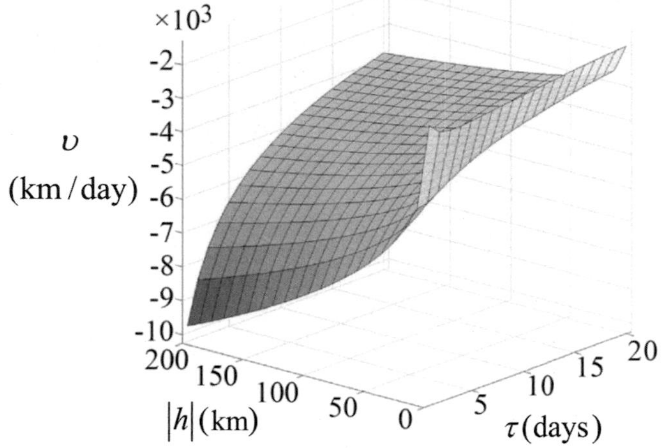

FIG. 2.4 Plot of the $PM_{2.5}$ spread velocity (in km/day) as a function of space lag $|\boldsymbol{h}|$ and time separation τ.

Step 2: Following the υ determination procedure above, it was found that (Christakos et al., 2018)

$$\upsilon = \frac{1}{\tau}\left[|\boldsymbol{h}| - \left(\frac{|\boldsymbol{h}|^{0.2562}}{(8.0215\tau^{0.8276}+1)^{0.0639}} + \frac{0.0709\ln(8.0215\tau^{0.8276}+1)}{0.0038}\right)^{\frac{1}{0.2562}}\right] \tag{2.8}$$

(in km/day), see Fig. 2.4.

Example 2.3 Hangzhou City is one of the biggest cities in China, located in its southeast coastal region (Fig. 2.5). According to the Chinese Cancer Registry Annual Report (2012), Hangzhou was also one of the cities with the highest *breast cancer* (*BC*) incidence, having 1700 newly diagnosed female *BC* cases in 2012, i.e., $X(\boldsymbol{p}) = BC(\boldsymbol{p})$, and the age-standardized incidence rate was about $33.63/10^5$ in 2009. The *BC* cases in Hangzhou during the period 2008–12 used in this study (Lou et al., 2017) were collected from the Center of Disease Control and Prevention. Specifically, there were 1643, 1727, 1820, 1812, and 1782 *BC* cases reported in 2008, 2009, 2010, 2011, and 2012, respectively.

Step 1: For data normalization purposes, the log(*BC* + 1)-transformed breast cancer incidence data values were detrended with a 100 km spatial radius and a 2 year time radius. Then, the de-trended *BC* data were used to calculate the empirical covariance values, and the theoretical separable (multiplicative) covariance model (Fig. 2.6)

$$c_X(\Delta\boldsymbol{p}) = c_0 e^{-\frac{3|\boldsymbol{h}|^2}{10000^2} - \frac{3\tau}{2}}, \tag{2.9}$$

was fitted to these values by minimizing the well-known Akaike Information Criterion, where $c_0 = 1$ (*BC* sill), with spatial *BC* correlation range $\varepsilon_s = \sqrt{3}a_s = \sqrt{3}\frac{10000}{\sqrt{3}} = 10000$ m, and temporal *BC* correlation range $\varepsilon_t = 3a_s = 3\frac{2}{3} = 2$ years.[c]

Step 2: By introducing the *BC* incidence covariance model of Eq. (2.9) above into Eq. (6a),

$$e^{-\frac{3|\boldsymbol{h}|^2}{10000^2} - \frac{3\tau}{2}} - e^{-\frac{3(|\boldsymbol{h}| - |\upsilon|\tau)^2}{10000^2}} = 0, \tag{2.10}$$

with the following speed physically consistent solution

[c] One may recall that a practical correlation range for a Gaussian covariance is $\varepsilon_s = \sqrt{3}a_s$, and for an exponential one it is $\varepsilon_t = 3a_t$ (Table 3.4, Section 3 of Chapter 5).

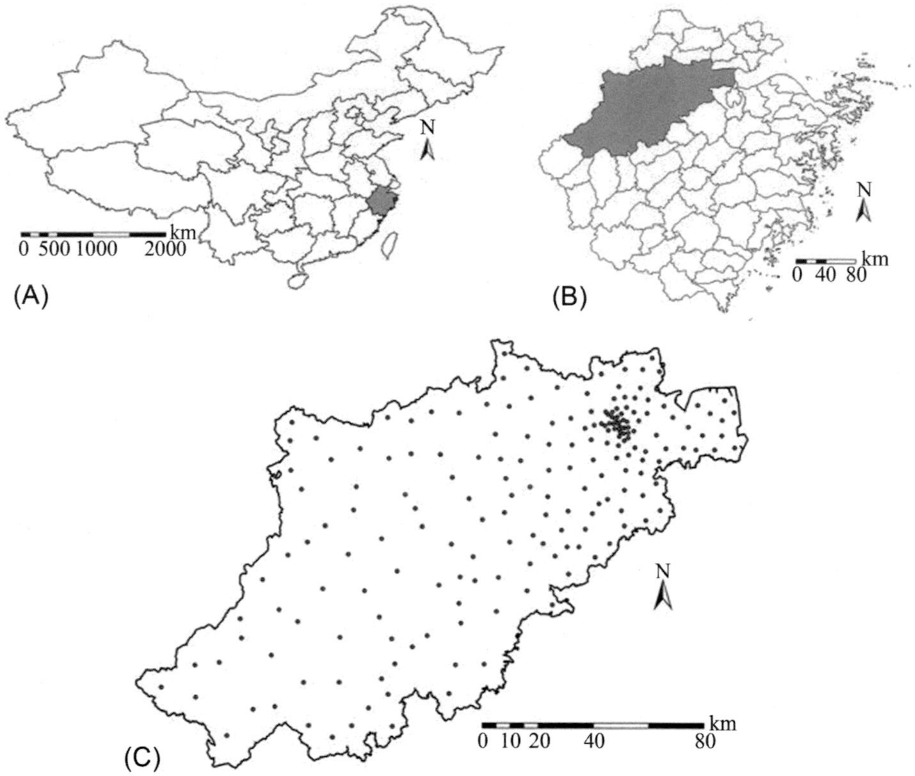

FIG. 2.5 *BC* data locations in Hangzhou City during the period 2008–12. (A) Chinese provinces boundaries; (B) city boundaries in Zhejiang Province; (C) Hangzhou City and *BC* incidence locations.

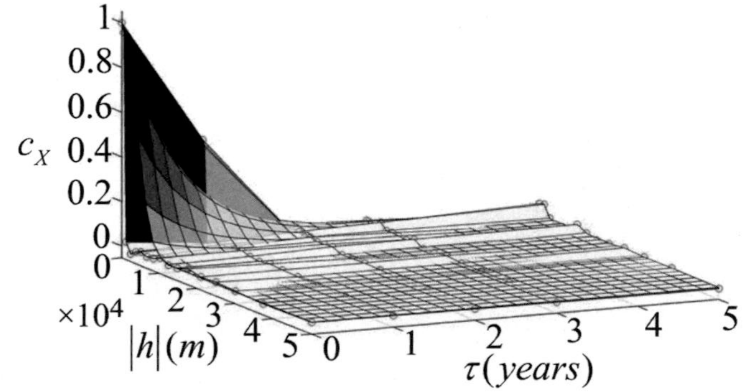

FIG. 2.6 Empirical covariance *(circles)* and theoretical covariance model *(continuous surface)* of the original space-time *BC* incidence distribution in $R^{2,1}$.

$$v = |\boldsymbol{v}| = \frac{|\boldsymbol{h}|}{\tau} - \frac{10000}{\tau}\left[\frac{|\boldsymbol{h}|^2}{10000^2} + \frac{\tau}{2}\right]^{\frac{1}{2}}, \quad (2.11)$$

which assures a positive $\widehat{h}$ value, $\widehat{h} = h - v\tau = 10000\left[\frac{|\boldsymbol{h}|^2}{10000^2} + \frac{\tau}{2}\right]^{\frac{1}{2}} > 0$, as is physically required. Interpretively, from a disease distribution perspective the *BC* incidence field "travels" in space with spread speed v along the direction of the spatial lag $\boldsymbol{h}$, i.e., v measures the correlation strength of *BC* incidences along this direction. The $\boldsymbol{v}$ distribution is plotted in Fig. 2.7 for better visualization. The northeast high *BC* incidence area provided the empirical means to calculate the *BC* velocity vector.

The following result is of wider interest as regards the significance of $\boldsymbol{v}$ determination in attribute mapping:

- Generally, a more adequate calculation of the velocity vector based on the available information leads to more **accurate** attribute interpolations and maps by the *STP* technique.

Table 2.1 provides a list of analytical v expressions for various space-time separable covariance models that can be readily implemented in practical applications.

Example 2.4 Let $X(\boldsymbol{p})$ be an attribute distribution in the original space-time domain $\boldsymbol{p} \in R^{2,1}$, and $\widehat{X}(\widehat{\boldsymbol{s}})$ be the transformed attribute distribution in the reduced dimensionality domain $\widehat{\boldsymbol{s}} \in R^2$ determined by Eqs. (2.3a) and (2.3b). The space-time dependency of $X(\boldsymbol{p})$ is quantitatively expressed by the covariance function

$$c_X(\Delta \boldsymbol{p}) = e^{-\left(|\boldsymbol{h}|^2 + 3\tau^2\right)^{\frac{1}{2}}}, \quad (2.12)$$

where $(|\boldsymbol{h}|, \tau) \in R^{1,1}$, with spatial range $\varepsilon_s = \sqrt{3}a_s = \sqrt{3}$, and temporal range $\varepsilon_t = 3a_s = 3\frac{1}{\sqrt{3}} = \sqrt{3}$. This is a space-time separable covariance consisting of Gaussian spatial and temporal components. The v expression is that of Case No. 5 of Table 2.1 ($a_s = 1, a_t = \frac{1}{\sqrt{3}}$) giving the physically consistent solution[d]

$$v = |\boldsymbol{v}| = \frac{h}{\tau} \pm \frac{a_s}{\tau}\left[\frac{h^2}{a_s^2} + \frac{\tau^2}{a_t^2}\right]^{\frac{1}{2}} = \frac{h}{\tau} - \frac{1}{\tau}\left[h^2 + 3\tau^2\right]^{\frac{1}{2}}. \quad (2.13)$$

Similarly, for the covariance of Eq. (2.9), Case No. 4 of Table 2.1 gives $v = |\boldsymbol{v}| = \frac{h}{\tau} - \frac{a_s}{\tau}\left[\frac{h^2}{a_s^2} + \frac{\tau}{a_t}\right]^{\frac{1}{2}} = \frac{|\boldsymbol{h}|}{\tau} - \frac{10000}{\tau}\left[\frac{|\boldsymbol{h}|^2}{10000^2} + \frac{\tau}{2}\right]^{\frac{1}{2}}$, i.e., Eq. (2.11).

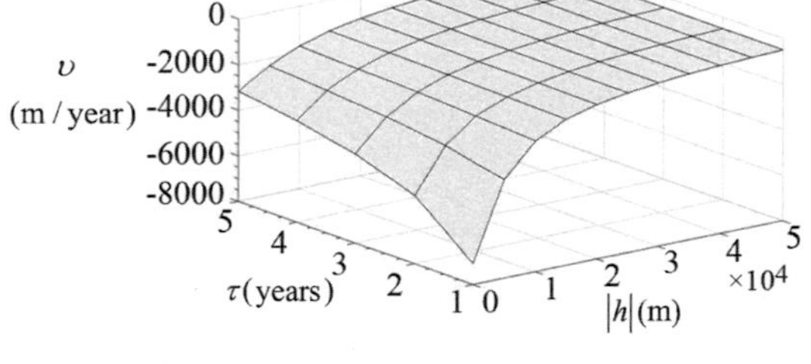

FIG. 2.7 Plot of the BC incidence spread velocity (in m/year) as a function of space lag, $|\boldsymbol{h}|$ and time separation τ.

[d] For confirmation purposes, Eq. (2.6a) yields $e^{-(|\boldsymbol{h}|^2+3\tau^2)^{0.5}} = e^{-((|\boldsymbol{h}|-v\tau)^2+3\times 0)^{0.5}} = e^{-(|\boldsymbol{h}|-v\tau)}$. The solution of this equation coincides with Eq. (2.13), as expected.

TABLE 2.1 Various space-time covariance models and the corresponding υ.

No.	Spatial component	Temporal component	Vector υ
1	Exponential	Exponential	$\upsilon=-\frac{a_s}{a_t}$
2	Exponential	Gaussian	$\upsilon=-\frac{a_s}{a_t^2}\tau$
3	Exponential	Spherical	$\upsilon=\frac{1}{3\tau}\left[0.6931-a_s\ln\frac{(\tau-a_t)^2(\tau+2a_t)}{a_t^3}\right]$
4	Gaussian	Exponential	$\upsilon=\frac{h}{\tau}\pm\frac{a_s}{\tau}\left[\frac{h^2}{a_s^2}+\frac{\tau}{a_t}\right]^{\frac{1}{2}}$
5	Gaussian	Gaussian	$\upsilon=\frac{h}{\tau}\pm\frac{a_s}{\tau}\left[\frac{h^2}{a_s^2}+\frac{\tau^2}{a_t^2}\right]^{\frac{1}{2}}$
6	Gaussian	Spherical	$\upsilon=\frac{h}{\tau}\pm\frac{a_s}{\sqrt{3}\tau}\left[0.6931+\frac{3h^2}{a_s^2}-\ln\frac{(\tau-a_t)^2(\tau+2a_t)}{a_t^3}\right]^{\frac{1}{2}}$

Example 2.5 As we saw in Example 3.9 of Section 3, Chapter 6, in the Jing-Jin-Ji region of North China the $PM_{2.5}$ concentration will reach a high level during winter because of burning fossil fuels, such as coal. Long- and short-term exposure to $PM_{2.5}$ is a great concern in China, due to its adverse population health effects. Therefore, for environmental health reasons it is important to evaluate the level of $PM_{2.5}$ in the entire North China region by collecting daily $PM_{2.5}$ monitoring data during October 1–31, 2015.

Step 1: The following spatiotemporal theoretical covariance model was fitted to the empirical covariance values of the $PM_{2.5}$ distribution (He and Christakos, 2018),

$$c_X(\boldsymbol{\Delta p})=e^{-\frac{3h}{250{,}000}-\frac{3\tau^2}{3^2}}, \tag{2.14}$$

with $a_s=\frac{250000}{3}=83{,}333.33$ m and $a_t=\sqrt{3}$ days. Physically, the distribution of the $PM_{2.5}$ concentrations exhibited the wave pattern of Eq. (2.2b) with $n=2$, which means that the $PM_{2.5}$ covariance satisfies Eq. (2.4b).

Step 2: Given the separable $PM_{2.5}$ covariance of Eq. (2.14), the υ was calculated from Table 2.1, Case No. 2, as

$$\upsilon=-\frac{a_s}{a_t^2}=-\frac{\frac{25\times10^4}{3}}{\sqrt{3}^2}\tau=-\frac{25\times10^4}{3^2}\tau=-27{,}778\tau. \tag{2.15}$$

(in m/day). As one can see, in this case, υ is a linear function of time separation.

2.3.2 *Differential υ-expressions*

In the case of a space and time-independent vector $\boldsymbol{\upsilon}$, another way to determine $\boldsymbol{\upsilon}$ is directly from the partial differential equation (Christakos, 2017),

$$\boldsymbol{\upsilon}=-\boldsymbol{H}_c^{-1}\frac{\partial}{\partial\tau}\nabla c_X, \tag{2.16}$$

where $\boldsymbol{H}_c$ is the Hessian matrix with regard to the spatial coordinates of the attribute covariance $c_X(\boldsymbol{h},\tau)$.

Example 2.6 In $R^{2,1}$, in particular, Eq. (2.16) gives

$$\boldsymbol{v} = \begin{bmatrix} v_1 \\ v_2 \end{bmatrix} = \frac{1}{\left(\frac{\partial^2 c_X}{\partial h_1 \partial h_2}\right)^2 - \frac{\partial^2 c_X}{\partial h_1^2}\frac{\partial^2 c_X}{\partial h_2^2}} \begin{bmatrix} \frac{\partial^2 c_X}{\partial h_2^2}\frac{\partial^2 c_X}{\partial h_1 \partial \tau} - \frac{\partial^2 c_X}{\partial h_2 \partial \tau}\frac{\partial^2 c_X}{\partial h_1 \partial h_2} \\ \frac{\partial^2 c_X}{\partial h_1^2}\frac{\partial^2 c_X}{\partial h_2 \partial \tau} - \frac{\partial^2 c_X}{\partial h_1 \partial h_2}\frac{\partial^2 c_X}{\partial h_1 \partial \tau} \end{bmatrix}, \tag{2.17}$$

which can be calculated anywhere on the covariance surface, apart from the locations where the determinant of the Hessian $Det\, \boldsymbol{H}_c$ vanishes. For illustration, let $c_X(\boldsymbol{\Delta p}) = e^{-(h_1^2+h_1^2)-\tau}$. Then, Eq. (2.17) gives

$$\boldsymbol{v} = \begin{bmatrix} v_1 \\ v_2 \end{bmatrix} = \frac{1}{(2h_1^2-1)(2h_2^2-1) - 4h_1^2h_2^2} \begin{bmatrix} h_1 \\ h_2 \end{bmatrix}, \tag{2.18}$$

i.e., $\boldsymbol{v}$ did not change with time separation in this case.

2.4 Coordinate calculation

An important observation concerning the consistency of $\boldsymbol{v}$ used on the one hand in Eqs. (2.2a) and (2.2b) in terms of attribute $X(\boldsymbol{p})$, and on the other hand in Eq. (2.4b) or Eq. (2.4c) in terms of its covariance $c_X(\boldsymbol{h},\tau)$ or variogram $\gamma_X(\boldsymbol{\Delta p})$ can be made as follows.

Since the vector $\boldsymbol{v}$ and the point vector $(\boldsymbol{s},t)$ are interdependent via Eqs. (2.2a) and (2.2b), to each $(\boldsymbol{s},t)$ we can associate a unique $\boldsymbol{v}$, and, since the same $\boldsymbol{v}$ that satisfies Eqs. (2.4b) and (2.4c) also satisfies Eqs. (2.2a) and (2.2b), one can use in Eqs. (2.2a) and (2.2b) the $\boldsymbol{v}$ determined by Eqs. (2.6a,b).

In the case, now, that Eqs. (2.6a,b) determine an isotropic $v = |\boldsymbol{v}| = [\sum_{i=1}^n v_i^2]^{\frac{1}{2}}$, each point $(s_1,\ldots,s_n;t)$ in $R^{n,1}$ can be associated with a point of the projected attribute distribution $\widehat{X}(\widehat{\boldsymbol{s}})$ with coordinates in R^n,

$$\widehat{\boldsymbol{s}} = \left(\widehat{s}_1,\ldots,\widehat{s}_n\right) = (s_1 - vt,\ldots,s_n - vt). \tag{2.19}$$

It should be noted that no information is lost during the transformation from the $R^{n,1}$ onto the R^n domain, since in the R^n domain the data analysis is actually "pseudo-spatial" and not "purely spatial" in the conventional sense. This is because the spatial coordinates, $\widehat{\boldsymbol{s}} = \boldsymbol{s} - \boldsymbol{v}t$, implicitly include temporal attribute information via the term $\boldsymbol{v}t$.

Let us start the investigation of the implementation of Eq. (2.19) by revisiting the simple Example 2.4.

Example 2.4(cont.): Given the covariance model of Eq. (2.12) and the v expression of Eq. (2.13), using Eq. (2.19) one finds that the original space-time coordinates $\boldsymbol{p} = (s_1,s_2;t)$ are associated with the projected coordinates as

$$\widehat{s}_i = s_i - vt = \left[s_i^2 + 3t^2\right]^{\frac{1}{2}}, \tag{2.20}$$

$i = 1, 2$. This simple example provides useful insight concerning the real-world calculation of the projected attribute coordinates.

The following real case study presents the *step-by-step* calculation of the **projected coordinates** $\widehat{\boldsymbol{s}}$ in practice.

Example 2.2(cont.): In the case of the $PM_{2.5}$ concentrations in Shandong Province (China) during the period January 1–31, 2014 (Example 2.2 above), the coordinates $\widehat{\boldsymbol{s}} = \left(\widehat{s}_1, \widehat{s}_2\right) = (s_1 - vt, s_2 - vt)$ were computed in R^2, as follows.

Step 3: Using the v expression of Eq. (2.8), the R^2-projected $\widehat{s}_i$ coordinates $(i = 1,2)$ for each one of the points with original coordinates $(s_1,s_2;t) \in R^{2,1}$ were calculated as follows:

$$\widehat{s}_i = s_i - v_i t = \left[\frac{s_i^{0.2562}}{(8.0215t^{0.8276}+1)^{0.0639}} + \frac{0.0709\ln(8.0215t^{0.8276}+1)}{0.0038}\right]^{\frac{1}{0.2562}}, \tag{2.21}$$

$i = 1, 2$.[e]

[e] When Eq. (2.8) was used in Eq. (2.21) the $|\boldsymbol{h}|$ and τ were replaced by s_i and t, respectively.

The next example provides some interesting visualizations of the coordinates in both the original *HD* domain and the projected *LD* domain in the case of the Hangzhou City study.

Example 2.3(cont.): Let us revisit the *BC* incidence study of Hangzhou City in Example 2.3 above. Given v, the coordinate pair $(\boldsymbol{s},t)$ in *HD* is related to a unique pair $\left(\widehat{\boldsymbol{s}},v\right)$ in *LD* through Eqs. (2.3a)–(2.3b).

Step 3: In practical terms this means that, given the coordinates of the original space-time *BC* points in $R^{2,1}$, $\boldsymbol{p}$, see Fig. 2.8A, the corresponding coordinates in R^2 of the projected space points can be calculated as

$$\widehat{s}_i = s_i - vt = 10000\left[\frac{s_i^2}{10000} + \frac{t}{2}\right]^{\frac{1}{2}}, \quad (2.22)$$

$i = 1, 2$, see Fig. 2.8B. Fig. 2.8A and B shows that the *HD* attribute points $\boldsymbol{p}$ at the same location s during various time instances t are projected along a line of the *LD*, and the projected locations $\widehat{s}$ associated with various time instants are close to each other. Also, the projected locations in *LD* are clustered into n separate groups (n is the number of time instants considered), each of which is derived at the same time instant of the original space-time *HD* realm. In each *LD* group, the distribution of locations among each other can be the same as in the original space-time domain. The readers are reminded of the following distinction:

- The R^2-domain of the Hangzhou *BC* study is "pseudo-spatial" rather than "purely spatial" in the conventional sense, since the spatial coordinates $\widehat{\boldsymbol{s}} = \boldsymbol{s} - \boldsymbol{v}t$ include temporal incidence information (via the term $\boldsymbol{v}t$), whereas the spatial coordinates $\boldsymbol{s}$ of the purely spatial analysis (e.g., spatial statistical regression) do not include temporal incidence information.

Example 2.5(cont.): The next steps in Example 2.5 above are as follows:

Step 3: Figs. 2.9A and B display the distribution of points before (s_1, s_2 at time t) and after dimensionality reduction $\left(\widehat{s}_1, \widehat{s}_2\right)$. Specifically, Fig. 2.9A displays the original monitoring stations (only hard data points are shown) in the *HD* domain with overlapping points across time; and Fig. 2.9B depicts the projected hard data points in the *LD* domain.

Step 4: The covariance of the projected $PM_{2.5}$ concentration distribution is given by

$$c_{\widehat{X}}\left(\widehat{h}\right) = e^{-\frac{3\left(h+27778\tau^2\right)}{250000}}. \quad (2.23)$$

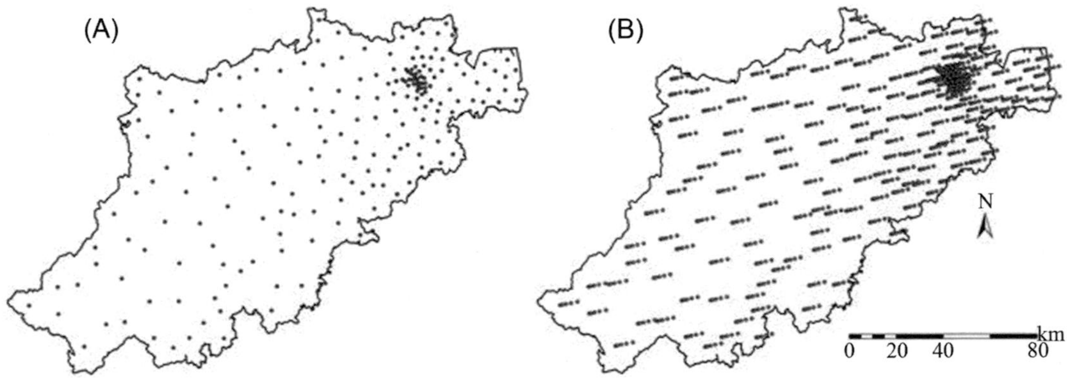

FIG. 2.8 (A) Distribution of original *BC* incidence points $\boldsymbol{p}$ in Hangzhou (the points are overlapped from various time instants), and (B) distribution of *BC* incidence location $\widehat{s}$ after coordinate transformation.

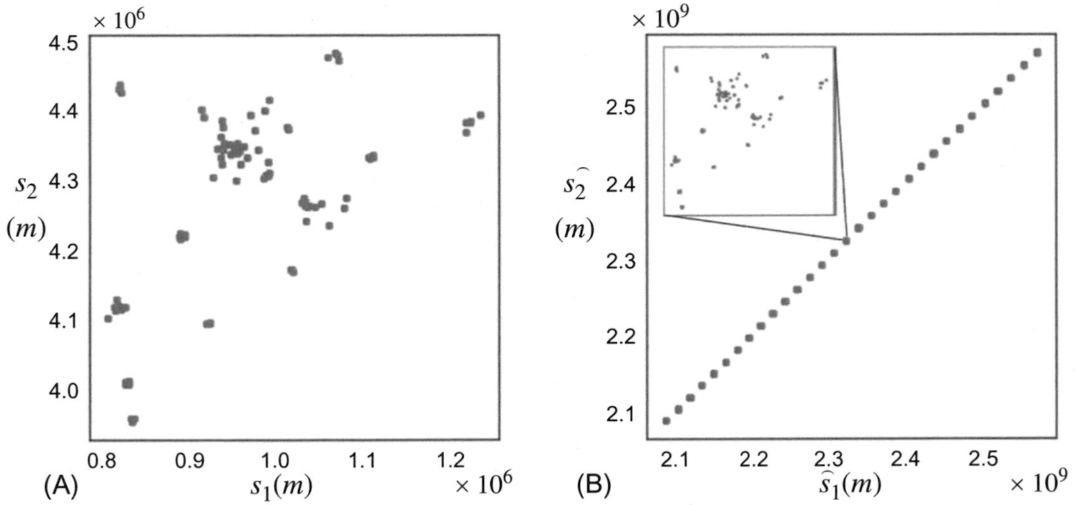

FIG. 2.9 Distribution of $PM_{2.5}$ monitoring points in the (A) *HD* and (B) *LD* domains.

2.5 Projected covariance and variogram models

Given the covariance or variogram models of the attribute $X(\boldsymbol{p})$ in the original *HD* domain, the corresponding covariance or variogram models of the projected attribute $\widehat{X}(\widehat{\boldsymbol{s}})$ in the *LD* domain can be determined. As usual, we start with a simple example.

Example 2.4(cont.): Given the covariance model of Eq. (2.12) and the υ expression of Eq. (2.13), the covariance of $\widehat{X}(\widehat{\boldsymbol{s}})$ in the domain is obtained from Eq. (2.4b) as.

$$c_{\widehat{X}}\left(\widehat{h}\right) = e^{-\widehat{h}}, \tag{2.24}$$

where $\widehat{h} = (|\,\boldsymbol{h}| - \upsilon\tau) = (h^2 + 3\tau^2)^{\frac{1}{2}} \in R^1$. As earlier, this simple example provides useful insight concerning the real-world calculation of the projected covariance or variogram models.

The following real-world example presents the *step-by-step process* of calculating the **projected covariance** or **variogram** model in practice.

Example 2.2(cont.): Continuing the Shandong Province study, the next step is as follows.

Step 5: Given the variogram model of Eq. (2.7) and the υ expression of Eq. (2.8), the projected empirical variogram values of the $PM_{2.5}$ distribution are obtained from Eq. (2.4b) and plotted in Fig. 2.10 together with the fitted variogram model

$$\gamma_{\widehat{X}}\left(\widehat{h}\right) = \begin{cases} 857.77 + 6428.4\left[1 - e^{-\frac{\widehat{h}}{5346.3}}\right] & \text{if } \widehat{h} \leq 3 \times 5346.3 \\ 857.77 + 6428.4 & \text{if } \widehat{h} > 3 \times 5346.3 \end{cases} \tag{2.25}$$

in the reduced dimensionality domain R^2 with argument $\widehat{h} \in R^1$, with nugget effect $\delta = 857.77$ $(\mu g/m^3)^2$, sill $\widehat{c}_0 = 6428.4(\mu g/m^3)^2$, and spatial correlation range $\widehat{\varepsilon}_s = 3\widehat{a}_s = 3 \times 5346.3 = 16{,}038.9$ km.[f] A noteworthy feature of the analysis above is that one can select more adequate

[f] The distance is estimated in the projected spatial domain.

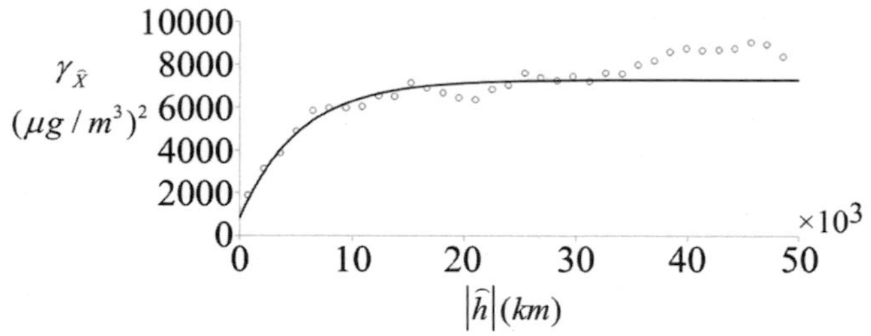

FIG. 2.10 Empirical variogram values *(dots)* and theoretical variogram model $\gamma_{\hat{X}}(\hat{h})$ *(continuous line)* of the projected $PM_{2.5}$ distribution.

$PM_{2.5}$ variogram models $\gamma_{\hat{X}}(\hat{h})$ in the reduced dimensionality domain than in the original space-time domain, in which it is often much harder to select variogram models $\gamma_X(\Delta \boldsymbol{p})$ that reflect adequately the composite space-time $PM_{2.5}$ variation structure.

Step 6: Following the standard rules of variogram interpretation presented in Section 3 of Chapter 5, the study of the shapes of the variogram models plotted in Figs. 2.3 and 2.10 revealed some interesting features of the $PM_{2.5}$ distribution in the original *vs.* the projected domains. In particular, the variogram model $\gamma_X(|\boldsymbol{h}|,\tau)$ plotted in Fig. 2.3 provided useful hints about the $PM_{2.5}$ distribution in the original Shandong Province domain.

① The variogram measured the space-time variability level among $PM_{2.5}$ concentrations. Higher variogram values represent a larger variance between $PM_{2.5}$ observations or a weaker dependence among pollutant concentration values and vice versa.

② The variogram accounted for space-time interactions among $PM_{2.5}$ concentrations and for the distinct features of the space and time components of $PM_{2.5}$ spread. The spatial coordinates are physically different from the time coordinate, and the effect of the spatial $PM_{2.5}$ concentration change differs for each time instance considered.

③ The rather fast variogram increase in Fig. 2.3 implies that strong $PM_{2.5}$ correlations occur at short space distances and time separations, but they are weaker at long distances and separations. At $\tau = 0$, the spatial correlation range is 200 km, whereas at $|\boldsymbol{h}| = 0$ the temporal correlation range is 20 days. These variogram features are consistent with the spatially homogeneous and temporally stationary distribution of the actual $PM_{2.5}$ concentrations in the Shandong Province.

④ It is visually obvious that the model $\gamma_{\hat{X}}(\hat{h})$ plotted in Fig. 2.10 offers a better fit to the projected $PM_{2.5}$ data than the $\gamma_X(\Delta \boldsymbol{p})$ does to the original data.

⑤ The $\gamma_{\hat{X}}(\hat{h})$ shape indicates that the range of the projected $PM_{2.5}$ concentrations is about 16,000 km, beyond which the $PM_{2.5}$ observations are independent (the correlation between two pollutant values is negligible beyond this range).

⑥ The variogram values measured the strength of the transformed dependence of the projected $PM_{2.5}$ concentrations. They also measured a special kind of dependence, i.e., the degree of linear relationship between $PM_{2.5}$ concentrations across space and time. The strength of this relationship was assessed by the variogram magnitude with the largest value, i.e., $7286.17(\mu g/m^3)^2$.

It is now time to revisit Example 2.3.

Example 2.3(cont.): Let us continue the *BC* incidence study of Hangzhou City in Example 2.3 above. The log(*BC* + 1)-transformed breast cancer incidence values in *LD* were detrended with a 100 km spatial radius.

Step 4: Given the covariance model of Eq. (2.9) and the υ expression of Eq. (2.11), the empirical covariance $\widehat{X}(\widehat{s}) = B\widehat{C}(\widehat{s})$ was calculated within a 50 km range, see Fig. 2.11. An exponential covariance model was fitted to the calculated projected BC covariance values, as follows (Lou et al., 2017):

$$c_{\widehat{X}}\left(\widehat{h}\right) = c_{\widehat{X}}(h - \upsilon\tau, 0) = c_0 e^{-\frac{\widehat{h}}{5}}, \quad (2.26)$$

with sill $c_0 = 1$ and range $\widehat{\varepsilon}_s = 3\widehat{a}_s = 3 \times 5 = 15\text{km}$, which is also plotted in Fig. 2.11. *BC* modeling becomes considerably easier and efficient as regards locational coordinate arrangements (Fig. 2.8) and covariance determination (Fig. 2.11), also compare Eq. (2.26) *vs*. Eq. (2.9). More specifically:

① Fig. 2.6 presents a two-dimensional plot of the *BC* incidence covariance as a function of two arguments, space and time, which have different effects on *BC* incidence variation. It is widely appreciated that physical distance in space differs drastically from "distance" in time, and the determination of composite space-time distances is usually a complicated process.

② There is an imbalance in the information content associated with the spatial *vs*. the temporal dimension (a common case is a geographically large study area with a short study period).

③ The above facts often make it much harder to select a covariance model that represents adequately the composite space-time variation structure of *BC* incidence. Complexity varies depending on the form of the selected theoretical covariance model to be fitted to the data. It is easier, e.g., to specify the parameters of a multiplicative (product) space-time model on the basis of the available data, and much more difficult to do the same for an additive (summation) space-time covariance model.

④ On the other hand, Fig. 2.11 is a unidimensional plot of the projected (transformed) *BC* covariance, the specification of which does not involve any of the complications mentioned above. Naturally, it is always easier to select an adequate covariance model and specify its parameters in the R^2 domain than in the $R^{2,1}$ domain.

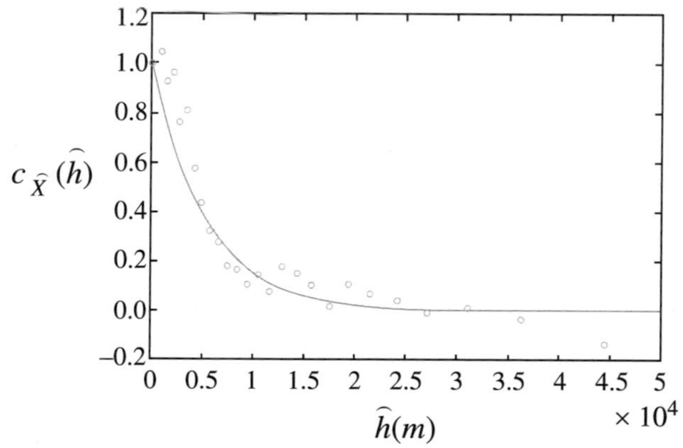

FIG. 2.11 Empirical covariance *(dots)* and theoretical covariance model *(continuous line)* of the projected *BC* incidence distribution in R^2.

2.6 Attribute interpolation and mapping

Naturally, the space-time interpolation points p_k are all projected from the *HD* on to the *LD* domain by using Eq. (2.2a) with the traveling vector v obtained from Eq. (2.6a,b) as

$$\widehat{s}_k = s_k - v t_k, \tag{2.27}$$

so that the interpolation points in the *HD* and *LD* domains are matched one-to-one. Given the projected attribute observations $X(\widehat{s})$ in *LD*, the attribute values $\widehat{X}(\widehat{s}_k)$ at points $\widehat{s}_k$ can be estimated by classical geostatistics techniques (e.g., spatial Kriging). Lastly, based on the one-to-one projection feature, the attribute interpolation values $\hat{X}(p_k)$ in the original domain are obtained as

$$\hat{X}(p) = \hat{\widehat{X}}\left(\widehat{s}\right). \tag{2.28}$$

The steps of the *STP* technique are presented in Table 2.2: Steps 5 and 6 are the two new steps here (Steps 1–4 were discussed in the preceding sections). In particular, the readers may notice the following methodological sequence:

- The $\hat{\widehat{X}}\left(\widehat{s}\right)$ map of Step 5 constitutes an intermediate step, which rather serves as a **modeling vehicle** that allows the attribute study to be carried out in the convenient R^n domain of reduced dimensionality and computational complexity, and then, in Step 6, the obtained results are back-transformed onto the original $R^{n,1}$ domain to obtain the real attribute map $\hat{X}(p)$.

Once more, one should be reminded that the time information is only temporarily compressed at Step 5 so that the attribute mapping problem is solved in the "compressed" domain (with the considerable modeling and practical advantages described earlier), and then the compressed time data information is released at the back-transformation (final) Step 6. In this setting, the "compressed" domain (R^n) is only a transitional one whose purpose is to simplify and improve the attribute data analysis. Lastly, the fact that the temporal neighboring points in the original spatiotemporal attribute field are converted to pseudo-spatial neighboring points in the *STP* projected attribute incidence domain implies that more sampling points are

TABLE 2.2 The *STP* mapping technique.

Step no.	Description
1	Using the available information, the attribute covariance model $c_X(\boldsymbol{h},\tau)$ (or variogram model $\gamma_X(\boldsymbol{h},\tau)$) is computed in the original domain $R^{n,1}$ as in Section 2.3 above
2	The velocity vector $\boldsymbol{v}$ of the attribute distribution is determined on the basis of the analysis of Section 2.3 above
3	Given the $(\boldsymbol{s},t)$ coordinates of the original space-time points in $R^{n,1}$, the corresponding $\widehat{\boldsymbol{s}} = \boldsymbol{s} - \boldsymbol{v}t$ coordinates (R^n) of the projected space points are calculated as in Section 2.4
4	Using the $\boldsymbol{v}$ values above, the values of the projected attribute covariance $c_{\widehat{X}}\left(\widehat{\boldsymbol{h}}\right)$ (or variogram $\gamma_{\widehat{X}}\left(\widehat{\boldsymbol{h}}\right)$) with $\widehat{\boldsymbol{h}} = \boldsymbol{h} - \boldsymbol{v}\tau$ (R^n) are computed as in Section 2.5
5	Estimates $\hat{\widehat{X}}\left(\widehat{\boldsymbol{s}}_k\right)$ are derived in R^n of the projected attribute values $\widehat{X}\left(\widehat{\boldsymbol{s}}_k\right)$ at the projected mapping points $\widehat{\boldsymbol{s}}_k$ of Eq. (2.27) using the attribute values at the projected data points (and other relevant information, if available)
6	In view of Eq. (2.28), the estimates $\hat{X}(\mathbf{p})$ of the attribute $X(\boldsymbol{p})$ in the original $R^{n,1}$ domain from the estimates $\hat{\widehat{X}}\left(\widehat{\boldsymbol{s}}\right)$ in the projected R^n domain are finally derived

established in the *STP* domain than by a conventional purely spatial technique (which uses only spatial neighboring points). This, in turn, implies that *STP* interpolation is numerically more accurate than mainstream purely spatial interpolation.

The following real-world examples present the *step-by-step* **STP attribute interpolation** and **mapping** in practice.

Example 2.2(cont.): Continuing with the Shandong Province study, its final steps are concerned with the interpolation and mapping of the $PM_{2.5}$ concentrations, as follows.

Step 7: Using the transformed $PM_{2.5}$ data and the variogram model $\gamma_{\widehat{X}}\left(\widehat{\boldsymbol{h}}\right)$, the projected $\widehat{\hat{X}}\left(\widehat{\boldsymbol{s}}\right)$ map was generated in R^2, see Fig 2.12A.[g]

Step 8: Lastly, the $\hat{X}_{stp}(\boldsymbol{p})$ interpolation map was generated in $R^{2,1}$ of the original $PM_{2.5}$ distribution in Shandong Province during the time period January 1–31 (2014), as shown in Fig. 2.12B. The $PM_{2.5}$ distribution characteristics move in the direction of the pollutant spread velocity $\boldsymbol{v}$ determined by the traveling model (e.g., one such characteristic is $PM_{2.5}$ concentration divergence from one location to another). The study of the $\hat{X}_{stp}$-map revealed the following:

- The mean $PM_{2.5}$ concentration was very high in Shandong during January 2014 (130.92 μg/m^3). About 73.47% of the locations-times experienced significant pollution (>75 μg/m^3) according to the National Ambient Air Quality Standard of China. Serious pollution clusters were detected during this month and the number of days with $PM_{2.5}$ exceeding 75 μg/m^3 differed with location.[h]

2.6.1 STP *vs.* STOK

It would be instructive to compare the *STP* technique *vs.* the widely used *STOK* technique of geostatistics (Chapter 8). This comparison is better achieved by means of a numerical example.

Example 2.2(cont.): We revisit the previous example and add one more step, as follows.

Step 9: For numerical comparison purposes, $PM_{2.5}$ concentration estimates in the Shandong Province were also generated by the mainstream *STOK* technique based on the same pollutant dataset and variogram model as those used by the *STP* technique. The $PM_{2.5}$ interpolation map $\hat{X}_{stok}(\boldsymbol{p})$ generated by the *STOK* technique is plotted in Fig. 2.12C.[i]

① A visual comparison of the *STP* and *STOK* maps of Fig. 2.12B and C, respectively, shows that the former provides a more detailed and realistic representation of the $PM_{2.5}$ distribution in the Shandong Province during the period January 1–31, 2014.

② The numerical accuracy of the two pollutant mapping techniques was tested by means of two commonly used accuracy indicators: the *MAE* and the *RMSE*. A total of $v = 433$

[g] At most 16 neighbor points located by a searching radius of 20 × 10^3 km were used to get an estimate at each one of the mapping (grid) points $\widehat{s} = \left(\widehat{s}_1, \widehat{s}_2\right)$ in R^2.

[h] The reasons for the serious $PM_{2.5}$ pollution during January were attributed to the facts that, first, during winter heating emits more harmful gases than during any other season and, second, meteorological factors (e.g., low wind speed, low relative humidity, scant rainfall) do not favor pollutant diffusion.

[i] A maximum of 16 neighboring points $\boldsymbol{p}$ located by a spatial searching radius of 1000 km and a temporal searching radius of 15 days were used to derive $PM_{2.5}$ estimates at each one of the mapping (grid) points in $R^{2,1}$ (the same as those used by the *STP* technique).

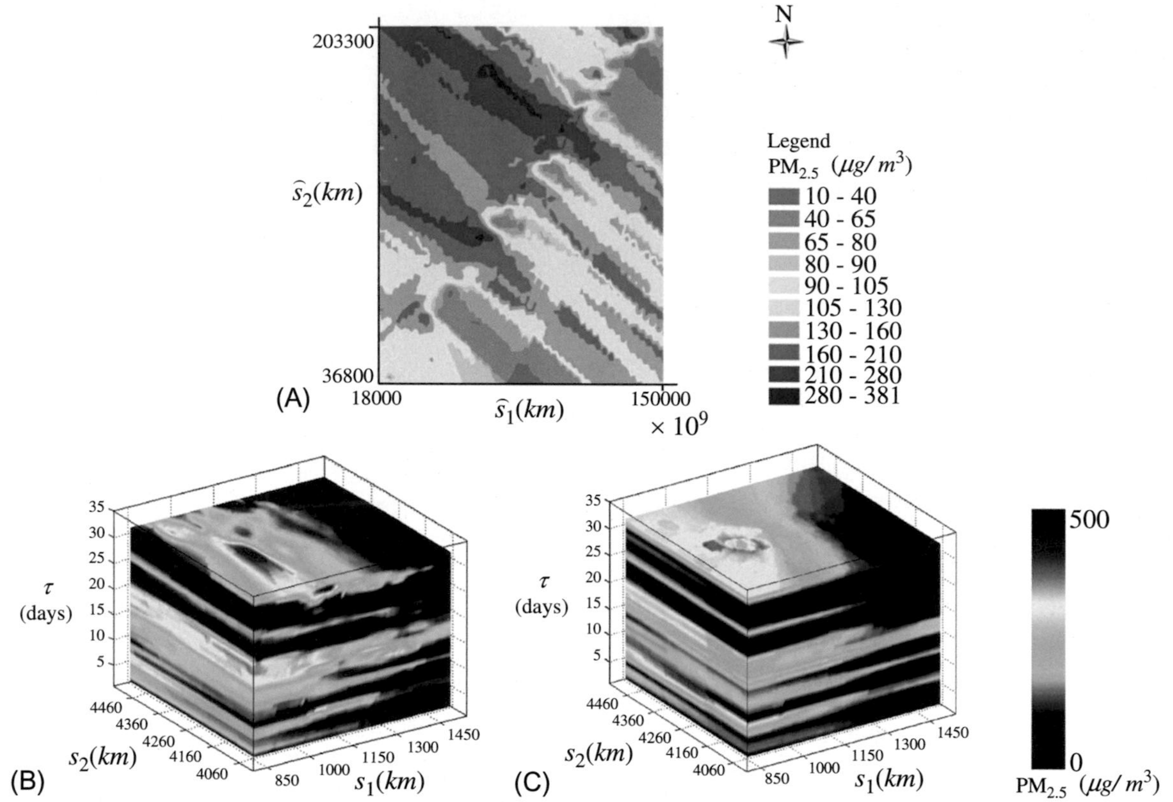

FIG. 2.12 (A) Projected $PM_{2.5}$ concentration map. (B) Space-time interpolation maps of the $PM_{2.5}$ distribution (μg/m^3) in Shandong Province during January 1–31 (2014) using the *STP* technique. (C) Space-time interpolation maps of the $PM_{2.5}$ distribution (μg/m^3) in Shandong Province during January 1–31 (2014) using the *STOK* technique.

validation[j] points were used. The obtained pollutant interpolation accuracy results for the two techniques are listed in Table 2.3. Clearly, the *STP* technique was more accurate than the *STOK* technique (26.2% and 28.5% more accurate, respectively, in terms of the *MAE* and the *RMSE* criteria).

③ As regards computational time, the results of Table 2.3 show that the *STP* technique again performed better than the *STOK* technique, i.e., about 19% less computational time was required in the implementation of the *STP* technique compared to that of the *STOK* technique.

④ Another technical difference between *STOK* and *STP* is that the *STOK* neighborhood search was based on the original coordinate set (s_1, s_2, t) whereas that of *STP* relied on the projected coordinate set $\left(\widehat{s}_1, \widehat{s}_2\right)$. In *STOK* the spatiotemporal distances were calculated using the Pythagorean metric (Example 3.3 of Section 3, Chapter 2)

[j] I.e., data points where the actual $PM_{2.5}$ concentrations were known and could be compared to the values estimated by the *STP* and *STOK* techniques.

TABLE 2.3 Accuracy indicators and computational time of *STP* *vs.* *STOK* techniques.

Technique	*MAE* (μg/m³)	*RMSE* (μg/m³)	Computational time (secs)
STOK	24.95	38.07	965
STP	18.41	27.23	759

$$|\Delta p|_P = \left(h_1^2 + h_2^2 + \varepsilon \tau^2\right)^{\frac{1}{2}},$$

where the coefficient ε had to be chosen properly. Often, this choice is not an easy matter. In the Shandong study, and in light of the $PM_{2.5}$ variogram plot (Fig. 2.3), the approximation

$$\varepsilon = \left(\frac{\text{spatial correlation range}}{\text{temporal correlation range}}\right)^2 = \left(\frac{200{,}000\ \text{m}}{20\ \text{days}}\right)^2 = 10^8\ \text{m/day}$$

was chosen so that all terms of $|\Delta \boldsymbol{p}|_P$ have the same units.[k] On the other hand, in the *STP* technique the projected distances of the traveling pollutant model were directly calculated using the formula $\left(\widehat{h}_1^{\,2} + \widehat{h}_2^{\,2}\right)^{\frac{1}{2}}$ that does not involve any approximations in terms of the ε coefficient, etc.

Lastly, our attention returns to the Hangzhou City case study (Lou et al., 2017) discussed in Example 2.3.

Example 2.3(cont.): Continuing with the Hangzhou City study, the final *STP* steps are concerned with the interpolation and mapping of *BC* incidences, as follows.

Step 5: Using *BC* data and its projected covariance model $c_{\widehat{X}}\left(\widehat{\boldsymbol{h}}\right) = c_{\widehat{BC}}\left(\widehat{\boldsymbol{h}}\right)$ of Eq. (2.4b), the corresponding projected $\hat{\widehat{X}}\left(\widehat{\boldsymbol{s}}\right) = \hat{\widehat{BC}}\left(\widehat{\boldsymbol{s}}\right)$ map was generated in R^2. Because of its reduced dimensionality, compared to working in the original $R^{2,1}$ domain, it is much easier and accurate to work in the R^2 domain, where one can calculate empirical *BC* covariance values that offer a valid representation of the *BC* incidence variation, select a projected *BC* covariance model (by choosing its parameters to best fit the empirical values), and implement a computationally much faster incidence interpolation technique.

Step 6: The $\hat{X}_{stp}(\boldsymbol{p}) = B\hat{C}_{stp}(\boldsymbol{p})$ interpolation maps were generated in the $R^{2,1}$ domain of the original *BC* distribution $X(\boldsymbol{p}) = BC(\boldsymbol{p})$ in Hangzhou City during 2008–12, as shown in Fig. 2.13. Based on these maps, the following conclusions were drawn:

- *BC* incidence distribution in Hangzhou City was temporally stable and spatially heterogeneous, revealing an increasing trend from SW to NE. This finding is consistent with previous studies of *BC* incidence heterogeneity in the time, the space, and the composite space-time domains using *Analysis of Variance*, *Poisson Regression*, and *Space-time Scan Statistics*. Many factors contributed to this heterogeneity, the key one being urban sprawl (Girgis et al., 2000; Fei et al., 2014). The recent economic development of towns and subdistricts made it easier for residents to get healthcare, so that more early-stage *BC* cases were diagnosed. On the other hand, economic development implies higher pollution (heavy metals and dioxins) that may lead to higher *BC* risk.

Step 7: In this case study too, *STP* demonstrated a superior performance compared to *STOK*, see Table 2.4—in addition to *MAE* and *RMSE* of Example 2.2 (cont.) above, the *ME* was also used to test the numerical *BC* interpolation accuracy of the two techniques. Lou et al. (2017) demonstrated the following:

[k] This procedure of defining space-time distances is used in the well-known *SEKSGUI* library, as well as in other space-time mapping libraries, when searching for neighboring points in the space-time domain.

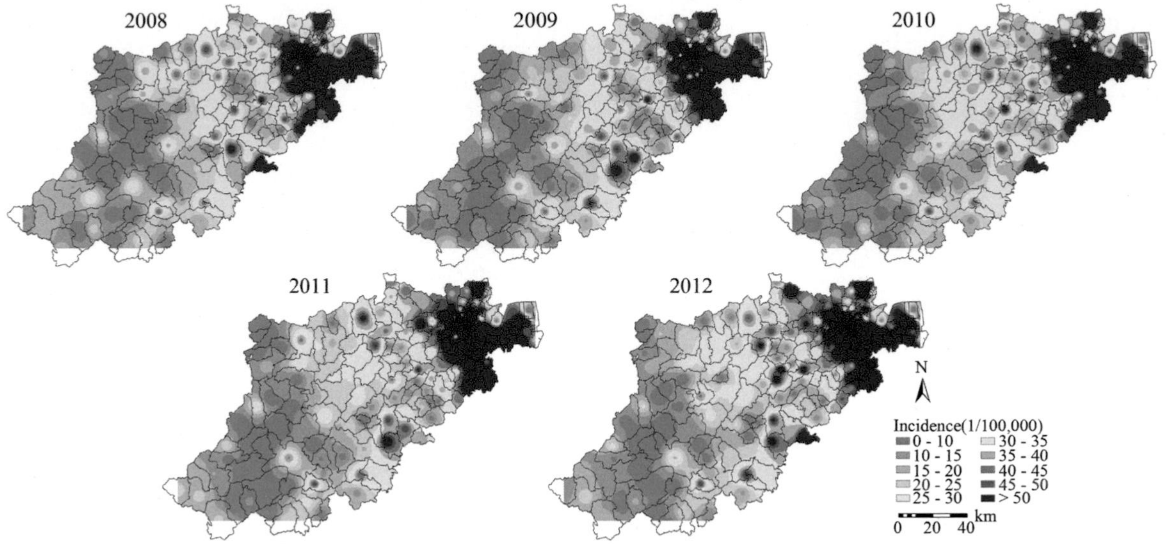

FIG. 2.13 *STP* interpolation maps of the *BC* incidence distribution in Hangzhou City during 2008–12.

TABLE 2.4 Comparison of *BC* incidence interpolation accuracy for *STP vs. STOK* techniques.

	ME (1/100,000)	*MAE* (1/100,000)	*RMSE* (1/100,000)
STOK	5.61	28.26	44.41
STP	−2.99	17.39	21.62

① *STP* yielded considerably more accurate *BC* results than *STOK*. The *ME*, *MAE*, and *RMSE* values of *BC* incidence interpolations obtained by *STP* were much lower (−2.99, 17.39, and 21.62/100,000 respectively) than those obtained by *STOK* (5.61, 28.26, and 44.41/100,000, respectively).

② This difference in accuracy in favor of the *STP* technique was also clearly observed in the corresponding *BC* incidence interpolation maps of Hangzhou City during the period 2008 to 2012.

③ Compared to the actual distribution of *BC* incidence, the *STOK* maps tended to overestimate the *BC* incidence in the southwest low incidence region and to underestimate it in the northeast high incidence region. The *STP* maps were closer to the actual *BC* distribution during the study period, and also much more stable. Specifically, these maps exhibited a definite trend from the southwest low incidence region to the northeast high incidence region, which is also the trend of the observed *BC* incidence distribution. Hence, *STP* provided a more informative and realistic representation of the actual *BC* distribution in Hangzhou City during the period 2008–12.

④ An additional accuracy-test shown in Fig. 2.14 (5-year averaged *BC* incidence of the actual data together with the *STP* and *STOK* interpolations in 200 townships) also suggested that *STP* performed considerably better than *STOK*. The *STOK* map tended to overestimate the *BC* incidence in low incidence regions and underestimate them in high incidence regions.

⑤ *STOK* generated unrealistic *BC* incidence interpolations in the middle and high incidence regions (seriously overestimated low *BC* incidence and underestimated high

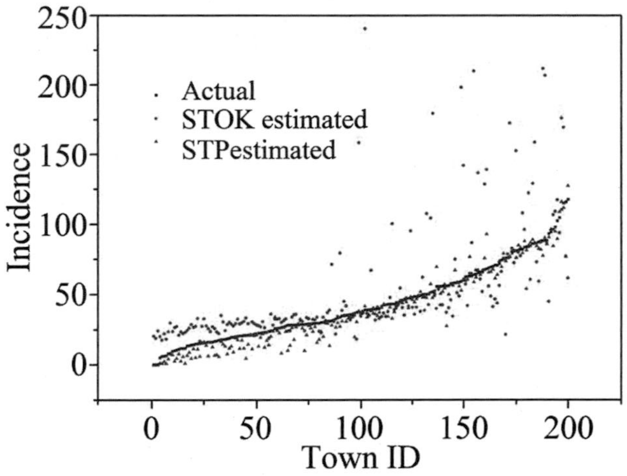

FIG. 2.14 Plots of the 5-year averaged *BC* incidence at 1000 town points for the period 2008–12: actual *BC* incidence values *(black squares)*, *STOK* estimated incidence values *(red points)*, and *STP* estimated incidence values *(blue triangles)*. Incidence per 100,000 people; town ID is denoted from 1 to 201 with ascending order of actual *BC* incidence data.

incidence). On the contrary, *STP* incidence interpolations provided an almost perfect fit to the actual *BC* incidences in the middle and high incidence regions, *albeit* they slightly underestimated the *BC* incidences in the low incidence region. An explanation may be that, as noted earlier, the velocity vector expresses average space-time *BC* spread. Another factor may be the considerable number of zero-incidence regions affecting data normalization and the *BC* incidence covariance. Nevertheless, compared to the *STOK* incidence maps, the *STP* maps provided much better representations of the actual *BC* situation, especially in the middle- and high-incidence regions, where it matters most.

⑥ Regarding the computational cost of the two techniques, a typical computer time of the *STOK* technique was 627 *s* and of the *STP* technique 463 *s* (i.e., an about 26.2% cost reduction).

Brevi manu, the above numerical studies showed that, compared to the *STOK* technique, the *STP* technique is more accurate, easier to implement, and also more workable with the software libraries available. Once one bears in mind the inherent constraints under which *STOK* operates, it is hard to escape the conclusion that its performance in a complex space-time environment could be challenged.

3 Noteworthy *STP* features

Putting the preceding comments together, we conclude that the *STP* technique has certain features that are noteworthy for their conceptual content as they are for their practical modeling and interpolation features.

① The main *STP* goal is to compute reduced models. As a result, a larger variety of attribute covariance and variogram models ($c_{\widehat{X}}(\widehat{h})$ and $\gamma_{\widehat{X}}(\widehat{h})$) are selected in the *LD* domain than in the original *HD* domain, in which it is often much harder to select covariance and variogram models ($c_X(\boldsymbol{\Delta p})$ and $\gamma_X(\boldsymbol{\Delta p})$) that reflect the composite space and time attribute variation structure.

Example 3.1 Difficulties in the original *HD* domain include model parameter specification, especially in the case of complex nonseparable covariance and variogram models. These complications are avoided by *STP* that focuses on the *LD* domain.

② In mainstream interpolation techniques the search for neighboring points typically involves the determination of certain metric parameters. This is often a difficult computational issue. The *STP* technique, however, does not need to calculate such metric parameters, thus avoiding the introduction of additional approximations and, at the same time, reducing the number of computations.

Example 3.2 The metric $|\Delta p|_P = \left(|h|^2 + \varepsilon\tau^2\right)^{\frac{1}{2}}$ commonly used by the *STOK* technique requires the calculation of the coefficient ε. No such calculation is required by *STP*.

③ The *STP* technique was shown to generate more accurate space-time attribute maps than the mainstream mapping technique.

Example 3.3 As was shown in the case studies of the previous section, the *STP* technique produced more accurate $PM_{2.5}$ concentration and *BC* incidence interpolation maps than the *STOK* technique.

④ Taking advantage of the lower dimensionality of its projected domain, the *STP* technique requires less computational effort than mainstream mapping techniques.

Example 3.4 The above *STP* feature could be particularly useful when big datasets are involved.

⑤ A large part of the uncertainty associated with mainstream mapping techniques is due to errors involved in the specification of the physical differences between spatial and temporal variations, and the determination of the space and time attribute cross-correlations. All these errors are eliminated by the coordinate transformation introduced by the *STP* technique.

Example 3.5 Traditional mapping techniques tend to ignore space and time cross-correlations in order to reduce the computational cost. However, the contributions of these space and time cross-correlations are important (Fournier et al., 2011).

4 Practice exercises

1. Prove the υ expression of Eq. (2.8) of Example 2.2 of Section 2.
2. Consider the data provided in Practice Exercise 9, Section 6 of Chapter 8: (*a*) use the *STP* technique to obtain the projected theoretical covariance model in the *LD* domain; and (*b*) compare the 10 fold cross-validation performance of the original Kriging (or *BME*) technique *vs.* the *STP*-based Kriging (or *BME*) technique.
3. Derive the υ expressions of Table 2.1 of Section 2 above.
4. Using Eq. (2.4b) of Section 2, confirms that the υ expression of Example 2.5 of Section 2 is $\upsilon = -27{,}778\tau$.
5. Given the attribute data $X(\boldsymbol{p}) \in R^{2,1}$ in Fig. 4.1, plot the corresponding $\widehat{X}(\widehat{s}) \in R^1$ for: (a) $\upsilon = 0.1$, and (b) $\upsilon = 0.1\frac{s}{t}$.
6. Calculate the υ according to Eq. (2.6a) of Section 2, assuming the following covariance models:

   (a) $c(\boldsymbol{\Delta p}) = \exp\left(-\frac{3h}{4} - \frac{3\tau^2}{16}\right)$

   (b) $c(\boldsymbol{\Delta p}) = \exp\left(-\frac{3h}{81} - \frac{3\tau}{8}\right)$

   (c) $c(\boldsymbol{\Delta p}) = \exp\left(-\frac{3h^2}{25} - \frac{3\tau^2}{64}\right)$

   (d) $c(\boldsymbol{\Delta p}) = \exp\left(-\frac{3h^2}{49} - \frac{3\tau}{7}\right)$
7. Consider the distribution of rainfall (in mm units) described by the dataset of Table 4.1 of Section 4, Chapter 9. The rainfall covariance is $c(\boldsymbol{\Delta p}) = \exp\left(-\frac{3h^2}{5.42} - \frac{3\tau^2}{5.77}\right)$, and

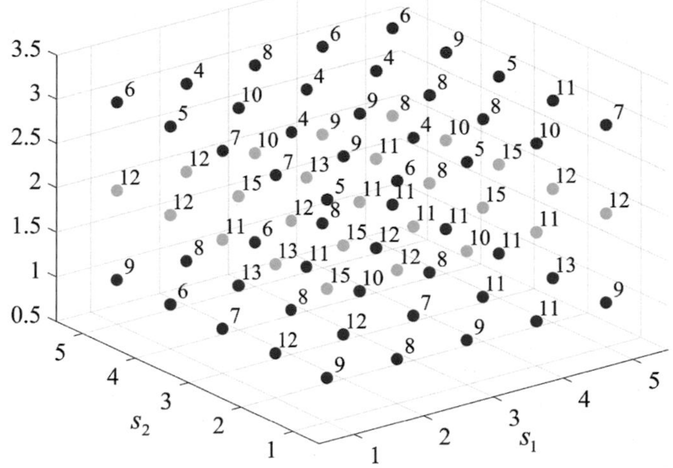

FIG. 4.1 Plot of the $X(\boldsymbol{p})$ data in $R^{2,1}$ (the numbers next to the dots denote data values).

the real rainfall value at point $\boldsymbol{p} = (7.5, 2; 9)$ is 3.54 mm. Using the standard *STOK* and the *STP* ordinary Kriging techniques calculate the rainfall value and its uncertainty at the above point using the data of Table 4.1 of Section 4, Chapter 9. Compare the results.

8. In Exercise 7 above assume the covariance model $c(\Delta\boldsymbol{p}) = \exp\left(-\frac{3h}{5} - \frac{3\tau^2}{5}\right)$, and calculate the rainfall value and uncertainty at point $\boldsymbol{p} = (7.5, 2; 9)$ using the same dataset. What are the differences between the current results and those obtained in Exercise 7? Based on these differences, what conclusions can you draw regarding the properties of the STP technique?

9. Consider a space-time dataset of your interest, and use the *STP* technique for mapping purposes by following the step-by-step process in Section 2.

10. Calculate the υ according to Eq. (2.6b) of Section 2, assuming the following variogram models:

(a) $\gamma(\Delta\boldsymbol{p}) = 100 + 50\left(1 - e^{-h}\right) + 30\left(1 - e^{-\tau^2/4}\right) + 10\left(1 - e^{-h}\right)\left(1 - e^{-\tau^2/4}\right)$

(b) $\gamma(\Delta\boldsymbol{p}) = 50h^3 + 10\tau$

11. Discuss the implementation of the *STP* technique in the chronotopologic interpolation and mapping of phenomena in the scientific field of your interest.

CHAPTER

12

DIA models

1 Introduction

As we saw in previous chapters, a typical *CTDA* study involves data collection, data analysis and modeling, and data interpretation. In other words:

> **Data** consists, generally speaking, of scientific measurements and/or observations that, once analyzed, modeled, and interpreted with the help of scientific theories, can be developed into evidence that can help address key questions about a phenomenon.

Scientific data collection ranges from a quick glance at a thermometer to the time-consuming operation of the Large Hadron Collider (*LHC*). Not surprisingly, data lie at the heart of scientific investigations, and all investigators collect data in one form or another, directly or indirectly. Let us consider an example *in actu*.

Example 1.1 Naturally, data types vary widely, including a series of profiles of soil properties obtained from boreholes, air pollutant distributions obtained at a network of monitoring stations, rock layer drawings of a mountainous region, annual temperatures generated by a global climate model, and the worldwide atlas of cancer cases. The possibilities offered by computers and the inflow of continuous satellite observations complementing ground-based records improve considerably the understanding of the geomagnetic field and the dynamics within Earth's core (Gillet, 2019).

The following assertion, then, seems rather obvious to anyone who takes a moment to examine it with an attentive mind:

- Investigators build both on their own work and that of others, which means that all investigators involved in this process must be **systematic** and **consistent** in their data collection procedures and make detailed records so that others can see and use the data they collect.

At the same time, the same investigators may find themselves in a position where they try to avoid a working environment in which the excessive use of *black-box* techniques can easily create human black boxes, although in several cases it may be the other way around.

1.1 Data constraints

Conventional data analysis techniques have been designed to extract insights from scarce, static, clean, and often poorly relational datasets,

https://doi.org/10.1016/B978-0-12-816341-2.00011-3

scientifically sampled and adhering to strict assumptions (such as independence, stationarity, and normality), and generated and analyzed with a specific question in mind. Yet, rather common data-related constraints exist:

① The substantial **gaps** in attribute domain coverage may raise questions regarding the ability of a dataset to adequately represent the phenomenon of interest.
② The insufficiency of many existing conventional data monitoring networks to correct for the **uncertainty** inherent in datasets exhibiting large natural variability.

Example 1.2 Real-world situations of the two limitations above include the following:

- As regards constraint ①, gaps in temperature coverage over the ocean can lead to a poor representation of the global system.
- As regards constraint ②, on any given day the weather in Antarctica is very different than that in the Sahara Desert.

Yet another kind of a data constraint may take the form of data *clusters* (this constraint is of considerable *CTDA* interest, and it has already been discussed in some detail in Section 4 of Chapter 4).

③ The basic goal of cluster **detection** is to properly detect attribute domains that are anomalous or unexpected in a certain sense that is of scientific interest.

Once we bear in mind the constraints under which *CTDA* often has to operate, it is hard to escape the conclusion that richer datasets should be welcomed that are, at the same time, of better quality and adequately distributed across space and time. Yet, one wonders whether this is always the case, and under what conditions this can be true.

1.2 Big data

In Chapter 3 the *Big Data* (*BD*) idea was introduced and some of its intricacies were discussed. It was stressed that with the development of new and powerful software tools of data analysis and processing the emerged *BD* era is characterized by the increasing sophistication and ability of these tools to potentially detect hidden patterns in vast amounts of data and draw inferences. The implications of these developments for science, engineering, and even culture are profound and, in some cases, unanticipated (e.g., studying the health records of hundreds of thousands of patients one can detect useful statistical correlations between specific drugs and their health effects). In certain disciplines, *BD* continues to generate massive, dynamic flows of diverse, fine-grained, relational, and scalable data. This is a fact with further implications:

- *BD* has led to the development of **Data-Intensive Analysis** (*DIA*) techniques, a chief concern of which is that as large amounts of data are continuously generated, a technology is needed to ensure that understanding is constantly updated.

In particular, there is no reason to doubt that the considerable evidential support that Earth's climate has been changing globally during the last century (IPCC, 2013) requires the development of a rigorous and effective monitoring of the Earth's climate at the global scale. In this case, *DIA* is characterized by the enormous increase (in size and variety) of the available information as a result of the massive datasets produced by satellite instruments, global and automated networks (*ARGO* floats and *AERONET*, Holben et al., 1998), the much denser global climate model resolutions, the advancement of computational tools and storage capabilities, and the huge volume of smartphone-based records provided by individuals worldwide (the so-called *Citizen Science*).

As a consequence of all the above, to the traditional *CTDA* challenge of acquiring the right data, it was added that of the large-scale analysis of the available data.

Example 1.3 The continuous gathering of thematic data using *remote sensing* (*RS*) techniques leads to the increasing availability of large amounts of chronotopologic information in *TGIS*. This information is valuable in the study of attribute anisotropies due to morphologic or tectonic structures (tectonic faults interrupting continuity, drainage systems inducing nested anisotropy patterns, etc.).

Naturally, then, a principal *DIA* component is computer science, in particular, branches of *Artificial Intelligence* (*AI*).

2 Machine learning

The particular need highlighted above to handle continuously generated huge amounts of data has led to the development of the following *AI* model of data management and high-performance computing:

> **Machine learning** (*ML*)[a] is a data-driven technique or algorithm that emerged from the need to develop innovative ways to handle large and continuous amounts of data in a way that new and improved levels of understanding are gained.

As a form of *AI*, it has been suggested that *ML* enables learning from data rather than through explicit programming. *ML* may detect correlations among huge numbers of variables (on the order of thousands, millions, or even billions) that may be beyond immediate human conception. Yet, it is rather common knowledge that collecting data is only one step of an investigation, and scientific knowledge consists of much more than a simple compilation of data points. Many observations can be generated, but not all of them provide a useful piece of information.

Example 2.1 A meteorologist could record air temperature values every second during a day, but this probably will not make weather forecasting more accurate than hourly temperature recordings.

Sine dubio, investigators routinely make choices about which dataset is most relevant to their objectives and what to do with it, how to turn a collection of measurements into a useful dataset through processing and analysis, and how to interpret those analyzed data in the context of what they already know. Thoughtful and systematic collection, analysis, and interpretation of data allow it to be developed into sound evidence that supports scientific ideas, arguments, and hypotheses.

- With such concerns in mind, *ML* algorithms (models) are mathematical expressions representing data in a specified context, i.e., *ML* generally is about moving from data to useful **insight**.

This insight is helped considerably by an understanding of the order lying behind the appearance that is provided by a scientific theory (see, also, our discussions in the preceding Chapters 6, 9, and 10, and in the following Chapter 13).

ML models are classified on the basis of their prime goals. Such a major classification is the one that distinguishes between:

① **Supervised** *ML* seeking to predict or explain a dataset (using previous input and output datasets to predict an output based on a new input).[b]

② **Unsupervised** *ML* seeking to relate and group observations that have similar features (evaluating data in terms of traits and using them to form clusters of entities that are similar to one another).[c]

Among the best-known supervised *ML* techniques are *linear regression* (Section 3 below),

[a] Recall discussion in Section 3 of Chapter 3.

[b] Supervised *ML* can be helpful when one seeks to predict the chronotopologic attribute distribution.

[c] Unsupervised *ML* can be helpful when one seeks to segment attribute values with similar features without the need to specify in advance which features to consider.

classification (e.g., logistic regression), and *ensemble* (e.g., random forest algorithm). The unsupervised *ML* techniques include *artificial neural networks* (Section 4 below), *clustering* (*K*-means), and *dimensionality reduction* (e.g., principal component analysis and spatiotemporal projection, Chapter 11). Using these algorithms, huge amounts of data are continuously collected and stored using advanced sensor technologies, powerful computing platforms and continuously improving online connectivity.

Before proceeding further, it is worth noticing that *ML* involves a variety of mechanistic models that are worth reviewing. The technical *ML* performance improves proportionally to the number of the model components. This means that the increasingly complex mechanistic models thus generated are also increasingly intelligible, and, therefore, potentially less explanatory, i.e., the models fail to explain scientifically why a certain phenomenon that occurs at a specified location and time adopts a certain physical value.

3 Linear regression techniques

Linear Regression (*LR*) is a concept at the center of a rather popular group of data-driven statistical techniques used in supervised *ML*. Supervised *ML* tasks are typically divided into classification and regression, the *LR* techniques focusing on the latter.

The goal of ***LR*** techniques is to describe the attribute of interest (*dependent* attribute) by introducing some related variables (*independent* or *explanatory* attributes) and then use linear combinations of the latter attributes to quantitatively obtain the former attribute.

Many of the *LR* techniques were originally developed in a space-free setting (mostly in statistics studies), and then considered in a space-dependent context (particularly in geographical sciences). *LR, ad usum*, has been used in two ways:

① To understand how changes in the independent attribute values are associated with changes in the dependent attribute.
② To generate dependent attribute interpolations based on the values of the independent attributes.

In the case ①, the independent attributes are typically used to explain changes in the dependent attribute. The case ② has already been discussed in Section 3.2 of Chapter 7, and will be explored further below.

Example 3.1 Let us consider two typical cases of the *LR* using ① and ② above:
- The investigators want to determine how changes in the soil moisture, the amount of fertilizer applied, and the amount of sunlight (independent attributes), are linked to plant growth change (dependent attribute).
- In environmental health studies, climate, demographic factors, and physical environment characteristics may be the independent attributes used to predict the human exposure effect, which is the dependent attribute.

In the following, we introduce the *multiple LR* notion in the *CTDA* setting, and then discuss some other techniques that are rather straightforward extensions of it.

3.1 Multiple linear regression

Multiple linear regression (*MLR*, Cook and Pocock, 1983; Fotheringham et al., 1998) is a data-driven statistical technique that models quantitatively the linear relationship between the independent (explanatory) attributes and a dependent (response) attribute. In spatial data analysis, arguably, *MLR* has been one of the most useful statistical models to identify the nature of relationships among variables.

In the *CTDA* setting, the ***MLR*** can be expressed mathematically as the series representation

$$X(\boldsymbol{p}_k) = \theta_0 + \sum_{j=1}^{m} \theta_j Y_j(\boldsymbol{p}_k) + \varepsilon(\boldsymbol{p}_k), \qquad (3.1)$$

where $X(\boldsymbol{p}_k)$ is the dependent attribute at unsampled points $\boldsymbol{p}_k$ $(k = 0,1,\ldots,n)$, $Y_j(\boldsymbol{p}_k)$ are the m independent attributes, θ_j $(j = 0,1,\ldots,m)$ are model parameters, and $\varepsilon(\boldsymbol{p}_k)$ are error terms (residuals).

Generally, *MLR* is widely used to determine how multiple independent variables are related to one dependent variable, i.e., the information on the independent variables is used to assess the effect they have on the dependent variable. Interpretationally, each θ_j $(j = 1,\ldots,m)$ measures a unit change in $X(\boldsymbol{p}_k)$ caused by a change in $Y_j(\boldsymbol{p}_k)$. The model creates a relationship in the form of a plane surface that best approximates all the individual data points.

MLR can be used to build models for inference or chronotopologic interpolation. These models usually rely on four assumptions: (i) A linear relationship exists between the dependent attribute and the independent attributes; (ii) the independent attributes are not too highly correlated with each other; (iii) the $X(\boldsymbol{p}_k)$ observations are selected independently and randomly within the domain of interest; and (*iv*) the error terms $\varepsilon(\boldsymbol{p}_k)$ should be independent and normally distributed random variables with zero mean and constant variance.

Observations of $X(\boldsymbol{p}_k)$ and $Y_j(\boldsymbol{p}_k)$ $(k = 0,1,\ldots,n)$ are obtained within the study domain, and estimates $\hat{\theta}_j$ of the parameters θ_j $(j = 0,1,\ldots,m)$ are derived by minimizing the sum of the squares in the difference between the observed and predicted values of the dependent attribute ($\hat{\theta}_j$ are unbiased estimators of the real values of θ_j). Hence, *MLR* generates attribute surfaces that are as close as possible to the data points in the least-squares sense.

Example 3.2 Below, we discuss a few of the numerous *MLR* applications.

- *MLR* has been used to link malaria cases in a study region (dependent attribute) with temperature, rainfall, population density, distance to ponds and streams, and normalized difference vegetation index (independent attributes).
- *MLR* is used in medical diagnostic and therapeutic studies in which the outcome is dependent on more than one factor.
- When an analyst needs to assess the effect of the financial markets on the stock price of an oil company, the independent attributes could be the price of oil, the interest rates, the price movement of oil futures, and the value of the S&P 500 index, whereas the stock price will be the dependent attribute.

3.1.1 *Simple linear regression*

Not surprisingly, several special cases of *MLR* have been considered in the literature with varying success. One of them is the *simple linear regression* (*SLR*) model:

SLR is a special case of *MLR* when a bivariate model is considered, i.e., a model in which there is only one independent attribute predicting a dependent attribute.

SLR chooses the parameters of the linear equation linking the dependent attribute and the independent attributes by the principle of least squares (minimizing the sum of the squares of the differences between the observed dependent attribute and those obtained by the linear equation).

Example 3.3 In the case of a single independent attribute, a bivariate regression may be used to fit a straight line to a scatterplot of observations on the independent and dependent attributes. The chosen straight line is the one that allows the least squared error between the predicted values of the dependent attribute and its actual values. Investigators are often interested about questions concerning the relationship between two attributes, such as if there is a relationship between age and baseline systolic blood pressure, how the *GNP*[d] and the life expectancy of different countries are linked, if female education and fertility are related in some way, whether or not countries with higher *GNP* levels have higher levels of life expectancy than countries with lower *GNP* levels, and if there is a positive relationship between employment opportunities and net migration?

The interpolation role of the *SLR* was examined in Section 3.2 of Chapter 7, in the form of the chronotopologic linear regression (*CTLR*) interpolator. Additional technical details can be found in the relevant literature (also cited in Chapter 7).

3.1.2 Land-use regression

Land use regression (*LUR*, Briggs et al., 1997, 2000; Gilliland et al., 2005; Hoek et al., 2008) is another data-driven technique focusing on the incorporation of site-specific variables in environmental health and epidemiology studies. It has been suggested by some authors (e.g., Ryan and LeMasters, 2007) that the incorporation of site-specific variables by *LUR* can detect small area variations more adequately than other data-driven interpolation techniques.

> ***LUR*** is a special formulation of *MLR* seeking to assess the relationship between an environmental dependent attribute and several independent geographic attributes.

Typically, in *LUR* the dependent attribute is an atmospheric pollutant or a health effect, and the independent attributes are traffic conditions, land use, topography, and other geographic factors, or air quality variables. *GIS* is used to construct independent attributes associated with land use and traffic conditions. Also, various modifications of the independent attribute supports are possible in the *LUR* setting. So, independent attributes based on land-use and population data are sometimes created in the form of buffer zones around the monitoring sites. On the other hand, independent attributes based on traffic data may be based on distance from the site to the road.

Example 3.4 *LUR* may use *MLR* modeling to describe the relationship between the monitored pollutant level (dependent attribute) and traffic conditions and land use (independent attributes), often employing *GIS* to collect measurements. This leads to an equation that can generate pollution level interpolations at unmonitored locations based on independent attribute data at a given set of locations.

The *LUR* technique may focus either on chronotopologic interpolation or on interpretation. In many applications, the *LUR* implementation is a combination of these two alternatives. Since the dependent attribute at each point is empirically linked to the independent attributes,[e] the *MLR* equation relating these attributes can generate dependent attribute values at unsampled points from the known values of the independent attributes and it can provide insights into the effects of the independent attributes on the dependent attribute as represented by the *MLR* equation. The interpretation of the effects of the independent attributes on the dependent attribute can be improved by maximizing the accuracy of the selected independent attributes and the regression coefficients.

Example 3.5 Land use and population density data can be used as the independent attributes of the *LUR* spatial interpolation technique to study the impact of social factors on the spatial Dengue Fever distribution viewed as the dependent attribute.

[d] Gross National Product.

[e] *LUR* requires some empirical association of the dependent to the independent attributes (especially environmental attributes that can affect pollutant emission intensity and dispersion efficiency).

The limitations of *LUR* modeling have been discussed extensively in the literature (e.g., Hoek et al., 2008; Isakov et al., 2011; Hankey and Marshall, 2015): the spatial *LUR* formulation performs inadequately when the spatiotemporal resolution of emission sources (e.g., traffic congestion) is poor, and generates interpolations that are not directly comparable to regulatory standards (temporal coverage is neglected); it models only long-term concentrations and one pollutant per model (although later efforts have attempted to address this limitation, at least in part, by considering multiple pollutants); extreme local variations close to emission source may be ignored; the models are location-specific and not applicable to other locations; and it is usually difficult to construct independent attributes illustrating land use and traffic conditions using *GIS*. Also, in many cases better alternatives to *LUR* exist, like the theory-based *BME* methods (Chapters 6, 9, and 10), which rigorously account for various core and site-specific knowledge bases, including physical laws, soft data, and secondary information.

3.2 Geographically weighted regression

Geographically weighted regression (*GWR*) is another *MLR* technique that was developed to address the concern that a regression model considered over the entire region of interest cannot adequately account for local spatial variations (e.g., Brunsdon et al., 1998).

Accordingly, ***GWR*** can be seen essentially as an extension of *MLR* (Section 3.1) that adds a level of modeling sophistication by allowing the relationships between the independent variables and the dependent variable to vary by locality.

GWR was originally developed in geography and other disciplines to analyze spatial point datasets and interpolate unsampled attribute values by assuming that the strength and direction of the relationship between a dependent attribute and the independent attributes may be modified by contextual factors. In this setting, *GWR* takes nonstationary[f] attributes into consideration and models the local relationships between these independent attributes and the dependent attribute of interest.

Example 3.6 *GWR* has been used to study a variety of issues, including the following:

- Determine the factors consistently influencing higher cancer rates within the chronogeographic domain of interest.
- Investigate the chronotopologic relationship between environmental exposure and certain illness or disease occurrences.
- Assess the relationship between educational attainment and income across the study region within a specific time period of regional growth.
- Identify habitats in need of protection and the required duration of the protection in order that the reintroduction of an endangered species may be encouraged.
- Discover at which county districts the children have achieved high science and mathematics test scores during the last decade.
- Generate $PM_{2.5}$ interpolations at unmonitored locations together with the associated variances, which can then be regarded as soft data to be used in chronotopologic interpolation (Xiao et al., 2018).

In the *CTDA* setting, the basic *GWR* idea is to study the chronogeographic change of the relationship between a dependent attribute and one or more independent attributes. Instead of fitting a unique model to the entire study domain, *GWR* is based on a *moving window* approach. This approach is briefly as follows:

[f] Nonstationarity here implies that locally weighted regression coefficients move away from their global values.

a search window is used to cover the available dataset by moving from one dataset point to another; as the search window rests on a sample point, all other points that are around it and within the search window are identified; and a regression model is then fitted to that subset of the data, giving most weight to the points that are closest to the one at the center. The issue of what area the search window should cover each time is addressed by a calibration process that selects an optimal search window size (termed as the optimal bandwidth).

In the case of ***GWR***, Eq. (3.1) of the *MLR* technique is replaced by the equation

$$X(p_k) = \beta_0(p_k) + \sum_{j=1}^{m} \beta_j(p_k) Y_j(p_k) + \varepsilon(p_k), \quad (3.2)$$

where the parameters $\beta_0(p_k), \beta_j(p_k)$ are assumed to be functions of the points on which the observations are obtained.

As noted above, *GWR* has been used both as an exploratory technique and as a prediction technique (although its usefulness in this respect is rather controversial). More specifically, as an exploratory technique *GWR* studies if and how the relationship between independent and dependent attributes varies chronotopologically. To achieve this, *GWR* constructs a separate *MLR* equation for every dataset point, which incorporates the dependent and independent attributes of points falling within a bandwidth of each target point. Among the known limitations of the *GWR* technique are multicollinearity and the computation of goodness of fit statistics (Wheeler and Tiefelsdorf, 2005). *Felix qui potuit rerum cognoscere causas*, yet, the GWR techniques do not necessarily establish any real cause-effect relationships.

4 Artificial neural network

For years, human modeling efforts have been trying to efficiently imitate biological processes at different scales. In this context, the notion of *Artificial Neural Network* (*ANN*, Haykin, 1994) or, simply, neural network, was conceived as a computational model mirroring one's central nervous system. Regarding *ANN* terminology, the motivation behind the use of the term "neural" is the human nervous system's basic functional unit "neuron" or nerve cells that are found in the human brain.

An ***ANN*** is an *ML* algorithm represented by a system of interconnected artificial neurons that can compute values from inputs in a way that simulates the behavior of biological systems composed of dendrites and synapses.

Otherwise said, the *ANN* models or algorithms are biologically inspired computer programs designed to simulate the way in which the human brain processes information. In practice, the *ANN* is a software implementation of a mathematical model that mimics the neuronal structure of the human brain and aims at enhancing existing *CTDA* technologies. Let us look at how this objective is achieved.

4.1 Human brain as a neural network

For our purposes, surely we do not need to address the complex biology of the human brain structure. Instead, it suffices to note that the human brain consists of millions of neurons that are like organic switches that send and process electrical and chemical signals (i.e., the switches can change their output state depending on the strength of their electrical or chemical input). These neurons are connected with synapses that allow neurons to pass signals. From large numbers of simulated neurons, neural networks are formed.

In particular, the typical nerve cell of the human brain has four parts with their corresponding functions described as follows (Fig. 4.1): the *dendrite* receives signals from other neurons, the *soma* (cell body) sums all the incoming signals to generate the input, so that when

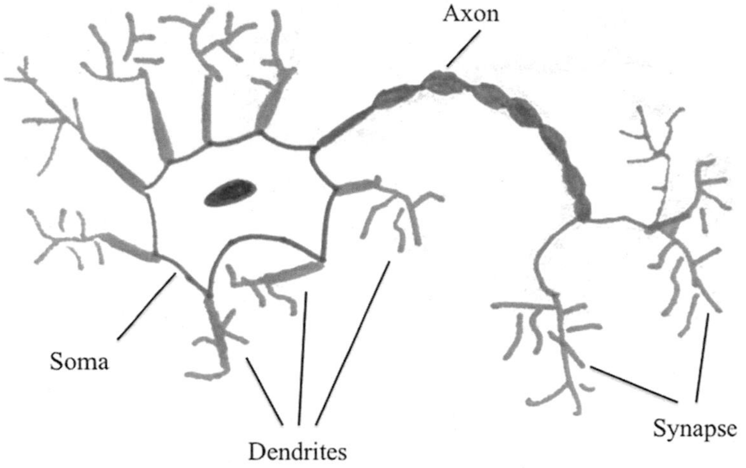

FIG. 4.1 An illustration of a biological neuron.

the sum reaches a threshold value the neuron fires and the signal travels down the *axon* to other neurons, and the *synapse*, which is the point of interconnection of one neuron with other neurons.

- The **biological neural network** (*BNN*) in a human's brain is a hugely interconnected network of neurons, where the output of any given neuron is the input to thousands of other neurons. Learning occurs by repeatedly activating certain neural connections over others, and involves feedback, i.e., when the desired outcome occurs given the specified inputs, the neural connections causing that outcome become strengthened.

Next, we examine how the various activities of a *BNN* are mimicked by the corresponding functions of an *ANN*.

4.2 From the *BNN* to the *ANN*

The *ANN* seeks to simplify and mimic the brain behavior above. It is represented by an oriented graph consisting of nodes (which in the biological analogy represent neurons) connected by arcs (which in the biological analogy represent synapses). Each arc is associated with a weight at each node, and it applies the value received as input by the node to define an activation function along the incoming arc, adjusted by the weight at each node. As such, an *ANN* is an information processing technique that works like the way the human brain processes information. *ANN* includes a large number of connected processing units that work together to process information and generate meaningful results. In this setting:

- *ANN* models gather their knowledge by detecting the patterns and relationships in data and learn (or are trained) through experience.

In light of Fig. 4.1, briefly, the following correspondence may be established between *BNN* and *ANN*: The *BNN* soma (cell body) may be compared to the *ANN* node, signals are received by the *BNN* dendrites that correspond to the *ANN* weighted inputs, and are sent down to the *BNN* axon that is analogous to the *ANN* output and its synapses (axon terminals) that connect axon to other neuron dendrites, i.e., the outgoing signals serve as inputs to other neurons, and the process is repeated.

As happens with the *ML* algorithms (Section 2 above), the *ANN* can be trained in two different ways:

❶ In a **supervised** *ANN* training takes place by providing input data to the network, which produces outputs that are compared with the desired outputs. An error signal is generated if there is a difference between the actual and the desired output, based on which the weights are adjusted until the actual outputs are matched with the desired outputs.

❷ In an **unsupervised** *ANN* learning takes place by combining input data of similar type to form clusters. When a new input pattern is applied, then the *ANN* gives an output response indicating the class to which the input pattern belongs.

Since there is no feedback from the environment as to what should be the desired output and if it is correct or incorrect, in unsupervised learning the *ANN* itself must discover the patterns and features from the input data, and the relation for the input data over the output: i.e., the *ANN* seeks to understand the structure of the provided input data on its own.

There are different types of neural networks, which can be generally classified into two prime types, as follows:

① **Feed-forward** networks, i.e., a nonrecurrent network with a structure that contains inputs, outputs, and hidden layers (Section 4.3), the signals can only travel in one direction, and the output is determined based upon a weighted sum of the inputs and a threshold transfer function.

② **Feed-back** networks, i.e., a recurrent neural network that has feed-back paths, all possible connections between neurons are allowed so that signals can travel in both directions using loops (continuously changing nonlinear dynamic system until a state of equilibrium is reached).

This chapter focuses on the first kind, i.e., on feed-forward networks.

4.3 ANN structure

The structure of a neural network is also referred to as its "architecture" or "topology." This structure can come in many different forms, but the most common, simple *ANN* structure consists of three kinds of layers as follows:

① **Input layer**. Its purpose is to receive as input the values of the independent attributes for each observation (the number of input nodes in an input layer is usually equal to that of independent attributes). Each node of the input layer receives a single value as its input, duplicates it by means of its many outputs, and sends it to the hidden nodes.

② **Hidden layer**. Its purpose is to determine the activity of each hidden unit. In the hidden layer, the processing is done via a system of weighted connections. The ability of the neural network to provide useful data manipulation lies in the proper selection of the weights.

③ **Output layer**. It receives connections from hidden or input layers. It returns an output value that corresponds to the prediction of the response variable. The active nodes of the output layer combine and change the data to produce the output values. The behavior of the output units depends on the activity of the hidden units and the weights between the hidden and output units.

Below, we discuss some essential *ANN* notions in more detail, with *CTDA* in mind.

4.3.1 *Activation or transfer functions*

In an *ANN*, the actual biological neuron is simulated by an *activation* or *transfer function* ϑ. The way ϑ simulates the "turning on" state of a biological neuron is that it has a "switch on" feature so that once the input χ is greater than a specified threshold, the output ϑ changes value (say, from 0 to 1). In this way noisy data can be identified for classification purposes.

The ϑ function translates the input signals to output signals. Four mathematical types of ϑ functions are commonly used:

1. **Unit step** function:

$$\vartheta(\chi) = \begin{cases} 1 & \text{if } \chi \geq 0 \\ 0 & \text{if } \chi < 0. \end{cases} \tag{4.1a}$$

2. **Sigmoid** function:

$$\vartheta(\chi) = \frac{1}{1 + e^{-a\chi}}, \tag{4.1b}$$

where a is a suitable coefficient that is activated (say, changing value from 0 to 1) when the χ is greater than a certain value.[g]

3. **Piecewise linear** function:

$$\vartheta(\chi) = \begin{cases} 1 & \text{if } \chi \geq \zeta_{\max} \\ ax + b & \text{if } \chi \in (\zeta_{\min}, \zeta_{\max}) \\ 0 & \text{if } \chi \leq \zeta_{\min}, \end{cases} \tag{4.1c}$$

where a and b are suitable coefficients.

4. **Gaussian** function:

$$\vartheta(\chi) = \frac{1}{\sqrt{2\pi}b} e^{-\frac{(\chi - a)^2}{2b^2}}, \tag{4.1d}$$

where a and b are suitable coefficients.

Example 4.1 Combinations of functions can also be used, such as the hyperbolic tangent function

$$\vartheta(\chi) = \tanh(\chi) = \frac{e^{\chi} - e^{-\chi}}{e^{\chi} + e^{-\chi}},$$

which is useful for multilayer networks.

The behavior of a neural network is determined by the activation or transfer function of its neurons, by the learning rule, and by the structure itself.

4.3.2 Nodes

The power of neural computations comes from connecting neurons in a network. Since biological neurons are connected hierarchical networks, with the outputs of some neurons being the inputs to others, these networks are represented in *ANN* as connected layers of *nodes* (or *perceptrons*). Each node receives multiple inputs χ_j weighted by $w_{ij}^{(l)}$ (i refers to the node number of the connection in layer $l+1$ and j refers to the node number of the connection in layer $l, l = 1, \ldots, L$), together with a bias weight denoted as $b_i^{(l)}$ (i is the node number in the layer $l+1$). The weights $w_{ij}^{(l)}$ are real-valued numbers (i.e., not binary 1 s or 0 s) that change during the learning process, and the bias term $b_i^{(l)}$ improves node flexibility (e.g., it is added so that the node can simulate a generic *if* function: "if $\chi > \zeta$ then 1, else 0"). Note that changing the weights $w_{ij}^{(l)}$ changes the slope of the ϑ output, which is useful if we want to model different strengths of relationships between the input and output variables. The bias $b_i^{(l)}$ plays a key role if we only want the output to change when $\chi > 1$. The $w_{ij}^{(l)}$ and $b_i^{(l)}$ values need to be calculated in the training phase of the *ANN*. *ANN* is a parameterized system, in the sense that the weights are the adjustable parameters.

The weights $w_{ij}^{(l)}$ at each node i are multiplied by the inputs χ_j, and then summed up in the node and the b term (b is the weight of the +1 bias element) is added to obtain the variable $v_i^{(l)} = \sum_{j=1}^{n} w_{ij}^{(l)} \chi_j + b_i^{(l)}$ ($i = 1, \ldots, n$; $l = 1, \ldots, L$). Otherwise said, the weighed sum of the inputs constitutes the activation of the neuron. Finally, each node applies the activation function ϑ to the summation $v_i^{(l)}$ to generate an *output* $\eta_j^{(l)}$ (i.e., the ϑ output), where j denotes the node number in the layer l of the *ANN*.

Example 4.2 *ANN* is typically organized in layers, which are made up of many interconnected nodes that contain an activation function. An example of such a structure is plotted in

[g] Note that $\vartheta(\chi)$ is not a step function. Its edge is soft and its value does not change instantaneously; thus, it is a derivable function, which is an important property for subsequent training purposes.

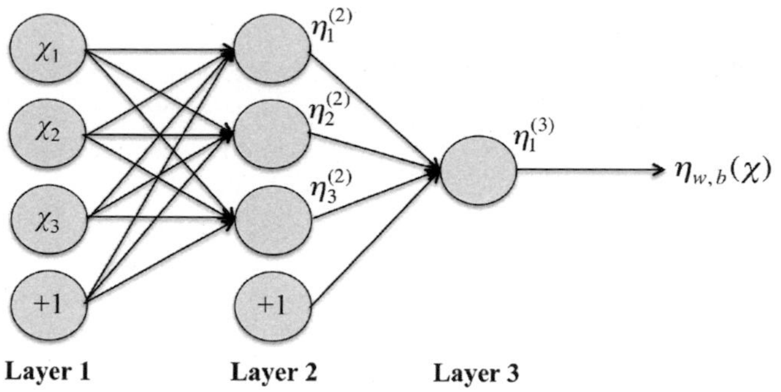

FIG. 4.2 An illustration of a three-layer *ANN*.

Fig. 4.2. The three layers of the network are: Layer 1 represents the *input layer*, where the external input data enters the network. Layer 2 is called the *hidden layer* as this layer is not part of the input or output. Note that a neural network can have many hidden layers, but in this case for simplicity, only one is included. For illustration, consider the connection between node 1 in the 1st layer ($l=1$) and node 2 in the 2nd layer ($l=2$), in which case the corresponding weight should be denoted as $w_{21}^{(1)}$. The weights between layers 1 and 2 can be represented as the matrix

$$W^{(1)} = \begin{bmatrix} w_{11}^{(1)} & w_{12}^{(1)} & w_{13}^{(1)} \\ w_{21}^{(1)} & w_{22}^{(1)} & w_{23}^{(1)} \\ w_{31}^{(1)} & w_{32}^{(1)} & w_{33}^{(1)} \end{bmatrix}.$$

Finally, the 3rd layer ($l=3$) is the *output layer*. Similarly, the weights between layers 2 and 3 can be represented as $W^{(2)} = [w_{11}^{(2)}\ w_{12}^{(2)}\ w_{13}^{(2)}]$. Each node in layer $l=1$ has a connection to all the nodes in the layer $l=2$, and each node in the layer $l=2$ is connected to the single output node of the layer $l=3$. Each of these connections has an associated weight. Moreover, the (+1) bias in Fig. 4.2 is connected to each of the nodes in the subsequent layer. So, the bias in layer 1 is connected to all the nodes in layer two. Because the bias is not a true node with an activation function, it has no inputs (it always outputs the value +1). For illustration, the weight on the connection between the bias in layer 1 and the second node in layer 2 is given by $b_2^{(l)}$. Each connection has an associated weight and the corresponding summations are $v_i^{(l)} = \sum_{j=1}^{3} w_{ij}^{(l)} \chi_j + b_i^{(l)}$ $(i,l = 1,2,3)$.

4.3.3 *Output values*

The feed-forward process (Section 4.2) computes the output values $\eta_j^{(l)}$ of the *ANN* given the values of the inputs and weights. To demonstrate how the output is computed from the input in the *ANN*, let us introduce the vector

$$\boldsymbol{v}^{(l+1)} = \mathbf{W}^{(l)} \boldsymbol{\chi}^{(l)} + \boldsymbol{b}^{(l)}, \tag{4.2}$$

where $\boldsymbol{\chi}^{(l)} = \begin{bmatrix} \chi_1 \\ \vdots \\ \chi_n \end{bmatrix}$, n is the number of nodes in layer l, $\mathbf{W}^{(l)} = \begin{bmatrix} w_{11}^{(l)} \cdots w_{1n}^{(l)} \\ \cdots \\ w_{n1}^{(l)} \cdots w_{nn}^{(l)} \end{bmatrix}$, $\boldsymbol{b}^{(l)} = \begin{bmatrix} b_1^{(l)} \\ \vdots \\ b_n^{(l)} \end{bmatrix}$,

and $\boldsymbol{v}^{(l+1)} = \begin{bmatrix} v_1^{(l+1)} \\ \vdots \\ v_n^{(l+1)} \end{bmatrix}$.

Then, the output vector of the nodes in the layer will be given by

$$\boldsymbol{\eta}^{(l+1)} = \boldsymbol{\vartheta}\left(\boldsymbol{v}^{(l+1)}\right), \tag{4.3}$$

where $\boldsymbol{\eta}^{(l+1)} = \begin{bmatrix} \eta_1^{(l+1)} \\ \vdots \\ \eta_n^{(l+1)} \end{bmatrix}, \boldsymbol{\vartheta}(\boldsymbol{v}^{(l+1)}) = \begin{bmatrix} \vartheta\left(v_1^{(l+1)}\right) \\ \vdots \\ \vartheta\left(v_n^{(l+1)}\right) \end{bmatrix}$,

and $\vartheta(\cdot)$ refers to the node activation function. The ith line in Eq. (4.3), $\eta_i^{(l+1)}$, is the output of the ith node in the $l+1$ layer, and its inputs are $w_{ij}^{(l)}\chi_j$ and $b_i^{(l)}$ $(i,j = 1,\ldots,n)$. An illustrative example follows.

Example 4.3 For illustration, consider the specific case of the three-layer neural network presented in Fig. 4.2. In this case, Eq. (4.3) yields the following *ANN* representation from layer 1 to layer 2[h]

$$\begin{aligned} \eta_1^{(2)} &= \vartheta\left(v_1^{(2)}\right) = \vartheta\left(\sum\nolimits_{j=1}^{3} w_{1j}^{(1)}\chi_j + b_1^{(1)}\right), \\ \eta_2^{(2)} &= \vartheta\left(v_2^{(2)}\right) = \vartheta\left(\sum\nolimits_{j=1}^{3} w_{2j}^{(1)}\chi_j + b_2^{(1)}\right), \\ \eta_3^{(2)} &= \vartheta\left(v_3^{(2)}\right) = \vartheta\left(\sum\nolimits_{j=1}^{3} w_{3j}^{(1)}\chi_j + b_3^{(1)}\right); \end{aligned} \quad (4.4a - c)$$

and from layer 2 to the single node of layer 3,

$$\eta_{w,b}(\chi) = \eta_1^{(3)} = \vartheta\left(v_1^{(3)}\right) = \vartheta\left(\sum\nolimits_{j=1}^{3} w_{1j}^{(2)}\chi_j + b_1^{(2)}\right), \quad (4.4d)$$

where the node activation function ϑ is the sigmoid function, in this case. Specifically, the 1st line in Eq. (4.4a) is the output $\eta_1^{(2)}$ of the 1st node ($i = 1$) in the 2nd layer ($l = 2$), where its inputs are $w_{1j}^{(1)}\chi_j$ ($j = 1,2,3$) and $b_1^{(1)}$ (see the three-layer connection in Fig 4.2). These inputs are then summed to get $v_1^{(2)}$, which is passed through ϑ to calculate the output $\eta_1^{(2)}$ of the 1st node ($i = 1$); similarly, for the other two nodes ($i = 2,3$) in the 2nd layer ($l = 2$), see $\eta_2^{(2)}$ and $\eta_3^{(2)}$ in the 2nd and 3rd lines in Eq. (4.4b and c). The final line in Eq. (4.46d) is the output of the only node in the 3rd and final layer ($l = 3$), which is the ultimate output of the *ANN*. As can be seen, rather than taking the weighted input variables (χ_j, $j = 1, 2, 3$), the final node takes as input the weighted output of the nodes of the second layer ($\eta_i^{(2)}$,$i = 1, 2, 3$), plus the weighted bias. Therefore, one can see in equation form the hierarchical nature of *ANN*.

- ***ANN*** models are similar to ***LR*** models in their structure and use: they consist of inputs (independent attributes) and output (dependent or outcome attribute), use connection weights (regression coefficients), bias weight (intercept coefficient) and estimation schemes to learn or train (parameter estimation) a model to be subsequently used for interpolation or interpretation purposes.

In sum, *ANN* consists of layers of computational units (neurons) with connections in different layers. Data are transformed from one layer to the next until it can be classified as an output. Each neuron multiplies the values by some weights, sums them up, adjusts the resulting number by the neuron's bias, and then the output is normalized by an activation function. The *ANN* is an iterative learning process during which the network trains by adjusting the weights so that the resulting outputs are sufficiently close to the desired outputs. During *ANN* training, the same dataset is processed several times during which the weights are continually refined. Once the network is trained and tested it can be given new input information to predict the output. Specifically, during training the inter-unit connections are optimized until the error in interpolations is minimized and the network reaches the specified level of accuracy.

4.3.4 *Pros and cons of ANN*

Despite its considerable popularity, some of the advantages and disadvantages of *ANN* are as follows (Tu, 1996; McGarry et al., 1999):

[h] For example, the output of the 2nd node in the 2nd layer is denoted as $\eta_2^{(2)}$.

① A common criticism of *ANN* is that it requires a large diversity of training in real-world operations, because any learning machine needs sufficient representative examples to capture the underlying structure that allows it to generalize to new cases (*ANN* model development is empirical, often requiring several attempts before an adequate model can be developed, thus they are challenging to use in the real world).
② In order to implement large and effective *ANN* software, huge processing and storage resources need to be committed.[i]
③ *ANN* algorithms learn from the analyzed data; they are referred to as "black box" models, and provide very little insight into what these models really do (the user just needs to feed it input and watch it train and await the output).
④ *ANN* models have limited ability in explicitly detecting possible causal associations.
⑤ *ANN* models are prone to overfitting.

Yet, although *ANN* is not comparable with the power of the human brain, still it is the basic building block of *AI*. In the end, instead of relying on a purely technical viewpoint coupled with arbitrary parameters, investigators may find it more profitable in the long run to take inspiration from the physical world to assess attribute variability across space and time. Primarily, the *ANN* should lead to inferences that are physically consistent. Given the noise and uncertainty associated with most real-world applications, technical inferences alone are generally insufficient; however, integrating physical meaning into *ANN* can provide a powerful framework.

4.4 ANN modeling in practice

The implementation of *ANN* in *CTDA* can take many forms. A rather typical outline of *ANN* implementation in practice is as follows:

> The *ANN* **operational process** starts with independent attribute datasets serving as *ANN* inputs, followed by the generation of dependent (output) attribute values, which are compared to a training set of values, tested using a test set and validated using a validation set, to be finally used for interpolation or interpretation purposes.

The quality, quantity, and structure of the input datasets are critical factors for the learning process and the chosen independent attributes must be relevant and measurable. *ANN* can be used in a variety of applications, including *image recognition* for agricultural usage (due to its ability to process large numbers of inputs to infer hidden and complex relationships); in *forecasting* involving many underlying factors, given *ANN*'s ability to model and extract unseen features and relationships and because, unlike traditional models, *ANN* doesn't impose any restriction on input and residual distributions; and in *soft data* generation (see application in Section 3 of Chapter 13).

The theoretical analysis presented in this chapter will be implemented in the real-world case studies that are discussed in the following Chapter 13, which is concerned with powerful syntheses of *CTDA* techniques with *DIA* models.

5 Practice exercises

1. Classify the type of *ANN* that is described as follows: The input data is passed onto a layer

[i] Simulating even a most simplified form on Von Neumann technology may compel an *ANN* designer to fill millions of database rows for its connections, which can consume vast amounts of computer memory and hard disk space.

of processing elements where it performs calculations. Each processing element makes its computation based upon a weighted sum of its inputs. The new calculated values then become the new input values that feed the next layer. This process continues until it has gone through all the layers and determines the output, which is quantified using a threshold transfer function.

2. Use the data presented in Table 6.2 of Section 6 in Chapter 6 to build a multiple linear regression (*MLR*) model, a geographically weighted regression (*GWR*), and an artificial neural network (*ANN*) model for depicting the relationship between the *AOD*, *PBLH*, wind, elevation, road length, farm area, and $PM_{2.5}$.
3. The *ANN* was designed according to a biological neural network (*BNN*). List some differences between *ANN* and *BNN*.
4. Assume that you need to evaluate the impact of the urban environment on taxi ridership in one city at various temporal instants, e.g., the morning peak (8:00–10:00 a.m.), afternoon peak (1:00–3:00 p.m.) and evening peak (9:00–11:00 p.m.). Provide an outline and specify some impact factors (e.g., road length, number of residential records, number of companies, education and government offices, number of hotels at specific locations) for building the *GWR* model. Explain your results.
5. Assume that you need to quantify the population density in a region during various years. Provide a modeling outline by considering some geographic impact factors (e.g., road length, elevation, normalized difference vegetation index, night light) and apply your model in a practical case. Compare the performance of *MLR, ANN,* and *GWR* in the population density modeling by using a cross-validation technique.
6. Compare the differences of the model structure among *MLR, ANN,* and *GWR*, and describe their specific characteristics.
7. Given the three-layer *ANN* shown in Fig. 4.2 of Section 4, and the weight matrices between various layers,

$$W^{(1)} = \begin{bmatrix} w_{11}^{(1)} & w_{12}^{(1)} & w_{13}^{(1)} \\ w_{21}^{(1)} & w_{22}^{(1)} & w_{23}^{(1)} \\ w_{31}^{(1)} & w_{32}^{(1)} & w_{33}^{(1)} \end{bmatrix} = \begin{bmatrix} 12.35 & 3.16 & -0.78 \\ 4.77 & 5.26 & 2.14 \\ 0.86 & 1.47 & 9.36 \end{bmatrix},$$

$$W^{(2)} = \begin{bmatrix} w_{11}^{(2)} & w_{12}^{(2)} & w_{13}^{(2)} \end{bmatrix} = [0.35\ 1.11\ 0.98],$$

calculate the *ANN* output in terms of the steps introduced in Example 4.2 of Section 4, under the conditions:

(a) $[\chi_1\ \chi_2\ \chi_3] = [1.46\ \ 2.98\ \ 3.33]$, and the Sigmoid activation function $\vartheta(\chi) = \frac{1}{1+e^{-\chi}}$.

(b) $[\chi_1\ \chi_2\ \chi_3] = [2.76\ \ -4.34\ \ 1.77]$, and the Sigmoid activation function $\vartheta(\chi) = \frac{1}{1+e^{-2\chi}}$.

(c) $[\chi_1\ \chi_2\ \chi_3] = [2.76\ \ -4.34\ \ 1.77]$, and the Unit step activation function $\vartheta(\chi) = \begin{cases} 1 \text{ if } \chi \geq 0 \\ 0 \text{ if } \chi < 0 \end{cases}$.

(d) $[\chi_1\ \chi_2\ \chi_3] = [3.47\ \ 2.22\ \ 7.10]$, and the Unit step activation function $\vartheta(\chi) = \begin{cases} 1 \text{ if } \chi \geq 0 \\ 0 \text{ if } \chi < 0 \end{cases}$.

8. Consider a four-layer *ANN* with one input layer, two hidden layers, and one output layer. The number of *nodes* in the four layers are 3, 3, 3, 1. The *ANN* structure is similar to Fig. 4.2 of Section 4, but with two hidden layers. The weights matrices between various layers are

$$W^{(1)} = \begin{bmatrix} w_{11}^{(1)} & w_{12}^{(1)} & w_{13}^{(1)} \\ w_{21}^{(1)} & w_{22}^{(1)} & w_{23}^{(1)} \\ w_{31}^{(1)} & w_{32}^{(1)} & w_{33}^{(1)} \end{bmatrix} = \begin{bmatrix} 12.35 & 3.16 & -0.78 \\ 4.77 & 5.26 & 2.14 \\ 0.86 & 1.47 & 9.36 \end{bmatrix},$$

$$W^{(2)} = \begin{bmatrix} w_{11}^{(2)} & w_{12}^{(2)} & w_{13}^{(2)} \\ w_{21}^{(2)} & w_{22}^{(2)} & w_{23}^{(2)} \\ w_{31}^{(2)} & w_{32}^{(2)} & w_{33}^{(2)} \end{bmatrix} = \begin{bmatrix} 0.38 & 1.10 & -0.88 \\ 2.37 & 1.26 & 0.14 \\ 0.16 & 3.43 & 1.36 \end{bmatrix},$$

$$W^{(3)} = \begin{bmatrix} w_{11}^{(3)} & w_{12}^{(3)} & w_{13}^{(3)} \end{bmatrix} = [0.15 \;\; 1.34 \;\; 0.75].$$

Calculate the *ANN* output subject to the following conditions:

(a) $[\chi_1 \; \chi_2 \; \chi_3] = [1.46 \;\; 2.98 \;\; 3.33]$, and the Sigmoid activation function $\vartheta(\chi) = \frac{1}{1+e^{-\chi}}$.

(b) $[\chi_1 \; \chi_2 \; \chi_3] = [2.76 \;\; -4.34 \;\; 1.77]$, and the Sigmoid activation function $\vartheta(\chi) = \frac{1}{1+e^{-2\chi}}$.

(c) $[\chi_1 \; \chi_2 \; \chi_3] = [2.76 \;\; -4.34 \;\; 1.77]$, and the Unit step activation function $\vartheta(\chi) = \begin{cases} 1 \text{ if } \chi \geq 0 \\ 0 \text{ if } \chi < 0 \end{cases}$.

(d) $[\chi_1 \; \chi_2 \; \chi_3] = [3.47 \;\; 2.22 \;\; 7.10]$, and the Unit step activation function

$$\vartheta(\chi) = \begin{cases} 1 \text{ if } \chi \geq 0 \\ 0 \text{ if } \chi < 0 \end{cases}.$$

9. In the field of remote sensing, the satellite remote sensing reflectance values (*Rrs*) at various spectra bands are used to retrieve the concentration of chlorophyll at the top layer of the sea. In other words, researchers are seeking algorithms relating the reflectance values with the sea surface chlorophyll, including *MLR*, and *ANN*. Given the data presented in Table 5.1, use various combinations of the *Rrs* from various bands to construct *MLR* or *ANN* models, and find the optimal band combinations of each algorithm to retrieve chlorophyll. The $Rrs_i (i = 1,\ldots,10)$ represent the *Rrs* values at 10 spectral bands, including 412, 443, 469, 488, 531, 547, 555, 645, 667, 678 nm. Given that the number of data is rather large, the complete Table 5.1 can be found on the website: https://www.dropbox.com/scl/fi/vzfre7yt1r7qdwly7wq5b/Exerciseof-Ch12.doc?dl=0&rlkey=l2836mm49e8r2wxwzbw1dr2b1.

10. Write a brief essay expanding on the argument that the huge amounts of data used by *DIA* should not be mistaken for confirmatory empiricism, and that what often counts is a small amount of data that falsifies the existing conditions or claims (*status quo*) about a phenomenon, i.e., disconfirmatory empiricism.

TABLE 5.1 Dataset.

$s_{i,1}$	$s_{i,2}$	t_i	*Chl*	Rrs_1	Rrs_2	Rrs_3	Rrs_4	Rrs_5	Rrs_6	Rrs_7	Rrs_8	Rrs_9	Rrs_{10}
−2.73E+05	5.36E+06	3364	0.143	1.16E-04	8.56E-04	1.29E-03	1.75E-03	1.85E-03	1.74E-03	1.55E-03	3.54E-04	2.51E-04	2.61E-04
−2.74E+05	5.39E+06	3364	0.326	8.87E-05	8.10E-04	1.28E-03	1.62E-03	1.67E-03	1.55E-03	1.33E-03	2.30E-04	1.79E-04	1.73E-04
−2.75E+05	5.42E+06	3364	0.376	−7.59E-04	3.91E-05	6.77E-04	1.08E-03	1.25E-03	1.14E-03	1.05E–03	2.36E-05	−2.27E-05	−2.33E-05
−8.92E+04	5.16E+06	3368	0.318	−1.56E-03	−6.06E-04	1.79E-04	5.51E-04	8.55E-04	8.02E-04	6.97E-04	−5.65E-05	−3.00E-06	1.17E-04
−4.27E+05	5.30E+06	3421	0.212	−8.80E-04	1.16E-04	7.52E-04	1.25E-03	2.30E-03	2.60E-03	2.53E-03	7.82E-04	5.96E-04	6.18E-04
−1.15E+05	5.17E+06	3426	0.954	−2.84E-04	6.02E-04	9.51E-04	1.34E-03	1.80E-03	1.76E-03	1.63E-03	3.31E-04	2.97E-04	3.09E-04
−4.50E+05	5.28E+06	3439	6.088	3.81E-04	1.12E-03	1.41E-03	1.66E-03	2.39E-03	2.57E-03	2.53E-03	1.24E-03	1.01E-03	1.01E-03
−4.48E+05	5.28E+06	3446	0.139	2.81E-03	3.41E-03	3.58E-03	3.69E-03	4.16E-03	4.26E-03	4.02E-03	2.47E-03	2.15E-03	2.09E-03
5.60E+04	5.13E+06	3453	0.478	1.63E-03	2.72E-03	3.30E-03	3.68E-03	3.42E-03	3.18E-03	2.66E-03	1.16E-04	2.39E-04	2.10E-04
5.63E+04	5.13E+06	3453	0.307	1.63E-03	2.72E-03	3.30E-03	3.68E-03	3.42E-03	3.18E-03	2.66E-03	1.16E-04	2.39E-04	2.10E-04

CHAPTER

13

Syntheses of *CTDA* techniques with *DIA* models

1 A broad synthesis perspective

Agility is a key issue for modern-day science and technology. That is, if a scientific or technological field fails to keep up with ongoing changes, then it is very probable that it will fall behind. This is why it is vital for *CTDA* to occasionally take a long view and think about what's coming down the line in the near future. In this respect, it has been generally claimed that "the total is larger than the summation of its parts." In the case of *CTDA*, this motto can be expressed as follows.

> **Theory-driven** techniques (Chapters 6, 9, and 10) and **data-driven** algorithms (Chapter 12)[a] can become more powerful when their individual strengths are combined in a proper manner, i.e., by acting **symbiotically** and complementing each other.

Inter alia, theory-driven techniques have the ability to estimate the accuracy, completeness, and efficiency of data-driven results, aid their interpretation, and continually update the understanding of a phenomenon by assimilating newly available data. Such syntheses of techniques, models, and data have been considered in the literature, which span length scales from the local to the global and time scales from a few tens of milliseconds to centuries (Karpatne et al., 2017).

Example 1.1 Some syntheses involve deterministic techniques (using, e.g., mathematical formulas involving weighted averages of nearby data, which, though, do not provide any interpolation error assessment), whilst some other syntheses involve stochastic techniques (using, e.g., statistical regression formulas, which can also provide an interpolation error assessment). Clearly, the latter usually offer a more realistic representation of the actual phenomenon.

There are types of syntheses that consider laws and models of the different components of a phenomenon, as well as different disciplines and natural attributes. For obvious reasons, this approach is often very useful in the real world.

[a] Including *MLR*, *GWR*, and *LUR*.

https://doi.org/10.1016/B978-0-12-816341-2.00007-1

An illustration of the essence of the approach is discussed next.

Example 1.2 He et al. (2020b) introduced a study of the COVID-19 spread that was essentially a fourfold synthesis: (a) a modified *susceptible-exposed-infected-removed* (*SEIR*) epidemic model represented mathematically the spread of the disease over time at geographical regions worldwide; (b) an *empirical law* described the evolution of the disease during its critical early phase; (c) *regression relationships* linked the environmental factors and the disease transmission rates; and (d) the effects of the measures taken at the various regions to control the disease were compared quantitatively in terms of the model and law *parameters*.

Often the idea behind synthesis is that instead of seeking a single new technique that has all the desired features but is necessarily rather complicated, one may get the desired result, but with fewer complications, by bringing advances from individual techniques and models of different areas and disciplines. Accordingly, the present chapter examines certain interesting cases of the synthetic perspective, and their comparative conceptual and implementation features.

- The syntheses to be considered in this chapter involve the fusion of the ***BME*** and ***STP*** techniques of modern geostatistics (Chapters 9 and 11) with ***DIA*** techniques (Chapter 12) and **artificial neural network** models (*ANN*).[b]

This synthesis involving *BME* theory seems quite appropriate given that *DIA* experience has brought to the fore some critical issues, such as whether data correlations can form the basis of a scientific method (see, also, Chapters 3 and 12). Indeed, although *DIA* may detect some statistical correlations in a massive dataset, the fundamental questions remain regarding how substantive (rather than merely numerical) these correlations are and whether they imply physical causation. Unfortunately, in both cases, the answer is usually disappointing: the correlations are merely numerical and statistical correlation does not represent causation.

- The syntheses of techniques discussed in this chapter are concerned about the chronotopologically varying attributes described in the previous chapters and mathematically indexed by space and time, i.e., $X(\boldsymbol{p})$, where $\boldsymbol{p} = (\boldsymbol{s}, t)$.

Let us refresh our memory regarding some practical space-time terminology. The chronotopologic attribute distribution is a function with values continuously varying across space and time. The term "location" or "place" refers to a set of geographical coordinates $\boldsymbol{s} = (s_1, \ldots, s_n)$. While the terms "concurrent" and "simultaneous" both mean values "occurring at the same time," the term "concurrent" is used only for values that occur over a period of time T, whereas the word "simultaneous" can also be used for values that occur at a time instance t.

Putting these various comments together, an adequate synthesis of theory, techniques, and models is essential in advancing our understanding of the complex links and mechanisms characterizing the chronotopologic variability and dependency structure of natural attributes and improving their realistic mapping.

1.1 *BME-ML* integration

It is, then, in the above-outlined setting that this chapter's subject unfolds. In particular, the following objective should be stressed:

> This chapter considers the augmentation of the theory-driven ***BME*** technique (Chapter 6) by integrating it with ***STP*** (dimensionality reduction; Chapter 11) and ***ML*** algorithms (*LUR* and *ANN*; Chapter 12).

[b] Haykin (1994).

BME is particularly suitable for this purpose, because of its epistemic underpinnings that distinguish between two knowledge bases (*KB*, Chapter 3):

- The core *G-KB* that is basically theory-driven, i.e., introduces **scientific consistency,** and the site-specific *S-KB* that is data-driven, i.e., introduces **empirical conditioning**.

As a little reflection shows, incorporating *G* alone or *S* alone is usually insufficient to assure a scientific study of complex real-world problems. This is because, on the one hand, the *G*-driven analysis uses theoretical models based on sound science (physical laws and ab initio principles) but the models may either under-represent the actual phenomena or their solution is very complicated; whereas, on the other hand, the *S*-driven analysis relies solely on information contained in the data and determines associative but not causative links among the data. By bringing *G* and *S* together, *BME* succeeds in combining the unique strengths and important features of both *KB*. In the process, an important strand tying both the *G*-based and *S*-based views of the phenomenon together is revealed that can improve its understanding.

Example 1.3 If the *CTDA* of a real-world phenomenon relies solely on site-specific (*S*) data and ignores core (*G*) knowledge, the resulting attribute models could be inadequate and even erroneous, regardless of the technologically advanced numerical algorithms and software that may be used. In many applications, in addition to the in-depth understanding and incorporation of the disciplinary core knowledge, the interdisciplinary knowledge sources (i.e., coming from different yet relevant disciplines) can be essential to improve physical modeling outcomes.

Traditionally, in many realistic cases, a *CTDA* study is *under-constrained*, in the sense that a limited number of data is available (i.e., the number of data is insufficient given the large number of the study variables). In recent years, the availability of big datasets and *ML* algorithms offered a possible solution to this problem. When we give this its due weight, we realize something crucial, in a *CTDA* context:

- The ***BME-ML*** synthesis may involve the assimilation of big datasets obtained by new technological advances in remote sensing, imaging, robotics, biotechnology, etc., which are necessary to adequately address the challenges of modern research and development.

Indeed, densely deployed wireless sensor networks with high-resolution sensing enable scientists to better understand the chronotopologic variability of natural attributes. Wireless sensing allows investigators to remotely view, enable, debug, and test *in-situ* sensor deployments from the safety of their laboratories. Practically, data from heterogeneous sources need to be integrated in a unified data management and exploration framework.

Example 1.4 The highly complex surfaces of mountainous regions require a massive amount of data (terabytes or higher) in order to understand land-atmosphere interactions that play a critical role in weather and climate prediction. This amount of data is not currently available in most cases and, in addition, it will be difficult to manage and analyze, if available. This reality has led to rapidly increasing efforts to develop big datasets and to establish collaborations between physical scientists and computer experts. Accordingly, data source arrangements of modern environmental sciences will rely on heterogeneous sensor deployment that may include (Lehning et al., 2009): (i) mobile stations, (ii) weather stations of various kinds (conventional, full-size, external), (iii) satellite imagery, (iv) weather radars, (v) mobile weather radar, (vi) stream observations, (vii) citizen-supplied observations, (viii) *LIDAR* (ground and aerial), (ix) nitrogen/methane measures, (x) snow hydrology and avalanche probes, (xi) seismic probes, (xii) distributed optical fiber temperature sensing, (xiii) water quality sampling, (xiv) stream gauging stations, (xv) rapid mass movements research, (xvi) runoff stations, and (xvii) soil research.

Certain notable aspects of the synthesis of techniques discussed in this chapter are summarized next:

① Investigators may need to derive and apply different models that **transform** sensor data into scientific and other practical forms.
② The synthesis of multisourced data (in some cases to be found over the Internet) with models and techniques is sometimes termed **data assimilation.**[c]
③ The **verification** and **assessment** of data-intensive synthesis results (e.g., assigning likelihood and confidence measures, and error bounds) should be carefully designed.[d]
④ Forecasting techniques (like *BME*) can generate short-term attribute predictions concerning hazardous phenomena (floods, landslides, avalanches, etc.), also known as **now-casting**.
⑤ Technique synthesis should be **domain-specific**, so that it can identify and address essential problems of major concern to the domain scientists.
⑥ The computational advances resulting from technique synthesis would provide effective representations for better understanding natural systems and essential tools for investigators to gain insight into **decision making**.

In view, then, of the above aspects, one can argue that this chapter is concerned with a theory-data *holistic* approach:

- An optimal synthesis should be sought of the theory-driven and the data-driven perspectives,[e] which implies an optimization of the **man–machine symbiosis**.

To achieve this objective, *a posse ad esse*, investigators need to fuse relevant advances in different fields. Cases in point include suitable syntheses of *BME*, *STP*, *LUR*, and *ANN* to be discussed in the remainder of this chapter.

2 A synthesis of the *STP* and *BME* techniques

Two prime aspects of such a synthesis will be considered, i.e., the methodological and empirical ones. Methodologically, the synthesis of the *STP* and *BME* techniques of modern geostatistics is rather straightforward involving three main steps as outlined in Fig. 2.1. In more words:

> ***STP*** is applied on the original attribute dataset to produce the projected dataset of reduced dimensionality; the dependency structure of the projected dataset is determined and the projected attribute maps are generated using ***BME***; lastly, the results are back-projected, ***STP***$^{-1}$, to generate maps of the original space-time attribute.

Empirically, the implementation of the *STP-BME* technique in the environmental health applications considered in this section would require close collaboration between physical, health, and computer sciences. This collaboration would involve high-resolution and high-fidelity data gathering, large-scale data management and sharing, data visualization, as well as realistic and systematic data modeling, and integrative reasoning (meaning the blending of different thinking styles).[f] Despite the occasional theoretical hardness, many real-world applications have

[c] As Brodie (2015) reports, 80% of the errors in *DIA* may arise during the data management processes.

[d] It has been argued that, in many cases, big data analysis is not adequately understood to allow quantification of the probability or likelihood that the analysis results will occur within estimated error bounds (Brodie, 2015).

[e] See, e.g., a theory-driven *DIA* (Section 3.1 of Chapter 3).

[f] That is, synthesis relies on the fusion of multisourced data and the simultaneous blending of the distinct (discipline-dependent) thinking modes of the participating investigators.

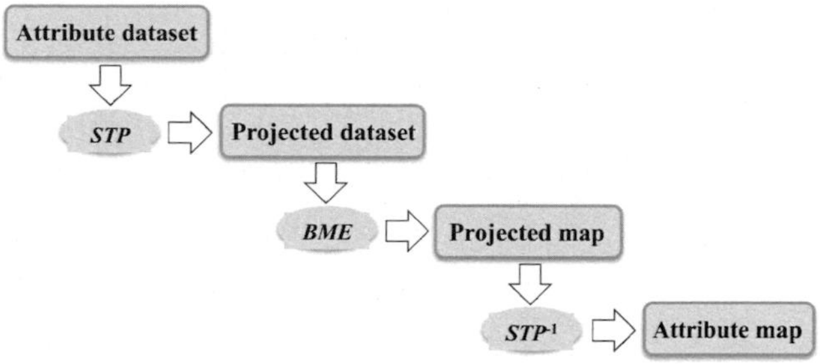

FIG. 2.1 An outline of the synthetic *STP-BME* framework. STP^{-1} denotes the backprojection process.

underlying structures that can be exploited to yield useful synthetic approaches of *CTDA*. This is the case in the following real-world study.

2.1 *HFRS* in Heilongjiang Province (China)

The topic of the Heilongjiang study was a rodent-borne zoonosis: *Hemorrhagic fever with renal syndrome* (*HFRS*). This disease is caused by hantavirus (Hantaviridae family). In China, the Hantaan and Seoul viruses dominate *HFRS* infection, the leading rodent hosts of which are *Apodemus agrarius* and *Rattus norvegicus*, respectively (Yan et al., 2007). The virus is transmitted from rodents to humans via inhalation of aerosols contaminated by rodents' urine, saliva, excreta, and dung, possibly through ingestion of contaminated food and by direct contact of contaminated materials with broken skin or mucous membranes, or by rodent bites (Liu et al., 2013; Chinikar et al., 2014). Once infected, the clinical manifestations include fever, headache, nausea, abdominal pain, adverse kidney effects, and subsequent pulmonary edema, shock, renal insufficiency, encephalopathy, hemorrhages, and cardiac complications (Simmons and Riley, 2002; Krautkrämer et al., 2012). The disease goes through five stages: febrile, hypotensive shock, oliguric, polyuric, and convalescent, which last, respectively, 1–7 days, 1–3 days, 2–6 days, 2 weeks, and 3–6 months (Jiang et al., 2016).

The chronotopologic domain of this case study was represented quantitatively by the symbolic arrangement,

$$\text{Chronotopologic domain } \boldsymbol{p}: \left\{ \begin{array}{l} \boldsymbol{s} = (s_1, s_2) \in \text{Heilongjiang Province (China)} \\ t \in \text{January 2005} - \text{December 2013 time period,} \end{array} \right. \quad (2.1)$$

where, as usual, $\boldsymbol{p} = (\boldsymbol{s}, t)$ denotes the geographical coordinates $\boldsymbol{s} = (s_1, s_2)$ and time instant t. Geographically, the Heilongjiang basin includes four major river systems: the Heilong, Songhua, Wusuli, and Suifen rivers (Fig. 2.2).

2.2 Datasets and modeling assumptions

Monthly *HFRS* reported cases (total 21,383 cases) were collected at 130 counties and districts in the chronotopologic domain of Eq. (2.1) by the *China Information System for Disease Control and Prevention* (*CISDCP*). The following clarifications should be made:

① Demographic data for each county came from the National Bureau of Statistics of China. A total of 14,040 chronotopologic records were included. The *HFRS* cases were population-standardized, and the averaged monthly *HFRS* incidences during 2005–13 are shown in Fig. 2.3. The highest

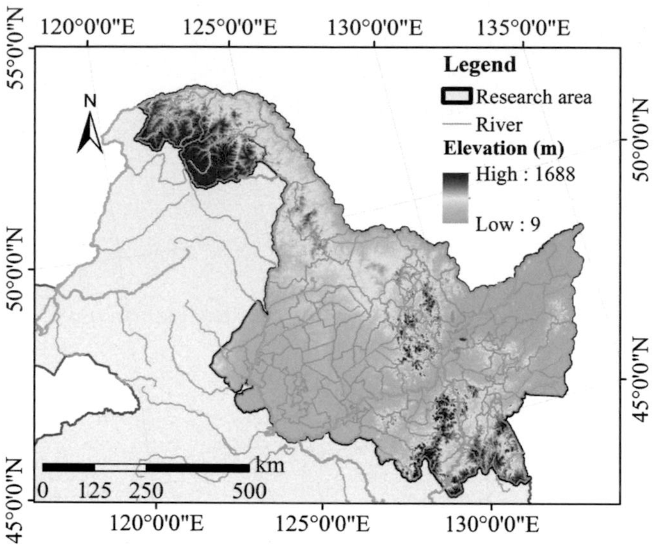

FIG. 2.2 Study area and delimiter of counties and districts.

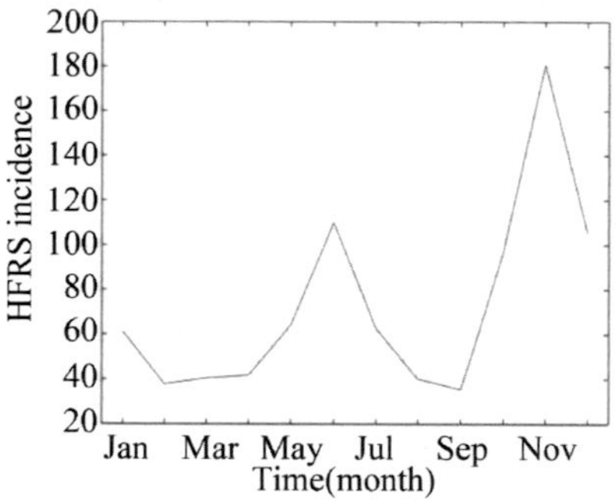

FIG. 2.3 Averaged monthly *HFRS* incidence during 2005–13.

incidences occurred during June and November, and the lowest during February and September.

② The standardized *HFRS* incidence was denoted as $X(\boldsymbol{p})$, with arguments shown in Eq. (2.1), where $\boldsymbol{s}=(s_1,s_2)$ denotes the centroid coordinates of each administration unit. In this domain, the *HFRS* incidence variability was measured by the isostationary *HFRS* covariance model plotted in Fig. 2.4A. This model had two components (an exponential model in space and a Gaussian model in time),

$$c_X(\boldsymbol{\Delta p}) = e^{-\frac{h}{72\times10^3}-\left(\frac{\tau}{2.6}\right)^2}, \tag{2.2}$$

where $\varepsilon_s = 3a_s = 3\times72\times10^3 = 216$ km and $\varepsilon_t = \sqrt{3}a_t = \sqrt{3}\times2.6 = 4.5$ mos are, respectively, the spatial and temporal correlation ranges of the *HFRS* incidence distribution.

③ Interpretations of Eq. (2.1) and the covariance shape (Fig. 2.4A and B) are phenomenological.[g] The resulting inference is that the *HFRS* case distribution in the domain of Eq. (2.1) was controlled by both spatial and temporal *HFRS* dependencies. In quantitative terms, the covariance value (around 0.3 $(\text{cases}/10^5)^2$) at time separation $\tau=4$ mos indicates a rather strong

[g] I.e., the meaning of any inference about the empirical world is to be analyzed into the experiences out of which it has been logically constructed (phenomenalism).

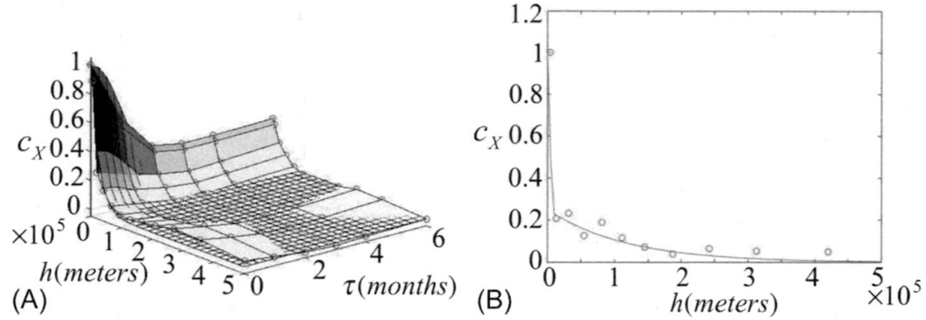

FIG. 2.4 Empirical *HFRS* covariance values (small circles) and fitted theoretical models (continuous surface and line) in (A) the $R^{2,1}$ domain, and (B) the R^2 domain.

temporal dependence among *HFRS* incidences. The influence of the spatial neighborhood extends to about 200 km, which also indicates a significant spatial dependence among the same incidences. In time, progressively the *HFRS* covariance becomes less sharp at the origin, with its peak (variance) reducing in size until it stabilizes around the value 0.3. The covariance's shape progressively smooths out in space until it becomes almost constant.

2.3 Step-by-step *STP-BME*

Subsequently, the synthetic *STP-BME* technique (Fig. 2.1) was implemented in this case study by means of the following step-by-step process.

Step 1: Following the rationale of the *STP* technique, the *HFRS* distribution was projected from the original space-time disease domain $R^{2,1}$ (Fig. 2.5A) onto a lower dimensionality domain R^2 (Fig. 2.5B) using Eq. (2.3b), Section 2 of Chapter 11.

Step 2: By inserting Eq. (2.2) into Eq. (2.6a), Section 2 of Chapter 11, the traveling coefficient was found to be $\upsilon = |\boldsymbol{\upsilon}| = -10{,}650.89\tau$. The *HFRS* incidence data points were projected from the $R^{2,1}$ domain (Fig. 2.5A) onto the reduced dimensionality R^2 domain (Fig. 2.5B).

Step 3: Following this projection process, the empirical *HFRS* covariance values and the fitted covariance model in the R^2 domain are plotted in Fig. 2.4B, where the theoretical *HFRS* covariance model is

$$c_{\widehat{X}}\left(\widehat{h}\right) = 0.75\left(1 - \frac{1.5\widehat{h}}{10^4} + \frac{0.5\widehat{h}^3}{10^{12}}\right) + 0.25e^{-\frac{\widehat{h}}{35\times10^4}}. \tag{2.3}$$

This covariance is of reduced dimensionality compared to that of Eq. (2.2). The theoretical

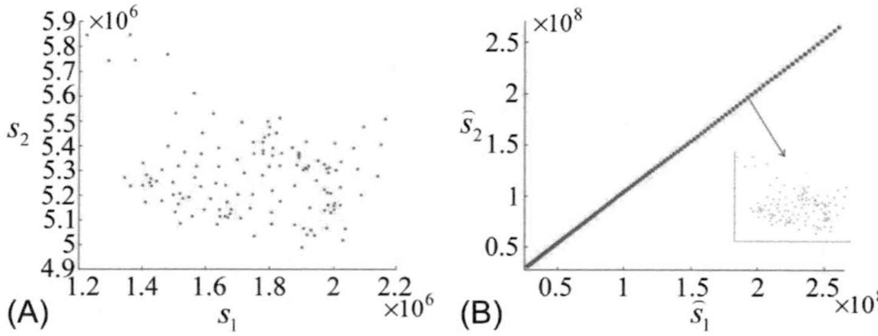

FIG. 2.5 Distribution of data points in (A) the $R^{2,1}$ domain and (B) the R^2 domain. Same geographic coordinate points from different time instants are overlapped in (A); Points from different times were gathered together, which seem like one point in (B).

covariance model of Eq. (2.3) provided a better fit to the empirical covariance values than the covariance model of Eq. (2.2).

Step 4: Several *HFRS* incidence maps were generated using the *STP-BME* technique. For illustration purposes, the monthly *HFRS* incidence maps for the year 2006 (January–December), as shown in Fig. 2.6.

Step 5: The following interpretations of the *HFRS* incidence maps were made in the present Heilongjiang study:

① Three areas with considerable *HFRS* incidences were identified in the maps of Fig. 2.6, particularly, in the eastern, western, and southern parts of the Heilongjiang

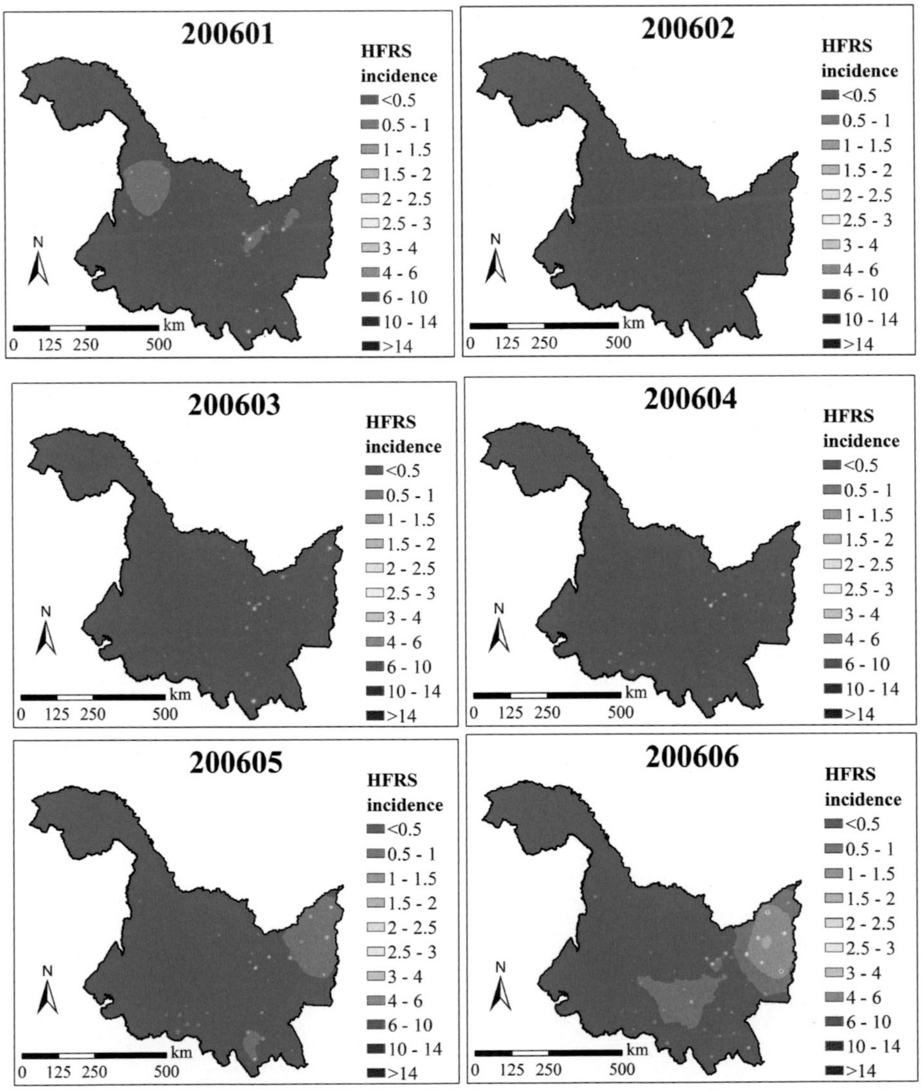

FIG. 2.6

(Continued)

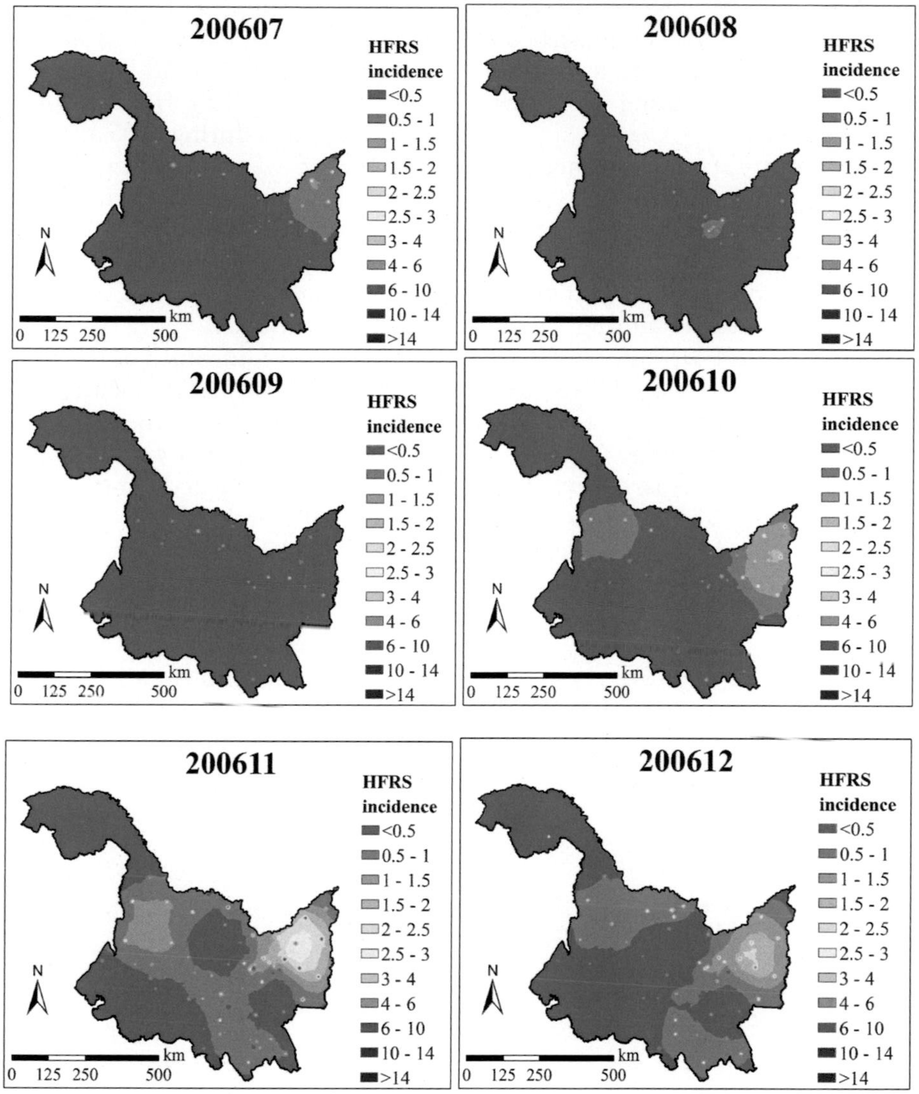

FIG. 2.6, CONT'D *HFRS* incidence maps during the period January–December 2006.

Province. The eastern part showed high incidences over a larger area than in the other two parts. As is shown in Figs. 2.6 and 2.3, the incidence begins to increase in April 2006, then a peak is reached in June 2006, and the incidence reduces significantly in September 2006. The next outbreak was observed from September 2006 to February 2007 with the peak occurring in November of 2006. Interestingly, the number of *HFRS* cases during the autumn-winter period was much larger than those during the spring-summer period. This phenomenon is probably due to the fact that the autumn-winter period coincides with the rice harvest season, i.e., after harvest the soil condition switches from a flood state to a dry state, which leads to rodents dispersal or migration causing a higher number of infected cases.

② In the western part of the Heilongjiang Province, low *HFRS* incidences were observed during February–September of 2006–13 (with the exception of the year 2005). Incidences at the southern part of the province remained high during June, October, November, and December of each year considered. Apparently, *HFRS* was transmitted to the southern part of the province from its eastern part. Overall, a declining incidence trend was observed in the maps for the period 2005–13, which may be due, at least in part, to improvements in medical conditions and disease prevention.

2.4 Comparative analysis

The cross-validation assessment of the *STP-BME* technique *vs.* the direct *BME* mapping is plotted in Fig. 2.7 for the years 2005–13: the *STP-BME* mapping was found to be more accurate in predicting *HFRS* at low incidence points, whereas the direct *BME* mapping was more accurate in predicting *HFRS* incidence at high incidence points during 2005 and 2007. Overall, the *STP-BME* was on average a better predictor of the *HFRS* incidence distribution than the direct *BME*: the *MAE* for *BME* over the entire domain was 0.524 cases/100,000 individuals, whereas that of *STP-BME* was 0.459 cases/ 100,000 individuals.

Using a central feature of the *STP-BME* synthesis, the study of *HFRS* incidence spread was technically transferred from the original three-dimensional domain (two space dimensions plus time, $R^{2,1}$) onto a reduced dimensionality domain (two space dimensions, R^2). In this way, the difficult to determine space-time metric is reduced to a much easier to define spatial distance. This means that the empirical space-time covariance of monthly incidences is accordingly transformed into a spatial covariance. As a result, it is technically much easier to fit a theoretical model to the spatial than to the spatiotemporal empirical covariance of monthly *HFRS* data.

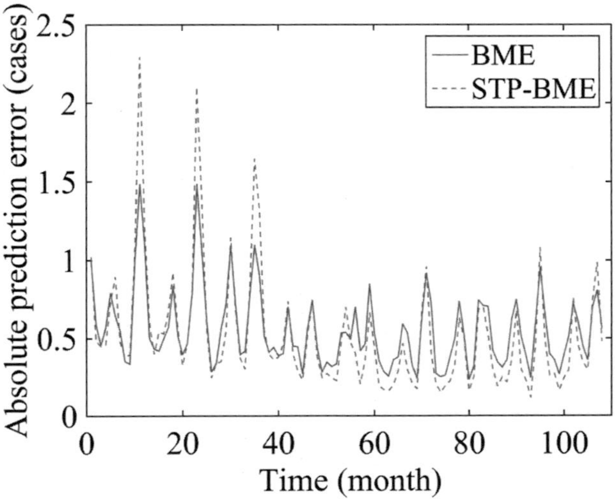

FIG. 2.7 Mean absolute prediction error of cross-validation in each month.

3 A synthesis of the *STP-BME* technique with the *LUR* and *ANN* models

The principal goal of this section is technically rather involved, since it combines techniques and models with different individual structures and objectives. The matter can be summarized as follows:

> Test the performance of a **synthesis** that leverages two different mapping techniques and assesses the impacts of models introducing soft data in the study of $PM_{2.5}$ pollution in North China.

With this overall aim in mind, the very important Jing-Jin-Ji Metropolitan Region of China (also known as Beijing-Tianjin-Hebei and as the Capital Economic Zone) was selected as the testing ground of the above synthetic approach. Fig. 3.1, then, presents a methodological framework that was designed so that it allowed the systematic quantitative blending of the *STP-BME* technique of space-time mapping with the *LUR* model that determines significant attribute-related environmental variables that serve, in turn, as inputs to the *ANN* model generating soft data.

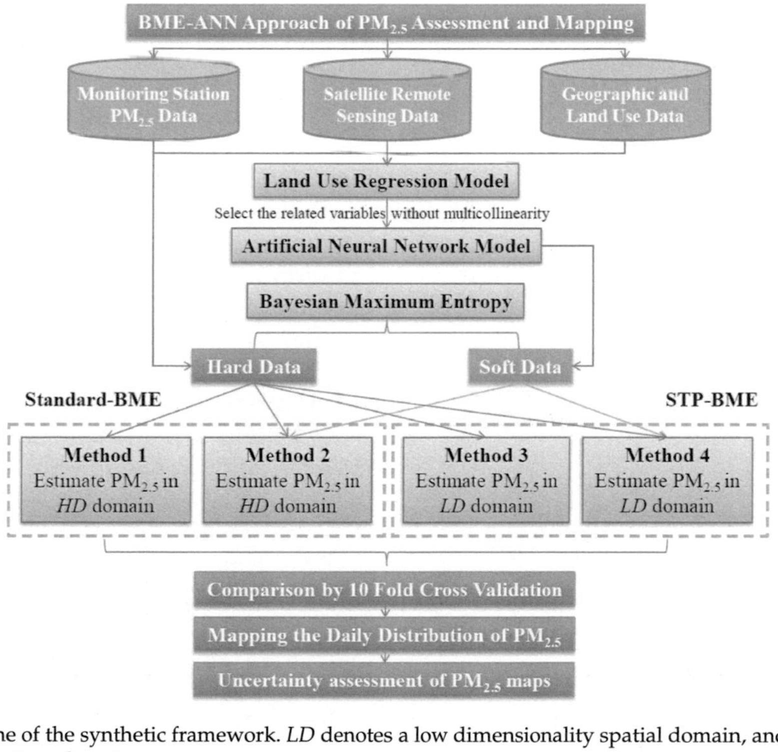

FIG. 3.1 An outline of the synthetic framework. *LD* denotes a low dimensionality spatial domain, and *HD* denotes a high dimensionality space-time domain.

3.1 The environmental Jing-Jin-Ji study

Jing-Jin-Ji is the National Capital Region of the People's Republic of China, being the biggest urbanized megalopolis region in North China. To demonstrate the empirical implementation of the synthesis of Fig. 3.1, let us continue the investigation of Example 3.9, Section 3 of Chapter 6, i.e., the $PM_{2.5}$ study in the domain represented quantitatively by the symbolic arrangement

$$\text{Chronotopologic domain } \boldsymbol{p}: \begin{cases} \boldsymbol{s} = (s_1, s_2) \in \text{Jing-Jin-Ji region (North China)} \\ t \in \text{October } 1-31,\ 2015 \text{ time period.} \end{cases} \tag{3.1}$$

3.2 Datasets and step-by-step synthesis

Two kinds of site-specific datasets were used in the Jing-Jin-Ji study during the period October 1–31 of 2015. *Hard* data denote the $PM_{2.5}$ observations collected from $PM_{2.5}$ monitoring stations, while *soft* data were generated by the *ANN* model depicting the relationship between $PM_{2.5}$ and related environmental attributes.

The subsequent investigation of the $PM_{2.5}$ pollution situation involved the following steps (see, also, Fig. 3.1):

Step 1: The *LUR* model was employed to define the significantly $PM_{2.5}$-related environmental variables, which were regarded as *ANN* inputs.

Step 2: In order to test the performance of the *STP* technique in the atmospheric pollution study and the impacts of introducing soft data, four methods were employed (Fig. 3.1):

① In method 1, space-time $PM_{2.5}$ concentrations were interpolated in the *HD* domain using the original $PM_{2.5}$ hard data only.

② By introducing soft data, method 2 combined them with hard data to estimate $PM_{2.5}$ concentrations in the *HD* domain.

③ In methods 3 and 4, respectively, hard and soft data were used, but the $PM_{2.5}$ concentrations were interpolated in the *LD* domain, following domain dimensionality reduction (*DDR*) transformation, as discussed above.

Step 3: A 10-fold *cross-validation* (*CV*) method (Rodriguez et al., 2010) was used to test potential model over-fitting and the performance of each of the four methods.

Step 4: Methods 1, 2, 3, and 4 were implemented using the *SEKSGUI* software library (Section 8 of Chapter 1, and Yu et al., 2007).

Step 5: The generalized additive model (*GAM*) and the *GWR* model were also considered for comparison purposes.[h]

Step 6: The *LUR* results indicated that the variables—aerosol optimal depth, relative humidity, planetary boundary layer height, precipitation surface temperature, wind speed, distance to coastline, elevation, road length, and farmland area—have individually contributed to $PM_{2.5}$ concentrations without multicollinearity (the R^2 of the *LUR* model was 0.529). The best performance network ($R^2 = 0.939$) had 4 layers: 1 input layer, 2 hidden layers (with 13 and 9 neurons, respectively), and 1 output layer. Based on *ANN* training that led to the best *ANN* model, 300 estimates were obtained at each soft data point (847 points, in total). The mean R^2 of the 300 *ANN* estimates was 0.899 and the mean range of the 95% confidence interval (*CI*) of soft data was 5.166 $\mu g/m^3$ (Fig. 3.2).

[h] *R* software was employed to implement *GAM* and *GWR*.

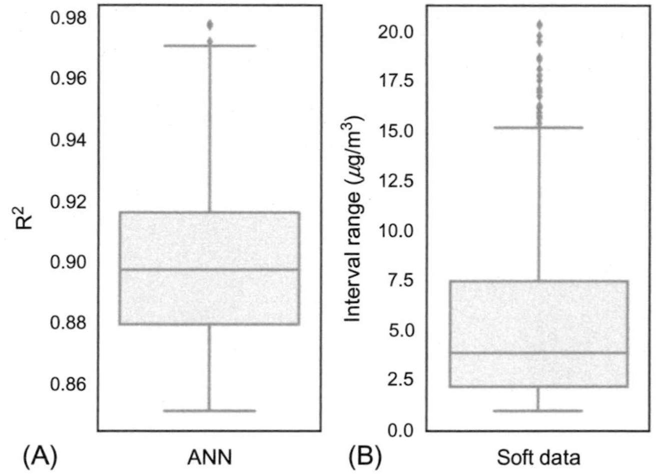

FIG. 3.2 Box plot of (A) 300 *ANN* training performance and (B) interval range at soft data points.

Step 7: The 10-fold cross-validation and computational cost results (time in seconds) demonstrated the validity of two points:

① The $PM_{2.5}$ mapping accuracy (measured in terms of the R^2 statistic) increased when soft data were involved. Compared to the *GAM* and *GWR* techniques, the improvement in R^2 accuracy gained by using method 4 (*STP-BME*, $R^2 = 0.944$) were found to be 48.47% and 9.34%, respectively.

② The mapping cost was reduced by 22.54% when dimensionality reduction was involved (*STP-BME* technique) compared to the direct *BME* technique with the same data. Additionally, compared to *GWR*, the method 4 allowed a 25.61% reduction of the computational time (Table 3.1).

The results in ① and ② facilitated the following finding:

- The use of soft data (enabled by *LUR* and *GWR*) and dimensionality reduction (enabled by *STP-BME*) had considerable positive effects on $PM_{2.5}$ mapping quality.

Step 8: Spatial daily maps (5 km × 5 km resolution, 8617 spatial points in total) of $PM_{2.5}$ concentration estimates with the associated standard mapping errors during the period October 1 to 31 were generated by the four methods. Given that method 4 offered the lowest standard $PM_{2.5}$ mapping error (i.e., the $PM_{2.5}$ maps of method 4 have the lowest uncertainty), it was employed for $PM_{2.5}$ concentration mapping purposes. Fig. 3.3 presents

TABLE 3.1 10-fold cross-validation results and computational time (cost).

Methods	R^2	*MAE* (μg/m³)	*RMSE* (μg/m³)	Time cost of prediction (s)[a]
Hard data with *DDR*	0.9420	9.236	15.232	2.44
Hard + soft data with *DDR*	0.9444	9.444	14.914	4.88
GAM	0.6361	24.564	32.724	0.06
GWR	0.8637	12.882	21.863	6.56

[a] *Time cost to predict per 100 points.*

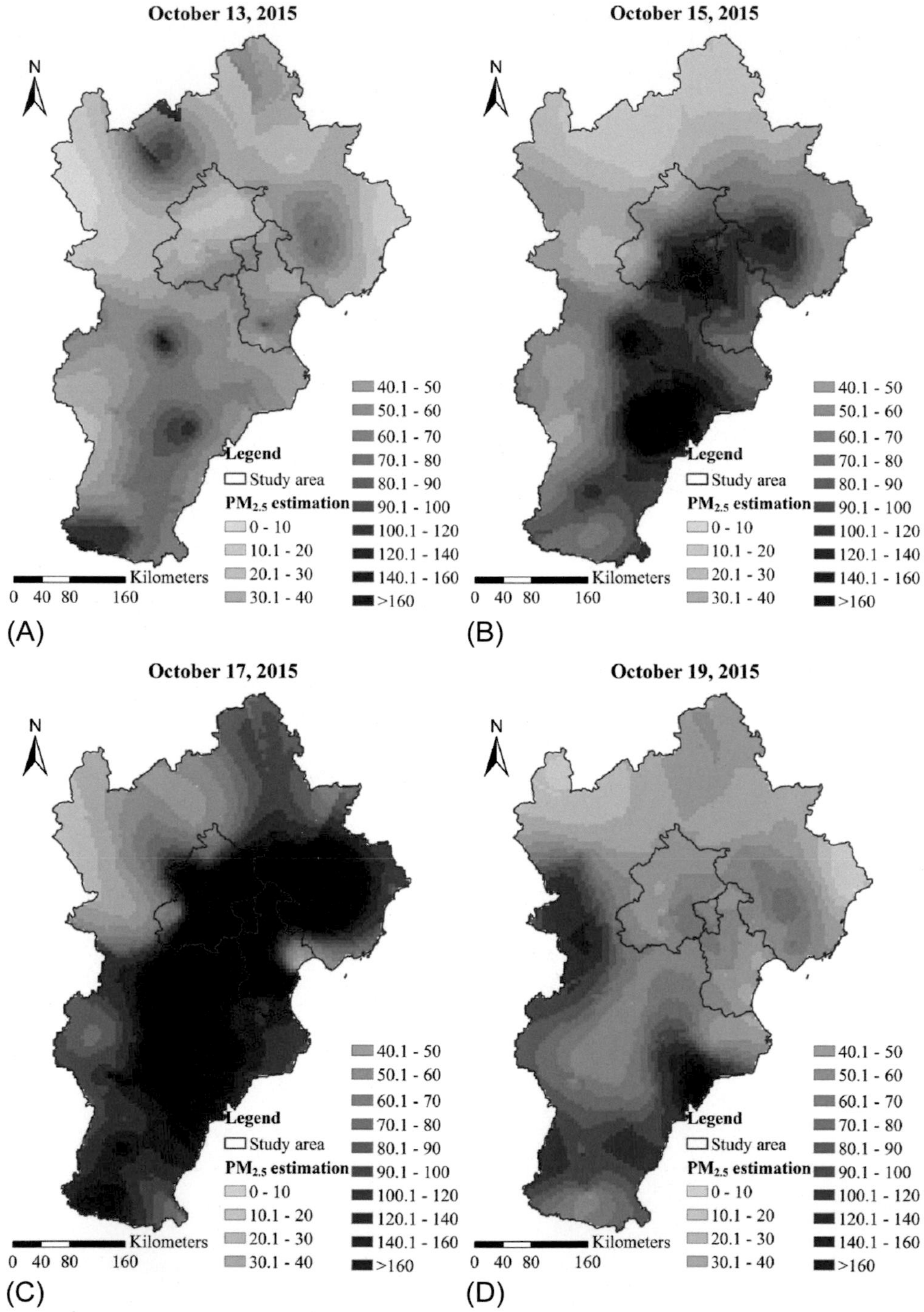

FIG. 3.3 The $PM_{2.5}$ distribution maps in the Jing-Jin-Ji region by using method 4. Selected maps (A)–(D) show a haze period (heavy pollution process) during October 13–19, 2015.

the $PM_{2.5}$ concentration maps of method 4, which clearly revealed a haze period. This severe pollution process began at the southern part of the Jing-Jin-Ji region on October 13, at which time Beijing was not polluted yet. During the next few days, the haze spread to the northern part of the region, and on October 17 it covered the entire Beijing City (the highest $PM_{2.5}$ concentration measurement was 403 $\mu g/m^3$). Finally, the pollution level decreased on October 19.

Step 9: Fig. 3.4 presents some important $PM_{2.5}$ mapping uncertainty plots derived by the four methods in terms of the temporal variation of the mean standard mapping error during October 2015 (8617 mapping points were used). Fig. 3.4 reveals that the two major peaks of the mapping error plots correspond to the two haze periods of heavy pollution, indicating that the $PM_{2.5}$ mapping uncertainty is much higher during the heavy pollution period than during other periods. Also, it is noteworthy that Fig. 3.4 confirms the cross-validation results (obtained earlier in this study) as follows:

- The decrease of $PM_{2.5}$ mapping uncertainty as a result of using soft data and dimensionality reduction was significant compared to other techniques, especially during the haze periods, thus producing more reliable $PM_{2.5}$ maps.

Step 10: The superior performance of the synthetic technique was also demonstrated in summary quantitative terms. Compared to the standard *BME* mapping, the mean uncertainty (error) of the $PM_{2.5}$ maps obtained by *STP-BME* was reduced by 50.83% (hard data were considered) and by 64.05% (when hard and soft data were assimilated). Lastly, by introducing soft data the mean uncertainty was reduced by 13.03% (standard *BME*) and 36.25% (*STP-BME*) compared to *BME* with hard data and *STP-BME* with hard data, respectively (Table 3.2). In summary terms, the following conclusion was drawn:

- The performance of the fourfold synthesis involving *STP* and *BME* (components of the $PM_{2.5}$ mapping part of the study) and *LUR* and *GWR* (components of the information processing part) was clearly superior to that of nonsynthetic approaches.

Step 11: Fig. 3.5 compares the maps of $PM_{2.5}$ mapping error (uncertainty) distribution obtained by the four methods over the Jing-Jin-Ji region in October 19. Lower mapping error values were consistently obtained by methods 3 and 4 compared to methods 1 and 2. Not surprisingly, the lower mapping uncertainty points were located near hard data points (monitoring stations) or near soft data points (Fig. 3.6). The introduction of soft data by method 2 led to a 6.55% (2.29 $\mu g/m^3$) reduction of the $PM_{2.5}$ mapping

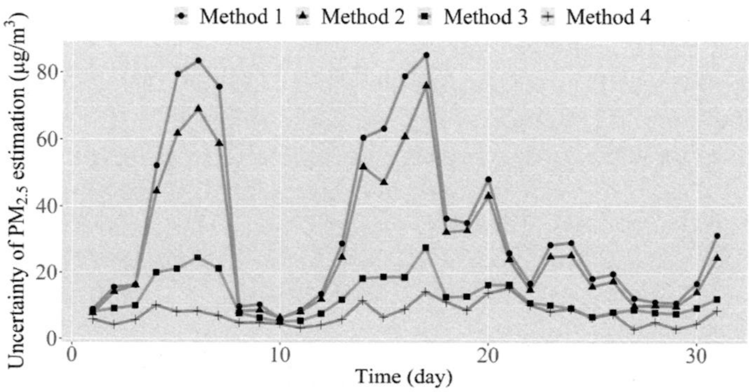

FIG. 3.4 Plots of the $PM_{2.5}$ mapping error or uncertainty (in $\mu g/m^3$ units) during October 1–31, 2015 by using the four methods.

TABLE 3.2 Uncertainty reduction of $PM_{2.5}$ mapping by comparing the daily mean map uncertainty during the period October 1–31, 2015.

Description	Min	Max	Mean
From method 1 to method 3	8.81%	76.31%	50.83%
From method 2 to method 4	23.48%	88.18%	64.05%
From method 1 to method 2	−3.00%	25.37%	13.03%
From method 3 to method 4	1.71%	67.28%	36.25%

error obtained by method 1, and a 32.15% (4.14 $\mu g/m^3$) error reduction obtained by method 4 compared to that by method 3. Similarly, the $PM_{2.5}$ mapping errors obtained by methods 3 and 4 were by 63.22% (22.13 $\mu g/m^3$) and by 73.30% (23.97 $\mu g/m^3$) lower than those obtained by method 1 and method 2, respectively. A comparison of the proportion of the 8617 mapping points exhibiting lower mapping uncertainty obtained by the four methods is presented in Fig. 3.7. Method 2 *vs.* method 1 represents the proportions of points with the mapping error of method 2 lower than that of method 1. Overall, method 4 *sine dubio* exhibited the best performance, leading to the following finding:

- The use of soft data allowed a very considerable decrease in the $PM_{2.5}$ mapping error compared to using hard data alone.

As the readers recall, a soft $PM_{2.5}$ datum is a datum characterized by nonnegligible uncertainty, in which case the study investigator needs to distinguish between appearance and reality. However, even uncertain $PM_{2.5}$ data may contain considerable levels of information (especially when the soft data points are chronotopologically near the interpolation points), which, if properly assimilated, can improve the accuracy of the resulting $PM_{2.5}$ maps.

Summarizing the results of the Jing-Jin-Ji case study, the following broad conclusion was drawn by the investigators:

- The type of **broad synthesis** of data and interpretation considered in the Jing-Jin-Ji study can be critical to the process of public health sciences, highlighting how individual scientists can assimilate different data sources as well as build on the work of others.

4 A synthesis of the *BME* technique with the *MLR* and *GWR* models

The synthetic perspective studied in this chapter has more than one facet, i.e., it seeks to combine different datasets, techniques, and thinking modes. In this context, the overall objective of this section is as follows:

> Investigate an approach that can combine **multisourced data** blending with **multitechnique** synthesis.

With this purpose in mind, Fig. 4.1 outlines the methodological framework of the synthesis of the *BME* mapping technique with the *MLR*[i] and *GWR* models as part of a case study that covered almost the entire China.

4.1 Public health concerns in China

Sine dubio, with the rapid development and urbanization, atmospheric pollution has become a considerable public health concern in the cities. This is the focus of the study by Xiao et al. (2018) that applied the *BME*

[i] In this case the *OLS* (*ordinary least squares*) scheme is used to estimate the parameters in the *MLR* model (there are also other schemes that can be used for the same purpose).

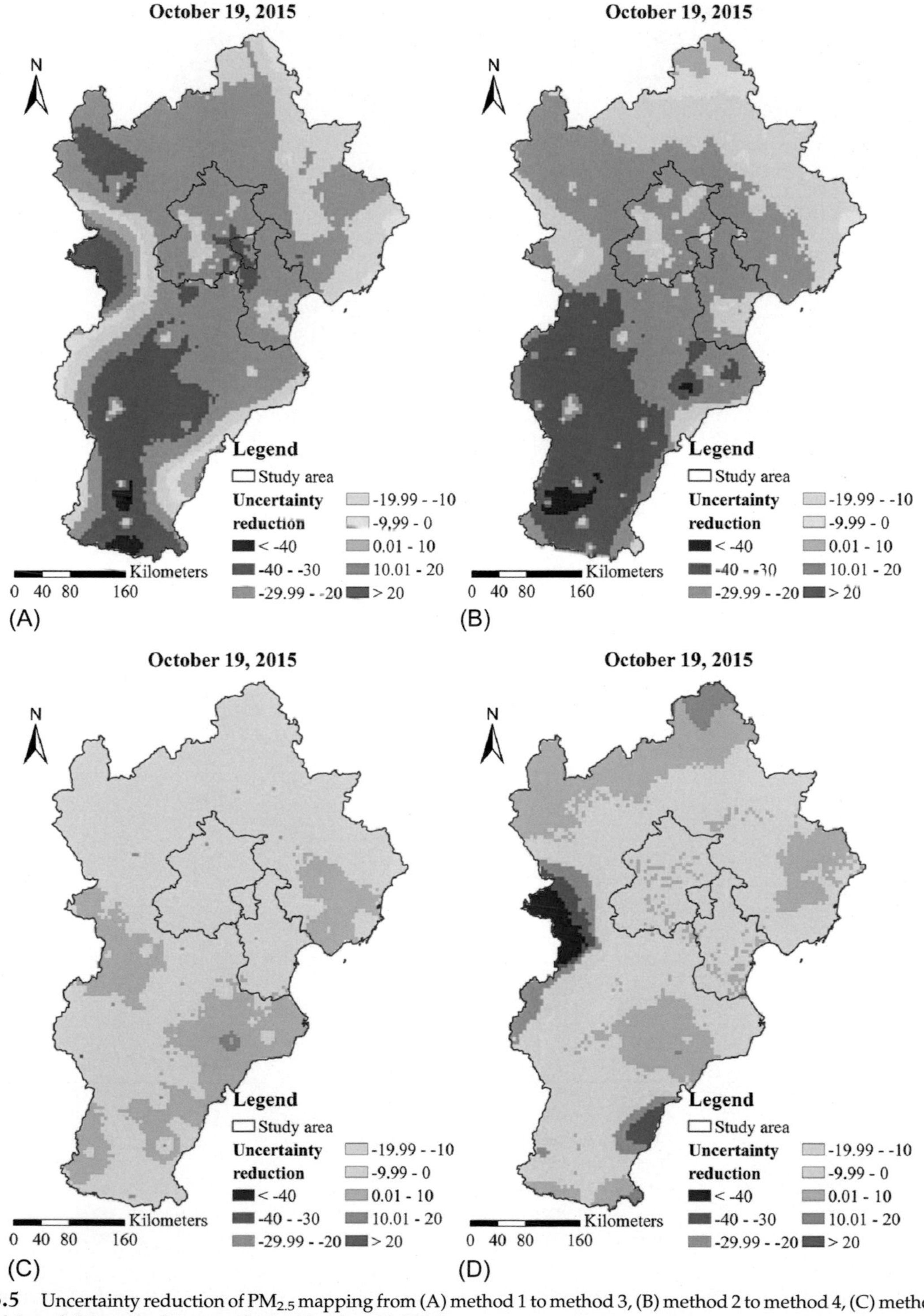

FIG. 3.5 Uncertainty reduction of $PM_{2.5}$ mapping from (A) method 1 to method 3, (B) method 2 to method 4, (C) method 1 to method 2, and (D) method 3 to method 4 (October 19, 2015).

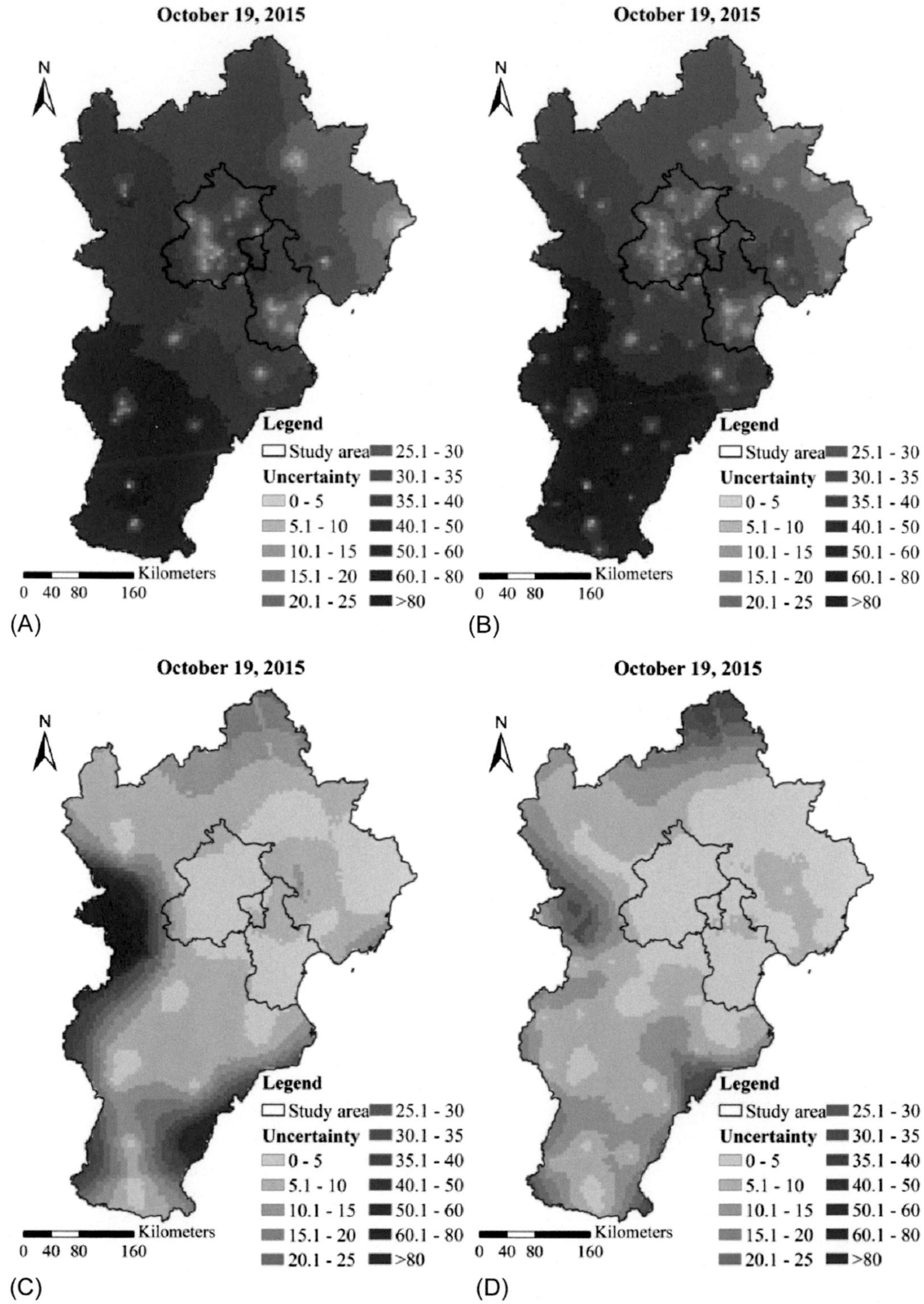

FIG. 3.6 Uncertainty maps of $PM_{2.5}$ concentration mapping on October 19 by using (A) method 1, (B) method 2, (C) method 3, and (D) method 4.

FIG. 3.7 Proportion of 8617 mapping points with lower uncertainty by comparing various methods. For example, the line with plus symbol shows that by using method 4, 86.48% of the mapping points in the 1st day have lower mapping error (uncertainty) than using method 2.

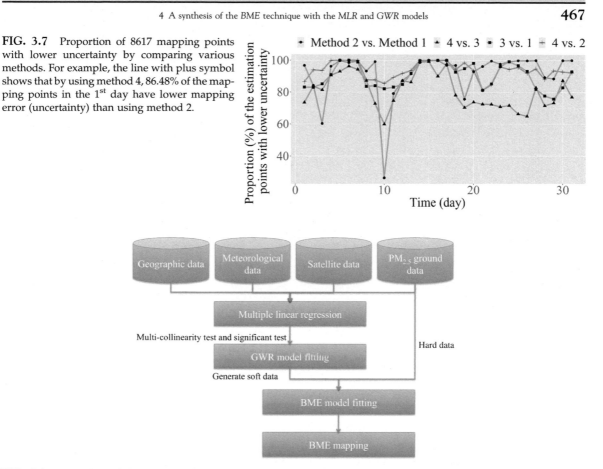

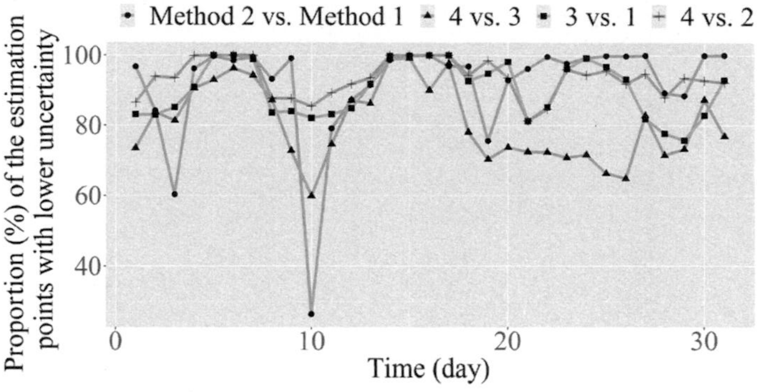

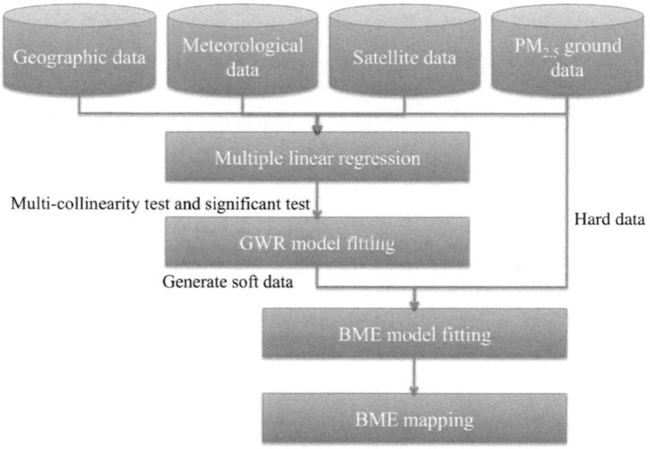

FIG. 4.1 An outline of the synthetic framework of the *BME* technique with *MLR*, *OLS*, and *GWR* models.

approach in atmospheric environmental research. It is important to have an accurate chronotopologic mapping of $PM_{2.5}$ concentrations in terms of high-resolution maps, which can benefit environmental control management and public health investigations. This study concentrated on the entire China with the exception of five provinces (Xinjiang, Tibet, Qinghai, Inner Mongolia, and Heilongjiang), see Fig. 4.2. These provinces were excluded because the monitoring stations were rather sparse, with only 89 stations in an area of 5.33 million km^2. The period of interest was from November 1, 2015 to February 28, 2016.

Similarly to the preceding case studies, the chronotopologic domain structure of the present Chinese case study was represented quantitatively by the symbolic arrangement

Chronotopologic domain $\boldsymbol{p}$:

$$\begin{cases} \boldsymbol{s} = (s_1, s_2) \in \text{China region (except four provinces)} \\ t \in \text{November 2015} - \text{February 2016 time period.} \end{cases} \tag{4.1}$$

Methodologically, this case study was characterized by two very different kinds of intersections.

4.2 Intersections of data

Multisourced data considerations can deliver insights. Accordingly, the first methodological concern of this case study was as follows:

- A prime feature of the synthesis narrative of the present pollution study is the anticipation that new possibilities open by **data intersection**, i.e., when one type of data crosses another.

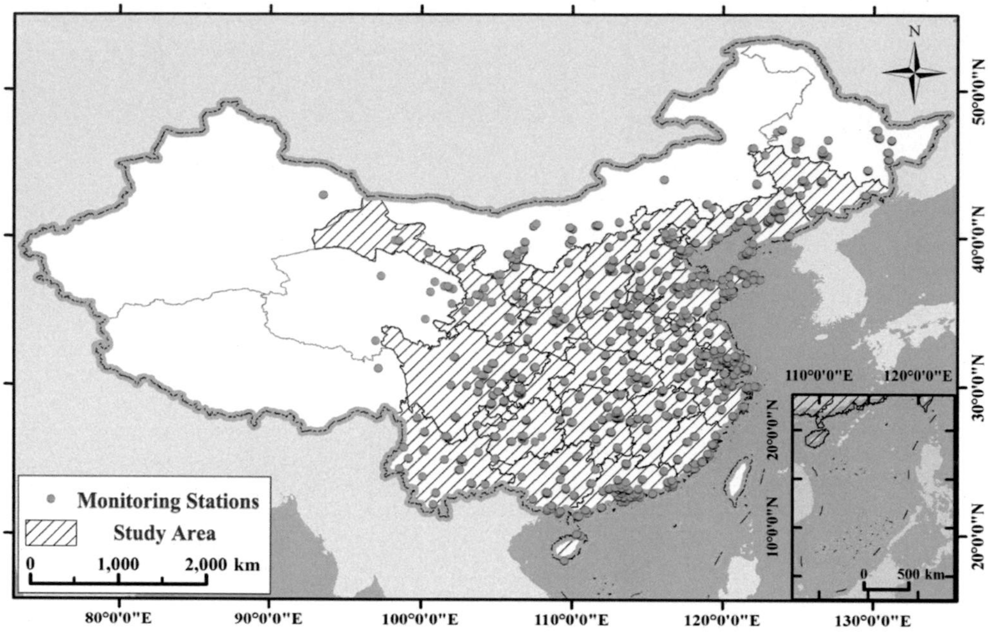

FIG. 4.2 The study region with the green dots representing the 1408 $PM_{2.5}$ monitoring sites within the region and the 43 sites in the neighboring provinces.

Otherwise said, this narrative is a vehicle for important information delivery. Given that the monthly atmospheric environment status was the study's main focus, taking advantage of data intersections was one of the key objectives in the present case study. Specifically, the initial three steps of this study were as follows:

Step 1: Daily mean $PM_{2.5}$, NO_2, and CO concentrations were collected from the China Environmental Monitoring Center (http://106.37.208.233:20035/, Table 4.1) at 1408 monitoring stations during the period November 1, 2015 to February 28, 2016. Concentrations less than 2 $\mu g/m^3$ (comprising 5.6% of the total records) were discarded because they were under the detection limit, i.e., these datasets were not reliable. Stations that operated less than 15 days per month were removed during the data preprocessing stage. The number of $PM_{2.5}$, NO_2, and CO observations during November, December, January, and February were 953, 707, 529, and 820, respectively. The monthly mean $PM_{2.5}$, NO_2, and CO concentrations were computed accordingly.

Step 2: Other $PM_{2.5}$-related environmental factors were taken into account in this pollution study, as described in Table 4.1.

Step 3: To assimilate all available data, a 3×3 km^2 grid resolution was used, including a total of 357,997 grid cells in the study domain, i.e., all data were arranged into these grid cells for further analysis. Monthly averaged values of all $PM_{2.5}$ related variables were obtained at each cell, and the $PM_{2.5}$ concentrations were arranged into the corresponding cells at the same points.

4.3 Integration of techniques

We now turn to the second major concern of the present case study, as follows:

- Another feature of the synthesis narrative adopted by the present pollution study is the anticipation that advances in chronotopologic modeling and mapping are

TABLE 4.1 $PM_{2.5}$ related variables.

Variable	Source	Original spatial resolution	Integration methods
Aerosol optical depth (*AOD*)	Moderate Resolution Imaging Spectroradiometer, MODIS, instruments on Aqua and Terra satellites (http://ladsweb.nascom.nasa.gov/)	3 km	None
NO_2 and CO	China Environmental Monitoring Center (http://106.37.208.233:20035/)	Monitoring station data	NO_2 and CO interpolated by inverse distance weighted (*IDW*) technique to match 3 km grid
Meteorologic variables (precipitation, temperature, relative humidity, air pressure, wind speed)	China Meteorological Data Sharing Service System (http://cdc.cma.gov.cn)	Monitoring station data	Meteorologic variables interpolated using relevant software for climatic data (*ANUSPLIN*)
Land use information (grassland, water, urban, city, forest)	Global Land Cover Facility-MODIS Land Cover (http://www.landcover.org/data/lc/)	500 m	Total area of five kinds of land use cover calculated on 3×3 km^2 AOD grid
Road network	OpenStreetMap (http://www.openstreetmap.org)	–	Total length calculated on same 3 km grid
Elevation	Shuttle Radar Topography Mission, *SRTM* (http://srtm.csi.cgiar.org/SELECTION/inputCoord.asp)	30 m	Elevations averaged at each 3 km grid
Population	Gridded Population of the World, Vers. 4 (*GPWv4*) (http://sedac.ciesin.columbia.edu/data/collection/gpw-v4)	1 km	Populations averaged at each 3 km grid

possible by **technique intersection**, i.e., when the important properties of the different techniques are combined.

Just as multisourced data integration can improve *CTDA*, techniques integration too is often necessary in order to optimize the solution of a problem. In this environmental study, a number of techniques were carefully combined, as follows:

Step 4: A technique based on the *MLR* model (with the *OLS* parameter estimation scheme) was used to select the significant $PM_{2.5}$ related variables and eliminate collinearities among them. When the *Variance Inflation Factor* (*VIF*, Miles, 2014) of a variable is larger than 10 or the tolerance is smaller than 0.1, there exists a collinearity problem; therefore, the corresponding variable should be excluded from the regression model. Following this procedure, the variables that finally remained in the model are listed in Table 4.2.

Step 5: After the collinearity problem was overcome, using the Koenker's studentized *Bruesch-Pagan* (*BP*, Gao and Li, 2011) statistic the relationship between the remaining independent attributes ($PM_{2.5}$-related variables)

TABLE 4.2 Variables of the regression model.

Month	Variables
November	*AOD*, precipitation, pressure, relative humidity, wind speed, temperature, NO_2, CO, road length, elevation, population, grassland, forest, city, water (15 variables)
December	*AOD*, precipitation, pressure, relative humidity, wind speed, temperature, NO_2, CO, road length, elevation, population, grassland, forest, city, water (15 variables)
January	*AOD*, precipitation, pressure, relative humidity, wind speed, NO_2, CO, elevation, forest (9 variables)
February	*AOD*, pressure, relative humidity, wind speed, NO_2, CO, city, forest, water (9 variables)

and the dependent attribute ($PM_{2.5}$) was found to be consistent (both in geographical and data space) and nonstationary (heteroskedastic).[j]

Step 6: The above results implied that the local sublinearity between environmental variables and $PM_{2.5}$ concentrations in a wide geographic area could be better interpreted by the *GWR* model,

$$PM_{2.5,i} = \beta_0(s_{1,i}, s_{2,i}) + \sum\nolimits_k \beta_k(s_{1,i}, s_{2,i}) x_{ik} + \varepsilon_i \tag{4.2}$$

($i = 1,2,\ldots n$), where $(s_{1,i}, s_{2,i})$ are the spatial coordinates of each sample point i, $\beta_0(s_{1,i}, s_{2,i})$ is the intercept of sample i, $\beta_k(s_{1,i}, s_{2,i})$ is the regression coefficient of sample point i, ε_i is the random error, x_{ik} is the value of the kth variable (Table 4.2), and $k = 15$, 15, 9, and 9 for November, December, January, and February, respectively. The *GWR* model provided the spatial variance of the relationship between $PM_{2.5}$ concentrations and other environmental variables.

Step 7: Given that the spatial arrangement of the $PM_{2.5}$ monitoring stations was still coarse for the region of interest, to make the $PM_{2.5}$ observation arrangement finer, the *GWR* model strength during each of the 4 months was used to generate $PM_{2.5}$ values at unmonitored locations. The *GWR* estimates and the associated variances were then regarded as soft data. The *GWR* model of each month was used to estimate $PM_{2.5}$ at a 30×30 km^2 grid (3582 grid cells) for generating soft data. Soft data combined with hard data (i.e., monthly mean $PM_{2.5}$ observations at monitoring stations) were integrated as site-specific (*S*) knowledge for *BME* modeling purposes.

4.4 Covariance modeling

Step 8: The empirical $PM_{2.5}$ covariance values[k] are shown in Fig. 4.3 together with the fitted theoretical (nonseparable across space and time) model

$$c_{PM_{2.5}}(\Delta \boldsymbol{p}) = c_1 \left(1 - \frac{3h}{2a_{s_1}} + \frac{h^3}{2a_{s_1}^3}\right)\left(1 - \frac{3\tau}{2a_{t_1}} + \frac{\tau^3}{2a_{t_1}^3}\right) + c_2 e^{-3\left(\frac{h}{a_{s_2}} + \frac{\tau}{a_{t_2}}\right)}, \tag{4.3}$$

where $c_1 = 0.8$, $c_2 = 0.2$, $a_{s_1} = 1$, $a_{s_2} = 14$ (°), $a_{t_1} = 3.8$, $a_{t_2} = 13$ (mos). This model succinctly captures the essential features of the chronotopologic $PM_{2.5}$ variation. The model plots are twofolded: a very sharp slope at the space origin combined with a very quick drop of the spatial covariance cross-section (indicating the irregular spatial variability of $PM_{2.5}$ concentrations and a very short spatial correlation range), and a rather smooth decline of the temporal covariance cross-section (indicating the long time correlation range of $PM_{2.5}$ concentrations).

[j] The *BP* tests were found to be statistically significant ($P < 0.01$) during the 4 months of the study.

[k] Notice that normalized $PM_{2.5}$ concentrations were used by the authors of this study (Xiao et al., 2018).

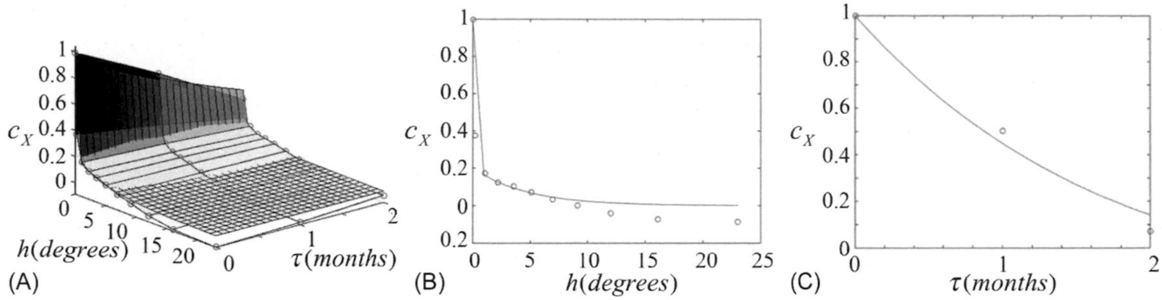

FIG. 4.3 Empirical $PM_{2.5}$ covariance values (small circles) and fitted theoretical model (continuous surface and continuous lines) (A) in the entire space-time domain, (B) in a spatial cross-section, and (C) in a temporal covariance cross-section.

4.5 *BME* assessment and mapping

Following $PM_{2.5}$ covariance modeling and interpretation, the two last computational stages of the study were as follows:

Step 10: The 10-fold cross-validation was used to evaluate the *BME* performance, which, for the 4 months of interest was numerically assessed in terms of the R^2 values 0.779, 0.828, 0.803, and 0.627, respectively, and in terms of the *MAE* values 10.89, 9.47, 11.14, and 12.64 $\mu g/m^3$, respectively (Fig. 4.4).

Step 11: The *BME* approach was used to estimate $PM_{2.5}$ concentrations at $3 \times 3\ km^2$ grid cells during each of the four study months. In particular, Fig. 4.5 presents ground-level $PM_{2.5}$ measurements and the corresponding monthly averaged $PM_{2.5}$ concentration values obtained using this approach.

Concluding our discussion, a few findings are worth pointing up concerning the present pollution study:

① Most of the $PM_{2.5}$ monitoring sites were clustered in urban areas (in fact, rural areas have little coverage in China).
② While the Beijing, Tianjin and Hebei Provinces were always highly polluted areas, the air quality has been much better in the southern provinces.
③ The geographical gradients of $PM_{2.5}$ concentrations showed a significant change during the four study months, and the overall concentration trend was high in the north and low in the south parts of the study region.
④ During the period November 2015 to February 2016, the temporal $PM_{2.5}$ concentrations in the study region showed a clear monthly variation, and there was a concentration trend from the northern coast toward inland.

5 Epilogue

In nuce, the syntheses of techniques and models presented in this chapter demonstrated that an effective methodological framework of chronotopologic modeling and mapping under uncertainty and partial observability can be achieved by a symbiosis of techniques and models from multiple areas of modern geostatistics, artificial intelligence, and machine learning. Empirically, each one of the presented syntheses of techniques and models was tested in the real-world conditions of properly selected case studies from the earth and environmental literature.

The results of such syntheses offer considerable promise in the study of complex and multidisciplinary real-world phenomena. Beyond the scientific and technical advances, the syntheses can also help to address the problem of establishing an intelligent coordination among investigators in different disciplines despite the availability of diverse and often limited *in-situ* information and the uncertainty in the study environment.

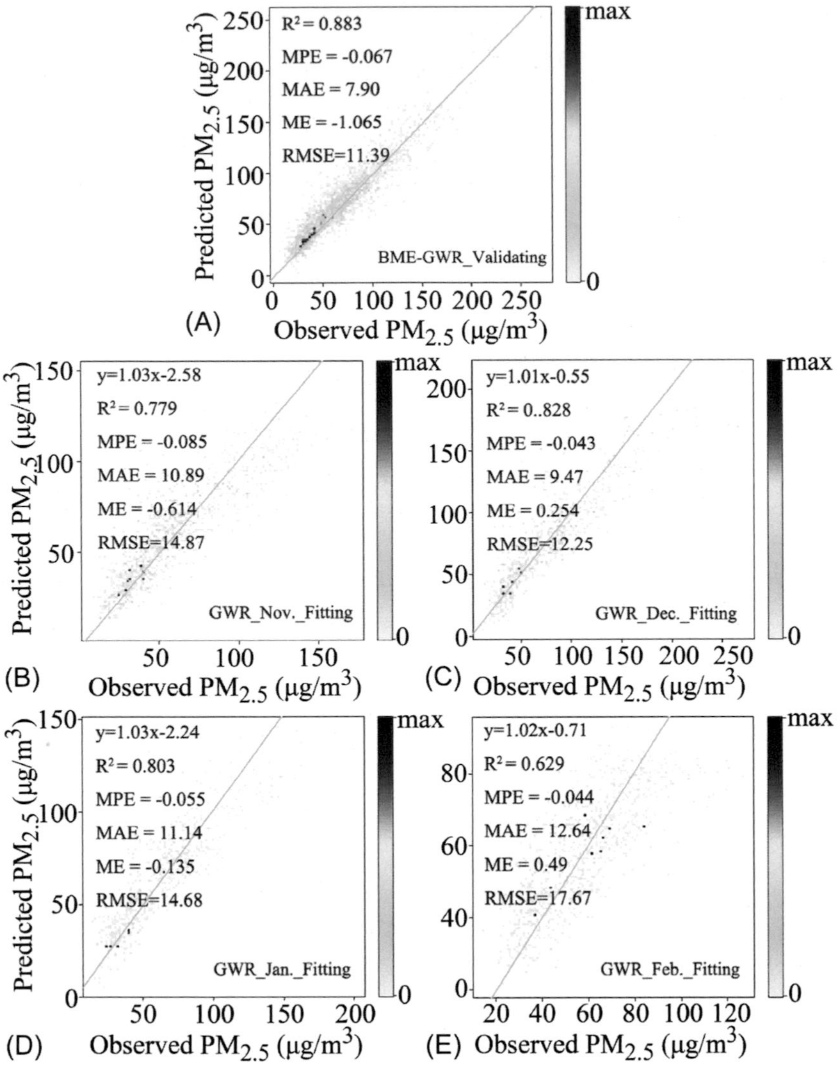

FIG. 4.4 Scatter plots of model fitting and validation result. The solid line denotes the trend line: (A) cross-validation results for the *BME-GWR* technique ($N = 3009$ mapping points, $t = 4$ months); (B)–(E) are *GWR* model fitting results during the 4 months considered.

6 Practice exercises

1. Using the hard and soft data listed in Tables 6.2–6.4 of Section 6 of Chapter 6, employ the *LUR* or *GWR* models to describe the relationship between AOD, PBLH, wind, elevation, road length, farm area, and $PM_{2.5}$. Then, use the STP-BME technique to study the spatiotemporal distribution of $PM_{2.5}$.
2. Comment on the following criticisms of big data-driven analysis:
   (a) More data often means more delusions, since human detection of nonexistent or false patterns is growing faster and

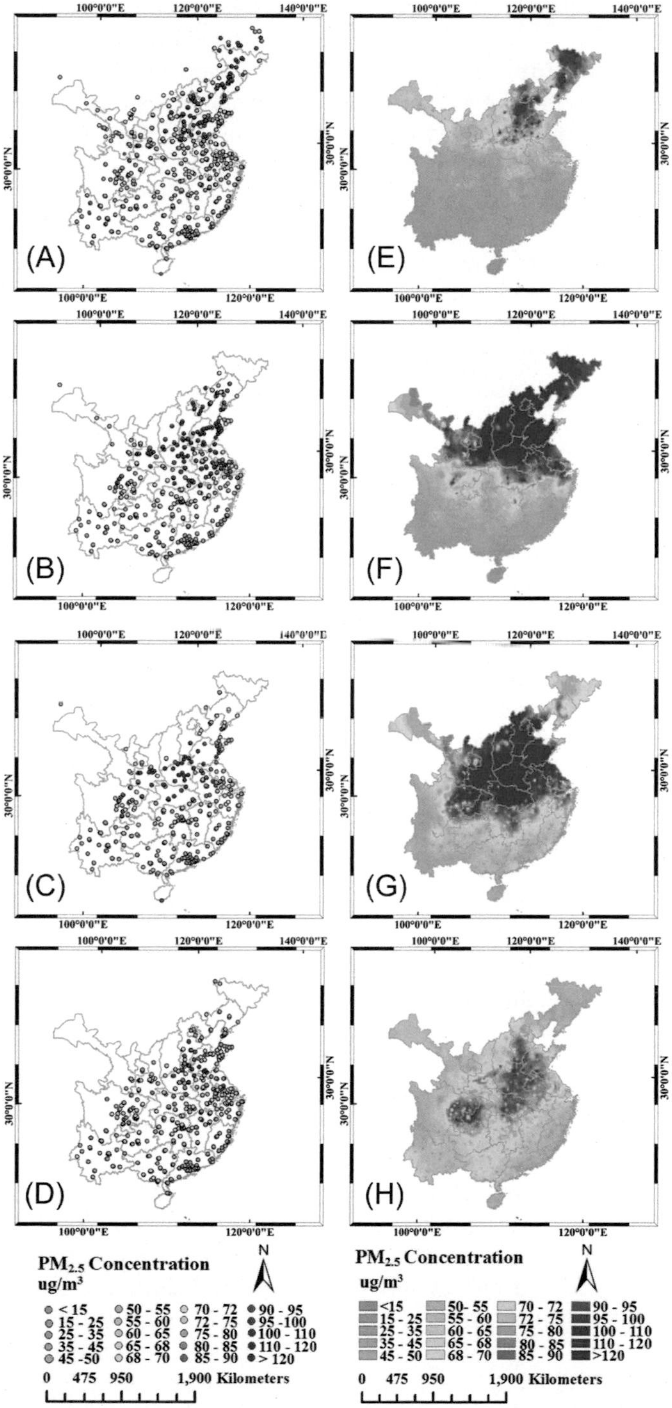

FIG. 4.5 (A)–(D) Ground measurements and (E)–(H) spatial distributions of monthly $PM_{2.5}$ concentrations generated by the *BME-GWR* space-time technique during November 2015, December 2015, January 2016, and February 2016.

faster as a side effect of the techno-information age.

(b) Our mental architecture is at an increased mismatch with the world in which we live, since the world becomes more and more complicated and our minds are trained for more and more simplification. That is, there is a mismatch between the messy randomness and noise of the info-rich modern world (with its complex interactions) *vs.* the human intuition of events, derived in our simpler mental architecture.

(c) The calamity of the information age is that the toxicity of data increases much faster than its benefits.

3. Consider the following viewpoints:

(a) The principal objective of the synthetic approach is to produce new knowledge and/or add to the existing body of knowledge in a specific field of science.

(b) The knowledge generated by the synthetic approach could be directly applicable for practice or the knowledge that needs further verification before application.

Do you agree or not with the above viewpoints, and to what extent?

4. Propose synthetic approaches that involve techniques from multiple areas of modern geostatistics, artificial intelligence, and machine learning.

5. You need to perform a *CTDA* of the distribution of chlorophyll-a concentrations in a coastal area during a given time period. A total of 50 cruise measurements across space and time are available (including chlorophyll-a data and sea temperature data), but they are not enough to cover the entire study area. A few additional dataset possibilities are listed below:

(a) Remote chlorophyll-a data.

(b) Remote temperature data

(c) Remote chlorophyll-a data and temperature data.

Based on these datasets, propose a synthesis of techniques to detect the chronotopologic characteristics of *chlorophyll a* concentrations.

6. Assume that the farmers need to know the heavy metal concentrations within the soil they grow their corn. Some relationships exist between soil copper concentration and several other variables, including soil moisture, soil temperature, and soil organic matter. These last three variables are much cheaper to measure than the soil copper concentration. Design a synthetic approach that can help the farmers understand the level and chronotopologic spread of contaminant concentration in the soil.

7. Suggest a synthesis of techniques through which a detailed direct assessment of chronotopologic variability and an informative attribute mapping can be achieved in a scientific field of your interest.

8. Assume you need to explore the chronotopologic distribution of chlorophyll-a concentration in a lake and you have 20 monitoring sites uniformly distributed throughout the lake. At each site, there is a buoy for monitoring several parameters, e.g., chlorophyll-a concentration, temperature, wind speed, pH, nitrogen, and phosphors. In addition, the variation dynamic of the chlorophyll-a is quite stable, i.e., it has stable cycles, but the largest and smallest values of chlorophyll-a concentration are different within the various cycles. These two values are related to the environmental conditions, as buoy measures. Design a methodological framework to combine the distributional characteristics and the dynamic behavior of chlorophyll-a concentration in the lake to map the chronotopologic distribution of the chlorophyll-a and also forecast the distribution of chlorophyll-a in the near future.

9. Suppose that a novel infectious disease occurs at one county of a region; design a methodological framework to map the chronotopologic transmission of the disease in the region over time. List all datasets you need and any infectious disease models you may consider.

10. Assuming that you need to investigate a spatiotemporal transmission pattern of a novel infectious disease, design a synthesis framework, including *SEIR* (Example 1.1 in Section 1 of Chapter 10) model and *BME*.

References

Aitchison, J., Silvey, S.D., 1958. Maximum-likelihood estimation of parameters subject to restraints. Ann. Math. Stat. 29 (3), 813–828.

Allen, L.J.S., Burgin, A.M., 2000. Comparison of deterministic and stochastic SIS and SIR models in discrete time. Math. Biosci. 163, 1–33.

Anderson C, 2008. The end of theory: the data deluge makes the scientific method obsolete, Wired, June 23, 2008, at http://www.wired.com/science/discoveries/magazine/16-07/pb_theory.

Anderson, J.L., Anderson, S.L., 1999. A Monte Carlo implementation of the nonlinear filtering problem to produce ensemble assimilations and forecasts. Mon. Weather Rev. 127 (12), 2741–2758.

Angulo, J., Yu, H.L., Langousis, A., Kolovos, A., Wang, J., et al., 2013. Correction: Spatiotemporal Infectious Disease Modeling: A BME-SIR Approach. PLoS One 8 (10). https://doi.org/10.1371/annotation/4adf8407-a5b8-4f4b-877d-e8b944f0e6ee.

Anselin, L., 1995. Local Indicators of Spatial Association-LISA. Geogr. Anal. 27 (2), 93–115.

Bishop, C., 2006. Pattern Recognition and Machine Learning. Springer-Verlag, Berlin, Germany.

Bohannon, J., 2015. Fears of an AI pioneer. Science 349 (6245), 252.

Bollier, D., 2010. (rapporteur). The Promise and Peril of Big Data. The Aspen Institute-Communications and Society Program, Washington DC.

Box, G., Jenkins, G., 1976. Time Series Analysis: Forecasting and Control. Holden-Day, Oakland, CA.

Boyd, D., Crawford, K., 2012. Critical questions for big data. Inf. Commun. Soc. 15 (5), 662–679.

Bracco, A., Clayton, S., Pasquero, C., 2009. Horizontal advection, diffusion, and plankton spectra at the sea surface. J. Geophys. Res.: Oceans 114 (C02001). https://doi.org/10.1029/2007JC004671.

Briggs, D.J., Collins, S., Elliott, P., Fischer, P., Kingham, S., Lebret, E., Pryl, K., Van Reeuwijk, H., Smallbone, K., Van der Veen, A., 1997. Mapping urban air pollution using GIS: a regression-based approach. Int. J. Geogr. Inform. Sci. 11, 699–718.

Briggs, D.J., de Hoogh, C., Gulliver, J., Wills, J., Elliott, P., Kingham, S., Smallbone, K., 2000. A regression-based method for mapping traffic-related air pollution: application and testing in four contrasting urban environments. Sci. Total Environ. 253 (1), 151–167. https://doi.org/10.1016/S0048-9697(00)00429-0.

Brodie, M.L., 2015. Understanding Data Science: An Emerging Discipline for Data-Intensive Discovery. In: Cutt, S. (Ed.), Chapter 4 in Getting Data Right. O'Reilly Media, Inc., ISBN: 9781491935361.

Brunsdon, C., Fotheringham, S., Charlton, M., 1998. Geographically weighted regression. J. Roy. Stat. Soc.: Ser. D (The Statistician) 47 (3), 431–443.

Buchan, I., Winn, J., Bishop, C., 2009. A unified modeling approach to data-intensive healthcare. In: Hey, T., Tansley, S., Tolle, K. (Eds.), The Fourth Paradigm: Data Intensive Scientific Discovery. Microsoft Research, Redmont, Washington, ISBN: 978-0-9825442-0-4, pp. 91–98.

Carr, M.-E., Friedrichs, M.A.M., Schmeltz, M., Noguchi, A. M., Antoine, D., Barber, R.T., Behrenfeld, M., Arrigo, K. R., Asanuma, I., Aumont, O., Bidigare, R., Buitenhuis, E., Campbell, J.W., Maria, C.A., Dierssen, H.M., Dowell, M., Dunne, J.P., Esaias, W.E., Gentili, B., Gregg, W., Groom, S., Hoepffner, N., Ishizaka, J., Kameda, T., Quere, C.L., Lohrenz, S.E., Marra, J., Melin, F., Moore, K., Morel, A., Reddy, T.E., Ryan, J., Scardi, M., Smyth, T., Turpie, K. R., Tilstone, G., Waters, K., Yamanaka, Y., 2006. A comparison of global estimates of marine primary production from ocean color. Deep Sea Res. Part II Top. Stud. Oceanogr. 53, 741–770.

Chainey, S.P., Reid, S., Stuart, N., 2002. When is a hotspot a hotspot? A procedure for creating statistically robust hotspot maps of crime. In: Kidner, D., Higgs, G., White, S. (Eds.), Socio-Economic Applications of Geographic Information Science—Innovations in GIS. vol. 9. Taylor & Francis, London, UK, pp. 21–36.

Chinikar, S., Javadi, A., Hajiannia, A., Ataei, B., Jalali, T., Khakifirouz, S., Nowotny, N., Schmidt-Chanasit, J., Shahhosseini, N., 2014. First evidence of Hantavirus in central Iran as an emerging viral disease. Adv. Infect. Dis. 4, 173–177.

Choi, K.M., Christakos, G., Serre, M.L., 1998. Recent developments in vectorial and multi-point BME analysis. In: Buccianti, A., Nardi, G., Potenza, R. (Eds.), Proceedings of 4th Annual International Association for Mathematical

Geology (IAMG) Conference. Vol. 1. De Frede Editore, Naples, Italy, pp. 91–96.
Christakos, G., 1985. Recursive parameter estimation with applications in earth sciences. Math. Geol. 17, 489–515. https://doi.org/10.1007/BF01032105.
Christakos, G., 1990. A Bayesian/maximum-entropy view to the spatial estimation problem. Math. Geol. 22 (7), 763–777.
Christakos, G., 1991. On certain classes of spatiotemporal random fields with applications to space-time data processing. IEEE Trans. Syst. Man Cybernet. 21 (4), 861–875.
Christakos, G., 1992. Certain results on spatiotemporal random fields and their applications in environmental research. In: Probabilistic and Stochastic Methods in Analysis, with Applications. Springer, Dordrecht, pp. 287–322.
Christakos, G., 1992a. Random Field Models in Earth Sciences. Acad Press, San Diego, CA.
Christakos, G., 1998a. Modern geostatistics in the analysis of spatiotemporal environmental data: the BME approach. In: Short Course Notes, IAMG98, Ischia, Italy.
Christakos, G., 1998b. Spatiotemporal information systems in soil and environmental sciences. Geoderma 85 (2–3), 141–179.
Christakos, G., 2000. Modern Spatiotemporal Geostatistics. Oxford Univ Press, New York, NY.
Christakos, G., 2010. Integrative Problem-Solving in a Time of Decadence. Springer-Verlag, New York, NY.
Christakos, G., 2017. Spatiotemporal Random Fields: Theory and Applications. Elsevier, Amsterdam, the Netherlands.
Christakos, G., Hristopulos, D.T., 1998. Spatiotemporal Environmental Health Modelling: A Tractatus Stochasticus. Kluwer Academic Publ, Boston, MA.
Christakos, G., Li, X., 1998. Bayesian maximum entropy analysis and mapping: a farewell to kriging estimators? Math. Geol. 30 (4), 435–462.
Christakos, G., Raghu, V.R., 1996. Dynamic stochastic estimation of physical variables. Math. Geol. 28 (3), 341–365.
Christakos, G., Serre, M.L., 2000. BME analysis of spatiotemporal particulate matter distributions in North Carolina. Atmos. Environ. 34, 3393–3406.
Christakos, G., Bogaert, P., Serre, M.L., 2002. Temporal GIS. Springer, New York.
Christakos, G., Olea, R.A., Serre, M.L., Wang, L.L., Yu, H.L., 2005. Interdisciplinary Public Health Reasoning and Epidemic Modelling: the Case of Black Death. p. 320, Springer, New York.
Christakos, G., Olea, R., Yu, H., 2007. Recent results on the spatiotemporal modeling and comparative analysis of Black Death and bubonic plgue epidemics. Public Health 121 (9), 700–720.
Christakos, G., Angulo, J.M., Yu, H.-L., 2011. Constructing space-time pdfs in geosciences. Bol. Geol. Miner. 122 (4), 531–542.
Christakos, G., Zhang, C., He, J., 2017. A traveling epidemic model of space–time disease spread. Stoch. Environ. Res. Risk Assess. 31, 305–314.
Christakos, G., Yang, Y., Wu, J., Zhang, C., Mei, Y., He, J., 2018. Improved space-time mapping of PM2. 5 distribution using a domain transformation method. Ecol. Indic. 85, 1273–1279.
Cobos, M., Lira-Loarca, A., Christakos, G., Baquerizo, A., 2019. Storm characterization using a BME approach. In: Valenzuela, O., Rojas, F., Pomares, H., Rojas, I. (Eds.), Theory and Applications of Time Series Analysis. ITISE 2018. Contributions to Statistics. Springer, Cham, https://doi.org/10.1007/978-3-030-26036-1_19.
Conan, R., Borgnino, J., Ziad, A., Martin, F., 2000. Analytical solution for the covariance and for the decorrelation time of the angle of arrival of a wave front corrugated by atmospheric turbulence. J. Opt. Soc. Am. A Opt. Image Sci. Vis. 17 (10), 1807–1818.
Cook, D.G., Pocock, S.J., 1983. Multiple regression in geographical mortality studies, with allowance for spatially correlated errors. Biometrics 39 (2), 361–371.
Darroch, J.N., Ratcliff, D., 1972. Generalized iterative scaling for log-linear models. Ann. Math. Stat., 1470–1480.
De Cesare, L., Myers, D.E., Posa, D., 2001. Product-sum covariance for space-time modeling: an environmental application. Environmetrics: Off. J. Int. Environ. Soc. 12 (1), 11–23.
Delaney, J.R., Barga, R.S., 2009. A 2020 vision for ocean science. In: Hey, T., Tansley, S., Tolle, K. (Eds.), The fourth Paradigm: Data Intensive Scientific Discovery. Microsoft Research, Redmont, Washington, pp. 27–38.
Della Pietra, S., Della Pietra, V., Lafferty, J., 1997. Inducing features of random fields. IEEE Trans. Pattern Anal. Mach. Intell. 19 (4), 380–393.
Dominy, S.C., Noppe, M.A., Annels, A.E., 2002. Errors and uncertainty in mineral resource and ore reserve estimation: The importance of getting things right. Explor. Min. Geol. 11 (1–4), 77–98.
Douaik, A., Van Meirvenne, M., Toth, T., 2005. Soil salinity mapping using spatio-temporal kriging and Bayesian maximum entropy with interval soft data. Geoderma 128, 234–248.
Drucker, R., Riser, S.C., 2014. Validation of Aquarius sea surface salinity with Argo: Analysis of error due to depth of measurement and vertical salinity stratification. J. Geophys. Res. Oceans 119 (7), 4626–4637.
Egger, A.E., Carpi, A., 2011. The Process of Science. Visionlearning, Inc., New Canaan, CT.
Emmanouil, G., Galanis, G., Kallos, G., 2012. Combination of statistical Kalman filters and data assimilation for improving ocean waves analysis and forecasting. Ocean Model. 59–60, 11–23.
Evensen, G., 1994. Inverse methods and data assimilation in nonlinear ocean models. Physica D 77, 108–129.

Fei, X., Yang, D., Kong, Z., Lou, Z., Wu, J., 2014. Thyroid cancer incidence in China between 2005 and 2009. Stoch. Env. Res. Risk 28, 1075–1082.

Fei, X., Lou, Z., Christakos, G., Liu, Q., Ren, Y., Wu, J., 2016a. A geographic analysis about the spatiotemporal pattern of breast cancer in Hangzhou from 2008 to 2012. PLoS One 11 (1), e0147866.

Fei, X., Christakos, G., Lou, Z., Ren, Y., Liu, Q., Wu, J., 2016b. Spatiotemporal co-existence of female thyroid and breast cancers in Hangzhou, China. Sci. Rep. 6, 28524.

Finlay, C., Lesur, V., Thébault, E., Vervelidou, F., Morschhauser, A., Shore, R., 2017. Challenges handling magnetospheric and ionospheric signals in internal geomagnetic field modelling. Space Sci. Rev. 206 (1–4), 157–189.

Fotheringham, A.S., Charlton, M.E., Brunsdon, C., 1998. Geographically weighted regression: a natural evolution of the expansion method for spatial data analysis. Environ. Plan. A 30 (11), 1905–1927.

Fournier, A., Aubert, J., Thébault, E., 2011. Inference on core surface flow from observations and 3-D dynamo modelling. Geophys. J. Int. 186 (1), 118–136.

Gao, J., Li, S., 2011. Detecting spatially non-stationary and scale-dependent relationships between urban landscape fragmentation and related factors using Geographically Weighted Regression. Appl. Geogr. 31 (1), 292–302.

Gershman, S.J., Horvitz, E.J., Tenenbaum, J.B., 2015. Computational rationality: A converging paradigm for intelligence in brains, minds, and machines. Science 349 (6245), 273–278.

Getis, A., Ord, K., 1992. The analysis of spatial association by use of distance statistics. Geographic. Anal. 24, 189–206.

Gillet, N., 2019. Spatial and temporal changes of the geomagnetic field: Insights from forward and inverse core field models. In: Mandea, M., Korte, M., Yau, A., Petrovsky, E. (Eds.), Geomagnetism, Aeronomy and Space Weather: A Journey from the Earth's Core to the Sun. Cambridge Univ. Press, Cambridge, UK, pp. 115–132. Special Publications of the Intern. Union of Geodesy and Geophysics.

Gilliland, F., Avol, E., Kinney, P., Jerrett, M., Dvonch, T., Lurmann, F., Buckley, T., Breysse, P., Keeler, G., McConnell, R., 2005. Air pollution exposure assessment for epidemiologic studies of pregnant women and children: Lessons learned from the Centers for Children's Environmental Health and Disease Prevention Research. Environ. Health Perspect. 113, 1447–1454.

Girgis, A., Boyes, A., Sanson-Fisher, R.W., Burrows, S., 2000. Perceived needs of women diagnosed with breast cancer: rural versus urban location. Aust. N. Z. J. Public Health 24, 166–173.

Graham, C., Kurtz, T., Méléard, S., Protter, P., Pulvirenti, M., Talay, D., 1996. Probabilistic models for nonlinear partial differential equations. In: Lecture Notes in Mathematics 1627. Springer, New York, NY.

Griffith, D.A., 1981. Interdependence in space and time: numerical and interpretative considerations. In: Griffith, D., McKinnon, R. (Eds.), Dynamic Spatial Models. Plenum Press, New York, NY, pp. 258–287.

Hägerstrand, T., 1975. Survival and arena: on the life-history of individuals in relation to their geographical environment. The Monadnock 49, 9–29.

Hand, J.L., Yang, C., Chopra, A.K., Moritz Jr., A.L., 1994. Ability of geostatistical simulations to reproduce geology: A critical evaluation. In: Presented at the SPE Annual Technical Conference and Exhibition, SPE paper 28414.

Hankey, S., Marshall, J.D., 2015. Land use regression models of on-road particulate air pollution (particle number, black carbon, PM2.5, particle size) using mobile monitoring. Environ. Sci. Technol. 49 (15), 9194–9202.

Hastie, T., Tibshirani, R., Friedman, J., 2001. The Elements of Statistical Learning-Data Mining, Inference, and Prediction. Springer-Verlag, Berlin, Germany.

Haykin, S., 1994. Neural Networks. MacMillan College Publ. Co., New York, NY.

He, J., Christakos, G., 2018. Space-time PM2.5 mapping in the severe haze region of Jing-Jin-Ji (China) using a synthetic approach. Environ. Pollut. 240, 319–329.

He, J., Kolovos, A., 2018. Bayesian maximum entropy approach and its applications: a review. Stoch. Environ. Res. Risk Assess. 32, 859.

He, J., Christakos, G., Zhang, W., Wang, Y., 2017. A space-time study of hemorrhagic fever with renal syndrome (HFRS) and its climatic associations in Heilongjiang province, China. Front. Appl. Math. Statist. 3, 16.

He, J., Christakos, G., Wu, J., Cazelles, B., Qian, Q., Mu, D., Wang, Y., Yin, W., Zhang, W., 2018. Spatiotemporal variation of the association between climate dynamics and HFRS outbreaks in Eastern China during 2005-2016 and its geographic determinants. PLoS Negl. Trop. Dis. 12 (6), e0006554.

He, J., Christakos, G., Jankowski, P., 2019a. Comparative performance of the LUR, ANN, and BME techniques in the multiscale spatiotemporal mapping of PM2.5 concentrations in North China. IEEE J. Select. Top. Appl. Earth Observ. Remote Sens. 12 (6), 1734–1747.

He, J., Christakos, G., Wu, J., Jankowski, P., Langousis, A., Wang, Y., et al., 2019b. Probabilistic logic analysis of the highly heterogeneous spatiotemporal HFRS incidence distribution in Heilongjiang province (China) during 2005–2013. PLoS Negl. Trop. Dis. 13 (1), e0007091.

He, M., He, J., Christakos, G., 2020a. Improved space-time sea surface salinity mapping in Western Pacific ocean using contingogram modeling. Stoch. Environ. Res. Risk Assess. 34, 355–368.

He, J., Chen, G., Jiang, Y., Jin, R., Shortridge, A., Agusti, S., He, M., Wu, J., Duarte, C., Christakos, G., 2020b. Comparative infection modeling and control of COVID-19 transmission patterns in China, South Korea, Italy and Iran.

Sci. Total Environ. https://doi.org/10.1016/j.scitotenv.2020.141447.

Healy, K., 2018. Data Visualization: A Practical Introduction. Princeton University Press.

Hey, T., Tansley, S., Tolle, K. (Eds.), 2009. The Fourth Paradigm: Data Intensive Scientific Discovery. Microsoft Research, Redmont, Washington, ISBN: 978-0-9825442-0-4.

Hoek, G., Beelen, R., Hoogh, K., Vienneau, D., Gulliver, J., Fischer, P., Briggs, D., 2008. A review of land-use regression models to assess spatial variation of outdoor air pollution. Atmos. Environ. 42, 7561–7578.

Holben, B.N., Eck, T.F., Slutsker, I., Tanre, D., Buis, J., Setzer, A., et al., 1998. AERONET—a federated instrument network and data archive for aerosol Characterization. Remote Sens. Environ. 66 (1), 1–16.

Horvitz, E., Mulligan, D., 2015. Data, privacy, and the greater good. Science 349 (6245), 253–255.

Hunt, J.R., Baldocchi, D.D., van Ingen, C., 2009. Redefining ecological science using data. In: Hey, T., Tansley, S., Tolle, K. (Eds.), The Fourth Paradigm: Data Intensive Scientific Discovery. Microsoft Research, Redmont, Washington, ISBN: 978-0-9825442-0-4, pp. 21–26.

IPCC, 2007. Climate change 2007: The physical science basis. Contribution of Working Group I to the Fourth Assessment Report of the Intergovernmental Panel on Climate Change. Cambridge University Press, New York, NY.

IPCC, 2013. Climate Change 2013: The Physical Science Basis. Contribution of Working Group I to the Fifth Assessment Report of the Intergovernmental Panel on Climate Change. Intergovernmental Panel on Climate Change https://www.ipcc.ch/site/assets/uploads/2018/03/WG1AR5_SummaryVolume_FINAL.pdf.

Isakov, V., Johnson, M., Touma, J., Özkaynak, H., 2011. Development and evaluation of land-use regression models using modeled air quality concentrations. Chapter 117, In: Steyn, D.G., Trini Castelli, S. (Eds.), NATO-Air Pollution Modeling and its Application, XXI. Netherlands, Series C. Springer, the Netherlands, pp. 717–722.

Jiang, H., Du, H., Wang, L., Wang, P., Bai, X., 2016. Hemorrhagic fever with renal syndrome: pathogenesis and clinical picture. Front. Cell. Infect. Microbiol. 6. https://doi.org/10.3389/fcimb.2016.00001.

Jones, P.D., Wigley, T.M.L., Wright, P.B., 1986. Global temperature variations between 1861 and 1984. Nature 322 (6078), 430–434.

Kao, H.Y., Lagerloef, G.S.E., Lee, T., Melnichenko, O., Meissner, T., Hacker, P., 2018. Assessment of aquarius sea surface salinity. Remote Sens. (Basel) 10 (9), 1341–1352.

Karpatne, A., Atluri, G., Faghmous, J., Steinbach, M., Banerjee, A., Ganguly, A., Shekhar, S., Samatova, N., Kumar, V., 2017. Theory-guided data science: A new paradigm for scientific discovery from data. IEEE Trans. Knowl. Data Eng. 29 (10), 2318–2331. https://doi.org/10.1109/TKDE.2017.2720168.

Kharif, C., Pelinovsky, E., Slunyaev, A., 2009. Rogue waves in the Ocean. Springer, Berlin, Germany.

Kirk, A., 2016. Data Visualisation: A Handbook for Data Driven Design. Sage.

Kitchin R, 2014. Big data, new epistemologies and paradigm shifts. Big Data Soc., April–June: 1–12. DOI: https://doi.org/10.1177/2053951714528481.

Klemas, V., 2011. Remote sensing of sea surface salinity: an overview with case studies. J. Coast. Res. 27 (5), 830–838.

Knaflic, C.N., 2015. Storytelling With Data: A Data Visualization Guide for Business Professionals. John Wiley & Sons.

Kolovos, A., Christakos, G., Serre, M.L., Miller, C.T., 2002a. Computational BME solution of a stochastic advection-reaction equation in the light of site-specific information. Water Resour. Res. 38 (12), 1318–1334.

Kolovos, A., Christakos, G., Serre, M.L., Miller, C.T., 2002. Computational Bayesian maximum entropy solution of a stochastic advection-reaction equation in the light of site-specific information. Water Resour. Res. 38 (12), 54-1–54-17.

Krautkrämer, E., Zeier, M., Plyusnin, A., 2012. Hantavirus infection: an emerging infectious disease causing acute renal failure. Kidney Int. 83, 23–27. https://doi.org/10.1038/ki.2012.360.

Krishnan, S., Journel, A.G., 2003. Spatial connectivity: from variograms to multiple-point measures. Math. Geol. 35 (8), 915–925.

Lagerloef, G.S.E., Colomb, F.R., Vine, D.L., Wentz, F., Yueh, S., Ruf, C., Lilly, J., Gunn, J., Chao, Y., Decharon, A., Feldman, G.C., Swift, C., 2008. The aquarius/SAC-D mission: designed to meet the salinity remote-sensing challenge. Oceanography 21 (1), 68–81.

Lang, Y., Christakos, G., 2018. Ocean pollution assessment by integrating physical law and site-specific sata. Environmetrics. https://doi.org/10.1002/env.2547.

Lazer, D., Kennedy, R., King, G., Vespignani, A., 2014. The parable of Google flu: Traps in Big Data analysis. Science 343, 1203–1205. 14 March.

Le Traon, P.-Y., Antoine, D., Bentamy, A., Bonekamp, H., Breivik, L.A., Chapron, B., Corlett, G., Dibarboure, G., DiGiacomo, P., Donlon, C., Faugère, Y., Font, J., Girard-Ardhuin, F., Gohin, F., Johannessen, J.A., Kamachi, M., Lagerloef, G., Lambin, J., Larnicol, G., Le Borgne, P., Leuliette, E., Lindstrom, E., Martin, M.J., Maturi, E., Miller, L., Mingsen, L., Morrow, R., Reul, N., Rio, M.H., Roquet, H., Santoleri, R., Wilkin, J., 2015. Use of satellite observations for operational oceanography: recent achievements and future prospects. J. Oper. Oceanogr. 8 (Suppl. 1). https://doi.org/10.1080/1755876X.2015.1022050. s12–s27.

Lebreton, L.C.-M., Greer, S.D., Borrero, J.C., 2012. Numerical modelling of floating debris in the world's oceans. Mar. Pollut. Bull. 64 (3), 653–661.

Lehning, M., Dawes, N., Bavay, M., Parlange, M., Nath, S., Zhao, F., 2009. Instrumenting the Earth: next-generation sensor networks and environmental science. In: Hey, T., Tansley, S., Tolle, K. (Eds.), The Fourth Paradigm: Data Intensive Scientific Discovery. Microsoft Research, Redmont, Washington, ISBN: 978-0-9825442-0-4, pp. 45–51.

Lermusiaux, P.F.J., Chiu, C.-S., Gawarkiewicz, G.G., Abbot, P., Robinson, A.R., Haley, P.J., Leslie, W.G., Majumdar, S.J., Pang, A., Lekien, F., 2006. Quantifying uncertainties in ocean predictions. Oceanography 19 (1), 80–94.

Li, A., Bo, Y., Zhu, Y., Guo, P., Bi, J., He, Y., 2013. Blending multi-resolution satellite sea surface temperature (SST) products using Bayesian maximum entropy method. Remote Sens. Environ. 135, 52–63.

Lindzen, R.S., 1990. Some coolness concerning global warming. Bull. Am. Meteorol. Soc. 71 (3), 288–299.

Liu, J., Xue, F.Z., Wang, J.Z., Liu, Q.Y., 2013. Association of haemorrhagic fever with renal syndrome and weather factors in Junan County, China: a case-crossover study. Epidemiol. Infect. 141, 697–705. https://doi.org/10.1017/S0950268812001434.

Lou, Z., Fei, X., Christakos, G., Yan, J., Wu, J., 2017. Improving spatiotemporal breast cancer assessment and prediction in Hangzhou City, China. Sci. Rep. 7 (1), 3188.

Maes, C., McPhaden, M.J., Behringer, D., 2002. Signatures of salinity variability in tropical Pacific Ocean dynamic height anomalies. J. Geophys. Res. 107 (C12), 8012. https://doi.org/10.1029/2000JC000737.

Mann, M.E., Bradley, R.S., Hughes, M.K., 1998. Global-scale temperature patterns and climate forcing over the past six centuries. Nature 392 (6678), 779–787.

Matérn, B., 1947. Metoden Att Uppskatta Noggranheten Vidlinje-Ochprovyte-Taxering. Meddelanden från Statens Skogsforskningsinstitut 36.

Matheron, G., 1965. Les Variables Regionalisees et Leur Estimation. Masson, Paris, France.

McGarry, K., Wermster, S., MacIntyre, J., 1999. Hybrid neural systems: From simple coupling to fully integrated neural networks. Neural Comput. Surv. 2, 62–93.

Melnichenko, O., Hacker, P., Maximenko, N., Lagerloef, G., Potemra, J., 2014. Spatial optimal interpolation of aquarius sea surface salinity: algorithms and implementation in the north atlantic. J. Atmos. Oceanic Tech. 31, 1583–1600.

Melnichenko, O., Hacker, P., Maximenko, N., Lagerloef, G., Potemra, J., 2016. Optimum interpolation analysis of Aquarius sea surface salinity. J. Geophys. Res. Oceans 121, 602–616. https://doi.org/10.1002/2015JC011343.

Mercier, R.S., 1985. The Reactive Transport of Suspended Particles: Mechanisms and Modeling. Doctoral Dissertation, Joint Program in Oceanography and Ocean Engineering, Massachusetts Institute of Technology and Woods Hole Oceanographic Institution WHOI-85-23.

Miles, J., 2014. Tolerance and Variance Inflation Factor. Wiley StatsRef, https://doi.org/10.1002/9781118445112.stat06593. Statistics Reference Online.

Miller, H.J., 2005. A measurement theory for time geography. Geographic. Anal. 37, 17–45.

Miller, R.N., Carter, E.F., Blue, S.L., 1999. Data assimilation into nonlinear stochastic models. Tellus 51A, 167–194.

Mitchell, T., 1997. Machine Learning. Mc Graw-Hill, New York, NY.

Morgan, C.J., 2011. Theoretical and practical aspects of variography. PhD thesis. University of Witwatersrand, Johannesburg, South Africa.

MSPH (Mailman School of Public Health), 2019. Hot Spot Detection. Columbia University, New York, NY. https://www.mailman.columbia.edu/research/population-health-methods/hot-spot-detection.

NIJ (National Institute of Justice), 2005. In: Eck, J.E., Chainey, S., Cameron, J.G., Leitner, M., Wilson, R.E. (Eds.), Mapping Crime: Understanding Hotspots. NIJ, Washington, DC.

Parkes, D.N., Thrift, N.J., 1980. Times, Spaces, and Places: A Chronogeographic Perspective. J. Wiley & Sons, New York, NY.

Poh, R., Kennedy, J., Blackwell, T., 2007. Particle swarm optimization. Swarm Int. 1 (1), 33–57.

Riser, S.C., Ren, L., Annie, W., 2008. Salinity in Argo: A modern view of a changing ocean. Oceanography 21 (1), 56–67.

Robinson, A.R., Lermusiaux, P.F.J., 2002. Data assimilation for modeling and predicting coupled physical-biological interactions in the sea. In: Robinson, A.R., McCarthy, J.R., Rothschild, B.J. (Eds.), The Sea: Biological-Physical Interactions in the Ocean. John Wiley & Sons, New York, NY, pp. 475–536. Ch 12.

Rockafellar, R.T., 1993. Lagrange multipliers and optimality. SIAM Rev. 35 (2), 183–238.

Rodriguez, J.D., Perez, A., Lozano, J.A., 2010. Sensitivity analysis of k-fold cross validation in prediction error estimation. IEEE Trans. Pattern Anal. Mach. Intell. 32, 569–575.

Rouhani, S., Hall, T.J., 1989. Space-time kriging of groundwater data. In: Geostatistics. Springer, Dordrecht.

Ruby, R., 1966. How the Scientist Thinks. In: Ohlsen, W., Hammond, F.L. (Eds.), From Paragraph to Essay. Charles Scribner's Sons, New York, N.Y.

Ryan, P.K., LeMasters, G.K., 2007. A review of land-use regression models for characterizing intraurban air pollution exposure. Inhal. Toxicol. 19 (Suppl 1), 127–133.

Sadoti, G., Pollock, M.G., Vierling, K.T., Albright, T.P., Strand, E.K., 2014. Variogram models reveal habitat gradients predicting patterns of territory occupancy and nest survival among vesper sparrows. Wildlife Biol. 20, 97–107.

SANLIB, 1995. Stochastic Analysis Software Library and User's Guide. Stochastic Research Group, Research Rept. No. SM/1.95, Dept. of Environmental Sci. and Engin., Univ. of North Carolina, Chapel Hill, NC. 63 p.

Serre, M.L., Christakos, G., 1999. BME studies of stochastic differential equations representing physical laws—Part II. In: 5th Annual Conference. vol. 1, pp. 93–98. Trodheim, Norway.

Shahid, R., Bertazzon, S., Knudtson, M.L., et al., 2009. Comparison of distance measures in spatial analytical modeling for health service planning. BMC Health Serv. Res. 9, 200. https://doi.org/10.1186/1472-6963-9-200.

Sibson, R., 1980. A vector identity for the Dirichlet tessellation. Math. Proc. Cambridge Philos. Soc. 87 (1), 151–155. https://doi.org/10.1017/S0305004100056589.

Silverman, B.-W., 1986. Density Estimation for Statistics and Data Analysis. Chapman & Hall, New York, NY.

Simmons, J.H., Riley, L.K., 2002. Hantaviruses: An overview. Comp. Med. 52, 97–110.

SMOS Team, 2010. SMOS L2 OS Algorithm Theoretical Baseline Document. IFREMER, 33–36.

Snepvangers, J.J.J.C., Heuvelink, G.B.M., Huisman, J.A., 2003. Soil water content interpolation using spatio-temporal kriging with external drift. Geoderma 112, 253–271.

Stephens, C., 2011. A Bayesian approach to absent evidence reasoning. Inform. Logic 31 (1), 56–65.

Succi, S., Coveney, P.V., 2019. Big data: the end of the scientific method? Phil. Trans. R. Soc. A 377. https://doi.org/10.1098/rsta.2018.0145.

Tobler, W.R., 1970. A computer movie simulating urban growth in the Detroit region. Econ. Geogr 46, 234–240.

Tøffner-Clausen, L., Lesur, V., Olsen, N., Finlay, C.C., 2016. In-flight scalar calibration and characterisation of the Swarm magnetometry package. Earth Planets Space 68 (1), 129.

Tu, J.V., 1996. Advantages and disadvantages of using artificial neural networks versus logistic regression for predicting medical outcomes. J. Clin. Epidemiol. 49 (11), 1225–1231.

Tufte, E.R., 2001. The Visual Display of Quantitative Information. vol. 2 Graphics Press, Cheshire, CT.

Vine, D.M.L., Lagerloef, G.S., Torrusio, S., 2010. Aquarius and the Aquarius/SAC-D mission. In: 11th Specialist on Microwave Radiometry and Remote Sensing of the Environment, pp. 33–36. Mar 1–4. Washington DC https://doi.org/10.1109/MICRORAD.2010.5559594.

Vyas, V., Christakos, G., 1997. Spatiotemporal analysis and mapping of sulfate deposition data over the conterminous USA. Atmos. Environ. 31 (21), 3623–3633.

Walker, G., 1931. On periodicity in series of related terms. Proc. Roy. Soc. Lond., Ser. A 131, 518–532.

Wang, J., Shen, Y., 2010. Modeling oil spills transportation in seas based on unstructured grid, finite-volume, wave-ocean model. Ocean Model. 35 (4), 332–344.

Wang, J., Zhang, T., Fu, B., 2016. A measure of spatial stratified heterogeneity. Ecol. Indic. 67, 250–256.

Wheeler, D., Tiefelsdorf, M., 2005. Multicollinearity and correlation among local regression coefficients in geographically weighted regression. J. Geogr. Syst. 7 (2), 161–187.

Wikle, C.K., Cressie, N., 1999. A dimensionality-reduced approach to space-time Kalman filtering. Biometrika 86, 815–829.

Wolter, K., Timlin, M.S., 2011. El Niño/Southern Oscillation behaviour since 1871 as diagnosed in an extended multivariate ENSO index (MEI.ext). Int. J. Climatol. 31 (7), 1074–1087.

Wu, Z., Phillips Jr., G.N., Tapia, R., Zhang, Y., 2001. A fast Newton algorithm for entropy maximization in phase determination. SIAM Rev. 43 (4), 623–642.

Xiao, X., He, J., Huang, H., Miller, T.R., Christakos, G., Reichwaldt, E.S., Ghadouani, A., Lin, S., Xu, X., Shi, J., 2017. A novel single-parameter approach for forecasting algal blooms. Water Res. 108, 222–231.

Xiao, L., Lang, Y., Christakos, G., 2018. High-resolution spatiotemporal mapping of PM 2.5 concentrations at Mainland China using a combined BME-GWR technique. Atmos. Environ. 173, 295–305.

Xiao, L., Christakos, G., He, J., Lang, Y., 2020. Space-Time Ground-Level PM2.5 Distribution at the Yangtze River Delta: A Comparison of Kriging, LUR, and Combined BME-LUR Techniques. J. Environ. Inform. 36 (1), 33–42.

Xie, Y., Deutsch, C., Tran, T., 2001. Preliminary Research Toward Direct Geostatistical Simulation of Unstructured Grids. The Annual Report of the Centre for Computational Geostatistics.

Yan, L., Fang, L.-Q., Huang, H.-G., Zhang, L.-Q., Feng, D., Zhao, W.-J., Zhang, W.-Y., Li, X.-W., Cao, W.-C., 2007. Landscape Elements and Hantaan Virus-Related Hemorrhagic Fever With Renal Syndrome. People's Republic of China, Landscape.

Yang, Y., Christakos, G., Yang, X., He, J., 2018. Spatiotemporal characterization and mapping of PM 2.5 concentrations in southern Jiangsu Province, China. Environ. Pollut. 234, 794–803.

Yu, H., Wang, C., 2013. Quantile-based Bayesian maximum entropy approach for spatiotemporal modeling of ambient air quality levels. Environ. Sci. Technol. 47 (3), 1416–1424.

Yu, H.-L., Kolovos, A., Christakos, G., Chen, J.-C., Warmerdam, S., Dev, B., 2007. Interactive spatiotemporal modeling of health systems: the SEKS-GUI framework. Stoch. Environ. Res. Risk Assess. 21 (5), 555–572.

Yule, G.U., 1927. On a method of investigating periodicities in disturbed series, with special reference to Wolfer's Sunspot numbers. Phil. Trans. Royal Soc. Lond., Ser. A 226, 267–298.

Index

Note: Page numbers followed by *f* indicate figures, *t* indicate tables, and *np* indicate footnotes.

Printed in the United States
by Baker & Taylor Publisher Services